T0156274

Smart Innovation, Systems and Technologies

Volume 39

Series editors

Robert J. Howlett, KES International, Shoreham-by-sea, UK
e-mail: rjhowlett@kesinternational.org

Lakhmi C. Jain, University of Canberra, Canberra, Australia, and
University of South Australia, Adelaide, Australia
email: Lakhmi.jain@unisa.edu.au

About this Series

The Smart Innovation, Systems and Technologies book series encompasses the topics of knowledge, intelligence, innovation and sustainability. The aim of the series is to make available a platform for the publication of books on all aspects of single and multi-disciplinary research on these themes in order to make the latest results available in a readily-accessible form. Volumes on interdisciplinary research combining two or more of these areas is particularly sought.

The series covers systems and paradigms that employ knowledge and intelligence in a broad sense. Its scope is systems having embedded knowledge and intelligence, which may be applied to the solution of world problems in industry, the environment and the community. It also focusses on the knowledge-transfer methodologies and innovation strategies employed to make this happen effectively. The combination of intelligent systems tools and a broad range of applications introduces a need for a synergy of disciplines from science, technology, business and the humanities. The series will include conference proceedings, edited collections, monographs, handbooks, reference books, and other relevant types of book in areas of science and technology where smart systems and technologies can offer innovative solutions.

High quality content is an essential feature for all book proposals accepted for the series. It is expected that editors of all accepted volumes will ensure that contributions are subjected to an appropriate level of reviewing process and adhere to KES quality principles.

More information about this series at http://www.springer.com/series/8767

Rui Neves-Silva · Lakhmi C. Jain
Robert J. Howlett
Editors

Intelligent Decision Technologies

Proceedings of the 7th KES International Conference on Intelligent Decision Technologies (KES-IDT 2015)

Editors
Rui Neves-Silva
FCT
Universidade Nova de Lisboa
Caparica
Portugal

Robert J. Howlett
KES International
Shoreham-by-sea
UK

Lakhmi C. Jain
Faculty of Education, Science, Technology
and Mathematics
University of Canberra
Canberra
Australia

ISSN 2190-3018 ISSN 2190-3026 (electronic)
Smart Innovation, Systems and Technologies
ISBN 978-3-319-36766-8 ISBN 978-3-319-19857-6 (eBook)
DOI 10.1007/978-3-319-19857-6

Springer Cham Heidelberg New York Dordrecht London

Softcover reprint of the hardcover 1st edition 2015

Printed on acid-free paper

Springer International Publishing AG Switzerland is part of Springer Science+Business Media
(www.springer.com)

Preface

The current volume includes the research results presented at the Seventh International Conference on *Intelligent Decision Technologies* (KES-IDT 2015) which took place during June 17–19, 2015, in Sorrento, Italy.

KES-IDT is a well-established international annual conference, an interdisciplinary conference in nature, and this edition consisted of keynote talks, oral and poster presentations, invited sessions, and workshops on the applications and theory of intelligent decision systems and related areas. It provided excellent opportunities for the presentation of interesting new research results and discussion about them, leading to knowledge transfer and generation of new ideas.

Sorrento is a town and commune in Campania, southern Italy, with approximately 16,500 inhabitants. It is a popular tourist destination overlooking the Bay of Naples and providing stunning views of Naples, Vesuvius, and the Isle of Capri. It forms an ideal base for visiting historic sites, such as Pompei, or for touring the Amalfi coast.

KES-IDT 2015 received many high quality submissions and all papers have been reviewed by at least two reviewers. Following a rigorous reviewing process, not all submissions could be accommodated for presentation at the conference. Of these, 57 papers were accepted for presentation and included in this proceedings. We are very satisfied with the quality of the program and would like to thank the authors for choosing KES-IDT as the forum for presentation of their work. Also, we gratefully acknowledge the hard work of the KES-IDT international program committee members and of the additional reviewers for taking the time to review the submitted papers rigorously and select the best among them for presentation at the conference and inclusion in its proceedings.

We are also grateful to the KES personnel for their great and efficient work in supporting KES-IDT 2015. Finally, we would like to thank the Springer personnel for their wonderful job in producing this volume.

The General and Program Co-chairs

Rui Neves-Silva, Universidade Nova de Lisboa, Portugal
Junzo Watada, Waseda University, Japan
Gloria Wren-Phillips, Loyola University, USA
Lakhmi C. Jain, University of South Australia, Australia

The Executive Chair

Robert J. Howlett, KES International & Bournemouth University, UK

The Program Committee Chair

Ana Rita Campos, UNINOVA, Portugal

International Programme Committee Members

Dr. Jair Minoro Abe, University of São Paulo, Brazil
Dr. Walid Adly Atteya, University of Bradford, UK
Dr. Ahmad Taher Azar, Benha University, Egypt
Dr. Dariusz Barbucha, Gdynia Maritime University, Poland
Prof. Alina Barbulescu, Ovidius University of Constanta, Romania
Dr. Laszlo Barna Iantovics, Petru Maior University, Romania
Dr. Márcio Basgalupp, UNIFESP, Brazil
Dr. Mokhtar Beldjehem, University of Ottawa, Canada
Prof. Valentina Emilia Balas, Aurel Vlaicu University of Arad, Romania
Prof. Monica Bianchini, Università di Siena, Italy
Dr. Gloria Bordogna, National Research Council, Italy
Dr. Janos Botzheim, Tokyo Metropolitan University, Japan
Dr. Ana Rita Campos, UNINOVA, Portugal
Dr. Anne Magály de Paula Canuto, UFRN, Brazil
Dr. Giovanna Castellano, University of Bari, Italy
Assoc. Prof. Hsing-Chung Chen, Asia University, Taiwan
Dr. Paolo Crippa, Università Politecnica delle Marche, Italy
Prof. Alfredo Cuzzocrea, University of Calabria, Italy
Prof. Dr. Irek Czarnowski, Gdynia Maritime University, Poland
Dr. Myriam Delgado, UFTPR, Brazil
Prof. Margarita Favorskaya, Siberian State Aerospace University, Russia
Dr. Wojciech Froelich, University of Silesia, Poland
Dr. Konrad Fuks, Poznań University of Economics, Poland
Dr. Mauro Gaggero, National Research Council, Italy
Dr. Petia Georgieva, University of Aveiro, Portugal
Dr. Christos Grecos, Independent Imaging Consultant
Prof. Jerzy W. Grzymala-Busse, University of Kansas, USA
Dr. Katarzyna Harezlak, Silesian University of Technology, Poland
Prof. Dr. Mohamed A. Moustafa Hassan, Cairo University, Egypt
Assoc. Prof. Ioannis Hatzilygeroudis, University of Patras, Greece
Prof. Dawn E. Holmes, University of California, USA

Dr. Milos Radovanovic, University of Novi Sad, Serbia
Dr. Ana Respício, Universidade de Lisboa, Portugal
Dr. John Ronczka, Australian Society of Rheology, Australia
Dr. Ewa Ratajczak-Ropel, Gdynia Maritime University, Poland
Prof. Marcello Sanguineti, University of Genova, Italy
Prof. Hirosato Seki, Kwansei Gakuin University, Japan
Prof. Udo Seiffert, Fraunhofer & University of Magdeburg, Germany
Dr. Urszula Stańczyk, Silesian University of Technology, Poland
Dr. Catalin Stoean, University of Craiova, Romania
Dr. Ruxandra Stoean, University of Craiova, Romania
Dr. Mika Sulkava, MTT Agrifood Research, Finland
Dr. Piotr Szymczyk, AGH University of Science and Technology, Poland
Prof. Shing Chiang Tan, Multimedia University, Malaysia
Prof. Mieko Tanaka-Yamawaki, Tottori University, Japan
Prof. Gheorghe Tecuci, George Mason University, USA
Dr. Renato Tinos, USP, Brazil
Prof. Claudio Turchetti, Università Politecnica delle Marche, Italy
Prof. Eiji Uchino, Yamaguchi University, Japan
Prof. Junzo Watada, Waseda University, Japan
Assoc. Prof. Fen Wang, Central Washington University, USA
Prof. Gian Piero Zarri, Sorbonne University, LaLIC/STIH Laboratory, France
Prof. Alfred Zimmermann, Reutlingen University, Germany

Contents

Propositional Algebra P_1

Jair Minoro Abe, Kazumi Nakamatsu, Seiki Akama and João I.S. Filho

Abstract In this semi-expository paper we investigate the propositional algebra P_1, some properties and its relationship with Curry algebra P_1. We show as in the classical case that both structures are equivalent in the sense of Bourbaki. Some results on the extension to the propositional algebras P_n ($1 \leq n \leq \omega$) are also made.

Keywords Propositional algebra • Curry algebra • Algebraic logic • Paracomplete logic

1 Introduction

It is well known that logic can be conceived as algebraic system. In this way, Boolean algebra is a version of classical logic, Heyting algebra is an algebraic version of some intuitionistic logics, and so on. In investigating non monotonic operators in lattices, Da Costa formulated the concept of Curry algebras, which their

J.M. Abe (✉)
Graduate Program in Production Engineering, ICET - Paulista University, R. Dr. Bacelar 1212, CEP, São Paulo, SP 04026-002, Brazil
e-mail: jairabe@uol.com.br

J.M. Abe
Institute for Advanced Studies, University of São Paulo, São Paulo, Brazil

K. Nakamatsu
School of Human Science and Environment/H.S.E., University of Hyogo, Kobe, Japan
e-mail: nakamatu@shse.u-hyogo.ac.jp

S. Akama
C-Republic, Tokyo, Japan
e-mail: akama@jcom.home.ne.jp

J.I.S. Filho
Santa Cecília University, Santos, Philippines
e-mail: jinacsf@yahoo.com.br

R. Neves-Silva et al. (eds.), *Intelligent Decision Technologies*,
Smart Innovation, Systems and Technologies 39,
DOI 10.1007/978-3-319-19857-6_1

fundamental relation is not the equality, but an equivalence relation. Through this expedient Da Costa has applied in the algebraization of his systems C_n ($1 \leq n \leq \omega$) [15], obtaining the Curry algebras C_n ($1 \leq n \leq \omega$). Since then such concept has been developed to furnish algebraic version of innumerous non-classical systems (even for classical logic, such pre-algebraic systems has also shown useful [16]). On the other hand, Rosenbloom [20] presents the concept of propositional algebra, which it is close to the notion of pre-algebra studied in Curry systems and it can also viewed as an algebraic version of classical propositional calculus. In this paper we apply these concepts to the study of paracomplete propositional systems P_n ($1 \leq n \leq \omega$) which are dual of the systems C_n ($1 \leq n \leq \omega$).

In [13], da Costa and Marconi have introduced a hierarchy P_n ($1 \leq n < \omega$) of paracomplete propositional calculi. Such calculi are duals of the propositional paraconsistent calculi C_n ($1 \leq n < \omega$).

For each n, $1 \leq n < \omega$, we have calculi symbolized by P_n. Such calculi were formulated with the following aim:

(1) The principle of the excluded middle in the form $A \vee \neg A$ is not valid in general;
(2) There are formulas A and $\neg A$ that are both false.
(3) The calculi should contain the most important schemes and inference rules of classical propositional calculus compatible with the conditions (1) and (2) above.

The language L of the calculi P_n ($1 \leq n < \omega$) is the same for all of them. The primitive symbols are the following:

- Propositional variables: a denumerable set of propositional variables;
- Logical connectives: $\rightarrow$ (implication), $\wedge$ (conjunction), $\vee$ (disjunction), $\neg$ (negation);
- Parentheses.

Formulas are defined in the usual manner.

Definition 1.1 Let A be any formula. Then $A^{\#}$ is shorthand for $A \vee \neg A$. We write, A^{n} for $A^{\#} \wedge A^{\#\#} \wedge A^{\#\#\#} \wedge \ldots \wedge A^{\#n}$, where #n means that the symbol # appears n times. Also, $A^{[n]} =_{\text{Def.}} A^1 \wedge A^2 \wedge \ldots \wedge A^n$ and $A \leftrightarrow B$ stands for $(A \rightarrow B) \wedge (B \rightarrow A)$.

The postulates (axiom schemes and inference rules) of P_n are: A, B, and C are formulas whatsoever.

(1) $A \rightarrow (B \rightarrow A)$
(2) $(A \rightarrow (B \rightarrow C)) \rightarrow ((A \rightarrow B) \rightarrow (A \rightarrow C))$
(3) $\frac{A, A \rightarrow B}{B}$
(4) $A \wedge B \rightarrow A$
(5) $A \wedge B \rightarrow B$
(6) $A \rightarrow (B \rightarrow (A \wedge B))$
7) $A \rightarrow A \vee B$
(8) $B \rightarrow A \vee B$

(9) $(A \to C) \to ((B \to C) \to ((A \vee B) \to C))$
(10) $\neg(\neg A \wedge A)$
(11) $A \to (\neg A \to B)$
(12) $A \to \neg\neg A$
(13) $A^{[n]} \to ((A \to B) \to ((A \to \neg B) \to \neg A))$
(14) $(A^{[n]} \wedge B^{[n]}) \to \left((A \wedge B)^{[n]} \wedge (A \vee B)^{[n]} \wedge (A \to B)^{[n]}\right)$

The calculus P_ω is obtained from P_n by dropping the axiom schemes (13) and (14).

In P_n, $A^{[n]}$ expresses intuitively that the formula A 'behaves' classically, so that the motivation of the postulates (13) and (14) are clear. Furthermore, in this calculus, the set of all well-behaved formulas together with the connectives $\to$, $\wedge$, $\vee$, and $\neg$ have all the properties of classical implication, conjunction, disjunction, and negation, respectively. Therefore the classical propositional calculus is contained in P_n, though it constitutes a strict sub-calculus of the former.

Theorem 1.1 In $P_n (1 \leq n < \omega)$ all theorems and inference rules of the classical positive implicative calculus are valid. In particular, the deduction theorem is valid in P_n.

Proof. Immediate consequence from the postulates of P_n.

Theorem 1.2 In P_n $(1 \leq n < \omega)$ the following schemes are valid:

(1) $\vdash A \wedge \neg A \to B$
(2) $\vdash A \vee (A \to B)$
(3) $\vdash A^{[n]} \to (\neg\neg A \leftrightarrow A)$
(4) $\vdash A^{[n]} \to ((A \to B) \wedge (A \to \neg B) \to \neg A)$
(5) $\vdash A \wedge \neg A \leftrightarrow B \wedge \neg B$

Proof. Immediate.

Theorem 1.3 In P_n $(1 \leq n)$ the following schemes are not valid:

(1) $A \vee \neg A$
(2) $\neg(A \vee B) \leftrightarrow \neg A \wedge \neg B$
(3) $\neg(A \wedge B) \leftrightarrow \neg A \wedge \neg B$
(4) $\neg\neg A \to A$
(5) $\neg\neg A \leftrightarrow A$
(6) $(A \to B) \to (\neg B \to \neg A)$
(7) $(A^{[n]})^{[n]}$

Proof. Immediate.

Definition 1.2 In P_n $(1 \leq n < \omega)$, we define the symbol $\neg^{\#}$ "strong negation" as follows: $\neg^{\#}A =_{\text{Def}} A \to (B \wedge \neg B)$, where B is a fixed formula.

Theorem 1.4 In P_n $(1 \leq n < \omega)$, the strong negation has all properties of the classical negation.

Proof. Immediate.

Corollary 1.4.1 In P_n $(1 \leq n < \omega)$ the following schemes are valid:

(1) $((A \to B) \to ((A \to \neg^{\#}B) \to {}^{\#}\neg A)$
(2) $A \to (\neg^{\#}A \to B)$
(3) $A \vee \neg^{\#}A$

Proof. Consequence of previous theorem.

Theorem 1.5 If P_0 is the classical propositional calculus, then each calculus of the sequence, starting from the 2nd one, $P_0, P_1, P_2, \ldots, P_n, \ldots, P_\omega$ is strictly stronger than each one of the preceding ones. P_i is the strongest calculus of the paracomplete calculus and P_ω the weakest one of the hierarchy.

Proof. Analogous to the calculi C_n.

2 Some Pre-Algebraic Structures

We begin with some basic concepts. For a detailed account see [11].

Definition 2.1 Suppose that in a non-empty set A is fixed an equivalence relation $\equiv$. We say that a n-ary operator φ on A is i-compatible with $\equiv$ if for any $x_1, \ldots, x_{i-1}, a, b, x_{i+1}, \ldots, x_n \in A$ and $a \equiv b$, imply $\varphi(x_1, \ldots, x_{i-1}, a, x_{i+1}, \ldots, x_n) \equiv \varphi(x_1, \ldots, x_{i-1}, b, x_{i+1}, \ldots, x_n)$. The operator is said to be compatible (or monotonic) with $\equiv$ if φ is i-compatible with $\equiv$ for all $i = 1, \ldots, n$. A relation R on A is said to be compatible with $\equiv$ if $(x_1, \ldots, x_n) \in R$ and $x_i \equiv x'_i$, $i = 1, \ldots, n$ then $(x'_1, \ldots, x'_n) \in R$.

Definition 2.2 A Curry system is a structure
$<A, (\equiv)_{i\in I}, (S)_{j\in J}, (R)_{k\in K}, (\varphi)_{l\in L}, (C)_{m\in M}>$ such that:

(1) $A \neq \varnothing$.
(2) $(\equiv)_{i \in I}$ is a collection of equivalence relations,
(3) $(S)_{j \in J}$ is a family of subsets of A,
(4) $(R)_{k \in K}$ is a finite collection of relations on A,
(5) $(\varphi)_{l \in L}$ is a family of operations on A,
(6) $(C)_{m \in M}$ is a finite collection of elements of A.

Normally we consider a unique equivalence relation $\equiv$ in a Curry system.

Definition 2.3 A Curry system is called a pre-algebra if the collection $(S)_{j \in J}$ is empty and all operations and relations regarding $\equiv$ are monotonic.

Definition 2.4 A Curry system is called a Curry algebra if there is at least one non-monotonic operation or relation on A.

Definition 2.5 A system $< A, \equiv, \leq >$ is called a pre-ordered system if

(1) For all $x \in A$, $x \leq x$.
(2) For all $x, y, z \in A$, $x \leq y$ and $y \leq z$ imply $x \leq z$.
(3) For all $x, y, x', y' \in A$, $x \leq y$, $x \equiv x'$, and $y \equiv y'$ imply $x' \leq y'$.
A pre-ordered system $< A, \equiv, \leq >$ is called a partially-ordered system if
(4) For all $x, y \in A$, $x \leq y$ and $y \leq x$ imply $x \equiv y$.
A partially-pre-ordered system $< A, \equiv, \leq >$ is called a pre-lattice if

(5) For all $x, y \in A$, the set of $\sup\{x, y\} \neq \emptyset$ and the set of $\inf\{x, y\} \neq \emptyset$ (sup abbreviates the ordinary supremum operation and inf abbreviates the ordinary infimum operation).
(We denote by $x \vee y$ one element of the set of $\sup\{x, y\}$and by $x \wedge y$ one element of the set of $\inf\{x, y\}$.
A system $< A, \equiv, \leq, \rightarrow >$ is called an implicative pre-lattice if $< A, \equiv, \leq >$ is a pre-lattice, and for all $x, y, z \in A$,

(6) $x \wedge (x \rightarrow y) \leq y$

(7) $x \wedge y \leq z$ iff $x \leq y \rightarrow z$.

$<A, \equiv, \leq, \rightarrow>$ is called classic implicative pre-lattice if it is an implicative pre-lattice and

(8) $(x \rightarrow y) \rightarrow x \leq x$ (Peirce's law).

With these definitions we can extend the majority of algebraic systems to pre-algebraic systems considering an equivalence relation $\equiv$ instead of equality relation.

Usual algebraic concepts, v.g., the ideas of Boolean pre-algebras, pre-filters, pre-ideals, etc. can be adapted to Curry systems.

Definition 2.6 A pre-Boolean algebra is a distributive pre-lattice with zero and unit such that each element has a complement, i.e. a system $< A, \equiv, \wedge, \vee, 0, 1, ' >$ with an maximum element 1, minimum element 0, and operators $\wedge$, $\vee$, and ' satisfying the conditions below:

(1) $x \vee y \equiv y \vee x$ and $x \wedge y \equiv y \wedge x$

(2) $(x \vee y) \vee z \equiv x \vee (y \vee z)$ and $(x \wedge y) \wedge z \equiv x \wedge (y \wedge z)$

(3) $x \wedge (x \vee y) \equiv x$ and $x \vee (x \wedge y) \equiv x$

(4) $x \vee (y \wedge z) \equiv (x \vee y) \wedge (x \wedge z)$

(5) $x \wedge (y \vee z) \equiv (x \wedge y) \vee (x \vee z)$

(6) $1 \wedge x \equiv x$

(7) $1 \vee x \equiv 1$

(8) $0 \wedge x \equiv 0$

(9) $0 \vee x \equiv x$

(10) $x \wedge x' \equiv 1$

(11) $x \wedge x' \equiv 0$

There is an order relation on A such that $x \leq y$ iff $x \equiv x \wedge y$ and $y \equiv x \vee y$

3 The Curry Algebra P_1

Definition 3.1 A Curry algebra P_1 (or a P_1-algebra) is a classical implicative pre-lattice $< A, \equiv, \leq, \wedge, \vee, \rightarrow, ' >$ with a greatest element 1 and operators $\wedge$, $\vee$, and ' satisfying the conditions below, where $x^{\#} =_{\text{Def.}} x \vee x'$:

(1) $x \leq x''$
(2) $(x \wedge x') \equiv 1$
(3) $x^{\#} \wedge y^{\#} \leq (x \rightarrow y)^{\#}$;
(4) $x^{\#} \wedge y^{\#} \leq (x \wedge y)^{\#}$;
(5) $x^{\#} \wedge y^{\#} \leq (x \vee y)^{\#}$;
(6) $x^{\#} \leq (x')^{\#}$
(7) $x^{\#} \leq (x \rightarrow y) \rightarrow ((x \rightarrow y') \rightarrow x')$
(8) $x \leq$
(9) $(x' \rightarrow y)$

Example 3.1 Let's consider the calculus P_1. A is the set of all formulas of P_1. Let's consider as operations, the logical connectives of conjunction, disjunction, implication, and negation. Let's define the relation on A:
$x \equiv y$ iff $\vdash x \leftrightarrow y$. It is easy to check that $\equiv$ is an equivalence relation on A.
$x \leq y$ iff $x \equiv x \wedge y$ and $y \leq x$ iff $y \equiv x \wedge y$. Also we take as 1 any fixed axiom instance.

The structure composed $< A, \equiv, \leq, \wedge, \vee, \rightarrow, ' >$ is a P_1-algebra.

Theorem 3.1 Let's $< A, \equiv, \leq, \wedge, \vee, \rightarrow, ' >$ be a P_1-algebra. Then the operator ' is non-monotone relatively $\equiv$.

Theorem 3.2 A P_1-algebra is distributive and has a greatest element, as well as a first element.

Definition 3.2 Let x be an element of a P_1-algebra. We put $x^* = x \rightarrow (y \wedge y')$, where y is a fixed element.

Theorem 3.3 In a P_1-algebra, x^* is a Boolean complement of x; so $x \vee x^* \equiv 1$ and $x \wedge x^* \equiv 0$. Moreover, in a P_1-algebra, the structure composed by the underlying set and by operations $\wedge$, $\vee$, and * is a (pre) Boolean algebra. If we pass to the quotient by the basic relation $\equiv$, we obtain a Boolean algebra in the usual sense.

Definition 3.3 Let $< A, \equiv, \leq, \wedge, \vee, \rightarrow, ' >$ be a P_1-algebra and $< A, \equiv, \leq, \wedge, \vee, \rightarrow, * >$ the Boolean algebra obtained as in the above theorem. Any Boolean algebra that is isomorphic to the quotient algebra of $< A, \equiv, \leq, \wedge, \vee, \rightarrow, * >$ by $\equiv$ is called Boolean algebra *associated with the* P_1-algebra.

Hence, we have the following representation theorems for P_1-algebras.

Theorem 3.4 Any P_1-algebra is associated with a field of sets. Moreover, any P_1-algebra is associated with the field of sets simultaneously open and closed of a totally disconnected compact Hausdorff space.

One open problem concerning P_1-algebras remains. How many non-isomorphic associated with the P_1-algebra are there?

By employing P_1-algebra we can get a completeness theorem of the calculus P_1.

Now we show a chain of Curry algebras beginning with the P_1-algebra.

Let $< A, \equiv, \leq, \wedge, \vee, \rightarrow, ' >$ be a P_1-algebra. If $x \in A$, x^1 abbreviates $x^{\#}$. x^n $(1 < n < \omega)$ abbreviates $x^{\#} \wedge x^{\#\#} \wedge \ldots \wedge x^{\#\#\ldots\#}$, where the symbol $^{\#}$ occurs n times. Also, $x^{(n)}$ abbreviates $x^1 \wedge x^2 \wedge \ldots \wedge x^n$.

Definition 3.4 A P_n-algebra ($1 < n < \omega$) is an implicative pre-lattice $< A, \equiv, \leq, \wedge, \vee, \rightarrow, ' >$ with a first element 1 and operators $\wedge$, $\vee$, and ' satisfying the conditions:

(1) $x \leq x''$
(2) $(x \wedge x') \equiv 1$
(3) $x^{(n)} \wedge y^{(n)} \leq (x \rightarrow y)^{(n)}$;
(4) $x^{(n)} \wedge y^{(n)} \leq (x \wedge y)^{(n)}$;
(5) $x^{(n)} \wedge y^{(n)} \leq (x \vee y)^{(n)}$;
(6) $x^{(n)} \leq (x')^{(n)}$
(7) $x^{(n)} \leq (x \rightarrow y) \rightarrow ((x \rightarrow y') \rightarrow x')$
(8) $x \leq (x' \rightarrow y)$

Usual algebraic structural concepts like homomorphism, monomorphism, etc. can be introduced for Curry algebras without extensive comments.

Theorem 3.5 Every P_n-algebra is embedded in any P_{n-1}-algebra ($1 < n < \omega$).

Corollary 3.5.1 Every P_n-algebra ($1 < n < \omega$) is embedded in any P_1-algebra.

If we indicate a P_n-algebra by P_n, the embedding hierarchy can be represented as $P_1 > P_2 > \ldots P_n > \ldots$

Definition 3.5 A P_ω-algebra is an implicative pre-lattice $< A, \equiv, \leq, \wedge, \vee, \rightarrow, ' >$ with a first element 1 and operators $\wedge$, $\vee$, and ' satisfying the conditions below:

(1) $(x \wedge x')' \equiv 1$
(2) $x \leq (x' \rightarrow y)$
(3) $x \leq x''$

By employing P_n-algebra we can get a completeness theorem of the calculus P_n.

4 Propositional Algebra P_1

In this session we present another study of the Curry algebra P_1 [1] named propositional algebra P_1.

Definition 4.1 A propositional algebra P_1 is a structure $< A, \tilde{A}, \rightarrow, \wedge, \vee, \neg >$ where $A \subseteq \tilde{A}$ (with $A \neq \varnothing$), $\rightarrow$, $\wedge$, and $\vee$ are binary operations, and $\neg$ is a unary operator on A satisfying the following conditions ($x^{\#}$ abbreviates. $x \vee \neg x$):

(1) $x, y \in A \Rightarrow x \rightarrow (y \rightarrow x) \in \tilde{A}$
(2) $x, y, z \in A \Rightarrow (x \rightarrow y) \rightarrow ((x \rightarrow (y \rightarrow z) \rightarrow (x \rightarrow z) \in \tilde{A}$
(3) $(x \in A \text{ and } x \rightarrow y \in \tilde{A}) \Rightarrow y \in \tilde{A}$
(4) $x, y \in A \Rightarrow x \wedge y \rightarrow x \in \tilde{A}$
(5) $x, y \in A \Rightarrow x \wedge y \rightarrow y \in \tilde{A}$
(6) $x, y \in A \Rightarrow x \rightarrow (y \rightarrow x \wedge y) \in \tilde{A}$

(7) $x, y \in A \Rightarrow x \rightarrow x \vee y \in \tilde{A}$
(8) $x, y \in A \Rightarrow y \rightarrow x \vee y \in \tilde{A}$
(9) $x, y, z \in A \Rightarrow (x \rightarrow z) \rightarrow ((y \rightarrow z) \rightarrow (x \vee y \rightarrow z)) \in \tilde{A}$
(10) $x \in A \Rightarrow \neg(\neg x \wedge x) \in \tilde{A}$
(11) $x, y \in A \Rightarrow x \rightarrow (\neg x \rightarrow y) \in \tilde{A}$
(12) $x, y \in A \Rightarrow x \rightarrow \neg\neg x \in \tilde{A}$
(13) $x, y \in A \Rightarrow x^{\#} \wedge y^{\#} \rightarrow (x \wedge y)^{\#} \wedge (x \vee y)^{\#} \wedge (x \rightarrow y)^{\#} \wedge (\neg x)^{\#} \in \tilde{A}$
(14) $x, y \in A \Rightarrow x^{\#} \rightarrow ((x \rightarrow y) \rightarrow ((x \rightarrow \neg y) \rightarrow \neg x) \in \tilde{A}$

Definition 4.2 (1) $x \sim y$ abbreviates $(x \rightarrow y) \wedge (y \rightarrow x)$
(2) $\vdash x$ abbreviates $x \in \tilde{A}$.

Theorem 4.1 In a propositional algebra P_1 we have:

(1) $\vdash x \rightarrow x$
(2) $\vdash x \wedge y \sim y \wedge x$
(3) $\vdash (x \rightarrow y) \wedge (y \rightarrow x) \sim (x \sim y)$
(4) $\vdash x \vee y \sim y \vee x$
(5) $\vdash x^{\#} \wedge x \wedge \neg x \rightarrow y$
(6) $\vdash y \rightarrow (x \rightarrow x)$
(7) $\vdash (x \wedge y) \wedge z \sim x \wedge (y \wedge z)$
(8) $\vdash (x \vee y) \vee z \sim x \vee (y \vee z)$
(9) $\vdash x \vee (y \wedge z) \equiv (x \vee y) \wedge (x \wedge z)$
(10) $\vdash x \wedge (y \vee z) \equiv (x \wedge y) \vee (x \vee z)$
(11) $\vdash (x \rightarrow y) \rightarrow (x \wedge z \rightarrow y \wedge z)$
(12) $\vdash x \vee (x \rightarrow y)$
(13) $\vdash ((x \rightarrow y) \rightarrow x) \rightarrow x$
(14) $\vdash (x^{\#})^{\#}$
(15) $\vdash (x \sim y) \wedge (y \sim z) \rightarrow (x \sim z)$
(16) $\vdash (x \wedge \neg x) \rightarrow y$;
(17) $\vdash x \vee (x \rightarrow y)$,
(18) $\vdash x^{\#} \rightarrow (\neg\neg x \leftrightarrow x)$;
(19) $\vdash x^{\#} \rightarrow ((x \rightarrow y) \rightarrow (\neg x))$;
(20) $\vdash (x \wedge \neg x) \leftrightarrow (y \wedge \neg y)$.

Definition 4.3 (1) $x \equiv y$ abbreviates $\vdash x \sim y$ and $x \leq y$ abbreviates $\vdash x \rightarrow y$.
We have $x \equiv y \Leftrightarrow x \leq y$ and $y \leq x$.

Theorem 4.2 We have:

(1) $\equiv$ is an equivalence relation
(2) $\leq$ is a quasi-order
(3) $x \leq y \Leftrightarrow x \wedge y \equiv x$
(4) $x \leq y \Leftrightarrow x \vee y \equiv y$
(5) $0 \leq x \leq 1$

Theorem 4.3 If $< A, \tilde{A}, \rightarrow, \wedge, \vee, \neg >$ is a propositional algebra P_1, if we define $\equiv$ and $0 =_{Def} x^{\#} \wedge \neg x^{\#}$ and if ' abbreviates $\neg$, then $< A, \equiv, \rightarrow, \wedge, \vee, 0, ' >$ is a Curry algebra P_1.

Theorem 4.4 If $< A, \equiv, \rightarrow, \wedge, \vee, 0, ' >$ is an algebra P_1, if we define $\tilde{A}$ as the set of elements x such that $x \equiv y \rightarrow y$ for some $y \in A$, and by representing the operator ' by $\neg$, the structure $< A, \tilde{A}, \rightarrow, \wedge, \vee, \neg >$ is a propositional algebra P_1.

Theorem 4.5 The concepts of Curry algebra P_1 and propositional algebra P_1 are equivalent.

Definition 4.4 A propositional algebra P_n is a structure $< A, \tilde{A}, \rightarrow, \wedge, \vee, \neg >$ where $A \subseteq \tilde{A}$ (with $A \neq \emptyset$), $\rightarrow$, $\wedge$, and $\vee$ are binary operations, and $\neg$ is a unary operator on A satisfying the following conditions ($x^{\#}$ abbreviates $x \vee \neg x$, x^1 abbreviates $x^{\#}$. x^n abbreviates $x^{\#} \wedge x^{\#\#} \wedge \ldots \wedge x^{\#\#\ldots\#}$, where the symbol $^{\#}$ occurs n times. Also, $x^{(n)}$ abbreviates $x^1 \wedge x^2 \wedge \ldots \wedge x^n$):

Conditions (1)–(12) of the Def. 4.1 plus:

(13') $x, y \in A \Rightarrow x^{(n)} \wedge y^{(n)} \rightarrow (x \wedge y)^{(n)} \wedge (x \vee y)^{(n)} \wedge (x \rightarrow y)^{(n)} \vee (\neg x)^{(n)} \in \tilde{A}$

(14') $x, y \in A \Rightarrow x^{(n)} \rightarrow ((x \rightarrow y) \rightarrow ((x \rightarrow \neg y) \rightarrow \neg x) \in \tilde{A}$

Theorem 4.6 If $< A, \tilde{A}, \rightarrow, \wedge, \vee, \neg >$ is a propositional algebra P_n, if we define $\equiv$ and $0 =_{Def} x^{(n)} \wedge \neg x^{(n)}$ and if ' abbreviates $\neg$, then $< A, \equiv, \rightarrow, \wedge, \vee, 0, ' >$ is a Curry algebra P_n.

Theorem 4.7 If $< A, \equiv, \rightarrow, \wedge, \vee, 0, ' >$ is an algebra P_n, if we define $\tilde{A}$ as the set of elements x such that $x \equiv y \rightarrow y$ for some $y \in A$, and by representing the operator ' by $\neg$, the structure $< A, \tilde{A}, \rightarrow, \wedge, \vee, \neg >$ is a propositional algebra P_n.

Theorem 4.8 The concepts of Curry algebra P_n and propositional algebra P_n are equivalent.

Theorem 4.9 In a propositional algebra P_n, $< A, \tilde{A}, \rightarrow, \wedge, \vee, \neg >$, the operator $\neg$ is not monotonic relatively to $\equiv$.

5 Conclusions

In this paper we've discussed some algebraic versions (namely trough concept of Curry algebra and propositional algebra) of the paracomplete propositional systems P_n ($1 \leq n \leq \omega$) which are dual of the systems C_n ($1 \leq n \leq \omega$) studied by Da Costa.

We've shown such 'algebraic' structures, which the fundamental relation is not the equality, but an equivalence relation (so perhaps instead of 'algebraic' structures, we can name them as 'pre-algebraic' structures). The existence of non monotonic operators regarding to the fundamental equivalence relation (which occurs in the majority of classical algebraic structures) naturally lead us to consider 'algebras' which their fundamental relation is an equivalence relation instead of equality. This is a root to follow to study algebraically some non-classical systems (among others). It is worth note that even for classical logic, the employment of pre-algebraic concept can be useful [16].

References

1. Abe, J.M.: A note on curry algebras. Bull. Sect. Logic Pol. Acad. Sci. **16**(4), 151–158 (1987)
2. Abe, J.M.: An introduction to the theory of curry systems. Scientiae Mathematicae Japonicae, Scientiae Mathematicae Japonicae, **76**(2), 175–194 (2013), :e-2013, 175–194, ISSN 1346–0862, publicado pela International Society for Mathematical Sciences, Osaka, Japão (2013)
3. Abe, J.M.: Curry algebras and propositional algebra C_1. Int. J. Knowl. Based Intell. Eng. Syst. ISSN: 1327–2314, **5**(2), 127–132 (2013)
4. Abe, J.M.: Curry algebras N_1. Atti Acc. Lincei Rend. Fis. **7**(9), 125–128 (1996)
5. Abe, J.M.: Curry algebras $P\tau$, Logique et Analyse, 161–162–163, 5–15 (1998)
6. Abe, J.M., Nakamatsu, K., Akama, S.: A Note on Monadic Curry System P_1. Lecture Notes in Computer Science, vol. 5712, pp. 388–394. Springer, Berlin/Heidelberg (2009)
7. Abe, J.M., Nakamatsu, K., Akama, S.: A Note on Monadic Curry System P_1. Lecture Notes in Computer Science—LNAI, vol. 5712, pp. 388–394, ISSN 0302–9743. Springer, Berlin/Heidelberg, Alemanha (2009). doi:10.1007/978-3-642-04592-9_49
8. Abe, J.M., Nakamatsu, K., Akama, S.: An Algebraic Version of the Monadic System C_1. New Advances In Intelligent Decision Technologies, Series: Studies in Computational Intelligence, vol. 199, pp. 341–349. Springer, Berlin/Heidelberg (2009)
9. Abe, J.M., Nakamatsu, K.: Curry systems for algebraization of some non-classical logics. Int. J. Reasoning-Based Intel. Syst. 59–69, ISSN: 17550556, , Inderściences Enterprize, UK (2011)
10. Abe, J.M., Akama, S., Nakamatsu, K.: Monadic Curry Algebras $Q\tau$, Lecture Notes in Computer Science, vol. 4693, pp. 893–900. Springer (2007)
11. Barros, C.M., da Costa, N.C.A., Abe, J.M.: Tópico de teoria dos sistemas ordenados: vol. II. Sistemas de Curry, Coleção Documentos, Série Lógica e Teoria da Ciência, IEA-USP **20**, 132p (1995)
12. Curry, H.B.: Foundations of Mathematical Logic. Dover, New York (1977)
13. da Costa, N.C.A., Marconi, D.: A note on paracomplete logic. Atti Acc. Lincei Rend. Fis. **80** (8), 504–509 (1986)
14. da Costa, N.C.A.: Logics that are both paraconsistent and paracomplete. Atti Acc. Lincei Rend. Fis. **83**, 29–32 (1990)
15. da Costa, N.C.A.: On the theory of inconsistent formal systems. Notre Dame J. Formal Logic **15**, 497–510 (1974)
16. Eytan, M.: Tableaux of Hintikka et Tout ça: un Point de Vue Algebrique. Math. Sci. Humaines **48**, 21–27 (1975)
17. Halmos, P.R.: Algebraic Logic. Chelsea Publishing Co., New York (1962)
18. Kleene, S.C.: Introduction to Metamathematics. Van Nostrand, Princeton (1952)
19. Mortensen, C.: Every quotient algebra for C_1 is trivial. Notre Dame J. Formal Logic **21**, 694–700 (1977)
20. Rosenbloom, P.C.: The Elements of Mathematical Logic, Dover (1950)

A Cloud-Based Vegetable Production and Distribution System

Alireza Ahrary and D.A.R. Ludena

Abstract The new paradigm of Big Data and its multiple benefits have being used in the novel nutrition-based vegetable production and distribution system in order to generate a healthy food recommendation to the end user and to provide different analytics to improve the system efficiency. As next step in this study, a new multidimensional matching algorithm was proposed in order to provide the end user with the best recommendation. The new multidimensional matching algorithm is 10 times faster than the standard algorithm based on the test with sample data. Also, different version of the user interface (PC and Smartphone) was designed keeping in mind features like: easy navigation, usability, ergonomics, etc.

Keywords Big data · Computer science · Data systems · IoT · Data analysis

1 Introduction

IoT share a common agreement regarding its definition, as we could express it as the seamlessly integration of internet-based sensors and devices in a wide area network that interact with a much more advanced Personal Area Network, allowing us to recognize in a much more detail manner the surround environment and interchange information with it, in an automatic manner. The future applications and research based on IoT will aver a profound impact in the user side, since most of its application will be in areas like: domotics, health, agriculture, intelligent services, etc.

A. Ahrary (✉) · D.A.R. Ludena
Faculty of Computer and Information Sciences, Sojo University, Kumamoto, Japan
e-mail: ahrary@cis.sojo-u.ac.jp

R. Neves-Silva et al. (eds.), *Intelligent Decision Technologies*,
Smart Innovation, Systems and Technologies 39,
DOI 10.1007/978-3-319-19857-6_2

Although IoT research is in its early stage in terms of development, it represents a challenge from the ethical and technological point of view.

Standardization is a mayor issue in IoT, different companies as well as independent organizations has tried to solve this problem with no success so far, or with no general agreement about a single methodology [1–5].

2 ICT Agriculture in Japan

Only industrialized countries were able to afford the high cost required to implement ICT solutions in agriculture. In that context Japan was one of the first countries to adopt and research about the introduction and use of ICT in Agriculture.

Japan represents a unique set of characteristics in the case of farming, among several we can extract the most important:

1. Difficult Geography

Japanese geography is one of the most challenging in the world due to the mountainous characteristic of the land. Japanese farmers had to create innovative methods in order to take the most of the restricted area useful to grow crops.

2. Farmers' aging society

One of the biggest problems for farmers and for the Japan in general is the fast peace growing age society. Specifically, farmers are used to pass their knowledge from generations, but the current scenario is leaving farmers with few or no generations, since their kids decide to migrate to the city on order to find a working position in manufacturing companies. This phenomenon is reducing at an alarming peace the number of young farmers putting in risk the continuity of the business.

3. Globalization

Japanese agriculture has key products in the year, their high quality means high prices in most of them and together with a strict quality control of the produce by the governmental regulator, creates a challenging scenario for farmers all around Japan. This scenario selects the best products at a high price but the ones that slightly do not accomplish its high standards are not selected and farmers have to find the way to commercialize these products.

The former scenario is becoming even more challenging since the Japanese government decided to sign different Free Trade Commerce (FTC) Agreements with different countries, specially in the Pacific Rim area, to provide Japan with different types of produce at lower prices than the local farmers. This increasing competition is making even more challenging the scenario for Japanese farmers.

3 The Academic Point of View

Japanese Academic institutions decided to use cutting edge technologies in order to provide the better solutions in the near future. Among those technologies are;

1. Sensor Networks

The use of different specific-application designed sensor in a network array connected to a high-speed networks, allow information exchange among different business partners. This sensor network is the core of multiple application layers that will be built over the information provided.

2. Cloud Computing

Many current applications, from storage to Software As a Service (ASA) are using this common affordable platform in order to reduce their hardware expenses. The information retrieved from the sensor networks could be processed or stored in the cloud for future purposes.

3. Augmented Reality (AR) Solutions

Using AR solutions, farmers could retrieve information on their lands effortless and in real-time. The crop information is retrieved from the sensor networks. A good example of this is the use on the field of Google Glasses, which can use the retrieved information previously stored in cloud-based databases, or to process it as well in the cloud.

4. Use of Unmanned Air Vehicles (UAV)

For supervisory/monitory purposes a new trend is the use of UAVs. Among the supervisory applications are: monitor plant growth, insect pests, plant and animal diseases, natural disasters, etc. the low-cost of this solution could serve as a good replacement for satellite-based applications.

5. Control Area Network (CAN)

The data generated by the different agriculture machines, e.g. yield of pesticides, production, etc., can be stored in the cloud as well.

The main disadvantage in the case of Academic ICT for Agriculture projects is the development cost and cost performance. For small size lands, the development cost is high compared to their large size counterparts and the cost performance is low. Another problem academic projects faced is the lack of long term research funding. Due that a highly politicized area like Agriculture in Japan is, research funding suffers the same issue.

4 Project Description

Farmers have been facing different challenges related to their business due to different reasons. In order to solve part of their problems and to create a new business platform the project "Novel nutrition-based vegetable production and distribution system" was created. The initiative of the project was to help farmers with their produce commercialization through the use of technology [7–10].

The creation of this innovative business model required the execution of different steps shown in this section.

Members and their functions and interconnections are defined as a first step.

1. Farmers
   In addition of representing the main project's benefits users, farmers represent the most important information provider. The information provided is vital in order to make the project work.

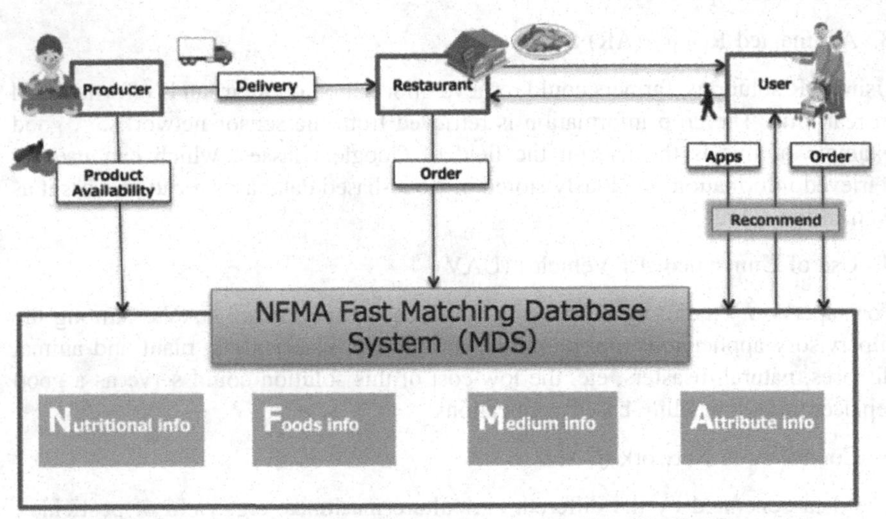

Fig. 1 Project perspective

2. End Users
   Although farmers are the ones that will benefit from the project, en users will be the ones taking most of the total benefits. Since the project is based on the use of the platform by end users.
3. Restaurants
   The "Ready meal" option represent here by the restaurants uses the vegetables provided by local farmers.

4. Knowledge Based Database Creators
   Two (2) databases were created during the project development.
   4.1. Nutritional requirement information
        User's nutritional information, e.g.: user's physical information, status information, physical condition, nutritional requirement, etc.
   4.2. Food information
        Food/vegetable information, like: nutritional calories, traceability, seasonal information, etc.
5. Project Integrator
   Sojo University represents the Project Integrator, meaning it will receive the different information presented previously and generate two additional databases:
   5.1. Platform information
        User's device details e.g.: mobile, tablet, Web, downloaded application, etc.
   5.2. Attribute information
        Information used at the registration procedure, e.g.: gender, age, family structure, etc.
   5.3. Recommendation generation algorithm
        A recommendation algorithm is designed in order to correlate the different created databases previously mentioned. Its objective is to generate a recommendation to relieve the user's symptoms based on the information provided by the user.

A project perspectivc is shown in Fig. 1.

5 Possible Threats Against IOT

For the specific purposes or the project we focus on the security analysis on two main aspects: authentication and data integrity.

Authentication and data integrity are two of the major problems related to IoT. Authentication is difficult in IoT due to its infrastructure requirements and servers that perform the authentication process through the information exchange with other nodes. In an IoT environment this could result not practical due to that RFID tags cannot exchange too many messages with the authentication servers because their issues related to energy management and messages standardization, this issue applies as well to sensor networks. Energy issues are one of the most difficult to overcome in RFID networks as well as sensor networks, due that there are scattered over a wide area and sometimes unmonitored, energy management is a key factor in order to ensure a long device life as well as usability. In the same manner, some authentication protocols could not be used due to their lack of standardization.

In this context, several approaches were developed for sensor networks [78 reference paper]. In these cases gateways that are part of the sensor networks are required to provide connection to the Internet. In the IoT scenario sensor nodes must be seen as nodes in the Internet, therefore authentication is required in order to differentiate them from sensors in the same area but not belonging to the same network.

In the case of RFID several approaches were presented, but most of them have serious issues, some of them mentioned in [6].

The "Man in the middle" attack is considered one of the biggest threats against wireless networks as well to IoT networks. Data integrity solutions should guarantee that the data in transaction cannot be modified and the system must be able to detect this situation. Data integrity as a issue has been extensively studied for standard network applications and communication systems and early results are related to sensor networks [11]. But, when a RFID networks with their own unique characteristics are included in the current Internet paradigm, different problems arise as well as unforeseen problems related to their use. Several approaches are developed or under research to solve the different new RFID related issues i.e. EPCGlobal Class-1 Generastion-2 and ISO/IEC 18000-3, both of them working in different process to protect the device memory. These approaches also consume large amount of the resources in encryption processes needed. The main used resources are: energy and bandwidth, both of them in the destination. Therefore, even using these approaches specific related problems with RFID still remain.

6 Matching Algorithm

Relational database (RDB) represent a one data management method that is most commonly available in the data warehouse (DWH) database. Relational databases (RDB) stores data in, such as with spreadsheet software Excel Table (Table) structure. The table is a "column pattern" specifying the data item. If, for example, food and beverage stores sales specification, sales date and store name, time stamp, membership number, customer name, primary contact, cashier, menu, quantity, tax amount, consumption tax, such as the total amount of money, etc. represent the data items. Each of the record in a form of a value represents a "row pattern".

Large amount of data is accumulated in time series in data warehouses (DWH). When data is analyzed, a pattern is often uses for that purpose. Focusing on the former characteristic, the speed of the aggregation process is an important issue. The proposed algorithm uses a space-filling curve version of the three-dimensional Hilbert curve (Fig. 2a). When storing a multidimensional value, it calculated whether the attribute value is the ordinal number on the Hilbert space-filling curve and store it as a key. Next, generated keys are stored in a three-dimensional space to perform a multi-dimensional search procedure (Figs. 3 and 4).

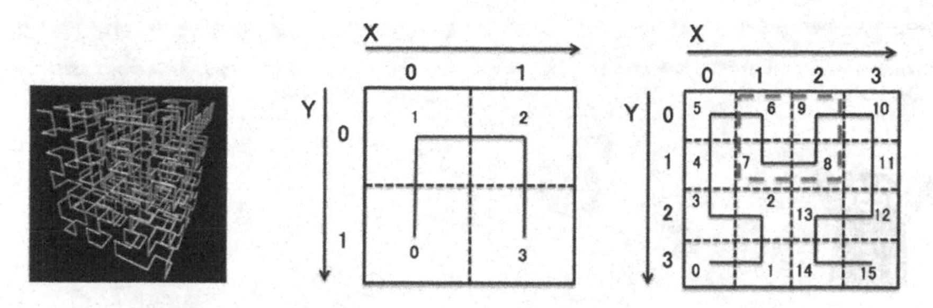

Fig. 2 (a) 3D Hilbert Curve (b) 2D Hilbert Curve (c) Search range example

Furthermore, close areas properties are defined and indexed in one table. Seeking the maximum key value in the search range, and the output is mapped into one dimension.

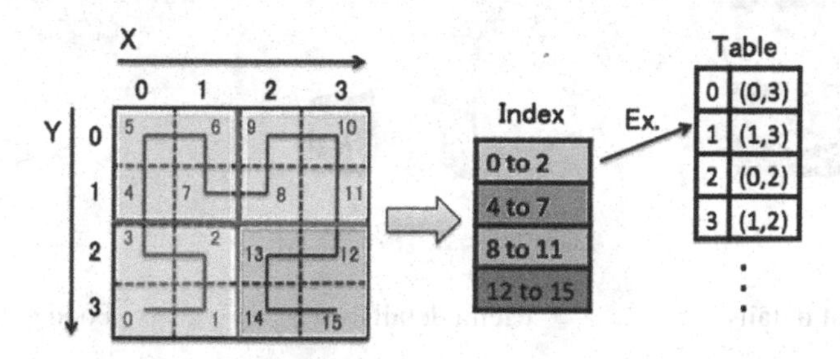

Fig. 3 Table registration

The proposed algorithm performed 10 times faster than the conventional method (KVS-HBase).

In addition, it is possible to dynamically register: restaurants, recipes (independent of the number), ingredients, and dynamically perform associations between each database (Independent of the number) to design the cloud-based system. Having an increase in the registered information will not require any modification in the system, making it elastic. 25 users, 20 recipes and its association with 15 food items are the target numbers in this stage of the project, The cloud-based system is designed to analyze the increment of the user (consumer) load, implementing load balancing procedures and a Web server.

For server administration purposes, a load balancing was implemented and the server specs increased.

Also, aware of the load balancing issue, a Master-Slave DB (Database) architecture was implemented. The Master DB performed the daily search and update procedures, meanwhile the Slave DB performed high-speed data retrieval. Implementing the MDS System in the server, the system was ready for the pilot stage.

Fig. 4 User interface designs

7 Interface Design

Because of the limited screen size on the Smartphone edition of the user interface, priority was given to the user's information input.

A high visibility and easy to navigate interface was designed with all navigation features like: search from symptoms list and nutrients. Extensive use of Java Script from the second navigation layer. For the PC version, it was designed in order to allow making an easy icon selection with a reduced directory structure. For display purposes, e.g.: food, nutrients, symptomology, etc. that contains large contents a matching program was built in order to provide specific information to the user's needs.

8 Conclusions and Future Work

Farmers have been facing some difficulties regarding their business because of weather change, price instability and constant financial problems due to foreign competition. Therefore, an expensive service like this could be only available to big producers in which case, markets are completely ensured long time contracts.

Academic and Company based research initiatives haven shown different issues that jeopardized their continuity, which affects directly to the farmer. Being the biggest issue in academic research: budget and availability of multidisciplinary human resources needed for the correct project development; and in the case of company based research, their cost, which could not be affordable by the farmer.

Big Data analytics allowed us to have a better understanding of the user needs and requirements regarding the project objectives.

The new IoT parading is becoming a part of the current and future Internet. The current Internet paradigm will drastically change into a much more personalized experienced with our surroundings that will lead to much more richer experience.

In the case of applications, is necessary to make a deep analysis of the requirements needed to be implemented, in order to specifically use what could result much more beneficial for a specific project purpose.

In the case of our project, IoT could bring not only automatic updates regarding key issues of the project, but allow us to drastically improve the user experience through different new services.

IoT Security as a new paradigm and new potentially useful research area must address its security concerns in order to show a more reliable platform for the user that, currently, could resist a functional initiative. Due that a single security incident could damage permanently the infinite benefits IoT could bring to the society.

A new multidimensional matching algorithm was proposed in order to provide the end user with the best recommendation. The new multidimensional matching algorithm is 10 times faster than the standard algorithm based on the test with sample data.

A redundant Master-Slave Database system is use in order to provide the maximum efficiency to the cloud-based system.

Different version of the user interface (PC and Smartphone) was designed keeping in mind features like: easy navigation, usability, ergonomics, etc.

Acknowledgments This work was supported by New Energy and Industrial Technology Development Organization (NEDO) in "IT Integration-based New Social System Development and Demonstration Projects", Grant No. 12102069.

References

1. Juels, A.: RFID security and privacy: a research survey. IEEE J. Sel. Areas Commun. **24**(2), 381–394 (2006)
2. Floerkemeier, C., Bhattacharyya, R., Sarma, S.: Beyond RFID. In: Proceedings of TIWDC 2009, Pula, Italy, September 2009
3. Sung, J., Sanchez Lopez, T., Kim, D.: The EPC sensor network for RFID and WSN integration infrastructure. In: Proceedings of the Fifth IEEE International Conference on Pervasive Computing and Communications Workshops, pp .618–621, March 19–23, 2007
4. Commission of the European Communities, Early Challenges Regarding the "Internet of Things" (2008)
5. Kushalnagar, N., Montenegro, G., Schumacher, C.: IPv6 over low-power wireless personal area networks (6LoWPANs): overview, assumptions, problem statement, and goals. In: IETF RFC 4919, Aug 2007
6. Weiser, M.: The computer for the 21st century. ACM SIGMOBILE Mob. Comput. Commun. Rev. **3**(3), 3–11 (1999)
7. Ludeña, D.A.R., Ahrary, A.: Big data approach in an ICT agriculture project. In: Proceedings for the 5th IEEE International Conference on Awareness Science and Technology (iCAST 2013), pp. 261–264. Aizu-Wakamatsu, Japan (2013)
8. Ludeña, D.A.R., Ahrary, A.: A big data approach for a new ICT agriculture application development. In: Proceedings for the 2013 International Conference on Cyber-Enabled Distributed Computing and Knowledge Discovery (CyberC 2013), pp. 140–143. Beijing, China (2013)
9. Ahrary, A., Ludeña, D.A.R.: Big Data approach to a novel nutrition-based vegetable production and distribution system. In: Proceedings of the The International Conference on Computational Intelligence and Cybernetics—CyberneticsCom 2013, pp. 131–135. Yogyakarta, Indonesia (2013)
10. Acharya, R., Asha, K: Data integrity and intrusion detection in wireless sensor networks. In: Proceedings of IEEE ICON 2008, New Delhi, India, Dec 2008
11. Research Trends: Special Issue on Big Data, Elsevier, Issue 30, Sept 2012

Capacity Planning in Non-uniform Depth Anchorages

Milad Malekipirbazari, Dindar Oz, Vural Aksakalli, A. Fuat Alkaya and Volkan Aydogdu

Abstract Commercial vessels utilize anchorages on a regular basis for various reasons such as waiting for loading/unloading, supply, and bad weather conditions. Recent increase in demand for anchorage areas has mandated a review of current anchorage planning strategies. In particular, current state-of-the-art anchorage planning algorithms assume that the anchorage areas are of uniform depth, which is quite unrealistic in general. In this study, we introduce an algorithmic modification to current anchorage planning methods that takes into account non-uniformity of anchorages. By exploiting the depth non-uniformity, our algorithm significantly improves the number of vessels that can be accommodated in an anchorage and it can easily be incorporated into existing anchorage capacity planning decision support systems.

Keywords Maritime transportation · Anchorage · Optimization algorithm · Decision support system

1 Introduction

Maritime transportation accounts for about 90 % of world's commerce and there are currently more than a hundred thousand commercial vessels operating in the world [1]. An important component in dealing with maritime traffic congestion is anchorages that serve as a temporary waiting area for vessels for various purposes such as land services (fueling, legal issues, repairs, etc.), loading/unloading of cargo, and as

M. Malekipirbazari · V. Aksakalli (✉)
Department of Industrial Engineering, Istanbul Sehir University, 34662 Istanbul, Turkey
e-mail: aksakalli@sehir.edu.tr

D. Oz · A.F. Alkaya
Department of Computer Engineering, Marmara University, 34722 Istanbul, Turkey

V. Aydogdu
Maritime Faculty, Istanbul Technical University, 34940 Istanbul, Turkey

R. Neves-Silva et al. (eds.), *Intelligent Decision Technologies*,
Smart Innovation, Systems and Technologies 39,
DOI 10.1007/978-3-319-19857-6_3

a refuge from bad weather conditions. As maritime trade routes proliferate and the demand for global seaborne shipping increases, efficient operation of anchorages has become a crucial task.

An important problem in anchorage planning is determining the optimal berth (i.e., anchorage) locations for incoming vessels such that the number of vessels that can be placed inside the anchorage is maximized. This capacity planning problem has been considered in previous studies, yet the prevailing underlying assumption in these studies is that the anchorage areas are of uniform depth, which is almost never the case in reality. In particular, portions of anchorages close to land are shallower and portions further away from the land are much deeper.

Our goal in this work is to contribute to anchorage planning research by exploiting depth non-uniformity for better utilization of anchorage areas. Specifically, we present a heuristic modification to an existing capacity planning algorithm that takes advantage of depth non-uniformity and drastically improves the number of vessels that can be placed inside the anchorage area. We evaluate our algorithm using real-world data from the Ahırkapı Anchorage in Istanbul as well as synthetic data obtained via Monte Carlo simulations. Ahırkapı Anchorage is located at the southern entrance of the Istanbul Strait and plays an important role in the overall efficiency of maritime traffic in the Istanbul Strait.

2 Problem Description and Previous Work

Environmental forces such as waves, winds, and sea currents play an important role in anchorages as non-moving vessels are much more susceptible to such forces. Thus, once a vessel drops an anchor, its exact location is largely determined by the prevalent environmental forces. On the other hand, such forces are are quite dynamic and they change frequently over time. Nonetheless, we can compute a safe maneuvering circle using the anchor position and depth of the anchorage via simple geometry as shown in Fig. 1. The radius of this circle is essentially a function of three components: the sea depth D at the anchor position, the vessel's length L, and anchor chain length. A practical formula for anchor chain length is $25\sqrt{D}$ [4], which yields the following formula for anchorage circle radius per the Pythagorean Theorem:

$$\text{Anchorage Circle Radius} = L + \sqrt{(25\sqrt{D})^2 - D^2}\ . \tag{1}$$

In order to prevent potential vessel collisions, anchorage circles must have no intersections with one another. Thus, the problem of maximizing capacity of an anchorage can be modeled as the classical NP-Hard problem of packing disks with different radii inside a polygonal area [2]. In this regard, previous studies concentrated on the improvement of capacity via a number of methodologies including an algorithm using a look-ahead method combined with beam search [3], an iterated tabu search [5], a coarse-to-fine quasi-physical optimization method [6], and an

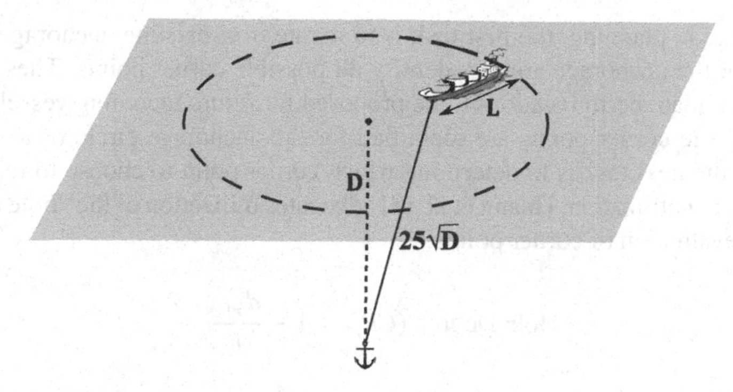

Fig. 1 Illustration of the anchorage circle associated with an anchored vessel

action-space-based global optimization approach for the problem of packing unequal circles into a square container [7].

The current state-of-the-art in anchorage planning is the MHDF Algorithm proposed in Huang et al. [8] based on the notion of corner points evaluated via Monte Carlo simulations. This algorithm is heuristic in nature without any optimality guarantees. The simulation tool and the algorithm introduced therein were later utilized in a marine traffic simulation system for hub ports [9]. Another simulation-based approach was presented in Shyshou et al. [10] where the authors investigated the optimal number of tug ships needed to move an oil rig with respect to equipment and weather related constraints. The algorithm we propose in this study is an extension of the MHDF Algorithm that exploits depth non-uniformity. Thus, our algorithm is also heuristic in nature. We are not aware of any previous studies on anchorage capacity planning that consider any depth non-uniformity aspects of the problem.

3 Corner Points and the MHDF Algorithm

A corner point CP_i associated with an anchorage circle i is defined as the point where the circle is tangent to at least two elements when its center is located at CP_i (see, e.g., [8]). These elements can either be the existing anchored vessels represented as circles or sides of the anchorage polygon. Corner points are classified into three categories with respect to the element types they are tangent to. These categories are illustrated in Fig. 2 and described below:

1. Side-and-Side (SS) Corner Points: The circle center is tangent to any two sides of the anchorage area.
2. Side-and-Circle (SC) Corner Points: The circle center is tangent to an existing anchorage and a side of the anchorage area.
3. Circle-and-Circle (CC) Corner Points: The circle center is tangent to two existing anchorage circles.

In anchorage planning, the first task is to iterate over existing anchorage circles and sides of the anchorage area to identify all possible corner points. These points serve as candidate berth locations to be proposed for future incoming vessels.

Once all the corner points are identified for the anchorage circle of an incoming vessel, the next task is to determine which corner point to choose to maximize anchorage area utilization. Huang et al. [8] advocates utilization of the "hole degree" metric for evaluation of corner points:

$$\text{Hole Degree } (CP_i) = 1 - \frac{d_{min}}{r} \tag{2}$$

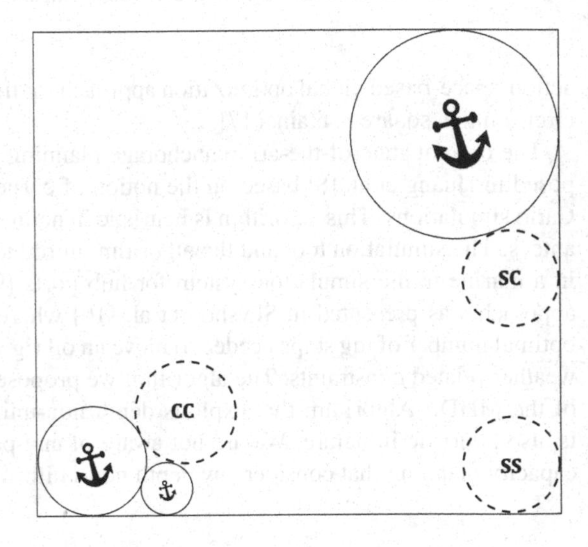

Fig. 2 The tree different types of corner points

where r is the radius of the circle associated with CP_i and d_{min} is the minimum distance from CP_i to the other circles and sides of the polygonal except the two closest item tangent to CP_i.

The Maximum Hole-Degree (MHDF) Algorithm of Huang et al. [8] starts with placing two discs at two corners of the anchorage area and puts the remaining discs one at a time by selecting the corner point with the highest hole degree. When no more corner points can be identified, the algorithm terminates and declares the anchorage area to be full. This algorithm operates under the assumption that the anchorage area is of uniform sea depth. In the next section, we show how this algorithm can be extended to account for non-uniform depth anchorages for better utilization of the anchorage area.

4 The Non-uniform MHDF Algorithm

4.1 A Model for Non-uniform Depth Anchorage Areas

We model non-uniform depth anchorage areas as a partitioning of the entire anchorage area into uniform-depth polygonal segments. This model is illustrated in Fig. 3 for the Dangerous Cargo Zone of Ahırkapı Anchorage. In a polygonal segment, radii of the anchorage circles are determined by the segment's sea depth as can be seen in Eq. 1. In addition, the segment depth determines the maximum length for vessels that can safely be placed inside the segment, as explained below.

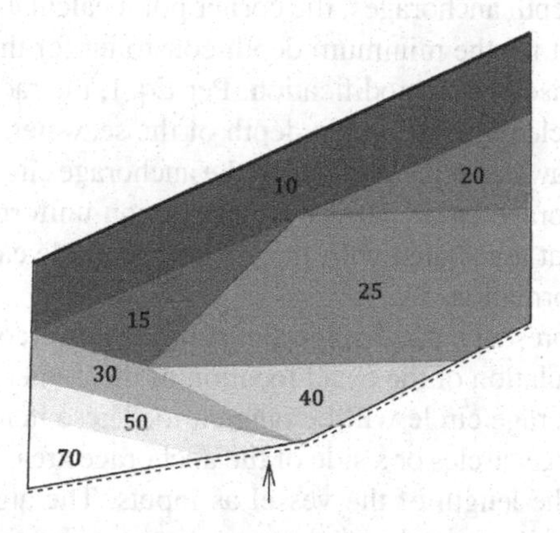

Fig. 3 The Ahırkapı Dangerous Cargo Anchorage Zone. The arrow shows the entry side of the zone. The numbers denote the sea depth in each polygonal sub-segment in meters

In seamanship terminology, *draft* refers to the depth of the ship from water line to the bottom of the ship which is typically equal to the length of the ship divided by 10.5 for tankers and 11.75 for bulk carriers and container ships [11]. The minimum sea depth that a vessel with a particular length can safely anchor can then be calculated as

$$\text{Minimum } D = \text{Draft} \times \text{UKCF} = \frac{L \times \text{UKCF}}{(10.5 \text{ or } 11.75)} \tag{3}$$

where UKCF stands for Under Keal Clearance Factor. A practical UKCF value for ships at anchorage is 1.5. For example, a tanker with a length of 100 m requires a sea depth of at least 14 m for safe anchorage. Thus, the longest vessel that can safely anchor at depth D can be calculated as follows:

$$\text{Maximum } L = \frac{D \times (10.5 \text{ or } 11.75)}{\text{UKCF}}. \tag{4}$$

Therefore, for each polygonal segment, there exist a maximum vessel length constraint as determined by Eq. 4. In our model, we take into account this constraint and we do not allow vessels anchor at segments that are not deep enough. Specifically, during the corner point identification process, we only consider corner points that are inside polygonal segments deep enough for the incoming vessel.

4.2 Corner Point Calculation in Non-uniform Depth Anchorages

In the MHDF Algorithm, all possible corner points are calculated and the one with the highest hole degree is chosen as the berth location for an incoming vessel. In the case of non-uniform depth anchorages, the corner point calculation process needs to be modified to account for the minimum depth constraint for the incoming vessels. A critical challenge arise in this modification: Per Eq. 1, the radius of the incoming vessel's anchorage circle depends on the depth of the seawater. Thus, we must first find the depth of the seawater at the position of the anchorage circle and then calculate the radius of the anchorage circle. However, due to non-uniforms depths, the exact position of corner point associated with the anchorage circle cannot be determined without the radius information.

In this section, we present a three-stage algorithm that overcomes the above challenge by iterative calculation of the exact location of the corner point given the pair of items that the anchorage circle will be tangent to. These items can be either one of the existing anchorage circles or a side of the anchorage area. The algorithm takes those two items and the length of the vessel as inputs. The algorithm is then executed for every pair of items in the anchorage area in order to calculate all possible anchorage circles for an incoming vessel. Once this three-stage algorithm is run and all the feasible corner points are identified, as in MHDF, we declare the corner point with the highest hole degree as the berth location for the incoming vessel. The MHDF Algorithm with this three-stage corner point calculation algorithm (in the case of non-uniform depths) is called the Non-uniform MHDF Algorithm, NU-MHDF Algorithm in short. The three-stage corner point calculation algorithm is described below.

1. **Initial Placement**: The algorithm starts with an initial placement of the corner point CP_i in the deepest polygonal segment in the anchorage area.
2. **Anchorage Circle Adjustment**: After the initial placement, the algorithm finds the depth segments that the anchorage circle at the corner point has intersections with, which we denote by X_i. Due to the minimum sea depth constraint, the algorithm takes the depth of the deepest segment in X_i and adjusts the radius of the anchorage circle accordingly. After the radius is updated, it repositions the corner point with the new radius and calculates the set X_i again. This loop continues until the deepest depth segment with which the anchorage circle has intersection with does not change in the next iteration or a depth value is obtained that has been previously calculated. The pseudo-code for this adjustment algorithm is given below.

Anchorage Circle Adjustment Algorithm

```
Function AdjustAnchorageCircle(Initial_Anchorage_Circle)
begin
    calculated_depth_list = empty_list
    c = Initial_Anchorage_Circle
    previous_depth = maximum_depth_in_anchorage_area
    while (!converged)
       depth = depth_of_deepest_segment_that_c_intersects
       if (depth == previous_depth)
          // Depth does not change any more
          converged = true
       else (depth in calculated_depth_list)
          // Loop detected. Stop.
          depth = maximum depth in calculated_depth_list
          converged = true
       else
          previous_depth = depth
          add depth to calculated_depth_list
          new_radius = calculateRadiusFromDepth(depth)
          c = reposition_anchorage_circle(new_radius)
       end if
    return c
    end while
end
```

3. **Anchorage Circle Validation**: The anchorage circle at the converged corner point needs to be checked for validity. In particular, the following conditions are checked:
   (a) **Boundary Condition**: Anchorage circle is inside the anchorage area.
   (b) **No Overlapping Condition**: The circle does not overlap with the existing anchorage circles.
   (c) **Maximum-Allowable-Length Condition**: As mentioned earlier, the longest vessel that can safely anchor at a particular polygonal segment can be calculated by Eq. 4. As opposed to radius calculation, this time we use the depth of the shallowest segment in X_i to calculate the maximum allowable vessel length to check for feasibility. We use the shallowest intersected segment for safety reasons, because if the length of the vessel satisfies the allowable length criteria for the shallowest segment, then it satisfies this criteria for all other segments in X_i.

4.3 Proof of Convergence

We now show that the Anchorage Circle Adjustment process above converges in a finite number of steps. Let $d_1, d_2, ...d_N \in X$ denote the finite set of depth segments in the anchorage A and d_{max} be the deepest segment in A. Let I_1 and I_2 be the two items for which the algorithm needs to find the anchorage circle. The algorithm computes the place of the anchorage circle tangent to I_1 and I_2 in A with the depth segments of X (i.e., A, X, I_1 and I_2 are the inputs).

Let $r_{initial}$ be the radius of the initial circle $C_{initial}$ calculated for I_1 and I_2 with depth d_{max}. Let C_i denote the anchorage circle calculated at the i–th iteration where $i > 0$ and r_i denotes its radius. Note that C_1 equals $C_{initial}$ and r_1 equals $r_{initial}$ according to this definition. Let X_i be the set of depth segments intersecting with C_i. Since the entire anchorage area is comprised by the depth segments in X, X_i cannot be empty. Let d_{max_i} denote the deepest segment in X_i. Finally, let CDL_i denote the calculated depth list at i–th iteration. CDL_i is the set of depth segments which are calculated as d_{max_j} at some j–th iteration where $j < i$. Notice that CDL_1 is simply the empty set as assigned in the algorithm.

At each iteration, the while loop has three branches listed below, two of which results in convergence.

1. If $d_{max_i} = d_{max_{i-1}}$, the algorithm converges. Note that $d_{max_0} = d_{max}$.
2. Else if $d_{max_i} \in CDL_i$, the algorithm converges.
3. Else
   $CDL_{i+1} = CDL_i + d_{max_i}$
   r_{i+1} = new radius calculated from d_{max_i}
   C_{i+1} = new circle calculated from r_{i+1}
   And proceed to $(i + 1)$-th iteration.

In order for the algorithm to diverge and execute indefinitely, the third branch listed above must be executed infinitely many times. To prove by contradiction, suppose the algorithm executes the third branch for infinitely many times for a particular instance of A, I_1, I_2, and X. Observe that CDL_i gets expanded by one element at each iteration when the loop falls into the third branch. Also note that CDL_i is the subset of X. If the algorithm falls into the third branch infinitely many times for this instance, then the set CDL_i must be expanded infinitely which is a contradiction since X is assumed to be a finite set. Thus, we conclude that the Anchorage Circle Adjustment algorithm converges in a finite number of iterations.

5 Computational Experiments

This section presents a comparison of NU-MHDF against regular MHDF via Monte Carlo simulations using three different vessel length distributions on three different real-world anchorage topologies. Throughout our experiments, the anchorage circle radii in MHDF are calculated with respect to the deepest sea depth inside the anchorage as any other depths would result in the dangerous situation of overlapping anchorage circles.

The vessel length distributions used in our experiments are (1) the historical vessel length distribution in the Ahırkapı Anchorage in 2013, (2) the uniform distribution between 25 and 250 m, and (3) Gamma distributed lengths between 25 and 250 m with shape parameters of 4 and 7 respectively (with a mean of about 100 m) that closely fits the historical Ahırkapı data.

The anchorage topologies we consider are the Harbor Approach, Departure Long Stay, and Dangerous Cargo anchorage zones inside the Ahırkapı Anchorage Area. We implemented a comprehensive and realistic simulation environment for comparison of the algorithms. In particular, we used common random numbers for both of the algorithms so that they were tested on the exact same sequences of vessel arrivals.

We conducted a total of 100 Monte Carlo simulations for each anchorage topology/vessel length distribution combination. In each simulation, we started with an empty anchorage and terminated the simulation when the anchorage became full. Table 1 summarizes the results of these experiments. The column *Improvement(%)* represents the percentage improvement of NU-MHDF over MHDF in terms of the total number of vessels accommodated at the end of the simulation. We observe that on the average, NU-MHDF outperforms regular MHDF by 74.1 % while this improvement can be as high as 210 %.

Table 1 Comparison of regular MHDF versus NU-MHDF

		Total number of vessels		
		MHDF	NU-MHDF	Improvement (%)
Historical data	Zone-A	63.0	**69.0**	**9.5**
	Zone-B	118.6	**153.0**	**29**
	Zone-C	40.4	**85.0**	**110.4**
Uniform	Zone-A	85.8	**111.7**	**30.2**
	Zone-B	128.2	**171.7**	**33.9**
	Zone-C	50.0	**154.8**	**210.1**
Gamma	Zone-A	84.3	**103.1**	**22.3**
	Zone-B	125.6	**162.7**	**29.5**
	Zone-C	50.0	**146.2**	**192.4**
Average		82.9	**128.6**	**74.1**

6 Summary and Conclusions

Efficient utilization of anchorage areas is critical in effective management of maritime traffic congestion. This study proposes an anchorage planning strategy, called NU-MHDF, with the goal of maximizing utilization by exploiting non-uniformity in sea depths inside the anchorages. The NU-MHDF Algorithm is an extension of the MHDF Algorithm that starts with the deepest polygonal segment inside the anchorage area and iteratively adjusts the corner point locations until convergence. We provide a proof that the algorithm converges in a finite number of steps. Our simulation results with three different vessel length distributions and three different anchorage zones show that NU-MHDF consistently yields much higher utilization levels compared to MHDF. The NU-MHDF Algorithm stands as a high-performing anchorage planning strategy that can be integrated into existing anchorage planning decision support systems in a straightforward manner.

Acknowledgments This work was supported by The Scientific and Technological Research Council of Turkey (TUBITAK), Grant No. 113M489. We would like to thank Turkish Directorate General of Coastal Safety for providing us the historical traffic data for the Ahırkapı Anchorage.

References

1. AGCS: Safety and shipping 1912–2012 from Titanic to Costa Concordia. Technical Report, Allianz Global Corporate & Specialty (2013)
2. Akeb, H., Hifi, M.: Algorithms for the circular two-dimensional open dimension problem. Int. Trans. Oper. Res. **15**(1), 685–704 (2008)
3. Akeb, H., Hifi, M.: Solving the circular open dimension problem by using separate beams and look-ahead strategies. Comput. Oper. Res. **40**(5), 1243–1255 (2013)
4. Danton, G.: The Theory and Practice of Seamanship. Routledge & Kegan Paul, London, UK (1996)
5. Fu, Z., Huang, W., Lu, Z.: Iterated tabu search for the circular open dimension problem. Eur. J. Oper. Res. **225**(2), 236–243 (2013)
6. He, K., Mo, D., Ye, T., Huang, W.: A coarse-to-fine quasi-physical optimization method for solving the circle packing problem with equilibrium constraints. Comput. Ind. Eng. **66**(4), 1049–1060 (2013)
7. He, K., Huang, M., Yang, C.: An action-space-based global optimization algorithm for packing circles into a square container. Comput. Oper. Res. **58**(1), 67–74 (2015)
8. Huang, S., Hsu, W., He, Y.: Assessing capacity and improving utilization of anchorages. Transp. Res. Part E Logistics Transp. Rev. **47**(2), 216–227 (2011)
9. Huang, S., Hsu, W., He, Y., Song, T.: A marine traffic simulation system for hub ports. In: Proceedings of ACM SIGSIM Conference on Principles of Advanced Discrete Simulation. pp. 295–304 (2013)
10. Shyshou, A., Gribkovskaia, I., Barcel, J.: A simulation study of the fleet sizing problem arising in offshore anchor handling operations. Eur. J. Oper. Res. **203**(1), 230–240 (2010)
11. Watson, D.: Practical Ship Design. Elsevier Science, Oxford (1998)

Automatic Image Classification for the Urinoculture Screening

Paolo Andreini, Simone Bonechi, Monica Bianchini, Alessandro Mecocci and Vincenzo Di Massa

Abstract Urinary Tract Infections (UTIs) represent a significant health problem, both in hospital and community–based settings. Normally, UTIs are diagnosed by traditional methods, based on cultivation of bacteria on Petri dishes, followed by a visual evaluation by human experts. In this paper, we present a fully automated system for the screening, that can provide quick and traceable results of UTIs. Actually, based on image processing techniques and machine learning tools, the recognition of bacteria and the colony count are automatically carried out, yielding accurate results. The proposed system, called AID (Automatic Infections Detector) provides support during the whole analysis process: first digital color images of the Petri dishes are automatically captured, then specific preprocessing and spatial clustering algorithms isolate the colonies from the culture ground, finally an accurate classification of the infection types and their severity is performed. Some important aspects of AID are: reduced time, results repeatability, reduced costs.

Keywords Artificial Neural Networks · Advanced Image Processing · Support Vector Machines · Urinoculture Screening

1 Introduction

Urinary Tract Infections (UTIs), together with those of the respiratory tract, are of great clinical relevance for the high frequency with which they are found in common medical practice[1] and because of the complications arising therefrom. They

[1] It is estimated that about 150 million UTIs occur world–wide yearly, giving rise to roughly 5 billion health care expenditures.

P. Andreini (✉) · S. Bonechi · M. Bianchini · A. Mecocci
Department of Information Engineering and Mathematics, University of Siena,
Via Roma 56, 53100 Siena, Italy
e-mail: paolo.andreini@yahoo.it
url: http://www.diism.unisi.it

V. Di Massa
Diesse Ricerche S.r.l., C/o TLS Via Fiorentina 1, 53100 Siena, Italy

R. Neves-Silva et al. (eds.), *Intelligent Decision Technologies*,
Smart Innovation, Systems and Technologies 39,
DOI 10.1007/978-3-319-19857-6_4

are mainly caused by Gram–negative microorganisms, with a high prevalence of *Escherichia coli* (E. Coli, 70 %), even if clinical cases frequently occur where complicated infections are caused by Gram–positive or multi–resistant germs, on which the common antimicrobial agents are inevitably ineffective, leading to therapeutic failures.

The urine culture is a screening test[2] in the case of hospitalized patients and pregnant women. In the standard protocol, the urine sample is seeded on a Petri dish that holds a culture substrate, used to artificially recreate the environment required for the bacterial growth, and incubated at 37 °C overnight. After the incubation, each dish must be examined by a human expert, adding some more time to the medical report output. This common situation significantly departs from the needs of the clinician to have results in quick time, to set a targeted therapy, avoiding the use of broad–spectrum antibiotics and improving the patient management.[3] Moreover, traditional analysis methods suffer from further problems, such as possible errors arising in the visual determination of the bacterial load—due to the skills and the expertise of the individual operator—, and difficulties in the traceability of samples and results [1].

Recent technological advancements in the biomedical field have allowed the development of innovative and complex systems like: biomedical image analysis systems [2], computer–based decision support systems [3], knowledge acquisition and management of medical imaging [4], artificial intelligence for healthcare [5–7]. Significant improvements in biology and medicine have been obtained by using hybrid methods, based on a combination of advanced image processing techniques, artificial intelligence tools, machine learning [8], expert systems, fuzzy logic [9], genetic algorithms, and Bayesian modeling [10]. In particular, the development of automated instruments for results assessment (screening systems), has attracted increasing research interest during the last decade, because of their higher repeatability, accuracy, reduced staff time (that are the main limiting factors of manual screening), and lower costs. *Automated urinalysis devices improve the capacity of the laboratory to screen more samples, producing results in less time than by manual screening. Moreover, the redeployment and lower grading of staff with the increased turnover and speed of urine screening, gave economic advantages of automated screening over manual screening* [11].

In this paper, we propose a tool called AID (Automatic Infections Detector) that provides a decision support system for biologists. The system automatically gets dish images from a color camera and, through a suitable preprocessing and spatial clustering, isolates the colonies from the culture ground, even in the presence of moderate ground disuniformities. Thereafter, the infection types and their severity are accurately established. Artificial neural networks and support vector machines (SVMs) have been used and comparatively tested for the classification of bacterial infec-

---

[2]The term *screening* is used here in its common meaning of a routine test performed on a large population, to identify those who are likely to develop a specified disease. Instead, in *in vitro* diagnostics it stands for preventive analyses aimed at establishing if a sample is positive or not.

[3]Rapid reporting is crucial especially when pediatric patients are involved since, in this case, the infection symptoms are not always specific, while it is urgent to decide if an antibiotic therapy is necessary or not and when to start it.

tions. The AID system allows a substantial speedup of the whole procedure, besides avoiding the continuous transition between sterile and external environments. The final outcomes are directly stored along with the related analysis records (the image, the type of infection and the colony count). The data used during experiments have been provided by DIESSE Ricerche Srl, Siena. Preliminary experiments show very promising results, both with respect to the classification accuracy and to the estimation of bacterial count.

The paper is organized as follows. The next section introduces the problem and describes the preprocessing steps. Section 3 presents the classification methods and the experimental results. Section 4 defines the procedure used to perform the bacterial count, whereas Sect. 5 specifically addresses the problem of Candida infections, whose classification and count is particularly demanding. Finally, conclusions are drawn in Sect. 6.

2 Data Preprocessing

To classify different types of infection, good quality images of the Petri dishes are fundamental. This is why a specific portable device has been developed to catch such images. It consists of a color camera controlled by a computer, on which the AID system runs (Fig. 1).

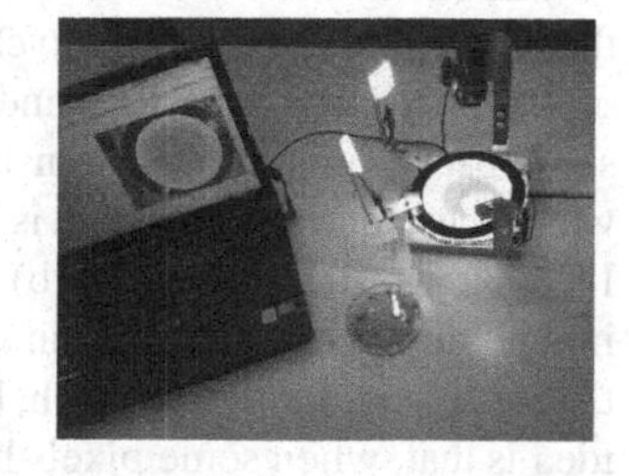

Fig. 1 The acquisition device connected to a PC

The automatic acquisition module employs a simple, nonetheless robust, motion detection algorithm based on frame differencing. The motion detection results are used to store a frame automatically, when the plate is correctly positioned, the scene is well illuminated, and no movements are observed. Subsequently, a Hough circle transform is used to detect the circular region comprising the Petri dish. Such a region is masked out to isolate it from the rest of the acquired scene. The acquisition system has been employed by DIESSE Ricerche to collect a set of 253 images, which has been used to train and validate the classifiers (actually the set has been divided into three subsets: a training, a validation, and a test set, comprising 154, 64, and 35 images respectively, see Table 1).

As a requirement, AID has to recognize eight different infections, namely: E. Coli, KES (Klebsiella, Enterobacter, Serratia), Enterococcus Faecalis, Streptococcus Agalactiae, Pseudomonas, Proteus, Staphylococcus Aureus, and Candida.

Table 1 Datasets' composition

Dataset	Coli	KES	Faecalis	Agalactiae	Proteus	Aureus	Pseudomonas	Candida
Training	34	16	12	15	12	12	7	46
Validation	16	7	8	5	5	5	5	13
Test	11	4	5	5	1	1	2	6

Since UriSelect 4, which is a chromogenic medium, is used as ground seed, the most important information to distinguish among the possible infections is represented by the chromatic features of the corresponding colonies. Moreover, to accurately evaluate their density, the colonies must be separated from the background, which means that the background must be localized and masked out, to preserve only the regions where bacteria are present. To this end, we used a chromatic description of the background, obtained by modeling its visual appearance, during a preliminary training phase. The chromatic information has been based on color opponent (a, b) of the CIE–Lab color space [12]. The lightness component (L) has been discharged because it turned out to convey little and somehow unstable information. To compensate for local background dishomogeneities, a Mean–Shift algorithm [13] has been used to associate each image pixel to the corresponding modal density value in the (a, b) space. The modal value so obtained, was compared with the background chromatic model, to establish if the pixel belongs to the background region or not.

The background segmentation step turned out to be harder than expected. In fact, the background color tends to change depending on the types of grown bacteria and on the proximity of background pixels to bacteria colonies. As a result, the use of a static background model is unsuitable (especially when yellow colonies are present, whose chromatic appearance is similar to that of some background disuniformities). In other words, a given (a, b) couple may actually represent a colony in certain images, and the background in others (Fig. 2). To solve this problem, we used a multistage segmentation approach, by introducing some auxiliary uncertain classes. The idea is that, when some pixels belong to a chromatic region that cannot be assigned with full certainty (for example, in case of yellow pixels that can be background or some infections), the final decision is postponed to a second stage, where specific processing steps are applied to the uncertain classes only. Moreover, from our experiments, it turned out that some spatial discontinuities (edges) are typically present between the colonies and the background. This observation has been used to define some of the uncertain classes. In particular, we have considered the following five regions for each dish:

(1) Colonies (E. Coli, KES, E. Faecalis, S. Agalactiae);
(2) Background without edges (blue in Fig. 2);
(3) Background with edges (pink in Fig. 2);
(4) Uncertainty region without edges (red in Fig. 2);
(5) Uncertainty region with edges (green in Fig. 2).

The previous regions are detected during the first step of our multistage approach. The edge presence has been revealed through a Sobel operator [14], due to its easy implementation and computational speed. Furthermore, according to our experiments, this operator gives good edge enhancement results, eliminating the need for more sophisticated and burdensome approaches. In the second stage of our multistage approach, regions (4) and (5) are assigned to the background or to the colonies. Here, instead of using the global information given by the background model, more localized information is used. The local information is taken into account by comparing, in each image, the H (hue) value of the HSV color space [15] of the uncertainty region with the H value of those pixels certainly belonging to the background. Actually, regions (4) and (5) are analyzed separately, by applying the same discriminating procedure, summarized in the following.

1. The Otsu method [16] is used to automatically perform a clustering–based image thresholding in (4) and (5);
   (a) Based on the computed threshold, the considered uncertainty region is divided into two sub–regions;
   (b) The histograms of the two sub–regions are computed and compared, to establish if a significant separation exists, i.e. if the peaks of the two sub–regions are far enough in the histogram of the H channel. In the last case, the two sub–regions are processed separately and identified with (6) and (7), respectively;
2. The peaks of the region (2) and of the analyzed regions ((6) and (7) separately or together, depending on the previous step) are compared;
3. If they are distant enough in the histogram of the H channel, the observed regions contain two different distributions and the Otsu method is employed again to calculate a threshold for deciding if the uncertainty region belongs to the background or to the colonies;
4. Instead, if the peaks are almost coincident, the uncertainty region shows a H value similar to the background, to which we suppose it belongs.

In other words, for both regions (4) and (5), we identify different sub–regions (if any) and, for each "homogeneous" zone, we establish its membership to the background or not.

Finally, further problems arise when considering Candida colonies, since their color is nearly the same as the culture ground. Actually, the above procedure is unable to isolate Candida in the acquired images (biologists too, anyway). To find out the presence of this infection, an expert usually rotates and moves the dish in different positions, to find out additional visual cues (reflections, thickness of the colony, etc.) to accomplish the task. Since a 2D image does not allow such a possibility, Candida was uncovered by detecting eventual edges inside the background (caused by the colony extrusions originating small reflections). Therefore, an ad hoc procedure was devised, particularly focused on analyzing segments belonging to region (3).

At the end of our proposed multistage segmentation, we have a reliable separation of the dish image into two sub–regions, namely: the background and the region comprising the whole set of colonies grown on the Petri dish. The region comprising the colonies must be further analyzed to correctly label the colonies and to evaluate their density.

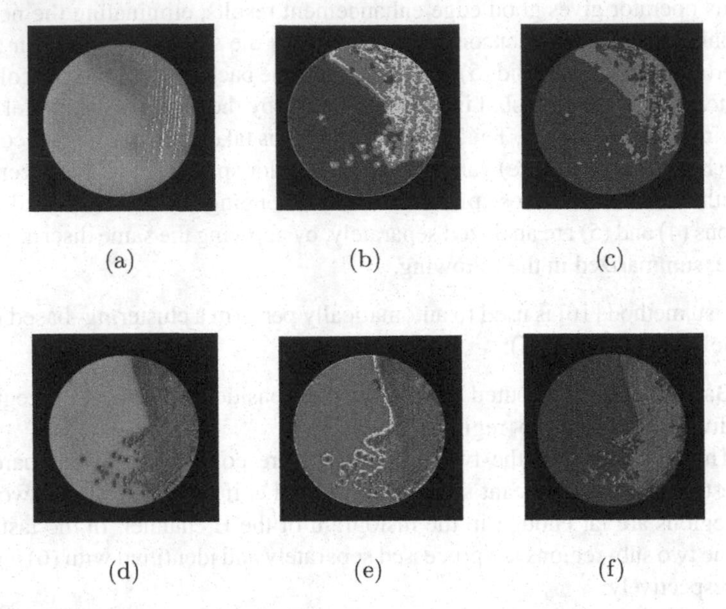

Fig. 2 Images (a) and (d) represent two dishes in which Pseudomonas and KES colonies are respectively grown; images (b) and (e) show the five different regions present on each Petri dish, and (c) and (f) the results obtained after the application of the described discriminating procedure

3 Urinary Bacterial Infections' Classification

The classification of bacteria grown on a Petri dish is a very difficult task. Actually, biologists are able to perform this operation using their *a priori* knowledge of the problem and thanks to the extraordinary ability of the human visual system. As previously stated, the culture ground used in this study is UriSelect 4, a chromogenic, milky, and opaque medium, used to isolate and enumerate the microorganisms of the urinary tract. UriSelect 4 is specifically designed to identify infections by color. In fact, the diverse colors taken by the colonies are due to the activity of different enzymes, and such a color information allows to directly recognize some macro–classes of infection, namely:

- E. Coli: red–pink, due to β–galactosidase and tryptophanase;
- Enterococcus Faecalis: turquoise blue, due to β–glucosidase;
- KES group: purplish blue, due to β–galactosidase and β–glucosidase;
- Proteus–Providencia–Morganella: brown–orange, due to tryptophane désaminase.

It is also possible to distinguish different types of infections within these macro–classes, but it is really difficult by using color information only.

Therefore, our approach to the problem of classifying different urinary bacteria again involves a multi–stage procedure. Firstly, infections are partitioned into groups which have similar chromatic features, producing a sort of pre–classification among colonies. After that, different classifiers that use both chromatic features and class–specific auxiliary information, are trained to distinguish within each group. Finally, a special procedure is applied to recognize the Candida infection.

Actually, the pre–classification phase aims at dividing infections into three main groups, that can be easily identified by their color: red for E. Coli, blue for Enterococcus Faecalis, KES and Streptococcus Agalactiae, and yellow for Pseudomonas, Proteus and Staphylococcus Aureus. Among the eight diverse types of infections, only E. Coli produces red colonies, so that it can be easily recognized as a result of this step. For the other classes, the pre–tag phase allows only infected/not–infected answers, whereas specific information is needed to identify each particular kind of infection. For example, in the blue class, the colony dimension is a useful feature to discriminate Enterococcus Faecalis (small colonies, 0.5–1.5 mm for the diameter) from KES (2–3 mm), whereas, in the yellow class, Proteus infections can be distinguished from the others by the presence of a halo surrounding the colonies.

To collect the data for the pre–tag classifier, we have first executed a Mean–Shift segmentation algorithm and then we have extracted the (a, b) color components of the colony in the CIE–Lab space. A multi–layer perceptron (MLP) network with two inputs, a single hidden layer with six neurons, and three output neurons, with sigmoid activations, was trained using BackPropagation. For comparison, we also trained an SVM with a Gaussian kernel, with $\gamma = 0.1$ and $C = 7.75$. The Weka software (http://www.cs.waikato.ac.nz/ml/weka/) was used to implement both MLPs and SVMs. All the architectural parameters were chosen via a *trial and error* procedure. Results for the pre–classification phase (accuracy and confusion matrices[4]) are reported in Table 2. We can notice that, based on the p–test[5] the two classification methods are statistically independent, but, in practice, they produce very similar numerical results, both using 50–fold crossvalidation or not.

After pre–classification, we tried to distinguish infections belonging to the blue class (Enterococcus Faecalis, KES, Streptococcus Agalactiae) and we found two main issues: sometimes the background extraction module was not able to eliminate all the background segments (mainly for E. Faecalis), badly affecting the classification performance; unpredictable chromatic variations can be produced when different species of bacteria are present and overlap on the same Petri dish (f.i., E. Coli and E. Faecalis on the same dish look like KES, in the overlapping regions).

To face the first problem, the information coming both from the background extractor and from the pre–tag classifier are used as input to a GrabCut algorithm [17], that allows us to remove the remaining background segments (Fig. 3).

[4]A confusion matrix allows the visualization of the performance of a supervised learning classifier. Each column represents the instances in the predicted class, while each row represents the instances in the actual class. Its name stems from the fact that it clearly shows if the system is confusing two classes (i.e. commonly mislabeling one as another).

[5]The p–test results for the Red, Blue and Yellow classifiers are respectively 0.0580, 0.0747, 0.0964.

Table 2 Classification results for the three main colors, for MLP and SVM classifiers

MLP/SVM classifiers for the three main colors		
Total Number of Segments	13292	Percentage
MLP Incorrectly Classified	23	0.179 %
MLP Correctly Classified	**13269**	**99.827 %**
SVM Incorrectly Classified	36	0.2708 %
SVM Correctly Classified	**13256**	**99.7292 %**

(a) Accuracy

Red	Blue	Yellow
4479/4479	2/0	8/10
0/0	8703/8691	0/12
13/14	0/0	87/86

(b) Confusion Matrices MLP/SVM

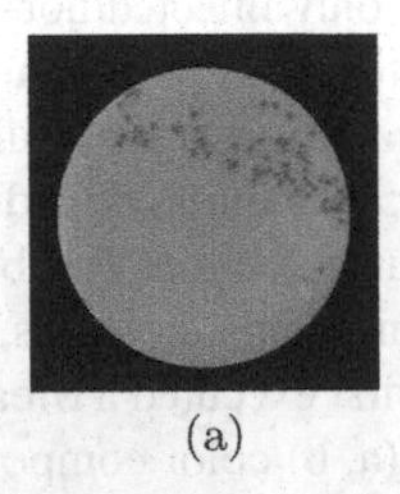

(a)

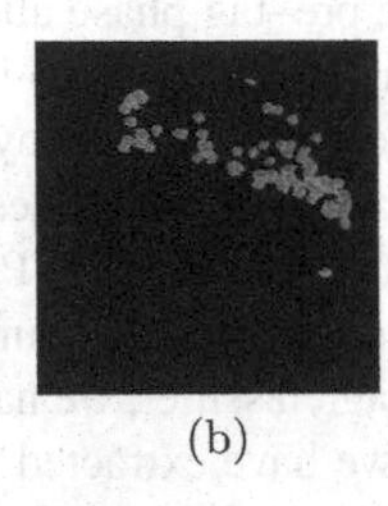

(b)

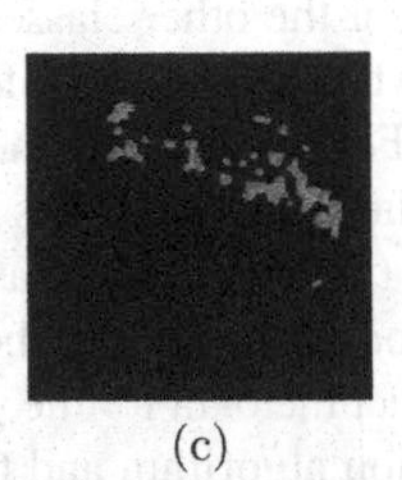

(c)

Fig. 3 In (a), the original image; (b) and (c) represent E. Faecalis colonies before and after the application of the GrubCut algorithm, respectively

The second problem is faced by first recognizing isolated colonies and then by using this information to forecast the kind of colonies possibly present in the overlapping regions. A relevant example is when E. Coli and E. Faecalis are present at the same time on a dish, producing an overlapping region very similar, in color, to a KES colony. This is a well–known situation and, therefore, we decided to use this a priori knowledge during the classification stage. When isolated colonies of E. Coli and E. Faecalis are contemporary detected on the dish, if a KES region is also detected, the latter is classified as a possible Coli/Faecalis overlapping region, which is also statistically more likely than having the simultaneous presence of E. Coli, E. Faecalis and KES.[6] Based on our experiments, such an assumption improves significantly the classification accuracy.

Results—obtained using an MLP (with two inputs, two hidden layers with three and six sigmoid neurons, respectively, and three sigmoid outputs) and an SVM (with a Gaussian kernel, $\gamma = 0.12$ and $C = 1$)—are described in Table 3. A similar procedure was implemented for the yellow class (Pseudomonas, Proteus and Staphylococcus Aureus)—using an MLP (with two inputs, three hidden and three output

[6] Actually, Petri dishes on which many different infections are simultaneously present are considered to be contaminated, and they are of little interest to biologists.

sigmoid neurons) and an SVM (with a Gaussian kernel, $\gamma = 0.12$ and $C = 0.88$)—showing very promising results (see Table 4). However, such results are very preliminary since they were achieved on a small image set; further experiments should be done to draw definitive conclusions.

Table 3 Classification results for the blue class

MLP/SVM classifier for KES, Faecalis and Agalactiae		
Total Number of Segments	7122	Percentage
MLP Incorrectly Classified	1234	17.3266 %
MLP Correctly Classified	**5888**	**82.6734 %**
SVM Incorrectly Classified	1240	17.4108 %
SVM Correctly Classified	**5882**	**82.5892 %**

(a) Accuracy

KES	Faecalis	Agalactiae
3003/3021	217/191	4/12
577/604	2619/2548	169/213
13/17	254/203	266/313

(b) Confusion Matrices MLP/SVM

Table 4 Classification results for the yellow class

MLP/SVM classifier for Proteus, Aureus and Pseudomonas		
Total Number of Segments	100	Percentage
MLP Incorrectly Classified	17	17 %
MLP Correctly Classified	**83**	**83 %**
SVM Incorrectly Classified	18	18 %
SVM Correctly Classified	**82**	**82 %**

(c) Accuracy

Proteus	Aureus	Pseudomonas
34/34	1/1	0/0
3/3	7/7	3/3
7/8	3/3	42/41

(d) Confusion Matrices MLP/SVM

4 Infection Severity Estimation

Besides the infection identification and classification, the AID system performs the bacterial count, giving an estimate of the number of microorganisms per milliliter of urine. The infection severity estimation is expressed in UFC/ml (Units Forming Colony per ml) by using a logarithmic evaluation scale, as shown in Fig. 4.

The actual severity value is obtained by multiplying the number of bacterial colonies counted on the dish by the inverse of the seeding dilution rate.

For the bacterial count, a multistage algorithm has been developed that, at first, searches for single, not overlapping colonies, and then tries to enucleate colonies belonging to slightly overlapping regions. Based on the ground seed identification phase, a binary image is constructed, in which the background is represented by

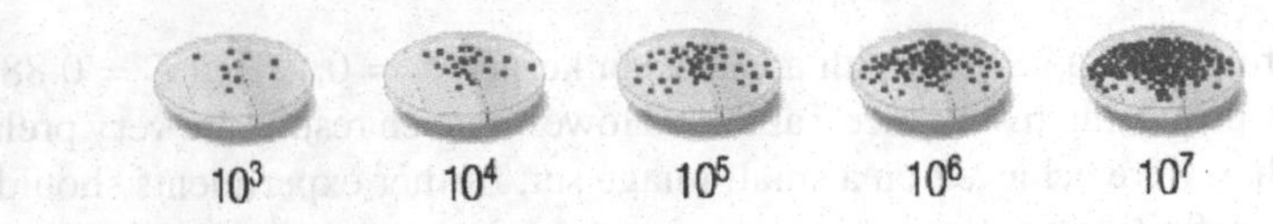

Fig. 4 Logarithmic scale for the colony count (UFC/ml)

the ground seed and the foreground by the colonies. Single colonies show a roughly circular shape and can be easily identified based on this feature. In particular, for each connected component in the binary image, we calculate the least enclosing circle and then, if the ratio between the circular area and the area of the given connected component is under a fixed threshold (chosen via a trial–and–error procedure), such a component is supposed to be a colony.

Obviously, this simple approach is not effective in massive overlapping regions, so that a different procedure has been developed to distinguish single colonies in this case. The convexity of each colony contour is calculated and, based on this information, sub–contours with convex shape are found, and the best ellipse (in the least square sense), that fits each sub–contour, is selected. Then a score matrix, that takes into account the axes rate and the ellipses points belonging (or not) to the contour, is constructed and used to remove non–relevant ellipses. Only single colonies discovered in this way and located in slightly overlapping regions, are counted and used to estimate the infection severity.

5 Identification of Candida Infections

The typical color of Candida colonies is nearly the same as that of the culture ground, making very difficult to identify and isolate this kind of infection from the background. A human expert can detect the presence of Candida by rotating the dish, in order to highlight the tridimensional structure of the grown colonies, which extrude small domes out of the ground surface. On the contrary, AID can only use a bidimensional representation of the Petri dish, since the camera catches just a frontal view of it. However, the presence of edges on the culture media—produced by reflection phenomena on the domes' surfaces—can actually give an indication of the probable presence of an undetected colony, even if noise can interfere by producing false edges. Therefore, to reduce the number of false positives, we used shape features (a colony tends to be circular), to distinguish colonies from noise. Finally, single non overlapping colonies were searched on the edge mask using the previously described algorithm. If a significant number of colonies can be detected, we assume that the corresponding edges are due to the presence of Candida and not to noise (see Fig. 5).

6 Conclusions

In this paper, an automatic tool, called AID, to detect, classify and count UTIs, was described. The system shows good accuracy when detecting the typical microorganisms present in humans. Preliminary promising results were reported, whereas AID is actually tested on real data by DIESSE biologists. It is a future matter of research to refine the classification approach with the aim of distinguishing colonies even if they exhibit very similar colors, based on a larger dataset of images. AID will also be extended to treat diverse types of culture grounds (possibly transparent), and based on different enzymes (producing different chromatic reactions).

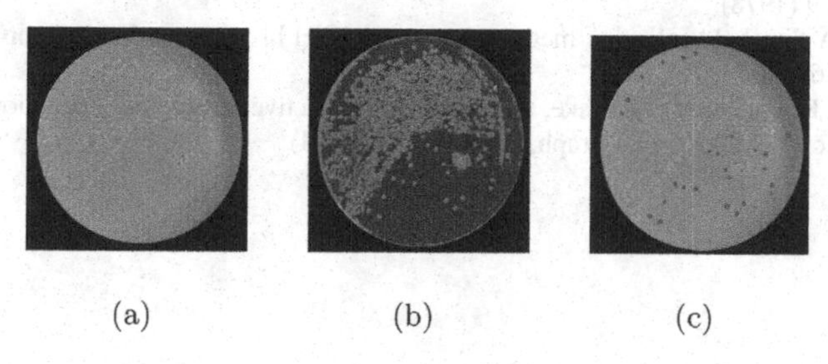

Fig. 5 In (a), the original image; in (b) the identified edges within the background, in pink; in c, Candida colonies found on the Petri dish (Color figure online)

Acknowledgments The authors want to thank the R&D team of DIESSE Ricerche Srl, Siena, for having stimulated this work and for the support given in collecting data and testing AID both in its prototype and final releases.

References

1. Ballabio, C., Venturi, N., Scala, M.R., Mocarelli, P., Brambilla, P.: Evaluation of an automated method for urinoculture screening. Microbiol. Med. **5**(3), 178–180 (2010)
2. Deserno, T.M. (ed.): Biomedical Image Processing. Springer, New York (2011)
3. Berlin, A., Sorani, M., Sim, I.: A taxonomic description of computer-based clinical decision support systems. J. Biomed. Inform. **39**, 656–667 (2006) (Elsevier)
4. Belazzi, R., Diomidous, M., Sarkar, I.N., Takabayashi, K., Ziegler, A., McCray, A.T., Sim, I.: Data analysis and data mining: current issues in biomedical informatics. Methods Inf. Med. **50**(6), 536–544 (2011) (Schattauer Publishers)
5. Agah, A. (ed.): Artificial Intelligence in Healthcare. CRC Press, Boca Raton (2014)
6. Heckerling, P.S., Canaris, G.J., Flach, S.D., Tape, T.G., Wigton, R.S., Gerber, B.S.: Predictors of urinary tract infection based on artificial neural networks and genetic algorithms. Int. J. Med. Inform. **76**(4), 289–296 (2007)
7. Bianchini, M., Maggini, M., Jain, L.C. (eds.): Handbook on Neural Information Processing, Intelligent Systems Reference Library, vol. 49. Springer, Berlin (2013)
8. Bandinelli, N., Bianchini, M., Scarselli, F.: Learning long-term dependencies using layered graph neural networks. In: Proceedings of IJCNN 2012, pp. 1–8 (2012)

9. Torres, A., Nieto, J.J.: Fuzzy Logic in Medicine and Bioinformatics. J. Biomed. Biotechnol. **91908** (2006)
10. Dey, D.K., Ghosh, S., Mallick, B.K.: Bayesian Modeling in Bioinformatics. CRC Press, Boca Raton (2010)
11. Automated urine screening systems. NHS Purchasing and Supply Agency, CEP 10031 Evaluation report, 2011
12. Hunter, R.S.: Photoelectric color difference meter. Proc. Winter Meet. Opt. Soc. Am. **38**(7), 661 (1948) (JOSA)
13. Comaniciu, D., Meer, P.: Mean-shift: a robust approach toward feature space analysis. IEEE Trans. Pattern Anal. Mach. Intell. **24**, 603–619 (2002)
14. Gonzalez, R., Woods, R.: Digital Image Processing, pp. 414–428. Addison Wesley, Boston (1992)
15. Smith, A.R.: Color gamut transform pairs. Proc. 5th Ann. Conf. Comput. Graph. Interact. Tech. **12**(3), 12–19 (1978)
16. Otsu, N.: A threshold selection method from gray-level histograms. IEEE Trans. Syst. Man Cybern. **9**, 62–66 (1979)
17. Rother, C., Kolmogorov, V., Blake, A.: GrabCut: interactive foreground extraction using iterated graph cuts. ACM Trans. Graph. **23**, 309–314 (2004)

Comparing Hybrid Metaheuristics for the Bus Driver Rostering Problem

Vítor Barbosa, Ana Respício and Filipe Alvelos

Abstract SearchCol is a recently proposed approach hybridizing column generation, problem specific algorithms and distinct well known metaheuristics (VNS, Tabu Search, Simulated Annealing, etc.). SearchCol allows to solve several combinatorial optimization problems by applying column generation to a given decomposition model, and using one of the available metaheuristics to search for an integer solution combining the previously generated columns, which are components of the problem. A new evolutionary algorithm (EA) was proposed as the first population based metaheuristic included in SearchCol. This EA uses a representation of individuals based on the generated columns and has been used to obtain integer solutions for a new model for the Bus Drivers Rostering problem (BDRP). Special features of this EA include local search and elitism. This paper presents a computational study evaluating the new population based heuristic (EA) versus two single solution heuristics: VNS and Simulated Annealing, exploiting different configurations of the framework on a set of benchmark instances for the BDRP.

Keywords Evolutionary algorithms • Metaheuristics • Hybrid methods • Rostering

V. Barbosa (✉)
Escola Superior de Ciências Empresariais do Instituto Politécnico de Setúbal, Setúbal, Portugal
e-mail: vitor.barbosa@esce.ips.pt

A. Respício
Universidade de Lisboa e CMAFIO, Lisbon, Portugal
e-mail: alrespicio@di.fc.ul.pt

F. Alvelos
Departamento de Produção e Sistemas/Centro Algoritmi, Universidade do Minho, Guimaraes, Portugal
e-mail: falvelos@dps.uminho.pt

R. Neves-Silva et al. (eds.), *Intelligent Decision Technologies*,
Smart Innovation, Systems and Technologies 39,
DOI 10.1007/978-3-319-19857-6_5

1 Introduction

Many real-life optimization problems map into combinatorial optimization problems which are NP-hard. Therefore, tight bounds on the value of optimal solutions are difficult to reach, even in the presence of medium sized real-life problem instances. The use of heuristics [1], metaheuristics [2] or hyper-heuristics [3] is justified whenever the time and computational resources required to reach a solution, albeit approximate, are more important than achieving the optimal solution, that may take too much time and too many computational resources to be achieved. Hence, employing heuristic methods is for those who need "good enough, soon enough, cheap enough" solutions to problems [3].

However, good trade-off between resolution speed and quality of the solution achieved is required, which has driven researchers to explore the combination of metaheuristics with exact algorithms. Combination approaches, as surveyed in [4], can be classified into collaborative or integrative, the latter being divided into two others, according to the type of integration: (1) the integration of exact algorithms in metaheuristics and (2) the integration of metaheuristics in exact algorithms. Combination approaches that integrate exact algorithms in metaheuristics are also surveyed in [5, 6].

SearchCol [7] is a recently proposed approach hybridizing column generation (CG) [8], which is an exact optimization method, with distinct metaheuristics (MH). These MH are used to explore the subproblem solutions generated during the CG procedure, in order to find integer solutions that are built from the combination of the subproblems solutions which are already integer. The SearchCol approach can address distinct problems, provided that a decomposition model exists to be solved by the column generation method, with a growing number of MH available to explore the subproblems solutions.

In previous work [9], a new decomposition model for the Bus Driver Rostering Problem (BDRP) was proposed to integrate the SearchCol++ framework, which is the computational framework implementing the SearchCol concept. In addition, a new MH was also included: an evolutionary algorithm (EA) tailored to integrate the framework. In this paper we compare an enhanced version of the EA proposed in [10] which considers elitism and local search with two single solution MH in the search, namely Variable Neighbourhood Search (VNS) and Simulated Annealing (SA).

Next section approaches rostering and presents the decomposition model for the BDRP. In Sect. 3, the concept of searching integer solutions from the column generation results is introduced, presenting an overview of the framework and its application to BDRP. The population based EA metaheuristic is presented as well as the two other metaheuristics with which EA is compared. Section 4 describes the computational tests, with details on the configurations used in the column generation stage and in each of the metaheuristics tested. Results obtained from all the metaheuristics are discussed. Finally, in the last section, the conclusions and future work are presented.

2 The Bus Driver Rostering Problem

Personnel scheduling or rostering [11] consists in defining the "work-schedule" for each of the workers in a company for a given period. A roster is a plan including the schedules for all workers. A work-schedule defines, for each day, if the worker is assigned to work or has a day-off and, in the first case, which daily task/shift has to be performed. The rostering problem arises because the company usually has diverse tasks to assign on each day, sometimes needing particular skills, and on the other hand, because the labour and company rules (days-off, rest time, etc.) restrict the blind assignment of tasks to workers. Rostering is addressed by many types of business, as surveyed in [12, 13].

The model for the BDRP is an integer programming formulation adapted from the one proposed in [14]. The model is only concerned with the rostering stage, assuming that the construction of the tasks was previously carried out by joining trips and rest times to obtain complete daily tasks ready to be assign to drivers. For each driver, the model considers a set of feasible schedules, represented by the columns assigned to that driver. The set of all the possible valid columns can be so vast that its enumeration becomes impossible. Therefore, we only consider a restricted subset of valid columns, leading to the following formulation of the restricted master problem (RMP) of the BDRP decomposition model proposed in [9].

RMP formulation:

$$Min \sum_{\nu \in V} \sum_{j \in J^{\nu}} p_j^{\nu} \lambda_j^{\nu} \tag{1}$$

Subject to:

$$\sum_{\nu \in V} \sum_{j \in J^{\nu}} a_{ih}^{j\nu} \lambda_j^{\nu} \quad \geq 1, i \in T_h^w, h = 1, \ldots, 28, \tag{2}$$

$$\sum_{j \in J} \lambda_j^{\nu} = 1, \nu \in V, \tag{3}$$

$$\lambda_j^{\nu} \in \{0, 1\}, j \in J^{\nu}, \nu \in V. \tag{4}$$

Where:

λ_j^v – Variable associated to the schedule j of driver v, from group of drivers V;
J^v – Set of valid schedules for driver v (generated by subproblem v);
p_j^v – Cost of the schedule j obtained from the subproblem of driver v;
a_{ih}^{jv} – Assumes value 1 if task i of day h is assigned in the schedule j of driver v;

In this model, the valid subproblem solutions are represented as columns, with cost p_j^v for the solution of the subproblem v at iteration j, and with the assignment of task i on day h, if $a_{ih}^{jv} = 1$; T_h^w is the set of work tasks available on day h.

The objective is to minimize the total cost of the selected work schedules. The first set of constraints, the linking constraints, assures that all tasks, in each day, are assigned to someone, and the last set of constraints, the convexity constraints, assures that a work schedule is selected for each driver/subproblem.

The BDRP, as most rostering problems, is a NP-Hard combinatorial optimization problem [14, 15], being computationally hard to obtain optimal solutions. To avoid the computational burden to achieve solutions by using exact methods, many authors approach the problem with heuristic methods which are usually faster for achieving good solutions. Examples of the use of non-exact methods can be found in [14, 16, 17]. The BDRP model was inserted in the SearchCol framework aiming to obtain valid solutions for this problem in short computational time.

3 Rostering by Metaheuristics

This section describes how the metaheuristics are used to improve valid rosters, the solutions to the BDRP, composed by the information resulting from the column generation optimization. The origin of the concept is presented in the first subsection and further, in the next subsection, its particular application to BDRP. The last subsection introduces the particular MH used in this study.

3.1 Concept Overview

The combination of column generation and metaheuristic search (SearchCol) to obtain approximate solutions of decomposable optimization problems was proposed in [7, 18].

Disregarding the details, the idea is to allow the use of diverse MH to build good solutions for a wide variety of problems. There are two conditions imposed by the framework [7]:
(1) the problems which may integrate the framework must have a decomposition model, allowing the optimization of their linear relaxation by the column generation method and,
(2) the MHs make use of the subproblems' solutions saved through the CG to improve global solutions.

When the framework is used in a new problem, after the definition of the decomposition model of the problem, CG is used to obtain an optimal linear solution composed by the combination of integer subproblems' solutions. Subproblems are solved as integer problems (with exact methods or heuristics) in each iteration. When the CG ends, the framework has a set of solution generators to build integer solutions with the information stored during the CG. A global solution for the complete problem is built with a solution from each subproblem. The available generators and their behaviour are described in [18].

As the global solution obtained by the generators is not guaranteed to be a good quality solution, or even a valid one, the solution must be evaluated according to its feasibility and infeasibility. The feasibility value is related with the cost of the global solution and the infeasibility value is related to the number of violated linking constraints. The function used to evaluate a global solution can be customized by each implementation of the problem decomposition model.

A global solution, or a population of those global solutions, is the raw material over which the MH are used to get a good global solution. The search space available to each MH is the union of the pools of solutions of all the subproblems.

3.2 Application to BDRP

The model presented in Sect. 1, which was implemented in the SearchCol + + (software implementation of the framework) in [9], is solved using column generation, after which a pool of valid schedules is available for each driver (represented by the subproblems in the model).

In each column generation iteration, the objective function of the subproblems are updated and each one is solved by using a heuristic or an exact method to obtain a valid work-schedule, which is added to the restricted master problem and to the pool of subproblems' solutions, if considered attractive by the CG.

To build a roster, which defines the schedules to all drivers, a solution from each subproblem needs to be selected in order to define for each driver the tasks he/she accomplishes in each day of the rostering period. A selection of those subproblems' solutions defines the global solution that the metaheuristics try to improve. Figure 1 show a roster where the selected schedule for driver 1 is the schedule represented by column 1, to the second driver is the seventh column and so on, until the last driver whose schedule corresponds to column 10.

C1	C7	C23	...	C10
Driver 1	Driver 2	Driver 3	...	Driver n

Fig. 1 A global solution (roster) composed by schedules selection

For the BDRP, the evaluation of a roster includes the infeasibility and feasibility proposed by the framework. The first one allows to measure the distance of a solution to be a valid roster, by counting the number of tasks that are not assigned to any driver. If the infeasibility value is zero, all tasks are assigned. The feasibility value represents the cost of the corresponding roster, and is obtained by adding the cost of each work-schedule. The solutions stored from the subproblems include the cost of the work-schedule, which is composed by a fixed cost, if the driver has tasks assigned, and the total cost of the tasks assigned. To obtain the cost (feasibility value) of a roster, the values of the entire set of selected subproblem' solutions are added.

When a new problem is included in the SearchCol++ framework, the evaluation function which calculates infeasibility and feasibility can be overwritten. We have decide to do this for the BDRP to evaluate each roster. Given the existence of the evaluation function related to the problem, the metaheuristics don't need to have particular knowledge about the problem. A roster can be changed by swapping the work-schedule for one or more drivers, by another work-schedule stored in the pool of subproblems' solutions (which assure that the work-schedule is valid to that driver).

3.3 Metaheuristics

Considering the inspiration of the genetic algorithm (GA) or of the evolutionary algorithm (EA), proposed by Holland [19], that is the evolution of the biological species, each individual is completely described by a chromosome allowing to depict all its characteristics by setting values in each gene. The similarities between the EA chromosomes and the global solutions representation in the SearchCol framework are considerable, which motivated the inclusion of an evolutionary algorithm as a new MH in the framework, simultaneously with the decomposition model for the BDRP [9]. The new EA algorithm included in the framework in [9] is the first MH based in a population of solutions, evolving all the population and not only a single solution as is done by other metaheuristics available in the framework.

When comparing two candidate solutions to be valid rosters, a feasible solution (infeasibility=0) is always better than an infeasible one. If both solutions are feasible (infeasible), the best solution is the one with lower feasibility (infeasibility).

The initial population to be used by the EA is very important because it includes most of the information to be explored by the EA along the generations. It should assure the equilibrium between having good quality solutions and a good diversity of solutions. The construction of the initial population uses the available generators in the framework, as described in [18], and the number of times each generator is used to include a new individual in the population is defined in run-time by a parameters file.

The selection operator currently implemented in the EA is the binary tournament [20]. The cost of the solution represented by the individual (feasibility value) is only compared for individuals with the same infeasibility value. Our priority is to obtain valid rosters (with all tasks assigned), and only after we try to reduce the cost.

The recombination operator is usually designated as crossover [20]. Considering that our solution representation has different content in each gene according to the locus of the gene, currently we have only implemented simple operators, with one and two crossover points. Given a chromosome, the mutation operator replaces a randomly chosen solution by other from the corresponding pool. The mutation operator is the only way to include new schedules into the individuals of the population, from the pool of solutions of each subproblem, that were never selected by the generators.

The elitism consists in, in each iteration of the EA, select the best individuals from the population and move them directly to the next generation. The number of individuals selected may vary, starting from a minimum value of one, and selecting only the best one. The increase of the number of individuals included in the elitism pool can originate a rapid stagnation of the population with identical individuals. The elitism used is configured by two parameters: the size of the elitism pool (parameter ε) and the percentage of individuals selected from the mating pool. The elitism is not used if the parameter ε is set to zero.

From the set of single-solution metaheuristics already implemented in the framework, we highlight here the ones used in the current research, which are the local search [21], variable neighborhood search (VNS) [22] and simulated annealing [23].

Local search [21] explores if neighboring solutions are better than the current solution by testing small changes in solutions. The neighborhood of a solution s is defined by the set of solutions obtained by changing the value of one of the variables. In our case, a neighbor of s is obtained by replacing the schedule of one driver by other schedule in the same pool. The search can stop when a better solution is found or only when all neighbors were explored, searching for the best possible one. Local search allows improving selected individuals and also helps the EA exploring the schedules that were not select by the generator, being available only in the pools of schedules of the subproblems.

The VNS uses cyclically the local search to explore neighbors, starting from a single solution and restarting the search every time a better solution is found from that new solution. The dimension of the neighborhood is increased by augmenting the number of changes allowed when no more improvements are achieved in the current neighborhood. Details about the VNS implementation are presented in [18].

Simulated annealing (SA) [23] is an optimization method based in the metal cooling physical process (the annealing process) which prevents the solutions from getting stuck in local optimum by allowing not only improvements, but also moves to worst solutions with a low, but dynamic, probability of occurrences.

4 Computational Tests

To perform the tests we used a subset of the instances for the BDRP, used in [14], designated as P80. All the instances have 36 drivers available which, applying the decomposition model from [9], results in 36 subproblems, each one responsible for building valid schedules for the corresponding driver/subproblem.

All the tests were run in the same computer, equipped with Intel Pentium G640 2.80 GHz CPU, 8 Gb of RAM, Windows7 Professional 64bits as operating system and with IBM ILOG CPLEX 12.5.1 64bits.

For all the tests, the configuration of the CG is the same. The time limit of the initial column generation was set to 30 s. All subproblems are solved in each iteration and the columns added, if attractive. The number of main cycles (CG and search) of the SearchCol metaheuristic was set to 20, with 3 as the maximum without improvement. The other CG (after the first) runs are limited to 10 s.

The parameters of the EA are the following: Initial population size: 240 (1/3 of individuals built by picking, totally at random, subproblem solutions; 1/3 of individuals built by picking random subproblem solutions, with the probability of being selected biased by the linear solution of the first CG; 1/3 of individuals built by picking random subproblem solutions with the probability of being selected biased by the linear solution of the last CG solved); elite population of 40 elements, and local search configured to stop at the first improvement found, applied when half of the limit of iterations without improvement is achieved; crossover and mutation probabilities: 80 % and 20 % (allows to explore the subproblems solutions not included in the individuals of the initial population and only improved solutions are kept); the stopping criterion is 200 generations without improvement.

The VNS configuration: the neighborhood size can grow from 1 to 6, with steps of 1; the local search has the same configuration of the one used in the EA; the maximum number of searches without improvement is 1.

The SA configuration: probability of a worse infeasible (infeasibility>0) solution to be accepted: 10 %; probability of a worse feasible (infeasibility=0) solution to be accepted: 90 %; temperature decrement (alpha):0.95; initial number of iterations in each temperature: 4, increased by 10 % in each temperature update.

For the VNS and SA MH, the initial solution is built by selecting, for each subproblem, the solution with the higher value from the optimal linear solution values in the last CG optimization.

4.1 Results

To test each of the MH, independent executions of the main SearchCol metaheuristic with each of the searchers, EA, VNS and SA, were run 30 times for each instance. Table 1 presents the average results for the EA, the VNS and the SA MH, respectively. Columns under "Time (s)" display the mean of the computational times in seconds; columns under "Solution Value" display the solution values mean and the best value found for each instance in the 30 runs of the corresponding MH. The line "*average*" displays the average value of the corresponding column heading for all the instances. The line "*#best found*" displays the total number of instances for which the corresponding MH was able to reach the best solution value, found in all the runs of the three MH. For each instance, the best value found is highlighted. Regarding the average results of the solutions' values, column "Mean", we can conclude that VNS was able to reach the best average results. In the opposite, the SA performed worst on average, followed by the EA. To evaluate the differences of the results, a paired samples t-test (using SPSS) was used, considering a level of significance of 5 %. The test indicates that the results of the EA are not statistically different from the ones obtained by the SA. The results obtained by the VNS are statistically different from the other MH.

Table 1 Results of the three MH for each instance: mean computational time; mean solution value; best solution found. Number of times the best solution was found by each MH

	SearchCol with EA			SearchCol with VNS			SearchCol with SA		
	Time (s)	Solution value		Time (s)	Solution value		Time (s)	Solution value	
Instance	Mean	Mean	Best	Mean	Mean	Best	Mean	Mean	Best
P80_1	98.0	3987.0	3819	96.6	3985.7	**3801**	75.3	3989.4	3819
P80_2	86.3	2906.5	**2873**	75.1	2905.8	**2873**	68.9	2907.4	**2873**
P80_3	105.5	5674.6	**5158**	100.1	5526.2	5309	92.7	5740.2	5341
P80_4	90.4	5148.8	**4380**	106.2	4872.7	4584	87.1	5057.8	4753
P80_5	100.3	3934.7	**3793**	98.7	3999.4	3815	72.8	4047.5	3905
P80_6	98.2	4588.6	4122	91.1	4212.7	**4026**	79.8	4346.5	4174
P80_7	90.6	5122.3	**4718**	118.6	5033.4	4723	94.0	5169.9	4814
P80_8	93.3	6028.2	5693	116.9	5987.8	**5671**	97.9	6135.6	5732
P80_9	90.4	4912.3	4489	111.1	4681.7	**4329**	89.7	4722.7	4542
P80_10	105.2	5152.3	**4869**	120.9	5229.2	4966	89.2	5251.8	4972
Average	95.8	4745.5	**4391**	*103.5*	***4643.5***	*4409*	***84.8***	*4736.9*	*4492*
#best found			*6*			*5*			*1*

Despite its low efficacy concerning the solutions quality, the SA was the MH with the faster convergence, presenting the lowest average computational running times.

The number of times each MH was able to reach best solution values allows assessing the ability of the MH to make a successful exploration of the SearchCol columns space. Concerning this metric, EA overcame both VNS and SA. It can be noted that SearchCol with the EA was able to attain the best solution value for six instances, while VNS reached the best solution for five instances, and SA for only one instance. Moreover, EA outperformed VNS in five instances – P80_3, P80_4, P80_5, P80_7 and P80_10 – against four for which VNS performed better – P80_1, P80_6, P80_8 and P80_9.

To summarize, for this set of BDRP instances, SearchCol with EA is the hybridization that was able to reach the best solution more times, which suggests that EA is better equipped to explore the space of columns generated by the CG, particularly by having a starting population of individuals with high diversity, some totally random, and other biased to the linear solution of the CG. This also justifies the higher accuracy of the other MH because they always start from a solution with the schedule with higher value in the linear solution of the CG for each subproblem, which is very similar in each run.

5 Conclusions and Future Work

In this study we assessed three metaheuristics as searcher component in the main metaheuristic SearchCol. As application problem, a NP-hard real-life rostering problem was used – the Bus Driver Rostering Problem. The metaheuristics under assessment comprise a recent EA, whose main features include: a special designed individual representation where each allele represents a possible schedule for the corresponding driver, a customizable evaluation function defined in the decomposition model in order to benefit from problem information, variation operators adapted to the solutions representation, and the data structures where the SearchCol stores the subproblems solutions. The EA offers the option of using an elite population to keep the best individuals and also the use of a local search procedure to explore the neighborhood of the best solution in some iterations of the EA. The other metaheuristics assessed were the VNS and the SA.

The computational results revealed that the hybridization of CG and the EA is a competitive metaheuristic. Albeit the good number of best solutions found by the EA in the tests performed, we need to assure a higher stability in the achievement of good solutions in every run. In the future we need to improve the generation of the initial population to include good starting solutions, without losing diversity. In addition, future work comprises the implementation of some diversity control, for the cases that use random generators, especially the ones which are biased by the CG linear solution values. We also need to test VNS and SA with solutions from other generators.

References

1. Pearl, J.: Heuristics: Intelligent Search Strategies for Computer Problem Solving. Addison-Wesley, Reading (1984)
2. Talbi, E.G.: Metaheuristics: From Design to Implementation. Wiley, New York (2009)
3. Burke, E., Kendall, G., Newall, J., Hart, E., Ross, P., Schulenburg, S.: Hyper-heuristics: an emerging direction in modern search technology. In: Glover, F., Kochenberger, G. (eds.) Handbook of Metaheuristics, vol. 57, pp. 457–474. Springer, US (2003)
4. Puchinger, J., Raidl, G.R.: Combining metaheuristics and exact algorithms in combinatorial optimization: a survey and classification. In: Mira, J., Álvarez, J.R. (eds.) First International Work-Conference on the Interplay Between Natural and Artificial Computation. Springer, Las Palmas, Spain (2005)
5. Dumitrescu, I., Stützle, T.: Combinations of local search and exact algorithms. In: Cagnoni, S., Johnson, C., Cardalda, J.R., Marchiori, E., Corne, D., Meyer, J.-A., Gottlieb, J., Middendorf, M., Guillot, A., Raidl, G., Hart, E. (eds.) Applications of Evolutionary Computing, vol. 2611, pp. 211–223. Springer, Berlin (2003)
6. Dumitrescu, I., Stützle, T.: Usage of exact algorithms to enhance stochastic local search algorithms. In: Maniezzo, V., Stützle, T., Voß, S. (eds.) Matheuristics, vol. 10, pp. 103–134. Springer, US (2010)
7. Alvelos, F., de Sousa, A., Santos, D.: SearchCol: metaheuristic search by column generation. In: Blesa, M., Blum, C., Raidl, G., Roli, A., Sampels, M. (eds.) Hybrid Metaheuristics, vol. 6373, pp. 190–205. Springer, Berlin/Heidelberg (2010)
8. Desaulniers, G., Desrosiers, J., Solomon, M.M.: Column Generation. Springer, New York (2005)
9. Barbosa, V., Respício, A., Alvelos, F.: A hybrid metaheuristic for the bus driver rostering problem. In: Vitoriano, B., Valente, F. (eds.) ICORES 2013—2nd International Conference on Operations Research and Enterprise Systems, pp. 32–42. SCITEPRESS, Barcelona (2013)

10. Barbosa, V., Respício, A., Alvelos, F.: Genetic algorithms for the SearchCol ++ framework: application to drivers' rostering. In: Oliveira, J.F., Vaz, C.B., Pereira, A.I. (eds.) IO2013—XVI Congresso da Associação Portuguesa de Investigação Operacional, pp. 38–47. Instituto Politécnico de Bragança, Bragança (2013)
11. Ernst, A.T., Jiang, H., Krishnamoorthy, M., Sier, D.: Staff scheduling and rostering: a review of applications, methods and models. Eur. J. Oper. Res. **153**, 3–27 (2004)
12. Ernst, A.T., Jiang, H., Krishnamoorthy, M., Owens, B., Sier, D.: An annotated bibliography of personnel scheduling and rostering. Ann. Oper. Res. **127**, 21–144 (2004)
13. Van den Bergh, J., Beliën, J., De Bruecker, P., Demeulemeester, E., De Boeck, L.: Personnel scheduling: a literature review. Eur. J. Oper. Res. **226**, 367–385 (2013)
14. Moz, M., Respício, A., Pato, M.: Bi-objective evolutionary heuristics for bus driver rostering. Public Transport **1**, 189–210 (2009)
15. Dorne, R.: Personnel shift scheduling and rostering. In: Voudouris, C., Lesaint, D., Owusu, G. (eds.) Service Chain Management, pp. 125–138. Springer, Berlin Heidelberg (2008)
16. Ruibin, B., Burke, E.K., Kendall, G., Jingpeng, L., McCollum, B.: A hybrid evolutionary approach to the nurse rostering problem. IEEE Trans. Evol. Comput. **14**, 580–590 (2010)
17. Respício, A., Moz, M., Vaz Pato, M.: Enhanced genetic algorithms for a bi-objective bus driver rostering problem: a computational study. Int. Trans. Oper. Res. **20**, 443–470 (2013)
18. Alvelos, F., Sousa, A., Santos, D.: Combining column generation and metaheuristics. In: Talbi, E.-G. (ed.) Hybrid Metaheuristics, vol. 434, pp. 285–334. Springer (2013)
19. Holland, J.H.: Adaptation in Natural and Artificial Systems. MIT Press, Cambridge (1992)
20. Mitchell, M.: An Introduction to Genetic Algorithms. MIT Press, Cambridge (1996)
21. Johnson, D.S., Papadimitriou, C.H., Yannakakis, M.: How easy is local search? J. Comput. Syst. Sci. **37**, 79–100 (1988)
22. Mladenović, N., Hansen, P.: Variable neighborhood search. Comput. Oper. Res. **24**, 1097–1100 (1997)
23. Kirkpatrick, S., Gelatt, C.D., Vecchi, M.P.: Optimization by simulated annealing. Science **220**, 671–680 (1983)

Effective Stopping Rule for Population-Based Cooperative Search Approach to the Vehicle Routing Problem

Dariusz Barbucha

Abstract The main goal of the paper is to propose an effective stopping criterion based on diversity of the population implemented in cooperative search approach to the vehicle routing problem. The main idea is to control diversity of the population during the whole process of solving particular problem and to stop algorithm when the stagnation in the population is observed, i.e. diversity of the population does not significantly change during a given period of time. Computational experiment which has been carried out confirmed that using the proposed diversity-based stopping criterion may significantly reduce the computation time when compared to using traditional criterion where algorithm stops after a given period of time (or iterations), without significant deterioration of the quality of obtained results.

Keywords Cooperative search · Metaheuristics · Stopping criteria · Vehicle routing problem

1 Introduction

Nowadays, metaheuristics are often used for solving computational difficult optimization problems. They basically try to combine basic heuristic methods in higher level frameworks aimed at efficiently and effectively exploring a search space [3]. Among many features of metaheuristics, which decide about the success of this class of methods in effective solving various problems, the following ones are worth mentioning: metaheuristics are strategies that "guide" the search process, techniques which constitute metaheuristic algorithms range from simple local search procedures to complex learning processes, during the whole process of search the dynamic balance between diversification and intensification is achieved, they may incorporate mechanisms to avoid getting trapped in local optima of the search space, and

D. Barbucha (✉)
Department of Information Systems, Gdynia Maritime University,
Morska 83, 81-225 Gdynia, Poland
e-mail: d.barbucha@wpit.am.gdynia.pl

R. Neves-Silva et al. (eds.), *Intelligent Decision Technologies*,
Smart Innovation, Systems and Technologies 39,
DOI 10.1007/978-3-319-19857-6_6

more advanced metaheuristics may use search experience (embodied in some form of memory) to guide the search [3].

Depending on the number of individuals on which the metaheuristics concentrate during the process of solving instances of a given problem one can distinguish *single solution based metaheuristics* (SBM), which concentrate on improving a single solution (individual), and *population-based metaheuristics* (PBM) [17], mostly inspired by biological or social processes, which handle a population of individuals (each individual represents particular solution of the problem) that evolves with the help of information exchange procedures. Whereas typical representatives of methods belonging to the first group are: Simulated Annealing (SA) [6], Tabu Search (TS) [10], or Greedy Randomized Adaptive Search Procedure (GRASP) [7], the *population-based metaheuristics* group, mostly inspired by biological or social processes, includes Evolutionary Algorithms (EA) [12, 16], Gene Expression Programming approaches [8], Scatter Search method (SS) [11], Particle Swarm Optimization (PSO) [14] and Ant Colony Optimization (ACO) [9] algorithms.

A common general framework describing the process of solving instances of the problem by PBM includes an initialization of population, generation of a new population of individuals by using the operators defined over the members of the population, and an integration this new population into the current one using some selection (integration) procedures. The search process is terminated when a given condition is satisfied [17].

Although all population-based methods share the main steps of the process of solving instances of the problem, their performance differs because of the many parameters, which should be tuned in the process of implementation and running of these methods and which can decide about success in using it for solving particular problem. Among them, the most characteristic include the *stopping criterion* which determines when the processes of search for the best solution should stop.

Most implementations of metaheuristics stop after performing a given maximum number of iterations, reaching a limit on CPU resources or performing a given maximum number of consecutive iterations without improvement in the best-known solution value [17]. However, in addition to the above traditional stopping criteria, the stopping criteria used in population-based methods may be based on some statistics of the current population or the evolution of a population.

The paper aims at proposing an effective stopping criterion based on diversity of the population. The main idea is to control diversity of the population during the whole process of solving particular problem, and to stop calculations when the stagnation in the population is observed. Keeping the execution of a population-based metaheuristic is useless in the situation when population diversity does not significantly change in a given period of time. It is expected that using the proposed stopping criterion the time needed by algorithm to solve the problem will decrease without losing the quality of obtained results when compare to using traditional stopping criterion where algorithm stops after a given period of time. The proposed stopping criterion has been implemented in Cooperative Search Approach to the Vehicle Routing Problem (VRP) and validated by a computational experiment.

The rest of the paper includes the following sections. Section 2 aims at presenting details of the proposed cooperative search approach to the VRP with a broad description of the suggested new stopping criterion. Section 3 presents assumptions, goal, and results of computational experiment carried-out in order to validate effectiveness of the approach with presence of diversity-based stopping criterion. Finally, main conclusions and directions of future research are included in Sect. 4.

2 Cooperative Search Approach to the VRP

Among many approaches designed for solving different classes of real-world hard optimization problems the dominant position enjoy algorithms which combine various algorithmic ideas, where population-based methods play important role. In the recent years technological advances enabled development of various parallel and distributed versions of the hybrid methods under the *cooperative search* umbrella for solving computationally difficult optimization problems. According to Toulouse et al. [18] "cooperative search is a parallelization strategy for search algorithms where parallelism is obtained by concurrently executing several search programs".

The idea of cooperative search has been used to design and implement Cooperative Search Approach (CSA) with diversity-based stopping criterion to well known combinatorial optimization problem - Vehicle Routing Problem. The main framework of the proposed approach has been based on e-JABAT multi-agent system implementation presented in [1].

2.1 Vehicle Routing Problem

VRP can be stated as the problem of determining optimal routes through a given set of locations (customers) and defined on an undirected graph $G = (V, E)$, where $V = \{0, 1, \dots, N\}$ is the set of nodes and $E = \{(i,j) | i,j \in V\}$ is a set of edges. Node 0 is a central depot with NV identical vehicles of capacity W. Each other node $i \in V \setminus \{0\}$ denotes customer characterized by coordinates in Euclidean space (x_i, y_i) and a non-negative demand d_i. Each link $(i,j) \in E$ denotes the shortest path from customer i to j and is described by the cost c_{ij} and the time t_{ij} of travel from i to j by shortest path $(i,j \in V)$. It is assumed that $c_{ij} = c_{ji}, t_{ij} = t_{ji} (i,j \in V)$.

The goal is to find a partition of V into NV routes of vehicles

$$R = \{R_1, R_2, \dots, R_{NV}\}$$

that covers all customers and minimize the total cost of travel, and satisfies the following constraints:

- each route R_k ($k \in \{1, 2, \ldots, NV\}$) starts and ends at the depot,
- each customer $i \in V \setminus \{0\}$ is serviced exactly once by a single vehicle,
- the total load on any vehicle associated with a given route does not exceed vehicle capacity ($\forall_{k \in \{1,2,\ldots,NV\}} \sum_{i \in R_k} d_i \le W$)),
- the total duration of any route R_k ($k \in \{1, 2, \ldots, NV\}$) should not exceed a preset bound T.

2.2 Model

The proposed cooperative search approach to the VRP (CSA-VRP) focuses on organizing and conducting the process of search for the best solution with using a set of *search programs* $SP = \{SP(1), SP(2), \ldots, SP(nSP)\}$ executed in parallel, where each search program is an implementation of a single-solution method. During their execution, the search programs operate on a population of individuals (trial solutions) $P = \{P_1, P_2, \ldots, P_{popSize}\}$, stored in a common, sharable *memory* (also called warehouse or pool of solutions). Execution of search programs is coordinated by a specially designed program, called *solution manager*, which acts as an intermediary between common memory and search programs.

The whole process of search is organized as a sequence of steps, including initialization and improvement phases, as it is shown in the pseudo-code in the Algorithm 1 and described as follows.

Algorithm 1 Cooperative Search Approach to the VRP

Require:
  $SP = \{SP(1), SP(2), \ldots, SP(nSP)\}$ - a set of nSP search programs,
  f - fitness function
Ensure:
  s^* - best solution found
1: Generate an initial population of solutions (individuals) $P = \{s_1, s_2, \ldots, s_{popSize}\}$, where *popSize* - population size, and store them in the common memory
2: $s^* \leftarrow \arg\min_{s_i \in P} f(s_i)$ {find the current best solution}
3: **repeat** {in parallel}
4: Select individual s_k ($k = 1, \ldots, popSize$) from the common memory
5: Select a search program $SP(i)$ ($i = 1, \ldots, nSP$)
6: Execute $SP(i)$ on selected individual s_k improving it and return s_k^* as a resulting individual
7: **if** (new individual s_k^* is accepted) **then**
8: Store s_k^* in the common memory
9: **end if**
10: **if** ($f(s_k^*) < f(s^*)$) **then**
11: $s^* \leftarrow s_k^*$
12: **end if**
13: **until** (stopping criterion is met)
14: **return** s^*

1. At first an initial population of solutions is generated randomly and stored in the memory (line 1 of Algorithm 1). The best solution from the population is recorded (line 2).
2. Next, at the following computation stages, individuals forming the initial population are improved by search programs until stopping criterion is met. Five search programs have been used in the proposed approach (see Table 1). They used dedicated local search heuristics operating on single ($R_i \in R$) or two ($R_i, R_j \in R$) randomly selected route(s) (R - set of m routes, $i \neq j$, and $i,j = 1, \ldots, m$). The process of improvement of individuals forming the population are performed in the following cycle (lines 3–13):

    (a) A single individual from the memory is selected randomly and sent to the selected search program, which tries to improve the received individual. In order to prevent such solution from being sent to the other search program, it is blocked for a period of time.
    (b) After execution, the search program returns an individual. If individual has been improved by search program, it is accepted and added to the population of individuals replacing the first found worse individual from the current population. Additionally, if a worse individual can not be found within a certain number of reviews (review is understood as a search for the worse individual after an improved solution is returned by the search program) then the worst individual in the common memory is replaced by the randomly generated one representing a feasible solution.
    (c) If currently improved individual is better than the best one it is remembered.

3. The process of search stops when the stopping criterion is met (line 13). The best solution stored in the population is taken as the final solution of the given problem instance (line 14).

2.3 Stopping Criterion

The proposed stopping criterion implemented in the CSA-VRP is based on population diversity, which is controlled during the whole process of search for the best solution. Measuring diversity of the population requires definition of the *distance* between solutions s_i and s_j in the search space - $dist(s_i, s_j)$. Having the distance, the *average distance* - $dmm(P)$ for the whole population P can be defined as $dmm(P) = (\sum_{s_i \in P} \sum_{s_j \in P, s_i \neq s_j} dist(s_i, s_j))/(popSize * (popSize - 1))$ [17]. It is easy to see, that if the average distance between individuals is large, the population diversity is high, otherwise the population has low diversity.

In order to control diversity of the population in the CSA-VRP, a measure partially based on the *Jaccard coefficient (distance)* (or Tanimoto distance, Marczewski-Steinhaus distance) - $J_\delta(A, B)$, which measures dissimilarity between sample sets A and B [13], has been defined. Jaccard distance is complementary to the *Jaccard simi-*

larity coefficient, measured as $J(A,B) = |A \cup B| / |A \cap B|$ and is obtained by subtracting the Jaccard distance from 1, or, equivalently, by dividing the difference of the sizes of the union and the intersection of two sets by the size of the union:

$$J_\delta(A,B) = 1 - J(A,B) = \frac{|A \cup B| - |A \cap B|}{|A \cup B|} \quad (1)$$

It is easy to see that if all elements in sets A and B are different then $J_\delta(A,B) = 1$, otherwise if A and B share the same elements then $J_\delta(A,B) = 0$.

Table 1 Agents and their characteristics [2]

Agent	Description
3Opt	An implementation of the *3-opt* procedure operating on a single route R_i. Three randomly selected edges are removed and next remaining segments are reconnected in all possible ways until a new feasible solution (route) is obtained
2Lambda	A modified implementation of the dedicated *local search method* based on *λ-interchange local optimization* method. At most λ customers are moved or exchanged between two selected routes R_i and R_j. In the proposed implementation, it has been assumed that $\lambda = 2$
2LambdaC	Another implementation of the dedicated *local search method* which operates on two routes, and is based on exchanging or moving selected customers between these routes. Here, selection of customers to exchange or movement is taken in accordance to their distance to the centroid of their original route. First, a given number of customers from two selected routes R_i and R_j for which the distance between them and the centroid of their routes are the greatest are removed from their original routes. Next, they are moved to the opposite routes and inserted in them on positions, which give the smallest distance between newly inserted customers and the centroid of this route
Cross1	An agent which is implementation of the one-point crossover operator. Initially one point is randomly selected on each route R_i and R_j, dividing these routes on two subroutes. Next, the first subroute of R_i is connected with the second subroute of R_j, and the first subroute of R_j is connected with the second subroute of R_i
Cross2	An implementation of the two-point crossover operator. Initially two points are selected randomly on each route R_i and R_j, dividing these routes on three subroutes. Next, the middle parts (between crossing points) of each route are exchanged between considered routes

Basing on the above Jaccard's similarity coefficient, and also being inspired by similarity measure for the TSP used by Boese et al. [4], a distance between two solutions s_i and s_j of the VRP has been defined. Let $E[i]$ and $E[j]$ be a set of edges in solution s_i and s_j, respectively. Similarity between two solutions s_i and s_j is measured as a number of edges shared by both solutions divided by the number of total edges used arising in both solutions. Hence, the distance ($distVRP(s_i, s_j)$) between these solutions is calculated as:

$$distVRP(s_i, s_j) = \frac{|E[i] \cup E[j]| - |E[i] \cap E[j]|}{|E[i] \cup E[j]|} \tag{2}$$

The average distance in population of solutions (*dmmVRP(P)*) is given by the formula:

$$dmmVRP(P) = \frac{\sum_{s_i \in P} \sum_{s_j \in P, s_i \neq s_j} distVRP(s_i, s_j)}{popSize * (popSize - 1)} \tag{3}$$

$dmmVRP(P) \in [0; 1]$ and if population consists of the same solutions (is not diversified) then $dmmVRP(P) = 0$, and if population of solutions includes completely different solutions (sets of routes including different edges) then $dmmVRP(P) = 1$.

Basing on the above notation the stopping criterion used in line 13 of Algorithm 1 has been defined in the following form

$$\boxed{((dmmVRP(P_t) - dmmVRP(P_{t-1})) \leq divLevel)}$$

where $dmmVRP(P_t)$ and $dmmVRP(P_{t-1})$ denote the average distances in population of solutions measured at time t and $t-1$, respectively, and *divLevel* is a predefined threshold measuring the minimal required difference between diversity of the population in two consecutive points of measure in order to say that the population is still diversified.

Proposed stopping criterion says that if difference between average distances in population defined on the left hand side of the stopping condition falls bellow a given threshold *divLevel* (the population stays *non-diversified* and the process of stagnation is observed), keeping the execution of the algorithm is useless and the algorithm stops.

3 Computational Experiment

3.1 Experiment Goal and Main Assumptions

Computational experiment has been carried out in order to validate the effectiveness of the proposed approach. The main goal of it was to determine to what extent (if any) the stopping criterion based on diversity of the population influences computation results produced by the proposed approach?

After the preliminary experiment, it has been decided that diversity of the population was measured every 300 iterations, and the threshold *divLevel* has been set and tested at three levels: $divLevel = 0.1, 0.01, 0.001$.

The benchmark set of Christofides et al. [5] including 14 instances has been involved in the experiment. The instances contain 50–199 customers with only capacity (vrpnc1-vrpnc5, vrpnc11-12) and, some of them, maximum length route restrictions (vrpnc6-vrpnc10, vrpnc13-14). Each instance was solved 10 times for each setting of *divLevel*. Additionally, each instance has been solved in presence of static and dynamic stopping criteria. In total 700 $((14*3+14*2)\times 10)$ test problems have been solved in the experiment.

Mean relative error - MRE (in %) from the optimal (or the best known) solution, reported in [15], and computation time (in sec.) have been chosen as measures of the quality of the results obtained by the proposed approach.

All computations have been carried out on PC with Intel Core i5-2540M CPU 2.60 GHz and 8 GB RAM running under MS Windows 7 operating system.

3.2 Results Presentation and Analysis

Results of the experiment are presented in Table 2. The first column of this table includes names of the instances, and for each instance, the remaining columns contain MRE and the computation time for the CSA-VRP with diversity-based stopping criteria applied with different *divLevel* values set.

Table 2 Results (MRE from the best known solution in % and time in s.) obtained by the CSA-VRP for all tested instances in presence of diversity-based stopping criterion

Instance	Diversity-based stopping criterion					
	divLevel = 0.1		*divLevel* = 0.01		*divLevel* = 0.001	
	MRE	Time	MRE	Time	MRE	Time
vrpnc1	0.23 %	6.90	0.00 %	11.53	0.00 %	23.49
vrpnc2	1.58 %	7.32	1.63 %	8.29	1.50 %	16.32
vrpnc3	0.89 %	12.55	0.84 %	14.47	0.82 %	39.38
vrpnc4	3.44 %	14.81	2.12 %	34.74	1.99 %	53.28
vrpnc5	4.03 %	22.69	3.39 %	64.86	2.79 %	117.76
vrpnc6	0.37 %	4.94	0.69 %	42.45	0.34 %	61.42
vrpnc7	2.24 %	9.33	1.90 %	18.89	1.25 %	39.69
vrpnc8	2.61 %	11.21	1.37 %	34.04	1.24 %	78.67
vrpnc9	4.14 %	14.85	2.91 %	39.75	2.80 %	59.46
vrpnc10	4.26 %	20.22	3.47 %	111.71	2.21 %	164.07
vrpnc11	2.27 %	14.83	0.36 %	14.15	0.12 %	44.37
vrpnc12	0.32 %	11.07	0.03 %	16.74	0.06 %	45.88
vrpnc13	2.43 %	14.33	2.36 %	46.52	2.03 %	53.11
vrpnc14	0.13 %	10.60	0.05 %	29.64	0.05 %	65.89
Average	2.07 %	12.55	1.51 %	34.84	1.23 %	61.63

Analysis of the obtained results allows one to conclude that the proposed CSA-VRP produces good results with average MRE equal to 1–2 %. Although, as it was expected, the best average MRE is observed for *divLevel* = 0.001 and the worst for *divLevel* = 0.1, the difference between these results does not exceed 1 %. Simultaneously one can observe that longer time has been required in order to obtain better results.

In order to discover benefits from applying diversity-based stopping criterion it has been also decided to carry out an experiment where typical static stopping criterion (algorithm stops after given period of time) has been used within the CSA-VRP instead of diversity-based one. Because of the fact that arbitrary choice of the amount of time after which the algorithm should stop is not obvious, it has been decided to stop algorithm after 3 min (3 times longer that average time for diversity-based stopping criterion with *divLevel* = 0.001). Results of this part of the experiment are presented in Table 3, where besides the name of the instance, MRE and computation time have been also included.

Looking at results in Table 3, it is easy to see that although the average MRE has been improved (0.97 %) in comparison with the best result from Table 2 (1.23 %), the difference in MRE is not significant (0.26 %), despite the significant increase of computation time (3 times longer).

Table 3 Results (MRE from the best known solution in % and time in s.) obtained by the CSA-VRP for all tested instances in presence of static stopping criterion (3 min)

Instance	Static stopping criterion	
	3 min	
	MRE	Time
vrpnc1	0.00 %	179.86
vrpnc2	0.01 %	181.56
vrpnc3	0.43 %	180.63
vrpnc4	0.85 %	182.10
vrpnc5	3.71 %	179.58
vrpnc6	0.00 %	178.56
vrpnc7	0.35 %	179.82
vrpnc8	0.00 %	180.54
vrpnc9	1.97 %	181.54
vrpnc10	4.09 %	180.02
vrpnc11	0.17 %	179.54
vrpnc12	0.00 %	180.41
vrpnc13	1.95 %	181.64
vrpnc14	0.00 %	181.68
Average	0.97 %	180.53

4 Conclusions

The paper focused on Cooperative Search Approach for the Vehicle Routing Problem, where the process of solving the problem is carried-out by a set of search programs operating on population of solutions stored in the common sharable memory. The main goal of the paper was to propose an effective stopping criterion based on diversity of the population. Diversification of the population is controlled during the whole process of solving the problem, and the algorithm stops when the stagnation in the population is observed, i.e. diversity of the population does not significantly change during a given period of time.

The experiment carried out on instances of the VRP showed that using the proposed diversity-based stopping criterion may significantly reduce the computation time when compare to using traditional criterion where algorithm stops after a given period of time (or iterations), without significant deterioration of the mean relative error.

Future research will focus on proposing different measures of distance between solutions and diversity of the population.

References

1. Barbucha, D., Czarnowski, I., Jędrzejowicz, P., Ratajczak-Ropel, E., Wierzbowska, I.: e-JABAT—an implementation of the web-based a-team. In: Nguyen, N.T., Jain, L.C. (eds.) Intelligent Agents in the Evolution of Web and Applications. Studies in Computational Intelligence, vol. 167, pp. 57–86. Springer, Berlin/Heidelberg (2009)
2. Barbucha, D.: Experimental Investigation of impact of migration topologies on performance of cooperative approach to the vehicle routing problem. In: Nguyen, N.T. et al. (eds.) ACIIDS 2015, LNAI 9011, pp. 250–259. Springer, Berlin/Heidelberg (2015)
3. Blum, C., Roli, A.: Metaheuristics in combinatorial optimization: overview and conceptual comparison. ACM Comput. Surv. **35**(3), 268–308 (2003)
4. Boese, K., Kahng, A., Muddu, S.: A new adaptive multistart technique for combinatorial global optimization. Oper. Res. Lett. **16**, 101–113 (1994)
5. Christofides, N., Mingozzi, A., Toth, P., Sandi, C. (eds.): Combinatorial Optimization. John Wiley, Chichester (1979)
6. Dorigo, M., Stutzle, T.: Ant Colony Optimization. MIT Press, Cambridge, MA (2004)
7. Eglese, R.W.: Simulated annealing: a tool for operational research. Eur. J. Oper. Res. **46**, 271–281 (1990)
8. Feo, T.A., Resende, M.G.C.: Greedy randomized adaptive search procedures. J. Glob. Optim. **6**, 109–133 (1995)
9. Ferreira, C.: Gene Expression Programming: Mathematical Modeling by an Artificial Intelligence. Springer, Heidelberg (2006)
10. Glover, F., Laguna, M.: Tabu Search. Kluwer, Boston (1997)
11. Glover, F., Laguna, M., Marti, R.: Fundamentals of scatter search and path relinking. Control Cybern. **39**, 653–684 (2000)
12. Holland, J.H.: Adaptation in Natural and Artificial Systems. The University of Michigan Press, Ann Arbor (1975)
13. Jaccard, P.: Etude comparative de la distribution florale dans une portion des Alpes et des Jura. Bulletin del la Societe Vaudoise des Sciences Naturelles **37**, 547–579 (1901)

14. Kennedy, J., Eberhart, R.C.: Particle swarm optimization. In: Proceedings of IEEE International Conference on Neural Networks, pp. 1942–1948. Piscataway, NJ (1995)
15. Laporte, G., Gendreau, M., Potvin, J., Semet, F.: Classical and modern heuristics for the vehicle routing problem. Int. Trans. Oper. Res. **7**, 285–300 (2000)
16. Michalewicz, Z.: Genetic Algorithms + Data Structures = Evolution Programs. Springer, Berlin-Heidelberg-New York (1994)
17. Talbi, E.G.: Metaheuristics: From Design to Implementation. Wiley, New York (2009)
18. Toulouse, M., Thulasiraman, K., Glover, F.: Multi-level cooperative search: a new paradigm for combinatorial optimization and an application to graph partitioning. In: Amestoy, P., et al. (eds.) Euro-Par99 Parallel Processing, LNCS, vol. 1685, pp. 533–542. Springer, Heidelberg (1999)

GEFCOM 2014—Probabilistic Electricity Price Forecasting

Gergo Barta, Gyula Borbely Gabor Nagy, Sandor Kazi and Tamas Henk

Abstract Energy price forecasting is a relevant yet hard task in the field of multi-step time series forecasting. In this paper we compare a well-known and established method, ARMA with exogenous variables with a relatively new technique Gradient Boosting Regression. The method was tested on data from Global Energy Forecasting Competition 2014 with a year long rolling window forecast. The results from the experiment reveal that a multi-model approach is significantly better performing in terms of error metrics. Gradient Boosting can deal with seasonality and auto-correlation out-of-the box and achieve lower rate of normalized mean absolute error on real-world data.

Keywords Time series · Forecasting · Gradient boosting regression trees · Ensemble models · ARMA · Competition · GEFCOM

1 Introduction

Forecasting electricity prices is a difficult task as they reflect the actions of various participants both inside and outside the market. Both producers and consumers use day-ahead price forecasts to derive their unique strategies and make informed decisions in their respective businesses and on the electricity market. High preci-

G. Barta (✉) · G. Borbely · G. Nagy · S. Kazi · T. Henk
Department of Telecommunications and Media Informatics, Budapest University of Technology and Economics, Magyar Tudosok Krt, Budapest 2. H-1117, Hungary
e-mail: barta@tmit.bme.hu
url: http://www.tmit.bme.hu

G. Nagy
e-mail: nagy@tmit.bme.hu

S. Kazi
e-mail: kazi@tmit.bme.hu

T. Henk
e-mail: henk@tmit.bme.hu

R. Neves-Silva et al. (eds.), *Intelligent Decision Technologies*, Smart Innovation, Systems and Technologies 39,
DOI 10.1007/978-3-319-19857-6_7

sion short-term price forecasting models are beneficial in maximizing their profits and conducting cost-efficient business. Day-ahead market forecasts also help system operators to match the bids of both generating companies and consumers and to allocate significant energy amounts ahead of time.

The methodology of the current research paper originates from the GEFCOM 2014 forecasting contest. In last year's contest our team achieved a high ranking position by ensembling multiple regressors using the Gradient Boosted Regression Trees paradigm. Promising results encouraged us to further explore potential of the initial approach and establish a framework to compare results with one of the most popular forecasting methods; ARMAX.

Global Energy Forecasting Competition is a well-established competition first announced in 2012 [1] with worldwide success. The 2014 edition [2] put focus on renewal energy sources and probabilistic forecasting. The GEFCOM 2014 Probabilistic Electricity Price Forecasting Track offered a unique approach to forecasting energy price outputs, since competition participants needed to forecast not a single value but a probability distribution of the forecasted variables. This methodological difference offers more information to stakeholders in the industry to incorporate into their daily work. As a side effect new methods had to be used to produce probabilistic forecasts.

The report contains five sections:

1. Methods show the underlying models in detail with references.
2. Data description provides some statistics and description about the target variables and the features used in research.
3. Experiment Methodology summarizes the training and testing environment and evaluation scheme the research was conducted on.
4. Results are presented in a the corresponding section.
5. Conclusions are drawn at the end.

2 Methods

Previous experience showed us that oftentimes multiple regressors are better than one [4]. Therefore we used an ensemble method that was successful in various other competitions: Gradient Boosted Regression Trees [5–7]. Experimental results were benchmarked using ARMAX; a model widely used for time series regression. GBR implementation was provided by Python's Scikit-learn [8] library and ARMAX by Statsmodels [9].

2.1 ARMAX

We used ARMAX to benchmark our methods because it is a widely applied methodology for time series regression [10–14]. This method expands the ARMA model with (a linear combination of) exogenic inputs (X). ARMA is an abbreviation of auto-regression (AR) and moving-average (MA). ARMA models were originally designed to describe stationary stochastic processes in terms of AR and MA to support hypothesis testing in time series analysis [15]. As the forecasting task in question has exogenic inputs by specification therefore ARMAX is a reasonable candidate to be used as a modeler.

Using the ARMAX model (considering a linear model wrt. the exogenous input) the following relation is assumed and modeled in terms of X_t which is the variable in question at the time denoted by t. According to this the value of X_t is a combination of AR(p) (auto-regression of order p), MA(q) (moving average of order q) and a linear combination of the exogenic input.

$$X_t = \epsilon_t + \sum_{i=1}^{p} \varphi_i X_{t-i} + \sum_{i=1}^{q} \theta_i \epsilon_{t-i} + \sum_{i=0}^{b} \eta_i d_t(i) \quad (1)$$

The symbol ϵ_t in the formula above represents an error term (generally regarded as Gaussian noise around zero). $\sum_{i=1}^{p} \varphi_i X_{t-i}$ represents the autoregression submodel with the order of p: φ_i is the i-th parameter to weight a previous value. The elements of the sum $\sum_{i=1}^{q} \theta_i \epsilon_{t-i}$ are the weighted error terms of the moving average submodel with the order of q. The last part of the formula is the linear combination of exogenic input d_t

.

Usually p and q are chosen to be as small as they can with an acceptable error. After choosing the values of p and q the ARMAX model can be trained using least squares regression to find a suitable parameter setting which minimizes the error.

2.2 *Gradient Boosting Decision Trees*

Gradient boosting is another ensemble method responsible for combining weak learners for higher model accuracy, as suggested by Friedman in 2000 [16]. The predictor generated in gradient boosting is a linear combination of weak learners, again we use tree models for this purpose. We iteratively build a sequence of models, and our final predictor will be the weighted average of these predictors. Boosting generally results in an additive prediction function:

$$f^*(X) = \beta_0 + f_1(X_1) + \ldots + f_p(X_p) \quad (2)$$

In each turn of the iteration the ensemble calculates two set of weights:

1. one for the current tree in the ensemble
2. one for each observation in the training dataset

The rows in the training set are iteratively reweighted by upweighting previously misclassified observations.

The general idea is to compute a sequence of simple trees, where each successive tree is built for the prediction residuals of the preceding tree. Each new base-learner is chosen to be maximally correlated with the negative gradient of the loss function, associated with the whole ensemble. This way the subsequent stages will work harder on fitting these examples and the resulting predictor is a linear combination of weak learners.

Utilizing boosting has many beneficial properties; various risk functions are applicable, intrinsic variable selection is carried out, also resolves multicollinearity issues, and works well with large number of features without overfitting.

3 Data Description

The original competition goal was to predict hourly electricity prices for every hour on a given day. The provided dataset contained information about the prices on hourly resolution for a roughly 3 year long period between 2011 and 2013 for an unknown zone. Beside the prices two additional variables were in the dataset. One was the Forecasted Zonal Load ($'z'$) and the other was the Forecasted Total Load ($'t'$). The first attribute is a forecasted electricity load value for the same zone where the price data came from. The second attribute contains the forecasted total electricity load in the provider network. The unit of measurement for these variables remain unknown, as is the precision of the forecasted values. Also, no additional data sources were allowed to be used for this competition.

Table 1 Descriptive statistics for the input variables and the target

	Price	Forecasted total load	Forecasted zonal load
count	25944	25944	25944
mean	48.146034	18164.103299	6105.566181
std	26.142308	3454.036495	1309.785562
min	12.520000	11544	3395
25 %	33.467500	15618	5131
50 %	42.860000	18067	6075
75 %	54.24	19853	6713.25
max	363.8	33449	11441

In Table 1 we can see the descriptive statistic values for the original variables and the target. The histogram of the target variable (Fig. 1) is a bit skewed to the left with a long tail on the right and some unusual high values. Due to this characteristic we decided to take the natural log value of the target and build models on that value. The model performance was better indeed when they were trained on this transformed target.

Fig. 1 Price histogram

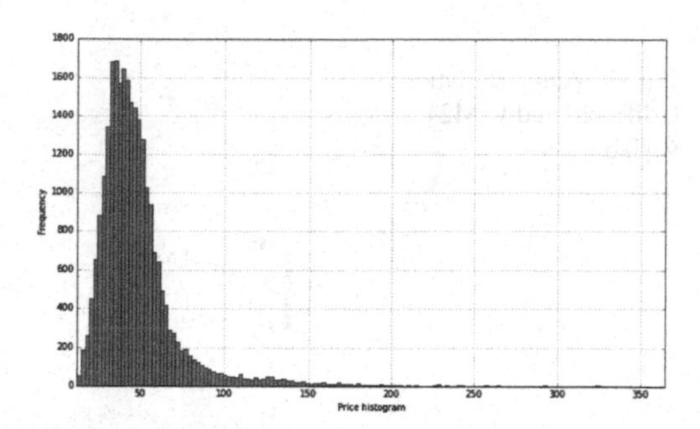

The distribution of the other two descriptive variables are far from normal as we can see on Fig. 2. As we can see the shapes are very similar for these variables with the peak, the left plateau and the tail on the right. They are also highly correlated with a correlation value of ~0.97, but not so much with the target itself (~0.5–0.58) (Table 2)

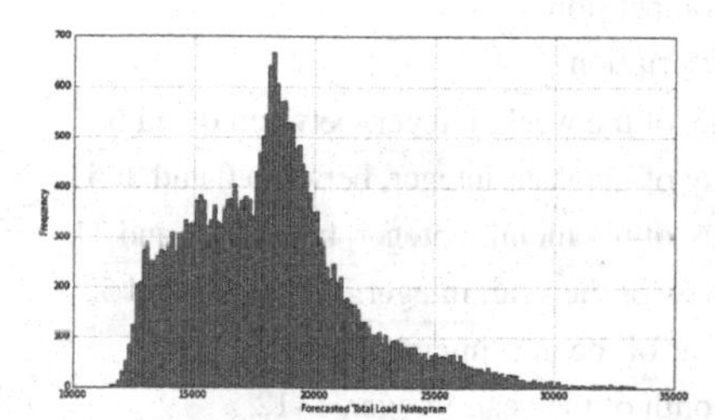

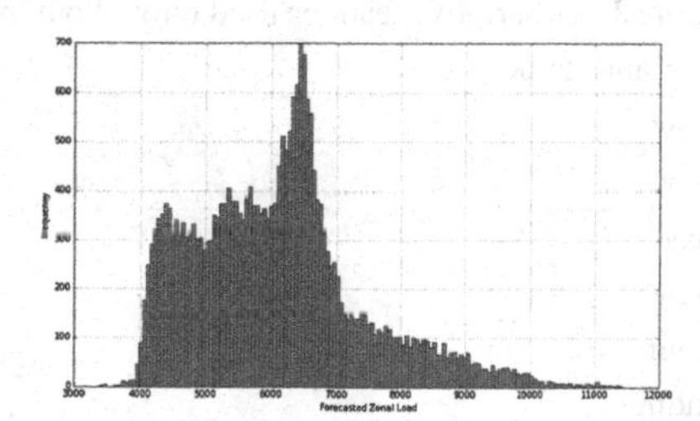

Fig. 2 Forecasted total load and forecasted zonal load histograms

Table 2 Correlation matrix of input variables

	Price	Forecasted zonal load	Forecasted total load
Price	1.0	0.501915	0.582029
Forecasted zonal load	0.501915	1.0	0.972629
Forecasted total load	0.582029	0.972629	1.0

Beside the variables of Table 1 we also calculated additional attributes based on them: several variables derived from the two exogenous variable $'z'$ and $'t'$, also date and time related attributes were extracted from the timestamps (see Table 3 for details).

During the analysis we observed from the autocorrelation plots that some variables value have stronger correlation with its +/− 1 h value, so we also calculated these values for every row. Figure 3 shows 3 selected variables to be shifted as the autocorrelation values are extremely high when a lagging window of less than 2 h is used.

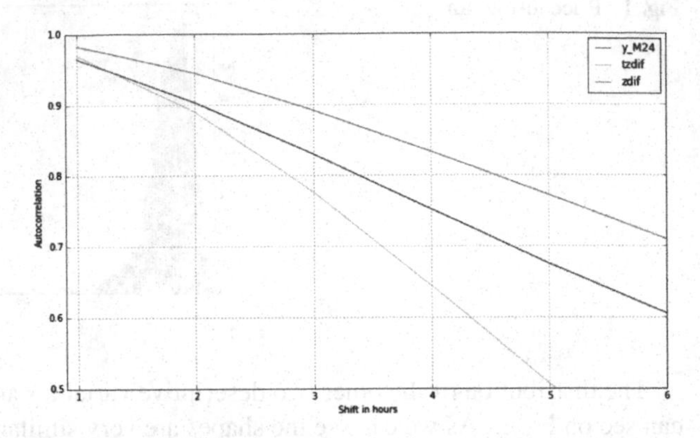

Fig. 3 Autocorrelation of tzdif, zdif and y_M24 variables

Table 3 Descriptive features used throughout the competition

Variable name	Description
dow	Day of the week, integer, between 0 and 6
doy	Day of the year, integer, between 0 and 365
day	Day of the month, integer, between 1 and 31
woy	Week of the year, integer, between 1 and 52
hour	Hour of the day, integer, 0–23
month	Month of the year, integer, 1–12
t_M24	t value from 24 h earlier
t_M48	t value from 48 h earlier
z_M24	z value from 24 h earlier
z_M48	z value from 48 h earlier
tzdif	The difference between t and z
tdif	The difference between t and t_M24
zdif	The difference between z and z_M24

In Fig. 4 figure we can see an autocorrelation plot of price values in specific hours and they are shifted in days (24 h). It is clearly seen that the autocorrelation values for the early and late hours are much higher than for the afternoon hours. That means it is worth to include shifted variables in the models as we did. Not surprisingly the errors at the early and late hours were much lower than midday and afternoon.

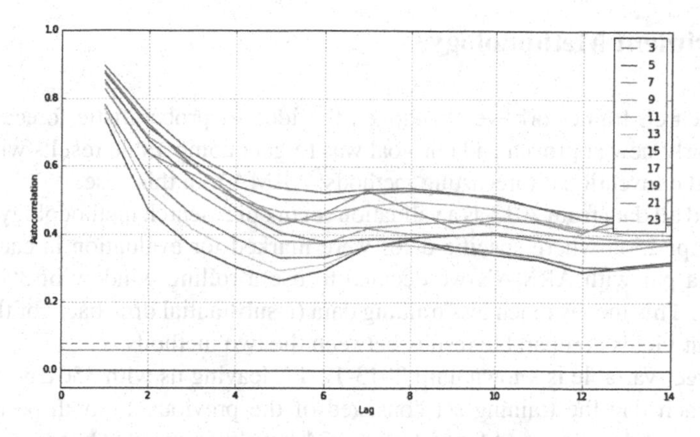

Fig. 4 Autocorrelation of price values at specific hours, shifted in days

Table 4 Attribute importances provided by GBR

Attribute	GBR variable importance
tzdif	0.118451
tdif	0.092485
zdif	0.090757
z	0.090276
hour	0.085597
t	0.078957
z_M48	0.078718
t_M48	0.076352
t_M24	0.069791
z_M24	0.069072
doy	0.067103
day	0.056018
dow	0.024973
month	0.001449

Gradient Boosting Regression Trees also provided intrinsic variable importance measures. Table 4 shows that (apart from the original input variables) the calculated differences were found to be important. The relatively high importance of the hour of day suggests strong within-day periodicity.

4 Experiment Methodology

In our research framework we abandoned the idea of probabilistic forecasting as this is a fairly new approach and our goal was to gain comparable results with well-established conventional forecasting methods; ARMAX in this case.

We used all data from 2013 as a validation set in our research methodology (unlike in the competition where specific dates were marked for evaluation in each task). To be on a par with ARMAX we decided to use a rolling window of 30 days to train GBR. This means much less training data (a substantial drawback for the GBR model), but yields comparable results between the two methods.

The target variable is known until 2013-12-17, leaving us with 350 days for testing. For each day the training set consisted of the previous 1 month period, and the subsequent day was used for testing the 24 hourly forecasts. On some days the ARMAX model did not converge leaving us with 347 days in total to be used to assess model performance. The forecasts are compared to the known target variable, we provide 2 metrics to compare the two methods: Mean Absolute Error (MAE) and Root Mean Squared Error (RMSE). Gradient Boosting and ARMAX optimizes Mean Squared Error directly meaning that one should focus more on RMSE than MAE.

5 Results

Figure 5 compares the model outputs with actual prices for a single day. While Table 5 shows the descriptive statistics of the error metrics: *mae_p_armax*, *rmse_p_armax*, *mae_p_gbr* and *rmse_p_gbr* are the Mean Absolute Errors and Root Mean Squared Errors of ARMAX and GBR models respectively. The average of the 24 forecasted observations are used for each day, and the average of daily means are depicted for all the 347 days. In terms of both RMSE and MAE the average and median error is significantly lower for the GBR model; surpassing ARMAX by approx. 20 % on average.

During the evaluation we came across several days that had very big error measures, filtering out these outliers represented by the top and bottom 5 % of the observed errors we have taken a t-test to confirm that the difference between the two models is indeed significant ($t = 2.3187, p = 0.0208$ for RMSE).

6 Conclusions and Future Work

The GEFCOM competition offered a novel way of forecasting; probabilistic forecasts offer more information to stakeholders and is an approach worth investigating in energy price forecasting. Our efforts in the contest were focused on developing accurate forecasts with the help of well-established estimators in the literature used

Table 5 Descriptive statistics for the error metrics

	mae_p_armax	rmse_p_armax	mae_p_gbr	rmse_p_gbr
count	347	347	347	347
mean	8.640447	10.395176	7.126920	8.496357
std	11.809438	13.822071	10.396122	11.627084
min	1.223160	1.781158	1.020160	1.302484
5.0 %	2.083880	2.673257	1.439134	1.785432
50 %	5.152181	6.088650	3.520733	4.144649
95 %	27.049138	31.339932	27.171626	31.122828
max	101.081747	106.317998	77.819519	83.958518

in a fairly different context. This approach was capable of achieving roughly 10th place in the GEFCOM 2014 competition Price Track and performs surprisingly well when compared to the conventional and widespread benchmarking method ARMAX overperforming it by roughly 20 %.

The methodology used in this paper can be easily applied in other domains of forecasting as well. Applying the framework and observing model performance on a wider range of datasets yields more robust results and shall be covered in future work.

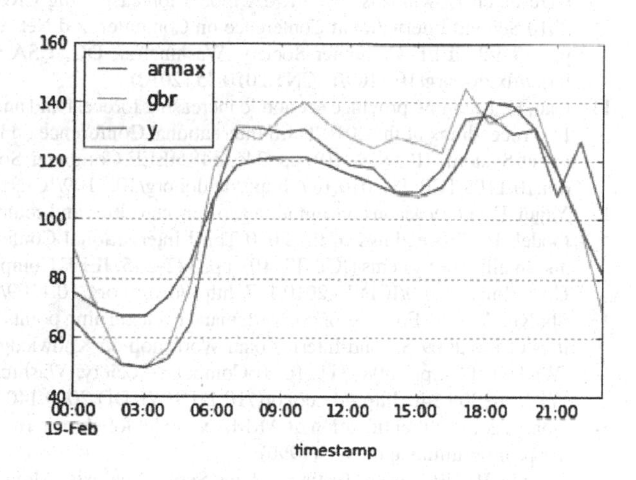

Fig. 5 Within-day price forecasts for 2013-02-19

During the competition we filtered the GBR training set to better represent the characteristics of the day to be forecasted, which greatly improved model performance. Automating this process is also a promising and chief goal of ongoing research.

References

1. Tao, H., Pierre, P., Shu, F.: Global energy forecasting competition 2012. Int. J. Forecast. **30**(2), 357–363 (2014)
2. Tao, H.: Energy forecasting: past, present and future. Foresight: Int. J. Appl. Forecast. (32), 43–48 (2014)
3. Weron, R.: Electricity price forecasting: a review of the state-of-the-art. Int. J. Forecast. **30**, 1030–1081 (2014)
4. Rokach, L.: Ensemble-based classifiers. Artif. Intell. Rev. **33**, 1–39 (2010)
5. Aggarwal, S.a.S.L.: Solar energy prediction using linear and non-linear regularization models: a study on AMS (American Meteorological Society) 2013–14 solar energy prediction contest. Energy (2014)
6. McMahan, H.B.e.a.: Ad click prediction: a view from the trenches. In: Proceedings of the 19th ACM SIGKDD International Conference on Knowledge Discovery and Data Mining, pp. 1222–1230 (2013)
7. Graepel, T.e.a.: Web-scale bayesian click-through rate prediction for sponsored search advertising in microsoft's bing search engine. In: Proceedings of the 27th International Conference on Machine Learning, pp. 13–20 (2010)
8. Pedregosa, F., et al.: Scikit-learn: Machine learning in Python. J. Mach. Learn. Res. Learn. **12**, 2825–2830 (2011)
9. Seabold, S., Perktold, J.: Statsmodels: Econometric and statistical modeling with python. In: Proceedings of the 9th Python in Science Conference (2010)
10. Tan, Ian K.T., Hoong, P.K., Keong, C.Y.: Towards Forecasting Low Network Traffic for Software Patch Downloads: An ARMA model forecast using CRONOS. In: Proceedings of the 2010 Second International Conference on Computer and Network Technology (ICCNT '10), pp. 88–92. IEEE Computer Society, Washington, DC, USA. doi:10.1109/ICCNT.2010.35, http://dx.doi.org/10.1109/ICCNT.2010.35 (2010)
11. Gao, F.: Liaoning province economic increasing forecast and analysis based on ARMA model. In: Proceedings of the 2010 Third International Conference on Intelligent Networks and Intelligent Systems (ICINIS '10), pp. 346–348. IEEE Computer Society, Washington, DC, USA. doi:10.1109/ICINIS.2010.107, http://dx.doi.org/10.1109/ICINIS.2010.107 (2010)
12. Yajun, H.: Forecast on consumption gap between cities and countries in China based on ARMA model. In: Proceedings of the 2010 Third International Conference on Intelligent Networks and Intelligent Systems (ICINIS '10), pp. 342–345. IEEE Computer Society, Washington, DC, USA. doi:10.1109/ICINIS.2010.137, http://dx.doi.org/10.1109/ICINIS.2010.137 (2010)
13. ShuXia, Y.: The Forecast of power demand cycle turning points based on ARMA. In: Proceedings of the 2009 Second International Workshop on Knowledge Discovery and Data Mining (WKDD '09), pp. 308–311. IEEE Computer Society, Washington, DC, USA. doi:10.1109/WKDD.2009.140, http://dx.doi.org/10.1109/WKDD.2009.140 (2009)
14. Hong-Tzer, Y.: Identification of ARMAX model for short term load forecasting: an evolutionary programming approach (1996)
15. Whittle, P.: Hypothesis Testing in Time Series Analysis. Almqvist & Wiksells boktr, Uppsala (1951)
16. Friedman, J.H.: Greedy function approximation: a gradient boosting machine. Ann. Stat. 1189–1232 (2001)

Nearest Neighbour Algorithms for Forecasting Call Arrivals in Call Centers

Sandjai Bhulai

Abstract We study a nearest neighbour algorithm for forecasting call arrivals to call centers. The algorithm does not require an underlying model for the arrival rates and it can be applied to historical data without pre-processing it. We show that this class of algorithms provides a more accurate forecast when compared to the conventional method that simply takes averages. The nearest neighbour algorithm with the Pearson correlation distance function is also able to take correlation structures, that are usually found in call center data, into account. Numerical experiments show that this algorithm provides smaller errors in the forecast and better staffing levels in call centers. The results can be used for a more flexible workforce management in call centers.

1 Introduction

Most organizations with customer contact have a call center nowadays, or hire specialized firms to handle their communications with customers through call centers. Current trends are towards an increase in economic scope and workforce (see [7]). Hence, there is an enormous financial interest in call centers that leads to the importance of efficient management of call centers. The efficiency relates to the efficient use of the workforce, since the costs in a call center are dominated by personnel costs.

The basis of efficient workforce management in call centers is the well-known Erlang-C model (see [6]). This model is used to compute the minimum number of service representatives needed to meet a target service level. Halfin and Whitt [10] suggest the square-root safety-staffing principle, recommending the number of representatives to be equal to the offered load with some additional safety staffing to compensate for stochastic variability. The Erlang-C model is perhaps the simplest

S. Bhulai (✉)
VU University Amsterdam, Faculty of Sciences, De Boelelaan 1081a,
1081 Hv Amsterdam, The Netherlands
e-mail: s.bhulai@vu.nl

R. Neves-Silva et al. (eds.), *Intelligent Decision Technologies*,
Smart Innovation, Systems and Technologies 39,
DOI 10.1007/978-3-319-19857-6_8

model in call center circles. For many applications, however, the model is an over-simplification due to the assumption of a constant arrival rate over the whole day. Common call center practice is to use the stationary independent period-by-period (SIPP) approximation (see [9]). The SIPP approximation uses the average arrival rate over a period of 15 or 30 min, based on historical data, and the number of service representatives in that period as input to the stationary Erlang-C model to approximate performance in that period. The pointwise stationary approximation (PSA, see [8]) is the limiting version of the SIPP approximation when the period length approaches zero.

The stationary models implicitly assume that the time required for the system to relax is small when compared to the length of the period. However, abrupt changes in the arrival rate, or overload situations during one or more periods lead to non-stationary behaviour that must be accounted for. Yoo [16] and Ingolfsson et al. [12] present methods to calculate the transient behaviour by numerically solving the Chapman-Kolmogorov forward equations for the time-varying $M_t/M/c_t$ queueing system. While non-stationary models perform well and can be used for workforce management, they assume that the overall arrival rate is known. In practice, the arrival rate is predicted from historical data and is not known with certainty in advance. The risk involved in ignoring this uncertainty can be substantial. A call center that is planning to operate at 95 % utilization, can experience an actual utilization of 99.75 % with exploding waiting times when the arrival rate turns out to be 5 % higher than planned (see [7]).

The rise in computational power and large call center databases bring forth new forecasting techniques that increase the accuracy of the forecasts. In this paper we present a K-nearest neighbour algorithm for forecasting the arrival rate function dynamically. The method is based on comparing arrival rate functions on different days with the observed pattern so far. The K arrival rate functions that are closest, with respect to some distance function, to the observed pattern are used to forecast the call arrival rate function for the rest of the planning horizon. The algorithm is able to take into account the correlations in call center data found by [3]. Moreover, the algorithm does not require a model and can be applied without pre-processing the historical data. The resulting forecast of the K-nearest neighbour algorithm can be used to create more accurate calculations of the number of representatives that is needed to meet the service level (e.g., by using the non-stationary models).

The outline of the paper is as follows. In Sect. 2 we give the exact problem formulation. We then continue to present the K-nearest neighbour algorithm, that will be used to solve the problem in Sect. 3. A case study with the results of applying this algorithm are presented in Sect. 4. Finally, Sect. 5 concludes the paper by summarizing the main results.

2 The Forecasting Problem

Consider a call center whose statistics are stored in a large database. Typically, the statistics contain a lot of information on each individual call, such as the starting time, the end time, the waiting time, the handling agent, and much more. In practice, many call centers store summarized historical data only, due to the historically high cost of maintaining and storing large databases even if these reasons are no longer prohibitive. Let us therefore assume that the data is aggregated over a period of length Δt minutes. Note that when Δt is sufficiently small, we get the information on individual calls back. For our purpose of forecasting the call arrival rate function, we concentrate on the number of calls in a period of length Δt.

The length of a period Δt together with the opening hours of the call center define n periods over the day, which we denote by $t_1, \ldots, t_n$. Assume that the database contains m records, i.e., data on aggregated call arrivals of m days. Then we can represent the database by an $m \times n$ matrix H, where entry $h_{d,i}$ represents the number of calls that occurred on day d in period t_i of length Δt for $d = 1, \ldots, m$ and $i = 1, \ldots, n$.

Data analysis (see, e.g., [3, 15]) shows that the shape of the call arrival rate function on a particular day of the week is usually the same over different weeks. In case of an unusual event, e.g., a holiday, the shape differs significantly. Therefore, in practice, the data is cleaned first by removing records for weeks containing unusual events, and the matrix H is divided into submatrices $H^{(1)}, \ldots, H^{(7)}$ per day of the week. Conventional forecasting algorithms base their forecast on this data by taking the mean number of call arrivals for that period based on the historical values of that day. Thus, the forecast F_i^{CV} for the number of calls in period t_i on day j of the week is given by

$$F_i^{\mathrm{CV}} = \frac{1}{k} \sum_{d=1}^{k} h_{d,i}^{(j)},$$

for $i = 1, \ldots, n$, where k is the number of records in the submatrix $H^{(j)}$.

The conventional algorithm ignores additional structure that is present in the call arrival pattern. From data analysis it is known that there is a significant correlation in call arrivals across different time periods in the same planning horizon. Typically, the first few hours of a day often provide significant information about the call volumes for the remainder of the day. The models suggested in [2, 13, 15] use this structure to determine the total number of calls over the day and distribute the calls according to the correlated structure over the different periods. These models, however, mostly focus on estimation rather than on prediction and also rely on the time-consuming data analysis and preparation to determine parameters of the models.

The discussion above shows that there is still a need for efficient computational algorithms that do not require the tedious data analysis and use the correlation structures in the call arrivals to yield accurate forecasts. Consequently, the algorithm should also be able to update the forecast as soon as new information on call arrivals is available. A possible way to do this is to compare arrival rate functions of different days with the observed pattern so far. The arrival rate functions that are closest, with respect to some distance function, to the observed pattern can then be used to forecast the call arrival rate function for the rest of the planning horizon. This leads to the class of nearest neighbour algorithms. The distance function can be used to capture the correlation structures in call arrivals. Moreover, the algorithm does not require a model and can be applied without pre-processing or analyzing the historical data.

3 The K-Nearest Neighbour Algorithm

The K-nearest neighbour algorithm (see, e.g., [4]) is a machine learning technique that belongs to the class of instance-based learners. Given a training set, a similarity measure over patterns, and a number K, the algorithm predicts the output of a new instance pattern by combining (e.g., by means of weighted average) the known outputs of its K most similar patterns in the training set. The training data is stored in memory and only used at run time for predicting the output of new instances. Advantages of this technique are the (implicit) construction of a local model for each new instance pattern, and its robustness to the presence of noisy training patterns (cf., e.g., [14]). Nearest neighbour algorithms have been successfully applied to many pattern recognition problems, such as scene analysis (see [5]) and robot control (see [1]).

Application of K-nearest neighbour algorithms to our forecasting problem in call centers translates into the prediction of the arrival rate function until the end of the planning horizon, based on matching K arrival rate functions to the pattern observed so far. More formally, suppose that we have observed the call arrival rates $r_1, \ldots, r_x$ on a specific day in periods $t_1, \ldots, t_x$ where $x < n$. Let us call this information the reference trace $\boldsymbol{r}_x = (r_1, \ldots, r_x)$. The nearest neighbour algorithm compares the reference trace to historical data $\boldsymbol{h}_{d,x} = (h_{d,1}, \ldots, h_{d,x})$ from the matrix H for $d = 1, \ldots, m$. The comparison is based on a distance function $\mathrm{D}(\boldsymbol{r}_x, \boldsymbol{h}_{d,x})$, which is defined by a norm on the space of the traces. The K nearest traces with respect to the distance function D are used to generate a forecast for the arrival rates $\hat{r}_{x+1}, \ldots, \hat{r}_n$ in periods $t_{x+1}, \ldots, t_n$. The result depends on the value of K as well as on the choice of the distance measure D.

In the next subsections we will describe two distance functions, the Euclidean distance and the Pearson correlation distance, that we will use in the numerical experiments. The former distance function can be seen as the conventional forecasting algorithm when restricted to the submatrices $H^{(1)}, \ldots, H^{(7)}$. The latter distance function is our novel approach to forecasting the call arrival rate function. Each subsection will motivate the choice of the distance function, and describe how to generate forecasts.

Euclidean Distance (ED)
The Euclidean distance is the most widely used and the most natural distance function to use. It is defined by the Euclidean norm

$$\mathrm{ED}(\boldsymbol{r}_x, \boldsymbol{h}_{d,x}) = \Big[\sum_{i=1}^{x}(r_i - h_{d,i})^2\Big]^{1/2}.$$

The effect of using this distance function in the nearest neighbour algorithm is that traces are considered to be near when the arrival rates in the historical data almost exactly match the observed arrival rates in the different periods. By selecting the K nearest traces, say on days $d_1, \ldots, d_K$, the forecast for period t_i is given by

$$F_i^{\mathrm{ED}} = \frac{1}{K}\big[h_{d_1,i} + \cdots + h_{d_K,i}\big],$$

for $i = x+1, \ldots, n$.

Note that by restricting to a specific day j of the week, i.e., to the submatrix $H^{(j)}$, we get the conventional forecast F_i^{CV}. Also observe that the distance function is sensitive to trends and days with special events. Hence, this distance function actually requires cleaned historical data without trends and days with special events. Therefore, the forecast needs to be adjusted for these situations.

Pearson Correlation Distance (PD)
The Pearson Correlation distance is based on the correlation coefficient between two vectors. A correlation value close to zero denotes little similarity, whereas a value close to one signifies a lot of similarity. Hence, the definition of the distance function is given by

$$\begin{aligned}\mathrm{PD}(\boldsymbol{r}_x, \boldsymbol{h}_{d,x}) &= 1 - |\mathrm{correlation}(\boldsymbol{r}_x, \boldsymbol{h}_{d,x})| \\ &= 1 - \left|\frac{(x-1)\sum_{i=1}^{x}(r_i - \mathrm{mean}(\boldsymbol{r}_x))\,(h_{d,i} - \mathrm{mean}(\boldsymbol{h}_{d,x}))}{\sum_{i=1}^{x}(r_i - \mathrm{mean}(\boldsymbol{r}_x))^2\,\sum_{i=1}^{x}(h_{d,i} - \mathrm{mean}(\boldsymbol{h}_{d,x}))^2}\right|,\end{aligned}$$

where $\mathrm{mean}(\boldsymbol{r}_x) = \frac{1}{x}\sum_{i=1}^{x} r_i$ and $\mathrm{mean}(\boldsymbol{h}_{d,x}) = \frac{1}{x}\sum_{i=1}^{x} h_{d,i}$. Note that the function looks at similarities in the shape of two traces rather than the exact values of the data. Consequently, this function answers the need to capture correlation structures in the call rates. Moreover, it does not require cleaned data, because it is less sensitive to trends and special events.

When the K nearest traces have been selected, say on days $d_1, \ldots, d_K$, the forecast for period t_i cannot be generated as in the case of the Euclidean distance. Since the distance function selects traces based on the shape, the offset of the trace for day d_j needs to be adjusted by

$$c_{d_j} = \frac{1}{x}\Big[\sum_{i=1}^{x}(r_i - h_{d_j,i})\Big].$$

Therefore, the forecast for period t_i is given by

$$F_i^{\mathrm{PD}} = \frac{1}{K}\big[(h_{d_1,i} + c_{d_1}) + \cdots + (h_{d_K,i} + c_{d_k})\big],$$

for $i = x + 1, \ldots, n$.

4 Numerical Experiments

In this section we illustrate the nearest neighbour algorithm using real call center data of an Israeli bank. The call center data with documentation are freely available from http://iew3.technion.ac.il/serveng/callcenterdata/. The data contains records during a period of a year. The call center is staffed from 7 am to midnight from Sunday to Thursday, it closes at 2 pm on Friday, and reopens at 8 pm on Sunday.

In our experiments we take for practical staffing purposes Δt to be equal to 15 min. Thus, taking the opening hours on Sunday until Thursday into account, we have 68 periods for which the number of arriving calls are aggregated per 15 min. Given these 5 days, we therefore have 5×52 records, which are used to construct the matrix H and the submatrices $H^{(1)}, \ldots, H^{(5)}$.

In order to evaluate the quality of the forecasts produced by the nearest neighbour algorithm, we pick a reference day and assume that the call arrival function $\boldsymbol{r}_x$ up to period $x \in \{1, \ldots, 68\}$ is known. The nearest neighbour algorithm then selects K traces that are nearest to $\boldsymbol{r}_x$ and generates forecasts as explained in the previous section. In principle, the nearest neighbour algorithm can be run every time new data of the reference trace is available. In practice, reacting to every new forecast might lead to many different staffing configurations which might not be manageable. Therefore, we choose to update the forecast only at specific moments of the day. Based on the data of the call center, see Fig. 1, we identify three moments over the day at which the forecast is updated. Figure 1 shows the mean arrival rates for

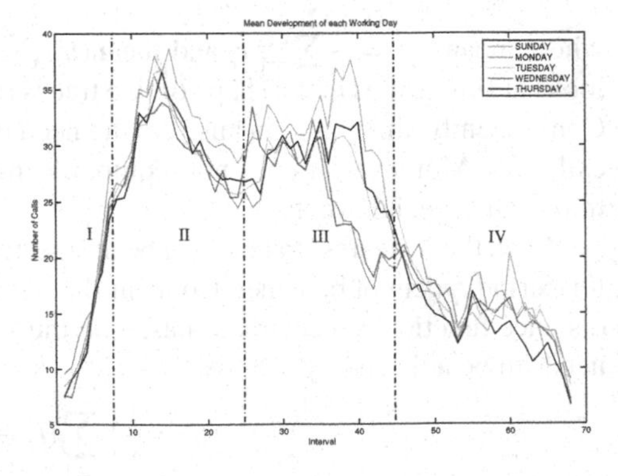

Fig. 1 Forecast update moments over the day

the different days of the week, and identifies 9 am (start of period 9), 1 pm (start of period 25), and 6 pm (start of period 45) as update moments.

Having described the setup of the experiments, we continue to describe the forecasting algorithms that we will compare in the experiments. We shall compare the conventional way of forecasting (CV), described in Sect. 2, and the K nearest neighbour algorithm with the Euclidean distance (ED) and the Pearson correlation distance (PD), described in Sect. 3. Note that the conventional algorithm CV is equivalent to ED when K is set equal to the total number of records available. These algorithms will also be compared by forecasting based on separated days of the week $H^{(1)}, \dots, H^{(5)}$ (SWD) and combined days of the week H (CWD).

Evaluation from a Statistical Perspective

In this subsection we will compare the algorithms CV, ED, and PD based on SWD and CWD, resulting in 6 algorithms. Since the presentation of these algorithms for 5 days of the week with three update moments in a day (giving 15 combinations) is too extensive, we only present the Wednesday with 1 pm as the update moment as a representative case. We evaluate the forecast at the start of period 25 and we define the *error of the forecast* as the average over the absolute differences between the forecasted number of calls and the actual realization in periods $25, \dots, 44$. It is not necessary to take the other periods into account, since at the start of period 45 the algorithm will update the forecast again.

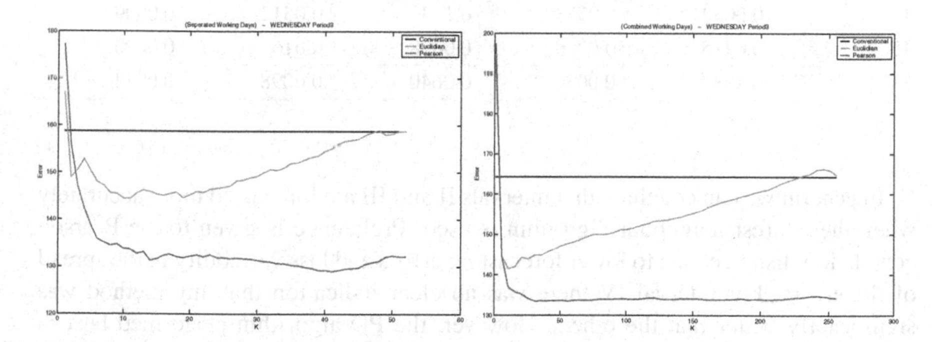

Fig. 2 Forecast errors for CV, ED, and PD under SWD and CWD as a function of K

We start by comparing the different algorithms by studying the error as a function of the size of the neighbourhood K. Figure 2 shows the error for the CV, ED, and PD algorithms under both SWD (left figure) and CWD (right figure) as a function of K. Both figures consistently show that the PD algorithm performs better than the CV and the ED algorithms. Also, the PD algorithm does not need data preparation and actually profits from working with the full dataset that is not split up for every day of the week. The results under CWD seem to be more stable for different values of K. Moreover, under CWD already moderate values for K suffice to run the PD algorithm.

The quality measure used to compare the performance of the considered algorithms is an average of the errors obtained by taking every Wednesday out of the 52 Wednesdays as a reference day. Therefore, it is interesting to analyze the variability in the errors of these forecasts. The variance for the nearest neighbour algorithms are denser around the lower error values. This strengthens the conclusion that the nearest neighbour algorithm outperforms the conventional forecasts. More formally, one can statistically test which of the algorithms performs better by using the Wilcoxon rank test. This test determines if for each pair of methods, with each using its best value of K, the median of the first method is significantly lower than that of the second one. By carrying out the test for every pair of methods (CV, ED, and PD), for every dataset (SWD $H^{(1)}, \ldots, H^{(5)}$, and CWD H), and for every update moment we obtained the significance values for all the data. For illustration we give the significance values in Table 1 for the comparison of ED and CV under SWD.

Table 1 shows, for each interval (see Fig. 1) and day combination, if the ED algorithm performs better than the CV algorithm. With a significance level of $\alpha = 5\,\%$, the table shows that for interval III the ED is significantly better than CV. With the exception of Tuesday and Wednesday, it is also better in interval II. Interval IV is somewhat unclear; Monday and Wednesday are predicted more accurately.

Table 1 Significance values for ED and CV under SWD

	Sunday	Monday	Tuesday	Wednesday	Thursday
II	0.0052	0.0373	0.1931	0.0512	0.0009
III	0.0215	0.0210	0.0008	0.0167	0.0000
IV	0.1988	0.0036	0.0640	0.0098	0.0934

In general we can conclude that intervals II and III are forecasted more accurately when the nearest neighbour algorithm is used. Preference is given to the Pearson correlation distance, due to lower forecasting errors and less variability in the spread of the errors. For interval IV there was no clear indication that any method was significantly better that the others. However, the PD algorithm performed best in most cases.

Evaluation from a Staffing Perspective

In this subsection we evaluate the results of the algorithms in the context of staffing in call centers. We first illustrate the effect of the nearest neighbour forecasting algorithm on staffing when the SIPP approach is used (see [9]). Next, we study the effect on staffing based on transient models (see [11]). We assume that the call center has a service level target specifying that 80 % of the customers should have a service representative on the line within 20 s. From the call center data, we can derive that the average service time of the representatives is 2.4 min per call. The staffing can now be based on the Erlang-C formula (see [6]), using the forecasts of the call arrivals, such that the service level is met in every period.

Table 2 Numerical experiments for SIPP staffing

h_i	n_i	SL_i	h_i^{CV}	n_i^{CV}	SL_i^{CV}	h_i^*	n_i^*	SL_i^*
23	6	0.846	28.706	7	0.940	20.821	6	0.846
23	6	0.846	33.627	8	0.979	22.089	6	0.846
20	6	0.914	30.176	8	0.991	22.309	6	0.914
21	6	0.894	29.137	7	0.962	25.016	7	0.962
19	5	0.812	28.333	7	0.977	23.431	6	0.931
18	5	0.845	30.608	8	0.995	26.041	7	0.983
25	7	0.911	29.941	8	0.966	25.382	7	0.911
21	6	0.894	28.510	7	0.962	24.187	6	0.894
24	6	0.817	27.922	7	0.927	23.455	6	0.817
24	6	0.817	30.333	8	0.973	26.065	7	0.927
20	6	0.914	30.961	8	0.991	27.480	7	0.970
27	7	0.873	27.196	7	0.873	22.577	6	0.708
15	5	0.921	22.980	6	0.976	18.675	5	0.921
20	6	0.914	23.039	6	0.914	18.138	5	0.775
11	4	0.912	23.196	6	0.994	18.528	5	0.976
14	4	0.813	22.333	6	0.982	18.211	5	0.939
14	4	0.813	19.549	6	0.982	14.797	5	0.939
8	3	0.871	18.431	5	0.994	13.333	4	0.969
16	5	0.899	20.549	6	0.968	15.870	5	0.899
17	5	0.874	19.843	6	0.957	15.163	5	0.874
	108	0.869		137	0.960		116	0.890

We analyze three scenarios in Table 2 for staffing on a specific Wednesday in interval III (i.e., periods 25 to 44). The first scenario uses the real call arrival rates h_i to staff the minimum number of service representatives n_i such that the attained service level SL_i is above 80 %. The second scenario uses the conventional forecasting method to derive estimates h_i^{CV} of the call arrival rate. Based on these forecasts, the number of representatives n_i^{CV} is determined such that the service level is met. However, in reality the service level experienced under the real call arrival rates is given by SL_i^{CV}. The last scenario is similar to the second with the exception that the forecast is based on the nearest neighbour algorithm with the Pearson correlation distance.

Table 2 shows that the conventional method systematically overestimates the call arrivals. This results in planning 137 representatives and an average weighted service level $\sum_i h_i \text{SL}_i / \sum_i h_i$ of 0.960. The nearest neighbour algorithm does not meet the service level in every period. It staffs 116 representatives with an average weighted service level of 0.890. It is close to the optimal staffing level of 108 with an average weighted service level of 0.869. In general, numerical experiments show that the nearest neighbour algorithm produces results closer to the optimal staffing levels and the optimal average weighted service level.

To illustrate the effect of the forecasting methods on staffing with transient models, we model the call center as an $M_t/M/c_t$ queueing system. The forecasts in scenarios 2 and 3 are updated at the beginning of interval II, III, and IV. As in the previous case, scenario 2 uses the conventional forecasting algorithm. In scenario 3, the best forecasting algorithm is used in each interval, i.e., the forecasting algorithm that gives the lowest errors in each interval for Wednesdays. Thus, the conventional forecast is used in interval I, the nearest neighbour forecasting algorithm PD is used in interval II and III, and for interval IV the nearest neighbour algorithm ED is used. Due to the computational complexity of this experiment, we take Δt to be 1 h, with an additional half hour at the end of the day for handling all remaining calls in the system. For the same reason, the service level is set to have an average speed of answer (ASA) of at most 30 s.

Table 3 shows the results for the three scenarios for several days (hence the index d instead of i). When we compare scenario 2 and 3 with each other we cannot draw firm conclusions. Out of the 50 Wednesdays that we examined, the nearest neighbour algorithm performed better in 26 cases, worse in 17 cases, and had similar performance in 7 cases. The variability in performance is caused by the estimate in interval I. The staffing based on this estimate can result in an ASA far from 30. This result has a big influence on the staffing in the next intervals, since the staffing levels will be adjusted to correct the average speed of answer. Hence, dynamic forecasting in combination with transient models should be done more carefully.

Table 3 Numerical experiments for staffing with transient models

n_d	SL_d	n_d^{CV}	$\mathrm{SL}_d^{\mathrm{CV}}$	n_d^*	SL_d^*
89	29.5783	91	30.9722	90.5	34.7052
84	27.8924	88	27.6744	105	32.7986
70	27.8958	86	12.1418	74	29.2221
102	29.0204	139	29.2838	134	29.8620
92	29.9510	95.5	30.0246	95.5	30.2343
117	29.7751	148	33.9807	142	34.6535

5 Conclusions

We have investigated the effectiveness of K-nearest neighbour algorithms for forecasting call volumes in call centers, based on the Euclidean and the Pearson correlation distance. From a statistical point of view, the nearest neighbour algorithm yields significantly more accurate predictions than the conventional method. Additionally, the nearest neighbour algorithm does not require data preparation and captures the correlation structures typically found in call center data. From a staffing perspective, the average weighted service level under the hybrid nearest neighbour algorithm performs very well. When the staffing is based on a transient model, however, no firm conclusion can be drawn. Hence, dynamic forecasting when using transient models warrants more research.

References

1. Atkeson, C., Moore, A., Schaal, S.: Locally weighted learning. Artif. Intell. Rev. **11**, 11–73 (1997)
2. Avradimis, A., Deslauriers, A., L'Ecuyer, P.: Modeling daily arrivals to a telephone call center. Manage. Sci. **50**, 896–908 (2004)
3. Brown, L., Gans, N., Mandelbaum, A., Sakov, A., Shen, H., Zeltyn, S., Zhao, L.: Statistical analysis of a telephone call center: a queueing-science perspective. J. Am. Stat. Assoc. **100**, 36–50 (2002)
4. Cover, T., Hart, P.: Nearest neighbour pattern classification. IEEE Trans. Inf. Theory **13**, 21–27 (1967)
5. Duda, R., Hart, P.: Pattern Classification and Scene Analysis. Wiley, New York (1973)
6. Erlang, A.: Solutions of some problems in the theory of probabilities of significance in automatic telephone exchanges. Electroteknikeren **13**, 5–13 (1917)
7. Gans, N., Koole, G., Mandelbaum, A.: Telephone call centers: tutorial, review, and research prospects. M&SOM **5**, 79–141 (2003)
8. Green, L., Kolcsar, P.: The pointwise stationary approximation for queues with non-stationary arrivals. Manage. Sci. **37**, 84–97 (1991)
9. Green, L., Kolesar, P., Soares, J.: Improving the SIPP approach for staffing service systems that have cyclic demands. Oper. Res. **49**, 549–564 (2001)
10. Halfin, S., Whitt, W.: Heavy-traffic limits for queues with many exponential servers. Oper. Res. **29**, 567–587 (1981)
11. Ingolfsson, A., Haque, M., Umnikov, A.: Accounting for time-varying queueing effects in workforce scheduling. EJOR **139**, 585–597 (2002)
12. Ingolfsson, A., Cabral, E., Wu, X.: Combining integer programming and the randomization method to schedule employees. EJOR **202**, 153–163 (2003)
13. Jongbloed, G., Koole, G.: Managing uncertainty in call centers using Poisson mixtures. Appl. Stoch. Models Bus. Ind. **17**, 307–318 (2001)
14. Mitchell, T.: Machine Learning. McGraw Hill (1997)
15. Steckley, S., Henderson, S., Mehrotra, V.: Service system planning in the presence of a random arrival rate. Submitted (2004)
16. Yoo, J.: Queueing models for staffing service operations. Ph.D. thesis, University of Maryland (1996)

A Rule Based Framework for Smart Training Using sEMG Signal

Giorgio Biagetti, Paolo Crippa, Laura Falaschetti, Simone Orcioni and Claudio Turchetti

Abstract The correctness of the training during sport and fitness activities involving repetitive movements is often related to the capability of maintaining the required cadence and muscular force. Muscle fatigue may induce a failure in maintaining the needed force, and can be detected by a shift towards lower frequencies in the surface electromyography (sEMG) signal. The exercise repetition frequency and the evaluation of muscular fatigue can be simultaneously obtained by using just the sEMG signal through the application of a two-component AM-FM model based on the Hilbert transform. These two features can be used as inputs of an intelligent decision making system based on fuzzy rules for optimizing the training strategy. As an application example this system was set up using signals recorded with a wireless electromyograph applied to several healthy subjects performing dumbbell biceps curls.

Keywords Surface electromyography (sEMG) · AM-FM decomposition · Muscle fatigue · Exercise cadence · Training · Fuzzy logic · Intelligent system

G. Biagetti · P. Crippa (✉) · L. Falaschetti · S. Orcioni · C. Turchetti
DII – Dipartimento di Ingegneria Dell'Informazione, Università Politecnica delle Marche, Via Brecce Bianche 12, 60131 Ancona, Italy
e-mail: g.biagetti@univpm.it

P. Crippa
e-mail: p.crippa@univpm.it

L. Falaschetti
e-mail: l.falaschetti@univpm.it

S. Orcioni
e-mail: s.orcioni@univpm.it

C. Turchetti
e-mail: c.turchetti@univpm.it

R. Neves-Silva et al. (eds.), *Intelligent Decision Technologies*, Smart Innovation, Systems and Technologies 39,
DOI 10.1007/978-3-319-19857-6_9

1 Introduction

Often sportive or training activities require the execution of repetitive movements. For some activities, such as cycling, running, the estimation of muscle metabolism is based on heart rate, oxygen uptake, lactate production, ventilatory threshold and other variables commonly used in sports medicine [15] requiring special instrumentation such as the cycloergometer. Additionally, recording forces produced by muscles during many physical performances is equally unpractical, often requiring use of special force or torque sensors, which can be bulky, expensive, and not user friendly.

The analysis of surface electromyography (sEMG) signals offers a simple alternative method for quantifying [9] and classifying [24] the muscular activity, and can be a practical tool when exercising on strength training machines or lifting freeweights, to set up ad-hoc training sessions, maximize efficiency, and even preventing injuries [16].

The spectral parameters derived from the EMG signal, such as mean frequency or median frequency, can be used to evaluate muscular fatigue [13, 21, 22, 25]. In fact, during a sustained isometric contraction, there is an increase in the amplitude of the low frequency band and a relative decrease in the higher frequencies, which is called EMG spectrum compression [27].

However, for dynamic or cyclic movements, or for contraction levels higher than 50 % of maximum voluntary contraction, the EMG is a non-stationary signal, thus the physical meaning of the overall spectrum is reduced since amplitude and frequency change over time. To deal with this variability, more sophisticated signal analysis techniques have been proposed [2, 5, 6, 17, 18, 26].

Sinusoidal AM-FM models are representations of signals that can be considered as resulting from simultaneous amplitude modulation and frequency modulation, where the carriers, amplitude envelopes, and instantaneous frequencies (IFs) need to be estimated [10–12]. Recent works have proposed different methodologies based on AM-FM models for evaluating fatigue from EMG signal in repetitive movements [1, 4, 14, 19].

The mean frequency of the amplitude spectrum (MFA) of the EMG signal, considered as a function of time, is directly related to the dynamics of the movement performed and to the fatigue of the involved muscles. If the movement is cyclic, MFA will display the same cyclic pattern, but its average will tend to decrease as the muscle becomes fatigued, due to the reduced conduction velocity of muscle fibers that cause a shift of the spectrum towards lower frequencies. These two effects have been simultaneously modeled by a multicomponent (two-component) AM-FM model [3].

More in detail we applied to the MFA of the EMG signal an AM-FM technique based on the Hilbert transform that is able to simultaneously extract features that are estimations of the cadence and of the resulting muscle fatigue. These features represent a simple and near real-time “summary” of the exercise and can be used by a fuzzy-rule-based decision making system to direct and try to maximize the effectiveness of the training.

Fuzzy-based decision making systems [20] are particular suitable for this kind of application as they can easily embed the vast amount of knowledge on training that can be collected from experts in the field. They are also well suited to *ad hoc* hardware and software implementation of their fuzzy membership functions (MFs), as reported in [7, 8, 23], and the parameters of their MFs can be estimated by using data obtained from previous exercises.

2 Exercise Evaluation

In order to evaluate the exercise execution and its potential effects on the training activity, we record an sEMG signal from the involved muscle and extract some fundamentals parameters according to the algorithm presented in [3] and briefly summarized in the following.

A flow-chart of the whole system is depicted in Fig. 1. First, a feature extractor simultaneously estimates both the cadence at which the exercise was performed and the resulting muscle fatigue. This data is then fed into a fuzzy engine, which ultimately gives suggestions as to how to proceed in the training, and whose rules are determined beforehand in accordance with the training objectives.

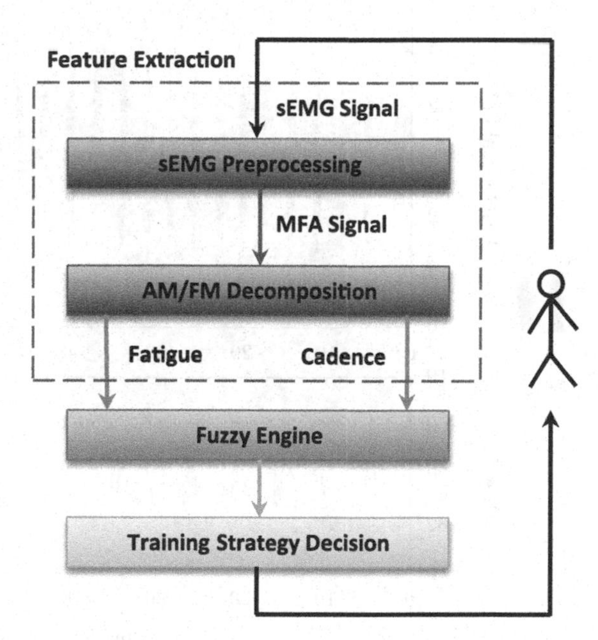

Fig. 1 Flowchart of the system

2.1 sEMG Feature Extraction

The recorded sEMG signal is first conditioned by passing it through a high-pass filter to remove some of the acquisition artifacts (e.g., due to cable movement), then the background noise during rest periods is cancelled. Let $y(t)$ be this cleaned signal. We compute the MFA trajectory from its sliding fast Fourier transform $Y(t, \omega)$, as

$$x(t) = \frac{1}{2\pi} \frac{\displaystyle\int_0^{f_H} \omega\, |Y(t,\omega)|\, d\omega}{\displaystyle\int_0^{f_H} |Y(t,\omega)|\, d\omega} \tag{1}$$

where f_H is the bandwidth of the electromyograph used to record the signals.

An example of such an MFA trajectory is shown in Fig. 2, together with its corresponding EMG signal. Its shape suggests that it can be well approximated by a simple two-component AM-FM model

$$x(t) \simeq \sum_{i=1}^{2} \hat{a}_i(t) \cos\left(\hat{\varphi}_i + \int_0^t \hat{\omega}_i(\tau)\, d\tau\right) \tag{2}$$

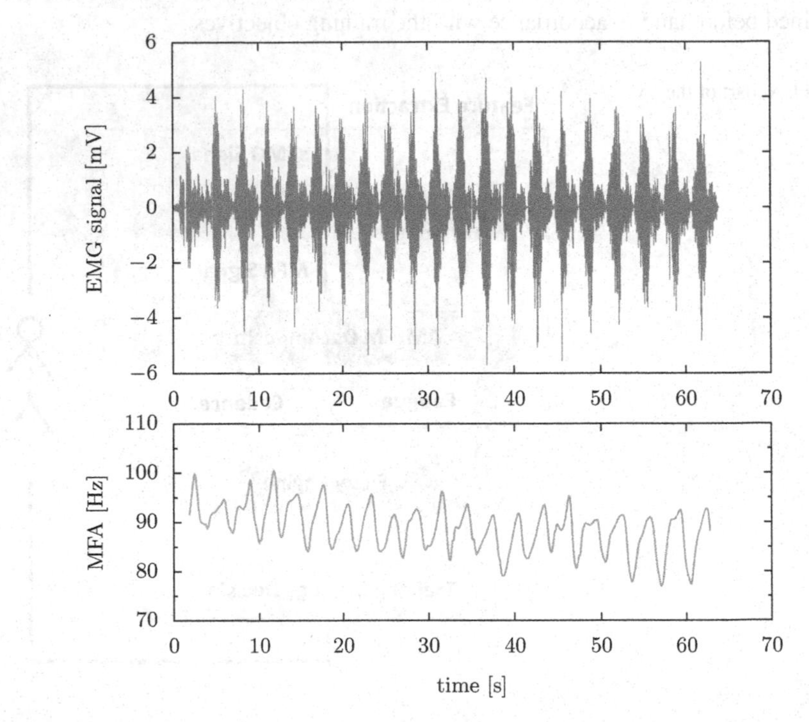

Fig. 2 EMG signal recorded from the *biceps brachii* muscle during a biceps curl exercise with a 3 kg dumbbell (top), and corresponding MFA trajectory (bottom)

where $\hat{a}_i(t)$, $\hat{\varphi}_i$, and $\hat{\omega}_i(t)$ are the estimated component's amplitude, initial phase, and frequency, respectively. The first component models the slowly time-varying average frequency decreasing trend due to fatigue, and the second component models the oscillations in frequency due to the various phases of performing the exercise, thus allowing the cadence to be extracted.

The trend of $x(t)$ is thus captured by the amplitude of the first component $\hat{a}_1(t)$ alone, with $\hat{\omega}_1(t) \simeq 0$ since the fatigue status is generally monotonic and not cyclic during a single exercise. On the other hand, the amplitude of the second component has little meaning (it's the difference between the "peak" and the "mean" MFA during one cycle of the movement), but its frequency $\hat{\omega}_2(t)$ corresponds to the cadence at which the exercise was performed. Figure 3 reports an example of these curves extracted by the algorithm in [3] from the signal of Fig. 2. The amplitude of the first component $\hat{a}_1(t)$ fits with the decreasing trend of $x(t)$, while $\hat{f}_2(t) = 2\pi\,\hat{\omega}_2(t)$ fits the (possibly varying) pace of the exercise.

A well-known fatigue index is the relative slope of the linear regression of $\hat{a}_1(t)$, that is, β/α if $\hat{a}_1(t) \approx \alpha + \beta\,t$. This value is reported, together with the mean cadence, in Table 1, which shows results obtained from 5 healthy subjects performing biceps curl exercises with different weights ranging from 2 kg to 7 kg, selected according to the level of fitness of each individual.

3 Fuzzy Engine for Training Strategy Decision

Once an exercise is performed, we have the two parameters called "fatigue" and "cadence" extracted as previously described. These are used as inputs to a fuzzy engine whose rules come from already available training experience.

An example is given below. We consider the problem of selecting the proper weight for training the *biceps brachii* muscles. The data reported in Table 1 can be used to help derive the shape of a few membership functions. For instance, the fatigue has been classified as "low", "medium", or "high" according to the weight selected by the subject. We assume that a 2 kg dumbbell produced a "low" fatigue status, and so on. The data from the 7 kg exercise was not used as there were not enough points. For each class we discarded the lowest and highest measurement and considered the remaining range of values as 100 % belonging to that class. Membership functions (MFs) are then tapered between these selected ranges, as shown in Fig. 4. The expected output of the system is a hint on the weight to use next, expressed as a percentage of increment, whose MFs are also shown in the same figure.

The set of rules used by the system are reported in Table 2, and the resulting input-output function in Fig. 5. For the inference algorithm, we used the mix/max functions for logic operations, product/sum functions for implication and aggregation, and centroid-based defuzzyfication.

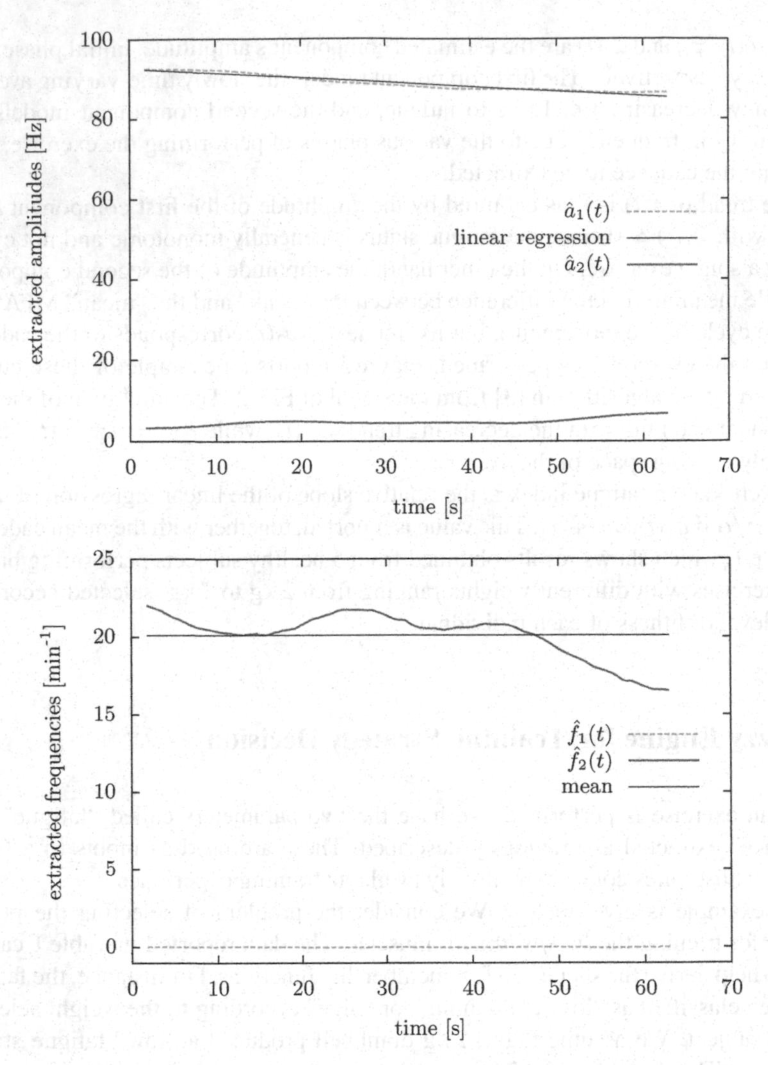

Fig. 3 Demodulated amplitudes and frequencies from the MFA signal of Fig. 2. The linear regression of $\hat{a}_1(t)$ and the mean of $\hat{f}_2(t)$ are also shown as they are used to obtain a compact representation of fatigue and cadence for the overall exercise

To validate the effectiveness of the set of rules previously enumerated, the outcome of the fuzzy engine was computed for each of the input corresponding to the data reported in Table 1. Results are shown in Table 3. The obtained results seems to be in good accordance to the level of fitness declared by the subjects that performed the exercise.

Table 1 Fatigue test performed on 5 healthy subjects: estimated rate of MFA variation and mean cadence, [%/ min] @ [reps/ min]. Data from [3]

Subject	Weight			
	2 kg	3 kg	5 kg	7 kg
subject 1	−7.9 @ 17.7	−4.2 @ 20.5	−12.4 @ 21.4	
subject 2	−0.6 @ 23.5	−9.8 @ 21.7	−30.3 @ 18.6	
subject 3	−1.2 @ 20.9	−7.5 @ 20.1	−19.1 @ 18.3	
subject 4		−7.4 @ 23.4	−16.8 @ 24.5	−23.7 @ 20.4
subject 5		−12.0 @ 22.4	−14.3 @ 24.9	−34.3 @ 22.6

Table 2 Inference rules used to select the next weight to use. The "=" symbol means keep the current weight, "−" decrease the weight, "+" increase the weight, "++" increase much the weight

		cadence		
		slow	right	fast
fatigue	high	−	=	=
	med	=	+	+
	low	=	+	++

4 Conclusion

In this paper an efficient framework to aid in the selection of a training strategy to improve muscular strength has been presented. The framework uses a lightweight wireless electromyograph applied on the involved muscles, and from the sEMG signal thus recorded both the muscular fatigue and the repetition frequency during the accomplishment of cyclic movements is estimated, by means of a two-component AM-FM decomposition based on the Hilbert transform.

These two features have been used as inputs of a fuzzy rule-based pattern recognition system whose outputs are the guidelines needed for optimizing and customizing individual training sessions. As an application example, some experimental data extracted from dumbbell biceps curls have been used to set-up the fuzzy system and obtain the input-output relationship for this kind of exercise employing inference rules written by exploiting knowledge on training techniques.

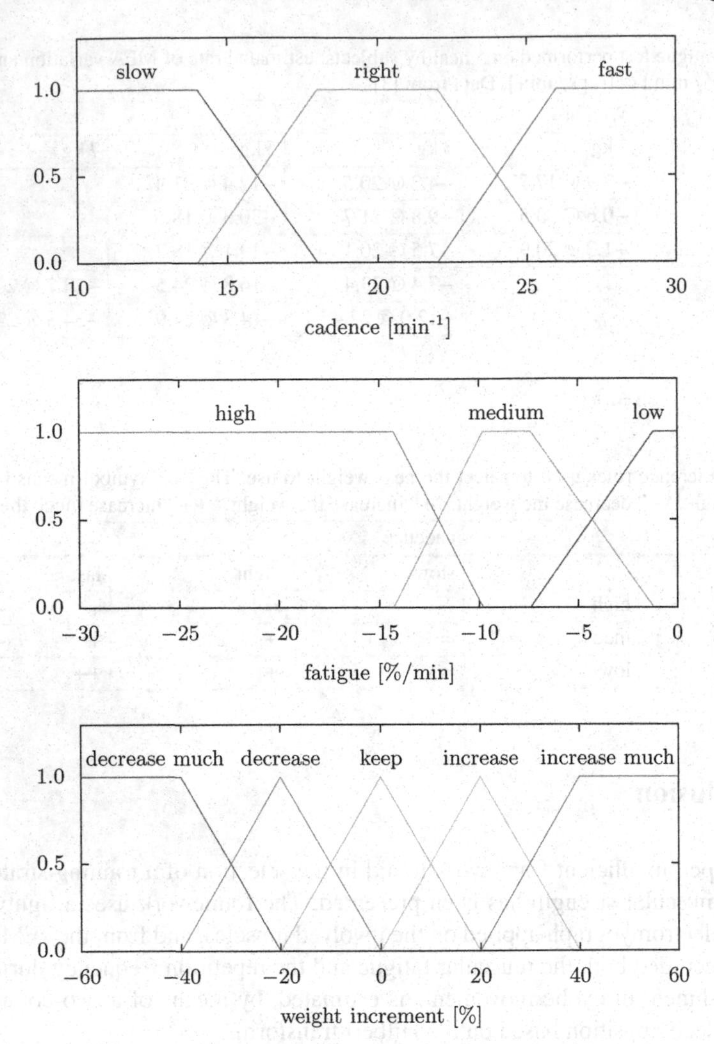

Fig. 4 Membership functions of the inputs and output of our fuzzy system

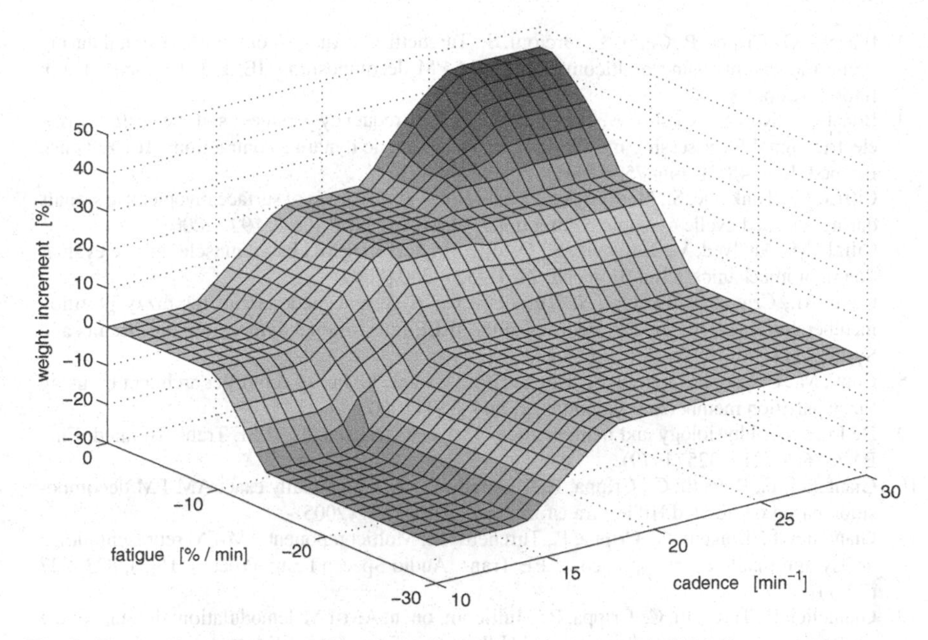

Fig. 5 Input-output relation resulting from the fuzzy rules reported in Table 2

Table 3 Results of the application of the fuzzy engine on the data of Table 1: suggested increase of the weight [%]

Subject	Weight			
	2 kg	3 kg	5 kg	7 kg
subject 1	+18.5	+20.0	+8.4	
subject 2	+31.9	+20.0	0.0	
subject 3	+20.0	+20.0	0.0	
subject 4		+20.0	0.0	0.0
subject 5		+10.2	0.0	0.0

References

1. Agostini, V., Knaflitz, M.: An algorithm for the estimation of the signal-to-noise ratio in surface myoelectric signals generated during cyclic movements. IEEE Trans. Biomed. Eng. **59**(1), 219–225 (2012)
2. Bai, F., Lubecki, T., Chew, C.M., Teo, C.L.: Novel time-frequency approach for muscle fatigue detection based on sEMG. In: IEEE Biomedical Circuits and Systems Conference (BioCAS). pp. 364–367 (2012)

3. Biagetti, G., Crippa, P., Curzi, A., Orcioni, S., Turchetti, C.: Analysis of the EMG signal during cyclic movements using multicomponent AM-FM decomposition. IEEE J. Biomed. Health Inform. (in press)
4. Bonato, P., Roy, S., Knaflitz, M., De Luca, C.: Time-frequency parameters of the surface myoelectric signal for assessing muscle fatigue during cyclic dynamic contractions. IEEE Trans. Biomed. Eng. **48**(7), 745–753 (2001)
5. Cifrek, M., Tonković, S., Medved, V.: Measurement and analysis of surface myoelectric signals during fatigued cyclic dynamic contractions. Measurement **27**(2), 85–92 (2000)
6. Cifrek, M., Medved, V., Tonković, S., Ostojić, S.: Surface EMG based muscle fatigue evaluation in biomechanics. Clin. Biomech. **24**(4), 327–340 (2009)
7. Conti, M., Crippa, P., Orcioni, S., Turchetti, C.: A current-mode circuit for fuzzy partition membership functions. In: Proceedings of the IEEE International Symposium on Circuits and Systems (ISCAS '99). vol. 5, pp. 391–394 (1999)
8. Conti, M., Crippa, P., Orcioni, S., Turchetti, C., Catani, V.: Fuzzy controller architecture using fuzzy partition membership functions. **2**, 864–867 (2000)
9. De Luca, C.: Physiology and mathematics of myoelectric signals. IEEE Trans. Biomed. Eng. BME **26**(6), 313–325 (1979)
10. Gianfelici, F., Biagetti, G., Crippa, P., Turchetti, C.: Asymptotically exact AM-FM decomposition based on iterated Hilbert transform. pp. 1121–1124 (2005)
11. Gianfelici, F., Biagetti, G., Crippa, P., Turchetti, C.: Multicomponent AM-FM representations: an asymptotically exact approach. IEEE Trans. Audio Speech Lang. Process. **15**(3), 823–837 (2007)
12. Gianfelici, F., Turchetti, C., Crippa, P.: Multicomponent AM-FM demodulation: the state of the art after the development of the iterated Hilbert transform. In: IEEE International Conference on Signal Processing and Communications, (ICSPC 2007). pp. 1471–1474 (Nov 2007)
13. González-Izal, M., Malanda, A., Navarro-Amézqueta, I., Gorostiaga, E.M., Mallor, F., Ibañez, J., Izquierdo, M.: EMG spectral indices and muscle power fatigue during dynamic contractions. J. Electromyogr. Kinesiol. **20**(2), 233–240 (2010)
14. Hotta, Y., Ito, K.: Detection of EMG-based muscle fatigue during cyclic dynamic contraction using a monopolar configuration. In: 2013 35th Annual International Conference of the IEEE Engineering in Medicine and Biology Society (EMBC). pp. 2140–2143 (2013)
15. Hug, F., Decherchi, P., Marqueste, T., Jammes, Y.: Emg versus oxygen uptake during cycling exercise in trained and untrained subjects. J. Electromyogr. Kinesiol. **14**(2), 187–195 (2004)
16. Izquierdo, M., Ibanez, J., Calbet, J.A.L., González-Izal, M., Navarro-Amézqueta, I., Granados, C., Malanda, A., Idoate, F., González-Badillo, J., Hakkinen, K., et al.: Neuromuscular fatigue after resistance training. Int. J. Sports Med. **30**(8), 614 (2009)
17. Karlsson, S., Yu, J., Akay, M.: Enhancement of spectral analysis of myoelectric signals during static contractions using wavelet methods. IEEE Trans. Biomed. Eng. **46**(6), 670–684 (1999)
18. Karlsson, S., Yu, J., Akay, M.: Time-frequency analysis of myoelectric signals during dynamic contractions: a comparative study. IEEE Trans. Biomed. Eng. **47**(2), 228–238 (2000)
19. Knaflitz, M., Bonato, P.: Time-frequency methods applied to muscle fatigue assessment during dynamic contractions. J. Electromyogr. Kinesiol. **9**(5), 337–350 (1999)
20. Lee, C.: Fuzzy logic in control systems: fuzzy logic controller. i. IEEE Trans. Syst. Man Cybern. **20**(2), 404–418 (1990)
21. Merletti, R., Conte, L.R.L.: Surface EMG signal processing during isometric contractions. J. Electromyogr. Kinesiol. **7**(4), 241–250 (1997)
22. Merletti, R., Knaflitz, M., De Luca, C.J.: Myoelectric manifestations of fatigue in voluntary and electrically elicited contractions. J. Appl. Physiol. **69**(5), 1810–1820 (1990)
23. Orcioni, S., Biagetti, G., Conti, M.: A mixed signal fuzzy controller using current mode circuits. Analog Integr. Circ. Sig. Process **38**(2–3), 215–231 (2004)
24. Ouyang, G., Zhu, X., Ju, Z., Liu, H.: Dynamical characteristics of surface EMG signals of hand grasps via recurrence plot. IEEE J. Biomed. Health Inform. **18**(1), 257–265 (2014)
25. Potvin, J., Bent, L.: A validation of techniques using surface EMG signals from dynamic contractions to quantify muscle fatigue during repetitive tasks. J. Electromyogr. Kinesiol. **7**(2), 131–139 (1997)

26. Ranniger, C., Akin, D.: EMG mean power frequency determination using wavelet analysis. In: Proceedings of the 19th Annual International Conference of the IEEE Engineering in Medicine and Biology Society. vol. 4, pp. 1589–1592 (1997)
27. Sakurai, T., Toda, M., Sakurazawa, S., Akita, J., Kondo, K., Nakamura, Y.: Detection of muscle fatigue by the surface electromyogram and its application. In: 2010 IEEE/ACIS 9th International Conference on Computer and Information Science (ICIS). pp. 43–47 (2010)

Distributed Speech Recognition for Lighting System Control

Giorgio Biagetti, Paolo Crippa, Alessandro Curzi, Laura Falaschetti, Simone Orcioni and Claudio Turchetti

Abstract This paper presents a distributed speech recognition (DSR) system for home/office lighting control by means of users' voice. In this scheme a back-end processes audio signals and transforms them into commands, so that they can be sent to the desired actuators of the lighting system. This paper discusses in detail the solutions and strategies we adopted to improve recognition accuracy and spotting command efficiency in home/office environments, i.e. in situations that involve distant speech and great amounts of background noise or unrelated sounds. Suitable solutions implemented in this recognition engine are able to detect commands also in a continuous listening context and the used DSR strategies greatly simplify the system installation and maintenance. A case study that implements the voice control of a digital addressable lighting interface (DALI) based lighting system has been selected to show the validity and the performance of the proposed system.

Keywords DSR · speech recognition · Vocal user interface · Feature extraction · MFCC · Cepstral analysis · Voice activity detection · Lighting control · DALI

G. Biagetti (✉) · P. Crippa · A. Curzi · L. Falaschetti · S. Orcioni · C. Turchetti
DII – Dipartimento di Ingegneria Dell'Informazione, Università Politecnica Delle Marche, Via Brecce Bianche 12, 60131 Ancona, Italy
e-mail: g.biagetti@univpm.it

P. Crippa
e-mail: p.crippa@univpm.it

A. Curzi
e-mail: a.curzi@univpm.it

L. Falaschetti
e-mail: l.falaschetti@univpm.it

S. Orcioni
e-mail: s.orcioni@univpm.it

C. Turchetti
e-mail: c.turchetti@univpm.it

R. Neves-Silva et al. (eds.), *Intelligent Decision Technologies*, Smart Innovation, Systems and Technologies 39,
DOI 10.1007/978-3-319-19857-6_10

1 Introduction

Voice is commonly viewed as one of the most easy-to-use methods for controlling home automation systems [4, 13]. In this work we aim to provide a speech interface to be the least obtrusive and combined with a robust automatic speech recognition (ASR) system [3]. Regarding the first objective, this is achieved by using an ad-hoc configuration that only requires at the user side, the installation of small and cheap audio front-ends (FEs), equipped with a panoramic microphone, and somehow connected to Internet. The home installation does not require other types of devices, as processing is performed by the speech recognition software running on a remote server in a distributed speech recognition (DSR) framework [7]. The FE extracts the representative parameters of speech (features), while the recognition of the transmitted message is performed by the back-end (BE) resident on the server, through the processing of the features stream. This configuration is extremely demanding for the ASR system, and requires solving a number of issues, namely:

- activation,
- speech capture,
- elimination of spoken text unrelated to the commands,
- user feedback,
- interpretation and execution of commands,
- system continuous listening.

With regard to problems related to the capture of the speech signal, the system must be able to distinguish voice from noise (as the microphone is always active), to recognize "distant speech", and to detect pronunciation errors. Problems related to the interpretation of commands arise since the user may not adhere perfectly to the grammar and the conversation might be confused with control commands. Thus, in order to achieve the second objective of realizing a robust ASR system, we focus our attention on two reasonably simple techniques that will help to deal with these problems: *(i)* an adaptive automatic voice activity detection (VAD) thresholding technique to enable the DSR system just when a spoken command is detected, and *(ii)* a mechanism that rejects non-command utterances, background noise, and unrelated sounds. This technique is based on the generation of a *garbage model* [8], which includes suitably placed decoy words [9] helping the ASR to identify out-of-vocabulary words, thus enabling it to discard non-command utterances. These strategies, together with an ad-hoc protocol to optimize the client/server communication, have resulted in a robust DSR system, as the experimental results show.

2 DSR Scheme

A DSR system is a speech recognition system with a client-server architecture [15]. The fundamental idea of such a system is to distribute the speech recognition processing between a client FE and a remote BE. The FE calculates and extracts representative parameters of speech (features) and discriminates commands from noise on the basis of an adaptive sound energy thresholding, also known as VAD. When the VAD detects an utterance, the FE sends a compressed stream of features to the BE, which replies with the recognised command, if any. Then, the FE can give suitable feedbacks to the user and can interact with the local home/office lighting control system (in this case a DALI command unit has been considered) to actually perform the desired command.

2.1 Front-End

The feature extraction algorithm we adopt in our voice control framework is based on the standard ETSI ES 201-212 [7], specifically defined for DSR. A block diagram of the algorithms involved in the DSR scheme is reported in Fig. 1.

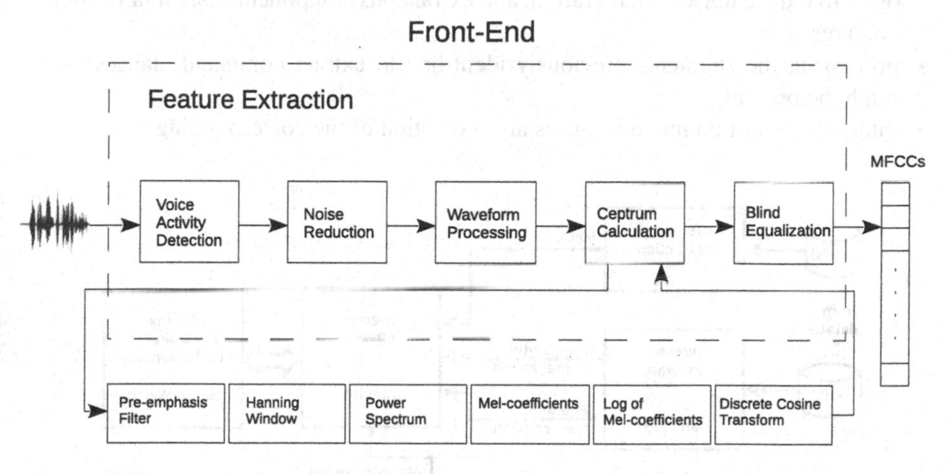

Fig. 1 ETSI standard scheme for feature extraction

The standard specification describes the FE algorithm for the computation of the Mel-cepstral features (MFCCs), from speech waveforms sampled at different rates (8 kHz, 11 kHz and 16 kHz). The feature vectors consist of 12 cepstral coefficients and a log-energy coefficient. In addition 26 dynamic features, the delta and delta-delta cepstral coefficients, are computed at the server-side, so that the final acoustic vector has 39 components. The MFCC coefficients are extracted from frames of 25 ms generated every 10 ms, thus consecutive 25 ms frames overlap by 15 ms. Before computing the cepstral coefficients, a noise reduction is performed and subsequently a waveform processing is applied to the clear signal. Then blind equalization of the resulting features is the final processing stage in the terminal front-end. In addition, algorithms for both pitch estimation and frame classification (voiced/unvoiced) have been implemented to suppress speaker mismatch.

This framework implements the ETSI feature extractor as a client module of the DSR in the recognition phase [10]. It has many advanced features also linked to the remote transmission of the features. The feature extraction module allows to calculate the features, of various types, given an input wave file or a list of files with an appropriate format. It supports various input and output formats, as well as various encodings of the features.

2.2 Back-End

The recognition of the message is performed at the BE by making extensive use of several Carnegie Mellon University Sphinx third-party libraries [14]. The recognition process steps are the following ones:

- identification of segments that potentially contain commands, their processing in order to reduce noise, reverberation, and extraneous components, and finally, their compression;
- processing the segments previously identified to extract command clauses that might be present;
- interpretation of command clauses and execution of the corresponding action.

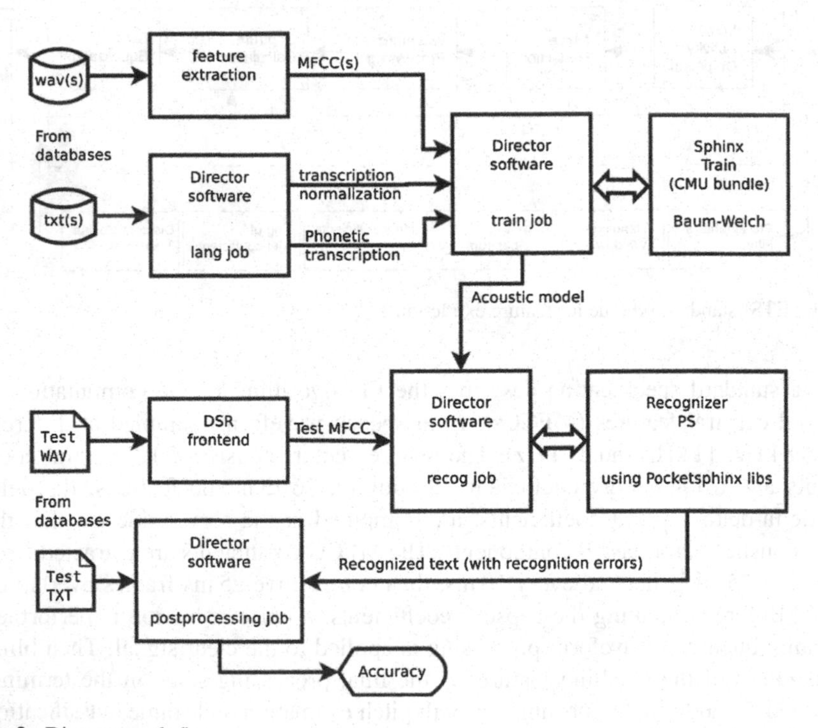

Fig. 2 Director work-flow

The BE returns the text corresponding to the spoken command, in case it satisfies the grammar rules. A device installed at the user site, typically the front-end itself, takes a decision on what action should be performed following the recognition of the command.

Our framework comprises also a special module, called director, that is set up as an assistant to the generation of acoustic and language models, generated almost exclusively through the tools included in Sphinx [12].

This module is responsible for:

- training of language models;
- training of acoustic models;
- recognition test;
- quality models analysis;
- adapting acoustic models.

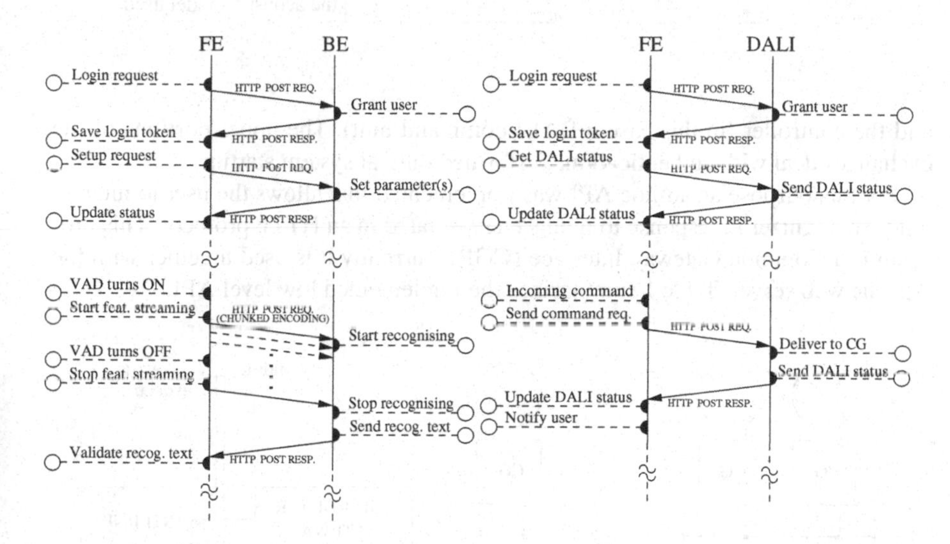

Fig. 3 DSR-dialogue

As it is shown in Fig. 2, all the operations of feature extraction, modeling, language processing, recognition and testing are coordinated by this module.

2.3 Communication Scheme

Figure 3 shows a scheme of the dialogue between the various entities that constitute the system. Data are exchanged between the audio FE and the recognition BE. After a command is spotted and recognized, another exchange takes place between the FE

Table 1 DSR HTTP API (low level API)

API Function	Method	Description
login	POST	obtains an access token (OTP) for authenticated requests
logout	POST	invalidates the current session on the server
echo	POST	tests the connection with the server
batch_adapt	POST	sends a file to the server in tar or compressed tar format
recog_frames	POST with query string	sends to the server a stream of frames for voice recognition
recog_utterance	POST with query string	sends the server a whole utterance for recognition
get_model_info	GET	obtains useful information on the acoustic model used

and the controller (in this case a DALI command unit). The upper portions of the exchanges deal with authentication, performed only at system startup.

For this purpose an ad-hoc API was implemented that allows the user to interact with a recognizer in response to inputs encapsulated in an HTTP protocol. This program is a Common Gateway Interface (CGI). Currently it is used together with the Apache web server. Table 1 summarizes the implemented low level API.

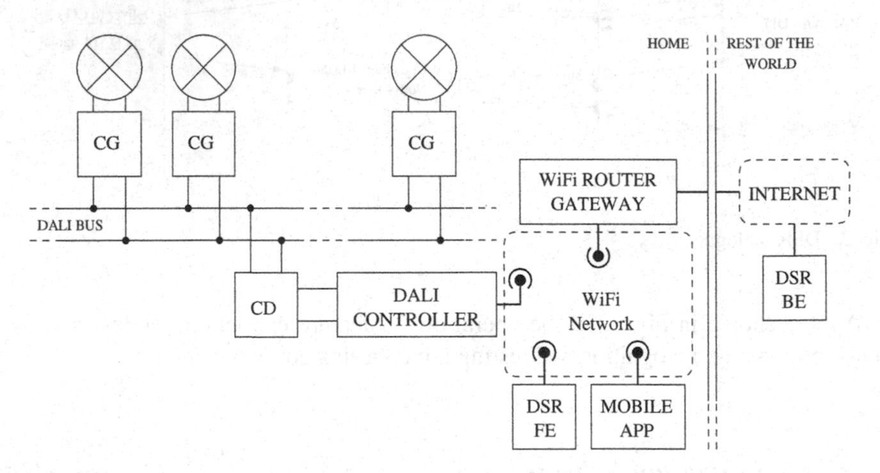

Fig. 4 Lighting system control

The high level API is a much more structured and easy-to-understand set of methods realized to allow the recognition and adaptation jobs without having to worry about the HTTP communication.

3 Case Study Architecture

Figure 4 shows the lighting system control we refer to in this paper [3]. It is a wireless system for digital control of lights, with a GUI control on a touch panel (DSR FE) and a speech recognition system (DSR BE), that contributes to send commands to the lamps; it is able to form a point-to-point wireless network consisting of several DALI devices (control gears (CGs) connected to the DALI bus) and includes:

- an audio FE; an inexpensive and power-efficient BeagleBoard®/Raspberry Pi® equipped with a standard USB microphone, or a smartphone equipped with a custom app;
- one or more DALI bridges. Each bridge can be connected to one or more CG through the DALI bus;
- a touch panel DALI master for control and configuration of the lighting network. The master is able to interoperate with different home automation systems, forwarding requests and commands to the home automation control bus;
- a custom Android application, that allows the voice control of lighting points and devices in the home automation network.

4 Experimental Results

4.1 System Setup

The experiments were carried out using a system setup including a panoramic USB microphone connected to the FE and the hidden Markov model-based recogniser running on the BE. The recognizer was configured for the Italian language and command-spotting usage; the acoustic model was trained for a generic large-vocabulary continuous speech recognition task [1, 11] and the language model is grammar-based with a hand-crafted grammar suitable to control a lighting system. The command-spotting system proficiency is achieved by the implementation of two different garbage models: a phone loop-based generic word model [5], essentially able to capture any out-of-grammar (OOG) sequence of phones, and a special decoy-based garbage model, also aimed at capturing OOG sequences. The decoys are obtained through a technique we devised [2], which semi-automatically finds the erroneous words that the ASR most often mistakenly substitutes for the correct words. It works by iteratively trying the ASR engine over the desired spoken words, each time adjusting the grammar so that the words that can most likely be confused with the target word are identified and removed.

The tests were carried out placing the microphone in an office room and recording different sessions of about 30 min: the first type of session involves two people doing their ordinary daily tasks (continuous speech), having asked them to speak commands from a given list, while the second type of session involves one speaker repeating one command in loop (round-robin). The first type of session was repeated for both on-line and off-line modes and for the following two different VAD types:

- The threshold VAD: is based on adaptive threshold energy. The VAD threshold parameter sets the speech sensitivity: a classification rule is applied to classify the acquired signal as speech or non-speech, by verifying if its energy exceeds or not a given threshold.
- The ETSI VAD: is based on ETSI standard.

4.2 Results

Several tests have been performed in order to verify the system performance (in both off-line and on-line modes), the system effectiveness in spotting commands, and the recognizer accuracy for different VAD types and garbage-model parameters. We evaluated the performance of the VAD in terms of VAD accuracy namely the number of speech signal segments (utterances) detected by the VAD in relation to the total number of utterances in the speech. Only for the threshold VAD we also considered the recognition performance as a function of the threshold value. The influence of the garbage model in speech recognition performance is tested for different values of the OOG probability.

The performance is evaluated by computing the sensitivity, specificity, precision, and accuracy, defined as follows:

Table 2 Threshold VAD performance for different threshold and OOG probability value, in off-line mode. The acquired trace has the following features: Duration: 0 h 36 m 58 s, rate: 8 kHz, SNR mean: 13.860 dB, SNR variance: 13.025 dB

OOG prob	Threshold	Sensitivity [%]	Specificity [%]	Precision [%]	Accuracy [%]	VAD accuracy
0.001	5	82	60	92	78	372/418
	7	85	42	85	77	461/418
	10	67	80	95	69	431/418
	13	57	57	84	57	289/418
0.01	5	67	60	90	66	376/418
	7	64	60	92	81	463/418
	10	64	80	94	66	436/418
	13	50	66	87	53	290/418
0.1	5	64	50	85	61	381/418
	7	82	75	95	81	470/418
	10	60	80	94	63	437/418
	13	46	66	86	60	291/418
1	5	79	75	95	68	383/418
	7	82	75	95	81	471/418
	10	50	100	100	60	438/418
	13	50	60	87	51	292/418

- Sensitivity = TP / (TP + FN)
- Specificity = TN / (TN + FP)
- Precision = TP / (TP + FP)
- Accuracy = (TP + TN) / (TP + TN + FP + FN)

where TP are the true positives (elements that belong to the grammar and are recognized as belonging to it), FN the false negatives (elements that belong to the grammar and are discarded), FP the false positives (elements that belong to garbage and are recognized as belonging to grammar), and TN the true negatives (elements that belong to garbage and are recognized as belonging to it).

Additionally we also considered the VAD segmentation, defined as the ratio of the number of utterances detected from VAD with respect to the total number of utterances.

As described in Subsect. 4.1 different types of recording sessions have been performed:

Continuous speech (off-line mode): The VAD accuracy is tested analyzing an audio trace containing continuous speech from multiple speakers interspersed with commands of the acceptable grammar and activating the OOG check option. The recognizer is in off-line mode, thus the processing is done entirely on the FE. Tables 2 and 3 show the values obtained.

The tables show that performance improves for both OOG probability greater than 0.1 and VAD threshold of 7; for this reason, the on-line test maintains this set-up of parameters.

Table 3 ETSI VAD performance for different OOG probability value, in off-line mode. The acquired trace has the following features: Duration: 0 h 36 m 58 s, rate: 8 kHz, SNR mean: 13.860 dB, SNR variance: 13.025 dB

OOG prob	Sensitivity [%]	Specificity [%]	Precision [%]	Accuracy [%]	VAD accuracy
0.001	82	66	92	79	383/418
0.01	71	100	100	75	390/418
0.1	85	66	92	87	397/418
1	75	80	95	75	403/418

Table 4 Threshold VAD performance in on-line mode. The audio track has the following features: Duration: 0 h 29 m 26 s, rate: 8 kHz, SNR mean: 19.673 dB, SNR variance: 22.580 dB

OOG prob	Th	Sensitivity [%]	Specificity [%]	Precision [%]	Accuracy [%]	VAD accuracy
0.1	7	60	83	98	61	138/122

Table 5 Round-robin performance. The audio track has the following features: Duration: 0 h 8 m 8 s, rate: 8 kHz, SNR mean: 17.918 dB, SNR variance: 15.409 dB

OOG prob	Th	Sensitivity [%]	Specificity [%]	Precision [%]	Accuracy [%]	VAD accuracy
1	7	89	0	98	87	135/123

Continuous speech (on-line mode): This test maintains the characteristics of the above except for the fact that the recognizer is enabled in on-line mode, and we test the performance of the DSR system. Results are reported in Table 4.

Round-robin (on-line mode): This test is realized in on-line mode with a single speaker repeating a loop of commands within the grammar and enabling the garbage model. This test does not include true negative checks; for this reason the scope *specificity* has null value. Results are reported in Table 5.

5 Conclusion and Future Works

In this paper we described a speech-operated system suitable to be unobtrusively installed in a user home to control the lighting system. Results show that, although further extended experimentation is needed to optimally tune the parameters, this system appears to be already able to offer an efficient voice speech interface for lighting control.

However the implementation of this system is evolving towards the creation of a high performance DSR system capable of integrating both voice synthesis and speaker recognition [6] besides speech recognition, and therefore providing a comprehensive service for home environments.

References

1. Alessandrini, M., Biagetti, G., Curzi, A., Turchetti, C.: Semi-automatic acoustic model generation from large unsynchronized audio and text chunks. In: Twelfth Annual Conference of the International Speech Communication Association (2011)
2. Alessandrini, M., Biagetti, G., Curzi, A., Turchetti, C.: A garbage model generation technique for embedded speech recognisers. In: Signal Processing: Algorithms, Architectures, Arrangements, and Applications (SPA), pp. 318–322 (2013)
3. Alessandrini, M., Biagetti, G., Curzi, A., Turchetti, C.: A speech interaction system for an ambient assisted living scenario. In: Ambient Assisted Living, pp. 233–239. Springer (2014)
4. Anastasiou, D.: Survey on speech, machine translation and gestures in ambient assisted living. In: Tralogy, Session 4—Tools for translators. Paris, France (2012)
5. Bazzi, I., Glass, J.R.: Modeling out-of-vocabulary words for robust speech recognition. In: Proceedings of the 6th International Conference on Spoken Language Processing (ICSLP 2000 / INTERSPEECH 2000), pp. 401–404. Beijing, China (Oct 2000)

6. Biagetti, G., Crippa, P., Curzi, A., Orcioni, S., Turchetti, C.: Speaker identification with short sequences of speech frames. In: International Conference on Pattern Recognition Applications and Methods (ICPRAM 2015), pp. 178–185 (Jan 2015)
7. ETSI ES 202 050 V1.1.5: Speech Processing, Transmission and Quality Aspects (STQ); Distributed speech recognition; Advanced front-end feature extraction algorithm; Compression algorithms (Jan 2007)
8. Hirota, S., Hayasaka, N., Iiguni, Y.: Experimental evaluation of structure of garbage model generated from in-vocabulary words. In: Proceedings of the 2012 International Symposium on Communications and Information Technologies (ISCIT 2012), pp. 87–92. Gold Coast, Australia (Oct 2012)
9. Levit, M., Chang, S., Buntschuh, B.: Garbage modeling with decoys for a sequential recognition scenario. In: Procedings of the 11th IEEE Workshop on Automatic Speech Recognition & Understanding (ASRU 2009), pp. 468–473. Merano, Italy (Dec 2009)
10. Pearce, D.: Enabling new speech driven services for mobile devices: An overview of the ETSI standards activities for distributed speech recognition front-ends. In: AVIOS 2000: The Speech Applications Conference, pp. 261–264 (2000)
11. Picone, J.: Continuous speech recognition using hidden Markov models. ASSP Mag. IEEE **7**(3), 26–41 (1990)
12. Seltzer, M., Singh, R.: Instructions for using the Sphinx3 trainer, http://www.speech.cs.cmu.edu/sphinxman/fr4.html
13. Vacher, M., Portet, F., Fleury, A., Noury, N.: Challenges in the processing of audio channels for ambient assisted living. In: Proceedings of the 12th IEEE International Conference on e-Health Networking Applications and Services (Healthcom'10), pp. 330–337. Lyon, France (Jul 2010)
14. Walker, W., Lamere, P., Kwok, P., Raj, B., Singh, R., Gouvea, E., Wolf, P., Woelfel, J.: Sphinx-4: A flexible open source framework for speech recognition. Technical Report SMLI TR2004-0811, Sun Microsystems Inc. (2004)
15. Zhang, W., He, L., Chow, Y.L., Yang, R., Su, Y.: The study on distributed speech recognition system. In: IEEE International Conference on Acoustics, Speech, and Signal Processing, vol. 3, pp. 1431–1434. IEEE (2000)

An Experimental Study of Scenarios for the Agent-Based RBF Network Design

Ireneusz Czarnowski and Piotr Jędrzejowicz

Abstract The paper focuses on the radial basis function neural design problem. Performance of the RBF neural network strongly depends on the network structure and parameters. Making choices with respect to the structure and value of the RBF network parameters involves both stages of its design: initialization and training. The basic question considered in this paper is how these two stages should be carried-out. Should they be carried-out sequentially, in parallel, or perhaps based on other predefined schema or strategy? In the paper an agent-based population learning algorithm is used as a tool for designing of the RBF network. Computational experiment has been planned and executed with a view to investigate effectiveness of different approaches. Experiment results have been analyzed to draw some general conclusions with respect to strategies for the agent-based RBF network design.

Keywords Neural networks · RBF network · Population learning algorithm

1 Introduction

Artificial Neural Networks are computational tools inspired by biological nervous systems with application in science and engineering, and are used to solve many different problems. Radial Basis Function Networks (RBFNs) can be placed among best-known neural networks types. They are successfully applied to multivariate nonlinear regression, classification and time series analysis [4]. The RBF networks are considered to serve as an universal approximation tool, similarly to the multilayer

I. Czarnowski (✉) · P. Jędrzejowicz
Department of Information Systems, Gdynia Maritime University, Morska 83,
81-225 Gdynia, Poland
e-mail: irek@am.gdynia.pl

P. Jędrzejowicz
e-mail: pj@am.gdynia.pl

R. Neves-Silva et al. (eds.), *Intelligent Decision Technologies*,
Smart Innovation, Systems and Technologies 39,
DOI 10.1007/978-3-319-19857-6_11

perceptrons (MLPs). However, radial basis function networks usually achieve faster convergence since only one layer of weights is required [10].

Radial basis functions have been first introduced by Powell to solve the real multivariate interpolation problem [4]. In the field of neural networks, radial basis functions were first used by Broomhead and Lowe [20]. *A radial basis function network is a neural network approached by viewing the design as a curve-fitting (approximation) problem in a high dimensional space. Learning is equivalent to finding a multidimensional function that provides a best fit to the training data, with the criterion for "best fit" being measured in some statistical sense* [20].

The RBF network is constructed from a three-layer architecture with a feedback. The input layer consisting of a set of source units connects the network to the environment. The hidden layer consists of hidden neurons with radial basis functions [10]. Each hidden unit in the RBFN represents a particular point in space of the input instances. The output of the hidden unit depends on the distance between the processed instances and the particular point in the input space of instances (the point is called an initial seed point, prototype, centroid or kernel of the basis function). The distance is calculated as a value of the activation function. Next, the distance is transformed into a similarity measure that is produced through a nonlinear transformation carried-out by the output function of the hidden unit called the radial basis function. RBFNs can use different functions in each hidden unit. The output of the RBF network is a linear combination of the outputs of the hidden units.

Performance of the RBF network depends on numerous factors. Among them is a task of optimizing values of the key parameters of the network. The basic problem with the RBFNs is to set an appropriate number of radial basis function, i.e. a number of hidden units. Deciding on this number results in fixing the number of clusters and their centroids. Another factor, called a shape parameter of radial basis function, plays also an important role from the point of view of accuracy and stability of the RBF-based approximations [9]. In numerous reported applications RBFs contain free shape parameters, which has to be tuned by the user. In [9] it is viewed as a disadvantage and somewhat ironic since the user is forced to make a decision on the choice of the shape parameter. In [12] it was suggested that the shape of radial basis functions should be changed depending on the data distribution. Such a flexibility should result in assuring better approximation effect in comparison with other approaches, where, for example, radial basis function parameters are set by some ad hoc criterion. Furthermore, connection weights as well as the method used for learning the input-output mapping have a direct influence on the RBFN performance, and both need to be somehow set. Thus, RBFN design is not a straightforward task and one of the main problems with RBF neural networks is lack of consensus on how to best implement them [27].

In general, designing the RBFN involves two stages:

- initialization – at this stage the RBF parameters, including the number of centroids with their locations and the number of the radial basis functions with their parameters, need to be calculated or somehow induced.

- training – at the second stage weights used to form a linear combination of the outputs of the hidden neurons need to be estimated.

The above stages are usually considered as being independent. This is consistent with the conventional strategy for the RBFN design. Some attempts to find specialized tools for automatically designing RBFNs have not, so far, produced a satisfactory results. Therefore, an algorithm that can automatically select network configuration for the RBFN would be beneficial. Most promising concepts leading towards automatic neural network design have been based on integrating neural networks and evolutionary approaches. A review of the evolutionary computation based approaches for the RBFN design can be found in [11].

The basic question considered in this paper is how the two stages of the RBFN design should be carried-out. Should they be carried-out sequentially, in parallel, or perhaps based on other predefined schema or strategy?

In the paper a framework for collaborative RBFNs design using the agent-based population learning algorithm is suggested. The approach extends earlier results of the authors proposed in [6]. To validate the framework an extensive computational experiment has been carried-out.

The paper is organized as follows. Section 2 reviews different schema for the RBF network design. The agent-based framework for designing RBFNs is presented in Sect. 3. Section 4 provides details on the computational experiment setup and discusses its results. Finally, the last section contains conclusions and suggestions for future research.

2 Schemas for RBF Network Design

As was shown in [16] the RBF network designing process may be conducted based on three schemas, i.e. one, two, and three-stage design. In the one-stage design, only the output weights are adjusted using some supervised method. The adjustment aims at minimizing the squared difference between the output of the RBF network and the target output. For this schema, at first, the structure of the RBFN, including the number of centroids with their locations and the number of the radial basis functions with their parameters, need to be set. This process is called an initialization which is independent from all other processes. The network parameters are fixed and have real values resulting from training data and supervised learning method.

Usually, the two-stage design is used for constructing RBF networks [16]. Two-stage schema is based on initializing the hidden layer using specialized strategy for determining centroids and other radial basis functions parameters. It is an important stage from the point of view of achieving a good approximation by the RBF network under development. Different approaches to initialization of the RBF hidden layer parameters have been proposed in the literature. Among classical methods used to RBFN initialization one should mention clustering techniques, such as the vector quantization or input-output clustering. Besides clustering methods, the support vector machine or the orthogonal forward selection approaches are used. In [14] several strategies for RBFN initialization were reviewed. For determining centroids they include random selection, clustering [18], sequential growing [19, 25], systematic seeding and editing techniques [26]. The discussion of the RBFN initialization approaches can be found in one of the earlier papers of the authors [5]. In [5] also a similarity-based approach as an editing technique has been proposed for cluster initialization. At the second stage, of the discussed schema, the weights of the output neurons are estimated. This scheme is based on the assumption that the stages are independent and don't overlap. In the two-stage approach computations are dominated by the forward construction.

In the three-stage design schema, parameters of the RBF network are adjusted through a feature optimization after having applied the two-stage design schema. The example of an approach is the locally regularized two-stage learning algorithm presented in [8]. The algorithm after having completed the second stage runs the process of recalculating of the RBF centers, which is called by the authors as backward network refinement process. In [8] it has been also suggested that it might be more advantageous to set values of RBNs parameters for both stages simultaneously. In [11] it has been concluded that it is desirable to combine the structure identification with parameter estimation within a single optimization problem.

In [29] the three-stage schema was based on a collaborative approach using the evolutionary computation tool for the RBFN design. In general, evolutionary computation tools can be used within the three-stage design schema, to evolve architecture and parameters (i.e. for determining the number of layers, hidden units, the activation function and its parameters such as learning rate, and etc.). In the literature collaborative approaches have been also discussed in context of the hybrid schema for the RBF network design (see, for example, [29]). Different combinations of evolutionary computation and neural networks techniques, as examples of hybrid and collaborative approaches, have been also discussed in [11, 16, 21, 29].

Another hybrid RBFN design schema has been discussed in [3]. The main feature of the approach is to apply some adaptation mechanisms and using different algorithms for setting values for each of the two sets of parameters. It has been also shown that an improvement in the hybrid learning schemas can be achieved by a growing structures. Finally in [3] the approach, which combines a set of different algorithms, including procedures for growing structures, to achieve a stable very fast determination of both the structural parameters and the network weights was proposed.

The integrated and customized approach to design of the RBF network, where the process of initialization and training are carried out in parallel, has been also proposed by the authors in [6]. It is a collaborative approach to RBFN design, where the agent-based population learning algorithm for selection of the set of parameters of the transfer function, determination of centroids of the hidden units and connection weights is applied. The approach is based on a hybrid training schema (see, for example, [29]). Important feature of the discussed approach is that the clusters, location of prototypes, type of the transfer function and its parameters, and the output weights are integrated into a single optimization problem and are determined jointly and in parallel using a set of dedicated agents. In [6] the following features making the RBFN design more flexible have been implemented:

- Structure of the RBNs is automatically initialized.
- Locations of centroids within clusters can be modified during the training process.
- Type of transfer function is determined during the training process.
- There is a possibility of producing a heterogeneous structure of the network.

In [6] several open questions has been formulated. Among them was the question on the impact of the selection of appropriate strategies and scenarios of agent cooperation on the RBF quality.

In the next section the framework for designing RBF networks using the agent-based population learning algorithm is presented in a more detailed manner. Further, the problem of choosing an appropriate schema for the RBFN design within the proposed agent-based framework is discussed.

3 A Framework for Designing RBFN Networks

Since the RBF neural network initialization and training belong to the class of computationally difficult combinatorial optimization problems [10], it is reasonable to apply to solve this task one of the known metaheuristics. This fact motivated implementation of the agent-based framework to solve the RBFN design problem. In [2] it has been shown that agent-based population learning search can be used as a robust and powerful optimizing technique.

3.1 Agent-Based Population Learning Algorithm

An agent-based population learning algorithm uses the idea of the A-Team approach. The A-Team concept was originally introduced in [22]. Concept of the A-Team was motivated by several approaches like blackboard systems and evolutionary algorithms, which have proven to be able to successfully solve some difficult combinatorial optimization problems. Within an A-Team agents achieve an implicit cooperation by sharing a population of solutions, to the problem to be solved.

An A-Team can be also defined as a set of agents and a set of memories, forming a network in which every agent remains in a closed loop. Each agent possesses some problem-solving skills and each memory contains a population of temporary solutions to the problem at hand. It also means that such an architecture can deal with several searches conducted in parallel. In each iteration of the process of searching for the best solution agents cooperate to construct, find and improve solutions which are read from the shared, common memory.

In the discussed population-based multi-agent approach multiple agents search for the best solution using local search heuristics and population based methods. The best solution is selected from the population of potential solutions which are kept in the common memory. Specialized agents try to improve solutions from the common memory by changing values of the decision variables. All agents can work asynchronously and in parallel. During their work agents cooperate to construct, find and improve solutions which are read from the shared common memory. Their interactions provide for the required adaptation capabilities and for the evolution of the population of potential solutions.

More information on the population learning algorithm with optimization procedures implemented as agents within an asynchronous team of agents (A-Team) can be found in [2]. In [2] also several A-Team implementations are described.

3.2 RBF Network Design

The main feature of the discussed framework with the agent-based population learning algorithm implementation is its ability to select centroids, determining the kind of transfer function for each hidden units and other parameters of the transfer function, and estimate values of output weights of the RBFN in cooperation between agents. Searches performed by each agent and information exchange between agents which is a form of cooperation, are carried-out in parallel. Most important assumptions behind the approach, can be summarized as follows:

- Shared memory of the A-Team is used to store a population of solutions to the RBFN design problem.
- A solution is represented by a string consisting of three parts:

  The first part contains integers representing numbers of instances selected as centroids. Clusters are generated using the procedure based on the similarity coefficients calculated in accordance with the scheme proposed in [26].
  The second part consists of real numbers for representing transfer functions parameters including the left and right slope.
  The third part consists of real numbers for representing weights of connections between neurons of the network.

- The initial population is generated randomly.
- Initially, potential solutions are generated through randomly selecting one or two centroids from each of the considered clusters.
- Initially, the real numbers representing slopes are generated randomly.
- Initially, the real numbers representing weights are generated randomly.

– Each solution from the population is evaluated and the value of its fitness is calculated. Solution fitness in this case could be the estimated classification accuracy or error approximation of the RBFN, assuming it is initialized using centroids, set of transfer function parameters and set of weights as indicated by the solution produced by the proposed approach.

The RBFN design problem is solved using three groups of optimizing agents:

- Agents executing procedures for centroid selection. These involves two local search procedures, one simple random search and another being an implementation of the tabu search for the prototype selection.
- Agents that are implementation of procedures for estimating the transfer function parameters.
- Agents with procedures dedicated for estimation of the output weights.

Optimizing agents improve solutions that are forwarded to them for the improvement. They may work together in parallel, sequentially, asynchronously or with using other strategy. Solutions for improvement are selected at random from the population of solutions. A returning individual replaces the current one solution in the population if it is evaluated as a better one. Evaluation of the solution is carried-out by estimating classification accuracy or error approximation of the RBFN.

4 Computational Experiment

This section contains results of the computational experiment carried out to evaluate the performance of the proposed approach, denoted further on as *ABRBF*, under different design schemas. Performance criterion was the classification accuracy of the RBFN-based classifier (*Acc*) and mean squared error (*MSE*) calculated as the approximation error over the test set in case of regression problems.

4.1 Computational Experiment Setting

In the reported experiment the following schemas have been considered:

– Case 1 – A sequential schema, where the beginning of search for the best solution is carried-out only by optimizing agents executing procedure for centroid selection. The search is carried-out for the predefined number of iteration. Afterwards agents of the first group are suspended. Next, also for the predefined number of iterations, the search is carried-out in the transfer function parameters dimension. After having carried-out their duty agents of the second group are also suspended. Finally, optimizing agents responsible for the output weights estimation are run - *ABRBFc1*.

- Case 2 – It is also a sequential schema, where, at first step, the process of searching for the best solution is carried-out by optimizing agents executing procedures for centroid selection and by optimizing agents responsible for estimating the transfer function parameters. After the predefined number of iteration the optimizing agents responsible for the output weights estimation are run and other agents are suspended - *ABRBFc2*.
- Case 3 – This case is similar to case no. 1, however agents are not suspended.— *ABRBFc3*.
- Case 4 – This case is similar to case No. 2, however different groups of agents are not suspended - *ABRBFc4*.
- Case 5 – This case assumes parallel processing. When the search process start, all optimizing agents are run. The process is carried-out by the predefined number of iteration - *ABRBFc5*.
- Case 6 – During the process of searching for the best solution, each optimizing agent monitors the value of the error function. When this value does not decrease after a predefined number of iterations the improvement procedure in a weight dimension is suspended. Then only a search for a better location of centroids within clusters and better transfer function parameters is carried-out by agents responsible for the centroid selection and slope parameters estimation. This search is carried-out for the predefined number of iteration. Subsequently, the weight searching process is resumed. These searches are carried-out (in parallel) for the predefined number of iterations. Subsequently, the weight searching process is resumed - *ABRBFc6*.

In the reported experiments the following RBFN initialization approaches have been also compared:

- The k-*means* clustering with the agent-based population learning algorithm used to locate prototypes (in this case at the first stage the *k-means* clustering has been implemented and next, from thus obtained clusters, the prototypes have been selected using the agent-based population learning algorithm) - k-*meansABRBFN*.
- The k-*means* algorithm used to locate centroids for each of the Gaussian kernels (in this case at the first stage the *k-means* clustering has been implemented and the cluster centers have been used as prototypes) - k-*meansRBFN*.
- The random search for kernel selection - *randomRBFN*.

Evaluation of the proposed approaches and performance comparisons are based on classification and regression problems. For both cases the proposed algorithms have been applied to solve respective problems using several benchmark datasets obtained from the UCI Machine Learning Repository [1]. Basic characteristics of these datasets are shown in Table 1.

Each benchmark problem has been solved 50 times, and the experiment plan involved 10 repetitions of the 10-cross-validation scheme. The reported values of the goal function have been averaged over all runs. The goal function, in case of the classification problems, was the correct classification ratio - accuracy (*Acc*). The

Table 1 Datasets used in the reported experiment

Dataset	Type of problem	Number of instances	Number of attributes	Number of classes	Best reported results
Forest Fires (FF)	Regression	517	12	–	–
Housing (Hous.)	Regression	506	14	–	–
WBC	Classification	699	9	2	97.5 % [1] (Acc.)
ACredit (AC)	Classification	690	15	2	86.9 % [1] (Acc.)
GCredit (GC)	Classification	999	21		77.47 % [13] (Acc.)
Sonar (So.)	Classification	208	60	2	97.1 % [1] (Acc.)
Satellite (Sat.)	Classification	6435	36	6	–
Diabetes (Diab.)	Classification	768	9	2	77.34 % [1] (Acc.)
Customer (Cust.)	Classification	24000	36	2	75.53 % [23] (Acc.)

overall goal function for regression problems was the mean squared error (*MSE*) calculated as the approximation error over the test set.

Parameter settings for computations involving different versions of the agent-based PLA algorithm applied to the RBFN design are shown in Table 2.

Table 2 Parameter settings for *ABRBF* in the reported experiment

Parameter	
Max number of iteration during the search	500
Max number of epoch reached in the RBF network training	1000
Population size	60
Number of iterations without an improvement until searching is stopped	100
Probability of mutation for the standard and non-uniform mutation (p_m, p_{mu})	20 %
Range values for left and right slope of the transfer function	[–1, 1]

4.2 Experiment Results

Table 3 shows performance comparison involving ABRBF, based on different designed schemas and some other approaches to RBF design including the k-means clustering with the agent-based population learning algorithm.

Table 3 Results obtained for different schemas for the RBFN design and their comparison with performance of several different competitive approaches

Problem	FF	Hous.	WBC	AC	GC	So.	Sat.	Diab.	Cust.
Algorithm	MSE				Acc. (%)				
ABRBFc1	3.82	37.4	86.31	80.46	67.01	73.15	76.11	68.52	68.47
ABRBFc2	3.27	37.58	89.9	81.7	67.17	75.91	74.87	72.41	63.87
ABRBFc3	2.14	36.82	90.5	82.61	67.22	75.94	77.25	73.71	68.59
ABRBFc4	2.09	36.4	93.84	83.33	70.09	78.6	82.4	75.42	68.34
ABRBFc5	**1.9**	35.22	96.9	84.91	**72.48**	84.72	85.1	77.51	70.0
ABRBFc6	1.92	**35.03**	**98.2**	**86.38**	72.15	**86.32**	**86.08**	**79.5**	**72.33**
k-*means*	2.29	35.87	95.83	84.16	70.07	81.15	83.57	73.69	70.8
k-*meansRBFN*	2.21	36.4	93.9	82.03	68.3	78.62	81.4	70.42	69.48
randomRBFN	3.41	47.84	84.92	77.5	67.2	72.79	74.84	62.15	65.4
MLP	2.11*	40.62*	96.7^	84.6^	77.2^	84.5^	83.75†	70.5^	71.5^
Multiple linear regression	2.38*	36.26*	–	–	–	–	–	–	–
SVR/SVM	1.97*	44.91*	96.9^	84.8^	–	76.9^	85.0“	75.3^	–
C 4.5	–	–	94.7^	85.5^	70.5	76.9^	–	73.82‡	73.32

* - source [28]; ^ - source [7]; † - source [15]; “ - source [24]; ‡ - source [5]

It can be observed that RBF networks designed using the agent-based approach assure a competitive performance in comparison to other approaches. The *AB-RBFc6* has improved accuracy for seven dataset, assuring better results than *ABRBF* for schemas numbered from 1 to 5, thus outperforming all other versions. The experiment results confirm that choosing an adequate RBFN design schema is a crucial factor from the point of view of the RBF performance. The experiment results also show that the schema for RBFNs designing based on case 6 performs better than k-meansABRBFN, k-*meansRBFN* and *randomRBFN*.

The results shown in Table 3 further demonstrate that the *ABRBF* can be superior to the other methods including MLP, multiple linear regression, SVM and C4.5 since in seven cases the proposed algorithm has been able to improve the respective generalization ability.

5 Conclusions

In the paper several scenarios for the agent-based RBFN design have been discussed and validated through the computational experiment. Its results clearly show that choice of the design scenario including schema and a level of integration of several involved optimization processes may influence the RBFN performance. It has been also shown that growing level of integration assures better performance of the network.

In the future it also is planned to extend the range of available optimization agents through incorporating agents responsible for the RBFN structure modification and implementation of new adaptation schemas.

References

1. Asuncion, A., Newman, D.J.: UCI Machine Learning Repository (http://www.ics.uci.edu/~mlearn/MLRepository.html). University of California, School of Information and Computer Science, Irvine (2007)
2. Barbucha, D., Czarnowski, I., Jędrzejowicz, P., Ratajczak-Ropel, E., Wierzbowska, I.: e-JABAT—an implementation of the web-based A-Team. In: Nguyen, N.T., Jain, I.C. (eds.) Intelligent Agents in the Evolution of Web and Applications. Studies in Computational Intelligence, vol. 167, pp. 57–86. Springer, Berlin (2009)
3. Borghese, N.A., Ferrari, S.: Hierarchical RBF networks and local parameters estimate. Neurocomuting **19**, 259–283 (1998)
4. Broomhead, D.S., Lowe, D.: Multivariable functional interpolation and adaptive networks. Complex Syst. **2**, 321–355 (1988)
5. Czarnowski, I., Jędrzejowicz, P.: Agent-based approach to the design of RBF networks. Cybern. Syst. **44**(2–3), 155–172 (2013)
6. Czarnowski, I., Jędrzejowicz, P.: Designing RBF networks using the agent-based population learning algorithm. New Gener. Comput. **32**(3–4), 331–351 (2014)
7. Datasets used for classification: comparison of results. In: Directory of Data Sets (2009). http://www.is.umk.pl/projects/datasets.html. Accessed 1 Sept 2009
8. Deng, J., Li, K., Irwin, G.W.: Locally regularized two-stage learning algorithm for RBF network centre selection. Int. J. Syst. Sci. **43**(6), 1157–1170 (2012)

9. Fasshauer, G.E., Zhang, J.G.: On choosing "optimal" shape parameters for RBF approximation. Numer. Algorithms **45**(1–4), 345–368 (2007)
10. Gao, H., Feng, B., Zhu, L.: Training RBF neural network with hybrid particle swarm optimisation. In: Weng, J., et al. (eds.) ISNN 2006. LNCS, vol. 3971, pp. 577–583. Springer, Berlin (2006)
11. Garg, S., Patra, K., Khetrapal, V., Pal, S.K., Chakraborty, D.: Genetically evolved radial basis function network based prediction of drill flank wear. Eng. Appl. Artif. Intell. **23**, 1112–1120 (2010)
12. Hanrahan, G.: Artificial Neural Networks in Biological and Environmental Analysis, Analytical Chemistry Series. CRC Press, Taylor & Francis Group (2011)
13. Jędrzejowicz, J., Jędrzejowicz, P.: Cellular GEP-induced classifiers. In: Pan, J.-S., Chen, S.-M., Nguyen, N.T. (eds.) ICCCI 2010, Part I. LNAI, vol. 6421, pp. 343–352. Springer, Berlin (2010)
14. Krishnaiah, P.R., Kanal, L.N.: Handbook of Statistics 2: Classification. Pattern Recognition and Reduction of Dimentionality. North Holland, Amsterdam (1982)
15. Liang, N.-Y., Huang, G.-B., Saratchandran, P., Sundararajan, N.: A fast and accurate online sequential learning algorithm for feedforward networks. IEEE Trans. Neural Networks **17**(6), 1411–1423 (2006)
16. Novakovic, J.: Wrapper approach for feature selections in RBF network classifier. Theor. Appl. Math. Comput. Sci. **1**(2), 31–41 (2011)
17. Powell, M.J.D.: Restart procedures for the conjugate gradient method. Math. Program. **12**, 241–254 (1977)
18. Qasem, S.N., Shamsuddin, S.M.H.: Radial basis function network based on multi-objective particle swarm optimization. In: Proceeding of the 6th International Symposium on Mechatronics and its Applications (ISMA09), Sharjah, UAE, 24–26 Mar 2009
19. Ros, F., Pintore, M., Chretie, J.R.: Automatic design of growing radial basis function neural networks based on neighboorhood concepts. Chemometr. Intell. Lab. Syst. **87**, 231–240 (2007)
20. Saliin, F.: A radial basis function approach to a color image classification problem in a real time industrial application. Master's Thesis, Virginia University (1997)
21. Sánchez, A.V.D.: Searching for a solution to the automatic RBF network design problem. Neurocomputing **42**(1–4), 147–170 (2002)
22. Talukdar, S., Baerentzen, L., Gove, A., de Souza, P.: Asynchronous teams: co-operation schemes for autonomous, computer-based agents. In: Technical Report EDRC 18-59-96, Carnegie Mellon University, Pittsburgh (1996)
23. The European Network of Excellence on Intelligence Technologies for Smart Adaptive Systems (EUNITE)—EUNITE World Competition in domain of Intelligent Technologies. http://neuron.tuke.sk/competition2 (2002). Accessed 1 Sept 2002
24. Wang, L., Yang, B., Chen, Y., Abraham, A., Sun, H., Chen, Z., Wang, H.: Improvement of neural network classifier using floating centroids. Knowl. Inf. Syst. **31**, 433–454 (2012)
25. Wei, L.Y., Sundararajan, N., Saratchandran, P.: Performance evaluation of a sequential minimal radial basis function (RBF) neural network learning algorithm. IEEE Trans. Neural Networks **9**, 308–318 (1998)
26. Wilson, D.R., Martinez, T.R.: Reduction techniques for instance-based learning algorithm. Mach. Learn. **33**(3), 257–286 (2000)
27. Yonaba, H., Anctil, F., Fortin, V.: Comparing sigmoid transfer functions for neural network multistep ahead streamflow forecasting. J. Hydrol. Eng. **15**(4), 275–283 (2010)
28. Zhang, D., Tian, Y., Zhang, P.: Kernel-based nonparametric regression method. In: Proceedings of the IEEE/WIC/ACM International Conference on Web Intelligence and Intelligent Agent Technology, pp. 410–413 (2008)
29. Zhao, Z.S., Hou, Z.G., Xu, D., Tan, M.: An evolutionary RBF networks based on RPCL and its application in fault diagnosis. In: Proceedings of International Conference on Machine Learning and Cybernetics, vol. 2, pp. 1005–1009 (2009)

Eye Movements Data Processing for Ab Initio Military Pilot Training

Emilien Dubois, Colin Blättler, Cyril Camachon and Christophe Hurter

Abstract French *ab initio* military pilots are trained to operate a new generation of aircraft equipped with glass cockpit avionics (Rafale, A400 M). However gaze scanning teachings can still be improved and remain a topic of great interest. Eye tracking devices can record trainee gaze patterns in order to compare them with correct ones. This paper presents experimentation conducted in a controlled simulation environment where trainee behaviors were analyzed with notifications given in real-time. In line with other research in civil aviation, this experimentation shows that student-pilots spend too much time looking at inboard instruments (inside the cockpit). In addition, preliminary results show that different notifications bring modifications of the visual gaze pattern. Finally we discuss future strategies to support a more efficient pilot training thanks to real-time gaze recording and its analysis.

Keywords Eye tracker · Gaze behavior · Military student pilots · Flight simulator · Automation issue

E. Dubois (✉) · C. Hurter
ENAC, Laboratoire D'Informatique Interactive (LII), Toulouse, France
e-mail: emilien.dubois@enac.fr

C. Hurter
e-mail: christophe.hurter@enac.fr

C. Blättler · C. Camachon
CReA, Facteurs Humains et Milieux Opérationnels (FHMO),
Salon-de-Provence, France
e-mail: colin.blattler@defense.gouv.fr

C. Camachon
e-mail: cyril.camachon@defense.gouv.fr

R. Neves-Silva et al. (eds.), *Intelligent Decision Technologies*,
Smart Innovation, Systems and Technologies 39,
DOI 10.1007/978-3-319-19857-6_12

1 Introduction

During summer 2012, the French Air Force academy replaced the analogical conventional cockpits of training aircraft by numerical glass-cockpits with sophisticated automation. The glass cockpit (see Fig. 1. for illustration) replaces the traditional electro-mechanical cockpit dials (altimeter, airspeed, turn and bank, vertical speed, altitude and heading) with two screens: the Primary Flight Display (PFD) and the MultiFunction Display (MFD). The PFD displays all of the information provided by the separate dials found in the traditional cockpit [22]. This change was an opportunity to study the impact of the modern cockpit environment in ab initio flight training [1]. Up to now, the transition to a modern cockpit environment occurred late in the French Air Force pilot training. However, teaching glass-cockpit earlier in pilot training raises the question of how the young pilots should be trained. Indeed, instructional techniques have been optimized over a long period of time for aircraft equipped with steam gauges [2]. Traditionally, pilots had to learn how to scan the six basic aircraft control steam gauges (attitude, altitude, airspeed, heading, climb rate and turn direction and rate) together with the outside environment. They were advised to look inside the cockpit no more than 4-5 s for every 16 s spent scanning the outside world [3]. Several studies stated that pilots had not achieved optimized visual scanning [4]. In addition, one of the main pitfalls is to spend too little time looking out the window: "too much head-down time" [5].

The modern glass cockpit technology was supposed to make the scan pattern easier and to help improve pilot's situation awareness [6]. To the best of our knowledge, there is no evidence that this is the case for novice pilots. Currently, there is no standardized "scan technique" training at the French Air Force Academy regarding glass cockpit management. The current teaching methods are based on flight instructors' experience, which is mostly acquired on conventional airplanes. According to French Air Force instructors, glass-cockpits really draw student pilot attention inside the airplane. In real flight, when they detect that a student pilot spends too much time looking inside the cockpit, instructors try to fix this incorrect gaze behavior with different methods. The less pervasive one is to orally notify the student, and the other one is to hide the information the student is focused on with an opaque paper. In this manner, the student understands that he or she has an incorrect behavior. However instructors do not have a tool to allow them to accurately analyze gaze behavior, so they cannot objectively detect deficient gaze behavior situations. Based on our observations and interviews, instructors use oral notifications as weaker warnings than visual notifications.

Furthermore, a major challenge is the growing use of Flight Training Devices (FTD) in pilot training in general and military aviation. FTD are largely used to train pilots at reduced costs [7]. There is also evidence that flight simulators are useful for ab initio flight training [8, 21] regarding instruments skills [7] even if the FTD is a low cost simulator (i.e. ordinary personal computer using commercial software [9, 10]). However, the use of flight simulators has some drawbacks. One example shown by Johnson, Wiegmann and Wickens [11] is the different gaze behavior induced by the analog cockpit and the glass cockpit on a simulator. In this experimentation, pilots using an analog cockpit (control group) spent approximately 40 % of their time looking out the window, instead of the 67 % to 75 % recommended allocation of attention to the outside world [12, 13]. Even more significantly pilots using a glass cockpit allocated only 10 % of their visual attention to the outside world. One of the most effective teaching devices (simulators) may actually increase the incorrect "too much head-down time" behavior.

In this paper, the "head-down time" issue regarding the glass cockpit environment in a training context will be addressed. A commercial eye tracking system will be adapted to be used in a glass cockpit flight training device. An apparatus has been developed to allow real-time collection and analysis of gaze behaviors in order to efficiently teach French military trainees the correct patterns.

2 Experiment

The first goal of our experimentation is to reduce the head-down time to approach the recommended standard of 30 % [12, 13]. More particularly the question here is to discover if real-time assistance based on notifications allow a reduction in head-down time. The other goal is to assess the effectiveness of the methodology currently used by the French Air Force in simulated flights. In this aim, we would like to validate our observations and assess if oral notifications are weaker warnings than visual ones.

In our experimentation, we will analyze gaze behavior in real-time, and send notifications in case of non-recommended gaze behaviors. Since the standard 30 % ratio of looking outside time is only applicable on the entire flight duration, we chose another criterion that allowed us to detect inappropriate gaze behavior in real-time. As instructors recommend not looking for more than 2 s at the inside of the cockpit we opted for this rule to trigger notifications. We call it the "2 s rule".

To conduct our experimentation, we designed an environment composed of a high fidelity, dynamic and interactive simulation [14] with a head mounted eye tracker. At this point, the real difficulty was to build a robust architecture to allow our eye tracker to interact in real-time with our simulation environment.

2.1 Eye Tracker Constraints

Recording, analyzing the location of the gaze and reacting to particular gaze behaviors in real-time, were the main challenges to address in this experiment. Moreover, the simulated environment setup brings too many constraints: five screens (3 for the outside world and 2 for the cockpit) and free head movements. This is why we opted for a head-mounted eye tracking solution. This eye tracker has two cameras: (1) one right eye focused camera responsible for the pupil position and size and (2) one for the recording of the environment (located between the eyes). Since this experimentation was not a study of usability or user experience, participant satisfaction was not tested. Our system does not have the vocation to be delivered to the French Air Force Academy or other training organizations. Participants in the experiment are subjects and under no circumstances final users.

To achieve our needs, the gaze location had to be known in real-time. We needed to find a robust and effective algorithm able to access and treat the adjusted gaze location thanks to head movements. The head movements could be detected thanks to a reference image which is captured at the beginning of the experimentation and tracked in real-time until the end of this experiment. Therefore a part of the solution was to use the OpenCV toolkit and the Surf algorithm [20] to track the location of the reference image in the picture provide by the environment camera.

However, the mounted eye tracking solution in our possession does not provide any suitable real-time gaze processing tools therefore we had to implement a specific module. The current gaze location is processed with a homographic computation between the gaze location in the reference image and the location of this reference image in the picture provide by the environment camera. This treatment takes time; nevertheless we achieved a suitable recording rate with 30 samples per second.

Furthermore, three static Areas of Interest (AOI) were defined (Fig. 1): one for the outside world, and two for the cockpit, the PFD and MFD respectively. The outside world area contains the three screens showing the flight simulator view, while the two others show a cockpit simulator view. In our experimentation, the "2 s rule" is only applicable to the time spent inside the PFD. In order to display visual notifications on the simulators screens (flight and cockpit), we developed another module with a windows overlay technique (shadow mask).

The first module which treats the gaze localization in real-time and analyzes behavior communicates with the other module responsible for showing visual notifications. We have made this possible thanks to the IVY data bus [15]. During the experimentation different data sources were collected and merged. The data is sent on this same logical bus IVY [15] and the data fusion is performed thanks to the time synchronization provided by a NTP server (Network Time protocol).

Our architecture is modular and handles the following constraints: temporal synchronization, modular communication, data logging and modular visual notifications.

2.2 Participants

Fifteen male students from the French Air Force (FAF) academy participated in the experiment. The students' total sailplane flight experience ranged from 5 to 100 h (mean = 37 SD = 22, 58) and their ages ranged from 21 to 27 (mean = 22, 4 SD = 1, 99).

2.3 Materials and Procedure

The participant's task was to navigate from Ajaccio to Solenzara (Corsica, France) in a Cirrus Perspective flight simulator. The glass cockpit of the Cirrus Perspective is a Garmin 1000. The left screen (Primary Flight Display, PFD) presents the airplane attitude information (speed, altitude, heading… see Fig. 1). The right screen presents engine information (MFD, see Fig. 1). No GPS navigation or any automation was allowed. Thus the student pilots had to fly with rudder and stick. The navigation duration was 31 min. Three horizontal screens showed the outside world.

Participants had to follow 31 altitude and heading instructions given orally while keeping constant speed (120 kt). Altitude and heading instructions were given every minute. Each student was told of the security recommendation not to look at instrumentation (inside the cockpit) for more than 2 s. They were divided into three different condition groups (5 students per condition). There was (1) a control condition which did not trigger notification when the 2 s rule was violated (2) and two experimental groups which each received audio and visual notifications. To match with reality, the design of notifications was made from the observation we performed. Therefore one audio notification which announced in a synthetic voice the "look outside" advice (sound condition) while the other visual notification (visual condition) hid all information on the PFD. In the sound condition, the audio notification was played continuously until the participant's gaze was moved away from the PFD. In the visual condition, a black screen was displayed and masked entirely the PFD until the participant's gaze had left the PFD.

To run our experimentation, we used Xplane 9.0 as a flight simulation and we recorded simulated aircraft locations. The head mounted eye tracker was the Pertech solution with an accuracy of 0.3° at 50 Hz. We computed the head movement and the gaze correction with a third computer: Core i7 2.2gh, 8Go ram. We developed our software with visual Studio and C#.

Two dependent variables were recorded. The first was the percentage of time spent looking at the PFD during the entire navigation. The expected effect of the notifications was to reduce the time spent looking inside at the PFD in the two experimental conditions compared to the control condition.

The second recorded dependent variable is the number of "2 s rule" violations. This number does not directly correspond to the number of notifications triggered because, in the control condition no notification was triggered. In order to allow a comparison, we analyzed in post-treatment the number of times they had exceeded the "2 s rule" for all participants. The expected effect was a lower number of "2 s rule" violations in the two experimental conditions compared to the control condition.

Fig. 1 Cirrus Perspective with Garmin 1000 simulator environment (CReA)

3 Results

3.1 Percentage of Time Spent Inside PFD

The Fig. 2 presents our results for the three groups of participants (control group, sound group, visual group) as a function the percentage of the time spent looking inside at the PFD during the entire navigation.

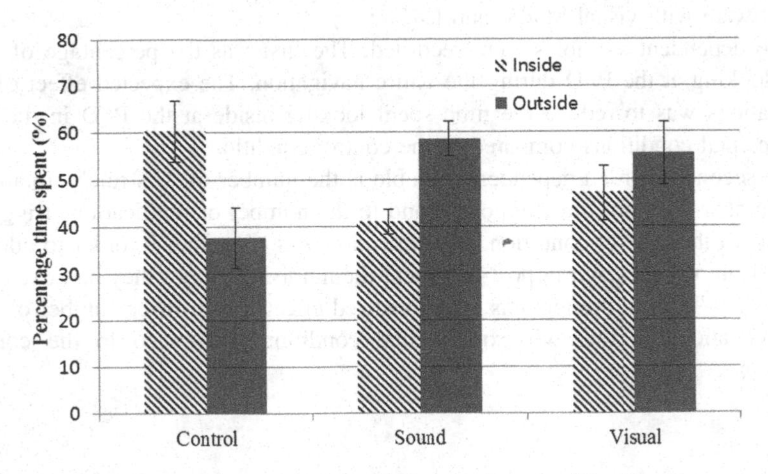

Fig. 2 Percentage of time spent inside and outside, averaged per condition. Error bars are standard errors

An ANOVA was conducted for the average percentage of time spent looking inside (control group = 60.41 %, SD = 14.8; sound group = 41.00 %, SD = 5.98; visual group = 47.09 %, SD = 12.97). The results show a difference between the three conditions which is close to being significant $F(2, 12) = 3.49$; MSE = 492.46; $p = 0.063$. T tests were conducted between these three means and show that the control group (1) is higher than the sound group $t(4) = 4.73$; $p < 0.01$ and (2) tends to be higher than the visual group $t(4) = 2.71$; $p = 0.053$. However, the difference between sound group and visual group is not significant $t(4) = 1.35$; $p > 0.05$.

Another ANOVA was conducted for the time spent looking outside percentage (control group = 37.62 %, SD = 14.31; sound group = 58.66 %, SD = 8.04; visual group = 55.45 %, SD = 14.95). The results show a significant difference between the three conditions $F(2, 12) = 3.91$; MSE = 642.78; $p < 0.05$. T test was conducted between these three means and shows that the control group (1) is higher than the sound group $t(4) = 7.33$; $p < 0.01$ and (2) tends to be significantly higher than the visual group $t(4) = 2.62$; $p = 0.058$. However, the difference between the sound group and visual group is not significant $t(4) = 0.51$; $p > 0.05$.

3.2 *Number of "2 s Rule" Violations*

Although the time spent looking inside and outside was analyzed over the whole duration of the experiment, the decision was taken to analyze this second dependent variable every minute. For every altitude and heading instruction, participants repeated the same task. Consequently this way of analyzing the data was more relevant.

An ANOVA was conducted for the number of "2 s rule" violations (control group = 5.29, SD = 0.83; sound group = 2.82, SD = 0.9; visual group = 2.49, SD = 1.11; see Fig. 3).

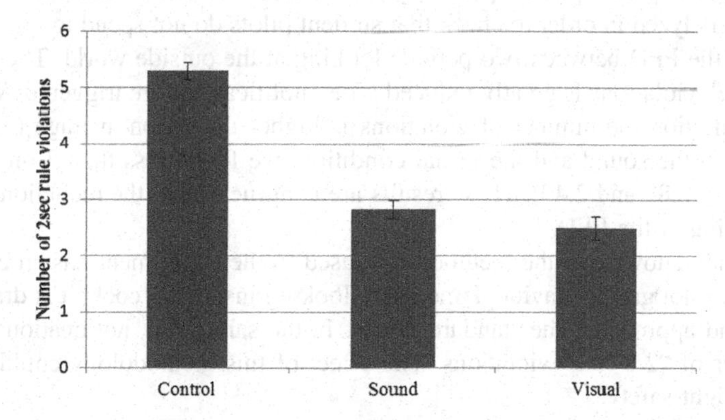

Fig. 3 Number of "2 s rule" violations per minute, averaged per condition. Error bars are standard errors

The results show a difference between the three conditions $F(2, 90) = 79.461$; $MSE = 72.612$; $p < 0.001$. T test was conducted between these three means and show that the control group is higher than the sound and the visual group $t(30) = 12.06$; $p < 0.001$ and $t(30) = 15.57$; $p < 0.001$ respectively. However, the difference between the sound group and visual group is not significant $t(30) = 1.58$; $p > 0.05$.

4 Discussion

In this paper, we investigated trainee pilot gaze behaviors during simulated flights thanks to an eye tracker device. Participants (French Air Force student pilots) were notified by a verbal message or by a visual black screen on the PFD each time they spent more than 2 s looking at the PFD. Two dependent variables were assessed (1) the percentage of time spent looking at the PFD and (2) the number of notifications presented to participants. The objectives of this study are both to assess the effectiveness of real-time warning notification in a flight simulator environment and to evaluate the relative impact of the oral and visual notifications.

In our experimentation, we first tried to observe the standard 30 % of flight time looking inside the cockpit [12, 13]. In the control condition, in which no notification was displayed, the trainees recorded a figure of 60 % of flight time looking at the PFD while in the sound and visual condition the figures were 41 % and 47 % respectively. The effects of our notifications are not enough to achieve the required standard. However this allows student pilots to produce behavior close to that of experienced pilots. Johnson, Wiegmann and Wickens [11] showed, with experienced pilots, the time spent looking inside the cockpit (analogical cockpit) is close to 40 %. The results also showed that the time spent looking elsewhere than at the PFD was transferred to the time spent looking outside. The control condition participants spent only 37 % of their time looking outside, the sound and visual condition participants spent respectively 58 % and 55 %. The obedience of the "2 s rule" was analyzed in order to check that student pilots do not spend too much time looking at the PFD between two periods looking at the outside world. The number of "2 s rule" violations is greatly reduced when notifications are triggered. While in control condition the number of violations is higher than 5 on average (5.29) per minute, with the sound and the visual condition, we found less than 3 on average (respectively 2.82 and 2.49). These results are congruent with the reduction of time spent looking at the PFD.

This study shows that the methodology used in the experiment has an effect on the student pilot gaze behavior. Time spent looking inside the cockpit is drastically reduced and approaches the standard figure. In the same way, notifications reduce the number of "2 s rule" violations. The effect of this methodology could lead to improve flight safety.

Regarding the second objective, there is no significant difference between audio and visual notifications either in terms of time spent looking (1) inside and

(2) outside, or of “2 s rule” violations. These results suggest, in a piloting assistance context,[1] that visual notifications bring no particular advantage compared to audio notifications (as expected from the military instructors’ interviews) in a simulated flight environment.

However in a learning assistance context, rather than in the context detailed in this paper, differences between these two kinds of notifications could be found. As future work, one could evaluate the “2 s rule” internalization using a post test methodology without any notification. This is a relevant question since audio notifications only inform whereas visual notifications force to look elsewhere. As mentioned by some psychology studies [16, 17], the degree of internalization (memorizing) of a behavior is linked to the deliberate choice of a behavior. A free choice behavior has a larger probability to be acquired than a forced choice. In our experiment the sound condition can be compared to a “free choice” and the visual condition can be compared to a “forced choice”. For example, the evaluation of the internalization of the “2 s rule” can be different, for these two conditions, in a post test (a few days after the experiment). This could be developed in another experimentation.

This study demonstrates a method of real-time warnings which allow an initial behavior to be changed to a targeted behavior. The methodology might help to create a more accurate scan pattern than the one used in this study (i.e., the ratio of time spent looking inside/outside the cockpit).

For instance, it may be useful to help novice pilots adopt practices similar to those of experienced pilots. In this aim it would be necessary to analyze more deeply the visual scan path of experienced pilots. We plan to analyze gaze data with interactive visualization tools [18] and processing algorithms [19]. Edge bundling algorithms have already proven to be an efficient tool to extract gaze patterns and thus will provide visual clues to assess the change of gaze behavior linked to the type of notification (audio or visual). As a future development, we also plan to use multiple eye tracker sources and to perform data fusion in order to improve gaze pattern detection. This gain of accuracy will require specific computation algorithms to perform the study in real-time and will be a technical challenge for future work.

5 Conclusion

This study shows that novice pilots fail to avoid the “too much head-down time” glass cockpit pitfall. However, with a real-time warning notification, this pitfall can be drastically reduced. Ocular behavior comes close to that of the official standard with real time warnings, while with no assistance it is still very far from this standard. The effect of this notification may increase flight safety.

[1]Treating only of the punctual effect of notifications on gaze behavior during a simulated flight.

Two practical consequences can be derived from these results. Firstly, this technology can be a good way to improve simulated flight pedagogy and reduce some of the drawbacks of the glass cockpit. However this statement can only be correct from a learning perspective. This current study does not enable us to reach such a conclusion and this point deserves further investigation as previously. And from a more general standpoint, it may be very beneficial to consider these results in any circumstance where people have to monitor a complex system (aircraft, drone system, nuclear plant etc.). Eye tracking devices can be considered as monitoring tools, in addition to being efficient teaching instruments.

Acknowledgments This study is supported by the "Direction Générale de l'Armement" (DGA).

References

1. Dahlstrom, N., Dekker, S., Nahlinder, S.: Introduction of technically advanced aircraft in ab-initio flight training. In: Technical Report, (2006-02) (2006)
2. Lindo, R.S., Deaton, J.E., Cain, J.H., Lang, C.: Methods of instrument training and effects on pilots' performance with different types of flight instrument displays. Aviat. Psychol. Appl. Hum. Factors **2**(2), 62 (2012)
3. Federal Aviation Administration (1998b). Scanning for Other Aircraft. Aeronautical Information Manual, 8-1-6-c. Oklahoma City, OK: Author
4. Colvin, K.W., Dodhia, R.M., Belcher, S.A., Dismukes, R.K.: Scanning for visual traffic: An eye tracking study. In: Proceedings of the 12th International Symposium on Aviation Psychology, pp. 255–260. Dayton, OH, USA: The Wright State University (2003)
5. Rudisill, M.: Flight crew experience with automation technologies on commercial transport flight decks. In: Human Performance in Automated Systems: Current Research and Trends, pp. 203–211. Hillsdale, NJ (1994)
6. Aircraft Owners & Pilots Association. Air Safety Institute. The Accident Record for TAA. Frederick: AOPA Foundation. In: Institute, A.S. (n.d.). The Accident ecord of the Technically Advanced Aircraft. Aircraft Owners and Pilots Association (2012)
7. Rantanen, E.M., Talleur, D.A.: Incremental transfer and cost effectiveness of groundbased flight trainers in university aviation programs. In: Proceedings of the Human Factors and Ergonomics Society Annual Meeting, vol. 49(7), pp. 764–768). SAGE Publications, London (2005)
8. Macchiarella, N.D., Arban, P.K., Doherty, S.M.: Transfer of training from flight training devices to flight for ab-initio pilots. Int. J. Appl. Aviat. Stud. **6**(2), 299–314 (2006)
9. Taylor, H.L., et al.: Incremental training effectiveness of personal computer aviation training devices (PCATD) used for instrument training. In: University of Illinois at Urbana-Champaign, Aviation Research Lab (2002)
10. Taylor, H.L., Talleur, D. A., Emanuel Jr, T.W., Rantanen, E.M.: Transfer of training effectiveness of a flight training device (FTD). In: Proceedings of the 13th International Symposium on Aviation Psychology, pp. 1–4 (2005)
11. Johnson, N., Wiegmann, D., Wickens, C.: Effects of advanced cockpit displays on general aviation pilots' decisions to continue visual flight rules flight into instrument meteorological conditions. In: Proceedings of the Human Factors and Ergonomics Society Annual Meeting, vol. 50(1), pp. 30–34. Sage Publications, London (2006)
12. AOPA Air Safety Foundation. How to avoid a midair collision. 16 Jan 1993
13. FAR/AIM. Aeronautical Information Manual/ Federal Aviation Regulations. McGraw-Hill (2003)

14. Eyrolle, H., Mariné, C., Mailles, S.: La simulation des environnements dynamiques: intérêts et limites. In: Cellier, J.M., De Keyser, V., Valot, C. (eds.) La gestion du temps dans les environnements dynamiques. PUF, Paris, pp. 103–121 (1996)
15. Buisson, M., Bustico, A., Chatty, S., Colin, F. R., Jestin, Y., Maury, S., Truillet, P.: Ivy: un bus logiciel au service du développement de prototypes de systèmes interactifs. In: Proceedings of the 14th French-speaking conference on Human-computer interaction (Conférence Francophone sur l'Interaction Homme-Machine), pp. 223–226. ACM (2002)
16. Kiesler, C.A.: The Psychology of Commitment: Experiments Linking Behavior to Belief. Academic Press, New York (1971)
17. Joule, R.V., Beauvois, J.L.: La soumission librement consentie: comment amener les gens à faire librement ce qu'ils doivent faire? Presses universitaires de France (1998)
18. Hurter, C., Tissoires, B., Conversy, S.: FromDaDy: Spreading data across views to support iterative exploration of aircraft trajectories. IEEE TVCG **15**(6), 1017–1024 (2009)
19. Hurter, C., Ersoy, O., Fabrikant, S., Klein, T., Telea, A.: Bundled Visualization of Dynamic Graph and Trail Data. (TVCG) Visualization and Computer Graphics (2013)
20. Bay, H., Ess, A., Tuytelaars, T., Van Gool, Luc: SURF: speeded up robust features. Comput. Vis. Image Underst. (CVIU) **110**(3), 346–359 (2008)
21. Stewart, J.E., Dohme, J.A., Nullmeyer, R.T.: US Army initial entry rotary-wing transfer of training research. Int. J. Aviat. Psychol. **12**(4), 359–375 (2002)
22. Wright, S., O'Hare, D.: Can a glass cockpit display help (or hinder) performance of novices in simulated flight training? Appl. Ergon. **47**, 292–299 (2015)

Massive-Scale Gaze Analytics Exploiting High Performance Computing

Andrew T. Duchowski, Takumi Bolte and Krzysztof Krejtz

Abstract Methods for parallelized eye movement analysis on a cluster are detailed. The distributed approach advocates the *single-core* job programming strategy, assigning processing of eye movement data across as many cluster cores as are available. A foreman-worker distribution algorithm takes care of job assignment via the Message Passing Interface (MPI) available on most high-performance computing clusters. Two versions of the MPI algorithm are presented, the first a straightforward implementation that assumes faultless operation, the second a more fault-tolerant revision that gives nodes an opportunity of communicating failure. Job scheduling is also briefly explained.

Keywords High-performance computing · Eye tracking · Gaze analytics

1 Introduction and Background

Eye movement data is voluminous. The large volume of data produced from eye tracking experiments stems not only from increasingly faster sampling rates (e.g., collecting an (x, y, t) data tuple every 16 ms when sampled at the modest rate of 60 Hz), or number of trials (e.g., dependent on number of stimuli viewed), but also from the need to visualize the data, either as temporally ordered *scanpaths* or aggregate *heatmaps* or as streaming x- or y-coordinate plots used to tune various digital filters employed during the process of fixation detection.

Consider a straightforward within-subjects experiment where each participant views three variants of a stimulus. Scanpath and heatmap visualizations yield $2 \times 3 = 6$ visualizations per participant. Given the rule of thumb of 10 participants per exper-

A.T. Duchowski (✉) · T. Bolte
School of Computing, Clemson University, Clemson, SC, USA
e-mail: duchowski@clemson.edu

K. Krejtz
National Information Processing Institute, University of Social Sciences and Humanities, Warsaw, Poland

R. Neves-Silva et al. (eds.), *Intelligent Decision Technologies*, Smart Innovation, Systems and Technologies 39,
DOI 10.1007/978-3-319-19857-6_13

imental design condition, suggesting 30 participants, produces 180 such visualizations. If each such visualization requires say 30 s to produce, the entire visualization catalog would then require about 90 min processing time (on a single CPU, without utilizing GPU hardware acceleration). Intermediate visualizations are beneficial for fine-tuning digital filtering parameters, but require plotting of additional graphs, e.g., raw, smoothed and differentiated data in 1D (separate x and y vs. t plots) and in 2D, increasing the number of plots per participant from 2 to 11 per stimulus.

(a) Clemson's Palmetto.

(b) Warsaw's Halo2.

(c) Lugano's Mönch

Fig. 1 Academic institution clusters for which the current MPI implementation is targeted

To alleviate the problem of processing of voluminous eye tracking data, Dao et al. [3] proposed a cloud-based architecture that collects data from multiple eye trackers, regardless of their physical location, and enables automatic real-time processing, evaluation and visualization of incoming eye-tracking streams. Experiments of their implementation of *EyeCloud* on Amazon's cloud showed advantages of cloud computing compared to a single PC in aspects of data aggregation and running time.

Although Dao et al.'s contribution is the only one we are aware of that proposes the use of a cluster for parallel processing of eye movement data (using Hadoop), they do not provide specific programming details beyond heatmap rendering.

In this paper, we focus on the parallelization of the uniprocessor approach using the Message Passing Interface, or MPI [9]. We provide programming details on how to organize the client/server architecture that Dao et al. omitted, however, we use the foreman/workers (or master/workers [2]) metaphor for job distribution. We also provide information on job scheduling.

The motivation for distributing computation of eye movement data, in this instance largely focusing on visualization, is to expose to the eye tracking community the use of High-Performance Computing (HPC) infrastructure which is becoming increasingly available (e.g., whether at local or national academic institutions, or via cloud computing). Although the application example used herein is straightforward, and falls under the category of *embarrassingly parallel*, it nevertheless provides an introduction to the use of MPI programming and HPC cluster scheduling. The triviality of the single-core visualization distribution to multiple nodes gives potential for further potentially more interesting computation, e.g., gaze transition entropy, variance of heatmap distributions, intercorrelation of Gaussian Mixture Models, etc.

Table 1 Selected cluster architectures

Cluster	Compute nodes	Cores	Interconnect	Queue	OS
Mönch	360	7,200	InfiniBand	SLURM	Linux
Halo2	512	8,192	InfiniBand	Torque (OpenPBS enhanced)	Linux
Palmetto	1,978	20,728	InfiniBand	PBS Pro	Linux

2 Selection of a Multi-core Cluster

The distributed programming approach to processing of multiple files, referred to as *single-core* job execution, scales to any number of cores available on a given cluster. Although the implementation is likely to port to cloud-based architectures that Dao et al. [3] promote, we target clusters that are likely to be available at academic institutions. A sample of academic institution clusters is shown in Fig. 1, with hardware and software platforms of these clusters given in Table 1, and includes: *Palmetto*, at Clemson University; *Halo2*, at the University of Warsaw; *Mönch*, at the Swiss National Supercomputing Center.

Some of these supercomputing centers house a number of other clusters (e.g., the CSCS in Lugano houses *Piz Daint* and *Blue Brain*, ranked 6 and 56, respectively, on the "Top500" list of supercomputers in the world), and it is important to select the appropriate infrastructure for the distributed paradigm outlined here, namely the single-core type of job. The single-core job would be discouraged from running on highly interconnected architectures such as *Piz Daint*'s.

The single-core job is a distributed SPMD (Single Program Multiple Data) programming approach and differs from other parallel programming approaches such as *multicore* and *many-core* paradigms. A key characteristic of the SPMD programming model is that the processors run asynchronously. Multicore programming generally refers to the use of multiple processing units found on a single chip [8]. Multicore programs are generally run on a single machine equipped with these chips, parallelizing code via an application program interface (API) such as OpenMP. Many-core programs, on the other hand, generally exploit the massively parallel SIMD (Single Instruction Multiple Data) "lock-stepped" style of instruction afforded by GPU architectures. GPUs can be used for general purpose programming through APIs such as OpenCL.

In general, high performance computing on a computational cluster can involve a mixture of programming models. For example, multicore programming can be effected by distributing jobs across multiple cores (housed on the same or different computational nodes, connected by high-speed interconnects). The same job can also be made to exploit the many-core model if each distributed version of the program makes use of the GPU. Designing programs to exploit different forms of parallelism is challenging. Here, we provide two examples: multicore parallelization by making use of multiple cores (e.g., essentially different computers connected by Ethernet) and a combination of multicore and many-core parallelization where each such core exploits the GPU for hardware-accelerated heatmap rendering.

3 Gaze Analytics Pipeline

Prior to describing approaches to parallelization, the gaze analytics pipeline is reviewed [4]. An individual's recorded eye movements, when exported as raw gaze data, is processed by the following steps:

1. denoising and filtering raw gaze data $g_i = (x_i, y_i, t_i)$, and classifying raw gaze into fixations $f_i = (x_i, y_i, t_i, d_i)$, where (x_i, y_i) coordinates indicate the position of the gaze point or centroid of the fixation, with t_i indicating the timestamp of the gaze point or fixation and d_i the fixation's duration,

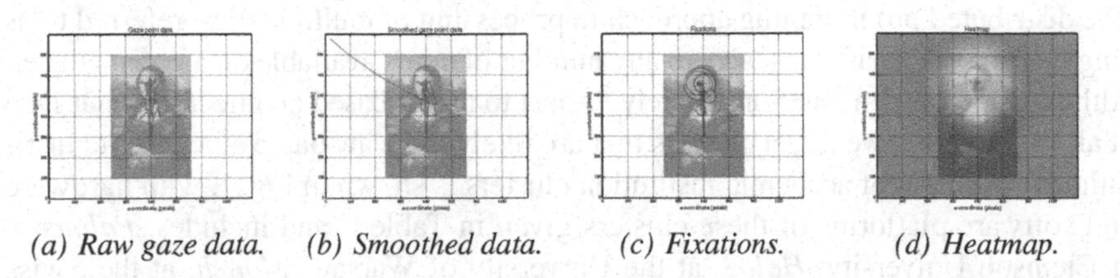

(a) Raw gaze data. *(b) Smoothed data.* *(c) Fixations.* *(d) Heatmap.*

Fig. 2 Representative (individual) gaze data processing pipeline showing progression from raw gaze data to smoothed data to visualizations

2. collating fixation-related information for its subsequent statistical comparison,
3. interpreting and visualizing statistical tests conducted on processed data.

The first step is applied to each data file, and can be performed either in sequence for all individuals' recorded data, or in parallel, where each data file is processed on its own core, hence the term single-core job. An often critical component accompanying data processing is visualization (see Fig. 2). Visualization in the form of scanpaths and heatmaps is useful for communicating results and for debugging purposes, e.g., examining filter performance. Unfortunately, graphical rendering of the data is computationally time-consuming. Heatmap rendering is particularly expensive [5], espeically if a many-core GPU card is unavailable.

Fixation Filtering. The eye tracker outputs a stream of gaze points (x_i, y_i, t_i). Typically, this data is noisy and requires smoothing. Smoothing (or differentiating) is achieved by convolving the gaze point input with a digital filter [11]. The filter used may be a Finite Impusle Response (FIR) or an Infinite Impulse Response (IIR) type [7]. Following Duchowski et al. [4], we chose a 2nd order Butterworth (IIR) filter to smooth the raw gaze data with sampling and cutoff frequencies of 60 and 5.65 Hz, (see Fig. 2b).

Following Andersson et al. [1] and Nyström and Holmqvist [10], a second-order Savitzky-Golay (SG) filter [6, 13] is used to differentiate the (smoothed) positional gaze signal into its velocity estimate. The Savitzky-Golay filter fits a polynomial curve of order n via least squares minimization prior to calculation of the curve's s^{th} derivative (e.g., 1st derivative ($s = 1$) for velocity estimation). We use a 7-tap (112 ms) SG filter with a threshold of $\pm$ 5 deg/s to produce fixations (see Fig. 2c).

Fine-tuning of the velocity threshold in degrees per second depends on viewing distance and screen resolution (e.g., in dots per inch). In the exemplar fixations of Fig. 2c, an 1280×1024 display (17″ diagonal) was viewed at 20″.

Filter Fine Tuning. Fine-tuning of both smoothing and differential filters is greatly aided by visualizations of the eye movement signal, especially by 1D plots of the x- and y-coordinates of gaze data versus time (see Fig. 3).

Visualization of velocity data, in particular, allows selection of the velocity threshold which is then used to identify saccades in the data streams.

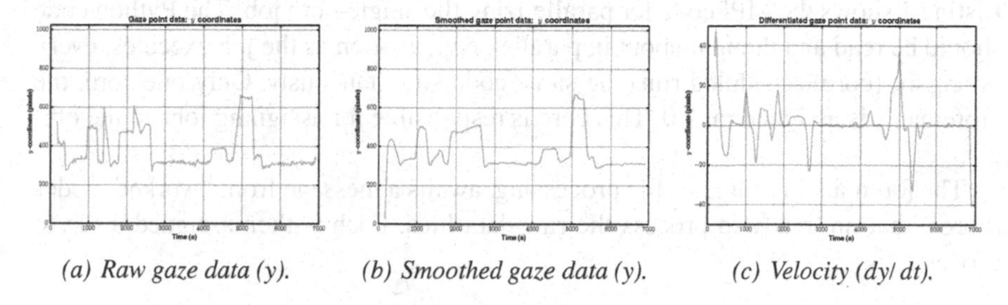

(a) Raw gaze data (y). *(b) Smoothed gaze data (y).* *(c) Velocity (dy/ dt).*

Fig. 3 Representative (individual) gaze data visualization (x- and y-coordinate versus time) showing progression from raw data to smoothed data to velocity

Heatmap Rendering. Heatmap rendering involves calculating pixel intensity $I(i,j)$ at coordinates (i,j), relative to the fixation at coordinates (x,y), by accumulating exponentially decaying "heat" intensity, modeled by the Gaussian point spread function (PSF): $I(i,j) = \exp\left(-((x-i)^2+(y-j)^2)/(2\sigma^2)\right)$. For smooth rendering, Gaussian kernel support should extend to image borders, requiring $O(n^2)$ iterations over an $n \times n$ image. With m gaze points (or fixations), an $O(mn^2)$ algorithm emerges [5]. Following accumulation of intensities, the resultant heatmap must be normalized prior to colorization.

4 Code Parallelization

Parallelization, the single-core approach advocated here, relies on applying the pipeline steps to each recorded data file:

1. parse and denoise raw data,
2. smooth the data,
3. differentiate smoothed data,
4. threshold differentiated data.

At each stage of the pipeline, render (plot) the intermediate results, e.g.,

1. 1D raw data (x, y vs. t)
2. 2D raw data
3. 1D smoothed data (x, y vs. t)
4. 2D smoothed data
5. 1D differentiated data (x, y vs. t)
6. 2D differentiated data
7. 2D fixations
8. 2D heatmap

Figure 3 shows select 1D plots. In practice, all steps are applied in sequence to each raw eye movement data file, e.g., written in Python, and can then be enclosed in an MPI script which sends assigns the input file to the next available machine on the cluster.

4.1 Multicore Parallelization with MPI

Listing 1 shows the MPI code for parallelizing the single-core job. The Python code should be read and thought about in parallel, e.g., as soon as the job executes, every processor (core) scheduled runs the same code simultaneously. Only one core, the "foreman", is assigned rank 0. This core is responsible for assigning jobs to the others.

The foreman, in charge of file processing, awaits a message from "worker" nodes (cores) that are ready to process the raw data files. Each is then assigned a file to process.

Listing 1 Foreman/workers code (courtesy of Ed Duffy @ Clemson)

```
from mpi4py import MPI

comm, stat = MPI.COMM_WORLD, MPI.Status()
rank, ncores = comm.Get_rank(), comm.Get_size()

TAG_WRK_NEED,TAG_WRK_TODO,TAG_WRK_DONE = range(3)

if rank == 0: # foreman
  # send out files for processing
  for file in list:
    comm.recv(source=MPI.ANY_SOURCE, tag=TAG_WRK_NEED, status=stat)
    comm.send((file,), dest=stat.source, tag=TAG_WRK_TODO)
  # shut down workers
  for i in range(1,ncores):
    comm.recv(source=MPI.ANY_SOURCE, tag=TAG_WRK_NEED, status=stat)
    comm.send((0,), dest=stat.source, tag=TAG_WRK_DONE)
else: # worker
  # keep working until no more work to do
  while True:
    comm.send((0,), dest=0,tag=TAG_WRK_NEED)
    file, = comm.recv(source=0,tag=MPI.ANY_TAG, status=stat)
    if stat.tag == TAG_WRK_DONE:
      break
    else:
      # process file
```

Once all files have been processed, the foreman then once again awaits a message from workers and then once each is ready to process, a message is sent indicating that no more work is required. Different MPI message tags are used to indicate whether the worker nodes have anything to process or whether they should terminate.

4.2 Fault-Tolerance

The MPI code in Listing 1 is a basic script for distributing the single-core job across multiple cores. This basic script has been successfully used when distributing the workload over 91 as well as 501 cores. The code can scale to any number of cores, depending on how many are available (see Job Scheduling, Sect. 5 below).

Although straightforward to understand, Listing 1 assumes that each scheduled worker node performs its function without fault. In other applications of this MPI script, when individual cores malfunction, the foreman node has no record of the failure. Incorporating fault tolerance, however, slightly increases the script complexity.

Listing 2 shows a possible fault-tolerant revision to the code in Listing 1. The key differences include:

1. the introduction of a new MPI tag which allows a worker node to communicate failure back to the foreman, and
2. the introduction of queues, which are used by the foreman to enqueue files to process as well as to record files processed by each worker.

Queues allow re-scheduling of files that could not be processed by workers that aborted processing for one reason or another. Queues then allow the foreman to complete processing of all files that may have been missed by the original code in Listing 1.

4.3 Many-Core Parallelization with OpenCL

SPMD parallelization can be augmented with many-core (SIMD) parallelism, provided that cluster nodes with GPUs are selected during execution (see Job Scheduling, Sect. 5 below). Each node can then use the GPU via an API such as OpenCL or GLSL.

One example of GPU usage is acceleration of heatmap rendering. Without a GPU, heatmap rendering can be sped up by truncating the Gaussian kernel beyond 2σ [12] during luminance accumulation, but this can lead to blocky image artifacts. Heatmap rendering can be written for the GPU, preserving the high image quality of extended-support Gaussian kernels while decreasing computation speed through parallelization [5]. Listing 3 shows an OpenCL implementation of the Gaussian kernel used

Listing 2 Fault-tolerant foreman/workers code

```
from mpi4py import MPI

comm, stat = MPI.COMM_WORLD, MPI.Status()
rank, ncores = comm.Get_rank(), comm.Get_size()

TAG_WRK_NEED,TAG_WRK_TODO,TAG_WRK_DONE,TAG_WRK_ABRT = range(4)

if rank == 0: # foreman
  # set up worker job list
  wrkr = defaultdict(deque)
  for i in range(ncores):
    wrkr[i] = deque()
  wrkrs = ncores
  # queue up jobs for processing
  jobs = deque()
  for file in list:
    jobs.append(file)
  # process job queue
  while len(jobs):
    comm.recv(source=MPI.ANY_SOURCE, tag=TAG_WRK_NEED, status=stat)
    if stat.tag == TAG_WRK_NEED:
      # record which job worker got
      wrkr[stat.source].append(jobs.pop())
      comm.send(wrkr[stat.source][-1],dest=stat.source,\
                                        tag=TAG_WRK_TODO)
    elif stat.tag == TAG_WRK_ABRT:
      # need to rerun job[wrkr[stat.source]]
      while wrkr[stat.source]:
        jobs.append(wrkr[stat.source].popleft())
      # subtract one worker from number of workers
      wrkrs -= 1
  # shut down workers
  for i in range(1,wrkrs):
    comm.recv(source=MPI.ANY_SOURCE, tag=TAG_WRK_NEED, status=stat)
    comm.send(None,dest=stat.source,tag=TAG_WRK_DONE)
else: # worker
  working = True
  # keep working until no more work to do
  while True:
    if working:
      comm.send((0,),dest=0,tag=TAG_WRK_NEED)
    else:
      comm.send((0,),dest=0,tag=TAG_WRK_ABRT)
    file, = comm.recv(source=0,tag=MPI.ANY_TAG, status=stat)
    if stat.tag == TAG_WRK_DONE:
      break
    else:
      # process file, if problem occurs, set working to False
```

Listing 3 Heatmap rendering using `OpenCL` kernel

```
__kernel
void gaussian(int2 dim, float2 pos, float sigma,
              __global float *gauss)
{
  // get the work-item's unique id
  int2 idx = {get_global_id(1), get_global_id(0)};
  float sigma2sq = -2.0 * sigma * sigma;

  float sx = idx.x-pos.x;
  float sy = idx.y-pos.y;
  float res = exp((sx*sx+sy*sy)/sigma2sq + (0.0));
  gauss[idx.y*dim.x+idx.x] += res;
}
```

for heatmap rendering. This OpenCL kernel has been used in a C++ version of the MPI script to compute heatmap entropy.

5 Job Scheduling on a Multi-Core Cluster

Parallel execution on a cluster depends on selection of a number of nodes followed by scheduling execution of the job on the cluster. This is accomplished by a queuing scheduler. In this example, PBS Pro is used. Job scheduling is largely a matter of syntax and is likely to be similar on different clusters, which may use other schedulers, such as OpenPBS, SLURM, MOAB, or some other variant.

A basic example requesting 91 instances of single cores on the cluster is:

```
qsub -l select=91:ncpus=1:mpiprocs=1 -l walltime=1:00:00
```

This `qsub` command specifies 1 core per CPU (`ncpus=1`), and 1 MPI process per core (`mpiprocs=1`) with expected total execution time of 1 h (`walltime =1:00:00`).

Other options, among others, that can be selected include number of nodes with GPUs (`ngpus`) as well as interconnect between nodes (e.g., `interconnect=1g` for 1 Gbps Ethernet; other interconnects include InfiniBand, Myrinet, etc.). For the single-core job described here, Ethernet or Myrinet will suffice.

The MPI job is then executed in parallel via the following:

```
mpiexec -n 91 python ./src/mpi_script.py
```

The `qsub` and `mpiexec` commands can be combined into a single shell script and then submitted to the `qsub` system. This allows batch processing of an MPI job, which is the intended way of utilizing the cluster. It may take some time before the requested resources are available, e.g., requesting 3,000 cores may take a long time to schedule, whereas requesting 91 may get the job running sooner. Different clusters may have different job queues, e.g., small, medium, or large, which may have different wait times based on the number of cores requested. Such considerations are likely to be installation-specific, hence familiarization with the local installation is beneficial.

6 Example Application Results

To report example timing performance metrics comparing execution of the MPI code versus code on a single processor machine, we use the experiment introduced at the outset, consisting of a straightforward within-subjects experiment where each of 30 participants viewed three variations of an image stimulus, namely the control image shown in Fig. 2 and two stylized variations (not shown).

The Python script executed produced 11 visualizations per image viewed, resulting in $11 \times 3 = 33$ visualizations per individual, yielding 990 files for 30 participants.

Running the MPI script on 94 cores, the job scheduler reported completion in a time of 00:01:17. The nodes selected via the scheduler used the Myrinet interconnect, which, on this particular cluster, heterogeneous in its makeup, would have been randomly selected from a pool manufactured by HP, Dell, Sun, or IBM, each with either Intel Xeon or AMD Opteron processors, with 12-16 GB RAM, and each with 8 cores.

Meanwhile, the same Python script (not wrapped in MPI) was run on a single machine (an Apple Mac Pro with 2 × 2.4 GHz 6-Core Intel Xeon CPUs running OS X Yosemite (10.10.2) with 12 GB of RAM) and, according to the `time` utility, completed in 1:07:37.

Direct comparison of execution times is somewhat problematic due to the heterogeneity of the cluster used and due to the multi-threaded nature of the Mac Pro used for the single machine used in this example (other processes were running at the time). Nevertheless, the advantage of using 94 CPUs over one is clear, resulting in a savings of about 1 h in this instance.

One may argue that producing 11 visualizations per image viewed by each participant is overkill. Granted not all visualizations may be useful in the long term, being somewhat transitory in their utility, but in the short term they do provide useful visual confirmation of participants' performance as well as of filter settings. In one instance, this type of visualization helped determine that the way data was being exported by eye tracking software resulted in duplication, e.g., every third or fourth data file was identical. Computing only statistics may not have identified this problem, and in fact, may have led to erroneous statistics being reported. Visualization clearly verified that several scanpaths were identical in appearance. The point is: if a high-performance computational cluster is available, it may as well be used. The present paper gives a primer on one possible strategy applicable to distributed processing of eye movement data.

7 Conclusion

Massive-scale parallelization of eye movement data processing and visualization was outlined via Message Passing Interface (MPI), executable on high-performance computing clusters. Two algorithmic variants were presented, one assuming fault-free execution, the other providing greater fault-tolerance should execution on a given computational core fail. A combined multi-core and many-core parallelization model was outlined and scheduling of the MPI job on the cluster was discussed.

References

1. Andersson, R., Nyström, M., Holmqvist, K.: Sampling frequency and eye-tracking measures: how speed affects durations, latencies, and more. J. Eye Mov. Res. **3**(3), 1–12 (2010)
2. Aoyama, Y., Nakano, J.: RS/6000 SP: Practical MPI Programming. IBM International Technical Supoprt Organization, Austin, TX (1999). http://www.redbooks.ibm.com/redbooks/pdfs/sg245380.pdf. Accessed Dec 2014
3. Dao, T.C., Bednarik, R., Vrzakova, H.: Heatmap rendering from large-scale distributed datasets using cloud computing. In: Proceedings of the Symposium on Eye Tracking Research and Applications, pp. 215–218. ETRA '14, ACM, New York (2014). http://doi.acm.org/10.1145/2578153.2578187
4. Duchowski, A.T., Babu, S.V., Bertrand, J., Krejtz, K.: Gaze analytics pipeline for Unity 3D Integration: signal filtering and analysis. In: Proceedings of the 2nd International Workshop on Eye Tracking for Spatial Research (ET4S), 23 Sept 2014
5. Duchowski, A.T., Price, M.M., Meyer, M., Orero, P.: Aggregate gaze visualization with real-time heatmaps. In: Proceedings of the Symposium on Eye Tracking Research and Applications, pp. 13–20. ETRA '12, ACM, New York (2012). http://doi.acm.org/10.1145/2168556.2168558
6. Gorry, P.A.: General least-squares smoothing and differentiation by the convolution (Savitzky-Golay) method. Anal. Chem. **62**(6), 570–573 (1990). http://pubs.acs.org/doi/abs/10.1021/ac00205a007
7. Hollos, S., Hollos, J.R.: Recursive digital filters: a concise guide. Exstrom Laboratories, LLC., Longmont, CO (April 2014), iSBN: 9781887187244 (ebook). http://www.abrazol.com/books/filter1/
8. Kirk, D.B., Hwu, W.M.W.: Programming Massively Parallel Processors: A Hands-on Approach. Morgan Kaufmann Publishers, Burlington (2010)
9. Message Passing Interface Forum: MPI: A Message-Passing Interface Standard. Version 3.0, University of Tennessee, Knoxville, TN (2012). http://www.mpi-forum.org/docs/mpi-3.0/mpi30-report.pdf. Accessed Dec 2014
10. Nyström, M., Holmqvist, K.: An adaptive algorithm for fixation, saccade, and glissade detection in eyetracking data. Behav. Res. Meth. **42**(1), 188–204 (2010)
11. Ouzts, A.D., Duchowski, A.T.: Comparison of eye movement metrics recorded at different sampling rates. In: Proceedings of the 2012 Symposium on Eye-Tracking Research and Applications. ETRA '12, ACM, New York. 28–30 March 2012
12. Paris, S., Durand, F.: A Fast Approximation of the Bilateral Filter using a Signal Processing Approach. Technical Report MIT-CSAIL-TR-2006-073, Massachusetts Institute of Technology (2006)
13. Savitzky, A., Golay, M.J.E.: Smoothing and differentiation of data by simplified least squares procedures. Anal. Chem. 36(8), 1627–1639 (1964). http://pubs.acs.org/doi/abs/10.1021/ac60214a047

A Simulation Approach to Select Interoperable Solutions in Supply Chain Dyads

Pedro Espadinha-Cruz and António Grilo

Abstract Business Interoperability has become an indisputable reality for companies that cooperate and struggle for competitiveness. Supply Chain Management is one kind of industrial cooperation which relies on large integration and coordination of processes. Though, supply chain operations are ruled and conditioned by interoperability factors, which until now misses a tool to identify and solve its problems. In this context, this article proposes a simulation approach to study the effects of interoperability solutions on the performance of supply chain dyads.

Keywords Business interoperability · SCM · Dyadic relationships · Simulation · Performance measurement

1 Introduction

Business interoperability (BI) is an organizational and operational ability of an enterprise to cooperate with its business partners and to efficiently establish, conduct and develop information technology (IT) supported business with the objective to create value [1]. In the context of supply chain management (SCM), business interoperability is an enabler that makes possible to execute the SC operations seamlessly, easing their alignment and the information flow, guaranteeing high performance and competitiveness [2]. However, lack of interoperability is an emerging issue in IT based cooperation [3]. Most of the existing research on interoperability areas concentrates in forms to classify and identify interoperability problems and barriers, and forms to measure and remove them.

P. Espadinha-Cruz (✉) · A. Grilo
UNIDEMI, Departamento de Engenharia Mecânica E Industrial,
Faculdade de Ciências E Tecnologia Da Universidade Nova de Lisboa,
Caparica, Portugal
e-mail: p.cruz@campus.fct.unl.pt

R. Neves-Silva et al. (eds.), *Intelligent Decision Technologies*,
Smart Innovation, Systems and Technologies 39,
DOI 10.1007/978-3-319-19857-6_14

On our research, we aim at the research question "How to achieve high levels of interoperability in supply chain dyads?", addressing one-to-one relationships in supply chains. To approach this issue, we address three topics: characterization and analysis of interoperability problems; cooperation re-design; and the study of the interoperability impact in the dyad performance. The present article proposes a method to study of interoperability impact on the dyad performance (in terms of SCM and interoperability performance), as a support to decision making in the dyad design and in the selection of suitable information systems to eliminate or mitigate interoperability problems.

The article is structured as follows: section two makes a brief review on the key topics (business interoperability, supply chain operations and performance); section three describes the methodology for analyzing and re-designing the supply chain dyadic cooperation; section four presents a case study on an automotive supply chain dyad; and section five presents the conclusions.

2 Business Interoperability

2.1 Business Interoperability Decomposition

BI is a concept that evolved from the technical perspective of interoperability incorporating several aspects of organization interactions. Frameworks and researches like IDEAS [4], INTEROP Framework [5, 6], ATHENA Interoperability Framework (AIF) [7], ATHENA Business Interoperability Framework (BIF) [7] and European Interoperability Framework (EIF) [8, 9] traced the evolutionary path that led to the exiting notion of business interoperability. In previous work from [10], several kinds of interoperability that contribute to the current definition of business interoperability were identified and related (see Fig. 1). In level 1 three interoperability types were suggested to contribute singly to the BI definition. Interoperability types shown in level 2 can provide input to more than one type of interoperability at level 1.

The different perspectives of interoperability reflect the issues that one must attend to achieve higher levels of interoperability or, as it was defined by [12], achieve "optimal interoperability".

2.2 Business Interoperability Measurement and Performance Metrics

Interoperability measurement and quantification is a branch of research dedicated to interoperability quantification in a qualitative or quantitative manner. Qualitative approaches to interoperability measurements are associated with subjective criteria

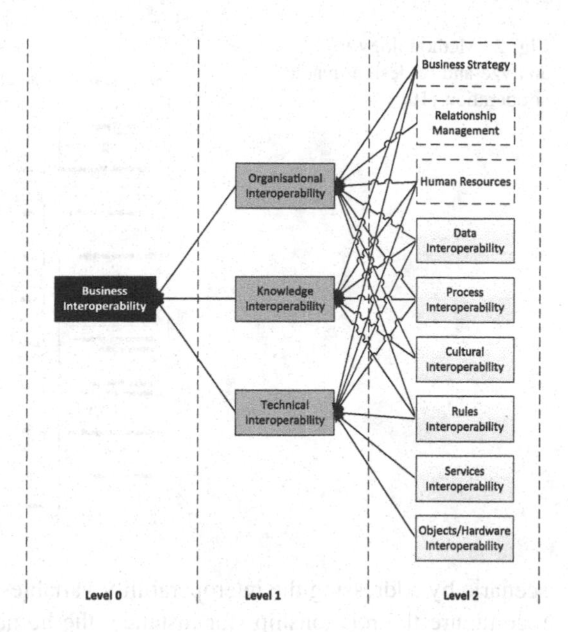

Fig. 1 Business interoperability components [10]

that permits to assign a certain level of interoperability (e.g. [13–15]), or a maturity level (e.g. [16, 17]), to a specific kind of interoperability.

On the other hand, quantitative approaches make an attempt to characterize the interoperations, proposing measurements (e.g. [18]) and scores [19] to convert interoperability issues into numeric values. The main problem with these approaches is that most of the numeric values that are obtained are as subjective as the interoperability issues that are analyzed.

Another branch of interoperability quantitative assessment is dedicated to performance measuring. Approaches to performance measurement as [7, 20–22] suggest ways to measure the impact of interoperability on metrics such as costs, time and quality. However, it is not known a direct way of relating interoperability issues, or the companies' decisions, with the interoperability metrics [7, 20–22].

3 Methodology to Analyze and Re-Design Dyadic Cooperation

The proposed method to analyze and re-design the supply chain dyads is depicted by Fig. 2.

In this method, the first phase is to analyze and model the dyad interoperability conditions in terms of the business interoperability components that represent the "as-is" situation. On the second stage, one simulates the "as-is" model and one identifies the various scenarios that may lead to a more interoperable situation. In this matter, we propose two kinds of approach: an improvement of the current

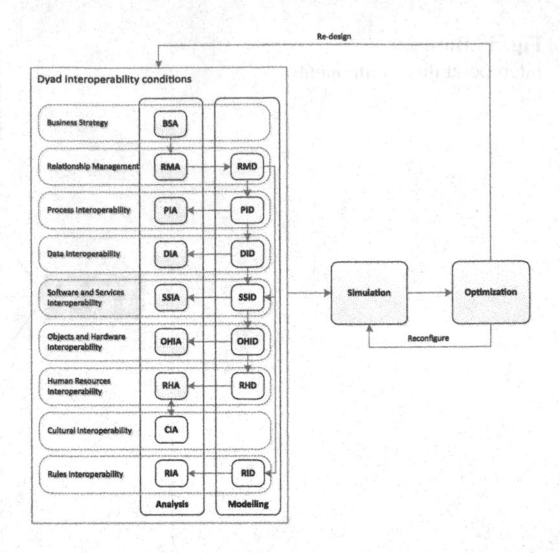

Fig. 2 Methodology to analyze and re-design dyadic cooperation [10]

scenario by addressing the interoperability variables that one can change in order to reconfigure the relationship (for instance, the human resources quantity on a specific process); or the re-design of certain aspects of interoperability, such as the process design or the selection of another information system that permits improving the dyad performance. In the last stage (optimization stage), one finds which one of those scenarios has the best performance in terms of interoperability and in terms of supply chain performance.

3.1 Stages of Analysis and Decomposition

As mentioned in the previous section, the first step of the method is to determine the dyad interoperability conditions. This is achieved by interleaving the interoperability and the performance analyses, and modeling the interoperability components in a process that we call analysis and decomposition stages (see Fig. 2). The sequence of these stages has to do with the relationship between the business interoperability components. On the top of the method are the managerial and governance aspects, such as the business strategy and the management of the relationships that impact subsequent components. For instance, in business strategy analysis (BSA), the cooperation objectives are addressed and the dyad is analyzed to verify if these ones are clear-cut to both companies and if the individual aspects are aligned into a cooperation business strategy. Managerial and governance aspects have impact in operations. Process interoperability decomposition (PID) and process interoperability analysis (PIA) are ruled by the prior aspects of interoperability, thus constituting the focus of this method. All the following stages are associated to the operations taken place in the dyad. For instance, data interoperability decomposition (DID) and data interoperability analysis (DIA) are stages acting on the exchange of data between the firms that perform the processes. Issues like semantic alignment, communication paths and data quality are addressed in this stage in order to ensure that the data is properly interpreted, that there are sufficient contact points to exchange data, and that data is usable.

In terms of interoperability, the process resources are the information technology assets (software and systems interoperability, as well as objects and hardware interoperability) and the human resources. These resources enable processes and data exchange. As in the case of data interoperability, these resources are connected to the process interoperability.

3.2 *Modeling and Measuring Interoperability Performance on Supply Chains*

Modeling supply chain processes derives from the concept of process integration and coordination [23]. The supply chain operations reference model (SCOR) [24] makes a link between performance measures, best practices and software requirements to business process models [25]. However, the SCOR model does not show how to proceed to achieve interoperability. In the application of the method portrayed in Fig. 2 we propose a systematic representation of the interoperability perspectives of the dyad. In this one, we address the supply chain operations that take place between the two firms. For instance, in [11] a buyer-seller interface was designed. To achieve this design, a mapping has been done since the strategic objectives to the process design decisions using Axiomatic Design Theory [26] combined with Business Process Notation [27] and Design Structure Matrix [28]. This procedure allowed to decompose the SC operations and to address the interoperability issues inherent to each activity. The interoperability impact study and the selection of the appropriate design is the contribution of this article, and allows to demonstrate how the findings from [10] and [11] are modeled using computer simulation.

The course between an actual ("as is") to a desired more interoperable state ("to be") is supported by the decisions taken place during the re-design and reconfiguration activities of Fig. 2. These decisions are formulated according to the identified interoperability barriers and tested through simulation. Here, in this part of the methodology the performance measurement becomes an essential aspect to achieve an interoperable dyadic relationship. Supply chain performance metrics and interoperability metrics portray a relevant part to strive, both, for a competitive and interoperable supply chain dyad.

In the next section we present a case study that is currently being developed on an automotive supply chain. Here is addressed the interaction between two firms in the context of purchase and delivery operations. These two operations were decomposed into interoperability aspects, and the business processes were modeled in order to help in the design of a simulation model. To evaluate the two companies three performance metrics were selected: order lead-time [29–33], time of interoperation and conversion time [7, 20, 21, 34, 35].

4 Case Study: Automotive Supply Chain Dyad

The present case study was implemented in a dyad constituted by a 2nd tier rubber parts supplier (company A) and a 1st tier automotive engine gaskets supplier (company B). The application of this method was made through several interviews in both companies and by analyzing companies' documentation. The internal and interface processes are presented in Fig. 3.

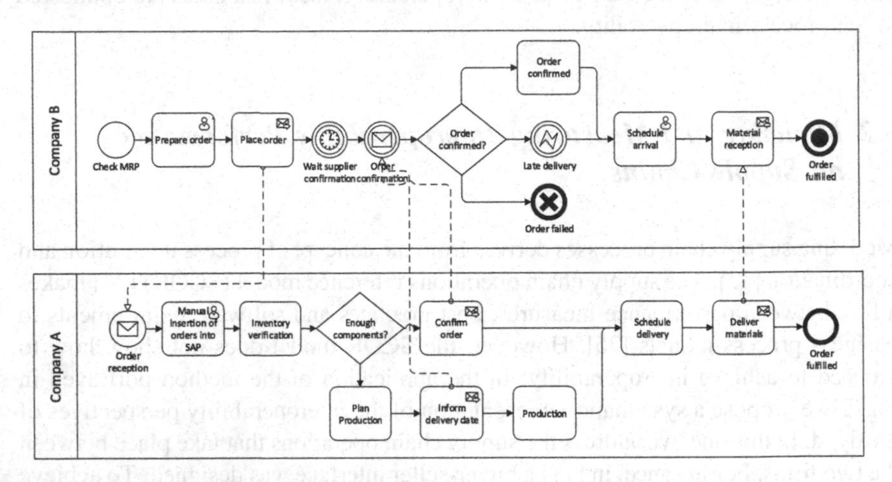

Fig. 3 Collaboration and internal activities business process model

The interoperability conditions for both are presented in Table 1.

Table 1 Interoperability conditions on the dyad

Interoperability aspect	Interoperability conditions
Business strategy	A contract was signed specifying the agreed lead-time of 7 days. The cooperation strategy was defined, but is not aligned with individual objectives
Relationship management	A long-term relationship was established
Human resources	Company A has 6 employees (5 responsible for inserting orders manually on SAP and 1 to validate orders) Company B has 2 employees to treat the orders
Process interoperability	In company A, 5 users insert manually orders into SAP. One HR verifies the inventory and confirms or calls for production. In company B, the ordering process is performed by 2 operators that check MRP data on SAP system and send the purchase orders to the supplier by e-mail and, then, wait for supplier response to validate the order and, then, wait for its fulfillment
Data interoperability	There are compatibility issues between the formats of the orders in both companies. Data must be treated manually in both cases
Software and systems interoperability	In both companies, SAP system and the E-mail system are not interoperable. This requires manual interaction between systems

The first improvement to test on the current approach for the collaboration is to study the use of the resources that enable cooperation. For simplification purposes, we only address the human resources quantity as variable to improve the "as-is" scenario. Other aspects featured on Fig. 1 should, if possible, be addressed in the performance analysis.

The results regarding the variation of human resources quantity are presented in Fig. 4.

Regarding order preparation from company B, currently there are 2 employees responsible for preparing, manually, the orders by accessing the Material Resource Plan on SAP system and send the needed orders by e-mail. On the "as-is" configuration, the average value of the order lead-time (OLT) is 163 h (7 days), which satisfies the agreed lead-time. Decreasing the number of employees to one permit reducing the OLT to 155 h (6 days) and the time of interoperation (TIP). However, the conversion time increases from 0,3 to 8,7 h for each order to be prepared. In counterpart, increasing the number of employees doesn't have effect on the metrics.

In respect to company A's activities, the number of employees on the manual insertion of orders on SAP could be decreased to a minimum of 3 in order to maintain the same OLT. Though, the minimum conversion time (Cv) is achieved with 4 employees.

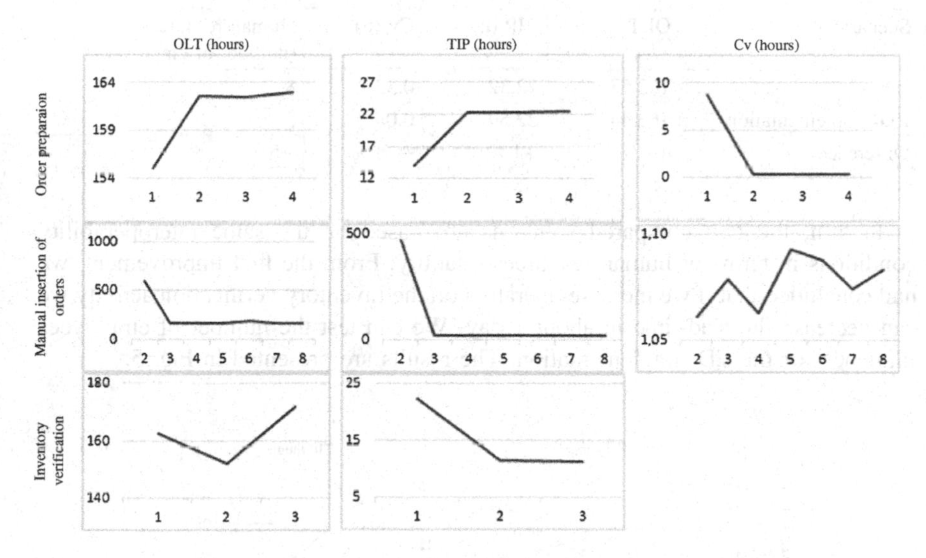

Fig. 4 Influence of human resources quantity on OLT, TIP and Cv for each process (obtained on Rockwell Arena Software in 20 replications with a confidence interval of 99 % and an error of 1,05 %)

Still in company A, increasing the employees to 2 permits to decrease the OLT to 152 h (6 days) and TIP to 11,54 h. This last improvement enhances the response time to the company B's requests. Instead of waiting 22 h to obtain the order confirmation, the increase of 1 employee permits to fulfill this in half of the time.

For this activity there are no Cv values because there is no conversion process involved.

The second improvement we propose is the implementation of an Electronic Data Interchange (EDI) system to replace the order placement communication path. This measure will enhance compatibility of data between the ICT and the order management system, reducing the time for order preparation in company B and eliminating the manual insertion process of company A. The obtained results are presented in Table 2.

Comparing the metrics for the "as-is" and the EDI implementation scenario, both OLT and TIP increase by 1 percent. In counterpart, there is a reduction of 76 % of the time to prepare the orders to send to company A.

In terms of human resources, the "as-is" scenario counts with 2 employees on company B and 6 employees (5 on manual insertion and 1 on inventory verification) on company A. The implementation of the EDI reduces the company A to 1 operator required to deal with company B's orders.

Table 2 Comparison between "as-is" and the implementation of EDI scenario (obtained on Rockwell Arena Software in 20 replications with a confidence interval of 99 % and an error of 1,05 %)

Scenario	OLT (h)	TIP (h)	Cv (h)	Human resources (number of employees)
"as-is"	162.58	22.32	0.32	8
EDI implementation	163.44	22.59	0.08	3
Difference	+1 %	+1 %	−76 %	−5

In turn, the two compared solutions are based on the same interoperability conditions in terms of human resources quantity. From the first improvement, we had concluded that if we increase operators on the inventory verification activity we can decrease the lead-time in about 1 day. We can test the number of employees influence for the EDI implementation. The results are presented in Fig. 5.

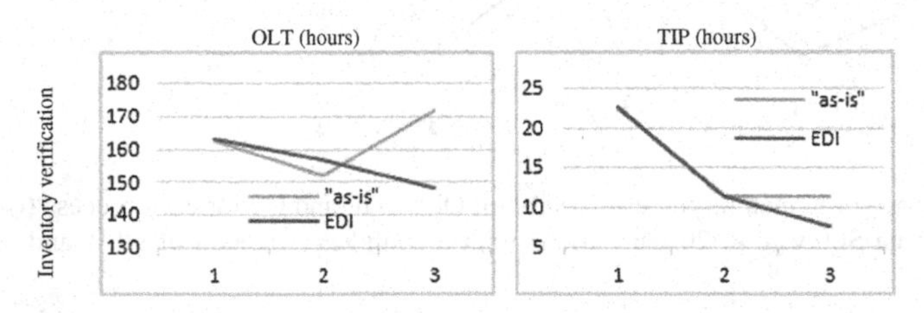

Fig. 5 Influence of human resources quantity on OLT and TIP for Inventory verification process for each scenario (obtained on Rockwell Arena Software in 20 replications with a confidence interval of 99 % and an error of 1,05 %)

If the companies decide to eliminate or mitigate the systems incompatibility (SAP and E-mail) by implementing an EDI, best results can be achieved if the number of employees on the inventory verification is increased to 3. However, if due to technical limitations the EDI implementation is not possible, the company A should add another employee to the inventory verification activity (by contracting a new employee) or remove one employee from manual insertion to inventory verification.

5 Conclusions

The presented research contributes to the development of an integrated framework to assess and re-design supply chain dyadic cooperation. It provides a method to study the interoperability impact on the performance of the dyad. This method allows one the test various scenarios without affecting the real system and providing the solution that may result in an improvement for the dyad.

Future work will concentrate on the integration of other interoperability aspects by implementing Design of Experiments and Taguchi methods. This will allow us to deal with the complexity of Business Interoperability by systematizing the influence of interoperability aspects on performance.

Acknowledgments The authors would like to thank Fundação para a Ciência e Tecnologia for providing a research grant to Pedro Espadinha da Cruz through the project PTDC/EME- GIN/ 115617/2009.

References

1. Legner, C., Wende, K.: Towards an excellence framework for business interoperability. In: 19th Bled eConference eValues, pp. 1–16 (2006)
2. Huhns, M.N., Stephens, L. M., Ivezic, N.: Automating supply-chain management. In: Proceedings of the First International Conference on Autonomous Agents MultiAgent System part 3—AAMAS'02, No. 483, p. 1017 (2002)
3. Elvesæter, B., Hahn, A., Berre, A., Neple, T., Ict, S., Blindern, P. O. B., Oslo, N.: Towards an interoperability framework for model-driven development of software systems, framework
4. IDEAS.: IDEAS project deliverables (WP1-WP7) (2003)
5. Chen, D.: Enterprise interoperability framework, EMOI-INTEROP (2006)
6. Chen, D., Doumeingts, G., Vernadat, F.: Architectures for enterprise integration and interoperability: Past, present and future. Comput. Ind. **59**(7), 647–659 (2008)
7. ATHENA: DA8.2 guidelines and best practices for applying the ATHENA interoperability framework to support SME participation in digital ecosystems (2007)
8. Vernadat, F.B.: Technical, semantic and organizational issues of enterprise interoperability and networking. Annu. Rev. Control **34**(1), 139–144 (2010)
9. IDABC: European interoperability framework—version 2.0 (2010)
10. Espadinha-Cruz, P., Grilo, A.: Methodology to analyse and re-design dyadic industrial cooperation. In: 21st International Annual EurOMA Conference, EurOMA 2014, No. Ieee 1990, pp. 1–10 (2014)

11. Espadinha-Cruz, P., Mourão, A.J.F., Gonçalves-Coelho, A., Grilo, A.: Business interoperability: dyadic supply chain process decomposition using axiomatic design. In: 8th International Conference on Axiomatic Design (ICAD 2014), pp. 93–99 (2014)
12. Legner, C., Lebreton, B.: Business interoperability research: present achievements and upcoming challenges. Electron. Mark. **49**(341), 176–186 (2007)
13. DoD: Levels of information systems interoperability (LISI) (1998)
14. Tolk, A., Muguira, J.A.: The levels of conceptual interoperability model, System, No. September. 2003 Fall simulation Interoperability Workshop, Orlando, Florida, pp. 1–11 (2003)
15. Sarantis, D., Charalabidis, Y., Psarras, J.: Towards standardising interoperability levels for information systems of public administrations, eJETA Special Issue, Interoperability Enterprise Admiral Worldwide (2008)
16. Guédria, W., Naudet, Y., Chen, D.: Maturity model for enterprise interoperability, Enterprise Information Systems No. October 2013, pp. 1–28 (2013)
17. Clark, T., Jones, R., Jones, L., Pty, C.: Organisational interoperability maturity model for C2. In: Proceedings of the 1999 Command and Control Research and Technology Symposium (1999)
18. Zutshi, A., Grilo, A., Jardim-Gonçalves, R.: The business interoperability quotient measurement model. Comput. Ind. **63**(5), 389–404 (2012)
19. Ford, T., Colombi, J.: The interoperability score. In: Proceedings of the Fifth Conference on Systems Engineering Research, pp. 1–10 (2007)
20. Chen, D., Vallespir, B., Daclin, N.: An approach for enterprise interoperability measurement. In: Proceedings of MoDISE-EUS 2008 Concepts, pp. 1–12 (2008)
21. Ducq, Y., Chen, D.: How to measure interoperability: concept and approach. In: 14th International Conference (2008)
22. ATHENA: D.B3.3 Interoperability Impact Analysis Model (2007)
23. Vernadat, F.: Enterprise modeling and integration: principles and applications, 1st edn, p. 513. Chapman & Hall, London (1996)
24. Supply Chain Council.: Supply Chain Operations Reference (SCOR) model—version 10.0, USA (2010)
25. Thakkar, J., Kanda, A., Deshmukh, S.G.: Supply chain performance measurement framework for small and medium scale enterprises. Benchmarking Int. J. **16**(5), 702–723 (2009)
26. Suh, N.P.: The Principles of Design, vol. 226, p. 401. Oxford University Press, Oxford (1990)
27. Fettke, P.: Business process modeling notation. Wirtschaftsinformatik **50**(6), 504–507 (2008)
28. Eppinger, S.D., Browning, T.R.: Design Structure Matrix Methods and Applications, p. 334. The MIT Press, Cambridge, Massachussetts (2012)
29. Azevedo, S.G., Carvalho, H., Cruz-Machado, V.: A proposal of LARG supply chain management practices and a performance measurement system. In: International Journal e-Education, e-Management e-Learning, vol. 1, no. 1, pp. 7–14 (2011)
30. Gunasekaran, C.P., McGaughey, R.E.: A framework for supply chain performance measurement. Int. J. Prod. Econ. **87**(3), 333–347 (2004)
31. Beamon, B.: Measuring supply chain performance. Int. J. Prod. Res. **39**(14), 3195–3218 (2001)
32. Otto, A., Kotzab, H.: Does supply chain management really pay? Six perspectives to measure the performance of managing a supply chain. Eur. J. Oper. Res. **144**, 306–320 (2003)
33. Chan, F.T.S.: Performance measurement in a supply chain. Int. J. Adv. Manuf. Technol. **21**(7), 534–548 (2003)
34. Camara, M.S., Ducq, Y., Dupas, R.: A methodology for the evaluation of interoperability improvements in inter-enterprises collaboration based on causal performance measurement models. Int. J. Comput. Integr. Manuf., no. October 2013, 1–17 (2013)
35. Razavi, M., Aliee, F.S.: An approach towards enterprise interoperability assessment. Enterp. Interoperability, Proc **38**, 52–65 (2009)

Probabilistic Weighted CP-nets

Sleh El Fidha and Nahla Ben Amor

Abstract This paper addresses the problem of combining uncertainty and precision in CP-nets. Existing approaches extended the network either with probability distribution to manage uncertainty i.e. PCP-nets or through weights like WCP-nets. This paper combines both concepts in order to make preference expression more flexible to the user. The experimental study shows that the proposed *Probabilistic Weighted CP-nets* (so called PWCP-nets for short), compared to WCP-nets and PCP-nets can really enhance the expressiveness of user preferences.

1 Introduction

Structuring user preferences has always been an arduous task, especially if they are expressed in natural languages. Although, several tools have been proposed to overcome this problem like Conditional Preference networks (CP-nets) initially proposed in [1]. Such models allow us to search for a non dominated alternative or simply to compare alternatives in a reasonable computation time [2–4]. Nevertheless, standard CP-nets or their variant TCP-nets [2] cannot represent all preference relations since qualitative comparison is not always feasible. This explains the appearance of various quantitative extensions for the standard model such as utility CP-nets [5], weighted CP-nets [6], probabilistic CP-nets [7] and dynamic probabilistic CP-nets [8].

In this paper we will mainly focus on two models: weighted CP-nets (WCP-nets) [6] allowing users to express fine grained preferences using a multiple levels scale of preference and the probabilistic version of CP-nets (PCP-nets) proposed in [7, 8]. In fact, we propose to enhance the expressiveness of CP-nets via a new model, that takes advantages of WCP-nets and PCP-nets, so-called Probabilistic Weighted CP-nets

S.E. Fidha (✉) · N.B. Amor
LARODEC, Institut Supérieur de Gestion Tunis, Université de Tunis, Tunisia, Africa
e-mail: sleh.fidha@yahoo.fr

N.B. Amor
e-mail: nahla.benamor@gmx.fr

R. Neves-Silva et al. (eds.), *Intelligent Decision Technologies*,
Smart Innovation, Systems and Technologies 39,
DOI 10.1007/978-3-319-19857-6_15

(PWCP-nets for short) allowing to include weights and probabilities on dependency relations and preference tables to express imprecision and uncertainty over the model which is, to the best of our knowledge, not yet addressed.

The paper is structured as follows. The next section introduces the basic concept of CP-nets, WCP-nets and PCP-nets. Section 3 presents probabilistic weighted CP-net (PWCP-nets), its semantic and inference task. Finally Sect. 4 reports and analyzes different comparative and experimental results.

2 Beyond CP-nets

CP-nets proposed by Boutilier et al. [1] are graphical models used for representing, compactly, qualitative preference relations over a set of variables $V = \{X_1, \ldots, X_n\}$. These models are based on the Ceteris Paribus principle defined as follows:

Definition 1 *Let X, Y, and* $Z \subseteq V$ *be nonempty sets that partition V and* $\succ$ *be a preference relation over the domain of V denoted D(V). X is said to be (conditionally) preferentially independent of Y given Z Ceteris Paribus (all else being equal) iff* $\forall x_1, x_2 \in D(X)$, $\forall y_1, y_2 \in D(Y)$, $\forall z \in D(Z)$:

$$x_1 y_1 z \succ x_2 y_1 z \textit{ iff } x_1 y_2 z \succ x_2 y_2 z.$$

A CP-net N is a directed graph $G = \{V, E, CPT\}$ where V stands for the problem variables, E for the set of edges and CPT for the set of CP-tables associated to the problem variables, encoding the preference rules of the user. An edge from a variable X_i to a variable X_j means that preferences over X_j depends on X_i. Formally:

Definition 2 *A CP-net N over* $V = \{X_1, \ldots, X_n\}$ *is a directed graph, whose variables are annotated with conditional preference table* $CPT(X_i)$ *associating a total order* $\succ_{X_i|u}$ *with each instantiation u of* X_i*'s parents denoted* $U(X_i)$*.*

Let $X \subseteq V$ be a subset of V, if X = V. An assignment to all variables of X is called an outcome denoted o. Given a CP-net N, two main reasoning tasks can be performed: (i) *Dominance* which is the task of deciding for two given outcomes o and o' whether N entails that o is preferred to o' denoted ($o \succ o'$). (ii) *Optimization* consisting in computing the best outcome according to N, that is the outcome which is undominated. This is using the forward sweep procedure where we assign for each variable its most preferred value starting from the root variable to the set of leaf variables.

Example 1 *Let V = {FilmGenre, Date, Language} be a set of variables, such that D(FilmGenre)= {Action, Drama} (D(Date) = {Recent, Classic} and D(Language) = {English, French}). Assuming user preferences defined by the CP-net N depicted in Fig. 1a where they are unconditional for the variable FilmGenre, while for Date (resp. for language) they are determined given FilmGenre (resp. Date). Moreover,*

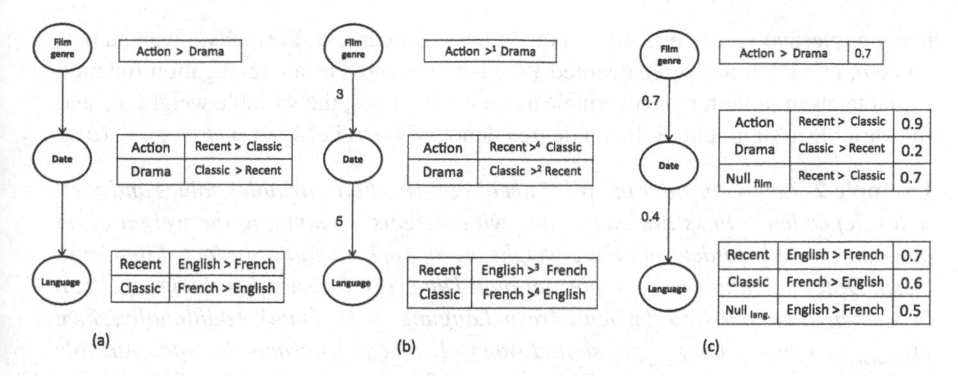

Fig. 1 **a** Standard CP-net, **b** WCP-net, **c** PCP-net

given $o = \{Action, Recent, French\}$ *and* $o' = \{Drama, Classic, French\}$, *dominance query entails that* $(o \succ o')$ *according to N, while optimization query identifies the outcome* $\{Action, Recent, English\}$ *to be the optimal outcome.*

2.1 Weighted CP-nets

Although CP-nets are simple and flexible tools used to express preferences qualitatively, they suffer from the inability of users to express their fine grained preferences. Specifically, they cannot indicate to what extent the preference would be. Moreover, the dependence relationships in CP-nets only indicate that parent variables are more important than children variables which results in many incomparable outcomes if their number increases.

To overcome the aforementioned issues, Wang et al. in [6] adopted the concept of multiple levels of relative importance by using a five levels scale i.e. *(1 (Equally preferred), 2 (Moderately preferred), 3 (Quite important), 4 (Demonstrably important) and 5 (Extremely preferred))* that can be applied to both variable values and variables. The extended model so called weighted CP-nets (WCP-nets) is defined as follows:

Definition 3 *A WCP-net N over* $V = \{X_1, \ldots, X_n\}$ *is a directed graph, whose variables are annotated with a weighted conditional preference table* $WCPT(X_i)$ *associating a relative importance denoted k to the total order* $\succ^k_{X_i|u}$ *with each instantiation u of* X_i*'s parents denoted* $U(X_i)$*. Furthermore, given two variables* X_i *parent of* X_j*, WCP-net assigns to the dependency link a relative importance k, used to compute the weight of each variable* X_i *denoted* w_{X_i} *where their summation should be 1.*

Though, WCP-nets increase the expressiveness of standard CP-nets by changing their basic reasoning task through the concept of violation degree. This allows for each outcome to compute its total weight which makes the comparison more feasible using numerical values instead of a qualitative comparison. Formally, given an outcome o, its violation degree denoted $V(o)$ is represented as an aggregation function F that takes in input for each variable a couple of values, the variable weight w_X and the variable assignment violation degree denoted $V_X(o)$ i.e. $V(o) = F(w_X, V_X(o))$.

Example 2 *Let us consider the WCP-net of Fig. 1b where variables values and variables dependency links are augmented with weights. Computing the weight of an outcome o needs to determine the weight w_X of each variable. In fact, FilmGenre (resp. Date) is 3 times more important than Date (res. 5 times than Language) can be written as Date = 1/3 FilmGenre (resp. Language = 1/5 Date). Additionally, since $w_{FilmGenre} + w_{Date} + w_{Language}$ should sum to 1, we can determine for each variable its weight w_X i.e. $w_{FilmGenre} = 0.714$, $w_{Date} = 0.23$ and $w_{Language} = 0.047$. Moreover, using the violation degree it is possible to determine the total weight of an outcome i.e. $o = \{Drama, Recent, French\}$ is equal to $(0.714 * 1) + (0.23 * 2) + (0.047 * 3) = 1.315$ where $V_{FilmGenre}(o) = 1$ (resp. $V_{Date}(o) = 2$ and $V_{Language}(o) = 3$) is the violation degree of FilmGenre (resp. Date and Language). Applying this process to all outcomes generates the following total order : $o_1 \succ o_2 \succ o_8 \succ o_7 \succ o_4 \succ o_3 \succ o_5 \succ o_6$, where o_1 (resp. o_2, o_3, o_4, o_5, o_6, o_7 and o_8) corresponds to the outcome (Action, Recent, English) (resp. (Action, Recent, French), (Action, Classic, English), (Action, Classic, French), (Drama, Recent, English), (Drama, Recent, French), (Drama, Classic, English) and (Drama, Classic, French)).*

2.2 Probabilistic CP-nets

A second recent extension of standard CP-nets named Probabilistic CP-nets (PCP-nets) were developed in a couple of works [7, 8]. In fact, authors in [7] were interested in optimization, while [8] was dedicated essentially to dominance testing by proposing an algorithm that enhance the complexity of this task in both, standard CP-nets and PCP-nets.

Definition 4 *A PCP-net N over $V = \{X_1, \ldots, X_n\}$ is a directed graph, whose variables are annotated with a probabilistic conditional preference table PCPT(X_i) associating a probability distribution p over the set of orders $\succ_{X_i|u}$: p with each instantiation u of X_i's parents denoted U(X_i). Moreover, given two variables X parent of Y in V, PCP-nets assign a probability p to its dependency link.*

Given a PCP-net, several reasoning tasks can be performed. The most natural are the probabilistic version of dominance given two outcomes, and the most probable optimal outcome [8]. Furthermore, given a PCP-net it is possible to determine the most probable CP-net as well as the most probable optimal outcome and the probability of a given outcome. The principle of such tasks is to transform the original PCP-net to a reduced version called Opt-net allowing the use of Bayesian Networks techniques [7].

Example 3 *Let us consider the PCP-net depicted in Fig. 1c where its PCP-tables associate each combination of parents assignment to a probability distribution over the set of total orderings and its dependency links are augmented with a probability of existence modeling the possible non existence of the dependency links between variables.*

3 A New Model to Manage Uncertainty and Weights in CP-nets

As mentioned above, WCP-nets proposed a new semantic based on relative importance between variable values and variables in order to enhance the expressiveness of the standard model and solve the incomparability resulting from qualitative comparison of outcomes. In addition, PCP-nets allow expressing uncertainty over user preferences by attaching probabilities to the possible assignments of a given variable in the context of its parents which lead to more than one order in its PCP-table. However, to the best of our knowledge none of existing works around CP-nets addressed both concepts in the same model. A typical situation is when expressing noisy preferences of a user. For instance, given $X = \{x_1, x_2\}$ the user *most often strongly prefers* x_1 *to* x_2. Similar sentences express at the same time the frequency and the strength of the preference relation between x_1 and x_2. However, existing models (i.e. PCP-nets or WCP-nets) partially describe this relation.

To overcome this problem, we propose to combine both concepts (probabilities and weights) in a new model so-called Probabilistic Weighted CP-nets (PWCP-nets for short) where the probability distributions and the weights are expressed over variable values and variables dependency links. The model is defined as follows:

Definition 5 *A PWCP-net N over a set of variables* $V = \{X_1, \ldots, X_n\}$ *is a binary directed acyclic graph G, where each variable* X_i *is annotated with a conditional probabilistic weighted preference table denoted* $PWCPT(X_i)$ *associating a probabilistic weighted total order* $\succ^k_{X_i|u} : p$ *with each instantiation u of* X_i*'s parents. Moreover, given two variables* X_i *parent of* X_j*, a probability of existence and a weight are assigned to their dependency link.*

Example 4 *Figure 2a shows an example of a PWCP-net where variable values and variable dependency links are augmented with probabilities and weights. As for PCP-nets, the presence of uncertainty over dependency links resulted in a set of statements in the PWCP-tables of children variables where parents are assigned null. Same interpretation holds for remaining preference statements and relations.*

Given a PWCP-net N, we are interested in the probabilistic weight of outcomes allowing users to determine the most probable weighted outcome or to compare the probabilistic weight of a set of outcomes. Indeed, the probabilistic weight of an outcome can be obtained via three main steps:

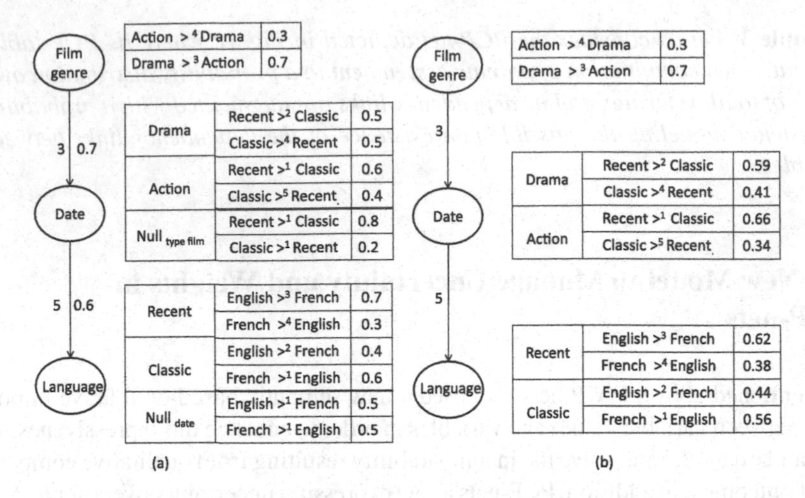

Fig. 2 **a** Probabilistic Weighted CP-net, **b** Opt-net with weights

Computing the Probability of Outcomes: Determining the probability of an outcome is based on the probabilistic information in PWCP-nets. Though, PWCP-nets doesn't allow us to compute directly the probability of a given outcome. For this, we propose to use the transformation of the PCP-net described in [8] so-called Opt-net sharing the same dependency relation as the original PWCP-net with probabilities only on its PWCP-tables using Eq. 1 where for a variable X_i in the context of its parents Y_i denoted $U(X_i)$, given an assignment of $U(X_i) = \{yi_1, \ldots, yi_n\}$, $y_i = \{g = (g_1 \ldots g_n) | g_i \in (y_i, null_i)\}$, $\alpha_i \neq 0$ (resp. $1 - \alpha_j$) is the probability that X_i depends on Y_i (resp. its complement) and p_{o_g} is the probability of ordering over the domain of X_i.

$$\sum_{(y_1 \ldots y_n) \in U(X_i)} (p_{o_g} . (\prod_{i.s.t. g_i \neq null_i} (\alpha_i). \prod_{i.s.t. g_j = null_j} (1 - \alpha_j))) \quad (1)$$

Furthermore, for each variable in the Opt-net, its domain is the set of values that are ranked first in the ordering of the PWCP-table. The CP-tables of the Opt-net in this case are obtained as follows: for each assignment of the parent variables, we consider the values of the dependent variable defined in the PWCP-table as most preferred in the set of orders according to that distribution.

In addition, since PCP-net can be obtained from a PWCP-net and it was demonstrated in [8] that there is a one-to-one correspondence between assignments with non zero probability in the Opt-net and the optimal outcomes in PCP-nets, we can then compute directly the joint probability of a given outcome. It is represented as the product of probabilities of the set of statements forming that outcome in the Opt-net.

Computing the Weight of Outcomes: In [6], Wang et al. adopted the concept of violation degree for each variable to compute the weight of outcomes. It refers to the relative importance of the variable value selected. For PWCP-net, computing the weight of outcomes is similar to WCP-nets. Yet, there are some differences. In fact, the CP-tables of both models differs in the definition of each statement in the context of its parents assignment since WCP-nets allow the user to express one certain order over the domain of a given variable, while in PWCP-nets the user is able to express a relative importance over several order in the context of the same parents assignment. For this, the weight assigned to each ordering in the PWCP-table will be associated to the value that is ranked first in that ordering. Next to that, given an outcome o, its weight is represented as an aggregation function F that takes in input for each variable a couple of values, the variable weight w_X and the variable assignment relative importance degree denoted $R_X(o)$ i.e. $W(o) = F(w_X, R_X(o))$.

Computing the Probabilistic Weight of Outcomes: The probabilistic weight of outcomes consists in computing for each one the product of its weight and probability which allows the integration of uncertainty expressed while constructing the PWCP-net.

Example 5 *In what follows we start explaining the reduction of the PWCP-net depicted in Fig. 2a to its corresponding Opt-net followed by the process of computing the probability of an outcome as well as its weight and we finish by computing its final probabilistic weight.*

- ***The Opt-net:*** *Figure 2b illustrates the transformation of the PWCP-net depicted in Fig. 2a where the Opt-net shares the same structure of the PWCP-net but statements with null parents assignment were combined in PCP-tables using Eq. 1 i.e. the probability 0.59 for the statement Recent $\succ$ Classic is obtained as follows: (0.7 * 0.5)+(0.3*0.8) where 0.7 (res. 0.3) is the probability of existence of the relation between variables Film Genre and Date (resp. its complement) and 0.5 (resp. 0.8) is the probability of the statement in the context of Drama Film (resp. the probability of the statement when null value is assigned to the parents).*
- ***The probability of outcomes:*** *The second step consists in computing the probability of each outcome. This is via the Opt-net where we sweep through the network to compute the probability of factors forming the outcome and ranked first in the set of variable orders i.e. the probability of outcome (Drama, Recent, French) (resp. (Drama, Classic, English)) is 0.7*0.59*0.38 = 0.15 (resp. 0.12). The total order obtained is as follows:* $o_5 \succ o_8 \succ o_6 \succ o_7 \succ o_1 \succ o_2 \succ o_4 \succ o_3$.
- ***The weight of outcomes:*** *Computing an outcome weight is based on two steps. (1) Determining the weight* w_i *of each variable where their summation should be 1 i.e. the PWCP-net of Fig. 2a where the weight of variables is the same of Fig. 1b. (2) For each outcome, we sweep through the network to compute its total weight* W_o *i.e. the weight of* o_6 *(Drama, Recent, French) is 2.79 = (0.714*3)+(0.23*2)+(0.047*4) where 0.714 (resp. 0.23 and 0.047) is the weight of variable Film Genre (resp. Date and language) and 3 (resp. 2 and 4) is the weight of the value Drama when ranked first (resp. Recent and French). The total order is then as follows :* $o_3 \succ o_4 \succ o_2 \succ o_1 \succ o_7 \succ o_8 \succ o_6 \succ o_5$.

– ***The probabilistic weight of outcomes:*** *The last step determines the final order of outcomes by computing the product of its probability and weight i.e. for the same statement (Drama, Recent, French) the probabilistic weight is 0.41 = 2.79 * 0.15. Applying this process to all outcomes generates the following order :* $o_5 \succ o_8 \succ o_6 \succ o_7 \succ o_1 \succ o_2 \succ o_4 \succ o_3$.

Considering the orders generated by WCP-net, PCP-net and PWCP-net, we notice that the optimal outcome of WCP-nets became the worst one in the PCP-net and the PWCP-net. This is due to the probabilistic component since according to user preferences this assignment is very uncertain whereas the outcome with the least weight in WCP-nets (outcome o_5*) has the highest probability which make it the more probable in the PCP-net and in the PWCP-net. Moreover, it is easy to identify a high resemblance between PCP-net and PWCP-net. In fact, this is explained by the closeness of the outcome weights that are varying between 2,81 and 4,14.*

4 Experimental Study

In order to evaluate the efficiency of the proposed PWCP-nets, we have developed a Java toolbox. We used *Spect Heart* dataset obtained from the University of California Irvine Machine Learning Repository,[1] composed of 267 records, each one having 22 binary variables.

4.1 Accuracy of Recommendation

The first set of experiments deals with the number of comparable outcomes obtained using different extensions of CP-nets. It consists in generating randomly a CP-net, a TCP-net, a WCP-net and a PWCP-net sharing all the same structure but with different preference tables configuration. Figure 3a details the result of incomparable outcomes according to the number of variables employed by different variants of CP-nets. In fact, since CP-nets and TCP-nets are purely qualitative, when their number of outcomes increases, several one appear to be incomparable i.e. outcomes (Action, Classic, English) and (Drama, Classic, French) in Fig. 1a. Indeed, (Action, Classic, English) violates the rules of all child variables, however, (Drama, Classic, French) violates the rule of the root variable only. In such situation, it is impossible to decide the dominant. On the other side, because of their quantitative components, WCP-nets, PCP-nets and PWCP-nets made the comparison feasible between all outcomes by assigning a specific weight to each one.

[1]https://archive.ics.uci.edu/ml/datasets/SPECT+Heart.

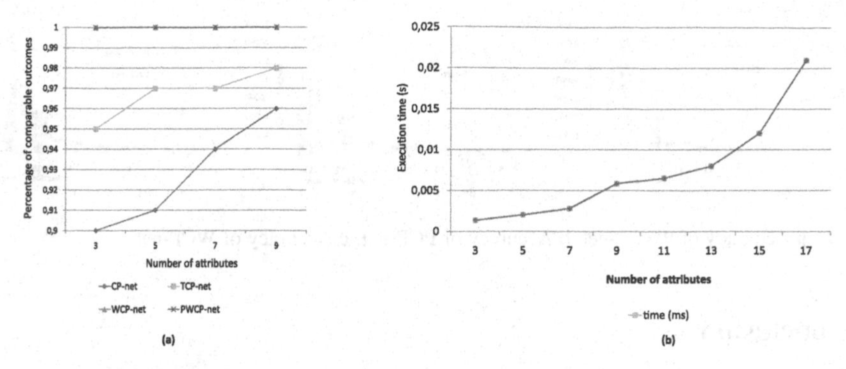

Fig. 3 **a** Percent of comparable outcomes, **b** Execution time

4.2 Execution Time

The second set of experiments is dedicated to evaluate the time needed to obtain the optimal outcome of PWCP-nets. Figure 3b shows that the time needed to find the optimal decision of PWCP-nets increases linearly with the number of variables. It starts with over 1 ms when using 3 variables while it exceeds 25 ms when the number of variables reaches its maximum (17).

4.3 Accuracy of Retrieved Outcomes

Our purpose from the third set of experiments is to evaluate the accuracy of the optimal outcome for PWCP-net (resp. PCP-net and WCP-net) compared to PCP-net and WCP-net (resp. PWCP-net).

To do this, we randomly choose 3, 5, 7, 9, 11, 13, 15, 17 variables to generate different graph structures from the dataset. Then, as measurement, for each model we count outcomes that are prior to the optimal one. The intuition behind is that the more accurate the preference model is, the less number of obtained outcomes is. All experiments are repeated 100 times and the result of the average performance are delineated in Fig. 4 where Figure a (resp. b and c) exposes the result when the optimal outcome of PWCP-net is selected (resp. PCP-net and WCP-net). Clearly, when few variables are employed (less than 7) to generate the graph, WCP-net holds a better performance. However if the number of variables increases, PWCP-nets outperforms WCP-net and PCP-net.

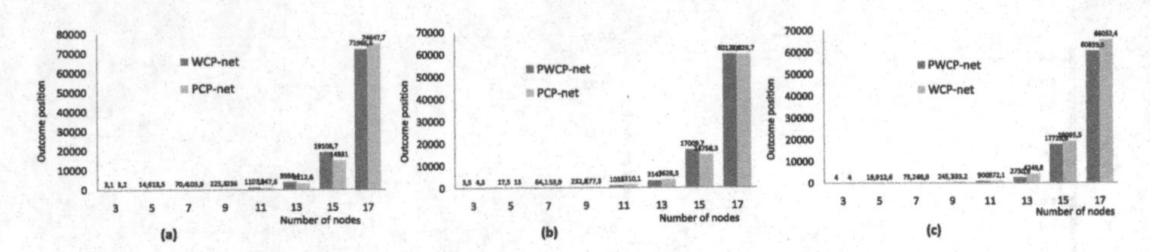

Fig. 4 **a** Accuracy of PWCP-net, **b** Accuracy of PCP-net, **c** Accuracy of WCP-net

5 Conclusion

This paper tackled the problem of combining uncertainty and imprecision of preferences expressed using CP-nets. It extended the standard model with probabilities presented in [7, 8] and weights detailed in [6]. Although, our proposed model handles only the binary case, which is a clear restriction over its expressiveness, it yields to good results. Consequently extending our model to a generic case will be a promising area of research.

References

1. Boutilier, C., Brafman, R.I., Domshlak, C., Hoos, H.H., Poole, D.: Cp-nets: a tool for representing and reasoning with conditional ceteris paribus preference statements. J. Artif. Intell. Res. (JAIR) **21**, 135–191 (2004)
2. Brafman, R.I., Domshlak, C.: Introducing variable importance tradeoffs into cp-nets, pp. 69–76 (2002)
3. Domshlak, C., Brafman, R.I.: Cp-nets, reasoning and consistency testing. In: Proceedings of the Eighth International Conference on Principles of Knowledge Representation and Reasoning, pp. 121–132 (2002)
4. Goldsmith, J., Lang, J., Truszczynski, M., Wilson, N.: The computational complexity of dominance and consistency in cp-nets. J. Artif. Intell. Res. (JAIR) **33**, 403–432 (2008)
5. Boutilier, C., Bacchus, F., Brafman, R.I.: Ucp-networks: a directed graphical representation of conditional utilities. In: Proceedings of the Seventeenth Conference on Uncertainty in Artificial intelligence, pp. 56–64. Morgan Kaufmann Publishers Inc., San Francisco (2001)
6. Wang, H., Zhang, J., Sun, W., Song, H., Guo, G., Zhou, X.: Wcp-nets: a weighted extension to cp-nets for web service selection. In: Service-Oriented Computing, pp. 298–312. Springer, New York (2012)
7. Cornelio, C., Goldsmith, J., Mattei, N., Rossi, F., Venable, K.B.: Dynamic and probabilistic cp-nets. CP Doctoral Program **2013**, 31 (2012)
8. Bigot, D., Fargier, H., Mengin, J., Zanuttini, B.: Probabilistic conditional preference networks. Septiémes Journées d'Intelligence Artificielle Fondamentale, p. 57 (2013)
9. Wilson, N.: Extending cp-nets with stronger conditional preference statements. AAAI **4**, 735–741 (2004)
10. Wang, H., Shao, S., Zhou, X., Wan, C., Bouguettaya, A.: Web service selection with incomplete or inconsistent user preferences. In: Service-Oriented Computing, pp. 83–98. Springer, New York (2009)
11. Xu, H., Hipel, K.W., Marc Kilgour, D.: Multiple levels of preference in interactive strategic decisions. Discrete Appl. Math. **157**(15), 3300–3313 (2009)

12. Wang, H., Liu, W.: Web service selection with quantitative and qualitative user preferences. In: Proceedings of the 2011 IEEE/WIC/ACM International Conferences on Web Intelligence and Intelligent Agent Technology—Volume 01, IEEE Computer Society, pp. 404–411 (2011)

Dealing with Seasonality While Forecasting Urban Water Demand

Wojciech Froelich

Abstract Forecasting water demand is required for the efficient controlling of pressure within water distribution systems, leading to the reduction of water leakages. For the purpose of this research, it is assumed that the control of water pressure is performed by pressure reduction valves (PRVs) working in the open loop mode. This means that water pressure is controlled on the basis of the daily water demand profile, a 24-step ahead forecasting of hourly time series. A key issue in such time series that affects the effectiveness of its forecasting is seasonality. Three different techniques to deal with seasonality are investigated in this paper: auto-regressive, differentiation, and the application of dummy variables. This paper details a comparative study of these three techniques with respect to water demand time series and different predictive models. We show that an approach based on dummy variables and linear regression outperforms the other methods.

Keywords Forecasting time series · Multiple seasonality · Water demand

1 Introduction

The objective of this research is the forecasting of water demand to enable the appropriate control of pressure within water distribution system (WDS). The adaptation of water pressure to the actual water demand leads to the reduction of leakages and thus decreases the cost of water delivery to consumers.

Two types of water leakages from a WDS can be distinguished; those occurring due to pipe bursts and those from background leakages [17]. Pipe bursts can be detected and eliminated by fixing the pipes. Background leakages occur due to other complex problems in WDS and are treated as unavoidable in the short term. Background leakages can be reduced by decreasing water pressure within WDS. However, water pressure cannot be reduced too much as it should be high enough to

W. Froelich (✉)
Institute of Computer Science, University of Silesia, Sosnowiec, Poland
e-mail: wojciech.froelich@us.edu.pl

R. Neves-Silva et al. (eds.), *Intelligent Decision Technologies*,
Smart Innovation, Systems and Technologies 39,
DOI 10.1007/978-3-319-19857-6_16

assure water delivery to all consumers, including those located at higher floors or in mountain districts.

For the purpose of this research, we assume that water pressure is controlled by the application of pressure reducing valves equipped with the appropriate programmable controllers. We also assume that the controller works in the open-loop mode [18]. In such a case, the so called time profile, or the pattern of water pressure with respect to time, must be uploaded to the controller. The time profile is prepared on the basis of the water demand profile, either separately for every day of the week or separately for working days and weekends. The time scale in which the water demand profile is prepared to enable efficient controlling over pressure should be at least hourly and is assumed as such in this study. This way, the addressed problem is actually a 24-step ahead forecasting of the water demand time series.

Forecasting of water demand has been already addressed by many researchers [3]. Different predictive techniques exploited to approximate multiple auto - regressive models for dealing with seasonality have been investigated [7]. A hybrid model consisting of daily and hourly forecasting has been proposed [19]. In that paper, the hourly module was designed to disaggregate the forecasted daily demand. A Predictor-Corrector approach for on-line forecasting of water usage was also investigated [9]. A maximum predictable time scale for the forecasting of hourly water consumption was reported [12]. Recently, forecasting hourly water demand time series focusing on the occurrence of multiple seasonal cycles has been addressed [6]. A comparative study regarding the real and forecasted water demand was performed [10]. Furthermore, reviews of existing methods of urban water demand forecasting can be found [2, 8].

In spite of the numerous papers on hourly water demand forecasting, to the best of our knowledge, there were no attempts to compare different methods of dealing with seasonality in such a time series. Especially when considering 24-h predictive forecasts, the best method to deal with seasonality has been not revealed. Moreover, it has been not clarified which predictive model coupled with which method of dealing with seasonality is the most effective.

The above mentioned limitations of existing works are addressed in this paper. First, we review three different methods for dealing with seasonality in time series: auto-regressive, differentiation, and the application of dummy variables. Second, we apply these methods to the considered problem of 24-step ahead, hourly forecasting of water demand. Afterward, we couple these methods to selected, well known state-of-the art forecasting methods and perform experiments to select the most efficient pair. As a reference point for this research, we assume the forecasting accuracy obtained by standard forecasting methods with built in mechanisms for dealing with seasonality. As a result of this study, it has been revealed that the application of dummy variables together with linear regression outperformed the other investigated methods.

The remainder of this paper is organized as follows. Section 2 introduces background knowledge on time series, seasonality analysis and forecasting models. In Sect. 3, the known methods of dealing with seasonality are adapted to the problem of 24-step ahead forecasting of hourly water demand time series. The experimental Sect. 4 presents the results of comparative experiments and selects the best methods of dealing with seasonality together with the corresponding forecasting model.

2 Time Series and Seasonality

Let $x \in \Re$ be a real-valued variable and let $t \in [1, 2, \ldots, n]$ be a discrete time scale, where $n \in \aleph$ is its length. The values of $s(t)$ are observed over time. A time series is denoted as a sequence $\{x(t)\} = \{x(1), x(2), \ldots, x(n)\}$.

The appropriate representation of trends and seasonal variation are key factors necessary for the effective prediction of time series. The trend is understood as the tendency of the time series to increase, decrease, or stay constant with respect to time. Seasonal variation is a component of a time series that describes its repetitive and predictable movement around the trend. We found that in our experiments, due to the lack of trend, the detection and exploitation of seasonality is the key tool for the effective forecasting of water demand time series. Seasonality in time series can be detected different ways [16]. First, it is possible to make graphic plots and visually recognize cycles in time series; multiple box plots can aid in visually detecting seasonality. Alternatively, assuming the additive or multiplicative model, time series can be decomposed to trend, seasonality, and irregular components. However, the most commonly used method for detecting seasonality is the autocorrelation plot. If there is a seasonality in a time series, the autocorrelation plot shows spikes at lags equal to the period of seasonality.

After detecting the seasonality it is possible to use several methods for exploiting it to improve the efficiency of forecasting [16]:

1. Autoregression - for auto-regressive models it is assumed that the current value of time series $x(t)$ is dependent on the value of $x(t-k)$, where the constant $k \in \aleph$ points to the time step in the previous period(season) of the detected seasonality. This way the seasonal variations are represented.
2. Seasonal adjustment (differentiation) - this is a method that relies on removing the seasonal component from time series before the forecasting is made. The seasonal component is recognized and then subtracted from the original time series by the calculation $x_a(t) = x(t) - x(t-k)$. The process is referred to as differentiation and removes the nonstationarity of the time series. The resulting seasonally adjusted series $x_a(t)$ are forecasted as series $x'_a(t)$. Afterward, the seasonal component is returned, resulting in the forecast of the original time series $x'(t) = x'_a(t)+x(t-k)$. That process is called integration.
3. Inclusion of dummy variables - the seasonality can also be reflected by adding to the predictive model the dummy variables. They play the role of explanatory variables where one variable is added for every of $n-1$ seasons, with n denoting the number of seasons. Each dummy variable is set to 1 if the value of time series

is drawn from the corresponding season and 0 otherwise. One of the seasons, e.g., the nth is reflected in the model by setting all dummy variables to 0.

There are a few state-of-the art forecasting methods equipped with built-in methods for dealing with seasonality. For the purposes of this paper, we selected those that are most representative and are known as the most effective [16]. The autoregressive integrated moving average (ARIMA) is a model that invokes the differentiation and integration processes in the case of nonstationarity in data. Especially in cases where seasonality is detected, the model is extended and denoted as SARIMA, involving components that reflect the seasonality within the time series. The Autoregressive Fractionally Integrated Moving Average (ARFIMA) model is dedicated to time series in which the autocorrelation is recognized for far lags. It is generalization of ARIMA model. The Holt-Winters method (HW) is the extended version of exponential smoothing forecasting. It takes into account both trend and seasonality within data. Its components refer to the estimated level, slope, and seasonal effects. The description of these state-of-the-art methods fall beyond the scope of the paper but can be readily found [4].

Besides the state-of-the-art forecasting models that deal with the integrated mechanisms for seasonality, it is possible to use general mathematical models that represent the dependencies between variables [16]. These models have to be appropriately supplemented when dealing with seasonality in time series. The linear regression model (LR) is used to relate a scalar dependent variable and one or more explanatory variables. Linear regression has been applied for forecasting [1]. Polynomial regression (PR) reflects the relationship between the independent variable and the dependent variables with a polynomial equation and has been used for forecasting [13]. Artificial neural network (ANN) is a system of interconnected units called neurons that can compute output values from inputs. ANNs have already been used for forecasting [15]. Fuzzy rule-based systems (FRBS) map the set of input variables to the dependent variable and have also been used for forecasting [1]. In addition to the above models, the naive approach is often used as a reference method for time series forecasting. It assumes that every prediction is the same as its previously observed value, i.e., $x'(t+1) = x(t)$.

The error of a single forecast (residual error) is calculated as $e(t) = x'(t) - x(t)$, where $x'(t)$, $x(t)$ denote the predicted and actual values of time series respectively. To calculate accumulated forecasting errors in this paper, we decided to apply only the absolute percentage error (MAPE), given as formula (1). The MAPE is a simple scale independent error measurement that relates forecasting errors to the values of the original time series.

$$MAPE = \frac{1}{n}\sum_{t=1}^{n}|\frac{e_t}{x_t}| \times 100\,\% \quad (1)$$

To compare statistically the time series of residuals generated by two different models, a Diebold-Mariano (DM) statistical test can be used [5]. The null hypothesis of the DM test H_0: Model 1/Model 2 is that Model 1 is more accurate than Model 2.

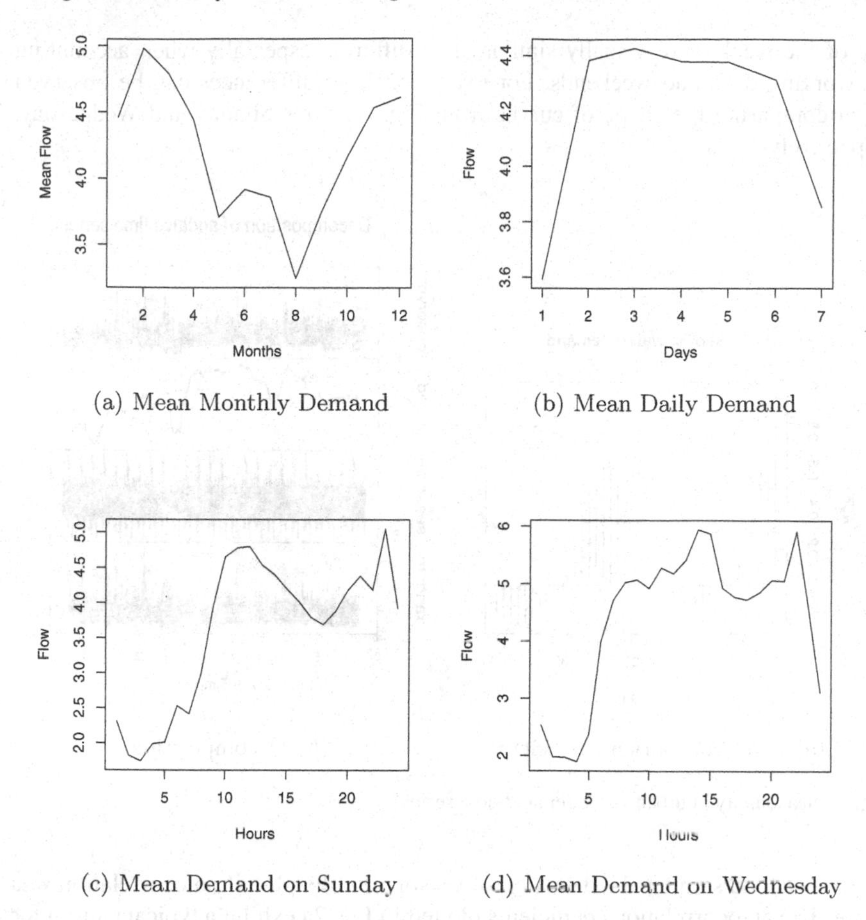

(a) Mean Monthly Demand (b) Mean Daily Demand

(c) Mean Demand on Sunday (d) Mean Demand on Wednesday

Fig. 1 Urban water demand in 'Kolejowa' district metered area of Sosnowiec, Poland

3 Dealing with Seasonality in Water Demand Time Series

For the validation of the proposed approach real-world data were acquired from the urban water distribution network of Sosnowiec, Poland. The data were gathered from the period of 12 September 2013 to 19 September 2014, limited by the time of installing appropriate sensors and data transmission channel. First, the mean monthly and mean weekly data were ploted and analyzed. As can be recognized in Fig. 1a, the data exhibit yearly patterns with decreased water demand during summer months of June, July, and August. Unfortunately, as the period of available data covers only one year it is too short to confirm the pattern on a yearly scale. For the same reason, the monthly seasonality cannot be exploited in our study.

A weekly pattern with lower water demand during the weekend can be recognized in Fig. 1b. When analyzing daily seasonality, we decided to plot the mean demand during every day with respect to hours of the day. Although the patterns for every day of the week were visually similar, they differed, especially when accounting for working days and weekends. For example, these differences can be observed when comparing the shape of curves from Fig. 1c, d for Sunday and Wednesday, respectively.

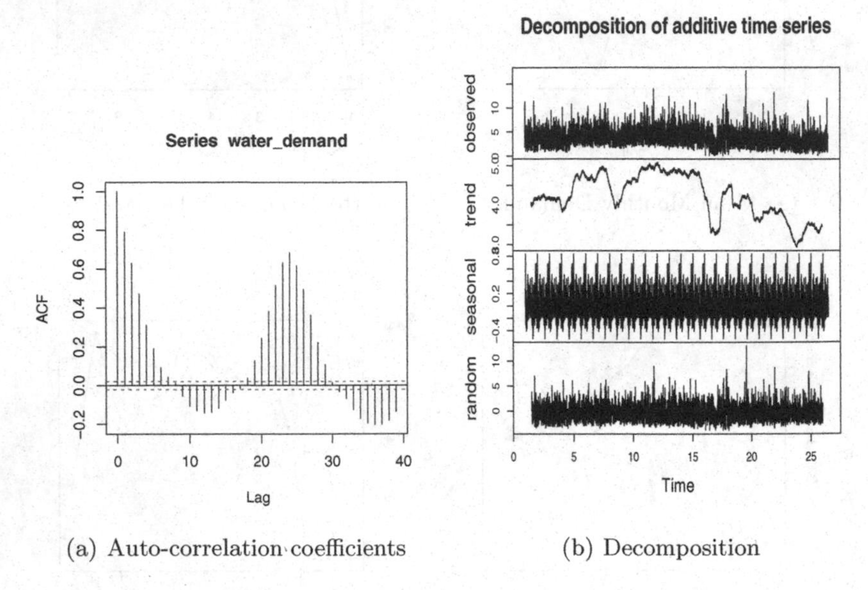

(a) Auto-correlation coefficients (b) Decomposition

Fig. 2 Seasonality in urban water demand time series

As a second step for the detection of seasonality the plot of autocorrelation was made. The autocorrelation coefficients plotted in Fig. 2a exhibit a typical pattern for seasonal time series without trend [3]. Also the decomposition plotted in in Fig. 2b confirms the existence of seasonality within the time series.

To exploit the recognized seasonality by the autoregressive and differentiation methods, appropriate time lags must be selected. Assuming an hourly time series (1 hour is a single step of time) and due to the detected weekly and daily seasonality, two corresponding time lags of $k = 168$ and $k = 24$ steps are assumed. For the third considered method, dummy variables are set consisting of two subsets: A - including 6 dummy variables for the first six days of the week, B - including 23 variables corresponding to the hours of the day.

In addition to the described settings, the considered time series had to be partitioned into learning and testing data. For that purpose, the idea of growing windows was applied. It assumes that the learning period begins at the first available observation of data and finishes at time $t - 1$. The minimum length of the learning period is set to 175 hourly steps equivalent to 7 days. The next 24-hours are assumed as the prediction horizon, i.e., the daily demand profile is forecasted every day, using all previous time steps for training the predictive model.

4 Experiments

The first issue encountered during the experiments were outliers found in the data. These outliers manifested as a rapid increase of water flow during the pipe burst. Due to the fact that the pipe burst occurred only once during the considered yearly period, we decided to manually remove the outliers and implement a linear approximation of the missing values for this paper. This led to the decrease of peaks in forecasting errors. To demonstrate the influence of the outliers on the accuracy of prediction, the residuals produced by the SARIMA model before and after the removal of the outliers are shown in Fig. 3a, b respectively.

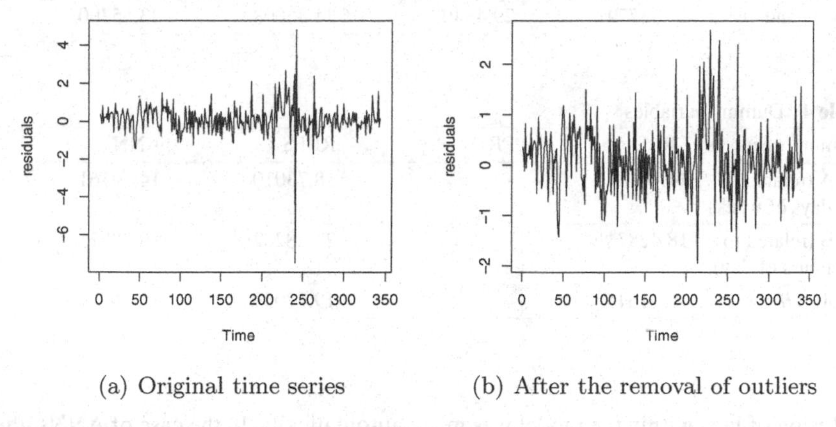

(a) Original time series (b) After the removal of outliers

Fig. 3 Residual errors generated by the ARIMA model

For the experiments with the first group of forecasting models with the integrated methods of dealing with seasonality 'auto.arima' , 'arfima' and 'ets' functions from the R package 'forecast' were used [14].

Table 1 Results

Naive	SARIMA	ARFIMA	HW
56.32478	37.29051	36.4749	48.10104

The results are shown in Table 1. As shown, all investigated methods were better than naive forecasting; however, even the best ARFIMA model demonstrated quite high forecasting errors.

In the second stage of experiments, the three different techniques of dealing with seasonality described in Sect. 2 were validated for every selected forecasting model. The experiments were implemented using the KNIME framework [11]. In every case, the parameters of the applied models were adjusted by numerous try-and-test experiments. Besides the lags related to the previously detected seasonality, the flow variable was lagged in time for up to 23 lags. In the case of LR, PR, and RBFS, the

Table 2 Autoregressive method

Seasonality	LR	PR	RBFS	ANN
lag 24	16.76324	17.26215	22.93245	18.44831
lag 168	17.05671	16.88909	23.08414	17.89695
lags 24 and 168	16.82078	17.05013	24.87860	17.43805

Table 3 Seasonal adjustment

Seasonality	LR	PR	RBFS	ANN
lag 24	21.12591	25.53785	29.01908	22.35639
lag 168	21.52052	22.31505	26.65640	21.87957
lags 24 and 168	27.83779	29.48400	34.86008	23.55300

Table 4 Dummy variables

Dummy variabels	LR	PR	RBFS	ANN
set A (related to the days of week)	**14.40186**	-	18.73010	14.79161
set B (related to the hours of day)	18.42874	-	27.28282	19.30515
set $A \bigcup B$	16.91645	-	19.97059	18.36589

inclusion of lags within the model was made automatically. In the case of ANNs, the importance of lags was adjusted by the weights assigned to the arcs of the network. When using dummy variables, due to the mutual dependence of attributes the implementation of polynomial regression in KNIME returned an error and could not be performed.

The results of the experiments are shown in Tables 2, 3, and 4. It is worthy of note that, independent of the predictive model, the method with seasonal adjustment produced higher errors that relied on the inclusion of dummy variables related to the days of week. When using dummy variables, the linear regression outperformed other methods. To confirm that result for the best two models, the Diebold-Mariano statistical test was performed. The attained p-value of 1 confirmed the superiority of linear regression over neural networks, assuming the involvement of hourly dummy variables (set A).

To confirm that result, for the best two models, the Diebold-Mariano statistical test was performed that confirmed by the p-value of 1 the superiority of linear regression over neural networks assuming the involvement of hourly dummy variables (set A).

5 Final Remarks

In the presented research, we investigated three methods of dealing with the seasonality that occurs in water demand time series together with the selected state-of-the art predictive models. The experiments reveal that linear regression coupled to the use of dummy variables corresponding to the days of week works the best. The direction of further research includes the investigation of other predictive models when coupling them with the known techniques of dealing with seasonality.

Acknowledgments The work was supported by ISS-EWATUS project which has received funding from the European Union's Seventh Framework Programme for research, technological development and demonstration under grant agreement no. 619228.

References

1. Bianco, V., Manca, O., Nardini, S.: Electricity consumption forecasting in Italy using linear regression models. Energy **34**(9), 1413–1421 (2009)
2. Billings, R.B., Jones, C.V.: Forecasting Urban Water Demand. American Water Works Association, Denver (2008)
3. Cortez, P., Rocha, M., Neves, J.: Genetic and evolutionary algorithms for time series forecasting. In: Proceedings of the IEA/AIE-2001: The 14th International Conference on Industrial and Engineering Applications of Artificial Intelligence and Expert Systems, pp. 393–402, 4–7 June 2001
4. Cowpertwait, P.S.P., Metcalfe, A.V.: Introductory Time Series with R, 1st edn. Springer Publishing Company, Incorporated (2009)
5. Diebold, F.X., Mariano, R.S.: Comparing predictive accuracy. J. Bus. Econ. Stat. **13**(3), 253–63 (1995)
6. Dudek, G.: Forecasting time series with multiple seasonal cycles using neural networks with local learning. In: Rutkowski, L., Korytkowski, M., Scherer, R., Tadeusiewicz, R., Zadeh, L., Zurada, J. (eds.) Artificial Intelligence and Soft Computing. Lecture Notes in Computer Science, vol. 7894, pp. 52–63. Springer, Berlin Heidelberg (2013)
7. Herrera, M., Torgo, L., Izquierdo, J., Prez-Garca, R.: Predictive models for forecasting hourly urban water demand. J. Hydrol. **387**(12), 141–150 (2010)
8. House-Peters, L.A., Chang, H.: Urban water demand modeling: review of concepts, methods, and organizing principles. Water Res. Res. **47**(5), n/a-n/a (2011)
9. Joby Boxall, C.M.C.P. (ed.): On-line Hydraulic State Estimation in Urban Water Networks Using Reduced Models (2009)
10. Kim, J., Shin, G., Choi, D.: Develpment of water demand forecasting simulator and its performance assessment. In: WDSA 2012: 14th Water Distribution Systems Analysis Conference, 24–27 Sept 2012 in Adelaide, pp. 1418–1423. South Australia. Barton, A.C.T.: Engineers Australia (2012)
11. KNIME: Professional open-source software. http://www.knime.org
12. Liu, J.Q., Zhang, T.Q., Yu, S.K.: Chaotic phenomenon and the maximum predictable time scale of observation series of urban hourly water consumption. J. Zhejiang Univ. Sci. **5**(9), 1053–1059 (2004)
13. Mavromatidis, L.E., Bykalyuk, A., Lequay, H.: Development of polynomial regression models for composite dynamic envelopes thermal performance forecasting. Appl. Energy **104**, 379–391 (2013)
14. Package, R.: http://www.r-project.org

15. Pulido-Calvo, I., Montesinos, P., Roldn, J., Ruiz-Navarro, F.: Linear regressions and neural approaches to water demand forecasting in irrigation districts with telemetry systems. Biosyst. Eng. **97**(2), 283–293 (2007)
16. Shmueli, G.: Practical Time Series Forecasting. Statistics.com LLC, 2nd edn. (2011)
17. Thornton, J., Sturm, R., Kunkel, G.: Water Loss Control, 2nd edn. The McGraw-Hill Companies, Inc., New York (2008)
18. Ulanicki, B., Bounds, P., Rance, J., Reynolds, L.: Open and closed loop pressure control for leakage reduction. Urban Water 2(2), 105–114 (2000), developments in water distribution systems
19. Zhou, S., McMahon, T., Walton, A., Lewis, J.: Forecasting daily urban water demand: a case study of melbourne. J. Hydrol. **236**(34), 153–164 (2000)

Short Term Load Forecasting for Residential Buildings—An Extensive Literature Review

Carola Gerwig

Abstract Accurate Short Term Load Forecasting is an essential step towards load balancing methods in energy systems. With the recent introduction of Smart Meters for residential buildings, load forecasting and shifting methods can be implemented for individual households. The high variance of the load demand on the household level requires specific forecasting methods. This paper provides an overview of the methods which have been applied and points out what results are comparable. Therefore a structured literature review is carried out. In the process, 375 papers are analyzed and categorized via a concept matrix. Based on this review it is pointed out, which methods achieve good results for which purpose and which publicly available datasets can be used for evaluation.

Keywords Short term load forecasting · Review · Time series · Residential buildings · Microgrid · Demand side management

1 Introduction

The integration of renewable energy resources in the existing energy systems comes along with several challenges: Energy generation becomes more decentralized and more volatile. In the ongoing development of smart grids, accurate short term load forecasting can significantly improve the micro-balancing capabilities of the energy systems [1]. On the basis of reliable forecasts, load shifting methods can be implemented to cushion peaks in demand and supply. Furthermore, with an increasing number of smart meters in residential buildings[1] a new potential for load shifting

[1] In Germany the installation of smart meters in new buildings has been forced since 2010 by law, cf. 21b Abs. 3a EnWG.

C. Gerwig (✉)
Institute for Information Systems, University Hildesheim,
Universit ätsplatz 1, 31141 Hildesheim, Germany
e-mail: gerwig@uni-hildesheim.de

R. Neves-Silva et al. (eds.), *Intelligent Decision Technologies*,
Smart Innovation, Systems and Technologies 39,
DOI 10.1007/978-3-319-19857-6_17

emerges. Therefore, specific short term load forecasting methods are required as the variance of the load is extremely high and established methods can fail easily.

On a larger scale, long-, mid- and short-term load forecasting have been on the research focus for quite a while. The methods are essential for the operation and planning of electricity systems. Reviews on general load forecasting can be found in [2] and [23]. Short Term Load Forecasting (STLF) is classified by the planning horizon's duration, which can be up to one day (cf. [2]). Usually the required forecast granularities are short time units between 15 min and 1 h. Used by energy providers to control operations, STLF provides a great saving potential (cf. [7]).

To apply load shifting to residential buildings, suitable methods have to be developed and evaluated. The standard methods for STLF like stochastic time series, artificial neural networks (ANN) and others have been applied and adapted. Thereby, exogenous parameters like weather parameters are used as input. The methods are then often evaluated on datasets which were recorded in individual research projects and are not publicly available. But the results strongly depend on the size of the demand and the available exogenous parameters. As a result, the methods and results are often not comparable.

Therefore, a literature review is needed, which delivers an overview of the applied methods ordered by the size of demand and a listing of publicly available datasets for residential load shifting. The guiding research questions are:

(1) What methods are used?
(2) What results are comparable with respect to the databasis?
(3) Are exogenous parameters included?
(4) Which datasets are publicly available?

To answer these questions, the literature review is structured as proposed in [6]. At first, a search string is developed and useful databases are identified in which the search is carried out (Chapter "A Cloud-based Vegetable Production and Distribution System"). After that, a concept matrix (cf. [26]) is designed, which provides dimensions and characteristics to categorize the findings in a useful manner (Chapter "Capacity Planning in Non-Uniform Depth Anchorages"). The outcome of the search is analyzed and the suitable articles are matched to the concept matrix. Methods, which are evaluated on similar sized datasets, are then compared and discussed (Chapter "Automatic Image Classification for the Urinoculture Screening"). Finally, the results are summarized pointing out next steps that should follow up in the research of STLF for residential buildings.

2 Research Design

2.1 Search String

The search string is derived from the research questions. First of all, "short term forecasting" needs to be part of the searchstring. The term "load" is added separately as not all papers use the term "short term load forecasting", but both terms should appear in the article. Furthermore, the focus of the research is residential STLF. Therefore, the term "residential" is added. The final search string combines the three terms with AND connectors:

"short term forecasting" AND "load" AND "residential"

By these terms the research is restricted to English written articles, which appears to be adequate with English as established scientific language.

2.2 Databases

The search is conducted in the IEEE xplore Digital Libary[2] and Google scholar.[3] IEEE xplore is a natural choice as it provides articles and papers about computer science, electrical engineering and electronics. Furthermore, today, most research papers are trackable by Google Scholar. Therefore, the search is conducted there as well. It searched in the full text and the meta index for IEEE xplore. Patents and Citations are excluded in Google Scholar. As mentioned before, the necessary Smart Meter technology for residential buildings has been introduced to the market about 2010. Thus the search is restricted to the time range 2010–2015. The search leads to a finding of 82 papers in IEEE xplore and 367 papers in Google scholar.[4] Most paper of IEEE xplore are also found in Google scholar.

2.3 Evaluation Method

The findings of the search are examined stepwise by article, abstract and in a final step by the whole paper. At first, all papers with titles obviously not related to the subject of STLF are sorted out. By reading the abstract of the remaining articles, suitable articles are identified and analyzed. The review is structured by the Concept Matrix, to which the contents of the papers are matched. The result of this procedure helps to discuss the contents and outcomes of the papers.

3 Concept Matrix

The Concept Matrix provides "the organizing framework" for a structured literature review according to Webster and Watson [26]. Therefore, the dimensions are directly derived from the research questions. One dimension is *methods* used for STLF. One

[2]http://ieeexplore.ieee.org/.

[3]https://scholar.google.de/.

[4]The search has been performed on Google Scholar between 27th and 29th *December* 2014 and on IEEE xplore between 7th and 10th *January* 2015.

dimension is *number of households*, which is a crucial factor for the comparability of methods. Furthermore, it is useful to know, if the paper itself provides a comparison of methods, if the dataset used for validation is publicly available and if exogenous parameters are used as input. Thus, the if-clauses are added as *additional artefacts*. The resulting framework is shown in Fig. 1.

3.1 Methods

The classification of the STLF methods is funded on a common structure: Five widely applied STLF techniques (cf. [17]) are multiple linear regression, stochastic time series, general exponential smoothing, state space models with Kalman filter and knowledge-based approaches. The latter one is not suitable for residential buildings as too much personal information of the inhabitants would be required. Instead, three additional methods from the field of artificial intelligence are added to the scope of the analysis: artificial neural networks (ANN), support vector machines for regression (SVR) and Clustering methods. At this point, only a brief overview of the selected approaches can be provided. Yet, it seems useful to give an insight of the advantages and disadvantages:

Multiple *linear regression* assumes that the value of the time series depends linearly on other variables. This is a quite simple model, which is easy to implement, when useful corresponding parameters are provided. *Stochastic time series* (referenced as autoregressive models) exploit the internal structure of a time series. The model is based on the autocorrelation within the time series and a combination of white noise terms. Therefore, they do not require a data collection of multiple variables, but a lot of historical data. As before, only linear relationships can be modeled. *Exponential Smoothing* is an averaging method of the recent past. Thereby, it assigns exponentially decreasing weights as the observation gets older. This method has a tendency to produce forecasts that lag behind. *Kalman Filter* also belong to the statistical methods. Transition matrices are constructed and updated by every new measurement. Unlike the other methods, it provides not only an estimate but also the variance of the estimation error. *Artificial neural networks* calculate the forecast by a complex system of hidden layers which determine the weights for a nonlinear model. Therewith, nonlinear relationships can be captured. For good results a sufficiently large input is required as training set. *SVR* can capture the non-linearity by mapping the regression problem in higher dimensions, in which the problem is assumed to be linear. The training data can be substantially smaller than for ANN. A disadvantage of SVR and ANN is the lack of transparency of results, which cannot be interpreted easily. *Clustering* partitions time series data into sequences based on similarities. Thus repeating patterns can be identified.

To summarize, the concept matrix will classify the methods of the analyzed papers in eight categories: *linear regression (LR)*, *autoregressive models (AR)*, *exponential smoothing (ES)*; *Kalman filter (KF)*, *ANN*, *SVR* and *Clustering*. Other methods are noted in the field *more*.

3.2 Number of Households

When results of different papers are compared, comparability of the underlying load variance needs to be assured. In accordance to the central limit theorem, the load variance decreases relatively with an increasing number of households. Therefore, three categories for the number of households are used for the Concept Matrix: (1) *individual households*, (2) *households up to 100* and (3) *small microgrid with less than 1000 households*. The number of households for microgrids is limited as the research focus of this review is on STLF methods which can cope with highly variable loads.

3.3 Additional Artefacts

Exogenous parameters can enhance the accuracy of forecasting methods, if they are correlated to the load demand. These can be time, temperature, solar irradiation, wind speed, humidity, day type (workday, weekend), etc. On the downside, less important feature can overfit the model. In every case, it needs to be assumed, that the exogenous data is available in the use cases. To compare the methods, the *use of exogenous parameters (ex.p.)* is noted in the concept matrix. The concept matrix also contains information whether the proposed approach of an article is *compared with other methods (comp.)*. The results are then included in the literature review. For further comparison, a method needs to be evaluated on a publicly available dataset. Therefore, the *availability of the used dataset (avail.)* is noted in the concept matrix.

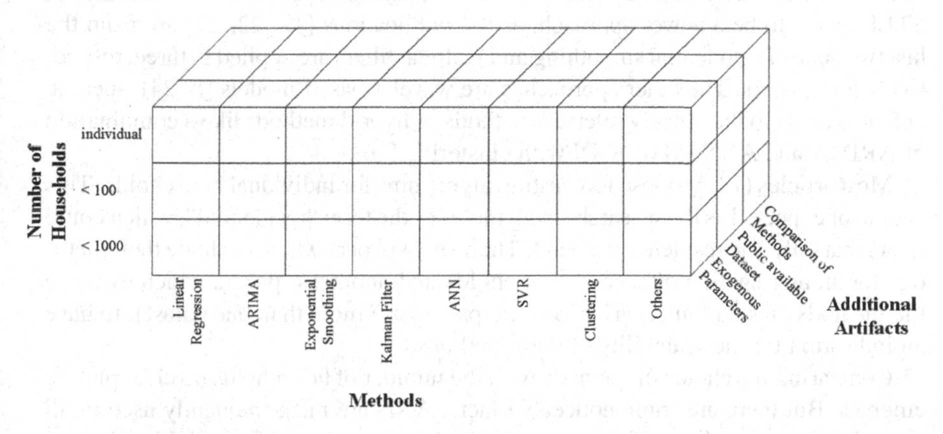

Fig. 1 Dimensions of the concept matrix

4 Outcome

4.1 Concept Matrix Augmented with References

The described keyword search lead to 375 results. Thereby, the articles proposed by both databases are only counted once. In the examining process all articles with titles which obviously are not related to the subject of STLF are rejected (166 articles). These are mainly articles about forecasting solar or wind generation, oil or water demand and some more exotic topics. By reading the abstract of the articles, another huge amount of articles can be sorted out. These are mainly articles about long term forecasting, algorithms for Demand Side Management and the management of Smart or Microgrids. This procedure leads to 50 remaining papers, which are organized and read as a whole. By doing this, even more articles are rejected. The remaining 18 articles are categorized by the concept matrix (Table 1).

4.2 Findings

Artificial neural networks are by far the most frequently used technique. Thereby, feed forward networks are used as well as recurrent back propagation networks. Further common applied methods are Linear Regression [12–14, 25, 27] and autoregressive methods [5, 22, 24, 25]. Also, clustering is a commonly used approach. In [8, 27] and [9] similar time sequences are matched. In [19] the load of several individual households are forecasted separately, but therefore the user patterns of all 127 households are compared and clustered. Applying Support Vector machines to STLF seems to be a newer approach, as the publications [12, 22, 27] are from the last two years. Exponential smoothing and Kalman filters are applied in three, respectively four papers. The other approaches are wavelet-based-models [8, 24], specific enhancements to the already referred methods or hybrid methods like a combination of ARIMA and ANN [5] or ANN with clustering [16].

Most articles (14) propose forecasting algorithms for individual households. This is kind of expected as the research string includes the term "residential", which could mean that only one residence is meant. There are 4 papers which evaluate their methods for an aggregation of several households and another 6 papers which evaluate the methods on small microgrids. Several papers use more than one dataset, to have an indication for the scalability of their methods.

Comparing the choice of methods with the number of households no clear picture emerges. But there are some noticeable facts: ANN are most frequently used in all sizes. Linear regression and clustering methods are often used for individual households. For more than 100 households, the variety of methods decreases and ANN, SVR and ARIMA are mainly used. These statistical statements must be treated with some caution due to the small number of underlying papers.

Table 1 Concept Matrx augemented with references

References	Methods								Households			Add. artefacts		
	LR	AR	ES	KF	ANN	SVR	Clust.	More	One	<100	<1000	Ex.p.	Comp.	Avail.
[5]		X			X			X			X	X	X	
[8]							X	X	X				(X)	X
[9]							X		X			(X)		
[10]				X					X			X		
[11]				X						X		X		
[12]	X				X	X			X		X	X	X	X
[13]	X								X			X		
[14]	X		X		X			X	X			X	X	
[16]					X			X			X	X	(X)	X
[19]							X		X			X		
[20]				X	X			X		X				
[21]			X						X	X				(X)

(continued)

Table 1 (continued)

References	Methods								Households			Add. artefacts		
	LR	AR	ES	KF	ANN	SVR	Clust.	More	One	<100	<1000	Ex.p.	Comp.	Avail.
[22]		X			X	X			X	X	X		X	
[24]		X			X			X	X		X	(X)	X	
[25]	X	X	X		X				X			X	X	X
[27]	X				X	X	X		X		X	X	X	X
[28]					X				X			X	(X)	
[29]				X					X			X		
sum	5	4	3	4	10	3	4	6	14	4	6	12(2)	7(3)	5

The research question about exogenous parameters can be answered clearly. Exogenous parameters are considered in almost all papers (14 of 18). These are weather variables (temperature and humidity) and information about the time (day of the week and time of day). The latter one is not necessarily needed for autoregressive methods and exponential smoothing, which is reflected in the findings: Two of the four autoregressive methods and one of three methods for ES does not use exogenous parameters. The same applies to the clustering methods. Again, these statements must be treated with some caution due to the small number of papers.

Several papers (10 of 18) provide a comparison of different methods in theirs evaluation. It is distinguished between papers which provide a comparison between different general approaches and papers which provide a comparison of a standard method and its tuned version. The first ones are marked with X, the latter ones with (X). An analysis of the evaluation is presented in the next subsection.

There are two publicly available datasets used for the evaluations: One is the CER Electricity Dataset[5] provided by the Irish Social Science Data Archive. It provides the half-hourly demand of 5000 Irish homes and businesses from 2009 to 2010. It is used by [8, 12, 16] and [27]. The other one is the Reference Energy Disaggregation dataset (REDD) for energy disaggregation research [15] provided by the Massachusetts Institute. The dataset contains the consumption data of six households for 18 days in the spring 2011. It is used in [25].

4.3 Comparison of the Evaluation Results

In the context of individual households and small groups the load variance is very high which leads to contradicting results when it comes to comparison of methods.

In [22] SARIMA beats an ANN method for individual users. This result is supported by [24], in which another autoregressive method (AR) is better than an artificial network (Echo State Network) and two other methods for individual users. Both articles recommend aggregating the load of more than 20 households as the error variance of one household is too high for good results. Contrary to this outcome, [25] states that ANN performs slightly better than ARIMA for individual, highly variable users. For stable consumption patterns they recommend persistent methods. Two other articles [12] and [27] state that linear regression works better than a Multi Layer Perceptron (ANN) and SVR for less than 32 households. Exponential Smoothing does not perform well on individual households [21].

Even for more households no clear picture can be drawn. Again [22] recommends SARIMA. The method can be used up to critical load of 16 MWh/h. The articles [5] and [24] both evaluate ANN und autoregressive methods on a dataset of Irish households provided by Irish Commision for Energy Regulation. Thereby, [5] favorizes a hybrid method of ANN and ARIMA. In [24] AR performs slightly better than the Echo State Network (ANN). SVR are seen as the best method for more than 32

[5]http://www.ucd.ie/issda/data/commissionforenergyregulationcer/.

Table 2 Comparable forecasting errors in different settings subdivided in number of households, time horizon and granularity of the sampling dataset and forecast

References	Setting			Error measures			Applied	Avail.
	# Households	Time horizon	Granularity	MAPE	NRMSE	RMSE	Method	Dataset
[10]	1	15 min	15 min	13 %	–	–	KF	–
[25]	1	15 min	15 min	48 %	–	–	AR	REDD
[28]	1	30 min	30 min	7.3 %	–	–	ANN	–
[10]	1	30 min	30 min	18 %	–	–	KF	–
[25]	1	30 min	30 min	49 %	–	–	ANN	REDD
[12, 27]	1	1 h	1 h	–	0.56	–	LR	CER
[24]	1	1 h	1 h	–	–	0.5 %kWh	KF	–
[10]	1	1 h	1 h	30.4 %	–	–	KF	–
[25]	1	1 h	1 h	47 %	–	–	AR	REDD
[25]	1	24 h	1 h	46 %	–	–	persistent	REDD
[21]	1	24 h	1 h	62.6 %	–	–	ES	–
[12, 27]	1	24 h	1 h	–	0.61	–	LR	CER
[21]	<100	24 h	1 h	7.4 %	–	–	ES	–
[11]	30	24 h	1 h	13 %	–	–	KF	–
[12, 27]	782	1 h	1 h	3.4 %	–	–	SVR	CER
[24]	150	24 h	1 h	–	–	21.4 kWh	AR	–
[16]	230	24 h	1 h	5 %	–	–	ANN	CER
[12, 27]	782	24 h	1 h	4.3 %	0.06	54.4 kWh	SVR	CER

households in [12], in which SVR outperforms linear regression and a Multi Layer Perceptron (ANN).

Comparing the results of the different papers is rather difficult as the evaluations do not vary only in the number of households but also in the length of the time horizon which is forecasted, in the granularity of the sampling dataset and the forecast and in the choice of error measures. In Table 2 the best evaluation results of each paper are presented for which one of the standard error measures is applied: Mean Absolute Percentage Error (MAPE), Root Mean Squared Error (RMSE) or the normalized RMSE (NRMSE) (cf. [27]). The results differ considerably. For further research one would have to reproduce the results on publicly available datasets.

5 Summary

By the extensive literature research which condensed down to 18 articles, it becomes obvious that STLF for residential buildings is a research topic still in progress. To compare the state-of-the-art methods, the articles were classified with a concept matrix and evaluated in different categories.

The evaluations within the articles lead to the conclusion that ANN, autoregressive methods and hybrid methods of both produce comparable results for individual households and up to 1000 households. For individual users, LR also seems to provide comparable results, while SVR works well for more than 32 households. Combining clustering methods with ANN or autoregressive methods could enhance the results as proposed in [16].

The presented subdivision of load size is a first step to make STLF methods for residential buildings more comparable. Nevertheless the results depend strongly on the variance of the individual dataset and the setting parameters of the forecast. For scientific benchmarking, there is strong need for publicly available datasets, which should then be used for an elaborated evaluation of the proposed method. The evaluation should always include the standard error measures, MAPE and NRMSE, even if a new error measure is proposed.

More available datasets besides the two datasets, which are used in few papers, are the Smart* Data [4], the UCI Data [3] and the Greenend dataset [18].

References

1. E-energy abschlussbericht: Ergebnisse und erkenntnisse aus der evaluation der sechs leuchtturmprojekte (2013). http://www.e-energie.info/documents/E-Energy_Ergebnisbericht_Handlungsempfehlungen_BAUM_140212.pdf
2. Alfares, H.K., Nazeeruddin, M.: Electric load forecasting: Literature survey and classification of methods. Int. J. Syst. Sci. **33**(1), 23–34 (2002)
3. Bache, K., Lichman, M.: UCI machine learning repository (2013). http://archive.ics.uci.edu/ml
4. Barker, S., Mishra, A., Irwin, D., Cecchet, E., Shenoy, P., Albrecht, J.: Smart*: An open data set and tools for enabling research in sustainable homes. SustKDD (2012)
5. Bennett, C., Stewart, R.A., Lu, J.: Autoregressive with exogenous variables and neural network short-term load forecast models for residential low voltage distribution networks. Energies **7**(5), 2938–2960 (2014)

6. vom Brocke, J., Simons, A., Niehaves, B., Riemer, K., Plattfaut, R., Cleven, A.: Reconstructing the giant: On the importance of rigour in documenting the literature search process. In: 17th European Conference on Information Systems, ECIS 2009, Verona, Italy, pp. 2206–2217 (2009)
7. Bunn, D.W.: Short-term forecasting: a review of procedures in the electricity supply industry. J. Oper. Res. Soc. 533–545 (1982)
8. Chaouch, M.: Clustering-based improvement of nonparametric functional time series forecasting: application to intra-day household-level load curves. IEEE Trans. Smart Grid **5**(1), 411–419 (2014)
9. Fujimoto, Y., Hayashi, Y.: Pattern sequence-based energy demand forecast using photovoltaic energy records. In: 2012 International Conference on Renewable Energy Research and Applications (ICRERA), pp. 1–6. IEEE (2012)
10. Ghofrani, M., Hassanzadeh, M., Etezadi-Amoli, M., Fadali, M.: Smart meter based short-term load forecasting for residential customers. In: North American Power Symposium (NAPS), pp. 1–5. IEEE (2011)
11. Hosking, J., Natarajan, R., Ghosh, S., Subramanian, S., Zhang, X.: Short-term forecasting of the daily load curve for residential electricity usage in the smart grid. Appl. Stoch. Models Bus. Ind. **29**(6), 604–620 (2013)
12. Humeau, S., Wijaya, T.K., Vasirani, M., Aberer, K.: Electricity load forecasting for residential customers: Exploiting aggregation and correlation between households. In: Sustainable Internet and ICT for Sustainability, SustainIT 2013. IEEE (2013)
13. Iwafune, Y., Yagita, Y., Ikegami, T., Ogimoto, K.: Short-term forecasting of residential building load for distributed energy management. In: 2014 IEEE International Energy Conference (ENERGYCON), pp. 1197–1204. IEEE (2014)
14. Javed, F., Arshad, N., Wallin, F., Vassileva, I., Dahlquist, E.: Forecasting for demand response in smart grids: an analysis on use of anthropologic and structural data and short term multiple loads forecasting. Appl. Energy **96**, 150–160 (2012)
15. Kolter, J.Z., Johnson, M.J.: Redd: A public data set for energy disaggregation research. In: Workshop on Data Mining Applications in Sustainability (SIGKDD), San Diego (2011)
16. Marinescu, A., Dusparic, I., Harris, C., Cahill, V., Clarke, S.: A dynamic forecasting method for small scale residential electrical demand. In: 2014 International Joint Conference on Neural Networks, IJCNN 2014, pp. 3767–3774 (2014)
17. Moghram, I., Rahman, S.: Analysis and evaluation of five short-term load forecasting techniques. Power Syst., IEEE Trans. **4**(4), 1484–1491 (1989)
18. Monacchi, A., Egarter, D., Elmenreich, W., D'Alessandro, S., Tonello, A.M.: GREEND: an energy consumption dataset of households in italy and austria. In: 2014 IEEE International Conference on Smart Grid Communications (SmartGridComm) (2014)
19. Mutanen, A., Repo, S., Järventausta, P.: Customer classification and load profiling based on amr measurements. In: Proceedings of the 21st International Conference on Electricity Distribution (CIRED 2011), Frankfurt, Germany, paper. vol. 277 (2010)
20. Qingfeng, T., Jianhua, Z., Zhengyong, X.: Short-term micro-grid load forecast method based on emd-kelm-ekf. In: 2014 International Conference on Intelligent Green Building and Smart Grid (IGBSG), pp. 1–4. IEEE (2014)
21. Rossi, M., Brunelli, D.: Electricity demand forecasting of single residential units. In: 2013 IEEE Workshop on Environmental Energy and Structural Monitoring Systems (EESMS), pp. 1–6. IEEE (2013)
22. Sevlian, R., Rajagopal, R.: Short term electricity load forecasting on varying levels of aggregation. arXiv preprint arXiv:1404.0058 (2014)
23. Suganthi, L., Samuel, A.A.: Energy models for demand forecastinga review. Renew. Sustainable Energy Rev. **16**(2), 1223–1240 (2012)
24. Tidemann, A., Høverstad, B.A., Langseth, H., Öztürk, P.: Effects of scale on load prediction algorithms (2013)
25. Veit, A., Goebel, C., Tidke, R., Doblander, C., Jacobsen, H.A.: Household electricity demand forecasting-benchmarking state-of-the-art methods. arXiv preprint arXiv:1404.0200 (2014)

26. Webster, J., Watson, R.T.: Analyzing the past to prepare for the future: writing a literature review. Manag. Inf. Syst. Quart. **26**(2), 3 (2002)
27. Wijaya, T.K., Humeau, S.F.R.J., Vasirani, M., Aberer, K.: Residential electricity load forecasting: evaluation of individual and aggregate forecasts. Technical report (2014)
28. Yang, H.T., Liao, J.T., Lin, C.I.: A load forecasting method for hems applications. In: PowerTech (POWERTECH), 2013 IEEE Grenoble, pp. 1–6. IEEE (2013)
29. Yoo, J., Park, B., An, K., Al-Ammar, E.A., Khan, Y., Hur, K., Kim, J.H.: Look-ahead energy management of a grid-connected residential pv system with energy storage under time-based rate programs. Energies **5**(4), 1116–1134 (2012)

Possibilistic Very Fast Decision Tree for Uncertain Data Streams

Mohamed Hamroun and Mohamed Salah Gouider

Abstract This paper addresses the classification problem with imperfect Data Streams. More precisely, it extends standard CVFDT to handle uncertainty in both building and classification procedures. Uncertainty here is represented by possibility distributions. The first part investigates the issue of building decision trees from Data Streams with uncertain attribute values by developing a non-specificity based information gain as the attribute selection measure which, in our case, is more appropriate than the standard selection measure based on Shannon entropy. The extended approach so-called Possibilistic Very Fast Decision Tree for Uncertain Data Streams (Poss-CVFDT) offers a more flexible building procedure. The second part addresses the classification phase. More specifically, it investigates the issue of predicting the class value of new instances presented with certain and/or uncertain attribute values.

Keywords Classification · Uncertainty · Possibility theory · Non specificity · Data streams · Decision tree

1 Introduction

Data Stream Classification represents an important task in machine learning and data mining applications. It consists in inducing a classifier from a set of historical examples with known class values and then using the induced classifier to predict the class value (the category) of new objects given known the values of their attributes (features). Classification is widely used in many real world applications including pharmacology, medicine, marketing, etc. However, classification with data streams is a very challenging task. Therefore, effective algorithms need to be designed in order to take into account temporal locality and the concept drift of the data.

M. Hamroun (✉) · M.S. Gouider
Institut Supérieur de Gestion, Laboratoire SOIE, Tunisia, Africa
e-mail: mohamed.hamrounn@gmail.com

M.S. Gouider
e-mail: ms.gouider@yahoo.fr

R. Neves-Silva et al. (eds.), *Intelligent Decision Technologies*,
Smart Innovation, Systems and Technologies 39,
DOI 10.1007/978-3-319-19857-6_18

To address these issues, many algorithms have been proposed, mainly including various algorithms based on ensemble approach and decision tree, e.g., [1] proposed algorithms to learn very fast decision trees (VFDT), [2] proposed a novel algorithm called concept-adapting Very Fast Decision Tree (CVFDT) which can learn decision trees incrementally with bounded memory usage, high processing speed, and detecting evolving concepts. In real world applications, the massive amounts of data are inseparably connected with imperfection. In fact, data can be imprecise or uncertain or even missing. These imperfections might result from using unreliable information sources. Standard classifiers, generally, ignore such imperfect data by rejecting them or replacing each imperfect data item by an arbitrary certain and precise value or by a statistical value such as a median, mode or a mean. This is not a good practice because it alters the real observed data. Consequently, ordinary Data Stream classification techniques such as CVFDT should be adequately adapted to take care of this problem.

Our idea is to treat different levels of uncertainty using possibility theory which is a non-classical theory of uncertainty [3]. More precisely, we will handle samples whose attribute values are given in the form of possibility distributions. We also adapt the attribute selection measure, used in the building phase, to the possibilistic framework by using a non-specificity based criterion instead of the Shannon entropy. In addition, we introduce a new sliding model based on samples's timestamp in order to improve the ability of CVFDT to cope with concept drift issue. Such possibilistic decision tree will be referred to by Poss-CVFDT. The paper is organized as follows: Sect. 2 presents a summary of related works. Section 3 proposes an extension of CVFDT, namely the Poss-CVFDT approach. This section defines the building procedure, then, it describes the method that we propose for the classification process. Before concluding, Sect. 4 presents and analyzes experimental results carried out on modified versions of commonly used data sets obtained from the U.C.I. machine learning repository.

2 Related Works

Uncertain Data Streams

Some previous work focused on building classification models on uncertain data examples. All of them assume that the Probability Density Function (PDF) of attribute values are known. [7] developed an Uncertain Decision Tree (UDT) for uncertain data. [6] developed another type of decision tree DTU for uncertain data classification. [8] proposed a neural network method for classifying uncertain data. Since the uncertainty is prevalent in data streams, the research of classification is dedicated to uncertain data streams nowadays. Unfortunately, only a few algorithms are available. [9] proposed two types of ensemble classification algorithms, Static Classifier Ensemble (SCE) and Dynamic Classifier Ensemble (DCE), for mining uncertain data streams. [10] proposed a CVFDT based decision tree named UCVFDT for uncertain data streams. UCVFDT has the ability to handle examples with uncertain attribute values by adopting the model described by [5] to represent uncertain nominal attribute values. In their work, uncertainty is represented by a probability degree on the set of possible values considered in the classification problem.

3 Poss-CVFDT

3.1 Data Streams Structure

Instead of rejecting instances having uncertain attributes values or adding a null attribute value to such instances, we use the possibility theory. More formally, we propose to represent the uncertainty on the attributes values of instances by a possibility degree on the set of possible values considered in the classification problem. Among the advantages for working under the possibility theory framework, we recall that the two extreme cases, total ignorance and complete knowledge, are easily satisfied. Given a stream data S= $\{S1, S2, \dots, St, \dots\}$ where St is a sample in S, arriving at time Ti for any i $\langle$ n , Ti $\langle$ Tn , we denoted by St = $\langle X^{uc}, y\rangle$.

- Here $X^{uc} = \{X_1^{uc}, X_2^{uc}, \dots, X_d^{uc}\}$ is a vector of d uncertain categorical attributes. Given a categorical domain Dom(X_i^{uc})= $\{v1, v2, \dots, vm\}$, X_i^{uc} is characterized by possibility distribution over Dom, where $0 \leq \pi(X_i^{uc}$=vj$) \leq 1$
- Yk $\in$ C denotes the class label of St, where C is set of class labels on stream data C= $\{y1, y2, \dots, yk, \dots, yn\}$.

we build and learn a fast decision tree model yk=$f(X^{uc})$ form S to classify the unknown samples.

The incoming sample will be passed down to a certain node from the root recursively according to its attribute value. Here we adopt a recursive process to partition the training samples into fractional samples based on [7, 10]. By giving a split attribute at node N, denoted by X_i^{uc}, sample st is divided into a set of samples $\{st1, st2, \dots, stm\}$ by X_i^{uc} where m=| $Dom(X_i^{uc})$ | and stj is a copy of st except for attribute X_i^{uc}.

3.2 Sliding Window Model

During the classification technique, we emphasis the concept drift issue which change the classifier result over time. The capture of such changes would help in updating the classifier model effectively. The use of an outdated model could lead to very low classification accuracy. That is why we need to focus only on the N recent records. Based on timestamp of samples, a new approach have been proposed to detect and identify outliers in the underlying data [11]. This model can be an interisting task in many sensor network applications.

3.3 Specificity Gain

Classical splitting measures such as Information Gain, Gini Index are not applicable in our case. In this section, based on possibility measures, we define new parameters which take into account the uncertainty encountered in the data flow. An interpretation have been proposed by [12, 13] based on the context of possibility theory assuming that the U-uncertainty measure of nonspecificities of possibility distribution, which is defined as

$$Nonspec(\pi) = \sum_{i=1}^{n} [(\pi(\omega_{(i)}) - \pi(\omega_{(i+1)}))log_2 i] + (1 - \pi_{(1)})log_2 n \quad (1)$$

can be justified as a proper generalization of Hartley Information [14] to the possibilistic setting. Non Specificity plays the same role to that of Shannon Entropy in probability theory.So we use it to construct a selection measure in the same way as information gain and Information gain ratio are constructed form Shannon Entropy. We calculate a Specificity Gain based on the nonspecificities on the possibility distributions $\pi_{X_i^{uc}}$ on the values of X_i^{uc}, π_C on the set of class labels and $\pi_{CX_i^{uc}}$ the joint distribution of the set of values of X_i^{uc} and the set of classes. The idea so suggests itself to construct these possibility distributions but we have to take into consideration the concept underlying possibility theory. The solution that we propose is the following : We will induce a representative possibility distribution that represents the marginal distribution of the different possibility degrees of the different values of X_i^{uc} and the set of class labels C. These marginal distributions are obtained not by summing values but by taking their maximum. Then, the average is applied as the case:

$$\pi_{X_i^{uc}}(vj) = Avg_{|C|}(max_{y \in C}(\pi_{CX_i^{uc}}(y, vj))) \quad (2)$$

$$\pi_C(y) = Avg_{|Dom(X_i^{uc})|}(max_{vj \in Dom(X_i^{uc})}(\pi_{CX_i^{uc}}(y, vj))) \quad (3)$$

We define the *Specificity gain* as:

$$S_{gain}(X_i^{uc}) = Nonspec(\pi_C) + Nonscpec(\pi_{X_i^{uc}}) - Nonspec(\pi_{CX_i^{uc}}) \quad (4)$$

We select for test, the attribute which yields the highest S_{gain}.

In order to collect statical data to Sgain, we only make a single pass over set Samples SN. So we propose, at each node of the nodes of the dynamic decision tree, for each possible value vj of each attribute $X_i^{uc} \in X^{uc}$, for each class yk, to associate it with a vector called Node Feature(NF) which is composed by four parameters ($PSijk$, N_{ijk}, N(S), t_s). Here,

- *PSijk* denotes the sum of possibility degrees of each coming sample in yk that X_i^{uc}= vj.
- N(S) defines the number of the samples contained in the NF.
- N_{ijk} denotes the ratio of the sum of possibility degrees of each sample in yk that X_i^{uc}=vj, $N_{ijk} = \frac{PS_{ijk}}{N(S)}$.
- t_s indicate the time of the recent sample incorporated in the NF.

→ for each new coming sample S^{new}, we only need to update PS_{ijk}, N_{ijk}, N(S) as follows:

$$PS_{ijk} = PS_{ijk} + \pi(X^{uc}_{i(S^{new})} = vj) \tag{5}$$

$$N(S) = N(S) + 1 \tag{6}$$

$$N_{ijk} = \frac{PS_{ijk}}{N(S)} \tag{7}$$

→ for removing sample S^{old} from a node, we only need to update PS_{ijk}, N_{ijk}, N(S) as follows:

$$PS_{ijk} = PS_{ijk} - \pi(X^{uc}_{i(S^{old})} = vj) \tag{8}$$

$$N(S) = N(S) - 1 \tag{9}$$

$$N_{ijk} = \frac{PS_{ijk}}{N(S)} \tag{10}$$

3.4 Poss-CVFDT Building

Based on [2], Algorithm 1 defines a pseudo code of the Poss-CVFDT building steps. In this study we use S_{gain} as G(.), X^{uc} denotes a vector of uncertain attributes, NF_{ijk}: the NF used to clollect statical data for computing specificity gain. t_s defines the arrival time of each sample used to enhance the ability of Poss-CVFDT to detect outdated samples.

Algorithm 1 illustrates the process of Poss-CVFDT learning in four steps : At the begining, as the classical technique CVFDT, our algorithm does some initializations and then process each sample (X^{uc},y) to the leaves (1–7). Next from line 8 to 15 each sample arrived is associated with a timestamp. Then it added to the sliding window. Old samples are forgetten from the tree as well as from the window. We detect and identify outdated samples according to their arrival time. The step 16 is for maitaining the tree (Algorithm 2). Finally, from line 17 to 20 we check the split validity of an internal node periodically.

Algorithm 2 details the growing of the uncertain decision tree. The node Feature's parameters are maintained between lines 3 and 7. From line 8 to line 18, a sample is split into a set of fractional samples. From line 19 to 29, the specificity gain is computed using the node feature at leaf nodes. Based on Hoeffding bound, split attribute is chose and the leaf node is split into an internal node.

Algorithm 1 illustrates the process of Poss-CVFDT learning:

Algorithm1: The learning algorithm for Poss-CVFDT

Inputs

S a stream of samples;
X^{UC} an uncertain categorical attribute vector;
$\bar{NF}_{ijk}$ a Node feature : a vector associated to each node ;
G(.) Specificity gain for split evaluation;
W the size of the Window ,based time;
Ti the arrival time of each sample ;
N_{min} the number between checks for growth;
N(s) defines the number of the samples contained in the node Feature;
f the number between checks for drift;

Outputs
HT a decision tree for uncertain samples

1 Begin
2 Let HT be a tree with a single leaf *l* , the root;
3 Let ALT(l) be an initially empty set of alternate trees for l;
4 For each each class yk do
5 For each possible value *vj* of each attribute X_i^{uc} ε X^{uc}
6 Let N_{ijk} = 0 , PS_{ijk}=0 , N(s) =0
//Initialize the Node Feature $\bar{NF}_{ijk}(l)$
7 End For
8 End For
9 For each sample (X^{uc} , y) in S do
10 Sort (X^{uc} , y) into a set of leaves L using HT of any node (X^{uc} , y) passes through
11 ts indicate the time of the recent sample incorporated in the node Feature ts = Tn
12 Add ((X^{uc} ,y), t(s)) to the beginning of W
13 Let ((X_w^{uc}, y_w, t(s)) be the last element of W.
14 Forget Samples (HT, ((X_w^{uc} ; y^w),t(s))) where $Ti \langle t(S) - W$

15 Let W with ((X_w^{uc}; y_w); t(s)) removed
16 Poss-CVFDTGROW(HT,G, (X^{uc} , y),n_{min})
// refer to Algorithm 2
17 If there have been f samples since the last checking of alternate trees
18 CheckSplitValidity(HT)
19 Return HT
20 End.

Algorithm 2 details the growing procedure of the uncertain decision tree.

Algorithm2: Poss-CVFDTGROW(**HT**,**G**, $(X^{uc}$, **y**), n_{min})

1 Let l be the root of HT
2 Let ALT(l) be an initially empty set of alternate trees for l;
3 For each class yk do
4 For each possible value vj of each attribute X_i^{uc} ε X^{uc}
5 Maintain the $\bar{NF}_{ijk}$ following formula (5) , (6) , (7)
6 End For
7 End For
8 For each T_{ALT} in ALT(l) do
9 Poss-CVFDTGROW(T_{ALT},G, (X_{uc} , y),n_{min})
10 End For
11 Label l with the majority class among the samples seen so far at l
12 If l is not a leaf
13 Split (X_{uc}, y) into a set of fractional samples FS
14 For each sample S_{tj} in FS do
15 Let l_j be the branch child for S_{tj}
16 Poss-CVFDTGROW(l_j,G, S_{tj} , n_{min})
17 End For
18 End IF
19 Else If $N(S)\rangle$ n_{min}
20 Compute G(X) for each attribute X_i^{uc} ε X^{uc} based on $\bar{NF}_{ijk}(l)$ and formula (4)
21 Let X_a^{uc} , X_b^{uc} be the attribute with highest and second-highest $\bar{G}$
22 Compute ε
23 If $\triangle\ \bar{G}_l = \bar{G}(X_a) - \bar{G}(X_b) \rangle\ \varepsilon$
24 Replace l by an internal node that splits on X_a^{uc}
25 For each class yk and each vj possible value of each attribute $X_i^{uc}\varepsilon$ X_j^{uc}
26 Initialize $\bar{NF}_{ijk}(l_j)$
27 End For
28 End If
29 End IF
30 Return HT

3.5 Prediction with Poss-CVFDT

Once a Poss-CVFDT is constructed, it can be used for predicting class types. The prediction process starts from the root node, the test condition is applied at each node in the tree and the appropriate branch is followed based on the outcome of the test. When the test sample S is certain, the process is quite straightforward since the test result will lead to one single branch without ambiguity. When the test is on an uncertain attribute, the prediction algorithm proceeds as follows:

Given an uncertain test sample st, $\langle X^{uc}, ?\rangle$, If the test condition is on a uncertain categorical attribute $X_i^{uc} \in X^{uc}$ and Dom(X_i^{uc})= $\{v1, v2, \ldots, vm\}$ a categorical domain which characterized by a possibility distribution where $0 <= \pi(X_i^{uc} = vj) <= 1$.

For the leaf node of Poss-CVFDT, each class yk ∈ C has a possibility degree $\pi_C(yk)$ which is the possibility degree for an instance to be in class yk if it falls in this leaf node. $\pi_C(yk)$ is computed based on the node feature and more especillay on Nijk in the node. Assume path L from the root to a leaf node contains t tests, and the data are classified into one class yk in the end,When predicting the class type for a sample S with uncertain attributes, it is possible that the process takes multiple paths. Suppose there are m paths taken in total, then the possibility degree of yk can be computed as the fraction of the total of possibility degree obtained at each path and the number of paths (m) : $\pi_C(yk) = \frac{\sum_{i=1}^{m} \pi_{yk}^{i}}{m}$.

Finally, the sample will be predicted to be of class yk which has the largest possibility degree $\pi_C(yk)$ among the set of class labes, where k=1,2,...,$|C|$ and poss-dist=$\{\pi_C(y1), \pi_C(y2), \pi_C(y3), \ldots, \pi_C(|C|)\}$ a possibility distribution over classes at a node leaf.

4 Expriment Study

In this section, we present the experimental results of the proposed decision tree algorithm Poss-CVFDT. A collection containing 3 real-world benchmark datasets were assembled from the UCI Repository. The evaluation of our Poss-CVFDT classifier was performed based mainly on three evaluation criterias, namely, the classification accuracy expressed by the percentage of correct classification (PCC), F1 Measure and the running time (t).

4.1 Artificial Uncertainty Creation in the Training Set

Due to a lack of real uncertain datasets, we introduce synthetic uncertainty into the datasets. We treated uncertainty in our approach by assigning a possibility distribution to attribute values. These possibility degrees are created artificially (Table 1).
Example:

- Education: Bachelors (B), Masters (M), Doctorate (D).
- Occupation: Technique support (TS), Craft repair (CR), Sales (S).
- Native_country: United States (US), Cambodia (C), England (E).

Table 1 The classical training set

S	Education	Occupation	Native_country
X1	B	TS	C
X2	D	CR	E

Table 2. details the new training set with uncertain attribute values.

Table 2 Training set based on possibility distributions

S	Education			Occupation			Native_country		
	B	M	D	TS	CR	S	US	C	E
X1	1	0	0	0.1	1	0.7	1	0.2	0.9
X2	0.5	1	0.6	1	0.7	0.2	1	0.3	0.4

4.2 Simulations on the Real Data Sets

We have modified UCI databases by introducing uncertainty in the attributes values of their instances as presented in Table 3.

Table 3 Description of datasets

Datasets	# Instances	# Attributes	# Classes
Solar-Flare	1389	10	3
Car evaluation	1728	6	4
Nursery	12960	8	5

4.3 Experimental Results: Poss-CVFDT VERSUS UCVFDT

We compare in this section the results obtained by our proposed approach and those obtained by the UCVFDT.

1. **Comparison of Classification Accuracy and F1 Measure**: Comparing two different methods which work under two differents frameworks seems to be not imperative. This section presents different results carried out from testing Poss-CVFDT and UCVFDT. We consider P as the percentage of attributes with uncertain values from the training set defined such that: $0 < P < 50$ or $50 < P < 100$. The percentage of P is useful to test the behavior of our classifier and its robustness in dealing with uncertainty. These uncertainty percentages represent the percentages of generated uncertain objects of one given dataset. For example, if we

fixe $P > 50\%$, it means that, for a given database which contains 10 attributes, more than 5 will be generated with uncertainty. These values of P allow us to generate the uncertain databases. We take g as the possibility degree defined for the different values of an attribute: $0 < g < 0,5$ or $0,5 < g <= 1$.
As seen in Figs. 1 and 2, both Accuracy and F1 of Poss-CVFDT are higher than that of UCVFDT. We can see that our approach achieves encouraging results when $P < 50$ and $g \in [0,0.5]$. As shown in Figs. 3 and 4, when the degree of uncertainty increases $P > 50$, both accuracy and F1 of Poss-CVFDT decline slowly. We can also observe that, both accuracy and F1 of Poss-CVFDT are quite comparable to those of UCVFDT. This demonstrates the robustness of our approach against the uncertainty is due to the possibility theory which handle the imprecise data efficiently by treating all the extreme cases.

2. **Running Time**: Figure 5 illustrates the comparison between UCVFDT and our approach in term of processing time. It is obvious that these encouraging results are obtained due to the sliding model which processes only the most recent records by discarding the obsolete ones. This mechanism allows obtaining good results in few times and with high accuracy.

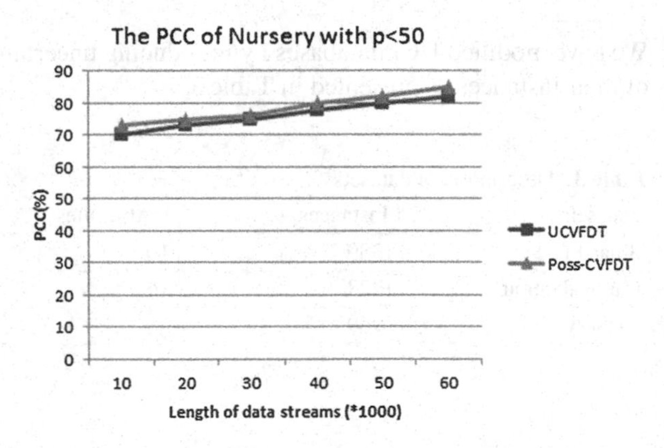

Fig. 1 PCC with $P < 50$ and $g < 0.5$

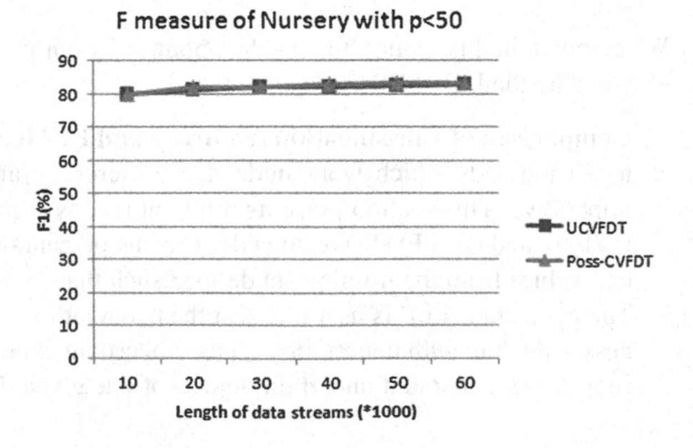

Fig. 2 F1 with $P < 50$ and $g < 0.5$

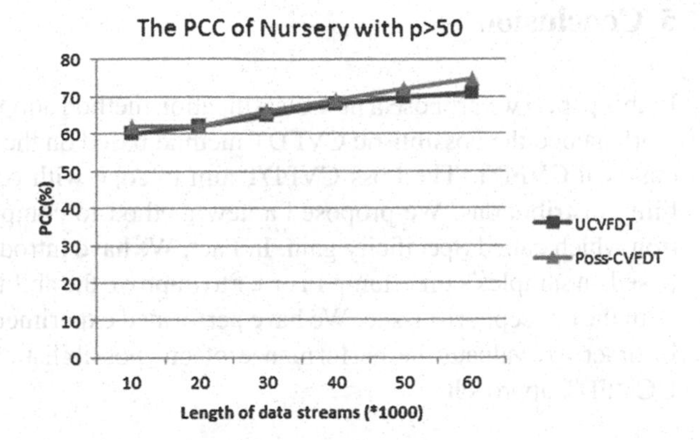

Fig. 3 PCC with $P < 50$ and $g < 0.5$

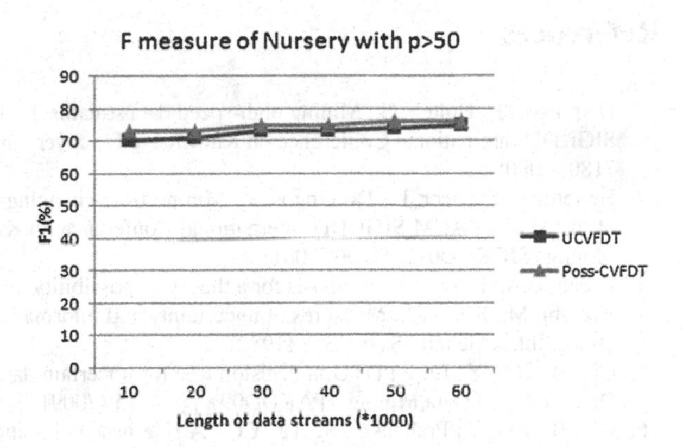

Fig. 4 F1 with $P < 50$ and $g < 0.5$

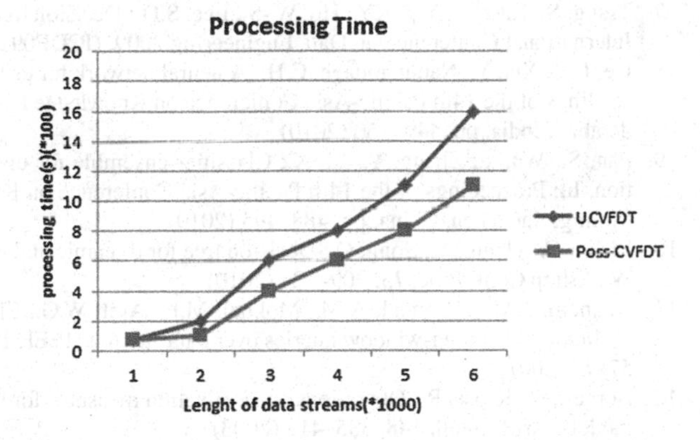

Fig. 5 Execution time

5 Conclusion

In this paper, we propose a new classification method adopted to an uncertain framework named the possiblistic CVFDT method based on the possibility theory and the classical CVFDT. The Poss-CVFDT aim to cope with objects described by possibility distributions. We proposed a new method to compute the measure of selection which called Specificity gain. In Fact, We have introduced a new sliding model based on samples's timestamp In order to improve the ability of Poss-CVFDT to cope with the concept drift issue. We have performed experimentations on UCI databases in order to evaluate the performance of our possibilistic CVFDT method and the UCVFDT approach.

References

1. Domingos, P., Hulten, G.: Mining high-speed data streams. In: Proceedings of the Sixth ACM SIGKDD International Conference on Knowledge Discovery and Data Mining (SIGKDD00), 7180 (2000)
2. Hulten, G., Spencer, L., Domingos, P.: Mining time-changing data streams. In: Proceedings of the Seventh ACM SIGKDD International Conference on Knowledge Discovery and Data Mining (SIGKDD01), 97106 (2001)
3. Zadeh, L.A.: Fuzzy sets as a basis for a theory of possibility. Fuzzy Sets Syst. **1**, 328 (1978)
4. Higashi, M., Klir, G.J.: Measures of uncertainty and information based on possibility distributions. Int. J. General Syst. 4358 (1982)
5. Qin, B., Xia, Y., Li, F.: DTU: a decision tree for uncertain data. In: Advances in Knowledge Discovery and Data Mining. (PAKDD09), pp. 4–15 (2009)
6. Qin, B., Xia, Y., Prabhakar, S., Tu, Y.C.: A rule-based classification algorithm for uncertain data. In: IEEE International Conference on Data Engineering, USA. (ICDE09), pp. 1633–1640 (2009)
7. Tsang, S., Kao, B., Yip, KY., Ho, W.-S., Lee, S.D.: Decision trees for uncertain data. In: IEEE International Conference on Data Engineering 2009. (ICDE09), pp. 441–444 (2009)
8. Ge, J.A., Xia, Y., Nadungodage, C.H.: A neural network for uncertain data classification. Proceedings of the 14th Pacific-Asia Conference on Knowledge Discovery and Data Mining, Hyderabad, India, pp. 449–460 (2010)
9. Pan, S., Wu, k., Zhang, Y., Li, X.: Classifier ensemble for uncertain data stream classification. In: Proceedings of the 14th Pacific-Asia Conference on Knowledge Discovery and Data Mining, Shenzhen, China, pp. 488–495 (2010)
10. Liang, C., Zhang, Y., Song, Q.: Decision tree for dynamic and uncertain data streams. JMLR: Workshop Conf. Proc. *13*, 209–224 (2010)
11. Ghanem, T.M., Hammad, A.M., Mokbel, M.F., Aref, W.G., Elmagarmid, A.K.: Incremental evaluation of sliding-window queries over data streams. IEEE Trans. Knowl. Data Eng. **19**(1), 57–72 (2007)
12. Borgelt, C., Kruse, R.: Operations and evaluation measures for learning possibilistic graphical models. Artif. Intell. **148**, 385–418 (2003)
13. Borgelt, C., Gebhardt, J., Kruse, R.: Concepts for probabilistic and possibilistic induction of decision trees on real world data. In: Proceedings of the 4th European Congress on Intelligent Techniques and Soft Computing (EUFIT96), Aachen, Germany, vol. 3, Verlag Mainz, Aachen, pp. 1556–1560 (1996)
14. Hartley, R.V.L.: Transmission of information. Bell Syst. Tech. J. **7**, 535–563 (1928)

15. Higashi, M., Klir, G.J.: Measures of uncertainty and information based on possibility distributions. Int. J. Gen. Syst. **9**, 43–58 (1982)
16. Aggarwal, C.: Data streams: models and algorithms. Advances in database systems. (2007) Proccedings of the 24th International Conference on Data Engineering, pp. 150–159 (2008)
17. Noga, A., Yossi, M., Mario, S.: The space complexity of approximating the frequency moments. J. Comput. Syst. Sci. **58**(1), 137–147 (1999)

Detection of Variations in Holter ECG Recordings Based on Dynamic Cluster Analysis

Matthias Hermann, Natividad Martinez Madrid and Ralf Seepold

Abstract The proposed approach applies current unsupervised clustering approaches in a different dynamic manner. Instead of taking all the data as input and finding clusters among them, the given approach clusters Holter ECG data (long-term electrocardiography data from a holter monitor) on a given interval which enables a dynamic clustering approach (DCA). Therefore advanced clustering techniques based on the well known Dynamic Time Warping algorithm are used. Having clusters e.g. on a daily basis, clusters can be compared by defining cluster shape properties. Doing this gives a measure for variation in unsupervised cluster shapes and may reveal unknown changes in healthiness. Embedding this approach into wearable devices offers advantages over the current techniques. On the one hand users get feedback if their ECG data characteristic changes unforeseeable over time which makes early detection possible. On the other hand cluster properties like biggest or smallest cluster may help a doctor in making diagnoses or observing several patients. Further, on found clusters known processing techniques like stress detection or arrhythmia classification may be applied.

Keywords Dynamic cluster analysis · ECG holter · Dynamic time warping · OPTICS clustering

M. Hermann (✉) · R. Seepold
Computer Science, HTWG Konstanz, Konstanz, Germany
e-mail: Matthias.Hermann@htwg-konstanz.de
url: http://uc-lab.in.htwg-konstanz.de

R. Seepold
e-mail: Ralf.Seepold@htwg-konstanz.de

N.M. Madrid
School of Informatics, Reutlingen University, Reutlingen, Germany
e-mail: natividad.martinez@reutlingen-university.de
url:http://iotlab.reutlingen-university.de

R. Neves-Silva et al. (eds.), *Intelligent Decision Technologies*,
Smart Innovation, Systems and Technologies 39,
DOI 10.1007/978-3-319-19857-6_19

1 Introduction

Through the rapid development in miniaturization there is a new wide area of possibilities in measuring and processing vital or biometric data. Holter ECG data i.e. long-term ECG recordings is often used as the first candidate in research. The use of analyzing this data is obvious. For analyzing this kind of data many different approaches are proposed. Many classification or clustering techniques can reveal certain cardiological conditions e.g. cardiac anomaly or stress. Clustering techniques are often used to reduce data dimensionality and ease complexity of diagnoses (unsupervised learning) whereas classification is used for pattern matching and detection of known cardiological arrhythmia (supervised learning).

Measuring vital signs aims mainly at two specific target groups. One group describes the so called self-optimizing people which try to keep track of their healthiness and fitness whenever possible. The other group describes medicals which need any data available to allow good diagnoses.

The proposed unsupervised approach tends to another direction – Grouping information which is unknown but similar allows at least detection of variations within the data over time. These variations in ECG signals may be because of continuous stress, overload or re-regeneration. Detection of variations allows both, user feedbacks on changes and doctors to keep track of different patients' healthiness in form of an alerting system.

2 Related Work

There has been a lot of work around ECG data in sense of applying modern machine learning techniques since the beginning of the 1970s. E.g. Willems et. al. showed in 1972 through the use of a computer that there are normal changes in ECG regarding amplitude and axis from day-to-day [1]. Simonson pointed out that patients with heart disease show a significant greater variability compared to healthy individuals (qt. in [2]).

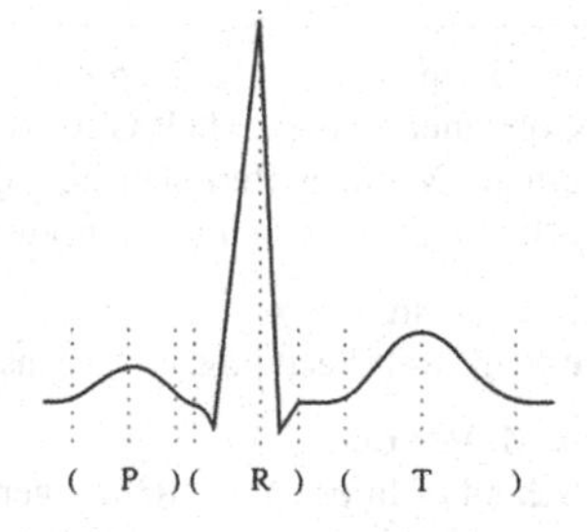

Fig. 1 ECG with P, QRS and T wave [3]

Beside examination of variations in the ECG signal there are other works that try to detect the different P, QRS and T waves, see Fig. 1, by using signal processing techniques [3]. In 2007 Vclav Chudek et. al. published a great comparison of different techniques for classifying ECG signals in discriminating between normal 'N' and unusual 'V' beats [4]. The compared methods reached from Decision Trees over Adaptive Neuronal Networks (ANN) to Support Vector Machines (SVM). A similar more sophisticated proposed approach classified ECG signals with respect to the heartbeat types recommended by AAMI (Association for the Advancement of Medical Instrumentation) [5].

Later Darin Dicheva et. al. proposed a technique for clustering ECG signals for visualization purposes by using Discrete Wavelet Transforms (DWT) [6]. Their idea was to reduce the time needed to manually inspect the whole Holter ECG. The newest techniques applied in clustering ECG data try to combine different approaches, reduce computational costs and even embed sensors in cloths [7]. A good example is the combination of Compressed Sensing (i.e. high undersampling), different transformations and a K-Nearest-Neighbours (K-NN) classifier [8].

3 Technical Background

This chapter gives a short overview on the algorithms used. In this case Dynamic Time Warping and OPTICS clustering.

3.1 Dynamic Time Warping

Because of lag between to signals Xiaopeng and many other authors demonstrated that using Dynamic Time Warping (DTW) for measuring similarity between time series is superior over Euclidian distance and PCA based measures [9–11].

Suppose there are two time series A and B of length n and m. To align these sequences using DTW, we construct an n-by-m matrix where the element $M_{i,j}$ of the matrix contains the distance between two points. The task is now to find a warping path through the distance matrix with minimal cost. There are arbitrary much such paths, but the only path that minimizes the warping costs at each element w_k is described by the following formula and can be determined by using dynamic programming technique.

$$DTW(A,B) = \min \sqrt{\sum_{k=1}^{K} w_k}$$

Additionally, this so called warping path is illustrated in Fig. 2. Adding another parameter *warping-window*, for instance the frequently used *Sakoe-Chiba-Band*, can restrict the path to stay within a given corridor [13]. It can be shown that this yields

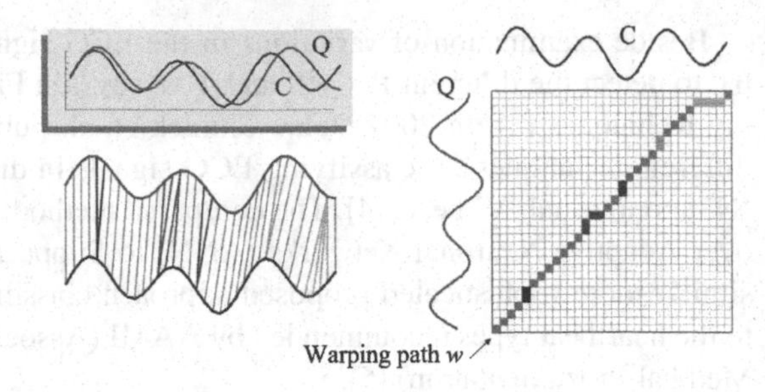

Fig. 2 Dynamic Time Warping algorithm with a warping-window of 4 [12]

better quality of similarity [14]. For further reading a detailed description of the algorithm can be found at [15].

3.2 OPTICS

One main advantage of density based clustering algorithms is that there is no technical need to specify how many clusters there will be. We are going to use an algorithm which is based on DBSCAN and is called *Ordering Points To Identify the Clustering Structure* (OPTICS) [16].

The biggest drawback of DBSCAN is that not specifying ε accordingly to the data, clusters with two different densities are contained in another cluster. OPTICS overcomes this drawback by adding three properties to each object: *order*, *core-distance* and *reachability-distance*.

- *Core-distance* is the distance to the *minPts*-nearest neighbour. So to speak that distance that would determine a core-point in DBSCAN.
- *Reachability-distance* of one point *p* to another point *o* is defined as the maximum of Core-distance and real distance.

The result of an OPTICS clustering is the so-called *reachability-plot* from where the clusters can easily identified by search for valleys as shown in Fig. 3.

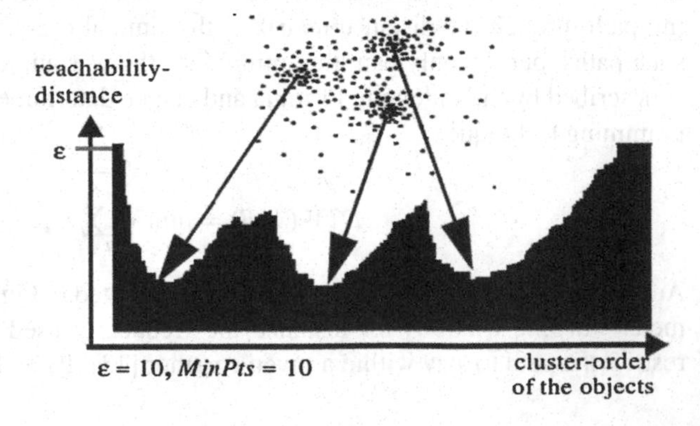

Fig. 3 Reachability-distance diagram (reachability-plot) [16]

4 Proposed Model

The proposed model takes the idea to apply unsupervised analysis techniques on the data collected on a permanent basis.

4.1 Architecture

This chapter describes the architecture i.e. the underlaying process of the proposed model illustrated in Fig. 4.

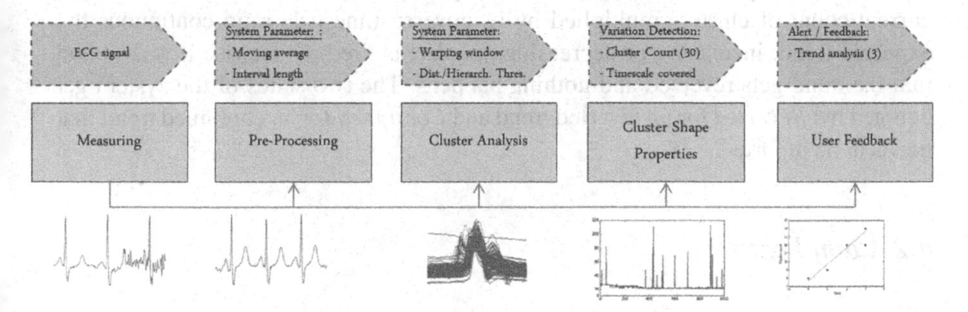

Fig. 4 Illustration of the underlaying process of the proposed model

Measurement and Preprocessing The first step is the *Measurement* of the signal. Next some *Preprocessing* to filter and smooth the signal needs to be applied. First the signal gets low-pass filtered by applying a simple moving average with an appropriate system parameter *window size* according to the sampling rate. One heartbeat consists at minimum of P, QRS and T wave plus offset, which makes $5 * 2 + 1 = 11$ linear approximations. As one heartbeat takes between 0.3 s and 2.0 s, the following frequencies are expected.

$$f = 1/T -> 1/0.3\,\mathrm{s} = 3.3\,\mathrm{Hz} * 11 = 36.6\,\mathrm{Hz} \text{ to } 5.5\,\mathrm{Hz}$$

Next the signal gets smoothed by an appropriate curve fitting technique. Afterwards the signal gets divided into pieces of the same length. The length needs to be specified through an system parameter called *interval length.*

Cluster Analysis Next step is the *Cluster Analysis*. There are two system parameters needed: *warping-window* for the DTW distance measurement between the signal pieces and the *distance threshold* i.e. that distance that describes the border to another cluster or geometrically free space between clusters. First the distance matrix for all signal pieces gets calculated. Afterwards OPTICS is applied to order the points with respect to their cluster structure. Result of the cluster analysis is the *reachability-plot*.

Cluster Shape Properties The cluster shape properties are defined in order to be able to compare dynamic clusters over time. This is needed because the process will run in regular periods of time.
The proposed architecture defines two properties:

- Cluster count of clusters having at least a specified *cluster size* to discard small noisy clusters.
- The timescale covered by the described clusters to weight periods with less noisy data more than noisy one.

The cluster properties can easily be extended to any desired measure.

User Feedback Having minimum cluster shape properties of three clustering intervals, a trend analysis can be applied. For simplicity a linear trend is assumed. If the current count of clusters multiplied by its covered timescale ratio continuous the expected trend, increasing or decreasing, an alert is fired. Otherwise it is assumed that the trend gets reversed and nothing happens. The two states of the system get denoted by: *Reverted* for an reverted trend and *Continued* for an continued trend that needs to be reported.

4.2 Complexity

When proposing such a model it is important to quick review the algorithmic complexity of the needed computations:

- Dynamic Time Warping of two signal intervals of size u: $\mathcal{O}(u^2)$
- Distance matrix of size *nxn*: $\mathcal{O}(n^2)$
- Optics clustering: $\varepsilon > maxd(o,p) \models \mathcal{O}(n^2)$
- Cluster size and trend analysis: $\mathcal{O}(n)$

This yields an overall complexity of $\mathcal{O}(n^2u^2 + n^2 + n)$. Restricting the warping-window yields an almost linear complexity and limiting epsilon allows spatial logarithmic indexing. This yields a complexity of $\mathcal{O}(un^2 + nlog(n) + n) \models \mathcal{O}(un^2)$ which is suitable.

5 Laboratory Prototype

This chapter describes the laboratory prototype. First the environment is introduced. Afterwards the experiment itself gets described. The laboratory environment consists of the E-Health Sensor Platform for Arduino and the scientific python stack (SciPy).

Measurement and Pre-Processing The data gets sampled at a rate of 250 samples per second, this yields a slightly adapted *window-size* of 5 and a cut-off frequency of 50 Hz. As we do not want linear but at minimum cubic approximation, the signal is undersampled with factor 2 which gives 125 samples per second. For the experiment the *interval length* is defined as 2 s. This results in 250 samples per piece of signal. Further we will use 5000 of such samples which yields a design matrix of size 5000×250 and a timescale of 168 min.

Cluster Analysis The *warping-window* defines how much warping is allowed. We define a value of 10. This value corresponds to a timescale of ≈ 0.1 s. As we have 5000 samples in the test data we define a *distance threshold* value of 500 which yields an estimation of 10 clusters. In Fig. 5 an arbitrary found cluster is shown. The result of the alignment of the DTW can easily be seen in the lag between the different samples.

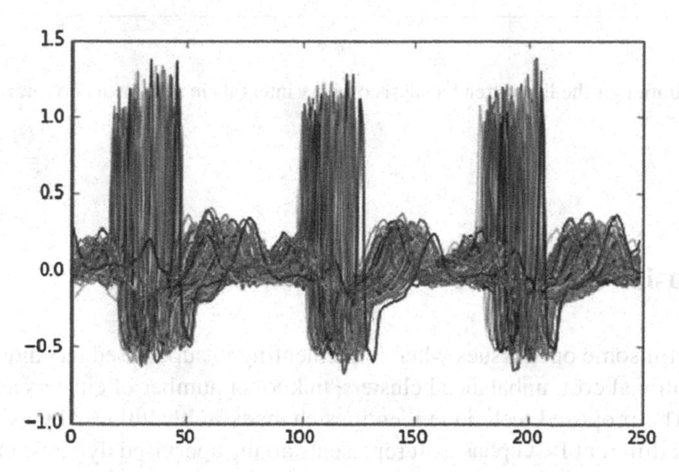

Fig. 5 Found cluster of size 92 covering 3 min

Cluster shape properties Before deriving the cluster shape properties we have to define the *hierarchical threshold* which can be seen as an horizontal line sweeping over the reachability-plot from top to bottom. It can be interpreted as an hierarchical clustering approach to OPTICS as the cluster count will raise during sweeping. As this threshold directly impacts cluster quality in defining the maximum dissimilarity of the included samples it must be well chosen. For the test set a value of 50 is defined.
Based on the resulting clusters the cluster shape properties can be derived.

- Taking candidates with at least 30 samples or 60 s results in 22 different clusters.
- These 22 clusters cover a timescale of 157 min of a total of 168 min which yields a ratio of 0.93.
- Combining the two yields a cluster shape property of 22.5 for the test set.

User Feedback The user feedback step is based on linear prediction of 3th order. This means we have to apply the process until first prediction at least four times.

The result of six runs is displayed in Fig. 6. It can be seen that the trend gets inverted in the first and second step, but gets continued in the last step. The measure in the last step would lead to an user feedback informing of an ongoing change in cluster shape properties.

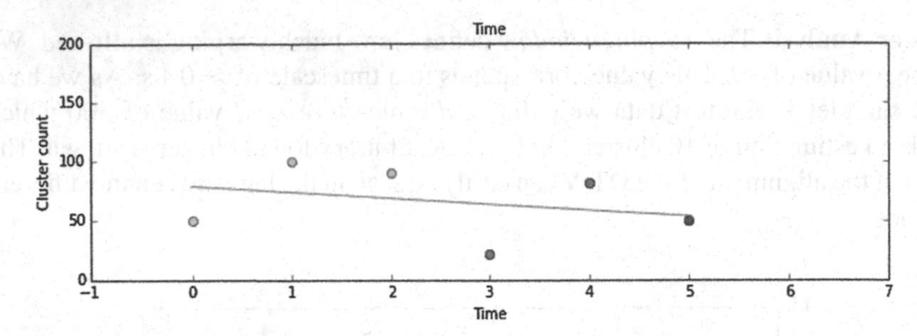

Fig. 6 Illustration of the linear trend analysis over six intervals in time with two *reverts* and one *continue* at t = 5

6 Conclusion and Future Work

There are still some open issues when implementing unsupervised techniques, such as computational cost, unbalanced clusters, unknown number of clusters and initial partition. The proposed technique identifies changes in healthiness by using commonness of different ECG phases. It represents an unsupervised dynamic clustering analysis methodology that can be implemented in oriented devices for analysis in real time. The considered model does not require prior training or heartbeat labeling by a doctor and can in principle be applied to any ECG signal.

Anyway, to improve accuracy and overall cluster quality further work could develop the proposed model from an unsupervised dynamic clustering approach to an supervised dynamic classifying approach. For instance, the model could be modified to classify ECG data into the AAMI ECG standard groups [5]. Cluster shape properties would be analogously defined by measuring the different classified group sizes. This would lead to a more medical backed prediction model.

References

1. Willems, J.L., Poblete, P.F., Pipberger, H.V.: Day-to-day variation of the normal orthogonal electrocardiogram and vectorcardiogram (1972)
2. Macfarlane, P.W., Lawrie, T.V.: The normal electrocardiogram and vectorcardiogram. In Macfarlane, P.W., van Oosterom, A., Pahlm, O., Kligfield, P., Janse, M., Camm, J. (eds.) Comprehensive Electrocardiology, pp. 485–546, 2nd edn. Springer (2010)

3. Vullings, E., Garca, J., Laguna, P.: Waveform detection in holter ECG using dynamic time warping. In: Comunicaciones del XV Congreso Anual de la Sociedad Espaola de Ingeniera Biomdica, Valencia, Spain, pp. 313–316 (1997)
4. Chudek, V., Petrk, M., Georgoulas, G., Epek, M., Lhotsk, L., Stylios, C.: Comparison of seven approaches for holter ECG clustering and classification. In: Proceedings of the 29th Annual International Conference of the IEEE EMBS, Lyon, France, IEEE, Aug 2007
5. Rodriguez-Sotelo, J., C.D. Acosta-Medina, G.C.D.: Recognition of cardiacarrhythmia by means of beat clustering on ECG-holter recordings. In Millis, R. (ed.) Advances in Electrocardiograms Methods and Analysis, InTech, pp. 225–250 (2012)
6. Vgner, A., Farkas, L., Juhsz, I.: Clustering and visualization of ECG signals. In: Dicheva, D., Markov, Z., Stefanova, E. (eds.) Third International Conference on Software, Services and Semantic Technologies S3T 2011. Advances in Intelligent and Soft Computing, vol. 101, pp. 47–51. Springer, Berlin (2011)
7. Van Laerhoven, K., Schmidt, A., Gellersen, H.W.: Multi-sensor context aware clothing. In: Proceedings Sixth International Symposium on Wearable Computers (ISWC 2002), pp. 49–56 (2002)
8. Balouchestani, M., Krishnan, S.: Fast clustering algorithm for large ECG data sets based on CS theory in combination with PCA and K-NN methods. In: Engineering in Medicine and Biology Society (EMBC), 2014 36th Annual International Conference of the IEEE, Chicago, USA, IEEE, pp. 98–101 (Aug 2014)
9. Xi, X., Keogh, E., Shelton, C., Wei, L., Ratanamahatana, C.A.: Fast time series classification using numerosity reduction. In: Proceedings of the 23rd International Conference on Machine Learning. ICML'06, pp. 1033–1040 (2006)
10. Kotsakos, D., Trajcevski, G., Gunopulos, D., Aggarwal, C.: Time-series data clustering. In: Data Clustering: Algorithms and Applications. CRC Press (2013)
11. Li, L., Prakash, B.A.: Time series clustering: complex is simpler! In: Getoor, L., Scheffer, T. (eds.) ICML, Omnipress, pp. 185–192 (2011)
12. Keogh, E., Ratanamahatana, C.A.: Exact indexing of dynamic time warping. Knowl. Inf. Syst. **7**(3), 358–386 (2005)
13. Sakoe, H., Chiba, S.: Dynamic programming algorithm optimization for spoken word recognition. IEEE Trans. Acoust. Speech Signal Process. (26/1), 43–49 (1978)
14. Ratanamahatana, C.A., Keogh, E.: Everything you know about dynamic time warping is wrong. In: KDD Workshop on Mining Temporal an Sequential Data, Seattle, US (2004)
15. Kruskall, J.B., Liberman, M.: The symmetric time warping algorithm: from continuous to discrete. In: Time Warps, String Edits and Macromolecules. Addison Wesley (1983)
16. Ankerst, M., Breunig, M.M., Kriegel, H.P., Sander, J.: Optics: ordering points to identify the clustering structure. In: Proceedings of the 1999 ACM SIGMOD International Conference on Management of Data, SIGMOD, pp. 49–60 (1999)

A Theoretical Framework for Trusted Communication

Dao Cheng Hong, Yan Chun Zhang and Chao Feng Sha

Abstract With the tendency of increasing ICT (information and communications technologies) applications, most existing literature examines trusted communication in decision making ignoring the effects of media inherent characteristics. However, according to the uncertainty reduction theory, media characteristics including media richness and media interactivity play a vital role in communication for decision making. To address the knowledge gap, this study proposes a novel framework for trusted communication (TCF) drawn on the media characteristics perspective and theory construction methodology. TCF theory, which is built on both a theory of behavioral trust and a theory of computational trust, provides a new lens for communication of decision making which currently plays a critical part in modern society.

1 Introduction

With the increasing variety of ICT applications, a growing number of social activities that have traditionally been conducted via physical mechanisms between people and/or objects are gradually becoming more and more virtual [1]. In particular, many decision makings are being conducted virtually via communication media, such as E-Commerce, E-Government, Smart Health and Wellbeing. Many decision making processes which would have been difficult to conduct with ICT only a few years ago are being undertaken by telecommunication systems. For example, the decision process of traditional shopping for acquiring goods from a salesman at a physical store is being conducted virtually on the Internet. Many retail banking processes that were handled by a human bank teller have migrated to

D.C. Hong · Y.C. Zhang · C.F. Sha (✉)
Shanghai Key Laboratory of Intelligent Information Processing, School of Computer Science, Fudan University, Shanghai, China

D.C. Hong · Y.C. Zhang
Centre for Applied Informatics, Victoria University, Melbourne, Australia

R. Neves-Silva et al. (eds.), *Intelligent Decision Technologies*,
Smart Innovation, Systems and Technologies 39,
DOI 10.1007/978-3-319-19857-6_20

online banking systems, or even mobile banking systems. Communications for decision making by means of various ICTs media continues to move forward at an ever-quickening pace, providing people with much greater convenience in work, study and social activities.

Researchers assert that some activities have proven more amenable to being conducted in mediated environments than others in industry practice because of a bad virtual reality experience [2]. Specifically, some business decision processes (such as purchasing a gift or a cell-phone over the web, especially in China) have proven popular, while others (such as buying a car or real estate) face many obstacles and have proven resistant to the Internet. Distance learning programs for experimental or internship processes are also resistant to mediated environments. Although social media (e.g., Facebook, Twitter, WeChat) have been a hot topic over the past several years, it is only a complementary tool for social networks in the physical world because of information privacy and security. These phenomena present increased challenges as they are not valid and reliable forms of communication in the decision making area. The foundations of trust in IT-based communication, as recognized by Gligor [3], are: participant preferences (risk aversion and betrayal aversion) and beliefs in the trustworthiness of other protocol participants. In relation to robust and trusted communication, the above practical observations lead to the main research questions: Why do different media applications with various characteristics have different communication trustworthiness in decision making? How do media characteristics influence the trusted communication?

In the relevant area, trust in communication has received a great deal of attention from researchers, particularly from different theoretical perspectives. However, there is very little literature on trusted communication of decision making from the perspective of media characteristics by integrating computer and social science. Therefore, it is necessary to conduct a deeper study and explore a novel theory for trusted communication of decision making from this new and integrated perspective. According to prior research, the uncertainty reduction theory suggests that humans reduce uncertainty and equivocality under the assistance of communication media, to make environments more predictable [4, 5]. Communication media has been applied to almost ICT applications (e.g., education, science, government, commerce) and plays a vital role in intelligent decision making. Hence, this research focuses on communication activities and proposes a robust and reliable framework for trusted communication (TCF), based on the media characteristic perspective [6] which aims to enhance communication security from user's perspective. In other words, this study explores media factors influencing trusted communication and develops the TCF theory for decision making drawn on media characteristics [6, 7] and the theory construction methodology [8, 9]. We aim to develop a new theoretical framework for trusted communication in decision making from both a theory of behavioral trust and a theory of computational trust.

2 Theoretical Fundamentals

Trusted Communication. The adoption of the Internet, the development of mobile internet and the emergence of the Internet of Things have increased ICT applications rapidly and has also resulted in heightened risk in communication security. Trusted communication has quietly descended on many fields, from governments and commerce to individual health and privacy. User participating in these mediated communications relies primarily on trust, since on-line verification of protocol compliance is often complex and impractical. Practical verification can lead to many problems, from co-NP complete test procedures to user inconvenience. Access management and architectures to protect services and devices in ubiquitous computing environments has been discussed in depth which allows access restrictions directly on services and object documents. Over the past three decades, there has been a vast body of work on trust for management in computer science and social science [3].

Trust can be divided into two broad areas: interpersonal trust and systems trust, both of which are critical in shaping participants' behaviour in ICT-mediated settings [10]. Trusted communication can also be divided into two broad areas: entities which are trustworthy; and media channels which are also trustworthy, as shown in Fig. 1 (source: [11]). These entities are primarily technologically mediated with each other, such as humans, forms of intelligence including mediated representations of remote humans via text, images, video, 3D avatars and artificial representations of humanoid or animal-like intelligence including virtual human agents, computers, smart devices and robots. A media channel is a broad concept which includes a variety of ICT applications, from the Internet to wireless communications. The reliability of entities and the relative attribute importance for the application's purpose have been integrated into the data anonymisation approach area [12]. Trusted communication is the credibility state of entities communicating with others via various media channels which is the perceived trustworthiness of

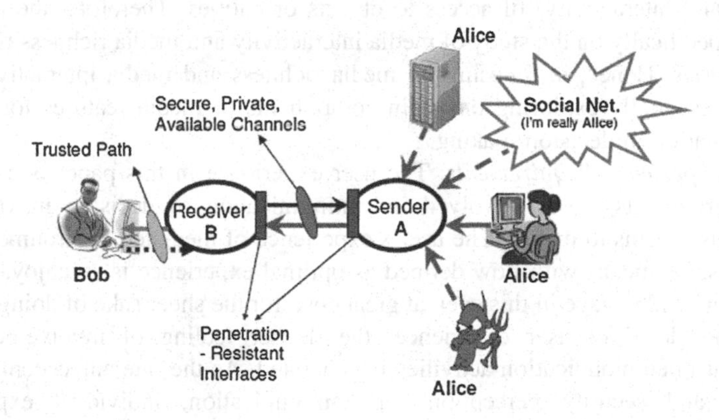

Fig. 1 Entities and media channels

communication by participants in decision making. Trusted communication involves comprehensive applications or systems of various ICTs and should be explored in depth from media perspective.

Media Characteristics. According to the uncertainty reduction theory [4, 5, 13], media from verbal to ICT-enabled types plays a vital role in contemporary society. The uncertainty reduction theory suggests that people reduce uncertainty and equivocality by means of communication media, to make their environments more predictable and reliable. Media vividness and interactivity which have been studied in certain areas (e.g. e-learning) are the two major characteristics [9, 14, 15]. The first dimension, media vividness referring to the ability of a medium to produce a sensorially rich mediated environment is also perceived as media richness by researchers. Media richness comes primarily from the literature on computer-mediated communications (CMC) and is most often associated with business communication [16]. In this paper, media richness is used to analyze communication media choice of decision making and to help reduce ambiguity and equivocality for mediated communication trustworthiness.

Interactivity refers to the extent to which individuals can participate in modifying the form and content of a mediated environment in real time, and is a rich research topic. Lombard and Snyder-Duch proposed that media interactivity should have five critical components: number of inputs acceptable; number and type of characteristics that are modifiable; the range of response possible; the speed of response; and the degree of correspondence between input and response [17]. Steuer construed media interactivity as having three components: speed, range, and mapping [6]. These correspond to the above-mentioned five components, as the first three of the five can be subsumed under "range". The potential for individuals to modify the mediated environment is also defined as user control, which has been conceptualized as the same as media interactivity by a number of researchers [1, 18]. Media interactivity describes the participants mutual communication either in real time (e.g., QQ, Skype) or in a store-and-forward basis (e.g., email), or to seek and gain access to objects an on-demand basis. Increasing network bandwidth, higher mobility, and more immersive designs promise to offer a better sense (much more richness and interactivity) of access to objects or entities. Therefore, the research focuses specifically on the study of media interactivity and media richness based on prior theories. Hence, in focusing on media richness and media interactivity, the study is essentially exploring the main components of media features for trusted communication in decision making.

User Experience Requirements. The user experience in this paper is a state in which participants are so involved in communication of decision making that nothing else seems to matter. The user's experience of mediated environments that have the same means with flow defined as optimal experience is so enjoyable that participants will engage in this even at great cost, for the sheer sake of doing it [19]. The higher level of user experience, the deeper feeling of involvement and enjoyment in communication activities is generated. As the one antecedent of user adoption and security perception for communication, individual experience requirements are about the need of involvement and enjoyment at the mediated communication state. Experience requirements include tasting, seeing, hearing, smelling, and touching other corresponding participants or objects, as well as the overall sensation that participants feel when engaging in a communication activity.

A focus on user perception and experience in the mediated environment has led to a proliferation of credible work in recent years. Researchers have studied how the mediated experience of users effects on communication and presented detailed guidelines for research [6, 7]. Overby described how the requirements of experience (sensory requirements, relationship requirements, synchronism requirements, and identification and control requirements) impact mediated communications [2]. A key premise of both prior work and this study is that the requirements of perception and experience can be used to make communication more resistant to be trustworthy for participants. The logical reasoning behind this premise that is straightforward is ICT-based communication lack of sensory connection and mutual interaction between entities. Evidently, participants with high experience requirements are critical to communication media and are reluctant to trust and adopt communication systems, because actual communication media can't create a virtual environment to satisfy all users' experience requirements. From this analysis, user requirements of experience have a negative effect on communication trustworthiness from user perception perspective.

Much more literature provides sufficient theoretical evidence for trust research in communication. Researchers proposed the guided foundations of the trust theory in computer and communication security through integrating a theory of behavioral trust and a theory of computational trust [3, 11]. Trusted communication can be used to increase the pool of services available to users, promote cooperation, and enable the network effect matter at an application level. However, the media characteristics have been ignored by most researchers which are necessary for reducing ambiguity and equivocality for participant's communication.

3 TCF Framework

As described in the above sections, people want to reduce uncertainty, ambiguity and equivocality in communication to make their environments more predictable and reliable. This section will focus on communication trustworthiness in decision making and will propose a robust and reliable model of trusted communication based on media characteristic perspectives. In this TCF theory, the dependent variable is trusted communication and the main constructs are media richness and media interactivity. Each of the main constructs (media richness and interactivity) is posited to have a positive effect on the dependent variable (communication trustworthiness). In other words, as each of the main constructs media richness and interactivity increases, the communication becomes amenable to be much more trusted. This does not mean that communication of decision making with a low degree of media richness and/or media interactivity can't be trustworthy. Rather, it means that it would be more amenable to being trusted by participants if the degrees of these media characteristics are high. As a dependent variable of TCF, trusted communication is continuous, not discrete, and should be thought of as a matter of degree, not of kind. This is a critical distinction for this study from other trusted communication research. The propositions of TCF theory that should not be interpreted as on/off are simply descriptions of the degree of trusted communication. In propositional terms of TCF, user requirements of experience for involvement and enjoyment have a negative moderating effect on the relations between the main constructs and trusted communication. Therefore we propose the research framework of TCF through integrating the research across relevant streams, as shown in Fig. 2.

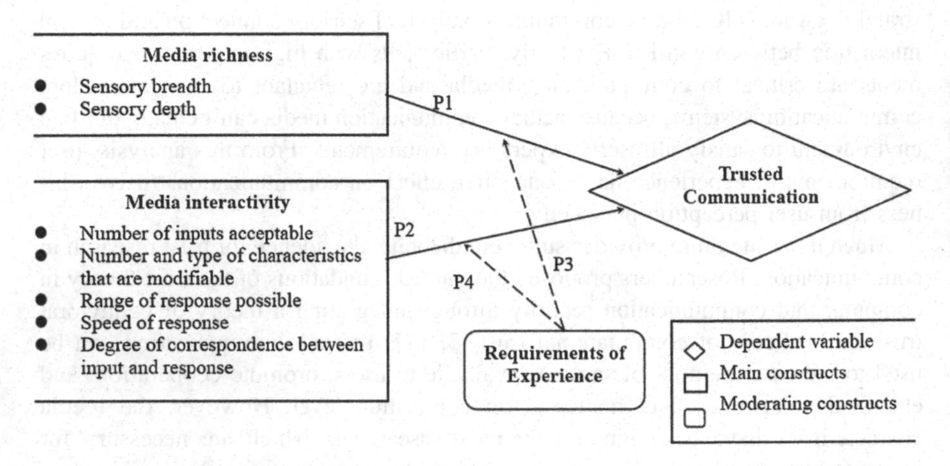

Fig. 2 TCF framework

3.1 Trusted Communication

The dependent variable in TCF theory is trusted communication, which presents the communication trustworthiness with a certain medium channel by participants in decision making. Researchers proposed a general trust theory in human-computer networks through integrating a theory of behavioral trust and a theory of computational trust [3, 11]. Trusted communication would increase the pool of services available to users, remove cooperation barriers, and enable the network effect matter at the application level. Practically, the state of trusted communication can be measured objectively either as adoption of the communication or the outcome quality of the communication systems [2]. For example, the adoption of e-commerce on Amazon or Alibaba over the past few years has indicated that certain shopping processes are amenable to be conducted by means of the Internet with high credibility perception. Considering the criteria of outcome quality, if the revenue of one commodity via the Internet is greater than via TV, this provides evidence that communication via the Internet media is highly reliable and trusted, at least under certain domains and for certain aims. As stated in the previous section, this study regards trusted communication as a continuous rather than discrete status, and a matter of degree, not of kind. Therefore, the dependent variable perception of trusted communication can be measured from two dimensions: adoption and revenue.

3.2 Media Richness

As indicated in the following, “the representational richness of a mediated environment as defined by its formal features; that is, the way in which an environment presents information to the senses” [6], media richness has been recognized as a critical component of media characteristics. The premise for research that the richness of a medium will influence individual perceptions of the environment lies in the ability of a media-rich environment to diminish user perceptions of mediation [20]. The logic is that the media can be used to simulate the sensory elements of the physical world for participants. This simulation of media richness will determine the level of user perception of trusted communication environments where participants are involved. Media richness is a better indicator to analyze communication media choice and to help participants reduce the ambiguity and equivocality of communication and increase trustworthiness in decision making.

Media richness embodies two core elements of the communication media for users: sensory breadth (number of communication channels) and depth (quality within each channel). Sensory breadth presents the number of different sensory channels that a medium utilizes (e.g., aural, olfactory, visual, tactile, gustatory), while sensory depth refers to the resolution within each of these perceptual channels. Thus, multimedia communications have greater breadth than single media communications (e.g., advertisements on flash or video on the web versus on TV). The user experience of a mediated representation will be related to the number of individual sensory channels that can be engaged. The greater the number and quality of sensory channels, the greater is the likelihood of user immersion in the mediated environment and the higher the perception level of communication credibility. Media richness is posited to have a positive relation to the trustworthiness of communication. As stated above, the first proposition describes the effect of media richness on trusted communication.

Proposition 1 (P1). Increased media richness promotes trusted communication.

3.3 Media Interactivity

Interactivity, which is viewed as a multidimensional construct in the literatures, is the extent to which the user can participate in modifying the form and content of a mediated environment with an immediate response. Researchers proposed that interactivity mainly includes five critical components: the number of inputs acceptable; the number and type of characteristics that are modifiable; the range of response possible; the speed of response; and the degree of correspondence between input and response [17]. In one experimental study, the researcher focused on certain constructs of interactivity for special situations which is an example of narrower aspects for user control [18]. Media interactivity describes a dynamic environment that affects everyone, from those directly involved to those who have to manage it and those who use it. This paper focuses on the interactivity of media feature which is pertinent for trusted communication.

Traditional studies of direct communication experience show that individuals have a high level of interactivity over how they interact with the objects what to look at, touch, smell, in what order, and for how long. In contrast with a highly mediated communication world, such as those transmitted via the Internet or wireless communication systems, the range of interactivity choices is limited. In such situations where fewer interactivity options are available, participants will thus perceive their experiences as more mediated because of the loss of media control. Generally, participants may perceive the environments in which they have greater interactivity as less mediated because of the association of interactivity with direct communication experience between entities. Therefore, media interactivity is posited to have a positive relation to the trustworthiness of communication. The second proposition presents the relation between media interactivity and trusted communication.

Proposition 2 (P2). Increased media interactivity promotes trusted communication.

3.4 Requirement of Experience

As shown in existing literature [6, 7, 19], participants' mediated experience of ICT applications including involvement and enjoyment, has an influence on the actual use or adoption of trusted communication. Different requirements of user experience have different effects on trusted communication of decision making. At single media channels situation with certain characteristics (media richness and interactivity), participants with high experience requirements are reluctant to trust and adopt communication systems, because practical communication media cannot create a virtual environment to satisfy all users' experience requirements. It is intuitive that with the same ICT-based media different communications were moderated heavily by user experience requirements of involvement and enjoyment. The logical reason for this is that different communication participants have different requirements of virtual reality experience for trusted communication. However, limited ICT applications cannot fully satisfy user requirements of experience. For example, the web-based purchasing if gifts and cell-phones is amenable to be adopted for high perception of trusted communication with low requirements of experience, but the Internet-based purchase of a car or a house has many obstacles to overcome before being adopted for low perception of communication trustworthiness with high requirements of user experience comparatively. In TCF theory, the moderating construct describes the degree of requirements for the user practical experience which negatively moderates the relations between the main constructs (media richness and interactivity) and the trusted communication. Through these discussions the third and fourth propositions can be drawn.

Proposition 3 (P3). The requirements of experience negatively moderate the relations between media richness and trusted communication.

Proposition 4 (P4). The requirements of experience negatively moderate the relations between media interactivity and trusted communication.

4 Discussion

4.1 Contributions

Our TCF theoretical framework explains what kinds of media characteristics affect trusted communication in decision making and how it works. In other words, this study describes why different communication applications for industry via ICT media have different user adoption and different levels of trustworthiness for applied systems. In this theory, if the main constructs (media richness and media interactivity) are low, the remote communication systems will be more difficult to adopted and trustworthy than if they were high. User requirements of experience (involvement and enjoyment) for ICT applications have a positive moderating impact on the relations between the main constructs (media richness and media interactivity) and trusted communication. A wider applied-promotion of the TCF framework can be conducted in the next research study.

Although TCF theory is mainly drawn from the literature, the research result is relevant to both academia and practice. The TCF framework provides researchers with a lens to explain and predict the media characteristics that influence whether a communication application is amenable to be adopted and trustworthy via a certain media. Practically, trusted communication occurs in nearly all aspects of society and business with ICT applications, which facilitates multiple field research including management, computer science, sociology, economics, and communication studies. TCF theory also presents the significance of media such as ICTs, which play a vital role in the decision making, communication and information systems field. This aspect of TCF theory will help researchers better understand why and how media characteristics continue to have profound impacts on trusted communication including adoption and revenue of applications.

TCF theory provides a useful framework for security evaluation and prediction in which communication is conducted and adopted when participants shall make decisions via various media. The analytical framework provided by TCF theory that will help engineers and managers with the communication process and channel design will become increasingly important as a combination of media factors and user requirements of experience in communication trustworthiness. For example, users can assess the degree of communication trustworthiness by considering media richness, media interactivity and user requirements of experience. If these components of media characteristics are low, engineers and managers will make more effort to introduce a viable communication system than if they are high. Especial in the same media environment condition, user experience requirements should be considered thoroughly for trusted communication.

4.2 Improvements

As a new theory, there is a certain amount of room to improve its development and practice. In the first place, TCF theory, which is mainly drawn on both a theory of behavioral trust and a theory of computational trust, explains communication trustworthiness in relation to media characteristics. This study discusses domain-specific factors of media in trusted communication of decision making which may hamper the general usage.

In the second place, TCF theory is not meant to be used to evaluate whether an ICT-based communication is better or worse than physical communication. Rather, it assesses whether communication is amenable to be trustworthy and adopted from the perspective of media characteristics. For example, there are many users of mobile banking, although the overall adoption and development is not good in China as the people prefer to transact via mobile payment, or have to adopt it according to regular or contract pressure. These types of preferences and pressures are not explicitly examined within the TCF theoretical framework.

Finally, to theorize TCF in decision making, this research draws on a theory construction methodology and media characteristics. All constructs are based on our interpretation from a media characteristics perspective and relevant streams. Therefore, this study is more or less constrained by theoretical biases which may engender limited promotion in the future.

5 Conclusions

This research focuses on media characteristics affecting trusted communication of decision making and proposes a TCF theory drawn on a theory of behavioral trust, a theory of computational trust and theory construction methodology. The TCF theoretical framework provides a new lens for researching communication trustworthiness which plays a central role but has been neglected in this digital age. The study identified the main constructs and propositions which offer guidance for trusted communication from user adoption and revenue perspectives. As with any newly proposed theory, the TCF framework can benefit from empirical and experimental testing which will improve and change some constructs. Hence, empirical and experimental research will be conducted in our future study.

There is little doubt that more and more remote communication will be conducted by means of various ICTs. However, it seems unlikely that society will abandon the physical world in favor of these advents of ICT-based communications especially in relation to trustworthiness, at least not in the near future. TCF theory is useful to understand and predict which media applications will continue to hinder communication from a trustworthiness aspect. It also may be used to help practitioners predict by which media based communication is trustworthy in the near future versus the long term. As a useful research perspective and framework, TCF theory provides a new lens and can explain the most trusted-communication phenomenon in decision making domain.

Acknowledgments The authors thank the anonymous referees for their valuable comments and suggestions. * Dr. Chaofeng SHA is the corresponding author of this study. This work was supported partly by CSC (201206105028), NSFC (61332013, 61472086, 61170095), Shanghai Pujiang Program, Senior Visiting Professor Project of Fudan University and IIPL (IIPL-2014-001).

References

1. Hong, D.C., Ling, H., Li, Y.M.: A novel theory of business process virtualization on E-commerce. ICIC Express Lett. **4**(2), 381–387 (2010)
2. Overby, E.: Process virtualization theory and the impact of information technology. Organ. Sci. **19**(2), 277–291 (2008)
3. Gligor, V.D.: On the foundations of trust in networks of humans and computers. In: ACM CCS 2012, p. 1, Raleigh, North Carolina, 16–18 Oct 2012
4. Berger, C.M.: Communicating under uncertainty. In: Roloff, M.E., Miller, G.R. (eds.) Interpersonal Processes: New Dimensions in Communications Research, pp. 39–62. Sage, Newbury Park (1987)
5. Berger, C.M., Bradac, J.J.: Language and social knowledge: uncertainty in interpersonal relations. Edward Arnold, London (1982)
6. Steuer, J.: Defining virtual reality: dimensions of determining telepresence. J. Commun. **42**(4), 73–93 (1992)
7. Klein, L.R.: Creation virtual product experiences: the role of telepresence. J. Interact. Mark. **17** (1), 41–55 (2003)
8. Eisenhardt, K.M.: Building theories from case study research. Acad. Manage. Rev. **14**(4), 532–550 (1989)
9. Wand, Y., Weber, R.: Research commentary: information systems and conceptual modeling—a research agenda. Inf. Syst. Res. **13**(4), 363–377 (2002)
10. Hsu, M.H., Chang, C.M., Yen, C.H.: Exploring the antecedents of trust in virtual communities. Behav. Inf. Technol. **30**(5), 587–601 (2011)
11. Gligor, V.D., Wing, J.: Towards a theory of trust in networks of humans and computers. In: Proceedings of the 19th International Workshop on Security Protocols. Cambridge, 28–30 March 2011
12. Sun, X.X., Wang, H., Li, J.Y., Zhang, Y.C.: Injecting purpose and trust into data anonymisation. Comput. Secur. **30**(1), 332–345 (2011)

13. Jiang, J.J., Klein, G., Carr, C.: Measuring information system service quality: SERVQUAL from the other side. MIS Quart. **26**(2), 145–166 (2002)
14. Bjorkman, E.A.A., Sarkani, S., Mazzuchi, T.A.: Test and evaluation resource allocation using uncertainty reduction as a measure of test value. IEEE Trans. Eng. Manage. **60**(3), 541–551 (2013)
15. Lombard, M., Ditton, T.: At the heart of it all: the concept of telepresence. J. Comput. Mediated Commun. **3**(2), 6–21 (1997)
16. Trevino, L., Lenge, R., Daft, R.: Media symbolism, media richness, and media choice in organizations. Commun. Res. **14**(5), 553–574 (1987)
17. Lombard, M., Snyder-Duch, J.: Interactive advertising and presence: a framework. J. Interact. Advertising, **1**(2), (2001)
18. Ariely, D.: Controlling the information flow: effects on consumers' decision making and preferences. J. Consum. Res. **26**(Sept), 233–248 (2000)
19. Konradt, U., Filip, R., Hoffmann, S.: Flow experience and positive affect during hypermedia learning. Br. J. Educ. Technol. **34**(3), 309–327 (2003)
20. Hoffman, L.J., Lawson-jenkins, K., Blum, J.: Trust beyond security: an expanded trust model. Commun. ACM **49**(7), 94–101 (2006)

Design Requirements to Integrate Eye Trackers in Simulation Environments: Aeronautical Use Case

Jean-Paul Imbert, Christophe Hurter, Vsevolod Peysakhovich, Colin Blättler, Frédéric Dehais and Cyril Camachon

Abstract Eye tracking (ET) provides various data like gaze position, pupil size, and eye movement events (blinks, fixations, saccades, etc.). These data can reveal users' cognitive/attentional state but also provide a worthwhile input to human-computer interfaces. Recording and processing such data is an issue especially when integrating ET systems within existing environments. The synchronization of events from the simulation environment and physiological measurement devices with ET data is also a concern. In this paper, we reflect upon a seamless integration process. We gather task fulfillment requirements and confront them with technical constraints. Based on this structured task analysis, we present architecture guidelines regarding efficient ET system integration. Finally, we provide a relevant use case of experimental environment where ET systems have been successfully integrated and discuss architecture solutions.

Keywords Eye tracking · Eye gaze analysis · Interaction · Human factors

---

J.-P. Imbert (✉) · C. Hurter
ENAC, Toulouse, France
e-mail: jean-paul.imbert@enac.fr

C. Hurter
e-mail: christophe.hurter@enac.fr

V. Peysakhovich · F. Dehais
ISAE, Toulouse, France
e-mail: vsevolod.peysakhovich@isae.fr

F. Dehais
e-mail: frederic.dehais@isae.fr

C. Blättler · C. Camachon
CReA, Salon-de-Provence, France
e-mail: colin.blattler@defense.gouv.fr

C. Camachon
e-mail: cyril.camachon@defense.gouv.fr

R. Neves-Silva et al. (eds.), *Intelligent Decision Technologies*,
Smart Innovation, Systems and Technologies 39,
DOI 10.1007/978-3-319-19857-6_21

1 Introduction

The analysis of user visual scanpaths gives insights about the way a Human-Machine Interface (HMI) is used. Eye tracking (ET) allows for the collection of information to infer user activity and his/her cognitive workload [10]. Nowadays, eye trackers are also used as an input device, providing system a new way to point to or interact with an [5]. Though the advantages of ET systems are known and unanimously accepted, ET integration within existing systems remains a challenging task. Firstly, ET devices are mostly off-the-shelf products and need to be integrated in existing systems by the customers themselves. Such integration can be a problem especially when existing environment (i.e., flight or drive simulators) does not allow communication with third party software. Secondly, eye trackers produce large amount of data which need to be stored and then processed. When an ET is used as a system input the data must be processed in real- or near real-time, thus adding complications. To the best of our knowledge, no literature has previously tried to reflect upon such ET integration. In this paper we gathered task fulfillment needs and countered them with technical constraints. From this structured task analysis, we extracted design requirements. The task analysis was completed based on user interviews and our own experience with ET systems in aeronautics. The presented taxonomy and design guidelines would help practitioners to better understand the challenges and the technical solution to the integration of ET with simulation systems. Our contributions are a taxonomy of tasks, technical challenges, and design architecture requirements for ET integration. The remainder of the paper is structured as follows: first we present a review of existing work using ET systems; then we detail our taxonomy with tasks and technical challenges; and lastly we present a use case where we fulfilled our identified design requirements. For each example, we detail and explain the technical challenges. Finally we summarize our architecture recommendations and conclude with future challenges for ET systems integration.

2 Experimental Process and Eye Tracking

In a multi-factorial approach, many data can be collected from different sources: various psychophysiological sensors (eye tracker, electroencephalography, electrocardiography, functional near-infrared spectroscopy, etc.) and the experimental environment (HMI events such as mouse/keyboard input, simulator events, interaction with other participants). The data collected from these sources have to be synchronized for further analysis, for example to verify if an event in the experimentation scenario is associated with a fixation over a moving object of the interface. Data merging from different sources can be complex and time consuming depending on the architecture of the experimentation. For example, centralized time synchronization is mandatory while receiving data from multiple computers.

Another issue, when conducting human factors research, is related to physical integration of ET device within the experimental set up. Integration of such a system is highly dependent of the experimental context/constraint. Indeed, in applied cognitive research, the Cognitive System Engineering (CSE) framework [20] proposes four stages which achieve a different balance between ecological validity and experimental control: Stage (1) cognitive processes testing (initial laboratory experiments); (2) functions testing (laboratory methods within a basic context); (3) functions testing within complex simulations; (4) and behavioral observation within an experimental operational setting. These stages describe different architectures in terms of complexity. The first stage is appropriate when new concepts are studied. Usually a new application must be developed in order to test these concepts. Therefore, the research engineer in charge will be able to integrate the sensors in the developed ad-hoc architecture. An example of the second station would be serious games or microworlds [12]. When initial concepts are validated, a microworld environment can help test more complex functions in a simplified setup compared to complex simulation. As with the first stage, a certain level of control on the development process may allow the definition of an architecture which will integrate physiological measurement devices. As for the last two stages, the architecture of the simulator is not always designed for human factors needs, thus seamless integration of eye trackers in the architecture is challenging. Usually in these cases, eye trackers will be used independently as standalone positions with their recording software and data analysis. Correlation with events in the simulation will require fastidious post-processing work. Despite these limitations, we explain how it is possible to facilitate the integration of eye trackers in complex systems by the choice of the software architecture.

2.1 Eye Tracker Systems Overview

Various systems are used to track eye movements [7, 11]. The setup can be head-mounted, table-based, or remote. These devices use video-cameras and processing software to compute the gaze position from the pupil/corneal reflection of an infrared emissive source. To increase data accuracy with table devices, it is possible to limit head movement with a fixed chin on the table. A detailed description of the experimentation setup with the apparatus, the screen(s), and the subject is mandatory. A calibration process is also mandatory to insure system accuracy. The calibration process usually consists of displaying several points in different locations of the viewing scene; the ET software will compute a transformation that processes pupil position and head location [9]. Table-based eye trackers are usually binocular and thus can calculate eye divergence and output raw coordinates of the Gaze Intersection Point (GIP) in x-y pixels applied to a screen in real-time. This feature allows integration of gaze position as an input for the HMI. Areas of Interest (AOIs) are then defined to interact with the user. When the gaze meets an AOI an event is generated and a specific piece of information will be sent. When an AOI is an element of the interface with some degree of freedom (a scrollbar, for instance), one is talking about a dynamic AOI (dAOI). Tracking of a dAOI is more challenging compared to a static one. Recently, Jambon [13] proposed a software architecture which allows the detection of a fixation in a dAOI. In this paper we propose a further generalization of this architecture. Another study proposed a tool for dynamic detection of AOI on a video or an animation [17].

There are two kinds of ET data collection methods. The first and the most common is to use the original software (for data recording and analysis) that is often provided by the device manufacturer. The second is to develop a specific software module (using a System Developer Kit (SDK), usually provided with the eye tracker) for data collection. Various parameters will impact the precision of raw data issued from the ET system. Among them, the video frame rate and the camera resolution are critical for the ET software. Existing systems use a video frame rate from 30 to 2000 Hz. For high precision ET, high frequency rate will improve data filtering but will also increase the data size and processing time which is critical for online processing.

2.2 Eye Tracker Data Analysis and Visualization

ET data collected during an experiment can be analyzed by statistical methods and visualization techniques to reveal characteristics of eye movements (fixations, hot spots, saccades, and scanpaths). A recent survey presents an overview of visualization techniques for ET data and describes their functionality [2]. Eye tracker data can either be processed offline for analysis purposes or online in order to adapt the HMI dynamically or to use gaze as a pointing device. Fixation, saccade, and smooth pursuit events [21] can be computed from raw data coordinates. To correlate these pieces of information with the HMI, some interface-related data have to be collected (i.e. object coordinates within the interface, HMI events like mouse hover, etc.). This information can be used to infer the user behavior:

- fixation (smooth pursuit) indicates visual encoding during overt orienting [8, 21];
- saccade is the process of visual research when the focus of attention is shifted;
- number of fixations on a specific object is often an indicator of the importance attached to a specific object [19];
- mean fixation duration or total dwell time can be correlated to the visual demand induces by the design of an object [14] or the associated task engagement.

Saccades are rapid eye movements that serve to change the point of fixation, and during which, as it is often considered, no information is encoded. Fixations occurs when the user fixate an object (usually during a 150 ms threshold) and encode relevant information. Sometimes shorter fixations are taken into account. Unlike long fixations that are considered to be a part of top-down visual processing, short ones are regarded as a part of bottom-up process. It is estimated that 90 % of viewing time is dedicated to fixations [7]. Other complex ocular events like glissades or retro-saccades could be considered. There exist numerous algorithms of eye movement event detection [11, 16]. Still, there is no general standard for these algorithms. The integration of such algorithms will be discussed in Sect. 4. The blink duration and frequency can be used to assess cognitive workload [3], both of which can be collected with an eyetracker. Variation of the pupil diameter can also be used as an indication of the cognitive workload [15, 18], defined by Beatty [1] as task-evoked pupillary response (TEPR). However, light sources (environment, electronic displays, etc.) must be strictly controlled since the pupil light reflex is more pronounced than the impact of the cognition on pupil size. Moreover, even the luminance of the fixation area (even when the luminance of the computer screen does not change) has an impact on the pupil size. Scanpaths can also provide insight on HMI usage. In general, collected and cleaned data can be analyzed to infer causal links, statistics, and user behavior [10].

3 Design Rationale

In this section, we present our taxonomy of tasks and a structured design space where every need regarding eye tracker data record and processing is gathered. This design rationale is the result of two brainstorming sessions with four human factors experts with expertise in ET, a research engineer, and one researcher in HMI and Information visualization. This section will not provide architecture solutions but rather questions that will help users to correctly define ET integration requirements:

What to Record?

ET systems can record many ocular features, all of which will be merged for future analysis. These features usually include gaze position, pupil size, head position and movements, eye divergence (in the case of binocular system). Some higher level complex features like scanpaths or activation of AOIs can be computed and recorded. A more detailed description of recording parameters can be found in [7]. In the case of a head mounted device, a reference image is often used and recorded to compute head movements and to correlate gaze and head position on the scene. In other cases, there is no need to know where the participant is looking at, therefore no correlation is necessary (e.g., if ET is used to estimate the cognitive load and/or attentional state by eye ballistic or pupil diameter). Additionally, contextual information must be (other physiological data, HMI events). Researchers usually record every piece of information available in the case of later analysis unanticipated at the moment of protocol development. Still, increasing the number of recorded features decreases the processing speed and requires extra storage space. If the ET system is used as a part of HMI, then the processing time becomes critical for (near) real-time interactions. In any case, recording unused parameters is undesirable. Thus, the question of relevant ET features choice is important.

What are the Environment Constraints?

Experiments can be conducted in various indoor (laboratory or simulator) or outdoor (airplane cockpit, car) environments. Environmental constraints directly define the possible ET features regarding the research objective. For example, ET usage can be an issue in high luminosity environments. Constricted pupils of 2–3 mm in diameter are difficult to track. If the ET is head-mounted and uses a field view camera, excessive light can saturate the image such that head movement calculation is be impossible. While recording during real flight [7], pilots usually wear sunglasses so that the tracking becomes quite delicate. Other problems such as vibration or sun power can occur in these environments. With remote ET devices, body and head movements are also a critical issue for accurate data recording; eyes can be lost with excessive head movement. During the calibration, the experimenter should verify that the participant's head has some degree of freedom but a slight head/body movement would not cause a loss of ET. If the data has to be of extreme precision (e.g., for use in psycho-physiological analysis), a chin rest is recommended. When users must be able to freely move their head (e.g., multi-display environment), head-mounted devices are preferable. Some environments have safety concerns; in such cases, the ET integration can be delicate and a specific care must be taken to monitor gaze without compromising safety. This concern is especially true with airline pilots. Little experimentation was done, and currently most of them appeared in training simulators. Even in a training simulator, some constrains remain. Simulation sessions are costly and participants, particularly professionals, have limited availability. The ET setup time must be as short as possible. The chosen ET system (head-/table-mounted, or remote using multiple cameras) is also constrained by the environment. Head-mounted devices lose tracking less often because of the proximity of the camera to the pupil. Still, head movements could have an impact on data recording, if the position of the device changes considerably due to participants knitting the brows, for example. In the latter case, the device would erroneously map the pupil position on the field camera. The head-mounted ET is also more intrusive and could interfere with participant activity. Calibration can require some time compared to other types, and light context and the physical environment can hinder the process. In the case of a head-mounted ET, a special care has to be taken for correct installation to avoid device shift and re-calibration. In the case of a head-mounted system, head motion tracking can be improved with physical markers (e.g. AR marker). Some calibration techniques can take into account multilayer depth (e.g. modern glass cockpit usage). Calibration can be problematic when the subject is wearing eyeglasses or contact lenses, or if there is a reflection in the eye, or if parts of the pupil are hidden by eyelashes or eyelids. Globally it takes more time to calibrate and setup head mounted devices than fixed one.

What are the Recording Requirements?

Sampling frequency should be decided with caution regarding the usage of the eye tracker. In the case of accurate time frame, the amount of recorded data can be huge. In this sense, special care must be taken to handle this information in term of size and extra computation requirement can be especially an issue when several eye trackers must be synchronized. Sampling frequency can be high for physiological experimentation, but a low sampling rate can be sufficient when doing general interface usage monitoring. In the case of complex experimentation platforms with multiple computers, time synchronization is crucial when correlation between various sources of data has to be made. For instance, if electroencephalography is used with an eye tracker during the experimentation, time synchronization has to be accurate to the millisecond. Finally, one must decide on the recorded data format. Usually human factors specialists will prefer comma-separated value (csv) files in order to process data in spreadsheet, statistical, or mathematical software. In case of high sampling rate like 500 Hz, file size can be an issue for text files. In order to assess the correctness of recorded information, one can monitor them with dedicated supervision of the HMI. For instance, camera accuracy, calibration quality, reference image, gaze position, reference image, and dynamic AOI all can be dynamically assessed. In case of incorrect data recording, it is possible to stop the experiment and to fix the problem. Supervision can also be used as a debugging tool, to retrieve a broad idea of the current experimentation validation. As a matter of fact, all recorded data can be monitored in real-time, but it can represent a technical challenge to visualize it.

What are the Processing Requirements?

Finally, processing and analyzing collected data can be time consuming, with important memory consumption and especially with big datasets. In some cases, real-time processing is necessary if ET is used as an input. Off-line processing can be used to manually adjust AOIs and to assess experiment results. Merging from different sources and data cleaning is always time consuming and is error prone. Thanks to the supervision, it is possible to qualitatively assess the recorded data. Data processing occurs with the mix of data cleaning and fusion, then finally, data can be segmented, processed and summarized.

4 Aeronautical Use Case

Contributors of this research paper come from the aeronautical domain and are human factors specialists and researchers in human computer interaction. We present a Stage 3 use case and the technical challenges addressed.

The main objective of this experiment was to verify in a complex setup the efficiency of a monitoring agent dedicated to assisting the controller. Three designs were selected for their saliency properties to be integrated in the radar image of a complex air traffic control simulator. This simulator is modular and constructed over the Ivy middleware [4]. Relevant information was detected by the monitoring agent and notified dynamically with the appropriate saliency on the radar image. A table-based eye tracker was chosen to measure the perception of the notifications.

Two technical challenges had to be addressed in this study: the integration of the eye tracker in this complex simulation platform and the complexity of the radar image; and the difficulty associated with the integration of ET data in complex software for the detection of dAOIs.

A specific module for fixation detection had been created to address the first challenge. Depending on the sampling frequency, it is possible to have a fixation module with a different algorithm without modifying the other modules. In this type of architecture, each function is a dedicated software agent in order to allow better flexibility in future experimentation. With our architecture, it is possible to add/remove a module depending on the study objectives, record relevant data for further analysis and even integrate high-level data like the workload index assessed dynamically by a mental state classifier. The architecture simplifies the post-treatment and the analysis phases for human factors specialists and allows HMI adaptation with ET data since raw data and high-level data can be collected dynamically on the software bus. Other optional modules could be integrated in this architecture like a dAOI module to correlate objects coordinates and fixation events or a TEPR module which analyze pupil data from the ET gateway and can produce outputs for a mental state classifier.

For the second challenge, two technical choices were possible. The first one was to improve he radar image code in order to output coordinates for meaningful objects and associate them in a dAOI module. The second one was to integrate fixations in the radar image to correlate them to objects inside the process. We chose the latter as our radar image was designed like Mozilla Firefox to integrate plugins to allow agile development of new functionalities without modifying the core application (dashed box in Fig. 1). Various plugins have already been created to support the use of a dynamic eye tracker like a gaze piloted cursor, an alarm validation plugin based on long fixations on notified objects, and detection of fixations on aircraft. Relevant information like graphical/data objects fixations were sent on the software bus and logged by a dedicated software agent. If the aim of the experiment is not to adapt the HMI but only record and correlate data between the eye tracker and HMI events or objects, an architecture without plugins and additional modules is more useful for reusability.

For this specific experiment, we have used the fixation module and a plugin included in the radar image for the detection of dAOI fixations and correlation with aircraft. Logging was done by a specific module and information send on Ivy (Fig. 1).

What to Record? Raw data and fixations on dAOI.

What are the Environmental Constraints? Controllers work indoors on a fixed screen; a table-based eye tracker is the better and simpler choice.

What are the Recording Requirements? In this simulator, we use a multi-computer architecture, as time synchronization NTP (Network Time Protocol) is mandatory. For data collection we used Ivy and a logging module.

What are the Processing Requirements? Detect and record fixations on aircraft (specific dAOI) to simplify the correlation process with events (HMI or monitoring).

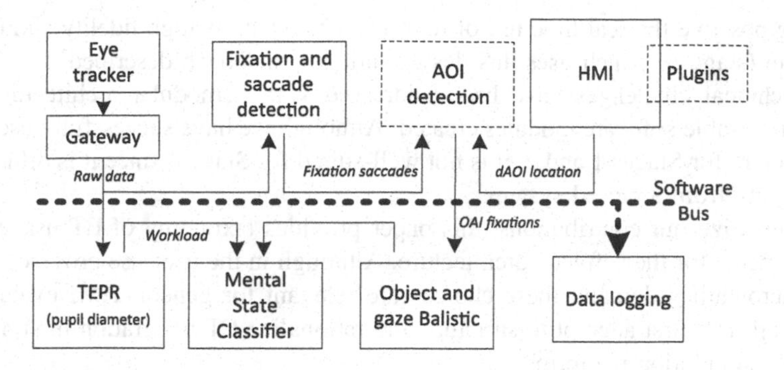

Fig. 1 A flexible and generic architecture

5 Conclusion

As a result, we proposed the following architecture recommendations:

- Use a software bus (middleware) to allow a simple and modular architecture and NTP protocol to address synchronization issues.
- Even in the case of laboratory evaluation (Stages 1 and 2 of the CSE framework) where research engineers have full control on the source code, using the more flexible architecture (use case 2) is the most relevant since it provides reusable components for future experimentation.
- It is mandatory to separate identified functions in a single software component and to connect them via a software bus. For instance, a fixation detection algorithm from raw coordinates or the measurement of TEPR (task-evoked pupillary response) amplitude can be considered as an optional and reusable module for future experimentation.
- The gateway between the eye tracker and the software architecture only outputs raw data from the eye tracker processed by another software component; no processing should be made. This allows for replacement of an eye tracker model with another by simply changing the eye tracker gateway.
- Raw coordinates coming from the eye tracker gateway should be stored in the log file even if high level eye-tracking data are already stored, since they can be replayed later for analysis purposes or processed by new filtering algorithms.
- Online calculation and correlations with HMI events must be performed when it is possible in order to store high level data (i.e. fixations on objects) and simplify the post-treatment phase for human factors specialists.
- Creating plugins for the HMI allows easier integration of high level ET data for HMI adaptability and permits a direct correlation between fixations and graphical objects.

This paper proposes a design rationale and architecture recommendations for their integration in the different stages of the CSE framework. These recommendations aim to improve the recording and analysis process for human factors specialist and

making possible the real-time use of high level ET data. A high fidelity simulation platform (stage 3) which uses this design rationale has been described.
The technical challenges have been addressed with a modular architecture and several reusable software modules created. Although we have successfully used this architecture for Stages 1 and 2, it is not well-suited for Stage 4 since it is difficult to gather data from a closed system.
To summarize our contributions, this paper provides taxonomy of ET usages and requirements for the software architecture. Although in the use case provided come from aeronautics domain, these choices are relevant for general HMI evaluation. This work is a first attempt to structure and rationalize ET integration in simple or complex simulation platforms.

References

1. Beatty, J.: Task-evoked pupillary responses, processing load, and the structure of processing resources. Psychol. Bull. **91**, 276–292 (1982)
2. Blascheck, T., Kurzhals, K., Raschke, M., Burch, M., Weiskopf, D., Ertl, T.: State-of-the-art of visualization for eye tracking data. In: Proceedings of EuroVis (2014)
3. Bruneau, D., Sasse, M.A. McCarthy, J.D.: The eyes never lie: the use of eye tracking data in HCI research. In: Proceedings of the CHI, vol. 2, p. 25 (2002)
4. Buisson et al.: Ivy: un bus logiciel au service du développement de prototypes de systèmes interactifs. In: IHM'02 (2002)
5. Bulling, A., Dachselt, R., Duchowski, A., Jacob, R., Stellmach, S., Sundstedt, V.: Gaze interaction in the post-WIMP world. In: CHI'12 Extended Abstracts on Human Factors in Computing Systems, pp. 1221–1224. ACM (2012)
6. Dehais, F., Causse, M., Pastor, J.: Embedded eye tracker in a real aircraft: new perspectives on pilot/aircraft interaction monitoring. In Proceedings from the 3rd International Conference on Research in Air Transportation. Federal Aviation Administration. Fairfax, USA (2008)
7. Duchowski, A.: Eye Tracking Methodology: Theory and Practice, vol. 373. Springer Science & Business Media (2007)
8. Goldberg, J.H., Kotval, X.P.: Eye movement-based evaluation of the computer interface. In: Kumar, S. (ed.) Advances in Occupational Ergonomics and Safety, pp. 529–532. ISO press, Amsterdam (1998)
9. Goldberg, H.J., Wichansky, A.M.: Eye tracking in usability evaluation: a practitioner's guide. In: Hyöna, J., Radach, R., Deubel, H., (Eds.), The Mind's Eye: Cognitive and Applied Aspects of Eye Movement Research, pp. 573–605. Amsterdam, Elsevier (2003)
10. Holmqvist, K., Andrà, C., Lindström, P., Arzarello, F., Ferrara, F., Robutti, O., Sabena, C.: A method for quantifying focused versus overview behavior in AOI sequences. Behav. Res. Meth. **43**(4), 987–998 (2011)
11. Holmqvist, K., Nyström, M., Andersson, R., Dewhurst, R., Jarodzka, H., Van de Weijer, J.: Eye Tracking: A Comprehensive Guide to Methods and Measures. Oxford University Press (2011)
12. Imbert J-P, Hodgetts H.M., Parise R., Vachon F., Tremblay S.: The LABY microworld: À Platform for Research, Training and System Engineering, HFES'2014
13. Jambon, F., Vanda L.: Analyse oculométrique « on-line » avec zones d'intérêt dynamiques: application aux environnements d'apprentissage sur simulateur. In: Proceedings of the 2012 Conference on Ergonomie et Interaction homme-machine (2012)

14. Just, M.A., Carpenter, P.A.: Eye fixations and cognitive processes. Cogn. Psychol. **8**, 441–480 (1976)
15. Matthews, G., Middleton, W., Gilmartin, B., Bullimore, M.A.: Pupillary diameter and cognitive load. J Psychophysiol. **5**, 265–271 (1991)
16. Nyström, M., Holmqvist, K.: An adaptive algorithm for fixation, saccade, and glissade detection in eyetracking data. Behav. Res. Meth. **42**(1), 188–204 (2010)
17. Papenmeier, F., Huff, M.: DynAOI: a tool for matching eye-movement data with dynamic areas of interest in animations and movies. Behav. Res. Meth. **42**(1), 179–187 (2010)
18. Pomplun, M., Sunkara, S.: Pupil dilation as an indicator of cognitive workload in human-computer interaction. In: Human-Centred Computing: Cognitive, Social, and Ergonomic Aspects. HCII 2003, pp. 542–546. Crete, Greece (2003)
19. Poole, A., Ball, L.J. Phillips, P.: In search of salience: a response time and eye movement analysis of bookmark recognition. In: Proceedings of HCI 2004 People and Computer XVIII, pp. 363–378. Springer, London (2004)
20. Rasmussen, J., Pejtersen, A.M., Goodstein, L.P.: Cognitive Systems Engineering. Wiley (1994)
21. Reimer, B., Sodhi, M.: Detecting eye movements in dynamic environments. Behav. Res. Meth. **38**, 667–682 (2006)

Human Friendly Associative Classifiers for Early Childhood Caries

Vladimir Ivančević, Marko Knežević, Ivan Tušek, Jasmina Tušek and Ivan Luković

Abstract Early childhood caries (ECC) is a widespread disease that may lead to serious complications and impact the whole society. For these reasons, we look for a predictive model that could be easily applied whenever and wherever necessary, especially in poor environments. As a result, we create human friendly classifiers for ECC that could be utilized in prevention programs. These classifiers are rule-based, with a few rules, easy to use even without computers, and without a loss in predictive performance. For this purpose, we mined association rules and clustered them by their contents. Next, we employed a genetic algorithm to assemble a classifier using dissimilar association rules. The proposed approach was tested on a data set about ECC in the South Bačka area (Vojvodina, Serbia). We compared the performance of the resulting classifiers to that of the logistic regression model built around the previously identified risk factors.

Keywords Early childhood caries · Associative classifiers · Data mining

V. Ivančević (✉) · M. Knežević · I. Luković
Faculty of Technical Sciences, University of Novi Sad, Novi Sad, Serbia
e-mail: dragoman@uns.ac.rs

M. Knežević
e-mail: marko.knezevic@uns.ac.rs

I. Luković
e-mail: ivan@uns.ac.rs

I. Tušek
Medical Faculty, University of Novi Sad, Novi Sad, Serbia
e-mail: ivantusek@gmail.com

J. Tušek
Palmadent, Petrovaradin, Serbia

R. Neves-Silva et al. (eds.), *Intelligent Decision Technologies*,
Smart Innovation, Systems and Technologies 39,
DOI 10.1007/978-3-319-19857-6_22

1 Introduction

Early childhood caries (ECC) has low to moderate prevalence throughout the developed countries including Serbia. However, the associated risks are not negligible, because ECC may lead to various other health problems, especially if left untreated. Moreover, ECC incidence is much higher in many other parts of the world, which further raises the need for preventive actions.

Researchers have identified numerous risk factors for ECC. Nonetheless, even with this knowledge, it may be difficult to understand the nature of ECC in a particular area. Certain strong risk factors may be absent, while several others may have a combined effect on the population. For this reason, each attempt at prevention should be first concerned with epidemiological data for the targeted area.

On the other hand, information technology may benefit both epidemiologists and general population. Epidemiologists may use computers first to analyse data, and later to track the application of the proposed measures and monitor the effects. With the proliferation of personal computers and smartphones, general population may easily access relevant epidemiological information. Unfortunately, many ECC-stricken areas tend to have suboptimal IT infrastructure and insufficient access to health-related information. As a result, the direct use of IT in such circumstances may be infeasible and mostly basic preventive methods may be employed.

Given all these issues, we decided to create classifiers that could be used to predict individual occurrence of ECC in a particular area without the use of IT. We present an approach to classifier creation that takes an ECC data sample for the analysed area as input. The output is a classifier with the following characteristics: (i) it is white-box and rule-based, so it could be applied without computers; (ii) it has only a few rules, so it would not be too demanding for manual use by humans; and (iii) its performance is comparable to that of other predictive models. In this manner, we obtain human friendly classifiers, i.e., readable and relatively simple classifiers with each rule targeting different risk factors.

The classifier creation process is divided into several phases. First, we perform association rule mining, filtering, and pruning on the data set with the goal of discovering interesting association rules that could be embedded into the classifier. Second, we perform divisive hierarchical clustering of the discovered association rules. Association rule with similar structure, i.e., a similar set of included potential risk factors, are put into the same or neighbouring clusters. Third, we use a genetic algorithm [1] to combine association rules from different clusters into a single associative classifier.

The clustering phase is important because association rule mining may produce a huge amount of large association rules, with subgroups in which association rules differ very little. In order to facilitate the application of the resulting classifier by human users, we look for combinations of association rules that clearly target different risk factors and have an empty intersection.

As input, we use a data set concerning ECC presence and relevant habits of children, as well as their parents, in the South Bačka area of Vojvodina, Serbia. In order to demonstrate the value of the obtained classifiers, we evaluate their predictive performance with respect to a baseline model. The baseline is a logistic regression model that is built around five key ECC risk factors for the South Bačka area [2].

Besides Introduction, the paper features five sections. In Sect. 2, we provide basic information about ECC, ECC research that relied on data mining, and associative classifiers. In Sect. 3, we introduce the ECC data set. In Sect. 4, we explain the employed methodology, while, in Sect. 5, we present the resulting classifiers and compare their performance to that of the logistic regression model. In Sect. 6, we give concluding remarks, together with some plans for future work.

2 Related Work

In this section, we cover relevant literature on ECC in the South Bačka area, how data mining was used in dental research, and associative classifiers in general.

2.1 ECC in South Bačka

According to the American Academy of Pediatric Dentists (AAPD), ECC is "the presence of one or more decayed (noncavitated or cavitated lesions), missing (due to caries), or filled tooth surfaces in any primary tooth in a child under the age of six" [3]. Criteria for ECC have varied among existing studies with respect to the location of dental decay and to the number of teeth affected, e.g. any labiolingual lesion in a maxillary incisors or DMFT score (the total number of decayed, missing, and filled teeth) of 5 or greater [4].

In Serbia, problems in oral health are officially recognized as particularly important in the national program of preventive dental protection [5]. In the same program, the prevalence of ECC in three-year-old children within the Province of Vojvodina was officially reported at 10.6 %.

Some other studies indicated substantially different values for the South Bačka area, which is located in Vojvodina. Vulović and Carević [6] pointed out that the ECC prevalence among three-year-old children was 22.07 %. In a more recent study by Tušek et al. [7], the ECC prevalence was reported at 30.5 %, with considerable variations between different ethnic groups. This number generally agrees with the ECC prevalence in some neighbouring countries, namely, Bulgaria [8] and Croatia [9].

2.2 Models in Dental Research

Powell conducted a large survey of caries models [10]. The results indicate that the majority of research relied on logistic regression analysis and, to a lesser extent, on linear discriminant analysis. There was only one use of classification trees [11].

However, data mining managed to find its way into some later studies. Tree models appear to be especially popular for caries models [12–14]. Moreover, artificial neural networks were used in various dental studies [15–17]. Nonetheless, association rule mining, which may be utilized both for descriptive and predictive purposes, was hardly, if at all, used in caries-related research.

2.3 Associative Classifiers

There are various approaches to associative classification [18, 19]. In our approach, we rely on some of the general steps in the creation of an associative classifier. On the other hand, we focus on selecting a relatively small number of association rules in order to make the classifier more human friendly.

We combine mined association rules into a single associative classifier that has two lists of association rules. Because the target variable about ECC is binary, we use one rule group for the prediction of ECC presence and the other for ECC absence. Although many proposed approaches include rules concerning both outcomes, in this paper, we also evaluate the cases when there are only individual rule groups within the classifier – either for ECC presence or for ECC absence.

3 ECC Data Set

The data set on ECC in South Bačka was collected in a study on ECC prevalence and risk factors by Tušek [2]. That study covered 341 children of various ethnicities who were examined at 15 different locations. Their age was between 18 and 64 months. The presence of ECC was also recorded using the DMFT (DMFS) score, which indicates the number of decayed, missing, and filled teeth (surfaces). Additional data about various potentially relevant habits were collected during interviews with the parents. The collected sample represents 10 % of preschool children in South Bačka.

In the present study, we focused on 36 categorical variables from the ECC data set. There are 35 variables describing various habits and socioeconomic factors related to the children and their families:

- **19 child-related variables** – ethnicity, age, gender, fluency in Serbian, birth order, birth weight, breastfeeding, breastfeeding frequency, breastfeeding during night, bottle feeding, use of infant formulas, additional food sweetening, use of fluoride supplements, use of fluoride toothpaste, oral hygiene, tooth brushing, diarrhea during infancy, use of medical syrups, and first dentist visit; and
- **16 family-related variables** – city, quality of housing, housing conditions, household monthly income, mother's age, marital status, mother's ethnicity, mother's fluency in Serbian, number of children, mother's education level,

mother's employment status, use of sweets during pregnancy, use of fluoride supplements during pregnancy, mother's oral health during pregnancy, mother's health awareness, and father's health awareness.

The one remaining variable is about ECC presence. It has two possible values: ECC present or ECC absent.

4 Methodology

In this section, we outline the methodology, while providing more details about the different phases in classifier creation and the evaluation scenarios.

4.1 Methodology Outline

There are three phases of classifier creation: (i) association rule discovery through association rule mining; (ii) association rule clustering; and (iii) combining association rules from different clusters into a single classifier. The result of this process is an associative classifier for manual use by humans.

4.2 Association Rule Introduction

Association rules are patterns that denote co-occurrence of some items/values within the analysed data set. For instance,

child gender = male, father health awareness = low => ECC = yes

is an association rule that indicates the simultaneous occurrence of low health awareness of the father and ECC in male children. The importance of association rules is evaluated using various measure of interestingness, e.g., support, confidence, and lift.

Rule support is the proportion of records in the data set that satisfy the rule, i.e., for which both sides of the rule are true. Rule confidence is the ratio of the number of records that satisfy both rule sides and the number of records that satisfy the left side of the rule. Rule lift is the ratio between the support for both rule sides, and the product of the support for the left rule side and the support for the right rule side.

4.3 Association Rule Mining

We use the well-known *Apriori* algorithm [20, 21] to mine for association rules within the ECC data set. The used implementation is available in the *arules* package [22] of the *R* environment for statistical computing [23]. There are two interesting groups of association rules: (i) rules denoting ECC presence (R1 rules) and (ii) rules denoting ECC absence (R0 rules). For both rule groups, we did filtering by setting the minimum confidence to 0.6. The minimum support for the R1 and R0 groups was set to 0.05 and 0.15, respectively.

The initial minimum support for the R1 rules was higher. However, since there are fewer cases of ECC, we initially obtained only few rules and had to decrease the support threshold. The length of the R1 and R0 rules was set to be between two and seven, i.e., there is exactly one condition in the right rule side and at least one condition (but not more than six) in the left rule side.

Due to the high number of generated rules, we performed rule filtering and pruning to reduce the number of available rules and facilitate identification of relevant rules. We removed all rules whose lift or odds ratio was below one. Moreover, we removed rules for which there were rules with less specific antecedents (more general conditions) but higher lift. A similar pruning method was briefly covered in [24]. In this manner, we did not attempt to remove trivial associations, as defined by Webb and Zhang [25], but to reduce potential overfitting by primarily focusing on more general rules. This resulted in pruned groups PR1 and PR0, obtained by pruning R1 and R0 respectively.

4.4 Association Rule Clustering

We clustered the pruned rule groups separately using the *diana* algorithm [26] for divisive hierarchical clustering. We relied on the implementation from the cluster package [27] of the *R* environment. We clustered rules by the variables from the data set excluding the target ECC variable.

One of the principal reasons for doing hierarchical clustering and choosing the *diana* algorithm is the possibility to first extract the hierarchy and later specify the exact number of clusters. This reduced the duration of our experiments with different cluster numbers and various classifier settings.

4.5 Associative Classifier

The associative classifier contains two pruned rule groups, one about ECC presence (PPR1) and one about ECC absence (PPR0). Each rule group contains N different association rules. The analyst is expected to set the default class (ECC presence or ECC absence). For a test case, the trained classifier operates in the following manner:

- PPR1 is searched for the first rule whose left side is satisfied by the test case;
- PPR0 is searched for the first rule whose left side is satisfied by the test case;

- if there is only one matching rule, the class from its right side is the output;
- if there are two matching rules, the class from the right side of the rule with higher confidence is the output (if confidence values are equal the default class is the output); and
- if there are no matching rules, the default class is the output.

In our case, the default class is ECC absence since it is more prevalent. The only rule interplay taken into account is the conflict arising when there are two matching rules, one for ECC presence and the other for ECC absence. Using multiple rules from the same group (PPR1 or PPR0) is not supported because a human friendly classifier should be as simple as possible to be used by humans. Moreover, for the used data set, lower N values are favoured because they are less demanding for human users.

4.6 Genetic Algorithm for Classifier Creation

We used the genetic algorithm to find associative classifiers of good predictive performance. At the beginning of the evolution process, the population contains randomly initialized individuals (solutions). Through selection, crossover (recombination), and mutation, characteristics of "good" individuals (chosen during selection) are exchanged (during crossover), randomly modified (during mutation), and passed to the next generation. In this manner, the population evolves over generations until some target fitness is achieved or population becomes homogenous.

An individual features enough information to define a single classifier. It contains a sequence of 2 * N integers, where each integer is the identifier of one association rule. The first N integers are different identifiers of the rules from the PPR1 group, while the second N integers are different identifiers of the rules from the PPR0 group. For instance, the individual with a sequence 3:2:1:2006:4892:2158 was used to create the classifier from Table 1 – the six integers correspond to the six classifier rules.

The fitness ("goodness") of an individual is measured in the following manner:

Fitness(i) = TSS(i) – RepetitionPenalty(i) – LengthPenalty(i) (1)

where TSS is True Skill Statistic [28] for the classifier represented by *i*, RepetitionPenalty is the penalty for the case of repeated risk factors in a single rule group represented within *i*, and LengthPenalty is the penalty for the case of long rules represented within *i*. TSS may be used to measure performance of different classifiers. We chose TSS over the oft-used kappa measure [29] since it is not sensitive to class prevalence while keeping the good properties of kappa.

During selection, pairs of individuals are formed. Individuals of higher fitness have a better chance of being selected into a pair. During crossover, individuals within each pair are combined to produce two individuals for the next generation. A prespecified number of integers are swapped between the two individuals, but with the restriction that only the values from the matching groups and positions may be exchanged.

During mutation, according to prespecified chance levels, an individual may remain the same or be modified in one of the two ways: (i) for each integer group, two neighbouring integers may switch places; and (ii) for each integer group, one integer may be substituted by some other valid integer that identifies an association rule.

As classifier performance depends on the contained rules and their order, these two mutation types are expected to introduce additional variation and help find a better solution. In order to support human friendliness of the classifiers, we implemented the restriction that, within each integer group, all rules need to be from different clusters.

4.7 Evaluation Scenarios

We prepared three evaluation scenarios: creation of a classifier that contains both the PPR1 and PPR0 groups; creation of a classifier that contains only the PPR1 group; and creation of a classifier that contains only the PPR0 group.

Each scenario was run 10 times. For each run, we recorded the TSS and Fitness values of the solution (the best discovered classifier). The settings for the genetic algorithm were the same for all three scenarios: population size was 300, number of generations 300, mutation chance 40 %, and target value for Fitness 0.95. The number of clusters was set to five for all scenarios. The RepetitionPenalty was set to 0.05 for all cases when at least one variable appears multiple times within a single rule group (either PPR1 or PPR0). For each scenario, the average performance over 10 runs was compared to the performance of the baseline model. The baseline is the logistic regression model (LOGR) that is built around five risk factors identified by Tušek [2]: city, child gender, birth order, child's body weight at birth, and child's use of medical syrups. When evaluation is done on the training set, the TSS value for LOGR is 0.325.

5 Results and Discussion

After pruning, we had 15 PR1 rules and 6306 PR0 rules. These rules were then used to build associative classifiers. The comparison between the three types of generated classifiers and the baseline model is given in terms of the TSS value in Fig. 1.

Once the number of rules within a group is three or higher (see Fig. 1), the generated classifiers start outperforming the baseline model, most notably the PPR0 and PPR1 + PPR0 classifiers. However, for the classifiers that include rules about ECC presence (PPR1 + PPR0 and PPR1), there was a decline in performance for N = 5. The decline may be attributed to the small number of PR1 rules, as each generated classifier was set to pick five rules from five different clusters. This most probably compromised the predictive qualities in favour of the classifier heterogeneity.

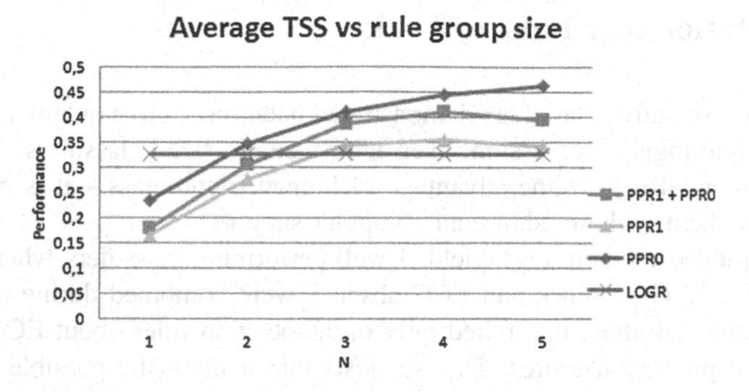

Fig. 1 The comparison of TSS values between the classifiers

When the RepetitionPenalty is taken into the account, the overall quality drops only for N values greater than three. This is most probably a consequence of having only five rule clusters to choose from. The predictive performance remains the same, although the human friendliness of those classifiers decreases. However, our policy for RepetitionPenalty was conservative as only one repetition of a factor within some rule group caused a 0.05 drop.

Overall, the most human friendly classifier in the scenarios could be obtained for N = 3. One such classifier, with the TSS and Fitness values of 0.41 is given in Table 1.

Table 1 An example of a human friendly classifier

Id	Rule	Support	Confidence	Lift
PPR1-1	child's age in years = 3 years, child's birth order > = 3rd => ECC = yes	0.06	0.66	2.15
PPR1-2	breastfeeding frequency > 8 times a day, use of fluoride supplements = no, use of medical syrups > 5 times a year => ECC = yes	0.05	0.72	2.36
PPR1-3	child's gender = male, father's health awareness = low => ECC = yes	0.07	0.64	2.09
PPR0-1	child's age in years = 3 years, mother's employment status = employed, use of infant formulas = no => ECC = no	0.17	0.83	1.19
PPR0-2	child Serbian language = understands and speaks, mother's ethnicity = Serbian, quality of housing = owns an apartment or a house, household conditions = comfortable, child's oral hygiene = after teeth eruption, mother's health awareness = medium => ECC = no	0.17	0.78	1.13
PPR0-3	child's gender = female, use of fluoride supplements = no, use of medical syrups = rarely => ECC = no	0.15	0.85	1.23

6 Conclusion and Future Work

The trained associative classifiers demonstrated better predictive quality as opposed to the baseline logistic regression, even with medium-sized classifiers. Moreover, these classifiers also have the advantage of human friendliness – it is possible to easily apply them without additional computer support.

The employed methodology yielded well-performing classifiers when association rules for ECC presence and ECC absence were combined during prediction. Moreover, the classifiers that relied only on association rules about ECC absence had the best prediction scores. This suggests that it might be possible to further simplify the classifier without sacrificing performance. A possible explanation is the availability of a large number of association rules concerning ECC absence.

However, validation on additional data would be necessary to check if there is a potential problem with overfitting. This might be a daunting task, because the collection of the ECC data for the analysed area was difficult. Moreover, data from other areas would probably feature location-specific patterns, so such data might not be very useful. In the future, we also plan to prepare experiments with human users who would test the human friendliness of the generated associative classifiers.

Acknowledgements The research presented in this paper was supported by the Ministry of Education, Science, and Technological Development of the Republic of Serbia under Grant III-44010

References

1. Holland, J.H.: Adaptation in Natural and Artificial Systems. MIT Press, Cambridge (1992)
2. Tušek, I.: The Influence of Social Environment and Ethnicity on Caries Prevalence in the Early Childhood (in Serbian). Ph.D. thesis, University of Belgrade (2009)
3. American academy of pediatric dentists policy on early childhood caries (ecc): Classifications, consequences, and preventive strategies. Ref. Manual **36**(6), 50–52 (2014)
4. Tinanoff, N., Berg, J., Slayton, R.: Use of fluoride. Early childhood oral health **1**, 92–109 (2009)
5. Program of preventive dental protection 2009–2015. Belgrade (2008)
6. Vulović, M., Carević, M.: The infectious nature of tooth caries (in Serbian). Zbornik referata i radova XII i XIII zdravstvenog vaspitanja u stomatologiji. Stom. Glas. S. **45**, 5–9 (1998)
7. Tušek, I., Carević, M., Tušek, J.: Prevalence of early childhood caries among members of different ethnic groups in the south bačka area (in Serbian). Vojnosanit. Pregl. **69**(12), 1046–1051 (2012)
8. Tinanoff, N., Kanellis, M., Vargas, C.: Current understanding of the epidemiology, mechanisms, and prevention of dental caries in preschool children. Pediatr. Dent. **24**, 543–551 (2002)
9. Pierce, K.M., Rozier, R.G., Vann, W.F.: Accuracy of pediatric primary care providers' screening and referral for early childhood caries. Pediatrics **109**, e82–e82 (2002)
10. Powell, L.: Caries prediction: a review of the literature. Commun. Dent. Oral Epidemiol. **26**, 361–371 (1998)

11. Stewart, P.W., Stamm, J.W.: Classification tree prediction models for dental caries from clinical, microbiological, and interview data. JDR **70**(9), 1239–1251 (1991)
12. Tamaki, Y. et al.: Construction of a dental caries prediction model by data mining. J. Oral Sci. **51** (2009)
13. Ito, A., Hayashi, M., Hamasaki, T., Ebisu, S.: Risk assessment of dental caries by using classification and regression trees. J. Dent. **39**, 457–463 (2013)
14. Li, H.F.: Data Mining and Pattern Discovery using Exploratory and Visualization Methods for Large Multidimensional Datasets. Ph.D. thesis, University of Kentucky (2013)
15. Goodey, R.D., Brickley, M.R., Hill, C.M., Shepherd, J.P.: A controlled trial of three referral methods for patients with third molars. Br. Dent. J. **189**, 556–560 (2000)
16. Amariti, M.L., Restori, M., De Ferrari, F., Paganelli, C., Faglia, R., Legnani, G.: A histological procedure to determine dental age. J. Forensic Odontostomatol **18**, 1–5 (2000)
17. Lux, C.J., Stellzig, A., Volz, D., Jager, W., Richardson, A., Komposch, G.: A neural network approach to the analysis and classification of human craniofacial growth. Growth Dev. Aging **62**(3), 95–106 (1998)
18. Liu, B., Hsu, W., Ma, Y.: Integrating classification and association rule mining. In: 4th International Conference on Knowledge Discovery and Data Mining (KDD), pp. 80–86 (1998)
19. Thabtah, F., Cowling, P., Peng, Y.: MCAR: Multi-class classification based on association rule. In: 3rd ACS/IEEE International Conference on Computer Systems and Applications (ICCSA), pp. 127–133. IEEE (2005)
20. Agrawal, R., Imieliński, T., Swami, A.: Mining association rules between sets of items in large databases. In: ACM SIGMOD Record vol. 2, pp. 207–216. ACM (1993)
21. Borgelt, C., Kruse, R.: Induction of association rules: Apriori implementation. Compstat, 395–400. Springer (2002)
22. Hahsler, M., Grün, B., Hornik, K.: A Computational Environment for Mining Association Rules and Frequent Item Sets (2005)
23. Team, R.C.: R: A Language and Environment for Statistical Computing. R Foundation for Statistical Computing. Vienna, Austria (2012)
24. Liu, B., Hsu, W., Ma, Y.: Pruning and summarizing the discovered associations. In: 5th ACM SIGKDD International Conference on Knowledge Discovery and Data Mining, pp. 125–134. ACM (1999)
25. Webb, G.I., Zhang, S.: Removing trivial associations in association rule discovery. In: 1st International NAISO Congress on Autonomous Intelligent Systems (ICAIS). (2002)
26. Kaufman, L., Rousseeuw, P.J.: Finding Groups in Data: An Introduction to Cluster Analysis. Wiley, New York (1990)
27. Maechler, M., Rousseeuw, P., Struyf, A., Hubert, M., Hornik, K.: Cluster: Cluster Analysis Basics and Extensions. R package version 2.0.1. (2006)
28. Allouche, O., Tsoar, A., Kadmon, R.: Assessing the accuracy of species distribution models: prevalence, kappa and the true skill statistic (TSS). J. Appl. Ecol. **43**, 1223–1232 (2006)
29. Fleiss, J.L.: Measuring nominal scale agreement among many raters. Psychol. Bull. **76**(5), 378–382 (1971)

QoS-Aware Web Service Composition Using Quantum Inspired Particle Swarm Optimization

Chandrashekar Jatoth and G.R. Gangadharan

Abstract Quality of Service (QoS)-aware web service composition is one of the challenging problems in service oriented computing. Due to the seamless proliferation of web services, it is difficult to find an optimal web service during composition that satisfies the requirements of an user. In order to enable dynamic QoS-aware web service composition, we propose an approach based on Quantum inspired particle swarm optimization. Experimental results show that the proposed QIPSO-WSC has effective and efficient performance in terms of low optimality rate and reduced time complexity.

Keywords Web service composition · Quality of service (QoS) · Particle swarm optimization · Quantum computing

1 Introduction

Web service composition is a process by which existing web services can be integrated together to create value added composite web services that satisfy the user's requirements [1]. Quality of Service (QoS), also known as non-functional properties of web services, differentiate those web services that are having same functionality. However, with the increasing adoption and presence of web services, it becomes more difficult to find the most appropriate web service during composition that satisfies the user's requirements.

C. Jatoth · G.R. Gangadharan (✉)
Institute for Development and Research in Banking Technology, Hyderabad, India
e-mail: grgangadharan@idrbt.ac.in

C. Jatoth
SCIS, University of Hyderabad, Hyderabad, India
e-mail: jcshekar@idrbt.ac.in

R. Neves-Silva et al. (eds.), *Intelligent Decision Technologies*,
Smart Innovation, Systems and Technologies 39,
DOI 10.1007/978-3-319-19857-6_23

In general, QoS-aware web service composition is solved at design time and invocation time using computational intelligence techniques [2–5]. Most Particle Swarm Optimization (PSO) based approaches do not have any major difficulties in representing QoS-aware web service composition and they consistently have faster convergence rates than genetic algorithms [6]. However, PSO approches require an initial service composition that needs to be optimized.

In this paper, we present and evaluate an approach based on quantum inspired particle swarm optimization (QIPSO) [7] to deal with QoS-aware web service composition (WSC), which combines the features of QIPSO and heuristic model. The experimental comparison of our proposed QIPSO-WSC shows effective and efficient performance.

The rest of the paper is organized as follows. Related work is described in Sect. 2. Theoretical Background related to our work is discussed in Sect. 3. The proposed approach on QIPSO-WSC is described in Sect. 4. Implementation and results are explained and analysed in Sect. 5 followed by concluding remarks in Sect. 6.

2 Related Work

QoS-aware web service composition is solved using different computational intelligence based approaches by several researchers. Yu et al. [8] used a tree based genetic algorithm to find an optimal solution for QoS-aware web service composition. Jianxin et al. [9] presented a niching particle swarm optimization for QoS-aware web service composition that supports multiple global constraints. Li et al. [10] discussed a novel approach for optimal web service selection by applying chaos particle swarm optimization. Xiangwei et al. [11] presented discrete particle swarm optimization algorithms and color petri nets in the context of QoS based web service composition. Zhao et al. [12] adopted an improved discrete immune optimization method based on PSO for QoS-aware web service composition. The said approaches use static workflows for QoS-aware web service composition.

Selecting the optimal web service that satisfies the user's requirements for web service composition during runtime (in a dynamic way) is a highly challenging problem. Parejo et al. [13] presented a hybrid approach combining greedy randomized search procedure (GRASP) and path re-linking algorithm to optimize the quality of service at runtime for web service composition. Zhang et al. [14] proposed ant colony optimization algorithm to solve QoS-based web service composition as a multi-objective optimal path execution in a dynamic environment. Guosheng et al. [15] discussed an efficent approach for optimal web service selection by applying PSO-GODSS (Global Optimization of Dynamic Web Service Selection Based on PSO). Liu et al. [16] proposed a hybrid QPSO algorithm for QoS-aware web service composition, formalizing the services composition as the combinatorial optimization problem.

In our study, we used a dynamic approach for QoS-aware web service composition using quantum inspired particle swarm optimization algorithm [7]. Our proposed approach performs global optimization by minimizing the QoS aggregation functions [8] and satisfying the user requirements in an effective and efficient way. [17–19], and [20] describe several applications of similar optimization algorithms.

3 Theoretical Background

Particle Swarm Algorithm: Particle Swarm Optimization (PSO) is a population based evolutionary technique inspired by birds flocking and fish schooling [21]. In this algorithm, the individual is called as particle and the population is called as swarm. Each particle represents the candidate solution and is associated with position, velocity, and memory vector. Each particle strengthens itself by using two optimal solutions. First one is finding the optimal solution in local search (itself) and second one is finding in global search (particle swarm). A standard particle swarm optimizer maintains a swarm of particles and each individual particle composes three D-dimensional vectors, where D represents the dimensionality of the search space. Each individual particle is composed with current position ($x^{t+1}_{i,d}$), previous best position ($x^{t}_{i,d}$), and velocity ($V^{t+1}_{i,d}$). After finding the two best solutions, velocity ($V^{t+1}_{i,d}$) and position ($X^{t+1}_{i,d}$) of each particles are updated by the following equations:

$$V^{t+1}_{i,d} = w.V^{t}_{i,d} + c_1.f_1(P^{t}_{i,d} - X^{t}_{i,d}) + c_2.f_2(P^{t}_{g,d} - X^{t}_{i,d}) \tag{1}$$

$$X^{t+1}_{i,d} = X^{t}_{i,d} + V^{t+1}_{i,d} \tag{2}$$

where $V_{i,d}$ represents the veocity of ith particle in cycle t; $X_{i,d}$ represents the position of ith particle in cycle t; c_1, c_2 are learning factors; f_1, f_2 are random factors uniformly distributed in the range of [0, 1]; w represents the inertia weight; $P^{t}_{i,d}$ represents the local best of particle i in cycle t; $P^{t}_{g,d}$ represents global best in cycle t.

Quantum Computing: Quantum computing is a technique that combines computer science and quantum mechanism [22]. The parallelism exhibited by quantum computing reduces the arithmetic complexity. Such an ability of parallellism by quantum computing is used to solve combinatorial optimization problems which require investigating of large solution spaces [7].

Quantum Inspired Particle Swarm Optimization: Quantum Inspired Particle Swarm Optimization (QIPSO) is mainly based on quantum representation of the search space and particle swarm optimization algorithm, trying to improve the optimization capacities of the PSO algorithm [7, 23]. The QIPSO algorithm depends on the representation of evaluation function, individual dynamics, and population dynamics. The QIPSO algorithm adopts a quantum representation that depicts the superposition of all potential solutions for a given problem [24].

4 Quantum Inspired Particle Swarm Optimization for Web Service Composition

Our proposed approach QIPSO-WSC (Quantum inspired particle swarm optimization for web service composition) is based on quantum representation of the search space incorporated with web service composition and on quantum operations used for exploring the search space.

4.1 Problem Formulation

- Let $S = \{s_1, s_2, \ldots, s_i, \ldots s_m\}$ be a composite web service that consists of m abstract services, where s_i is the ith abstract service of S.
- An abstract service $s_i = \{(c_{i,1}, c_{i,2}, \ldots, c_{i,k})\}$ consists of k concrete services, where $c_{i,k}$ is the kth concrete service of ith abstract service.
- The composite web service S can be formulated as $S = \{(c_{1,1}, c_{1,2}, \ldots c_{1,k_1}), (c_{2,1}, c_{2,2}, \ldots c_{1,k_2}), \ldots, (c_{m,1}, c_{m,2}, \ldots, c_{m,k_m})\}$, where $c_{i,j}$ represents j^{th} concrete service of an abstract service c_i.
- Let $Q = \{q_1, q_2, q_3, \ldots, q_k\}$ be the set of values for the QoS properties of $c_{i,j}$, where q_k is the value for the kth criteria of $c_{i,j}$.
- QoS properties are bifurcated as positive properties and negative properties. Positive properties denote the higher value with higher user utility, while negative ones represent the lower user utility with higher values. For example, availability and reliability are positive properties and cost and response time are negative properties.
- Let $C = \{co_1, co_2, co_3, \ldots, co_k\}$ be a set of QoS constraints given by the user, where each $co_i \in C$ is a constraint on q_i. If q_i is a positive QoS property then co_i imposes a lower bound. Isobilaterally, if q_i is a negative QoS property, then co_i imposes a upper bound.
- Let $W = \{w_1, w_2, w_3, \ldots, w_n\}$ be a set of weights given by the users. Each $w_i (1 \leq i \leq n)$ corresponds to each QoS property. For each q_k, a user assigns a qos weight w_k, such that all weights satisfy $\sum_{k=1}^{n} w_k = 1$ and $0 \leq w_k \leq 1$. The utility of value x for a QoS property $q_{i,j}$ is defined as follows [13]:

$$q'_{i,j}(x) = \begin{cases} \frac{q_{max}(j) - q_{i,j}}{q_{max}(j) - q_{min}(j)} & \text{if } q_{i,j} \text{is negative} \\ 1 & \text{if } q_{max}(j) = q_{min}(j) \\ \frac{q_{i,j} - q_{min}(j)}{q_{max}(j) - q_{min}(j)} & \text{if } q_{i,j} \text{ is positive} \end{cases} \quad (3)$$

The fitness function/objective function is defined as follows [12]:

$$fitness\ function(f_i) = \sum_{k=1}^{n} w_k \times q'_{i,j} \quad (4)$$

To evaluate multi-objective QoS-aware web service composition we used the objective function or fitness function mapped with the following four QoS attributes (Response time, Availabilty, Reliability, and Cost). We model optimization problem as follows:

$$Min \cdot QoS(WSC) = \sum_{k=1}^{n} QoS(f_i) \tag{5}$$

subject to constraints:

$$f_i \leq ResponseTime_{max}; f_i \leq Cost_{max}; f_i \geq Availabilty_{min}; f_i \geq Reliability_{min} \tag{6}$$

where $\{ResponseTime, Cost, Availabilty, Reliability\} \in Q$.

4.2 Quantum Representation of Web Service Composition

To apply quantum principles over web service composition, we design possible web services compositions into quantum representation that can be easily used by quantum operators. All possible web service compositions are represented as a binary vector of size M, where M is the number of web services. Each element of the vector represents a possible web service composition. In quantum computing representation, each solution (composition) is represented as a quantum register of length M and each quantum register contains a superposition of all possible solutions. Each column in the quantum register $\begin{pmatrix} p_i \\ q_i \end{pmatrix}$ represents a single qubit and corresponds to a binary digit 0 or 1. The probabilities of (p_i, q_i) should satisfy $| p_i |^2 + | q_i |^2 = 1$. For each qubit, $| p_i |^2$ and $| q_i |^2$ are determined as 0 or 1 according to their possibilities [7]. The value 1 represents that the candidate web service is present at the position i of the current composition solution and 0 represents that the candidate web service is not present at the position i of the current composition solution.

4.3 QIPSO-WSC Framework

Our proposed QIPSO-WSC framework (Fig. 1) consists of a synergy module, a web services repository, and a web service composition (wsc) and optimization module. Initially, the user enters his preferred QoS parameters, values and constraints through the Synergy module and sends to the web service composition and optimization module to find the best web service composition. The web service composition and optimization module plays three roles: generation of initial population, evaluation of solutions with fitness function and selection of optimal solution with

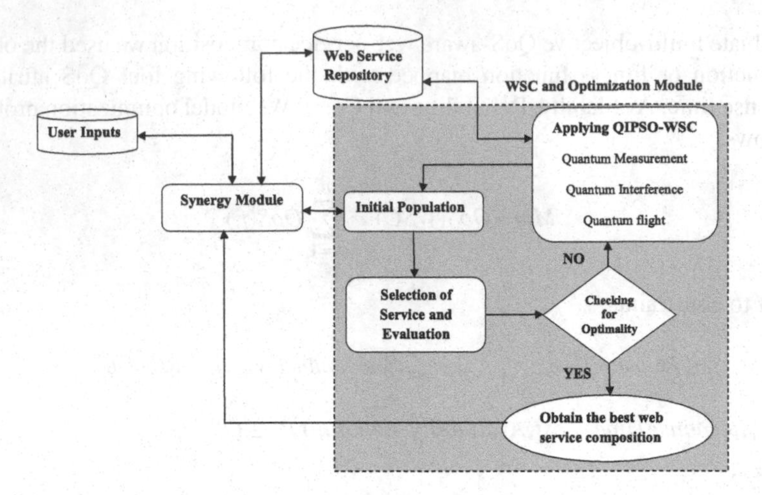

Fig. 1 QIPSO-WSC Framework

selection operator (elitism strategy) used in genetic algorithms [25], and application of quantum inspired particle swarm optimization.

4.4 Heuristic Algorithm for Initial Population

Initialization of population in web service composition is one of the pivotal phases in our proposed approach. We follow an efficient and scalable stochastic heuristic approach (See Algorithm 1) to generate an initial population, based on [26]. The proposed heuristic starts from finding web services based on user inputs and terminates by finding the best web service composition or output which meets the user's requirements.

Algorithm 1 Heuristic for initial population (Adopted from [26])

Input: Input, Output, Existing web services
Output: The Optimal web service compositon
1: **begin**
2: Initialize the web service composition solution to zero ;
3: Find the random services S_i such as Input = $S_{i_{input}}$;
4: $S_{candidate} = S_i$;
5: Set the value 1 in the vector's solution at position i ;
6: **while** Output not found **do**
7: Find S_i such as $S_{i_{input}} = S_{candidate_{output}}$;
8: $S_{candidate} = S_i$;
9: Set the value 1 in the vector's solution at position i ;
10: **end while**
11: **end**

4.5 *Quantum Inspired Particle Swaram Optimization Operators*

Application of quantum inspired particle swarm optimization is performed by the following operators: Quantum measurement, Quantum interference, Quantum flight. These operators are used in entire swarm through generations.

Quantum Measurement This operator allows us to acquire binary particles from quantum particles, and transforms by projection the quantum vector into a binary vector [7]. By using this operator, we get a feasible solution. If we did not get a feasible solution that is not corresponding to the requirements of a user, then we use an improved heuristic as mentioned in [26] for web service composition.

Quantum Interference This operator increases the optimal composition amplitude and decreases the bad composition. The main aim of quantum interference is to move the state of each qubit towards optimal composition (solution). The rotation angle is achieved by using unit transformation whose angle is a function of amplitude (p_i, q_i). To avoid premature convergence, the rotation angle value $\delta\theta$ is firmly set. In our approach, we use rotation angle $\delta\theta = \pi/20$.

Quantum Flight This operator allows to move from the current position $(X^t_{i,d})$ to the next position $(X^{t+1}_{i,d})$ of its neighbor to enhance the capacity of search space. It generates the new solution $(X^{t+1}_{i,d})$ by using the updation of velocity (Eq. 1) and position of each particle (Eq. 2), as explained in Sect. 3.

4.6 *QIPSO Algorithm for QoS-Aware Web Service Composition*

Initially, the population of quantum particle is created at a random position to represent all possible solutions and the best solutions are generated with the proposed heuristic model. For each particle, we calculate the fitness function (f_i) using Sect. 4.1. During each iteration, we perform the following operations: (i) evaluate quantum particles in current population by using measure operation in order to get binary solution. (ii) apply interference operation on some particles. Then, we modify velocity based on its global and local fitness of the swarm. We update the position of each particle by using quantum flight operation. Finally, we update the global best solution. This whole process is repeated until reaching an optimal solution (See Algorithm 2).

Algorithm 2 QIPSO algorithm for QoS-aware web service composition

Input: Problem data
Output: Problem solution
1: **begin**
2: Initialize the population, positions, and velocities randomly;
3: **while** Non stop criterion **do**
4: **for** each particle (P_i) **do**
5: Calculate fitness function (f_i) by using Equation 4;
6: **end for**
7: Apply the measure operation on each particle;
8: Apply the interference operation randomly on each particle;
9: Save the highest fitness of each individuals;
10: Modify the velocity based on particle ($P_{i,d}$) and ($P_{g,d}$) positions;
11: Update the particle position by using quantum flight operations;
12: Update the local best of particle i ($P_{i,d}$) ;
13: Update the global best ($P_{g,d}$) ;
14: **end while**
15: Go to step 4;
16: **end**

5 Implementation and Results

QoS-aware web service composition using QIPSO (described in Sect. 4) is implemented in Java and tested on a computer system with a processor of Intel (R) core (TM) i5 2.60 GHZ and 8 GB of memory. We evaluate computational time and optimality rate and compare our proposed approach with PSO and IDPSO. In this study, we used the real time QWS dataset given by [27], where each service has four QoS properties: Response time (Rt), Availability (A), Reliability (R), and Cost (C). The values for QoS properties, their bounds values, and their weights are specified in Table 1. Table 2 presents QIPSO-WSC parameters and their values. We evaluate the optimality rate for PSO, IDPSO, and QIPSO-WSC algorithms using the following equation based on [26] (as we are minimizing the QoS aggregated function).

$$Optimality\ \ Rate = 1 - \frac{Optimal\ \ solution}{Initial\ \ solution} \tag{7}$$

Table 1 QoS properties, bounds values, and weights

QoS properties	Bounds values	Weights
Response time	0–2000 (ms)	0.4
Availability	0.5–1.0	0.2
Reliability	0–1.0	0.2
Cost	0–20 (Dollar)	0.2

Table 2 QIPSO-WSC Parameters and values

Parameter name	Values
Learning factor (c_1)	1.8
Learning factor (c_2)	2.0
Rotation angle	$\Pi/20$
Number of iterations	[50, 1000]
Population size	20

The optimality rate and computation time are mentioned in Table 3. From Table 3, we observed that the optimality rate of QIPSO-WSC is better than PSO and IDPSO approaches. In terms of computation time QIPSO-WSC is efficient than other two approaches. The optimality rate of PSO, IDPSO, and QIPSO are shown in Fig. 2. From Fig. 2, we observed that the optimality rate of QIPSO and IDPSO algorithms in solving services composition are comparatively better than PSO, and with the higher number of iterations QIPSO performs better than other two approaches. The computation time of PSO, IDPSO, and QIPSO algorithms are shown in Fig. 3. From Fig. 3, we observe that the computation time of our approach is better than PSO and IDPSO approaches.

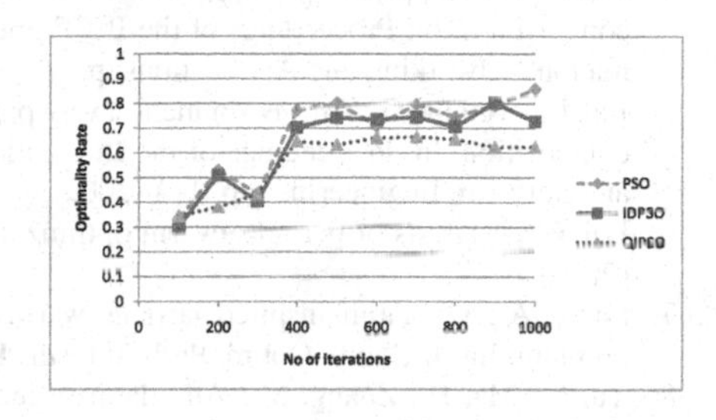

Fig. 2 Optimality Rate of PSO, IDPSO, and QIPSO algorithms

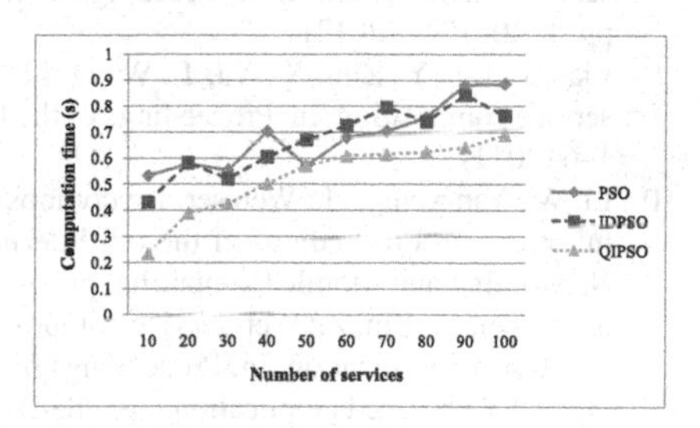

Fig. 3 Computation time of PSO, IDPSO, and QIPSO algorithms

Table 3 Results comparing PSO, IDPSO, and QIPSO-WSC

Algorithm	Optimality rate	Computation time (s)
PSO	0.8581390	0.8848251
IDPSO	0.7248137	0.7648123
QIPSO-WSC	0.6227350	0.6881229

6 Concluding Remarks

QIPSO based QoS-aware web service composition combines quantum computing principles and PSO to optimize the QoS for web service composition. We compared our approach with the existing PSO approaches and found that the optimality rate and computation time of QIPSO-WSC are better than other approaches. In our future work, we plan to apply quantum represenation approach to other meta-heuristic methods.

References

1. Sheng, Q.Z., Qiao, X., Vasilakos, A.V., Szabo, C., Bourne, S., Xu, X.: Web services composition: a decade's overview. Inf. Sci. **280**, 218–238 (2014)
2. Strunk, A.: Qos-aware service composition: a survey. In: Proceedings of the IEEE 8th European Conference on Web Services, pp. 67–74 (2010)
3. Amiri, M., Serajzadeh, H.: Effective web service composition using particle swarm optimization algorithm. In: Proceedings of the Sixth International Symposium on Telecommunications, pp. 1190–1194 (2012)
4. Ludwig, S.: Applying particle swarm optimization to quality-of-service-driven web service composition. In: Proceedings of the IEEE 26th International Conference on Advanced Information Networking and Applications, pp. 613–620 (2012)
5. Jun, L., Weihua, G.: An environment-aware particle swarm optimization algorithm for services composition. In: Proceedings of the International Conference on Computational Intelligence and Software Engineering, pp. 1–4 (2009)
6. Bai, Q.: Analysis of particle swarm optimization algorithm. Comput. Inf. Sci. **3**(1), 180–184 (2010)
7. Layeb, A.: A quantum inspired particle swarm algorithm for solving the maximum satisfiability problem. Int. J. Comb. Optim. Prob. Inform. **1**(1), 13–23 (2010)
8. Yu, Y., Ma, H., Zhang, M.: An adaptive genetic programming approach to qos-aware web services composition. In: Proceedings of the IEEE Congress on Evolutionary Computation, pp. 1740–1747 (2013)
9. Liao, J., Liu, Y., Zhu, X., Xu, T., Wang, J.: Niching particle swarm optimization algorithm for service composition. In: Proceedings of the IEEE Global Telecommunications Conference, pp. 1–6 (2011)
10. Li, W., Yan-xiang, H.: Web service composition based on qos with chaos particle swarm optimization. In: Proceedings of the 6th International Conference on Wireless Communications Networking and Mobile Computing, pp. 1–4 (2010)
11. Xiangwei, L., Yin, Z.: Web service composition with global constraint based on discrete particle swarm optimization. In: Proceedings of the Second Pacific-Asia Conference on Web Mining and Web-based Application, pp. 183–186 (2009)
12. Zhao, X., Song, B., Huang, P., Wen, Z., Weng, J., Fan, Y.: An improved discrete immune optimization algorithm based on pso for qos-driven web service composition. Appl. Soft Comput. **12**(8), 2208–2216 (2012)
13. Parejo, J.A., Segura, S., Fernandez, P., Ruiz-Cortes, A.: Qos-aware web services composition using grasp with path relinking. Expert Syst. Appl. **41**(9), 4211–4223 (2014)
14. Zhang, W., Chang, C., Feng, T., yi Jiang, H.: Qos-based dynamic web service composition with ant colony optimization. In: Proceedings of the IEEE 34th Annual Conference on Computer Software and Applications, pp. 493–502 (2010)

15. Kang, G., Liu, J., Tang, M., Xu, Y.: An effective dynamic web service selection strategy with global optimal qos based on particle swarm optimization algorithm. In: Proceedings of the IEEE 26th International Symposium Workshops Ph.D. Forum Parallel and Distributed Processing, pp. 2280–2285 (2012)
16. Liu, Y., Miao, H., Li, Z., Gao, H.: Qos-aware web services composition based on hqpso algorithm. In: Proceedings of the First International Conference on Computers, Networks, Systems and Industrial Engineering, pp. 400–405 (2011)
17. Bastos-Filho, C.J., Chaves, D.A., e Silva, F., Pereira, H.A., Martins-Filho, J.F.A.: Wavelength assignment for physical-layer-impaired optical networks using evolutionary computation. J. Opt. Commun. Networking **3**(3), 178–188 (2011)
18. Precup, R.-E., David, R.-C., Petriu, E., Preitl, S., Paul, A.: Gravitational search algorithm-based tuning of fuzzy control systems with a reduced parametric sensitivity. In: Proceedings of the Soft Computing in Industrial Applications, vol. 96, pp. 141–150 (2011)
19. Mota, P., Campos, A.R., Neves-Silva, R.: First look at mcdm: Choosing a decision method. Adv. Smart Syst. Res. **3**(2), 25–30 (2013)
20. El-Hefnawy, N.: Solving bi-level problems using modified particle swarm optimization algorithm. Int. J. Artif. Intell. **12**(2), 88–101 (2014)
21. Poli, R., Kennedy, J., Blackwell, T.: Particle swarm optimization. Swarm Intell. **1**(1), 33–57 (2007)
22. Williams, C.P., Clearwater, S.H.: Explorations in Quantum Computing, vol. 1. Springer (1998)
23. Sun, J., Feng, B., Xu, W.: Particle swarm optimization with particles having quantum behavior. In: Proceedings of the Congress on Evolutionary Computation, vol. 1, pp. 325–331 (2004)
24. Xi, M., Sun, J., Xu, W.: An improved quantum-behaved particle swarm optimization algorithm with weighted mean best position. Appl. Math. Comput. **205**(2), 751–759 (2008)
25. Layeb, A.: A novel quantum inspired cuckoo search for knapsack problems. Int. J. Bio-Inspired Comput. **3**(5), 297–305 (2011)
26. Boussalia, B., Chaoui, A.: Optimizing qos-based web services composition by using quantum inspired cuckoo search algorithm. In: Proceedings of the Mobile Web Information Systems, vol. 8640, pp. 41–55. Springer (2014)
27. Al-Masri, E., Mahmoud, Q.H.: Investigating web services on the world wide web. In: Proceedings of the 17th International Conference on World Wide Web, pp. 795–804 (2008)

Fitness Metrics for QoS-Aware Web Service Composition Using Metaheuristics

Chandrashekar Jatoth and G.R. Gangadharan

Abstract Quality of Service (QoS)-aware composition of web services is significant field of research, moving towards heuristic solutions based on soft computing techniques. However, the existing solution are specific to one or two QoS factors mainly response time and reliability. According to the real-time practices, the assessment of services by one or two QoS factors is impractical. Moreover these soft computing approaches have the computational complexity as $O(n^2)$, due to the exponential increase in the number of web services that are available to select. In this context we propose two fitness metrics using metaheuristics that help in assessing web services and composition fitness, resulting in the computational complexity of web service composition as $O(n * log(n))$. Our experiment results indicate the significance of the proposed approach towards scalable and robust QoS-aware service composition.

Keywords Web service composition · Quality of service (QoS) · Local fitness value · Global fitness threshold · Fitness metrics · Metaheuristics

1 Introduction

Web service composition is a process that loosely connects the existing web services and defines a sequence to execute them [1], which creates a value added composite web services to fulfill the desired task. The task of web service composition involves the search to select optimal services from the service repository; the optimality is a factor that varies from one context of service composition to another [2–4]. The importance and priority given to the QoS factors (availability, accessibility, cost,

C. Jatoth (✉) · G.R. Gangadharan
Institute for Development and Research in Banking Technology, Hyderabad, India
e-mail: jcshekar@idrbt.ac.in

G.R. Gangadharan
e-mail: grgangadharan@idrbt.ac.in

C. Jatoth
SCIS, University of Hyderabad, Hyderabad, India

R. Neves-Silva et al. (eds.), *Intelligent Decision Technologies*,
Smart Innovation, Systems and Technologies 39,
DOI 10.1007/978-3-319-19857-6_24

integrity, throughput, response time, reliability, and security) of web services are the definition aspects of service optimality. For example, the response time of an independent service can be optimal. However, the particular service may fail to fit in composition [5–7].

In general, several services from different providers exist for performing a single business process/task. Choosing one of these services to fulfill the task completion is generally done by ranking them, based on their QoS factors. The prioritizing the specific QoS factor towards ranking the services meant for a specific task is purely based on the context of composition. For instance, one specific context of composition may demand the reliability of the service but no matter how much costly it would be. The other context of composition might have a barrier about the service cost. Another context may expect the balance in QoS factors such as cost and reliability. Hence the search for an optimal web service from a given repository is complex and plays a key role to determine scalability and robustness of the service composition strategy. Hence the selecting optimal web services for service composition under the given user constraints is a most challenging task.

The web service composition helps to define the dynamic solution for applications with combination of independent tasks. In regard to this the composition process should adapt suitable service among the available, which is based on the QoS factors and their priority order defined by the composition [8].

1.1 Illustrative Example

Zheng et al. [9] and Parejo et al. [10] described a goods ordering process as an example to illustrate the issues of the QoS-aware web service composition. A customer initially registers the details of the goods, and initiates the payment transaction through his/her credit card. Once the card is verified (task t_1), then the payment transaction (task t_2) is done successfully, followed by the stock checking (task t_3) and reserving the registered goods for pickup (task t_4). If the stock is not available, then the customer is updated with delay in goods delivery and tasks t_3 and t_4 of the composition waits for some time and repeats. This wait and repeat continues till the process is completed.

According to the said example, task t_1 is accomplished by services s_{1_A} and s_{1_B}, and t_2 is accomplished by services s_{2_A} and s_{2_B}. s_{1_A} and s_{2_A} are owned by a provider A and s_{1_B} and s_{2_B} owned by a provider B. According to the choice of composition context, accomplish the tasks t_1 and t_2 either the A or B can be selected. If the choice of composition context exclusively focus on the cost then A is selected, assuming that the cost of s_{1_A} and s_{2_A} is cheaper than the services provided by B. If response time is the choice of the composition context, then s_{1_B} and s_{2_B} are preferable assuming that the response time s_{1_B} and s_{2_B} is lower. If cost and response time are the choice of the composition context, then selecting services becomes a challenging task.

In case of task t_4, the service should invoke if the ordered good in stock (verified by the task t_3), else it waits till the stock is available. The service that opted to accomplish task t_4 may wait and repeat, which is based on the response of the service that opted for task t_3. Hence it is obvious to choose services for tasks t_3 and t_4 from the same provider.

The other composition context of the depicted goods ordering process example is, accomplishing two tasks parallel. Once goods are available in stock, the task t_5 (pickup and delivery) and task t_6 (sending digitally signed invoice to the client through an email) can be performed parallel. The failure of the service related to t_5 must not be tolerable against successful completion of the service related to t_6. Once the successful completion of the services related to t_5 and t_6, the service in schedule, to accomplish the task t_7 (taking feedback from the customer) initiated. Further, other non functional constraints such as maximum time limit for the process completion or maximum time limit for the composition completion would also play a vital role.

The exploration of the service composition in the context of the goods ordering process example illustrates that the QoS-aware service composition is highly challenging if several QoS factors are considered and is NP-hard [11, 12] if increased number of services for each task. Hence to simplify the service composition in the context of challenges explored, we propose a novel meta-heuristic approach for service selection towards QoS-aware service composition by introducing two assessment metrics: local fitness value and global fitness threshold, for acessing web services and their composition using meta-heuristics.

The reminder of the paper is organized as follows. Related work is described in Sect. 2. The proposed metrics are presented in Sect. 3. Section 4 discusses experiments and results followed by conclusions in Sect. 5.

2 Related Work

A considerable research towards defining metaheuristic models for QoS aware service composition is found in recent literature. Yu et al. [5] used a greedy approach to incorporate QoS features for composite services and applied an adaptive strategy to get efficient performance with in the minimum search time to find the solution. Xiangbing et al. [6] proposed a web service modeling ontology based web service composition method to solve QoS based service composition and applied a genetic algorithm to find the optimal solution. Li et al. [13] discussed a novel approach for optimal web service selection by applying chaos particle swarm optimization. Xiangwei et al. [14] presented a discrete particle swarm optimization algorithms and color petri nets in the context of QoS based web service composition. Mao et al. [15] presented different meta-heuristic algorithms (particle swarm optimization, estimation of distribution algorithm, genetic algorithm) for efficient performance in web service composition. Zhao et al. [16] used immune algorithm to improve the local best strategy and applied particle swarm optimization algorithm to find the global optimization value and reduce the search capability and high scalability. Parejo et al. [10] presented a hybrid approach GRASP and path re-linking algorithm to optimize the quality of service for QoS-aware web service composition at runtime.

The said approaches consider one or two QoS factors to assess services in composition which is impractical in reality. Moreover the computational complexity of service composition in said approaches is $O(n^2)$, considering the seamless increase in number of services per each task and the number of services required for composition.

3 Proposed Method

Let us consider an application with a set of m tasks and each task t can be fulfilled with any individual service among available services $S = \{st_1 = \{s_{1,1}, s_{1,2}, \cdots, s_{1,i}\}, st_2 = \{s_{2,1}, s_{2,2}, \ldots s_{2,j}\}, \ldots, st_m = \{s_{m,1}, s_{m,2}, \ldots s_{m,p}\}$. The services in the set $\{st_i = \{s_1, s_2, \ldots s_x\}$ are the similar services performing the task t_i of a given application. Hence the solution to the given application is the composition of the services such that only one service among the x similar services of each task should be considered for the composition. Thus, the objective of our proposal is to select a service from each set of x similar services. As the selected services for composition could influence the resulting QoS of composition, it is essential to pick the optimal services. Following are the characteristics of services that could impact the resulting composition. (i) A service can be rated best by considering the independent performance of itself. But the same service might fail to deliver the same level of performance as a dependent service during composition. (ii) A service can be ranked divergently with respect to its various QoS factors. As an example, a service can be best with respect to uptime, but the service might be moderate in terms of cost, and worst in the context of execution time. (iii) The importance of the QoS factors might vary from one composition requirement to other.

According to the characteristics of the services described, it is evident that the best ranked independent service is not always the optimal in the composition. However, verification of composition with all possible services of a task is also not scalable and robust. The services under a composition that perform well under some prioritized QoS factors need not be the best fit for service composition under other prioritized QoS factors. With respect to this, the said meta-heuristic model in its first stage, finds the fitness of the independent services, based on primary QoS factor opted. This process is termed as local fitness evaluation of the services. Further services are ranked according to their fitness and are used in the same order to finalize a service towards composition. In the second stage, the composition update is scaled against best fit threshold determined under a heuristic strategy.

3.1 Evaluation Strategy of Local Fitness of Services

Let F be a set of QoS factors $\{f_1, f_2, \ldots, f_n\}$ of each service in the given service set $S = \{st_1 = \{s_{1,1}, s_{1,2}, \ldots s_{1,i}\}, st_2 = \{s_{2,1}, s_{2,2}, \ldots s_{2,j}\}, \ldots, st_m = \{s_{m,1}, s_{m,2}, \ldots s_{m,p}\}$. A QoS factor f_{opt} is said to be the anchor factor to rank the services. The QoS factors of services are classified as positive and negative factors. The factors that are having highest values as optimal values are positive factors and the factors that are optimal with minimal values are negative factors. The normalization of positive and negative factors is described in Algorithm 1. Then the services of each service set are ranked by their normalized values in descending order, such that each service gets different ranks for different factors. Further these ranks will be used as inputs to measure the local fitness.

Algorithm 1 Positive and Negative Factors Normalization.

1: **for** each service set $[st_i \exists st_i \in S]$ **do**
2: **for** each service $[s_j \exists s_j \in st_i]$ **do**
3: **for** each factor $[f_k \exists f_k \in F_{sj}]$ **do** // here F_{sj} is the set factors of service s_j
4: **if** f_k is positive factor **then**
5: $norm(f_k) = 1 - \frac{1}{val(f_k)}$
6: **else if** f_k is negative factor **then**
7: $norm(f_k) = \frac{1}{val(f_k)}$
8: **end if**
9: **end for**
10: **end for**
11: **end for**

Let a rank set of a service $[s_j \exists s_j \in st_i \wedge st_i \in S]$ be $rs(s_j) = [r(f_1), r(f_2), \cdots, r(f_n)]$. Then, the local fitness value (lfv) of this service is measured as follows.

$$lfv(s_j) = \left[\frac{\sqrt{\left[\sum_{k=1}^{n} \left[\frac{\sum_{i=1}^{n} r(f_i \exists f_i \in F_{sj})}{n} \right] - r(f_k \exists f_k \in F_{sj}) \right]^2}}{n} \right]^{-1} \quad (1)$$

Equation 1 is derived from the statistical approach of calculating variance between the given number of attribute values. In this equation, $\left[\frac{\sum_{i=1}^{n} r(f_i \exists f_i \in F_{sj})}{n} \right]$ represents the mean of ranks of all features in the feature set F_{sj}. Then the local fitness of the services are sorted based on the rank of f_{opt} $\left([f_{opt} \exists f_{opt} \cong f_i \exists f_i \in F] \right)$. Then the set of services $[pst_{ij} \subseteq st_{ij}]$ is considered, which is based on the maximum rank threshold (mrt) given. The processed service set pst_j is sorted in ascending order of their local fitness value and the same order is preferred to select services for composition.

3.2 Evaluation Stragegy of Global Fitness Threshold of Service Compositions

This model derives a heuristic process that estimates the global fitness value of each composition of a given optimal set of compositions. Further, it estimates the global fitness threshold and its upper and lower boundaries. Then these upper and lower bounds of the global fitness threshold are used to assess the current fitness state of the composition. This strategy leads to direct the best fit service selection for composition.

Let C be a set of compositions $C = \left[c_1, c_2, \ldots, c_n\right]$. Let S be a set of services $S = \left\{s_1, s_2, \ldots, s_m\right\}$, which are involved to prepare the compositions opted. Let s_i and s_j be two services. s_i is connected with s_j if and only if $\left(s_i, s_j\right) \in \left[c_i \exists c_i \in C\right]$. We build an undirected weighted graph with services as vertices and connection between services as edges. An edge between the two services s_i and s_j is weighted as follows:

$$w_{(s_i \leftrightarrow s_j)} = \frac{\sum_{k=1}^{|S|} \left\{1 \exists \left[\left(s_i, s_j\right) \subseteq c_k \wedge i \neq j\right]\right\}}{|\, C\,|}, for\ each \left\{s \forall s \in S\right\} \tag{2}$$

In the process of building a weighted graph, we consider that an edge between any two services exists if and only if the edge weight $w > 0$. To represent each composite service with their candidate services, we build a duplex graph (Fig. 1) between C and S. If a service s_i is a part of a composition c_j, then the weight of the connection between s_i and c_j is measured as follows:

$$cw_{(s_i \leftrightarrow c_j)} = \frac{\sum_{k=1}^{|c_i|} \left\{ w_{(s_i \leftrightarrow s_k)} \exists \left[i \neq k \wedge \left(s_i, s_k\right) \in c_j\right] \right\}}{|\, c_j\,|} \tag{3}$$

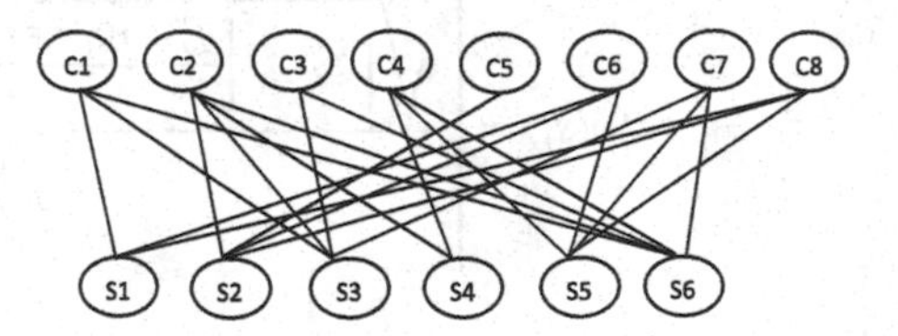

Fig. 1 Duplex graph between compositions C and services S

Equation 3 considers the sum of all edge weights from the undirected graph such that there exists an edge between s_i and other services of the composition c_j. The $|\, c_j\,|$ indicates the total number of services in a composition c_j. The process of identifying global fitness values for composition using duplex graph is explored as follows:

Consider a matrix having the connection weights of the edges between services as prerogatives and compositions as pivots in given a duplex graph (The underpinning association of web service compositions and web service descriptors is that of association between pivots and prerogatives.). The initial pivot value in the duplex graph is initialized as 1, which we represent as matrix in Table 1.

Table 1 Initializing the initial pivot value in duplex graph with 1 and representing them as a matrix *pw*

1	1	1	1	1	1

The matrix in Table 3 is M', which is the transpose of matrix M in Table 2. We find prerogative weights *prw* of the duplex graph as follows.

$$prw = M' \bigotimes pw \tag{4}$$

Table 2 Matrix M, representation of connection weights between pivots and prerogatives from Fig. 1

$wc_{(s_1 \leftrightarrow c_1)}$	$wc_{(s_1 \leftrightarrow c_2)}$	$wc_{(s_1 \leftrightarrow c_3)}$	$wc_{(s_1 \leftrightarrow c_4)}$	$wc_{(s_1 \leftrightarrow c_5)}$	$wc_{(s_1 \leftrightarrow c_6)}$	$wc_{(s_1 \leftrightarrow c_7)}$	$wc_{(s_1 \leftrightarrow c_8)}$
$wc_{(s_2 \leftrightarrow c_1)}$	$wc_{(s_2 \leftrightarrow c_2)}$	$wc_{(s_2 \leftrightarrow c_3)}$	$wc_{(s_2 \leftrightarrow c_4)}$	$wc_{(s_2 \leftrightarrow c_5)}$	$wc_{(s_2 \leftrightarrow c_6)}$	$wc_{(s_2 \leftrightarrow c_7)}$	$wc_{(s_2 \leftrightarrow c_8)}$
$wc_{(s_3 \leftrightarrow c_1)}$	$wc_{(s_3 \leftrightarrow c_2)}$	$wc_{(s_3 \leftrightarrow c_3)}$	$wc_{(s_3 \leftrightarrow c_4)}$	$wc_{(s_3 \leftrightarrow c_5)}$	$wc_{(s_3 \leftrightarrow c_6)}$	$wc_{(s_3 \leftrightarrow c_7)}$	$wc_{(s_3 \leftrightarrow c_8)}$
$wc_{(s_4 \leftrightarrow c_1)}$	$wc_{(s_4 \leftrightarrow c_2)}$	$wc_{(s_4 \leftrightarrow c_3)}$	$wc_{(s_4 \leftrightarrow c_4)}$	$wc_{(s_4 \leftrightarrow c_5)}$	$wc_{(s_4 \leftrightarrow c_6)}$	$wc_{(s_4 \leftrightarrow c_7)}$	$wc_{(s_4 \leftrightarrow c_8)}$
$wc_{(s_5 \leftrightarrow c_1)}$	$wc_{(s_5 \leftrightarrow c_2)}$	$wc_{(s_5 \leftrightarrow c_3)}$	$wc_{(s_5 \leftrightarrow c_4)}$	$wc_{(s_5 \leftrightarrow c_5)}$	$wc_{(s_5 \leftrightarrow c_6)}$	$wc_{(s_5 \leftrightarrow c_7)}$	$wc_{(s_5 \leftrightarrow c_8)}$
$wc_{(s_6 \leftrightarrow c_1)}$	$wc_{(s_6 \leftrightarrow c_2)}$	$wc_{(s_6 \leftrightarrow c_3)}$	$wc_{(s_6 \leftrightarrow c_4)}$	$wc_{(s_6 \leftrightarrow c_5)}$	$wc_{(s_6 \leftrightarrow c_6)}$	$wc_{(s_6 \leftrightarrow c_7)}$	$wc_{(s_6 \leftrightarrow c_8)}$

Table 3 Transpose matrix M' of matrix M

$wc_{(s_1 \leftrightarrow c_1)}$	$wc_{(s_2 \leftrightarrow c_1)}$	$wc_{(s_3 \leftrightarrow c_1)}$	$wc_{(s_4 \leftrightarrow c_1)}$	$wc_{(s_5 \leftrightarrow c_1)}$	$wc_{(s_6 \leftrightarrow c_1)}$
$wc_{(s_1 \leftrightarrow c_2)}$	$wc_{(s_2 \leftrightarrow c_2)}$	$wc_{(s_3 \leftrightarrow c_2)}$	$wc_{(s_4 \leftrightarrow c_2)}$	$wc_{(s_5 \leftrightarrow c_2)}$	$wc_{(s_6 \leftrightarrow c_2)}$
$wc_{(s_1 \leftrightarrow c_3)}$	$wc_{(s_2 \leftrightarrow c_3)}$	$wc_{(s_3 \leftrightarrow c_3)}$	$wc_{(s_4 \leftrightarrow c_3)}$	$wc_{(s_5 \leftrightarrow c_3)}$	$wc_{(s_6 \leftrightarrow c_3)}$
$wc_{(s_1 \leftrightarrow c_4)}$	$wc_{(s_2 \leftrightarrow c_4)}$	$wc_{(s_3 \leftrightarrow c_4)}$	$wc_{(s_4 \leftrightarrow c_4)}$	$wc_{(s_5 \leftrightarrow c_4)}$	$wc_{(s_6 \leftrightarrow c_4)}$
$wc_{(s_1 \leftrightarrow c_5)}$	$wc_{(s_2 \leftrightarrow c_5)}$	$wc_{(s_3 \leftrightarrow c_5)}$	$wc_{(s_4 \leftrightarrow c_5)}$	$wc_{(s_5 \leftrightarrow c_5)}$	$wc_{(s_6 \leftrightarrow c_5)}$
$wc_{(s_1 \leftrightarrow c_6)}$	$wc_{(s_2 \leftrightarrow c_6)}$	$wc_{(s_3 \leftrightarrow c_6)}$	$wc_{(s_4 \leftrightarrow c_6)}$	$wc_{(s_5 \leftrightarrow c_6)}$	$wc_{(s_6 \leftrightarrow c_6)}$
$wc_{(s_1 \leftrightarrow c_7)}$	$wc_{(s_2 \leftrightarrow c_7)}$	$wc_{(s_3 \leftrightarrow c_7)}$	$wc_{(s_4 \leftrightarrow c_7)}$	$wc_{(s_5 \leftrightarrow c_7)}$	$wc_{(s_6 \leftrightarrow c_7)}$
$wc_{(s_1 \leftrightarrow c_8)}$	$wc_{(s_2 \leftrightarrow c_8)}$	$wc_{(s_3 \leftrightarrow c_8)}$	$wc_{(s_4 \leftrightarrow c_8)}$	$wc_{(s_5 \leftrightarrow c_8)}$	$wc_{(s_6 \leftrightarrow c_8)}$

In Eq. 4, *prw* represents the matrix representation of the prerogative weights and M' is the transpose matrix of the matrix M representing connection weights between compositions as pivots and services as prerogatives in the duplex graph. Then the actual pivot weights pw is measured as follows.

$$pw = M \bigotimes prw \tag{5}$$

The matrix multiplication between matrix M and matrix prw results the actual pivot weights. Then the service weight sw of a service s is measured as follows.

$$sw_s = \frac{\sum_{i=1}^{|C|} \left\{ pw_{c_i} \exists cw_{s \leftrightarrow ci} \neq 0 \right\}}{\sum_{i=1}^{|c|} pw_{c_i}} \tag{6}$$

The global fitness value gfv of each composition is found as follows.

$$gfv_{c_i} = 1 - \frac{\sum_{j=1}^{|S|} \left\{ sw_{s_j} \exists s_j \in c_i \right\}}{| S |} \tag{7}$$

In Eq. 7, $| S |$ indicates the total number of services involved to create the compositions of C. Then the global fitness threshold gft is measured as follows.

$$gft = \frac{\sum_{i=1}^{|C|} gfv_{c_i}}{| C |} \tag{8}$$

In Eq. 8, $| C |$ indicates the total number of compositions considered. Then, the variance v of the derived all global fitness values from gft is measured as follows.

$$v = \sqrt{\frac{\left[\sum_{i=1}^{|C|} \left(gfv_{c_i} - gft \right)^2 \right]}{| C | - 1}} \tag{9}$$

The global fitness threshold minimal and maximal boundaries are explored as follows: Lower bound of global fitness threshold (gft_l) is $gft_l = gft - v$.

Upper bound of global fitness threshold (gft_u) is $gft_u = gft + v$.

The fitness of a composition at current state can be best if $gfv_{c_i} > gfv_u$, good if $gfv_{c_i} > gft$, and safe if $gfv_{c_i} > gfv_l$.

4 Experiments and Results Exploration

The dataset (in Table 4) for analyzing the performance of our proposed model is collected from a software organization. We have not opted the existing data sets such as [17, 18] because these datasets are not considering maximum number of QoS factors to assess the service quality and they limit to a few QoS factors mainly response time and throughput. But in reality, service selection could be influenced by several QoS factors including availability, accessibility, cost, integrity, throughput, response time, reliability, regulatory and security. The implementation of the application that recommends the composition using the proposed strategy is built using Java. The total number of input composition enabled applications of size 500 was partitioned

Table 4 Description of the dataset used for experiments

Number of tasks	4500
Range of services available for each task	7–16
No of qualified compositions	500
Range of tasks in each composition	6–25

into 73 % and 27 %. Here, in this partition 73 % is used to estimate the global fitness threshold and the remaining 27 % is used for the statistical assessment of the proposal. The global fitness threshold (*gft*) and its lower bound (gft_l) and upper bound (gft_u) found in this experiment are 0.5095930, 0.3509121, and 0.6682740 respectively. The observations from the compositions for the 27 % tasks of the given dataset are discribed in Table 5. The valid compositions are taken from the given set as input, which is 27 % of the total set. The low scaled compositions are partly failed compositions, which are not included in the given set of valid compositions. The total number of compositions used as input is 135. The average number of services evaluated for each task is 3 and the mean time taken for composition of services with the range between 10 to 21 services is less than 500 milliseconds. The time taken to evaluate global fitness threshold and its lower and upper bounds from 400 valid compositions of 3150 services is less than 30000 milliseconds. The precision, recall and F-Measure from experiment results are 0.837837838, 0.788135593, and 0.812227074 respectively. The Precision and F-Measure of the statistical analysis indicate the significance of the proposed model. The recall is bit low due to the variation between the count of qualified and unqualified compositions used as input (see Table 5).

Table 5 Details of test data and the results obtained

Total valid compositions considered for Testing	100
Total low scale compositions considered for testing	35
True positives	93
False Positives	24
False Negatives	18

5 Conclusion

In this paper we proposed two metrics for web service composition using meta-heuristic for acessing web services and thier composition fitness. Unlike other bench marking models (see the Sect. 2), the proposed metrics are not specific to one or two particular QoS factors. The services are selected under the consideration of vast number of combination of QoS factors and prioritization of these combinations is done through an anchor factor defined by the composition initiator. The best fit services for composition are selected by their local fitness value, which is one of the metric proposed. This local fitness value metric helps to avoid the process of verifying each available service of a specific task towards service composition by assessing and ordering the services by their metric value. Further the composition initiator uses only those services that are having local fitness value than the given threshold. Hence the NP-hard problem observed in the existing models is significantly handled in this proposed approach. The other metric, global fitness value plays a vital role towards minimizing the computational complexity towards service composition.

These results open different directions for future research. One of that is estimating the correlation between QoS factors towards assessing local fitness and the same way assessing correlation between services towards finding the global fitness threshold and its boundaries. Another interesting approach would be the usage of fuzzy logic to assess the local fitness value metric, which will be our next direction of research.

References

1. Sheng, Q.Z., Qiao, X., Vasilakos, A.V., Szabo, C., Bourne, S., Xu, X.: Web services composition: a decade's overview. Inf. Sci. **280**, 218–238 (2014)
2. Anane, R., Chao, K.-M., Li, Y.: Hybrid composition of web services and grid services. In: Proceedings of the IEEE International Conference on e-Technology, e-Commerce and e-Service, pp. 426–431 (2005)
3. El Hadad, J., Manouvrier, M., Rukoz, M.: Tqos: Transactional and qos-aware selection algorithm for automatic web service composition. IEEE TSC **3**(1), 73–85 (2010)
4. Strunk, A.: Qos-aware service composition: a survey. In: Proceedings of the IEEE 8th European Conference on Web Services, pp. 67–74 (2010)
5. Yu, Y., Ma, H., Zhang, M.: An adaptive genetic programming approach to qos-aware web services composition. In: Proceedings of the IEEE Congress on Evolutionary Computation, pp. 1740–1747 (2013)
6. Xiangbing, Z., Hongjiang, M. Fang, M.: An optimal approach to the qos-based wsmo web service composition using genetic algorithm. In: Proceedings of the International Conference on Service Oriented Computing, pp. 127–139 (2013)
7. Li, W., Yan-xiang, H.: A web service composition algorithm based on global qos optimizing with mocaco. In: Proceedings of the Algorithms and Architectures for Parallel Processing, pp. 218–224 (2010)
8. Ardagna, D., Pernici, B.: Adaptive service composition in flexible processes. IEEE TSC **33**(6), 369–384 (2007)
9. Zheng, H., Zhao, W., Yang, J., Bouguettaya, A.: Qos analysis for web service compositions with complex structures. IEEE TSC **6**(3), 373–386 (2013)
10. Parejo, J.A., Segura, S., Fernandez, P., Ruiz-Cortes, A.: Qos-aware web services composition using grasp with path relinking. Expert Syst. Appl. **41**(9), 4211–4223 (2014)
11. Bonatti, P.A., Festa, P.: On optimal service selection. In: Proceedings of the 14th International Conference on World Wide Web, pp. 530–538 (2005)
12. Ardagna, D., Pernici, B.: Global and local qos guarantee in web service selection. In: Proceedings of the Business Process Management Workshops, pp. 32–46 (2006)
13. Li, W., Yan-xiang, H.: Web service composition based on qos with chaos particle swarm optimization. In: Proceedings of the 6th International Conference on Wireless Communications Networking and Mobile Computing, pp. 1–4 (2010)

14. Xiangwei, L., Yin, Z.: Web service composition with global constraint based on discrete particle swarm optimization. In: Proceedings of the 2nd Pacific-Asia Conference on Web Mining and Web-based Application, pp. 183–186 (2009)
15. Mao, C., Chen, J., Yu, X.: An empirical study on meta-heuristic search-based web service composition. In: Proceedings of the IEEE 9th International Conference on e-Business Engineering, pp. 117–122 (2012)
16. Zhao, X., Song, B., Huang, P., Wen, Z., Weng, J., Fan, Y.: An improved discrete immune optimization algorithm based on pso for qos-driven web service composition. Appl. Soft Comput. **12**(8), 2208–2216 (2012)
17. Zheng, Z., Zhang, Y., Lyu, M.: Investigating qos of real-world web services. IEEE TSC **7**(1), 32–39 (2014)
18. Zhang, Y., Zheng, Z., Lyu, M.: Wspred: A time-aware personalized qos prediction framework for web services. In: Proceedings of the IEEE 22nd International Symposium on Software Reliability Engineering, pp. 210–219 (2011)

Distance-Based Ensemble Online Classifier with Kernel Clustering

Joanna Jędrzejowicz and Piotr Jędrzejowicz

Abstract In this paper an on-line distance-based classifier is considered. The approach extends earlier proposed idea where a family of the online distance-based classifiers based on fuzzy C-means clustering followed by calculation of distances between cluster centroids and the incoming instance for which the class label is to be predicted, were suggested [8]. Now, instead of fuzzy C-means clustering we use kernel-based clustering method. The proposed algorithm works in rounds, where at each round a new instance is given and the algorithm makes a prediction. A portfolio of similarity or distance measures used to construct the ensemble of classifiers predicting the class of coming instances. The proposed approach is validated experimentally.

Keywords Kernel functions · Kernel based fuzzy clustering · Online learning

1 Introduction

Online learning is considered to be of increasing importance to deal with never ending and usually massive stream of received data such as sensor data, traffic information, economic indexes, video streams, etc. [16]. Online approach is, as a rule, required when the amount of data collected over time is increasing rapidly. Processing streams of continuously incoming data implies requirements concerning a limited amount of memory and a short processing time, especially when data streams are large [14]. Common feature of such streams is a high possibility that data may evolve over time. Reviews of algorithms and approaches to data stream mining can

J. Jędrzejowicz (✉)
Institute of Informatics, Gdańsk University, Wita Stwosza 57, 80-952 Gdańsk, Poland
e-mail: jj@inf.ug.edu.pl

P. Jędrzejowicz
Department of Information Systems, Gdynia Maritime University,
Morska 83, 81-225 Gdynia, Poland
e-mail: pj@am.gdynia.pl

R. Neves-Silva et al. (eds.), *Intelligent Decision Technologies*,
Smart Innovation, Systems and Technologies 39,
DOI 10.1007/978-3-319-19857-6_25

be found in [6] and [14]. Online classifiers are induced from the initially available dataset as in case of the static approach. However, in addition, there is also some adaptation mechanism providing for a classifier evolution after the classification task has been initiated and started. In each round a class label of the incoming instance is predicted and afterwards information as to whether the prediction was correct or not, becomes available. Based on this information adaptation mechanism may decide to leave a classifier unchanged, or modify it, or induce a new one. Usual approach to deal with the online classification problems is to design and implement an online classifier incorporating some incremental learning algorithm [9]. According to [13] an algorithm is incremental if it results in a sequence of classifiers with different hypothesis for a sequence of training requirements. An ideal incremental learning algorithm should be able to detect concept drifts, to recover its accuracy, to adjust itself to the current concept and to use past experience when needed [17]. Examples of some state of the art online classifiers include approaches proposed in [16] and [2]. In this paper an approach based on kernel clustering is proposed. Kernel method is used to produce clusters from available training instances and to induce an ensemble of simple distance-based or similarity-based classifiers. The approach works in rounds, where at each round a new instance is arriving and the algorithm makes a prediction. After the true class of the instance is revealed, the learning algorithm updates its internal hypothesis. The paper is organized as follows. Section 1 contains introduction. Section 2 contains the description of the kernel method used. Section 3 gives a detailed description of the proposed online classification process and discusses computational complexity of the approach. Section 4 presents validation experiment settings and experiment results. Section 5 contains conclusions and suggestions for future research.

2 Kernel Based Fuzzy Clustering

Basic classification of clustering methods assigns them into two groups: crisp and fuzzy. One of the most used fuzzy methods is fuzzy C-means clustering (FCM) [3]. FCM considers each cluster as a fuzzy set with membership function measuring the possibility for each data to belong to a given cluster. The method proved useful overcoming some drawbacks of crisp C-means clustering. Recently, to deal with over-lapping and noisy data kernel methods have been applied to fuzzy clustering ([4, 10, 18]) and the kernelized version of FCM is referred to as kernel-based fuzzy C-means clustering. Graves and Pedrycz [7] give a broad report on kernel methods applied to clustering. The basic idea of kernel-based fuzzy C-means clustering is to transform the input data (feature space) into a higher dimensional kernel space via a non-linear mapping Φ which increases the possibility of linear separability of the patterns in kernel space and then allows for fuzzy C-means clustering in the feature space. As noted by [7], the kernel method uses the fact that product in the kernel space can be expressed by a Mercer kernel K given by

$$K(\mathbf{x}, \mathbf{y}) \equiv \Phi(\mathbf{x})^T \Phi(\mathbf{y}).$$

This allows to replace the computation of distances in the kernel space by a Mercer kernel function, which is known as a 'kernel trick'. Two variants of most common kernel functions are:

- gaussian kernel function $K(\mathbf{x}, \mathbf{y}) = \exp \frac{-dist^2(\mathbf{x},\mathbf{y})}{\sigma^2}$ for $\sigma^2 > 0$,
- polynomial kernel function: $K(\mathbf{x}, \mathbf{y}) = (\mathbf{x}^T\ \mathbf{y} + \Theta)^p$ where $\Theta \geq 0$, and p - natural.

Following the terminology in [7], two classes of kernel-based fuzzy clustering algorithms are considered:

1. KFCM-F - kernel-based fuzzy C-means clustering with centroids (or, prototypes as denoted in [7]) in feature space, and
2. KFCM-K - kernel-based fuzzy C-means clustering with centroids in kernel space.

In what follows, it is assumed that $D = \{x_1, \ldots, x_M\} \subset R^N$ is a finite collection of data to be clustered into nc clusters. Clusters are characterized by centroids c_i and fuzzy partition matrix $U = (u_{ij})$ of size $nc \times M$, where u_{ik} is the degree of membership of x_k in cluster i. Parameter $m \in N$ is a fixed natural, most often equal 2. The standard constraints for fuzzy clustering are assumed:

$$0 \leq u_{ik} \leq 1, \quad \sum_{i=1}^{nc} u_{ik} = 1 \; for\ each\ k, \; 0 < \sum_{k=1}^{M} u_{ik} < M \; for\ each\ i. \tag{1}$$

The partition of data D into clusters is performed in accordance with the maximal value of membership degree:

$$clust(x_k) = \arg max_{1 \leq j \leq nc} u_{jk} \tag{2}$$

2.1 KFCM-F Algorithm

In case of KFCM-F algorithm centroids are constructed in the feature space (along the data) and the algorithm minimizes the following objective function:

$$J_m = \sum_{i=1}^{nc} \sum_{k=1}^{M} u_{ik}^m \; dist^2(\Phi(x_k), \Phi(c_i))$$

For the gaussian kernel this leads to the following expressions for the fuzzy partition matrix U and centroids c_i (the details are in [7]):

$$u_{ik} = \frac{1}{\sum_{j=1}^{nc} (\frac{1-K(x_k,c_i)}{1-K(x_k,c_j)})^{\frac{1}{m-1}}} \tag{3}$$

$$c_i = \frac{\sum_{k=1}^{M} u_{ik}^m K(x_k, c_i)\, x_k}{\sum_{k=1}^{M} u_{ik}^m\, K(x_k, c_i)} \tag{4}$$

The algorithm of kernel based fuzzy C-means clustering with prototypes in feature space is shown as **Algorithm 1**. In this paper what is meant by stopping criteria is no change in centroids or, the maximal change in U not exceeding the predefined threshold ϵ.

Algorithm 1 Kernel-based fuzzy C-means clustering, version KFCM-F

Input: data D, kernel function K, number of clusters nc,
Output: fuzzy partition U, centroids $\{c_1, \ldots, c_{nc}\}$
1: initialize U to random fuzzy partition satisfying (1)
2: initialize centroids as random data from D,
3: **repeat**
4: update centroids according to (4)
5: update U according to (3)
6: **until** stopping criteria satisfied or maximum number iterations reached

2.2 KFCM-K Algorithm

For KFCM-K it is assumed that prototypes $\boldsymbol{v}_i$ are located in the kernel space and further centroids need to be approximated by an inverse mapping $c_i = \boldsymbol{\Phi}^{-1}(\boldsymbol{v}_i)$ to the feature space. In this case the objective function to be minimized is:

$$J_m = \sum_{i=1}^{nc} \sum_{k=1}^{M} u_{ik}^m\, dist^2(\boldsymbol{\Phi}(x_k), \boldsymbol{v}_i)$$

Optimizing J_m with respect to $\boldsymbol{v}_i$ gives:

$$\boldsymbol{v}_i = \frac{\sum_{k=1}^{M} u_{ik}^m \boldsymbol{\Phi}(x_k)}{\sum_{k=1}^{M} u_{ik}^m}$$

which leads to the formula for partition matrix

$$u_{ik} = \frac{1}{\sum_{j=1}^{nc} \left(\frac{dist(\boldsymbol{\Phi}(x_k), \boldsymbol{v}_i)}{dist(\boldsymbol{\Phi}(x_k), \boldsymbol{v}_j)}\right)^{\frac{2}{m-1}}} \tag{5}$$

where

$$dist^2(\Phi(x_k), v_i) = K(x_k, x_k) - 2\frac{\sum_{j=1}^{M} u_{ij}^m K(x_k, x_j)}{\sum_{j=1}^{M} u_{ij}^m} + \frac{\sum_{j=1}^{M}\sum_{l=1}^{M} u_{ij}^m u_{il}^m K(x_j, x_l)}{(\sum_{j=1}^{M} u_{ij}^m)^2} \tag{6}$$

Centroids c_i are approximated in the feature space using the minimization of

$$V = \sum_{i=1}^{nc} dist(\Phi(c_i), v_i)$$

which in case of gaussian kernel gives:

$$c_i = \frac{\sum_{k=1}^{M} u_{ik}^m \cdot K(x_k, c_i)\, x_k}{\sum_{k=1}^{M} u_{ik}^m\, K(x_k, c_i)} \tag{7}$$

and for the polynomial kernel:

$$c_i = \frac{\sum_{k=1}^{M} u_{ik}^m (x_k^T c_i + \Theta)^{p-1} x_k}{(c_i^T c_i + \Theta)^{p-1} \sum_{j=1}^{M} u_{ij}^m} \tag{8}$$

The algorithm of kernel based fuzzy C-means clustering is shown as **Algorithm 2**.

Algorithm 2 Kernel-based fuzzy C-means clustering, version KFCM-K

Input: data D, kernel function K, number of clusters nc,
Output: fuzzy partition U, centroids $\{c_1, \ldots, c_{nc}\}$
1: initialize U to random fuzzy partition satisfying (1)
2: **repeat**
3: update U according to (5) and (6)
4: **until** stopping criteria satisfied or maximum number iterations reached
5: **repeat**
6: update centroids according to (7) for gaussian kernel and (8) for polynomial kernel
7: **until** maximum number iterations reached

2.3 Fixing the Number of Clusters

As noted by [19] an additional advantage of using kernel functions is the possibility to determine the number of clusters based on significant eigenvalues of a matrix determined by the kernel function applied to data rows. In the sequel it is assumed that each kernel-based fuzzy clustering is preceded by **Algorithm 3** to fix the number of clusters.

Algorithm 3 Number of clusters estimation

Input: data D, kernel function K, threshold δ
Output: number of clusters nc
1: let $K_{ij} = K(x_i, x_j)$ be the quadratic matrix of size $M \times M$,
2: calculate the eigenvalues of matrix (K_{ij})
3: nc← number of eigenvalues exceeding δ
4: **return** nc

2.4 Clustering for Classification

For the online algorithms considered in the paper two measures are applied to classify unseen data rows. The first one is connected with clustering partition. Suppose that the fuzzy algorithm with nc -number of clusters, was applied to a set of data rows $X = \bigcup_{c \in C} X^c$. As a result one obtains, for each $c \in C$, centroids set Cnt^c of size nc and using the partition matrix with application of (2), the partition into clusters $X^c = \bigcup_{j=1}^{nc} X_j^c$.

To classify new data row $r \notin X$ the distance from r to any element from cluster X_j^c is calculated. Then the distances are sorted in a non-decreasing order so that l_1^{cj} points to the element nearest to r, l_2^{cj} - to the second nearest etc. The coefficient S_{cj}^x measures the mean distance from x neighbours:

$$S_{cj}^x = \frac{\sum_{i=1}^x dist(r, l_i^{cj})}{x} \tag{9}$$

For each class $c \in C$ and each cluster $X_j^c \subset TD^c$ the coefficient S_{cj}^x is calculated. Using the partition, the row r is classified as class c, for which the value (9) is minimal:

$$class(r) = arg \min_{c \in C, j \leq nc} S_{cj}^x \tag{10}$$

For the second measure one makes use of incremental mode of considered algorithms and distances from centroids from previous rounds are used.

3 Online Algorithms

Our online algorithms work in rounds. In each round a chunk of data is classified by an ensemble of classifiers. All the classifiers in the ensemble use **Algorithm 4** but differ in distance metric *dist*. In each round, for each classifier separately, kernel-based fuzzy clustering is performed on a fixed chunk of data (learning step). In the next step (classification) for each data row from the next chunk of data class value

Algorithm 4 Classification based on fuzzy clustering partition

Input: partition of learning data $X = \bigcup_{c \in C} \bigcup_{j \leq nc} X_j^c$, set of centroids *cnAll*, data row $r \notin X$, number of neighbours x
Output: class for row r
1: calculate S_{cj}^x for all $c \in C, j \leq nc$ according to (9)
2: calculate $dist(r, cnt)$ for all $cnt \in cnAll$,
3: **if** $\min_{c \in C, j \leq nc} S_{cj}^x < \min_{cnt \in cnAll} dist(r, cnt)$ **then**
4: $\quad cl = arg \min_{c \in C, j \leq nc} S_{cj}^x$
5: **else**
6: $\quad cl =$ class of centroid *cnt* for which $dist(r, cnt)$ is minimal
7: **end if**
8: **return** class *cl*

is assigned relying on the clustering information as well as using centroids from previous rounds. After all the ensemble members have made their decisions majority vote is performed for each row, and next round is performed. The number of rounds is proportional to the size of data divided by the size of the chunk. The details are in **Algorithm 5**. In line 22 where precision is calculated, the testing data size is data size diminished by the chunk size since first *ch* rows are used for training.

Algorithm 5 Online algorithm

Input: data D, chunk size ch, number L of ensemble classifiers
Output: qc - quality of online classification
1: initialize $correctClsf \leftarrow 0$
2: $dataClust \leftarrow$ first ch rows from D
3: $dataClassif \leftarrow$ next ch rows from D
4: $centroidsAll \leftarrow \emptyset$
5: **while** not all rows in D considered yet **do**
6: $\quad$ **for** l=1 to L **do**
7: $\quad\quad$ perform kernel-based clustering on *dataClust* according to Algorithm 1 or Algorithm 2
8: $\quad\quad$ **for all** row $r \in dataClassif$ **do**
9: $\quad\quad\quad$ assign class label c_l^r to r according to Algorithm 4, using current clustering and *centroidsAll*
10: $\quad\quad$ **end for**
11: $\quad\quad$ update *centroidsAll* by adding centroids from the current clustering
12: $\quad$ **end for**
13: $\quad$ **for all** row $r \in dataClassif$ **do**
14: $\quad\quad$ perform majority vote on $\{c_1^r, \ldots, c_L^r\}$ to assign class c to r
15: $\quad\quad$ **if** c is a correct label **then**
16: $\quad\quad\quad$ $correctClsf \leftarrow correctClfs + 1$
17: $\quad\quad$ **end if**
18: $\quad$ **end for**
19: $\quad$ $dataClust \leftarrow dataClassif$
20: $\quad$ $dataClassif \leftarrow$ next ch rows from D
21: **end while**
22: $qc \leftarrow \frac{correctClfs}{|D| - ch}$

Remark. In **Algoritms 1, 2, 3, 4** data rows stand for subsets of R^N, where N is the number of attributes and in Algorithm 5 data also contains class value, in order to check the classification quality.

As far as computational complexity is concerned observe that

- the complexity of both kernel based clustering (Algorithm 1, Algorithm 2) is $O(t \cdot M \cdot nc)$, where t is the number of iterations of the C-means algorithm, M is the number of data rows and nc is the number of clusters,
- estimation of number of clusters needs $O(p \cdot M^2)$ steps where p is the number of iterations of eigenvectors computation used by Apache Common Mathematics Library,
- classification based on calculating the classification coefficient demands sorting of M values which sums up to complexity $O(M^2)$.
- online algorithm **Algorithm 5** is performed in M/ch rounds, where ch is the chunk size; each round requires clustering thus the the computational complexity is $O(t \cdot M^3 \cdot nc)$.

4 Validating Experiment Results

To test accuracy of the proposed approach to the online classification problem, computational experiment has been carried-out. In the experiment we used a set of publicly available benchmark datasets including those often used to test incremental learning algorithms. The experiment involved the following datasets from the UCI Machine Learning Repository [1]: Bank Marketing (4522/17), Breast (263/10), Heart (303/14), Image (2086/19), Ionosphere (351/35), Magic (19020/11), P-H (141179/11), Sonar (208/61), Spam (4601/58), Thyroid (7000/21), Waveform (5000/41), the following sets from [11]: Banana (5300/16), SEA (50000/4), Twonorm (7400/21), the sets from [20]: Chess (503/9), Luxembourg (1901/32) and Electricity (44976/6) from [15]. In Table 1 classification accuracy averaged over 20 runs obtained by two variants of the proposed ensemble KFCM-K and KFCM-F, are shown. It can be observed that the proposed variants of the ensemble kernel-based classifiers perform, statistically, equally well. Comparison with performance of some state of the art online classifiers shown in Table 2 suggests that the proposed approach could be a useful tool for the online classification. This statement is supported by a relatively low computational complexity of the approach as mentioned in Sect. 3.

Table 1 Classification accuracy of the compared online classifiers

	KFCM-K		KFCM-F	
Dataset	Chunk	Accuracy	Chunk	Accuracy
Banana	15	87,9 +-0,4	10	**88,6 +- 0,5**
Bank	10	85,8 +-0,6	4	**88,5 + - 0,8**
Breast	50	**74,06 +-0,5**	5	74,4 + - 0,6
Chess	1	75,2 +-1,1	1	**75,9 + - 1,2**
Electricity	1	**94,01 +-0,7**	4	88,6 + - 0,4
Heart	5	7,2 +-0,4	5	**80,5 + - 0,5**
Image	1	**96,4+-0,9**	1	96,3 + -1,6
Ionosphere	3	91,7+-0,6	3	**91,9 + - 0,5**
Luxemburg	4	87,4 +-0,8	4	**87,6 + - 0,9**
Magic	1	**1 +-0**	1	75,8 + - 0,6
P-H	1	51,2 +-0,2	3	**54,4 + - 0,2**
Sea	1	77,1 +-0,7	4	**78,5 + - 0,5**
Sonar	1	89,7 +-0,6	1	**90,8 + - 0,7**
Spam	1	**1 +-0**	5	99,9 + - 0,0
Thyroid	1	**92,1 +-0,4**	10	89,9 + - 0,4
Twonorm	1	93,9 +-0,5	15	**95,5 + - 0,1**
Waveform	3	76,4 +-0,7	40	**80,9 + - 0,9**

Table 2 Accuracy of the OL-Kernel-E classifier versus some recently reported results

Name	Best of KFCM-K and KFCM-F	Recently reported
Banana	88,6	89,3 (Inc SVM [16])
Bank M.	**88,5**	86,9 (LibSVM [15])
Breast	**74,6**	72,2 (IncSVM [16])
Chess	**75,9**	71,8 (IncL [5])
Electricity	**94,1**	88,5(IncSVM [5])
Heart	80,5	83,8 (IncSVM [16])
Image	**96,4**	94,2 (SVM [12])
Poker-Hand	**54,4**	52,2 (LWF [12]
Sea	78,5	96,6 (Kaoginc [2])
Spam	**1**	85,5 (K-a graph [2])
Thyroid	92,1	95,8 (LibSVM [15])
Twonorm	95,5	97,6 (FPA [16])
Waveform	80,9	86,7 (Inc SVM [16])

5 Conclusions

The paper contributes through proposing the online (incremental) ensemble classifier. The idea is to cluster reference examples from the training set using a kernel function. Next, majority vote based on several distance/similarity measures us used to predict class label of the newly arrived instance. After this step a set of the reference instances called here a chunk, is updated using the sliding window concept, and the cycle is repeated with the next incoming set of attributes, representing an instance which class is to be predicted.

Computational complexity analysis has shown that the approach assures classification in polynomial time. Proposed approach have been validated experimentally. The reported computational experiment proves that online KFCM classifiers perform well in terms of the classification accuracy outperforming some state-of-the-art incremental classifiers reported in the literature.

Future research will focus on finding more sophisticated approach to predicting class label than majority voting. It also expected to extend the range of similarity measures or even integrating them with other approaches through incorporating simple base classifiers.

References

1. Asuncion, A., Newman, D.J.: UCI machine learning repository. http://www.ics.uci.edu/mlearn/MLRepository.html, University of California, School of Information and Computer Science (2007)
2. Bertini, J.R., Zhao, L., Lopes, A.: An incremental learning algorithm based on the K-associated graph for non-stationary data classification. Inf. Sci. **246**, 52–68 (2013)
3. Bezdek, J. C.: Pattern Recognition with Fuzzy Objective Function Algorithms. Kluwer Academic Publishers (1981)
4. Chiang, J.H., Hao, P.Y.: A new kernel-based fuzzy clustering approach: support vector clustering with cell growing. IEEE T. Fuzzy Syst. **11**(4), 518–527 (2003)
5. Ditzler, G., Polikar, R.: Incremental learning of concept drift from streaming imbalanced data. IEEE Trans. Knowl. Data Eng. **25**(10), 2283–2301 (2013)
6. Gaber, M.M., Zaslavsky, A., Krishnaswamy, S.: Data stream mining. In: Maimon, O., Rokach, L. (eds.) Data Mining and Knowledge Discovery Handbook, pp. 759–787, Part 6 (2010)
7. Graves, D., Pedrycz, W.: Kernel-based fuzzy clustering and fuzzy clustering: A comparative experimental study. Fuzzy Sets Syst. **161**(4), 522–543 (2010)
8. Jędrzejowicz, J., Jędrzejowicz, P.: A family of the online distance-based classifiers. In: Nguyen, N.T. et al. (eds.) ACIIDS 2014, LNAI, vol. 8398, pp. 177–186. Springer, Heidelberg (2014)
9. Last, M.: Online classification of nonstationary data streams. Intell. Data Anal. **6**, 129–147 (2002)
10. Li, Z., Tang, S., Xue, J., Jiang, J.: Modified FCM clustering based on kernel mapping. Proc. SPIE **4554**, 241–245 (2001)
11. Machine learning data set repository. http://mldata.org/repository/tags/data/IDA_Benchmark_Repository/ (2013)
12. Moreno-Torres, J.G., Sáez, J.A., Herrera, F.: Study on the impact of partition-induced dataset shift on k -fold cross-validation. IEEE Trans. Neural Netw. Learn. Syst. **23**(8), 1304–1312 (2012)

13. Murata, N., Kawanabe, N., Ziehe, A., Muller, K.R., Amari, S.: On-line learning in changing environments with application in supervised and unsupervised learning. Neural Netw. **15**, 743–760 (2002)
14. Pramod, S., Vyas, O.P.: Data stream mining: A review on windowing approach. Global J. Comput. Sci. Technol. Softw. Data Eng. **12**(11), 26–30 (2012)
15. Waikato.: http://moa.cms.waikato.ac.nz/datasets/ (2013)
16. Wang, L., Ji, H.-B., Jin, Y.: Fuzzy passive aggressive classification: A robust and efficient algorithm for online classification problems. Inf. Sci. **220**, 46–63 (2013)
17. Widmar, G., Kubat, M.: Learning in the presence of concept drift and hidden contexts. Mach. Learn. **23**, 69–101 (1996)
18. Zhang, D., Chen, S.: Clustering incomplete data using Kernel-based fuzzy C-means algorithm. Neural Process. Lett. **18**(3), 155–162 (2003)
19. Zhang, D., Chen, S.: Fuzzy clustering using kernel method. In: Proceedings of the International Conference on Control and Automation ICCA, pp. 162–163. Xiamen, China (2002)
20. Žliobaite, I.: Controlled permutations for testing adaptive classifiers. In: Proceedings of the 14th International Conference on Discovery Science, Springer LNCS, vol. 6926, pp. 365–379 (2011)

A Survey on Expert Systems for Diagnosis Support in the Field of Neurology

Mirco Josefiok, Tobias Krahn and Jürgen Sauer

Abstract In this paper we present a survey of expert systems in neurology. Especially in the field of neurology diseases occur in varying combinations of symptoms. Supporting medical staff in the process of finding a correct diagnosis to a hypothesis in a timely manner is very desirable to improve a patients outcome. We searched Google Scholar, Mendeley and PubMed and compared the found systems regarding their performance, precision and applicability in practice. Moreover we highlight gaps and point out possible further research directions.

Keywords Experts in neurology · Decision support systems in neurology · AI in neurology · Survey · Meta analysis

1 Introduction

Neurological diagnoses occur in different and varying combinations of symptoms. Differential diagnosis is a challenge for doctors, especially for general practitioners or those with less experience. For example the stroke is one of the most common diseases in Germany. About six billion Euros are spent each year for the treatment of the most frequent types.[1] A rapid and early diagnosis significantly improves patient

[1] See diagnosis code I60, I61, I63, I65 at http://www.gbe-bund.de - last visited 2014-12-14.

M. Josefiok (✉) · T. Krahn
OFFIS—Institute for Information Technology, Escherweg 2, Oldenburg, Germany
e-mail: mirco.josefiok@offis.de
url: http://www.offis.de

T. Krahn
e-mail: tobias.krahn@offis.de

J. Sauer
Carl von Ossietzky Universität Oldenburg, Ammerländer
Heerstr. 114-118, 26129 Oldenburg, Germany
e-mail: juergen.sauer@uni-oldenburg.de
http://www.uni-oldenburg.de

R. Neves-Silva et al. (eds.), *Intelligent Decision Technologies*,
Smart Innovation, Systems and Technologies 39,
DOI 10.1007/978-3-319-19857-6_26

prognosis and allows for applying appropriate measures. It is therefore desirable to offer as much support to medical staff as possible.

Especially in the healthcare domain, various attempts of applying expert systems exist and the possible benefits are widely recognized [1, 5, 11]. Research and development of medical expert systems started in the 70s and systems are still evolving. Software systems supporting medical staff in the diagnostic process have been developed for nearly every medical domain. While at present computer systems are used in various steps of a patients treatment, e.g. in detecting adverse drug events [17], diagnosis related expert systems are rarely used in practice [2, 14, 21]. Expert systems are software solutions utilizing specialists' knowledge in order to support the decision-making process [19]. Contrary to other knowledge-based systems, to which expert systems belong, the knowledge base is derived from human experts. However, expert systems seek to reduce the need for human experts in practice. This is one main advantage, because human experts are often extremely costly, scarce or unavailable [4]. Expert systems can be distinguished by their representation of the underlying knowledge base. In general there are rule-based systems and systems based on artificial intelligence. Rule-based systems can be subdivided further according to their individual representation of knowledge [4].

This paper is organized as follows. In Sect. 2 we present our research method. In Sect. 2.3 we give an overview of the surveyed papers and present our findings. The paper concludes with Sect. 4 in which we discuss our findings, give a brief conclusion and point out possible research directions.

2 Method

Our previous research revealed an ongoing, if not increasing, demand for computer aided support in finding a correct diagnosis [15, 26]. Through expert interviews we were able to develop a theoretical concept and a corresponding business process to properly integrate computer based decision support in the daily routine of medical staff [26]. For enhanced orientation, we carried out a systematic literature review by searching PubMed, Mendeley and Google Scholar for the terms *neurology*, *expert systems*, *diagnosis support systems*, *decision support systems* and *healthcare* in various combinations. To conduct the review we used the PRISMA (Reporting Items for Systematic Reviews and Meta-Analyses) statement [18]. The PRISMA statement describes an approach on how to conduct systematic reviews, surveys and meta-analyses. It consists of a flow diagram which describes the process of filtering and selecting references, a checklist for each section and corresponding literature. While no restrictions were placed on publication date, we focused on research from the past decade. We considered only literature in English and German. The last search has been conducted on December 09, 2014. Results from Mendeley were hard to process, because even with multiple filters and numerous search terms large amounts of unrelated results were returned.

Our research questions were:

- What approaches for expert systems in neurology do exist?
- Are the expert systems applicable in daily practice and what are possible gaps and further research directions?

2.1 Literature Selection

A first selection of the found literature was performed based on the title and the abstract. We screened the belonging references and sorted out the unrelated ones. For our survey we did not include comprehensive theoretical approaches, reviews and editorials. In addition, we considered expert systems supporting the analyses of medical images or laboratory values as out of scope for this work and skipped them as well. Moreover, we left out works that deal with diagnosis support for animals.

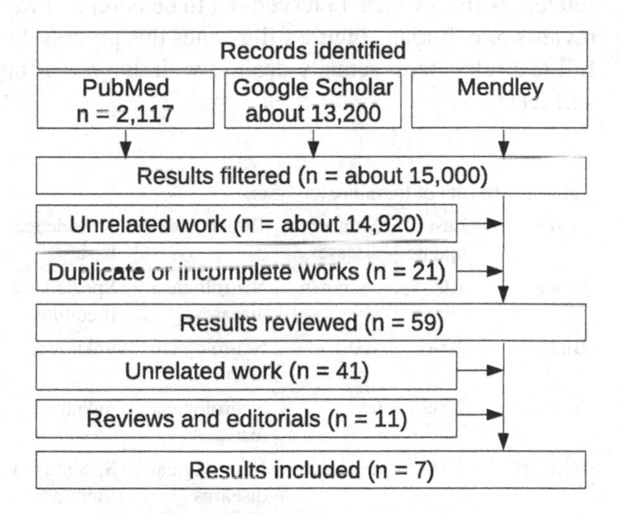

Fig. 1 Literature selection process

Figure 1 shows a flow diagram of the survey process. In Sect. 2.3, the survey process is explained in more detail. Most expert systems in the field of neurology seem to deal with supporting medical staff in analyzing images or medical data (e.g. Electroencephalography (EEG) results) and are therefore not related to support the process of finding an appropriate diagnosis.

2.2 Criteria for Our Analysis

For our analysis, the following criteria were included: Year of initial publication, year of the last update, number of considered neurological diseases, and foundation of knowledge base (e.g. literature study, expert interviews etc.), implementation of knowledge base, interface and applicability in practice. In addition we tried to find out how the individual systems were validated and how they performed in practice. This includes information about correctly identified diagnoses.

2.3 Overview of Surveyed References

Following the described search strategy, we identified more than 15,000 references. After further filtering, we were able to reduce this set to 14,920 references. From those, we identified and removed 21 duplicates. Finally, we were left with 59 full text references from which 43 turned out to be unrelated work and 11 were identified as reviews or editorials. Figure 1 illustrates this process. Subsequent to this, only eight full text references, actually dealing with diagnosis support systems in neurology, still remained.

Table 1 Results of literature analysis

Name	Last update	Number of diseases	Target diseases	Knowledge base	Method	Reference
Bunke	1988	unknown	Neurological diseases	Specialized literature	Database, rule-based	[9]
Bickel	2006	400	Neurological diseases	unknown	Database, statistical	[6]
Roses	2009	6	Neurological diseases	unknown	Rule-based	[23]
Reimers	2009	unknown	Neurological diseases	Specialized literature	Database, rule-based	[20]
Borgohain	2012	4	Neuromuscular disorders	Expert interviews	Rule-based	[7]
Borgohain	2012	1	Cerebral Palys Diagnosis	Expert interviews	Rule-based	[8]
Ghahazi	2014	1	Multiple Sclerosis	Spreadsheet	Fuzzy logic	[13]

3 Results

Table 1 gives an overview of our findings, which we present in more detail in this section. While we discovered different approaches for expert systems in neurology, we found no references which show or hint at a routine application of any expert system. This emphasizes our previous findings [26]. We initially started the survey with the objective to compare the different expert systems. However, because of the lack of information (e.g., unclear information how the specific knowledge base is build, number of targeted diseases etc.), different approaches and incomplete and inconsistent validation, make a comparison nearly impossible. Therefore, in the following sections, we would like to discuss the works in more detail.

3.1 Presentation and Discussion of Individual References

In the following we will present a short overview of the individual reviewed references.

Bunke et al.: Bunke et al. [9] discuss two different approaches for expert systems in neurology. Their main intention was to compare a rule-based approach with a database approach. Two prototypes were developed and compared. The authors decided to only implement their expert systems focusing on diagnoses related to the median nerve. They created the necessary rules by using statements like *if condition then conclusion* and a commercial expert system shell called EXSYS. The authors implemented 163 rules and 17 possible diagnoses. The database was created in a similar fashion but using relations. Subsequently the two approaches have been compared regarding their response time and accuracy. The authors found out that the database oriented prototype responded faster but was less accurate. They assumed this was because no probabilities were taken into account. No details are given on how the knowledge base was created or its validation.

Bickel et al.: Bickel and Grunewald [6] present an expert system for the field of neurology developed on top of the commercial program Filemaker-7.0. They added about 400 diagnoses with their corresponding symptoms and created the necessary dialogs for accessing the data. The authors evaluated their approach in two steps. First, the performance was tested with 15 predefined cases. The test persons had minor neurological experience. In this case the program was able to find the correct diagnosis in each case. Second, real cases were used. With real cases a correct diagnosis was found in approximately 80 % of the trials. The authors are both practising physicians and thereofore derived the data necessary from the expert system from their daily work and corresponding specialized literature.

Roses: Roses is the name of an expert system developed by Vaghefi and Isfahani [23] with the objective to support the diagnostic process of six neurological diseases in children. The knowledge base is modeled by using Java Expert System Shell (JESS)

[12]. In its last incarnation the system suffered from low reliability. Only about 65 % correct diagnoses could be made. Another pressing issue identified during testing, was the need for a graphical user interface, enabling the modification of the knowledge base by the users.

Reimers et al.: Reimers and Bitsch [20] developed and distributed an expert system which aims at supporting medical staff in finding a correct diagnosis to a hypothesis. The authors created a comprehensive knowledge base by evaluating specialized literature. Neither information on how the knowledge base is represented nor validation results are given.

Borgohain et al.: Borgohain and Sanyal [7, 8] present two expert systems which share the same approach but a different knowledge base. Both systems are created using JESS and therefore do not contain any probability. The main source for creating the knowledge base for the systems were interviews with doctors and postgraduate students of a nearby hospital. The knowledge base was thereafter extended by using specialized literature and internet research. The systems were evaluated by testing its rule base against a few test cases which have not been further specified. Both systems provided accurate diagnoses when the symptoms were fully given. However, the authors described no practical application experiences or tests with real cases.

Ghahazi et al.: Ghahazi et al. [13] present an approach for an expert system based on fuzzy logic for diagnosing Multiple Sclerosis. They perform the diagnosis an base of basic patient data, symptoms and signs. They work with crisp inputs which lead to uncertain results. For storing their data they use a spreadsheet. The system was evaluated with multiple test cases from specialized literature and performed well. The integration of a spreadsheet based solution in the daily workflow of medical staff is perceived as problematic.

3.2 General Findings

In view of the progress in soft computing, or to be more specific in the development of expert systems, the small amount of systems or even only approaches we found is surprising [21]. In the healthcare domain with MYCIN [22, 24] and CADUCEUS [3] early attempts exist regarding expert systems with an inference engine. The experts systems of the CADIAG family are an enhancement of those approaches. In a first iteration the system's knowledge base was represented by if-then rules and interferences were performed using propositional logic. To overcome the shortcomings of imprecise or incomplete information, a second iteration was developed which was able to work with graded inputs and weighted rules. A yet unsolved problem with CARDIAG-like systems concerns the interpretation of the results provided by an expert system. Usually information is provided to the system and the result is displayed as a numeric value associated to a fact [10]. A shortcoming, from which all reviewed expert systems suffer, is the lack of dealing with probability, imprecise information and lack of proper validation. Therefore, we would like to emphasize these topics in more detail in the following subsections, namely the foundation of knowledge base, the implementation of knowledge base as well as the validation of expert systems.

Foundation of Knowledge Base The knowledge base of an expert system can be considered as its most important part. Therefore the choice of a correct source for the information leading to the knowledge and their validation are of the utmost importance.

Most systems reviewed in this work do not evaluate the underlying knowledge. The main source of knowledge for expert systems is directly acquired from experts. In order to be considered an expert, certain criteria must be fulfilled. These are for example: experts posses above average abilities in their domain, experts use their experience to solve problems, experts can solve problems even although incomplete and imprecise information is given, experts are rare and expensive etc. This means in particular that, if an expert system is created, one must prove, that the knowledge is acquired from human experts [4]. In some cases it was not even possible to find out how the knowledge base was created [6, 13, 23]. In other cases, the main source for the knowledge base was specialized literature [9, 20]. In the two cases, in which expert interviews were mentioned as a primary source for creating the knowledge bases, no evidence was given whether experts were interviewed or not [7, 8].

Implementation of Knowledge Base For implementing a knowledge base, different approaches exist. According to a recent survey [25] in healthcare, early systems used Bayesian approaches. They were mostly replaced by rule-based approaches with an interference engine because they proved to be better suited for modeling correlations of symptoms and diseases. With an increasing rule base it became obvious that they were hard to maintain. Moreover, probability could not be modeled. Therefore, fuzzy rules, fuzzy sets and neuro fuzzy approaches were introduced. In addition, artificial neuronal networks (ANN) were established. ANNs are a mathematical model which simulate a biological neural network. While systems using ANNs perform well, the results are not interpretable and they strongly depend on the used training data. Most of the reviewed systems in this paper use a rule-based approach [7–9, 20, 23]. However, one system is based on a fuzzy based approach [13] and one used a heuristic approach [6].

Validation of Expert Systems A proper validation of expert systems is necessary [16]. Whilst most systems are evaluated regarding their accuracy [6–9, 13, 23], only a few are evaluated in practice. For one system, no validation was executed at all [20]. In Table 2, an overview of the validation of the different expert systems is given. Whilst most systems are evaluated regarding their accuracy [6–9, 13, 23] few are evaluated in practice. For one even no validation is given [20]. And even if an expert system is evaluated there is few information available regarding the used cases. In Table 2 an overview of the validation of the different expert systems is given.

In summary, no statement can be made whether the performance of practitioners and therefore patient care can be improved when using one of the proposed expert systems.

Table 2 Validation of expert systems

Name	Count of cases used for validation	Percentage of correct diagnoses	Reference
Bunke	unknown	unknown	[9]
Bickel	15	81 %	[6]
Roses	unknown	65 %	[23]
Reimers	no validation		[20]
Borgohain	some	100 %	[7]
Borgohain	some	100 %	[8]
Ghahazi	unknown	unknown	[13]

3.3 Limitations

We performed the systematic literature search as thoroughly as possible, some articles could have been missed. Especially articles that are written in other languages than English or German. Moreover, there exists some solely commercial expert systems for which no or only few information is available [20]. Because of the limited available information and the great variety in targeted diseases, methods applied, knowledge base and validation no statistical comparison of the expert systems is possible.

4 Conclusion and Further Research Questions

Concerning our research questions, it happens that there isn't any system in practical use. The main reasons are:

- Insufficient validation of the knowledge base.
- Unclear origin of the knowledge base.
- The absence of transparency and accountability of results and data.

Regarding the question, which approaches exist for expert systems in neurology, we found that most of the reviewed systems in this paper use a rule-based approach and only one is based on fuzzy logic [13].

We found three different aspects to be most important, when executing the review. Those are foundation of knowledge base, implementation of knowledge base and its validation. We covered the different aspects in Sect. 3.2. Regarding the reviewed articles, none was found to be fully convincing. Moreover, in all cases no considerations regarding user interface, expandability and customizability of knowledge are published. For practitioners in healthcare it is very important that the results of the experts are comprehensible.

We were able to identify the following starting points for further research:

- How can one validate expert systems in a correct manner?
- How can a potential influence of a patients outcome be measured?
- How can medical staff be empowered to expand and customize the knowledge base?
- What requirements must an expert system fulfill that its knowledge base can be expanded and customized from domain experts?

Considering the demand for computer assistance in finding a correct diagnosis and the potential positive effects for medical staff further research will hopefully resolve the questions that have arisen [1].

Our next steps include performing further expert interviews, drawing up and implementing a corresponding technical concept and finally creating a usable prototype for an expert system. Our overal aim ist to provide a sensible set of rules with which we can produce comprehensible results. For medical staff it is utmost important the the results provided by an expert system are comprehensible and understandable. Thereby we aim to continously evaluate our work with the support of practionaires from a local hospital.

References

1. Adhikari, N.K.J., Beyene, J., Sam, J., Haynes, R.B.: Effects of computerized clinical decision support systems on practitioner performance. JAMA **293**(10), 1223–1238 (2014)
2. Angeli, C.: Diagnostic expert systems: from expert's knowledge to real-time systems. In: Sajja, P., Akerkar, R. (eds.) Advanced Knowledge Based Systems: Model, Applications & Research, vol. 1, Chap. 4, pp. 50–73 (2010)
3. Banks, G.: Artificial intelligence in medical diagnosis: the INTERNIST/CADUCEUS approach. Crit. Rev. Med. Inf. **1**, 23–54 (1986)
4. Beierle, C., Kern-Isberner, G.: Methoden wissensbasierter Systeme: Grundlagen, Algorithmen, Anwendungen (Computational Intelligence) (German Edition). Springer Vieweg, 5, überar edn. (2014), http://amazon.com/o/ASIN/3834818968/http://link.springer.com/content/pdf/10.1007/978-3-8348-9517-2.pdf
5. Berner, E.: Clinical decision support systems: state of the art. AHRQ Publication (09) (2009), http://scholar.google.com/scholar?hl=en&btnG=Search&q=intitle:Clinical+Decision+Support+Systems+:+State+of+the+Art#0
6. Bickel, A., Grunewald, M.: Ein Expertensystem für das Fachgebiet Neurologie: Möglichkeiten und Grenzen. Fortschr. Neurol. Psychiatr. **74**(12), 723–731 (2006)
7. Borgohain, R., Sanyal, S.: Rule Based Expert System for Cerebral Palsy Diagnosis. arXiv preprint arXiv:1207.0117, pp. 1–4 (2012), http://arxiv.org/abs/1207.0117
8. Borgohain, R., Sanyal, S.: Rule Based Expert System for Diagnosis of Neuromuscular Disorders. arXiv:1207.2104, pp. 1–5 (2012)
9. Bunke, H., Flückiger, F., Ludin, H.P., Conti, F.: Diagnostic expert systems in neurology—a comparison between a rule based prototype and a database oriented approach. In: Rienhoff, O., Piccolo, U., Schneider, B. (eds.) Expert Systems and Decision Support in Medicine, pp. 209–212. Springer (1988)
10. Ciabattoni, A., Picado, D., Vetterlein, T., El-zekey, M.: Formal approaches to rule-based systems in medicine: the case of CADIAG-2. Int. J. Approx. Reasoning **54**, 132–148 (2013). http://www.sciencedirect.com/science/article/pii/S0888613X12001612

11. Dinevski, D., Bele, U., Šarenac, T., Rajkovič, U., Šušteršič, O.: Clinical decision support systems. Studies in health technology and informatics, pp. 217–238 (2013)
12. Friedman-Hill, E.: JESS in Action. Manning Greenwich, CT (2003)
13. Ghahazi, M.A., Fazel Zarandi, M.H., Harirchian, M.H., Damirchi-Darasi, S.R.: Fuzzy rule based expert system for diagnosis of multiple sclerosis. In: 2014 IEEE Conference on Norbert Wiener in the 21st Century (21CW), pp. 1–5. IEEE (2014)
14. Ishak, W., Siraj, F.: Artificial intelligence in medical application: an exploration. Health Informatics Europe Journal (2002). http://www.researchgate.net/publication/240943548_ARTIFICIAL_INTELLIGENCE_IN_MEDICAL_APPLICATION_AN_EXPLORATION/file/60b7d5292d45d674ab.pdf
15. Josefiok, M.: Measuring and Monitoring the Rehabilitation of Patients on Monitoring Stations via the Analyses of poly-structured Data. In: Plödereder, E., Grunske, L., Schneider, E., Dominik, U. (eds.) 44. Jahrestagung der Gesellschaft für Informatik, Informatik 2014, Big Data—Komplexitnät meistern. pp. 2305–2310. LNI, GI (2014)
16. Kaplan, B.: Evaluating informatics applications–clinical decision support systems literature review. Int. J. Med. Inf. **64**(1), 15–37 (2001). http://linkinghub.elsevier.com/retrieve/pii/S1386505601001836, http://www.sciencedirect.com/science/article/pii/S1386505601001836
17. Krahn, T., Eichelberg, M., Gudenkauf, S., Laleci, E., Gokce, B., Hans-Jürgen, A.: adverse drug event notification system-reusing clinical patient data for semi-automatic ADE detection. In: Proceedings 27th International Symposium on Computer-Based Medical Systems (IEEE CBMS 2014) (2014)
18. Moher, D., Liberati, A., Tetzlaff, J., Altman, D.G., Group, T.P.: Preferred Reporting Items for Systematic Reviews and Meta-Analyses: the PRISMA Statement. PLoS Med **6**(7), e1000097 (2009). http://dx.doi.org/10.1371/journal.pmed.1000097
19. Puppe, F.: Einführung in Expertensysteme (Studienreihe Informatik) (German Edition). Springer (1988). http://amazon.com/o/ASIN/3540194819/
20. Reimers, C.D., Bitsch, A.: Differenzialdiagnose Neurologie. Kohlhammer, Stuttgart (2008)
21. Sharma, T., Jain, S.: Surv. Expert Syst. **2**(1), 343–346 (2012)
22. Shortliffe, E.H. (ed.): Computer-Based Medical Consultations: Mycin. Elsevier (1976)
23. Vaghefi, S.Y.M., Isfahani, T.M.: ROSES: an exper system for diagnosing six neurologic diseases in children. In: International Conference Intelligent Systems & Agents, pp. 259–260 (2009)
24. Van Remoortere, P.: Computer-based medical consultations: MYCIN (1979)
25. Wagholikar, K.B., Sundararajan, V., Deshpande, A.W.: Modeling paradigms for medical diagnostic decision support: a survey and future directions. J. Med. Syst. **36**(5), 3029–3049 (2012). http://link.springer.com/article/10.1007/s10916-011-9780-4/fulltext.html
26. Weiß, J.P., Josefiok, M., Krahn, T., Appelrath, H.J.: Entwicklung eines Fachkonzepts für die klinische Entscheidungsunterstützung durch Analytische Informationssysteme. In: Proceedings of the 12th International Conference on Wirtschaftsinformatik (WI2015) (2015), to be published

Towards a New Approach for an Effective Querying of Distributed Databases

Abir Belhadj Kacem and Amel Grissa Touzi

Abstract Nowadays, with the evolution of data and their geographical distribution, Distributed Database Management Systems (DDBMS) have become undoubtedly a need for Information Systems (IS) users. Unfortunately, query optimization remains a handicap for existing DDBMS, given the high cost of network traffic caused by the access to geographically distributed data in different sites. To remedy this problem, we propose a new effective approach of querying Distributed Database (DDB) based on the definition of relevant sites to the query knowing fragmentation and /or duplication of distributed data. This approach allows us to minimize the volume of transferred data via network and consequently reduces the query execution cost. This approach has been validated by implementing a layer "effective-query" on Oracle DDBMS.

Keywords Distributed database • Query • DDBMS • Query processing

1 Introduction

The end of the last century was marked by a rapid evolution of the distributed context in several areas such as industry, science, and commerce that often need to store and manipulate data at geographically distributed sites using the computing power of distributed systems. This led to the rise of distributed databases (DDB) and Distributed Database Management Systems (DDBMS). Thus, many DDBMS have been commercialized such as Oracle [1], MySQL [2], Ingres [3], Cassandra [4], etc.

A.B. Kacem (✉) · A.G. Touzi (✉)
National School of Engineers of Tunis, University of Tunis El Manar,
LR-SITI, Bp.37, Le Belevedere, 1002 Tunis, Tunisia
e-mail: belhadj.kacem.abir@gmail.com

A.G. Touzi
e-mail: amel.touzi@enit.rnu.tn

R. Neves-Silva et al. (eds.), *Intelligent Decision Technologies*,
Smart Innovation, Systems and Technologies 39,
DOI 10.1007/978-3-319-19857-6_27

Although these DDBMS don't automatically ensure the transparency of data distribution and do not directly represent the basic concepts of DDB such as fragmentation and replication, query optimization requiring high cost of network traffic remains their major handicap.

Indeed, global queries addressed to these DDBMS are decomposed into sub-queries. These sub-queries are sent to the available systems on the local sites where they are executed. Local responses are then reconstituted to elaborate query response [5].

This process is faced with an imminent problem caused by the solicitation of sites whose fragments are useless to elaborate the query response increasing then the query execution cost.

Several approaches have been proposed for the evaluation and optimization of distributed queries. In [6], Hevner et al. have proposed to apply a method to extend the optimal algorithms to derive efficient distribution strategies for general query processing in distributed database systems. Besides Parallel execution strategies for horizontally fragmented databases is treated in [7].

Unfortunately, these approaches don't take in account execution strategies for replicated fragments in distributed databases. In this paper, we propose an intelligent approach to evaluate queries in DDB. This approach is based on the concept of query rewriting by restricting search to relevant sites and data knowing fragmentation and/or duplication of tables:

1. If a table is vertically fragmented, we access only to the sites where relevant fragments to our query are stored.
2. If a table is replicated on different sites, we ask the most useful site for our request.

Besides this introduction, this paper includes four sections. Section 2 presents the basic concepts of DDB. Section 3 presents our motivations for this work. Section 4 presents our new approach for querying DDB. Section 5 presents the validation of our approach to simple queries. Finally, we conclude by evaluating this work and proposing some future perspectives.

2 Basic Concepts of DDB

2.1 Definition

We define a distributed database (DDB) as a collection of multiple, logically interrelated databases, distributed over a computer network [8].

A distributed database management system (DDBMS) is then defined as the software system that permits the management of the distributed database and makes the distribution transparent to the users [9]. As examples of DDBMS, we can mention: Oracle [1, 9], MySQL [2], Ingres [3] and Cassandra [4].

According to ANSI/SPARC architecture, there are three views of data: the external view, which is that of the end user, the internal view, that of the system or machine and the conceptual view. For each of these views, an appropriate schema definition is required [10].

2.2 DDB Design

To design a DDB, we must master the following techniques:

- Fragmentation: a relation can be divided into a number of sub-Relations, called fragments, allocated to one or more sites. There are two types of fragmentation: horizontal and vertical. The horizontal fragments are subsets of tuples and vertical fragments are subsets of attributes of relations. Fragmentation also has two other types: mixed or derived. The derived fragmentation is a type of horizontal fragmentation where the fragmentation of a relationship is based on cutting another relation.
- Replication: It's the reproduction of a subset of the main base on a different site. The duplication of data items is mainly due to reliability and efficiency considerations [11].

Example. Three institutions of the University of Tunis El Manar: National Engineering School of Tunis (ENIT), Faculty of Mathematical, Physical and Biology Sciences of Tunis (FST) and Faculty of Economics and Management of Tunis (FSEGT) have decided to pool their libraries and service loans, to enable all students to borrow books in all the libraries of the participating institutions. Joint management of libraries and borrowing is done by a distributed database over 3 sites (Site1 = ENIT, Site2 = FST and Site3 = FSEGT). The global schema of this database is as follows:

EMPLOYEE (IDEMP, NAME, ADRESS, STATUS, ASSIGNMENT)

STUDENT (IDSTU, LASTNAME, FIRSTNAME, ADRESS, INSTITUTION, CURSUS, NBBORROW)
DEPARTMENT (IDDEP, NAMEDEP)
BOOK (IDBOOK, TITLE, EDITOR, YEAR, AREA, STOCK, SITE)
AUTHOR (IDAUTH, NAMEAUTEUR)
LOAN (IDBOOK, IDSTU, BORROWINGDATE, RETURNDATE)

The table STUDENT is solicited by two separate entities: the administration of the institution and the library. The library needs a fast student identification to manage the loans and control of the number of borrows, while the administration will not need the information on the number of borrows. The administration of each university is responsible for the management of its students, but the library must have a list of all students in order to distribute and manage the loans properly. Then, we apply a horizontal fragmentation to table STUDENT based on university. The books are managed by the responsible of each library. The AUTHOR table is fragmented on different sites depending on the foreign key IDBOOK. Similarly for LOAN table. The table DEPARTMENT and $STUDENT_{LIB}$ are replicated to all sites because they are common to all sites and are much in demand as any student can borrow through the country and are not limited to the context of the university. Then, the fragmentation is as follows:

$$STUDENT_i = \Pi_{IDSTU,\, LASTNAME,\, FIRSTNAME,\, ADRESS,\, CURSUS}(\sigma_{STATEMENT=i}(STUDENT))$$

$$STUDENT_{LIB} = \Pi_{IDSTU,\, LASTNAME,\, FIRSTNAME,\, NBBORROW}(STUDENT)$$

$$EMPLOYE_i = \Pi_{IDEMP,\, NAME,\, ADRESS,\, STATUS}(\sigma_{ASSIGNMENT=i}(EMPLOYEE))$$

$$BOOK_i = \Pi_{IDBOOK,\, TITLE,\, EDITOR,\, YEAR,\, AREA,\, STOCK,\, SITE}(\sigma_{affectation=i}(BOOK))$$

with $i = ENIT, FST, \text{FSEGT}$

2.3 Query Processing in DDB

A query refers to the action of retrieving data from the database. We present in what follows the principle and steps of query processing in DDB.

In a distributed environment, the global query is decomposed into sub-queries. These sub-queries are sent to the local sites where they are executed. Local responses are then reconstituted to elaborate global query response [5]. The distributed query processing includes mainly the steps shown in Fig. 1.

Query Decomposition. Query decomposition is the first phase of query processing that transforms a relational query into a query described in relational algebra. Both input and output queries refer to global relations, without knowledge of the distribution of data. The successive steps of query decomposition are (1) normalization, (2) analysis, (3) elimination of redundancy, and (4) rewriting.

Data Localization. The localization layer translates a query on global relations described in relational algebra into a query expressed on physical fragments.

Query Optimization. Query optimization refers to the process of producing a query execution plan which represents an execution strategy for the query [12].

Sub-queries Execution. The distributed query execution step is very similar to centralized query execution. However, we should consider the need (1) to send and receive operators while exchanging fragments of queries to local sites and shipping return results, and (2) to execute dedicated algorithms, especially for the treatment of join, semi-join, hash joins etc. [13].

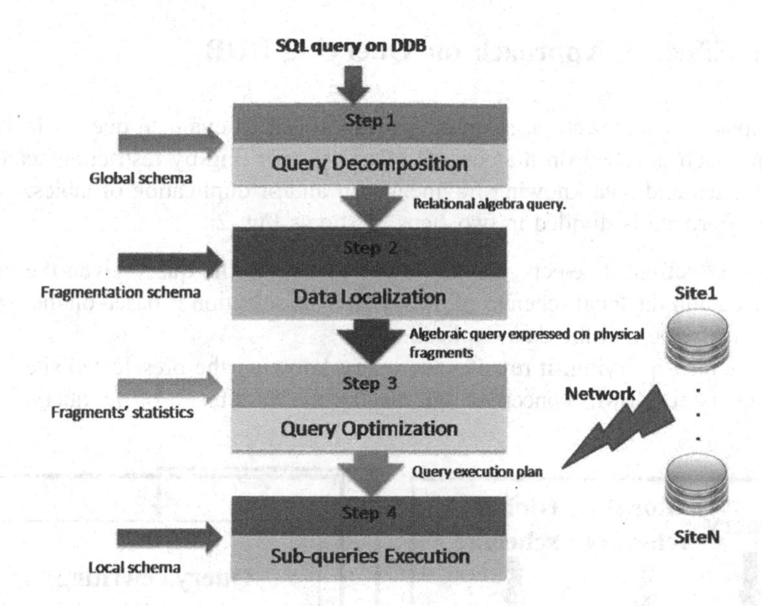

Fig. 1 Distributed query processing

3 Motivations

Several approaches have been proposed for the evaluation and optimization of distributed queries. In [6], Hevner et al. have proposed to apply a method to extend the optimal algorithms to derive efficient distribution strategies for general query processing in distributed database systems. Besides Parallel execution strategies for horizontally fragmented databases is treated in [4]. Unfortunately, these approaches don't take in account execution strategies for replicated fragments in distributed databases. Thus, the distributed query processing has prominent problems:

- High response time: At any given time, one of the sites may be unavailable for querying, or the network may be affected. This increases the response time and also the total execution time, or worse, may lead to the failure of query execution even if the query is independent of this site.
- High cost of network traffic: As the network throughput varies from day to day and even from hour to hour, optimization solutions must be found to avoid the transfer of useless data for our query.

The following section presents an effective approach to evaluate queries in DDB in order to remedy to these limitations.

4 An Effective Approach for Querying DDB

We propose, in this section, an intelligent approach to evaluate queries in DDB. This approach is based on the concept of query rewriting by restricting search to relevant sites and data knowing fragmentation and/or duplication of tables.

This approach is divided in two steps as shows Fig. 2:

- Data collection: it selects relevant sites and data to the query given the global schema and the local schemas of the DDB. This selection is based on the type of fragmentation and the data duplication.
- The step of querying: it rewrites the query knowing the preselected sites, send the query to DBMS concerned and display the final result of the query.

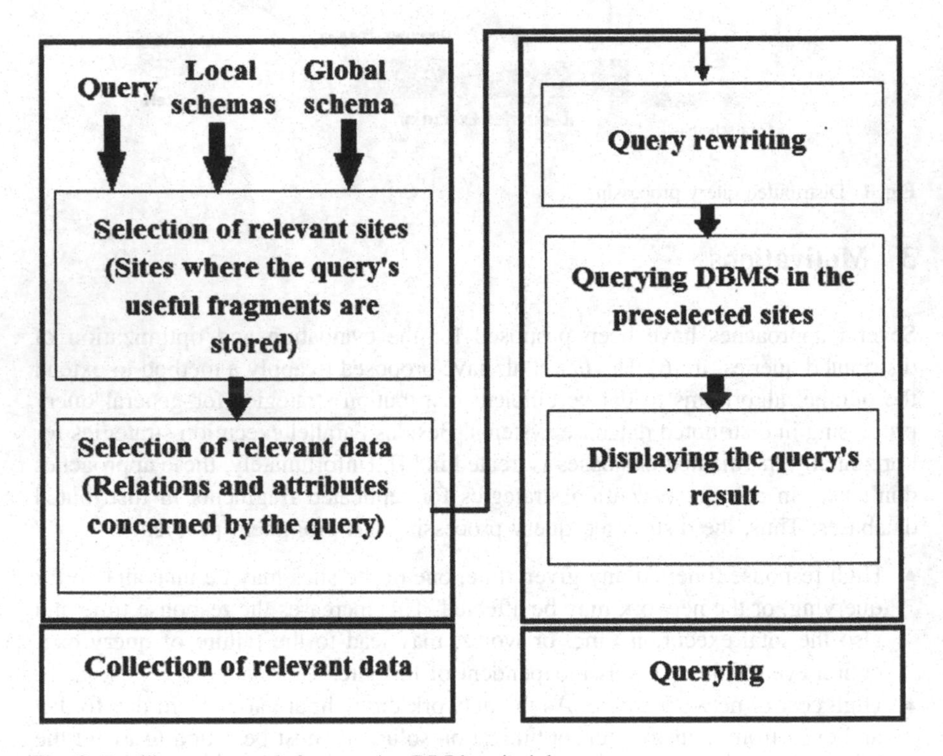

Fig. 2 Intelligent approach for querying DDB's principle

4.1 Collection of Relevant Data

The definition of relevant data is the first phase of this approach. It aims to prepare sites where relations and attributes are related to the inserted query in order to optimize distributed querying. This choice of relevant sites and data is based on the type of fragmentation and/or duplication of data.

This phase proceeds in two steps:

- Selection of relevant sites: It chooses DDB's sites where the query's useful fragments are stored given the local schema of each site and the DB's global schema.
- Selection of relevant data: It selects the entities (tables/attributes) concerned by the inserted query stored in the preselected sites.

The process of relevant data collection can be summarized by the following algorithm.

Algorithm: Collection of relevant data

INPUT:

- DB's global schema
- Local schemas $SI = \{ S_i = BD_i = \{ Table\ j\ (a_{j1}, \dots, a_{jk})\ ; 1 \leq j \leq N \} \}$
- Query
  $R = \{ Table1(a_{11}, \dots, a_{1l}), Table2(a_{21}, \dots, a_{2m}), \dots, Table\ N(a_{N1}, \dots, a_{Np}) \}$

N: number of tables concerned by the query R
l, m, p: number of attributes of $Table1$, $Table2$ and $Table\ N$

OUTPUT: Set of relevant data $SP = \{ S_i = BD_i = \{ Table\ j\ (a_{j1}, \dots, a_{jk})\ ; 1 \leq j \leq N \} \}$

Begin
$SP = SI$
For i from to N

    For j from $i + 1$ to N

    $S'_i = \{ a_{ij} \}$; i: table index ; j: attribute index
    $S'_j = \{ a_{ij} \}$; i: table index ; j: attribute index
    $R' = \{ a_{ij} \}$; i: table index ; j: attribute index
    If $S'_i \cap S'_j = E \neq \emptyset$ (vertical fragmentation case) then
        If $S'_i \cap R' \setminus E = \emptyset$ then $SP = SP/S_i$
        else
            If $S'_j \cap R' \setminus E = \emptyset$ then $SP = SP/S_i$
            End If.
        End If.
    End If.
    If $S_i \cap S_j = E \neq \emptyset$ (duplication case) then
        If $S_i \cap R = \emptyset$ then $SP = SP/S_i$
        else
            If $S_j \cap R = \emptyset$ then $SP = SP/S_i$
            End If.
        End If.
    End If.

    End For
End For

End

4.2 Querying

The querying is the second phase of the approach. This phase aims to rewrite the query with the preselected sites and entities prepared by the first phase. This phase proceeds in three steps:

- Rewrite the query knowing the relevant data defined in advance in order to optimize its cost.
- Querying the preselected sites
- Display the final result.

This process can be summarized by the following algorithm.

Algorithm: Querying DBs

INPUT: : Set of relevant data $SP = \{S_i = BD_i = \{Table\ j\ (a_{j1}, \ldots, a_{jk})\, ; 1 \le j \le N\}\}$

OUTPUT: The query result Res

Begin

1. While ($SP \neq \emptyset$)
   a) Select a site $S_i = BD_i$
   b) Extract appropriate data from S_i
   c) Rep_i = response of site S_i
   d) $SP = SP/S_i$
2. $Res = \bigcup_{i=1}^{n} Rep_i$

End

---

5 Experimental Results

Our effective approach is mainly about limiting the querying to relevant sites and data. To test this approach, we have developed a simple application which permits to enter a query, view relevant sites and display the result of the query and its execution time. This application deals with the distributed database described and fragmented in Sect. 2.2. Table 1 shows some queries addressed to this database and the geographic sites invoked by each query in the case of its execution by the DDBMS Oracle without and with adding our "efficient-layer".

Table 1 Queries examples

Query	Sites invoked by Oracle without our "efficient-layer"	Sites invoked by Oracle without our "efficient-layer"
Select all ENIT's students' names, borrows number and assigned departments	Site 1(ENIT)	Site 1(ENIT)
Select all students' names, borrows number and assigned departments	Site1, Site2, Site3	Site1 or Site2 or Site3
Select all students' names and assigned departments	Site1, Site2, Site3	Site1 or Site2 or Site3

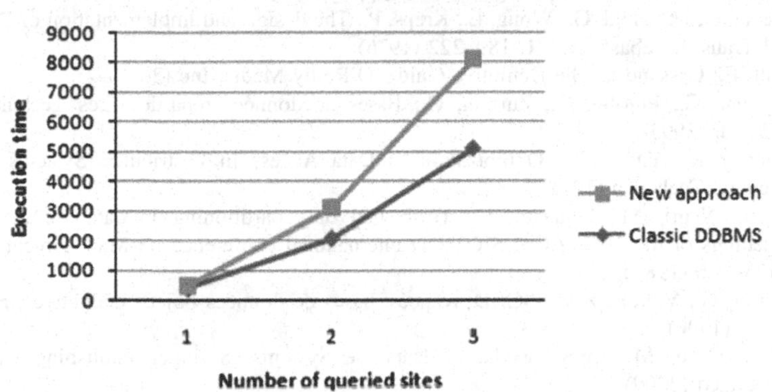

Fig. 3 Variation of execution time depending on the approach

We have executed these queries requiring 1, 2 and 3 sites with a classic DDBMS (Oracle). These queries are then executed with Oracle having our "efficient-layer" and their execution times are shown in Fig. 3

This curve shows that for simple queries, our approach minimizes up to half the execution time comparing to the classic DDBMS.

6 Conclusion

In this paper, we are interested to query optimization in a distributed environment which remains the major handicap for existing DDBMS, given the high cost of network traffic caused by the access to geographically distributed data in different sites.

To remedy this problem, we have presented a new approach of querying DDB. It is summed up in limiting the querying to the relevant site by rewriting the query knowing fragmentation and/or duplication of tables.

We proved that this approach reduces the cost of network transfer and optimizes the query execution time. This approach has been validated by implementing a layer "effective-query" on Oracle DDBMS. As perspective to this work, we plan to extend our new approach to distributed big data.

References

1. Dye, C., Russell, D.: Oracle distributed systems. O'Reilly (1999)
2. Franke, C., Morin, S., Chebotko, A., Abraham, J., Brazier, P.: Distributed semantic web data management in HBase and MySQL cluster. In: Proceedings of 2011 IEEE International Conference on Cloud Computing (CLOUD), pp. 105–112 (2011)
3. Stonebraker, M., Held, G., Wong, E., Kreps, P.: The design and implementation of INGRES. ACM Trans. Database Syst. **1**, 189–222 (1976)
4. Hewitt, E.: Cassandra: The Definitive Guide. O'Reilly Media, Inc (2010)
5. Chrisment, C., Pujolle, G., Zurfluh, G.: Bases de données réparties. Les Techniques de l'Ingénieur (1993)
6. Hevner, A.R., Yao, S.B.: Optimization of Data Access in Distributed Systems. Purdue University, Tech. rep (1978)
7. Ceri, S., Negri, M., Pelagatti, G.: Horizontal data partitioning in database design. In: Proceedings of the 1982 ACM SIGMOD international conference on Management of data, pp. 128–136 (1982)
8. Gardarin, G., Valduriez, P.: SGBD avancés: bases de données objets, déductives, réparties. Eyrolles (1990)
9. Liu, L., Özsu, M.: Encyclopedia of Database Systems. Springer Publishing Company, Incorporated (2009)
10. Özsu, M., Valduriez, P.: Principles of Distributed Database Systems. Springer Science and Business Media (2011)
11. Polychroniou, O., Sen, R., Ross, K.A.: Track join: distributed joins with minimal network traffic. Proceedings of the 2014 ACM SIGMOD International Conference on Management of Data, pp. 1483–1494 (2014)
12. Iacob, N.: Distributed query optimization. Ann. Econ. Ser. **4**, 182–192 (2010)
13. Kemme, B., Jiménez-Peris, R., Patiño-Martinez, M.: Database Replication. Morgan & Claypool Publishers

Strengthening the Rationale of Recommendations Through a Hybrid Explanations Building Framework

Andreas Charissiadis and Nikos Karacapilidis

Abstract In recommender systems research, there is growing awareness of the need to make the recommendation process more transparent and persuasive. This can be achieved by providing users with explanatory information about the recommended items. In this way, recommendation generating algorithms do not act like black boxes, but provide a rationale for each recommendation. In this paper, we briefly discuss and comment on existing approaches, and propose a novel framework for explanations building in recommender systems. The proposed framework builds explanations by following a hybrid approach that incorporates collaborative filtering and sentiment analysis features into classical multi-attribute based ranking. The enhanced trustworthiness of the explanation patterns produced by our approach is demonstrated through a series of illustrative examples.

Keywords Recommender systems • Explanations • Collaborative filtering • Sentiment analysis • Multi-attribute decision making

1 Introduction

The popularity of recommender systems has increased in the last decade with many commercial applications adopting them. The mainstream of research in these systems has been focused on the development and evaluation of new algorithms that can provide more accurate recommendations. However, it is broadly admitted that the most accurate systems may not be those that provide the most useful recommendations [1]. The assumption associating the accuracy of a recommendation algorithm with the quality of the system has been challenged, since other factors seem to also play a crucial role, such as transparency and persuasion. It has been also recognized that many recommender systems act like "black boxes", providing

A. Charissiadis · N. Karacapilidis (✉)
IMIS Lab, MEAD, University of Patras, 26504 Rio Patras, Greece
e-mail: nikos@mech.upatras.gr

R. Neves-Silva et al. (eds.), *Intelligent Decision Technologies*,
Smart Innovation, Systems and Technologies 39,
DOI 10.1007/978-3-319-19857-6_28

no transparency into the rationale of the recommendation process [2]. In order to increase their quality, recommender systems must be able to explain what they do and justify their actions in terms that are understandable to the user.

Explanations can play a crucial role in influencing a user's acceptance of a recommender system. An explanation can be considered as any type of additional information accompanying a system's output, having as ultimate goal to achieve certain objectives [3]. Besides assisting the user in understanding the output and rationale of the system, explanations can also improve the acceptance and the effectiveness of the system. The term explanation was first introduced in the field of Expert Systems, where the user could ask how the system reached the conclusion presented, and an explanation in the form of reasoning trace from the system was returned. Nowadays, the need of justification becomes even more crucial due to "shilling attacks" by malicious web robots. In a system that is under such an attack and does not provide justifications, users cannot understand why they receive incorrect advices [4].

The majority of existing recommendation algorithms and associated systems pay no or limited attention to the provision of explanations in their recommendations. In any case, their explanations simply reflect the classical multi-attribute based ranking of available items, and do not take into account the explicit preferences of people with similar profiles as well as their implicit opinion already expressed in reviews. Arguing that a meaningful orchestration of the above information augments a recommender system's transparency and persuasion, this paper proposes a hybrid framework for explanations building that combines multi-attribute based ranking, collaborative filtering and sentiment analysis. The remainder of the paper is structured as follows: Sect. 1 reports on background work and provides hints about our motivation; Sect. 2 describes in detail the proposed approach; Sect. 3 provides three examples to demonstrate the enhanced credibility and trustworthiness of the associated explanation pattern; finally, Sect. 4 concludes by discussing final remarks and sketching future work directions.

2 Related Work

Work performed in the context of recommender systems has actually led to the regeneration of research on explanation frameworks. Among other things, explanations could help increase user trust, satisfaction, make it faster and easier to find what a user wants and persuade them to try or purchase the recommended item. We can categorize explanations, based on a set of explanation goals, keeping in mind that the goals an explanation has to fulfill vary according to the domain, the user and the system. The main goals an explanation aims to achieve are [5]:

- *Transparency:* An explanation should give user an understanding of how the system reached a conclusion.

- *Scrutability:* Explanations should be part of a cycle, where the user understands what is going on in the system and exerts control over the type of recommendation made [6].
- *Trust:* Trust is usually associated with transparency and increase of a user's confidence in the system.
- *Persuasiveness:* This aim refers to convincing users to buy an item. It has been shown that users can be manipulated to give a rating closer to the system's prediction, whether this prediction is accurate or not [7].
- *Effectiveness:* Besides helping users to purchase an item, an explanation should also assist a user to make better decisions. This is why effectiveness is directly associated with the accuracy of the recommendation algorithm.
- *Efficiency:* Explanations could make it easier and faster for users to find the item which matches their needs best. This can be achieved by allowing the user to understand the relation between competing options.
- *Satisfaction:* Explanations may increase a user's satisfaction or on the contrary decrease a user's interest when he/she is given poorly designed explanations.

As shown in the next section, our explanations building framework explicitly addresses the goals of transparency, trust and effectiveness. Future enhancements, discussed in Sect. 4, aim to improve scrutability, persuasiveness, efficiency and satisfaction.

As far as access to explanations is concerned, the literature distinguishes two types, namely *user-invoked* and *automatic*. User-invoked explanations are requested by the user, while automatic ones are proactively provided by the system. Experiments have shown that the more difficult it is to access the explanations, the more the users perceive the explanation as a separate tool and become less aware of its informational value. Our approach enables automatic explanations, which provide a way to add trustworthiness to the conclusion presented to the user.

It is also noted that the explanation component can be provided before or after the advice given by the system. Explanations provided before advice are called *feedforward* and provide the user with a means to find out why a question is being asked during a consultation. On the other hand, *feedback* explanations present a trace of the rules that were invoked during a consultation and display intermediate inferences in order to achieve a conclusion [8]. In our approach, we apply feedback explanations aiming to make the decision process and the conclusion we reach each time transparent to the user.

Finally, related research has recently focused on the association between the justification provided from the users and possible *explanation patterns* they may have adopted in order to choose an option. The main idea here is to collect data from justifications expressed in natural language, analyze them in order to extract quantitative information and design possible patterns in which a justification may be expressed. The study described in [9] introduced a new issue in explanation research area, concerning explanations provided from the users to justify their choices (and not from the recommender system). However, this work faces some limitations: First, these patterns take into consideration only an item's attributes and do not make any use of data originating from users with similar tastes; our framework incorporates collaborative filtering features, thus exploiting information about related users' past behavior. Second, most of these patterns can serve only expert users who have a high knowledge level of the domain; instead, our framework provides simple and transparent explanations, thus being able to serve even novice users.

3 Our Approach

Our approach aims to remedy drawbacks of existing works on explanations building and enhance the overall quality of recommendations in terms of transparency, trust and effectiveness. We have developed a novel mechanism which, beyond classical ranking based on multiple attributes, incorporates collaborative filtering and sentiment analysis techniques for short texts. To our knowledge, this is the first attempt to jointly consider all the above in the area of explanations building

It is well-known that multi-attribute decision making, on which many recommender mechanisms are based, concerns the structuring and solving of decision and planning problems that involve multiple criteria (attributes). Cost is usually one of the main criteria, while some measure of quality is typically another criterion that is in conflict with cost. In many cases, there does not exist a unique optimal solution for such problems and it is necessary to use decision makers' preferences to differentiate between solutions [10].

On the other hand, the main idea of collaborative filtering (CF) is based on the assumption that people with similar tastes may share the same preferences [11]. Exploiting information about users' past behavior, CF-based recommender systems try to predict the interest of the active user. CF consists of two steps: first, it predicts the likelihood for an item of a particular user and then, it provides a recommendation list of *top-N* items. Collaborative filtering builds a database including user's preferences for items and represents them into a user-item matrix. This matrix consists of a list of n users $U = \{u_1, u_2, u_3, \ldots, u_n\}$ and a list of m items $I = \{i_1, i_2, i_3, \ldots, i_m\}$.

Finally, sentiment analysis (also known as opinion mining) refers to the use of natural language processing, text analysis and computational linguistics to identify and extract subjective information in source materials [12]. Generally speaking, sentiment analysis aims to determine the attitude of a speaker or a writer with respect to some topic or the overall contextual polarity of a document. A basic task in sentiment analysis is classifying the polarity of a given text to decide whether the expressed opinion in a document is positive, negative, or neutral.

We have developed a hybrid explanation pattern, which takes advantage of the three research fields mentioned above. Our pattern concerns the recommendation of an option o_c from a set of available options *Opt*. Each $o_i \in Opt$ is described through a set of attributes *Att*, where each $a_i \in Att$ is associated with a domain D_i that establishes the values allowed for that attribute.

In our approach, the decision function $DF(o_i) \rightarrow [1,0]$ is based on a weighted sum of three factors:

$$DF(o_i) = b{*}AttCost(o_i) + c{*}CBF(o_i) + d{*}SA(o_i), \tag{1}$$

where *b, c* and *d* correspond to the importance given to each of these factors ($b+c+d=1$). In the experiments described in the next section, we assume that $b=0.5$ and $c=d=0.25$, thus assigning double importance to the first factor.

The first factor is the attribute cost function, $AttCost(o_i) \rightarrow [0,1]$, which is actually the weighted sum of the scores of a particular option's individual attributes (this is applicable in our approach since all scores are expressed in the same scale; in addition, we assume that all the attributes are benefit attributes, that is, the higher the values are, the better it is). It is: $AttCost(o_i)=\sum_{j=1}^{k} w_j s_{i,j}$, where w_j denotes the weight of an attribute *j* and $s_{i,j}$ the score of option o_i with respect to attribute *j* ($w_j \in [0,1]$, $\sum w_j = 1$).

The second factor, $CBF(o_i) \rightarrow [0,1]$ is actually an implementation of collaborative filtering theory. It is:

$$CBF(o_i) = k(o_i){*}z(n), \tag{2}$$

$$\text{where } k(oi) = \begin{cases} \frac{1}{120}, 0 \le z < 10 \\ \frac{1}{60}, 10 \le z < 20 \\ \frac{1}{30}, 20 \le z \le 30 \end{cases} \text{ and } z(n) = \textstyle\sum_1^{30} n_i \tag{3}$$

We use a neighborhood size of 30, based on the results from a related research (see [2]). In Eq. (3), n_i is the variable describing if a user had chosen a particular option in the past. This variable takes the value 1 if the user had chosen the option; otherwise, it takes the value 0.

We also use the *Pearson correlation similarity* of two users *x, y* in order to find users with the highest rate of similarity. We use this measure to find the neighborhood of the active user and then examine how many of these neighbors had chosen an option in the past. The Pearson correlation similarity of two users *x, y* is defined as:

$$simil(x,y) = \frac{\sum_{i \in Ixy}(r_{x,i} - \bar{r}_x)(r_{y,i} - \bar{r}_y)}{\sqrt{\sum_{i \in Ixy}(r_{x,i} - \bar{r}_x)^2 \sum_{i \in Ixy}(r_{y,i} - \bar{r}_y)^2}}, \tag{4}$$

where I_{xy} is the set of items rated by both user *x* and user *y*, $r_{x,i}$ is the rating of user *x* for item *i*, $\bar{r}_x$ is the average rating of user *x* (for all the items rated by that user), $r_{y,i}$ is the rating of user *y* for item *i*, and $\bar{r}_y$ is the average rating of user *y* (for all the items rated by that user). This measure takes values from -1 to +1, where 1 corresponds to *total positive correlation*, 0 to *no correlation*, and −1 to *total negative correlation*. In our approach, two users *x* and *y* are considered as *similar* when $0.75 \le simil(x,y) < 0.85$, and as *highly similar* when $simil(x,y) \ge 0.85$.

The third factor, $SA(o_i) \rightarrow [0,1]$, reflects sentiment analysis techniques. It is:

$$SA(oi) = \frac{w_i - min}{max - min}, \quad (5)$$

where $w_i = (\sum_{0}^{x} SA_i)/x$, SA_i is the sentiment analysis score for each review of o_i, x is the number of reviews for the particular object, *min* is the minimum possible sentiment score for a review, and *max* is the maximum possible sentiment score for a review.

We use an open source sentiment analysis tool, namely *Semantria*, which exploits Natural Language Processing techniques to estimate the strength of positive, negative and neutral sentiment in short texts (https://semantria.com). Semantria allows for gaining valuable insights from unstructured text content by extracting categories, topics, themes, facets and sentiment. The results of sentiment analysis are presented in a scale of -2 to 2. Semantria analyzes each document and its components based on sophisticated algorithms developed to extract sentiment from content in a similar manner as a human. In our approach, we use a free Excel add-in provided from Semantria that permits content processing directly from Excel without any coding or integration work required. Semantria's cloud-based software extracts the sentiment of a document and its components through the following steps:

- A document is broken in its basic parts of speech, which identify the structural elements of a document, paragraph, or sentence (i.e. nouns, adjectives, verbs and adverbs);
- Sentiment-bearing phrases, such as "terrible service", are identified through the use of specifically designed algorithms;
- Each sentiment-bearing phrase in a document is given a score based on a logarithmic scale that ranges between -10 and 10;
- Finally, the scores are combined to determine the overall sentiment of the document or sentence.

In our approach (taking into account Eq. (5)), sentiment analysis scores in the interval [0, 0.25) are characterized as *very negative*, scores in the interval [0.25, 0.5) are characterized as *negative*, scores in the interval [0.5, 0.75) are characterized as *positive* and scores in the interval [0.75, 1] are characterized as *very positive*.

Based on the above, our framework is able to produce recommendations of the following explanation pattern template:

[Option_X] is recommended to you because it has been highly rated for its **[k_attribute]** from **[Y out_of Z]** users with **[similar /highly similar]** tastes to yours, who have already chosen it. In addition, these users have expressed **[very negative /negative /positive /very positive]** reviews for **[Option_X]**.

In this template, *K_attribute* represents the highest rated attribute of the recommended option, *Y* represents the number of active user's neighbors, and *Z* the total number of users. It is noted that our explanation pattern is produced in its entirety in cases where information about all three factors of Eq. (1) is available (see Example 3); in cases where some information is missing (see Examples 1 and 2), the pattern is produced partially.

4 Examples

In our initial set of experiments, we used a dataset containing hotel reviews from *TripAdvisor* (http://www.tripadvisor.com). This dataset contains 235,793 hotel reviews crawled from TripAdvisor in about one month period (from 14 February 2009 to 15 March 2009). We chose this data set because in addition to the overall rating, reviewers have also provided their ratings for seven hotel attributes, namely: value, room, location, cleanliness, front desk, service, and business service (ratings are expressed in the "1-5 stars" scale; these data is available at http://times.cs.uiuc.edu/~wang296/Data).

As mentioned above, we use a decision function based on a weighted sum of three factors. For the attributes used in our experiments, we have considered the weights shown in Table 1. We have also normalized the rating scale (1-5 stars) to a scale from 0 to 1 as follows: 1 star → 0; 2 stars → 0.25; 3 stars → 0.5; 4 stars → 0.75; 5 stars → 1.

Table 1 Hotel attributes and their weights

Attribute	Value	Room	Location	Cleanliness	Front desk	Service	Business service
Weight	0.25	0.20	0.15	0.15	0.10	0.10	0.05

After finding the results of our decision function for all available options, we use the best option's attributes and information about related users to construct our explanation.

Example 1

Hotel A, which has the higher decision function score, has been rated from 10 users. However, after the calculation of similarities between them and the active user, we consider in this example the case where these similarities are low (i.e. <0.75). In other words, the active user has no similar users who had visited this hotel in the past. In addition, we consider that there were not any reviews available for this hotel. The ratings for the attributes of hotel A are summarized in Table 2.

The average rating of each attribute (last row of Table 2) is calculated by adding the normalized ratings of the 10 users who have already rated hotel A and then dividing the result with the number of users. In this example, the highest rated attribute is location. The explanation is based only on the attribute cost factor and contains information about the location attribute of the recommended hotel. Thus, the explanation provided is:

Table 2 Hotel attributes' ratings – first example

	Value	Room	Location	Cleanliness	Front desk	Service	Business service
Active User	1	1	1	1	1	1	1
User1	5	5	5	5	5	5	5
User2	4	4	4	5	5	5	5
User3	5	5	4	4	5	4	4
User4	5	4	5	5	5	5	4
User5	4	4	5	4	5	4	5
User6	5	5	5	5	5	5	5
User7	4	5	5	5	3	5	1
User8	5	3	4	5	5	5	1
User9	5	5	5	4	4	5	5
User10	2	2	3	2	1	1	1
Average	0.85	0.80	**0.875**	0.85	0.825	0.85	0.65

```
Hotel A is recommended to you because it has been highly
rated for its location from 10 users, who have already
chosen it.
```

Example 2

As in the previous example, Hotel A has been rated from 10 users. We now consider the case where the active user has 7 highly similar users who have visited hotel A in the past (see Table 3).

Table 3 Hotel attributes' ratings – second example

	Value	Room	Location	Cleanliness	Front desk	Service	Business service
Active User	5	5	5	4	4	5	5
User1	5	5	5	5	5	5	5
User2	4	4	4	5	5	5	5
User3	5	5	4	4	5	4	4
User4	5	4	5	5	5	5	4
User5	4	4	5	4	5	4	5
User6	2	1	1	1	1	1	1
User7	4	5	5	5	5	5	1
User8	5	3	4	5	5	5	1
User9	1	1	1	1	1	1	1
User10	2	2	3	2	1	1	1
Average	0.675	0.60	0.675	0.675	**0.70**	0.65	0.45

To compute the similarities between the active user and the 10 other users who have rated hotel A, we use the Pearson correlation similarity measure (see Eq. (4)). The similarities between the active user and the other 10 users are summarized in Table 4 (one can easily observe that for seven users the similarity score is ≥0.85).

Table 4 User Similarities

Similarities	User 1	User 2	User 3	User 4	User 5	User 6	User 7	User 8	User 9	User 10
Score	0.94	0.91	0.87	0.94	0.87	0.12	0.89	0.85	0.05	0.15

In this example, where we also have ratings from similar users, the explanation provided is:

> Hotel A is recommended to you because it has been highly rated for its **front desk** from **7 out of 10** users with **highly similar** tastes to yours, who have already chosen it.

Example 3

In addition to the information given in Example 2, each of the 10 users that have rated hotel A has also written a review to describe their experience (see Appendix). In this example, we exploit both collaborative filtering and sentiment analysis data. Through Semantria's Excel add-in, we calculate the scores given in Table 5. Using these scores, we compute the sentiment analysis factor of our decision function according to Eq. (5).

Table 5 Sentiment analysis scores

Reviews	User 1	User 2	User 3	User 4	User 5	User 6	User 7	User 8	User 9	User 10
Score	0.70	0.68	0.80	0.90	0.69	0.68	0.83	0.71	-0.14	-0.05

It is: $w_A = (0.70 + 0.68 + 0.80 + 0.90 + 0.69 + 0.68 + 0.83 + 0.71 - 0.14 - 0.05)/10 = 0.58$, and $\mathrm{SA(A)} = \frac{0.58 - (-2)}{2 - (-2)} = 0.645$. Since $0.5 \leq \mathrm{SA(A)} < 0.75$, the sentiment score is characterized as *positive* and the explanation provided is:

> Hotel A is recommended to you because it has been highly rated for its **front desk** from **7 out of 10** users with **highly similar** tastes to yours, who have already chosen it. In addition, these users have expressed **positive** reviews for **Hotel A**.

5 Discussion and Conclusion

The use of explanations in recommender systems is gaining increasing importance as they are able to augment the adoption of these systems and the options proposed. In this paper, we presented a novel explanation building framework that incorporates collaborative filtering and sentiment analysis features in the classical multi-attribute decision model. A preliminary evaluation of this framework, exploiting the TripAdvisor dataset and aiming to compare our approach with previous ones, has demonstrated very encouraging results in terms of increasing transparency, trust and effectiveness. However, we admit that more work and studies are needed to engineer the various parameters of the proposed framework and draw more conclusive findings.

Towards this, we have already set a series of future work directions. First, we plan to develop an online questionnaire to compare our approach against previous ones that build explanations by taking into account only one factor of our decision function (e.g., works described in [9] and [13]). Such a comparison will be performed through the users' ranking of the goals an explanation aims to achieve (see Sect. 1). Second, we plan to verify our approach with other existing and rich datasets (e.g. the Amazon product co-purchasing network metadata, provided by Stanford University). Third, we will develop a web application for testing our explanation building framework in various domains and through different user groups. This application will be able to both use legacy datasets and maintain new ones. It will serve the fulfilment and evaluation of additional explanation goals such as those of scrutability, persuasiveness, efficiency and satisfaction. Particular emphasis will be given to making this application as interactive as possible, thus enabling the user to easily perform a 'what-if' analysis of the recommended options and associated recommendations by considering alternative decision parameters. Fourth, we plan to further investigate existing sentiment analysis tools to better align their capabilities and features with our overall framework. In this direction, we are already working on tuning Semantria's lexicon to strictly consider the attributes taken into account each time in the first factor of our framework's decision function. Fifth, we plan to go beyond textual recommendations and provide alternative visual explanation interfaces that users are most likely to enjoy and use towards making faster and informed decisions. This includes, for instance, histograms and infographics about how similar users had rated the recommended option, with appropriate clustering and visualization of the top attributes' ratings and related reviews. Finally, we intend to integrate user-centric principles about the decisive pros and cons for an option, in line with the work proposed in [13].

As a concluding note, we mention that our overall approach to explanation building adopts the guidelines proposed in [5, 9] for: (i) the qualities and forms of explanation needed to best meet user expectations, and (ii) patterns for explanations to be given under different circumstances. Also, we argue that our approach is able to mine and exploit similarities among both users and options, and meaningfully integrate explanation styles influenced by content-based, collaborative, demographic and knowledge-based algorithms [14, 15].

Appendix: User Reviews

User1	Great Hotel! My fiancée and I stayed at the Hotel Monaco over the recent Thanksgiving holiday. The hotel is fantastic. The rooms are very comfortable and clean. The management is very responsive. During our second night, the people in the room next to us were extremely loud. When I told the manager this in the morning, I was immediately given 1/2 off the night's stay and moved to suite at no additional cost. The hotel location is ideal, within walking distance of the fish market, shops, bars and restaurants. I highly recommend staying here.
User2	Very Impressed! My husband and I stayed in the Monaco for 3 nights and had a wonderful time. Seattle is an amazingly clean city and the Hotel Monaco was also spotless. Our room and bathroom were quite small and quite dark but, as we were really only sleeping there, this didn't bother us. The room however, is perfectly equipped and the bed was really comfortable. The lobby of the hotel is quite impressive and the staff was very friendly. We didn't eat at the hotel restaurant but it was very nice and always very busy. The hotel is ideally situated for seeing all of Seattle although, everyone should be aware that Seattle is a very hilly city in parts. We had a great stay at the Monaco and I would highly recommend it and indeed Seattle.
User3	Excellent Stayed in Hotel Monaco this past w/e, found it a delight. The reception staff is friendly and professional; our room was smart with a very comfortable bed. We particularly liked the reception our small dog received, all the staff guests spoke to him, he loved it. My only mild negative is the distance (uphill) from PPMarket and restaurants on 1st. Overall a great experience.
User4	Very Nice Hotel and Staff The hotel is in a good location. My room was very comfortable and a nice size. The staff was extremely accommodating. I would definitely stay there again.
User5	Wonderful relaxing stay. Everything was a breeze, from check-into check-out. The beds were incredibly comfortable and the room (a double queen room) was spacious. I especially liked the l'occitane shampoo and conditioner; it's a nice treat when the toiletries are nicer than what I use on a normal day. I would recommend this hotel to anyone who wants an upgrade to the other nearby downtown hotels.
User6	Great Experience. Much more interesting than staying at the bigger chains, and I've had good experiences with the Seattle Sheraton. Staff is very friendly, as are most people in Seattle, and rooms have a lot of cool features, including the optional gold fish and bathrobes. I'd definitely stay again.
User7	Excellent Stay! What a delightful surprise to stay at the Monaco. We thoroughly enjoyed our stay. The room was very comfortable, lovely amenities, and friendly staff. Especially enjoyed the hour of indulgence!! Definitely will come back!
User8	Fantastic Service! Stayed here for my girlfriend's birthday over a weekend in June. While the rooms were on the smallish side, they were nicely decorated and the bed was extremely comfortable.. During our stay, the staff was extremely friendly, polite, and helpful. We found the concierge staff to be especially helpful. We are looking forward to another trip to Seattle and stay at the Monaco!

(continued)

User9	Horrible Customer Service. My friend and I picked the Hotel Monaco because of its appealing website and online package which included champagne, late checkout at 3, free valet and a gift from the spa for the weekend … Sunday morning we called the front desk to speak to management about the sheets and they became very aggravated and rude with us. The young man we spoke to said that they would cover the food, adding that the person that changed the sheets said that it was fresh blood, in a very rude tone. Our checkout was at 3 pm because of the package that we booked. I finally packed all my things up and went downstairs to check out. Quickly signed the paper and took off. When on the way, I took a closer look at the room bill. Unfortunately they never covered the food, which they had offered and we were charged for the valet. I called the front desk to ask about the charges, the lady that answered snapped back saying that they were aware of the problem we experienced. I have never in my life been treated like this at any hotel anywhere. I am not sure if this hotel constantly has problems or if we were the lucky ones, but I would never recommend it to anybody I know.
User10	Disappointing Hotel Monaco has very little going for it except location. In fact, I would hazard to say they go out of their way to stay away from guest. We had 4 in the room including our children and staff kept leaving only 3 sets of towels. Not a big deal but for $320/night I expect better. This hotel has some management problems that need to be addressed. The restaurant was nothing but ordinary and way overpriced. The rooms are dull but spacious. The evening wine reception is a nice touch but the way the lobby is arranged makes for a poor place to try and meet other guest. Really not worth the time. At the end of the day I was sorely disappointed and would not go back.

References

1. McNee, S., Riedl, J., Konstan, J.: Being accurate is not enough: how accuracy metrics have hurt recommender systems. In: Proceedings of CHI ’06, pp. 1097–1101. ACM, New York (2006)
2. Herlocker, L., Konstan, J., Riedl, J.: Explaining collaborative filtering recommendations. In: Proceedings of the 2000 ACM Conference on Computer Supported Cooperative Work (CSCW’00), pp. 241–250. Philadelphia, Pennsylvania (2000)
3. Tintarev, N., Masthoff, J.: Effective explanations of recommendations: user-centered design. In: Proceedings of the 2007 ACM Conference on Recommender Systems (RecSys’07), pp. 153–156. ACM, MN, Minneapolis, USA (2007)
4. Gunes, I., Kaleli, C., Bilge, A., Polat, H.: Shilling attacks against recommender systems: a comprehensive survey. Artif. Intell. Rev. **42**(4), 767–799 (2012)
5. Tintarev, N., Masthoff, J.: Designing and evaluating explanations for recommender systems. In: Ricci, F., Rokach, L., Shapira, B., Kantor, P.B. (eds.) Recommender Systems Handbook, pp. 479–510. Springer, Berlin (2011)
6. Czarkowski, A.: Scrutable Adaptive Hypertext. Ph.D. thesis, University of Sydney (2006)
7. Cosley, D., Lam, S., Albert, I., Konstan, J., Riedl, J.: Is seeing believing? how recommender interfaces affect users’ opinions. In: Proceedings of CHI 2003, pp. 585–592. ACM, New York (2003)
8. Darlington, K.: Aspects of intelligent systems explanation. Univers. J. Control Autom. **1**, 40–51 (2013)
9. Nunes, I., Miles, S., Luck, M.: Investigating explanations to justify choice. In: Masthoff, J., Mobasher, B., Desmarais, M., Nkambou, R. (eds.) UMAP 2012. LNCS, vol. 7379, pp. 212–224. Springer, Heidelberg (2012)

10. Figuera, J., Greco, S., Ehrgott, M. (eds.): Multiple Criteria Decision Analysis: State of the Art Surveys. Springer, New York (2003)
11. Burke, R.: A case-based approach to collaborative filtering. In: Proceedings of the Fifth European Workshop on Case-Based Reasoning, vol. 1898. Springer, LNAI (2000)
12. Pang, B., Lee, L.: Opinion mining and sentiment analysis. Found. Trends Inf. Retrieval **2** (1–2), 1–135 (2008)
13. Nunes, I., Miles, S., Luck, M., Barbosa, S., Lucena, C.: Pattern-based explanation for automated decisions. In: Proceedings of ECAI, pp. 669–674 (2014)
14. Lazanas, A., Karacapilidis, N.: Augmenting transportation-related recommendations through data mining. Int. J. Adv. Intell. Paradigms **2**(1), 78–89 (2010)
15. Karacapilidis, N., Hatzieleftheriou, L.: A hybrid framework for similarity-based recommendations. Int. J. Bus. Intell. Data Min. **1**(1), 107–121 (2005)

The Eye Tracking Methods in User Interfaces Assessment

Katarzyna Harezlak, Jacek Rzeszutek and Pawel Kasprowski

Abstract Acquiring basic information about the user's needs, is one of the most important problem which a user interface designer has to face. It influences the selection of the design patterns which match the user's requirements. Most frequently lots of possible solutions could be found and the appropriate choice has to be done. The results from some previously conducted research regarding human–computer interaction proved that collecting and analysing the eye movement data may be useful in the user interfaces assessment as well. The aim of the preliminary studies presented in this paper was to analyse to what extent the eye tracing methods and eye movement metrics can support the process of user interfaces' assessment and how this process can be automated.

1 Introduction

Many people believe that a computer program is all they see and therefore assess its usability based on a user interface impression. This fact, in conjunction with observations of current trends in increasing software influence on current human life, has made UI design an important task. Every day, millions of people use the Internet in order to find out necessary information. Finding information more efficiently and having a nicer user experience while browsing for it can significantly improve the day-to-day life of Internet users.

However, designing a useful user interface is not a simple task, which is strengthened by the wide range of user interface technologies. Using a desktop application differs a lot from using a mobile one and the areas and purposes of the software usage change all the time. Nowadays, software is used not only for

K. Harezlak (✉) · J. Rzeszutek · P. Kasprowski
Institute of Informatics, Silesian University of Technology, Gliwice, Poland
e-mail: katarzyna.harezlak@polsl.pl

P. Kasprowski
e-mail: pawel.kasprowski@polsl.pl

R. Neves-Silva et al. (eds.), *Intelligent Decision Technologies*,
Smart Innovation, Systems and Technologies 39,
DOI 10.1007/978-3-319-19857-6_29

managing invoices or accounting, but also in complex data analysis, every-day entertainment or managing an online university. Nevertheless Web search tools belong to the most heavily used applications. That's why it is very important to make the human-computer interaction really efficient.

The key aspect in designing a user interface is to understand the user and his/her needs. Acquiring basic information about the user's needs, is one of the most important problem which a user interface designer has to face. It influences the selection of the design patterns which match the user's requirements. Most frequently lots of possible ones can be found and the appropriate choice has to be done. The results from some previously conducted research regarding human–computer interaction proved that collecting and analysing the eye movement data may be useful in user interfaces assessment as well.

The aim of the preliminary studies presented in this paper was to analyse to what extent the eye tracing methods and eye movement metrics can support the process of user interfaces' assessment and how this process can be automated.

2 The State of the Art

The problem of developing user interfaces of high quality is broadly discussed in literature. There are items considering the design patterns that deal with the most common UI interaction problems [1, 2] as well as those that are addressed to the assessment of their usability [3–6]. In [3] the author, based on nearly 200 studies, reviewed and validated current practices in measuring usability.

There were also some experiments conducted, which used eye trackers to find patterns of people's behaviour while interacting with user interfaces. They were based on an extraction of the main eye movement components (fixations and saccades), made during an observation of the given area of interest (AOI). A fixation is recognized when the point-of-gaze remains within a small area for a given time. Fixations are interlaced with saccades - quick movements made to reach another point-of-regard [7]. In the experiments described in [8] the participants were to visit a set of websites within a given time period to answer questions asked by an experimenter. The tasks were presented to a participant in a fixed order. The efficiency of the search was analysed in three aspects: scope of the task, the screen and the areas of interest. A number of fixations, a total and mean fixation duration related to various screens and AOIs as well as sequence information in the form of scan paths were taken into consideration. The authors of the studies presented in [9] analysed an influence of search results organisation on the performance of informational and navigational tasks. The experiments used various ways of the search result presentation including short, medium and long result description and a changed position of the outcome in the a results list. The analysis of the results was based on fixations regarding particular AOIs, the total time spent on a search task realisation and accuracy in finding a result most fitted to the search condition.

Another research [10] concentrated on four factors (a number of fixations, total and mean time of a fixation duration and a fixation dispersion) to assess the difficulty of task realization, whereas in [11] a correlation between eye movement patterns and usability problems was searched for. In the latter case the "think-aloud" and retrospective protocols were used to enrich information gathered from the eye tracker. Together, with eye movements, the input from a webcam, audio from the microphone, mouse and keyboard recordings were collected. The results were studied in terms of the number of fixations, fixations duration, fixations order and numbers of vertical and horizontal saccades. An interesting research was presented in [12], where authors, on the basis of three web sites of a similar type, considered the influence of web elements layout on the search performance. The participants of the experiment were expected to complete four tasks for which scan paths and a number of fixations within particular parts of the websites were analysed.

The studies presented in this paper are similar to that described in [12], however they differ in several respects. The first distinction regards the web sites used. We prepared exactly the same content to be displayed and placed it in three various types of layouts (Fig. 2). Another difference was using an extended set of eye movement metrics to assess these interfaces. Finally the aim of the research was different than those described earlier. The purpose of the studies was not to prove that one layout is better than another one, yet to provide the guidelines, which can be valuable while developing a user interface and collecting feedback by applying eye tracking methods for measuring the user interface ergonomics. Additionally we attempted to introduce an automatic process for verifying interfaces usability. There were reference ratios calculated for this purpose.

It is worth mentioning that the experiments were carried out using our own developed eye tracker and our own implemented data analysis software. These features make the experimental setup repeatable in many ordinary computer environments being commonly used. Detailed information about the aforementioned issues will be provided in the subsequent chapters.

3 The Experimental Setup

The experimental setup of the tracking system used in the research is shown in Fig. 1. Its main component was a video-oculography (VOG) head-mounted eye tracker developed with single CMOS camera with USB 2.0 interface (Logitech QuickCam Express) with 352 × 288 sensor and lens with IRPass filter. The camera was mounted on an arm attached to a head and was pointing at the right eye. The eye was illuminated with a single IR LED, placed off the axis of the eye that causes the "dark pupil" effect, which was useful during pupil detection. The image obtained from the camera was converted to grayscale. Subsequently it was thresholded to convert the image into black and white scale. For such processed image the contour detection algorithm was applied to find the smallest convex polygon surrounding the white shape. The centre of gravity of this polygon was the estimated centre of the pupil. The system generated 20–25 measurements of the centre of the pupil per second.

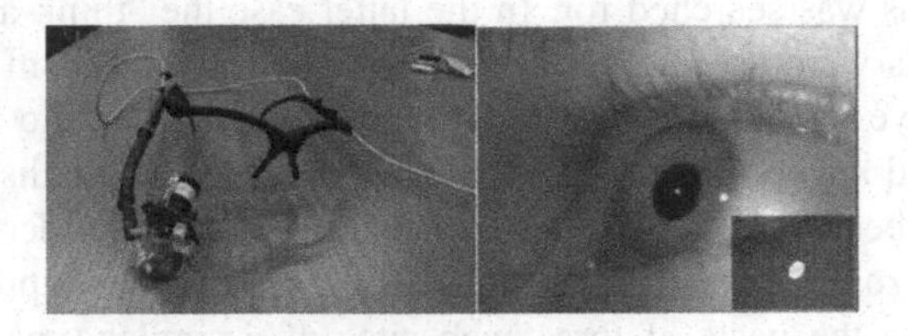

Fig. 1 The own developed eye tracker and an eye image

The tool chosen for the eye movement tracking uses an artificial neural network (ANN) for mapping an eye position to screen coordinates. During the calibration, the mapping function (neural network) is trained using a back propagation learning algorithm. The network is trained for 2000 epochs or until the moment when a network error is below 0.1. Error on level 0.1 ensures good work for the eye tracking mapping purposes. The learning ratio used for training is equal to 0.5. The validation error was calculated as an average distance (in degrees) between the compared points. Before the experiments started, the calibration, based on 9 points evenly distributed over the screen, was performed. Each of the calibration points was represented on the screen with a small, black dot. When the user was supposed to focus on the particular calibration point, it was highlighted using a circular marker around it. The calibration was performed based on studies presented in [13, 14].

4 The Method of the Experiments

Real-life scenarios concerning the booking of flight tickets were used for the purpose of the experiment The participants were presented with three test user interfaces, which provided exactly the same functionalities. They differed from each other in the way of content presentation (Fig. 2). All of them were displayed on a screen of size 52.04 cm × 32.6 cm and a display resolution set to 1920 × 1200 pixels.

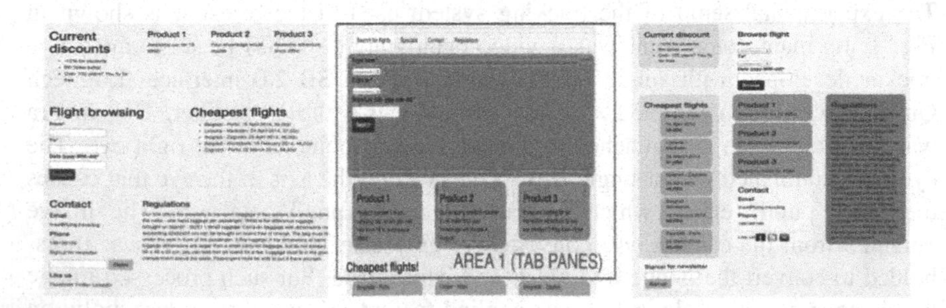

Fig. 2 The example web site layouts

The participants were supposed to fulfil a few simple usability scenarios:

1. Analyse the cheapest flights and find the cheapest one from Belgrade.
2. Search for the cheapest flight from Lisbon to Belgrade around 1st April 2014.
3. Analyse rules for the luggage transportation. Is it allowed to travel with luggage of size 30 cm × 30 cm × 50 cm?
4. Find a phone number and an email address for contacting a given airline.

The participants.
A group of four participants took part in the experiments. There were representatives of people of various ages – the teenager, two young men and the middle-age women. All the users were supposed to fulfil the scenarios in random order (for both all user interfaces and all scenarios) to make the results more reliable. There were 48 sessions recorded, each of them was preceded by a calibration process. A session was started when calibration error was lower than 0.5 deg.

Metrics used for the results assessment.
Recorded data were analysed using a subset of metrics described in [15]. These metrics were analysed and chosen in term of their usefulness in the presented studies:

- number of all fixations – a bigger number of fixations indicates longer search,
- time to first fixation on-target – shorter time may indicate that a user was able to search more efficiently,
- ratio of on-target fixations to the number of all fixations – a low number may indicate that a user spent a lot of time on non-relevant parts of the screen, which means that the search was less efficient,
- verification whether expected (mentioned in scenarios) points were visited – to check whether a user really visited expected parts of UI,
- scan path length – length of a whole path that the user's eye movement covered on the tested user interface (during the recording phase). The shorter the path was, the more efficiently a user was searching for elements.

Additionally, the average fixation duration and heap maps – a graphical representation of fixations positions density, showing which parts of UI were catching most of the user's attention – were used.

The values obtained for a specific scenario and for a particular user were compared to other users' recordings and against the reference values of metrics used. These values, among which 'ideal path' was defined, were obtained from the experienced web site user. Similarly, the verification whether the expected points were visited was developed. It creates the opportunities for an immediate discovering the most crucial usability issues.

Fixation detection.
The main role in the previously discussed analysis plays a set of fixations extracted from an eye-movement signal. The fixations were determined using the I-DT algorithm [16, 17]. For the analysis purposes, the procedure parameters were

initially set to values suggested in papers [16–18] - the dispersion threshold equal to 1° (0.028 in the screen coordinates system) and the duration threshold equal to 150 ms. The sampling frequency parameter was set to 25 Hz (frequency of the eye-tracker), which was used for calculating fixation durations. The dispersion window was determined as a difference between maximal and minimal values for x and y axis.

Target area.
One of the metrics used during the analysis was verifying whether a fixation is registered on-target or not. For this purpose the target areas, described using rectangle with minimal and maximal values for x and y axis specified, were defined. These areas were identified for each scenario and for each user interface.

Time to first fixation on-target.
The time to first fixation on-target was calculated as a sum of the durations of all fixations made until the user focused his/her gaze on the target area. After the first fixation was found, the total time (in seconds) of the recordings registered before that fixation was returned.

Ratio of on-target fixations to the number of all fixations.
The ratio of on-target fixations to the overall number of fixations was calculated by checking how many of fixations were made within the target area. A value returned by this calculation is within range [0.0, 1.0].

Verification whether expected points were visited.
This metric checked whether the user visited a set of expected points, which meant there was a fixation lying within 3° distance (0.084 in the screen coordinates system) from the expected point. The metric was calculated as the Euclidean distance between the fixation and the closest point belonging to the set of expected points. As it was described earlier, they were defined based on recordings obtained for the user experienced in the usage of test interfaces.

Scan path length.
The scan path length is a sum of the distances between subsequent fixations for the single recording determined using the Euclidean distance.

It should be emphasized that studies of only one of metrics is not sufficient and may skew the results. Therefore, to obtain valuable and reliable results all the aforementioned metrics should be taken into consideration. For example a scan path length can be similar to the ideal one, yet may concern other parts of an interface. It is easy to check while the test results are analysed one by one on the screen. However, in case of their automatic verification, a confirmation by verifying visited points is required.

5 Results and Discussion

The results obtained during the experiment were analysed in terms of their applicability in the developed method of automatic user interfaces assessing. There were conclusions drawn in each paragraph, based on values obtained for the particular metrics.

Number of fixations.
Primarily, the number of all fixations registered for particular interfaces and scenarios was considered, taking into account the less fixations were made the quicker the task was done, assuming the task realization was successful.

The results slightly differed for various scenarios. For three of them (scenario 2, scenario 3, scenario 4), the third user interface served better than the two other ones. During the scenario 1, the users made more fixations for the third interface (average value of 34 fixations, when for the second interfaces this value was only 20 fixations).

The second user interface was also working well with respect to the third interface, apart from the scenario 4, when the users made on average twice as many fixations (30 fixations) as for the third user interface (15 fixations).

The first user interface was significantly less efficient during the scenario 2 and scenario 3. The average numbers of fixations for these scenarios was 88 and 70 fixations, which was much more than the results of the other user interfaces (46 and 56 for the second interface and 41 and 46 for the third one).

Scan path length.
The analysed metric values for the recorded scenarios varied a lot. The third user interface performed the best apart from the scenario 1, when its average scan path length (3.3 in the screen coordinate system) was almost three times higher than the best result (average length 1.3).

The first user interface turned out to be much worse than the other ones for the scenario 2 (average length 5.8 while the third user interface had average of 2.5). The second user interface was much worse than its competitors during the scenario 4, where its average scan path length was more than twice longer than for the other ones.

All the user interfaces achieved very close results during the scenario 3 (all results within 4.5 and 6.0).

Verification whether expected points were visited.
The analysis regarding the average ratios of expected points visited is supposed to provide information to what extent the solved task was successful. The value of 1.0 indicates that all the expected points were visited.

In most of the cases, the obtained results were satisfactory (on average 75 % visited points). The worst performance regarded the second user interface during the scenario 1, when the participants managed on average to visit only 0.52 (52 %) of the expected points.

During the scenario 2 (all user interfaces), the participants visited almost all the expected points, which means that they found all the important places on the user interface. The third user interfaces served the best considering all the scenarios. Its result has never gone below 0.7 (visiting at least 70 % of the expected points).

Time to first fixation on-target.
The results regarding the average times to first fixation on-target varied a lot for each scenario, yet all the results were fitting acceptable values. The worst result was obtained for the first user interface during the scenario 3 and was equal to 2.2 s, however spending roughly 2 s on finding the target area is still a reasonable amount of time, so the user interface was usable.

During the scenario 1, for both the second and third interfaces the participants immediately focused their gaze on the target area. Results obtained during the scenario 2 (all the user interfaces) also indicated almost an instant focus on the target area after the recording was started (all the results there are within 0.5 and 0.6 s).

During three of all the recordings, one tested user has not managed to make any fixation on-target, which indicated that he/she has not done the task. Earlier evaluated metrics - the scan path and the number of fixations - have not revealed this problem because the participant fixated eyes in various points and moved his/her eye over the screen. The application of the *Time to first fixation on-target* metric made discovering the problem possible.

Ratio of on-target fixations to number of all fixations.
Analysing the results regarding the ratios of on-target fixations to the number of all fixations, the assessment of the effective time of a task realization was achievable. The higher value of this metric indicated that the user focused more times on the relevant parts of the interface (so the design was more efficient). The highest possible and the best value is equal to 1.0. In case of the experiments presented in the paper, this values was achieved twice and value within range [0.9, 1.0] four times. On the other hand there were two trials with ratio equal 0.0.

While testing all the scenarios, the third user interface turned to be the best or was very close to being the best. The result for the scenario 1, indicated that the participants made on average 85 % of their fixations within the target areas, which is a very good result. The first user interface proved to be worse than its competitors during the scenario 1, when the user made only 20 % of his/her fixations on the target area, whereas 50 % on the second interface and 85 % on the third one. The second user interface had its poorest usage during the scenario 4, when the user made only 25 % of his/her fixations on the target area (while the user achieved 50 % result on the other interfaces).

6 The Scoring System

The diversity of the previously-presented outcomes, concerning various scenarios and participants, make their analysis aimed at identifying the best interface rather complicated. It can be expected that in case of involving more participants it would be even more complex. Conducting it manually studying the results presentation on the screen could be wearisome, take of lot of time and finally could turn out to be too difficult to accomplish. To facilitate this task, by automatically realized assessment, the scoring system based on the comparison of the results obtained for a particular web site against both the expected (reference) values and the outcomes of competing user interfaces was proposed. In the former case, every time a participant has managed to perform as expected, the user interface obtained 3 points. When the user has performed at acceptable level, then 1 point was granted.

The chosen results (for the first interface, the first and the second scenarios) were presented in Table 1 to clarify the idea.

Table 1 Example results of the carried out experiments

Interface	Scenario	User	Ratio of points visited	Time to first fixation on target [s]	Ratio of fixations on target area	Scan path length (in screen coordinate system)
1	1	1	0.79	1.88	0.35	1.36
		2	0.57		0.00	1.43
		3	0.86	1.32	0.46	1.55
		4	0.86	1.92	0.05	1.13
	2	1	1.00	0.28	0.94	6.37
		2	1.00	0.16	0.66	5.39
		3	1.00	1.40	0.28	9.30
		4	1.00	0.32	0.95	2.37

Values were coloured so that their background expresses a result of the comparison with the reference data. The green colour points out the results that matched (or were better than) the reference values, the orange colour highlights average (but still acceptable) ones and the red colour marks the worst ones, where the user failed to meet expectations.

Comparing the results between the user interfaces, each time the average value of analysed metric (for the single scenario) was the best for particular user interface, it was granted 10 points. If two of the user interfaces had the same average value of the analysed metric, both of them obtained 10 points. Points were granted only for first place in the comparison. Using such *the scorings system* the further analysis can be done to summarize detailed outcomes and point out the best solution.

7 Conclusion

The aim of the research was to develop the method based on eye-tracking data, for the automatic assessment of the user interfaces usability. This goal has been achieved. The research was conducted using the simple web application. The results obtained during the experiment proved the eye-tracing may play an important role in GUIs designing and development, pointing out their weak and strong features.

However, it must be emphasized that, to get the meaningful assessment, many metrics have to be used. As it was shown in the paper, some metrics provide data on visited parts of UI (heat maps, verification whether expected points were visited) and other concern time spent either on searching an appropriate information or within the chosen area (time to first fixation on-target, ratio of on-target fixations to number of all fixations, fixations scan path).

The analysis of each group of metrics may reveal only a part of information, but studied together it may provide a complete information how an interface serves in the particular scenario. However, when user interfaces are verified by several people realizing several tasks and their cooperation with this interfaces provides ambiguous assessments, reaching a conclusion may be a complex or even unrealizable task.

The proposed scoring system seems to be a useful tool facilitating such undertakings. The scores used in the experiments should be considered as exemplary. They can be adjusted to a particular experiment and calculated for instance on the basis of a scenario difficulty or its duration. The important issue is to define reference values serving as a baseline for assessing results.

Finally, the usage of own developed eye tracer should be mentioned once again. It demonstrates the opportunities for developing an own, not very expensive testing set-up, possible to be built by any computer user.

References

1. Tidwell, J.: Designing Interfaces: Patterns for Effective Interaction Design. O'Reilly Media, Dec 2010
2. Bahadur, S., Sagar, B.K.: User interface design with visualization techniques. Int. J. Res. Eng. Appl. Sci. **2**(6) (2012)
3. Hornbæk, K.: Current practice in measuring usability: challenges to usability studies and research. Int. J. Hum. Comput. Stud. **64**(2), 79–102 (2006)
4. Badashian, A.S., Mahdavi, M., Pourshirmohammadi, A., Monajjemi nejad, M.: Fundamental usability guidelines for user interface design. In: Proceedings of the 2008 International Conference on Computational Sciences and its Applications (ICCSA'08). IEEE Computer Society, pp. 106–113. Washington, DC, USA
5. Hilbert, D.M., Redmiles, D.F.: Extracting usability information from user interface events. ACM Comput. Surv. **32**(4), 384–421 (2000)
6. Ivory, M.Y., Hearst, M.A.: The state of the art in automating usability evaluation of user interfaces. ACM Comput. Surv. **33**(4), 470–516 (2001)

7. Holmqvist, K., Nyström, M., Andersson, R., Dewhurst, R., Jarodzka, H., van de Weijer, J.: Eye Tracking A Comprehensive Guide to Methods and Measures. Oxford University Press, Oxford (2011)
8. Goldberg, J.H., Stimson, M.J., Lewenstein, M., Scott, N., Wichansky, A.M.: Eye tracking in web search tasks: design implications. In: Proceedings of the 2002 symposium on Eye tracking research & applications (ETRA '02), pp. 51–58. ACM, New York, NY, USA
9. Cutrell, E., Guan, Z.: What are you looking for?: an eye-tracking study of information usage in web search. In: Proceedings of the SIGCHI Conference on Human Factors in Computing Systems (CHI'07), pp. 407–416. ACM, New York, NY, USA
10. Cowen, L., Ball, L.J., Delin, J.: An eye-movement analysis of web-page usability. In: People and Computers XVI—Memorable yet Invisible: Proceedings of HCI 2002, pp. 317–335. Springer, London (2002)
11. Ehmke, C., Wilson, S.: Identifying web usability problems from eyetracking data. In: Paper presented at the British HCI conference 2007. University of Lancaster (3-7 Sept 2007)
12. McCarthy, J.D., Sasse, M.A., Riegelsberger, J.: The geometry of web search. In: Proceedings of HCI2004. Leeds, UK, 6–10 Sep 2004
13. Kasprowski, P., Harężlak, K., Stasch, M.: Guidelines for the eye tracker calibration using points of regard. Information Technologies in Biomedicine, vol. 4. Advances in Intelligent Systems and Computing, vol. 284, pp. 225–236. Springer (2014)
14. Harężlak, K., Kasprowski, P., Stasch, M.: Towards accurate eye tracker calibration—methods and procedures. Procedia Comput. Sci. **35**, 1073–1081 (2014) (Elsevier)
15. Poole, A., Ball, L.J., Poole, A., Ball, L.J.: Eye Tracking in Human-Computer Interaction and Usability Research: Current Status and Future Prospects in Encyclopedia of Human Computer Interaction. IGI Global, Hershey (2005)
16. Salvucci, D.D., Goldberg, J.H.: Identifying fixations and saccades in eye-tracking protocols. In: Proceedings of the 2000 Symposium on Eye Tracking Research and Applications, pp. 71–78. ACM Press, NY (2000)
17. Shic, F., Chawarska, K., Scassellati, B.: The incomplete fixation measure. In: Proceedings of the 2008 Symposium on Eye Tracking Research and Applications, pp. 111–114. ACM Press, NY (2008)
18. Blignaut, P.: Fixation identification: the optimum threshold for a dispersion algorithm. Attention Percept. Psychophysics **71**(4), 881–895 (2009)

The Eye Movement Data Storage – Checking the Possibilities

Katarzyna Harezlak, Tomasz Wasinski and Pawel Kasprowski

Abstract The research currently conducted in various fields of interest base their experiments on the eye movement signal analysis. Among these areas cognitive studies on people's willingness, intentions or skills and advanced interface design can be enumerated. There are also some studies attempting to apply the eye tracking in security tasks and biometric identification. During these experiments a lot of data is registered, which has to be stored for further analyses. This research purpose is to find efficient structures for managing data gathered in such a way. Four different storages were chosen: Oracle, Excel, MongoDB and Cassandra systems that were examined during various task realizations.

1 Introduction

The research currently conducted in various fields of interest uses the eye movement signal to obtain information about people's willingness, intentions or skills [1, 2] and for supporting GUIs development [3, 4]. There are some studies attempting to apply eye tracking in security tasks [5, 6] and biometric identification [7–12].

Such studies require a preparation of an experimental set-up based on a specialized device, called eye tracker (ET), responsible for providing eye movement data. There are several types of eye trackers available, some of them are shown in Fig. 1.

K. Harezlak (✉) · T. Wasinski · P. Kasprowski
Institute of Informatics, Silesian University of Technology, Gliwice, Poland
e-mail: katarzyna.harezlak@polsl.pl

P. Kasprowski
e-mail: pawel.kasprowski@polsl.pl

R. Neves-Silva et al. (eds.), *Intelligent Decision Technologies*,
Smart Innovation, Systems and Technologies 39,
DOI 10.1007/978-3-319-19857-6_30

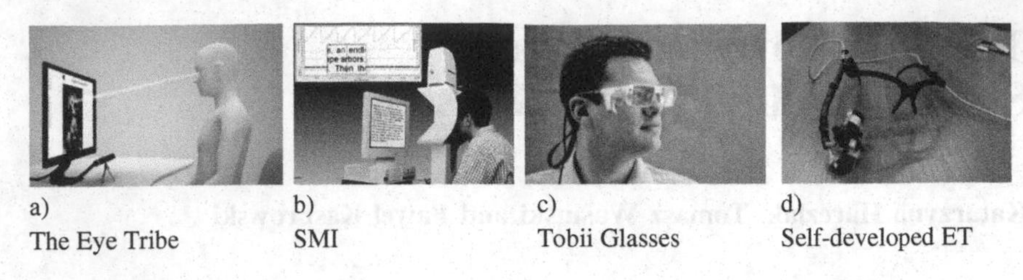

Fig. 1 Example eye tracker types

It is noteworthy that it is possible to build a simple eye tracker using a single CMOS camera with IRPass filter. Such a kind of video-oculography (VOG) eye-tracker, developed by the authors of the paper, is presented in Fig. 1d.

Dependent on recording frequency of an eye tracker used there may exist a necessity of registering between 25 to even 2000 samples per second. For such data to be useful, they should be stored and accessible for further cognitive analysis.

The aim of the research presented in the paper was to find the most efficient structures to support the aforementioned task. To our knowledge there was not such studies conducted so far, whereas much research dealing with processing and analyzing of eye movement data type were carried out. It can be expected that there are sets of data collected during such experiments, which may be used repeatedly for various purposes. Thus the question is how to store that data to use it efficiently.

Providing an answer for this question was the task of the experiments described in the subsequent sections. They were devoted to the problems of moving data from primary to the chosen structures, an efficiency of particular recording search and delivering data to the commonly used algorithms for extracting events from an eye movement signal - like I-DT (Dispersion-Threshold Identification) or I-VT (Velocity-Threshold Identification) [16, 17].

2 The Test Environment

The research was conducted with usage of the previously presented self-developed VOG eye tracker. The system generated 20–25 measurements per second. At the beginning of the experiments all data obtained from the eye tracker was saved in text files formatted according to TSV (Tab Separated Values) standard.

The test set consisted of 20 files including altogether 30 282 samples. Each sample was described by a type, its registration time eye gaze coordinates. A content of an exemplary file is presented in Table 1. Point type represents a position of a stimuli on the screen (SS) or an eye movement sample (R) containing information about the current gaze point.

Table 1 Data on simple recordings

Point type	Registration time	X coordinate	Y coordinate
R	1383154553160	0.5630817	0.5127366
R	1383154553225	0.5637367	0.5132968
R	1383154553281	0.5638777	0.5136657
R	1383154553355	0.5637559	0.5130093
SS	1383154553664	0.05	0.05

It could be expected that using storages dedicated to and supporting data processing may improve overall analysis of the eye movement signal. Thus, there were four type of storages chosen to study in terms of the research purposes:

- Oracle Database 11 g Express Edition,
- MongoDB 2.6 Standard [14, 15] with MongoDB 1.5.2 PHP driver,
- Apache Cassandra 2.0.7 [13],
- Microsoft Excel - Microsoft Office 2010.

The computer serving as a workstation was equipped with:

- Processor: Intel(R) Core(TM) i5-3210 M CPU @ 2.50 GHz,
- RAM: 8 GB,
- Windows 7 operating system.

Ensuring the same conditions for all chosen data storages was an important aspect of the carried out studies. To avoid burdening results by tests previously done, only one database system was operating at a time and RAM memory was cleaned before each test.

3 Moving Data to New Storages

As it was discussed above – in most cases – the eye movement data is at first being saved in text files (like for example in the previously presented format). To take advantage of the other storage types it has to be moved there from an original location. The assessment of this process efficiency was the first research step of the studies. Additionally, to check its scalability the experiment was repeated tenfold for different amount of data – 50 000, 500 000 and 1000 000 samples.

The obtained results, described in detail in subsequent paragraphs, were averaged and summarized on the chart presented in Fig. 2. A logarithmic scale was used to make this presentation possible.

Oracle.
In the case of Oracle database, before the migration process started, one additional step had to be done. Tables in a relational database are expected to have a primary key for each record. This fact entailed necessity of a sequence object creation to

ensure generating primary key automatically. After that the migration process became possible. The average results of the tenfold each test execution are presented in Fig. 2. Moving 50 000 samples from the text file to the Oracle database took 26.977 s, 500 000–324.407 s and for one million samples it lasted 783.153 s. These results have presented almost linear relationship between amount of data and its migration duration.

Excel.
Test undertaken using MS Excel in case of processing 50 000 samples finished without any problems and the average duration of that task was 55.031 s. However, the trial regarding 500 000 samples has revealed a problem with RAM memory usage, influencing the migrating data duration - averaging on 4 085.048 s (Fig. 2). The analysis of reasons has indicated that the problem was the library used (PHPExcel - http://www.codeplex.com/PHPExcel). It stored all analysed data in an inner table being a type of cache, heavily using RAM memory. For this reason experiments with 1 000 000 samples have been omitted.

MongoDB.
The results of migrating the eye movement recordings to MongoDB database turned out to be interesting, especially in case of the set of 1 000 000 samples (Fig. 2). While the time of data migration increased linearly for two first amounts of data (9.787 s for 50 000 and 94.042 s for 500 000 samples), in case of 1 000 000 the duration was only about one and a half longer than for 500 000 elements.

Cassandra.
The outcomes obtained for Cassandra database – 52.625 s for 50 000; 503.986 s for 500 00 and 1 027.915 s for 1 000 000 samples – have shown, similarly to Oracle system, almost a linear relationship between them. However, in this instance the execution time was two times longer for each case comparing to Oracle.

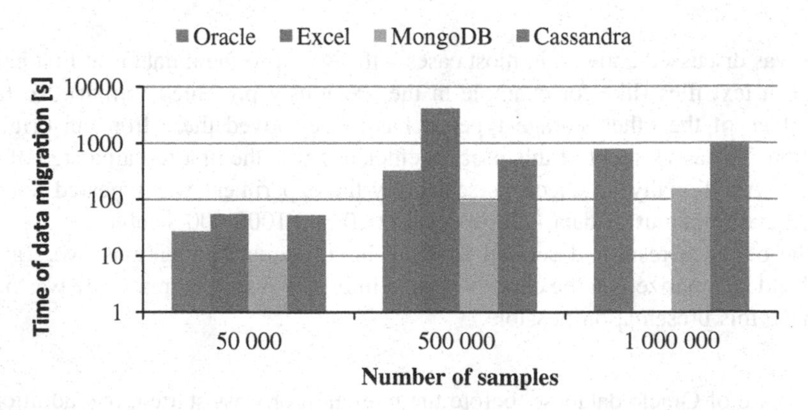

Fig. 2 Duration of moving data from text file to different storages

The analysis of the chart clearly indicates that filling MongoDB database is related to the most efficient migration process execution. Additionally, its outcome allows to have an expectation that MongoDB efficiency will increase along with the growth of the amount of data. On the other hand, exceptionally weak results of Excel incline to the opinion that this program is not a good solution for managing big sets of data, in this case, concerning eye movements. Yet, it should be checked using other programing languages environments.

4 The Efficiency of Data Access

Gathering of an eye moment signal makes sense if it is accessible for a further processing and analysis – for example determining gaze points or finding areas attracting user's attention. For this purpose two algorithms – I-DT and I-VT [16, 17] - extracting sets of fixations from eye movements have been implemented. During the experiments data structures of the algorithms were filled with data from the previously described storages. This operation was repeated tenfold for each of the structures and the results were averaged. Due to the fact that both algorithms data structures were similar, the analysis is relating to only one of them.

Once again the tests were conducted using sets with 50 000, 500 00 and 1 000 000 samples. The schema of the test setup is presented in Fig. 3.

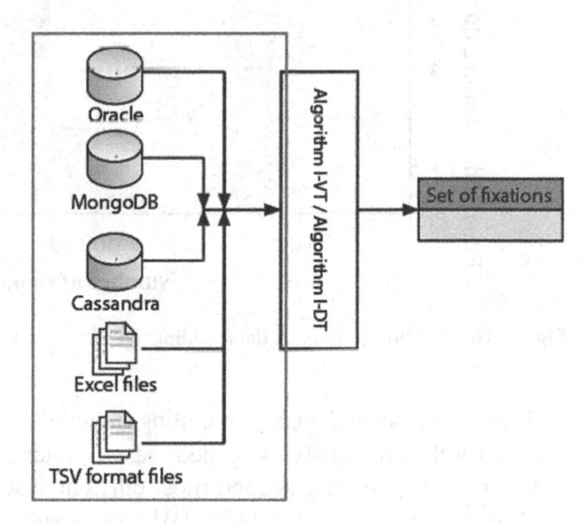

Fig. 3 The schema of the test setup

Each experiment concerning one data storage has been divided into three parts:

- the task of the first one was to provide to the algorithms data from a particular eye movement recording – which means that data was filtered with a given condition;

- subsequently, the tests aimed at counting all samples in a particular eye movement recording were carried out;
- finally, during the last part of the experiment, the time required to load all data available in the storage to the algorithms' data structure, were studied.

Examining the efficiency of providing data for further processing, the files in TSV format were taken into consideration as well. The results obtained for the first task realization showed that TSV format, with average duration equal 0.048 s (independently of the number of recordings), and Oracle (Fig. 4) were the most efficient storages. The stable outcome for text file resulted from the fact that each recording was saved in a separate file with the known location on the disk. Thus, access to the recording was tantamount to an access to a single file. The good results for Oracle system can be easily explained taking well developed functionality of the search engine into account.

The tests were also carried out for Excel storage, however their duration for 50 000 samples (16.789 s) caused the abandonment of further studies in this environment. The result is omitted on the chart to make the analysis of other values convenient.

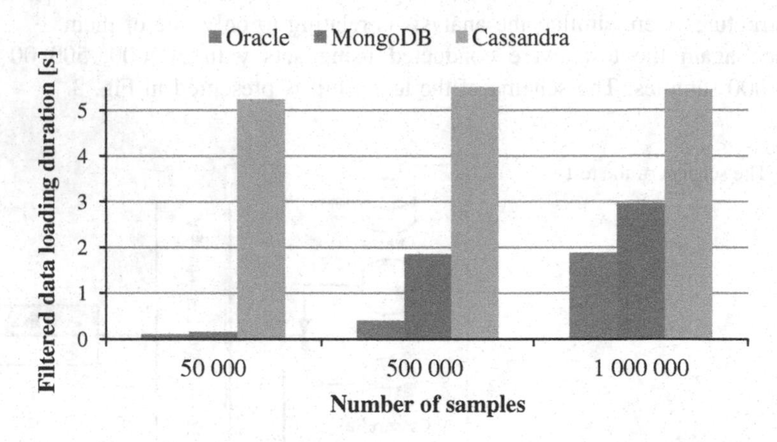

Fig. 4 The duration of filtered data loading

During the second task – counting samples in a chosen recording – the best results, for the previously explained reasons, once again were obtained for TSV file (0.012 s). However, the second most efficient system was MongoDB with almost the same average task duration (0.013 s), whereas Oracle database outcome was twice worse (0.0251 s) than both the previously mentioned storages types. Yet, significantly the least efficient system was Cassandra one, reaching outcome between two and five hundred times worse than other systems (5.551 s).

In the last performance tests (loading all samples into algorithms' data structures) Cassandra surprisingly turned out to be the best one for two bigger sets, realizing its task in about 7 s, while the task duration was from two to four times worse in case of other systems (Fig. 5). The difference for files in TSV format was even bigger (eight times worse result – not presented on the chart).

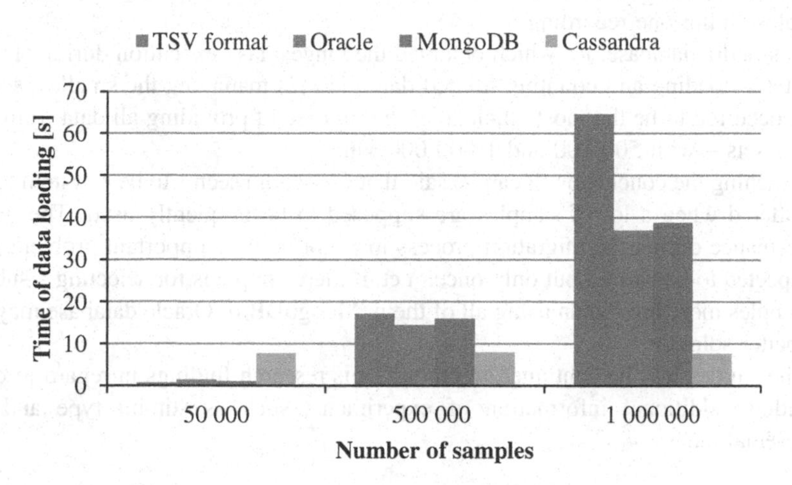

Fig. 5 All data loading time

5 Conclusion

The variety of the tasks, which were studied during the experiments, have revealed that performance of a particular storage depend on an operation being executed.

The analysis of the obtained results showed that Excel was not a good solution when big sets of samples had to be taken into account. It was perceived in both migration and loading data tasks. MongoDB database, in the former operation, featured by the good performance, especially in case of big sets of data. The task duration did not increased proportionally to size of data yet was lower than expected. The Oracle system was slightly worse than MongoDB, which may be explained by the necessity of generating primary key value for each sample. The poorest performance was observed for Cassandra. Generating the row identifier required while inserting a new data, like in the case of Oracle database, may be pointed out as one of the reasons. Yet, more important influence on making migration process more costly had data structures used in Cassandra database.

Different conclusions may be drawn when studying the outcomes of the rest of the experiments. The files with TSV format turned out to be the most efficient structure while working with filtered data – counting and providing data from one recording. However, it has to be once again emphasized that in case of this structure, samples of a particular recording were stored in one file in the known location. Much worse results were achieved when all data, from all files, had to be taken into account. In this case the worst performance from all of the analysed storages, with exception of Excel one, were observed.

Oracle system was the most efficient in two tasks – providing filtered and all recordings to I-DT and I-VT algorithms. However the results were close to outcome of MongoDB database, which in turn was more efficient in checking the number of samples within one recording.

Cassandra database, for which obtained the longest task execution during (1) the first tests (loading and counting filtered data) and (2) managing the smallest set of data, occurred to be the most efficient system in case of providing all data from the bigger sets – with 500 000 and 1 000 000 samples.

Reaching the conclusion, it can be said that Cassandra seems to be a system to be considered when a lot of samples are supposed to be frequently used. The weak performance during the migration process may not be very important problem if it is expected to be carried out only once. Yet, if there are plans for selecting a subset of samples more often than using all of them, MongoDB or Oracle database may be the better solution.

The studies will be continued to confirm this research findings in regard to data including additional information of experiments such as stimuli type and its representation.

Bibliography

1. Bednarik, R., Vrzakova, H., Hradis, M.: What do you want to do next: a novel approach for intent prediction in gaze-based interaction. In: Spencer, S.N. (ed.) Proceedings of the Symposium on Eye Tracking Research and Applications (ETRA'12), pp. 83–90. ACM, New York, USA (2012)
2. Jarodzka, H., Scheiter, K., Gerjets, P., van Gog, T.: In the eyes of the beholder: How experts and novices interpret dynamic stimuli. Learn. Instr. **20**, 146–154 (2010)
3. Goldberg, J.H., Stimson, M.J., Lewenstein, M., Scott, N., Wichansky, A.M.: Eye tracking in web search tasks: design implications. In: Proceedings of the 2002 symposium on Eye tracking research & applications (ETRA'02), pp. 51–58. ACM, New York, USA. doi:10.1145/507072.507082, http://doi.acm.org/10.1145/507072.507082
4. Jacob, R.J.K.: Eye tracking in advanced interface design. In: Barfield, W., Furness, T.A. III (eds.) Virtual Environments and Advanced Interface Design, pp. 258–288. Oxford University Press, Inc., New York, USA
5. Kasprowski, P., Harężlak, K.: Cheap and easy PIN entering using eye gaze. In: 3rd Conference on Cryptography and Security Systems. Annales UMCS Informatica, vol. 14, issue 1, pp. 75–84, Lublin (2014)
6. De Luca, A., Denzel, M., Hussmann, H.: Look into my eyes!: can you guess my password?. In: Proceedings of the 5th Symposium on Usable Privacy and Security (SOUPS'09), pp. 1–12. ACM, New York, USA, Article 7 (2009)
7. Kasprowski, P., Ober, J.: Eye movement in biometrics. In: Proceedings of Biometric Authentication Workshop, European Conference on Computer Vision in Prague 2004, LNCS, vol. 3087, Springer (2004)
8. Bednarik, R., Kinnunen, T., Mihaila, A., Fränti, P.: Eye-movements as a biometric. In: 14 Scandinavian Conference on Image Analysis, Lecture Notes in Computer Science, vol. 3540, pp. 780–789. Springer (2005)

9. Kinnunen, T., Sedlak, F., Bednarik, R.: Towards task-independent person authentication using eye movement signals. In: Proceedings of the 2010 Symposium on Eye-Tracking Research & Applications, ACM, New York, USA (2010)
10. Komogortsev, O., Jayarathna, S., Aragon, C.R., Mahmoud, M.: Biometric identification via an oculomotor plant mathematical model. In: Proceedings of the 2010 Symposium on Eye-Tracking Research & Applications, ACM, New York, USA (2010)
11. Rigas, I., Economou, G., Fotopoulos, S.: Biometric identification based on the eye movements and graph matching techniques. Pattern Recognit. Lett. **33**(6), 786–792, ISSN 0167-8655 (2012)
12. Kasprowski, P., Komogortsev, O.V., Karpov, A.: first eye movement verification and identification competition at BTAS 2012. In: IEEE Fifth International Conference on Biometrics: Theory, Applications and Systems (BTAS) (2012)
13. http://cassandra.apache.org/ (dostęp wrzesień 2014)
14. http://www.mongodb.org/ (dostęp wrzesień 2014)
15. http://php.net/manual/en/mongo.tutorial.php (dostęp wrzesień 2014)
16. Salvucci, D.D., Goldberg, J.H.: Identifying fixations and saccades in eye-tracking protocols. In: Proceedings of the 2000 Symposium on Eye Tracking Research and Applications, pp. 71–78. ACM Press, NY (2000)
17. Shic, F., Chawarska, K., Scassellati, B.: The incomplete fixation measure. In: Proceedings of the 2008 Symposium on Eye Tracking Research and Applications, pp. 111–114. ACM Press, NY (2008)

Using Non-calibrated Eye Movement Data to Enhance Human Computer Interfaces

Pawel Kasprowski and Katarzyna Harezlak

Abstract Eye movement may be regarded as a new promising modality for human computer interfaces. With the growing popularity of cheap and easy to use eye trackers, gaze data may become a popular way to enter information and to control computer interfaces. However, properly working gaze contingent interface requires intelligent methods for processing data obtained from an eye tracker. They should reflect users' intentions regardless of a quality of the signal obtained from an eye tracker. The paper presents the results of an experiment during which algorithms processing eye movement data while 4-digits PIN was entered with eyes were checked for both calibrated and non-calibrated users.

1 Introduction

The usefulness of the eye movement analysis was confirmed in research in many areas of interests. It may be used for example in advertisements developing, sociology, medicine and cognitive studies [3, 6]. Recently, a lot of attention has been focused on possibilities to use eye movement for enhancing human-computer interfaces. Using gaze information as a new input device in a way similar to mouse seems to be the promising technique, making cooperation with computers even easier for unexperienced users. Nevertheless, the eye movement processing still faces a lot of usability problems so a lot of effort must be done to make this

P. Kasprowski (✉) · K. Harezlak
Institute of Informatics, Silesian University of Technology, Gliwice, Poland
e-mail: pawel.kasprowski@polsl.pl

K. Harezlak
e-mail: katarzyna.harezlak@polsl.pl

R. Neves-Silva et al. (eds.), *Intelligent Decision Technologies*,
Smart Innovation, Systems and Technologies 39,
DOI 10.1007/978-3-319-19857-6_31

technique really user friendly. One of the examples is a so called Midas touch problem [7], which addresses the difficulty to decide when user looks at a button if he wants to click it or just to read its caption.

One of the most important obstacles in making eye movement based interfaces robust and convenient is the necessity to calibrate an eye tracker for each user before any usage [11]. The aim of the studies presented in this paper was to check whether it is possible to utilize information obtained from an eye tracker without prior calibration done by the user that is being measured. One of the simplest tasks - the PIN entering - was taken into consideration and these studies are based on the research discussed in [9].

The main contribution of the research presented in the paper is the introduction of the idea to shorten eye tracking sessions by carrying out the same calibration for various users. Thus, the novel so called *regression based* algorithm was implemented and compared to an intuitive *distance based* algorithm. The correctness of the analysed task realization for both cases and factors influencing the results were analysed.

All classic experiments based on eye tracking methods are conducted in accordance with a commonly used schema. Each trial starts with a calibration process. The aim of that step is to find the correlation between coordinates of user's gaze point and coordinates represented in an eye tracker system. During a calibration users are asked to move their eyes over a screen as a reaction to a presented stimulus. Dependent on an experiment type there may be various stimuli used, yet the most popular is a point jumping over a screen. Each change of the point position triggers eyes movement. Recorded eye movement samples correspond to a given point on a screen and a user is expected to keep the focus in the same point for a while, long enough to collect such set of samples, which ensure good adjustment of a screen and an eye tracking system coordinates. The time of a single point presentation is usually set within scope of 2–3 s [4]. A number of point's locations and their dispersion on the screen are other issues [1, 8]. It is obvious that the higher quality may be achieved for more points, yet taking user's convenience into consideration the lowest possible number is better solution. The reason is time required to perform an eye tracker calibration – too long can be wearisome for participants, discouraging them for the involvement in core experiments.

The main problem of the calibration is that it must be repeated before every trial as it depends on an environment used in experiments and on characteristic features of a particular user. As the calibration process is rather cumbersome for a user, the idea of the paper was to check if, for some human-system interactions, which do not require highly accurate eye tracker's adjustment, it is possible to omit calibration step and still achieve satisfactory results. Tasks regarding entering a PIN, in which focusing eyes on a specific area is sufficient to determine a digit, can be taken as an example. Such task may be used to lock and unlock computer screen with eyes or to enter PIN at ATM [10]. According to [2] it prevents shoulder surfing attack and is generally more difficult to forge.

2 The Experiment

The eye movements were registered using the Eye Tribe - an eye tracking system working with sampling rates 60 Hz. The accuracy and spatial resolution declared by manufacturer equals 0.5° – 1° and 0.1° respectively. The eye tracker was placed below the screen. A PC computer was used to control the experiment (show stimulus, control the eye tracker and save recordings). The users were sitting centrally at a distance of 60 cm.

The experiment consisted of two parts. During the first part one participant was calibrated using a classic scenario with 9 points evenly distributed on the screen. It lasted for about 20 s. Then a screen with evenly distributed circles with digits from 0 to 9 was displayed and the participant was asked to look at a digit and press a trigger button, then look at subsequent digit and press the trigger and so on. The PIN was a four digits sequence for which every two subsequent digits were always different. After four trigger clicks the attempt (a trial) was saved.

The second (and more interesting) part of the experiment started when some number of different participants were asked to enter their PINs with eyes in the same manner, but this time without any calibration. These participants' eye movements were registered using a calibration function built for the first user. Eye movement data for every PIN entering was saved (later referred to as a *trial*).

There were 802 trials collected including 204 with own calibration (called 'calibrated') and 598 without own calibration (called 'non-calibrated'). 41 participants took part in the experiment.

To examine samples gathered during the experiments two own developed methods were used. Both of them are based on sets of fixations extracted from eye movement signal and are described in details in the subsequent section.

3 Method

The purpose of the PIN extraction algorithm described in this section was to obtain information about a sequence of digits pointed with eyes from the recorded eye movement data. In the algorithm two phases may be distinguished:

- Extraction of fixations,
- Assigning fixations to digits on the screen.

3.1 Extraction of Fixations

Typical eye movement signal consists of two events: fixations - when eye is relatively still and the brain acquires information from the scene; and saccades - a rapid movement when an eye position changes to another fixation. The extraction of fixations from a raw eye movement signal may be done using different algorithms [12, 13]. It was our own implementation of one of the most popular dispersion-threshold algorithm (I-DT) used in this work.

At first the algorithm classifies each eye movement sample according to a simple rule: if the distance among this sample and five previous samples is less than a specified threshold (Th) the sample is classified as a potential part of a fixation (F) and it is classified as a potential part of a saccade (S) otherwise. In the next step all neighboring F-points are gathered together as potential fixations. Every fixation has four attributes: its start time, x and y coordinates of its center and the fixation duration. The subsequent steps convert this preliminary list of fixations into the final list using different techniques for fixation merging and removing. All details of the algorithm are presented in [5].

The value of threshold parameter (Th) started from 0.2° and was increased by 0.2° until one of following conditions was met: I-DT algorithm returned exactly four fixations or threshold value reached 8°.

In the latter case the trial was rejected and no further analyses were done.

3.2 Discovering Chosen Digits

After extracting the four most dominant fixations, it was assumed that these fixations occurred while a person was looking at specific digits. The next task was to discover which digits were pointed with eyes. There were two different methods used in the research to pair fixations with proper digits.

Distance based approach.
The first – and the most obvious - method divides a screen into regions of interest (ROIs) using digits locations as points in Voronoi diagram. As a result every fixation is classified as a digit which is the closest one for this fixation. It may happen that a fixation location is almost in the middle between two digits (near a boundary of a Voronoi cell). As a result one of the digits must always be chosen but we may expect that the choice is somehow random in such case. Therefore, an additional step and an additional parameter: proximity coefficient (PCF) were introduced. After finding the closest digit for a fixation, it is checked whether a distance between that digit and the fixation multiplied by PCF is still lower than distances between the fixation and all other digits:

$$\forall_{0 \leq i \leq 9} \|F - D_i\| > \min_{0 \leq i \leq 9}(\|F - D_i\|) * PCF \quad (1)$$

where F is fixation's location and D_i is a location of digit i.

If this condition does not hold for any fixation in a sequence, the whole trial is rejected. Obviously, for PCF equal to one there are no rejections (there is always one minimal distance) and as its value increases the number of rejections increases as well. For instance PCF equal to 2 means that the distance between the closest digit and the fixation must be twice lower than the distances between that fixation and all other digits.

To check whether a simple, one point calibration is enough to improve results, the additional assumption was added that the first digit of PIN is known. Therefore, the extended version of the algorithm introduces the additional step: before any classification all fixations are shifted in space so that the first fixation is positioned exactly in a location of the first digit. The confirmation of usefulness of such activity may be subsequently used for one point calibration displayed for example in the middle of the screen.

Regression based approach.
The next method was using slightly different approach. The basic assumption was that the regression model for a correctly adjusted PIN should provide the lowest error. Thus, the algorithm starts with building regression models that map 4 fixations into 4 digits for every possible combination of PIN digits. There are 7290 possible combinations when assumption that subsequent digits are always different holds. For every such model new fixations' positions are calculated and Mean Square Error between these new and correct positions of PIN's digits are calculated. At the end there is a list of possible PIN numbers with MSE for every PIN available. The PIN with the lowest error is chosen as a correct one.

The only problem in the algorithm described above is which regression function should be chosen to obtain reliable results. Usually, second degree polynomial function is used for eye trackers calibration [1] but such model is too precise. For 4 points it is able to build a function that maps given fixations to any sequence of digits with almost no errors. Therefore, it was a first degree polynomial function used to evaluate new values for X and Y independently:

$$X_{new} = A_x * X + B_x \tag{2}$$

$$Y_{new} = A_y * Y + B_y \tag{3}$$

The Levenberg Marquardt algorithm was used to calculate coefficients for each 4 fixations – 4 digits pair. Because we were not interested in mirror mapping of PIN numbers, an additional assumption that coefficients A_x and A_y must be positive numbers was made.

Similarly to the proximity coefficient in the distance based method, it was necessary to add possibility to allow for rejection of trials for which values found are not reliable. Therefore, an additional *min_error* (MER) parameter was introduced. If the lowest value of MSE for the trial is higher than MER, the trial is rejected as unreliable.

To make a fair comparison to the distance based approach that uses information about the first digit, there was also a version of the algorithm evaluated that calculates models only for PINs starting with a known digit.

4 Results

The most obvious results that may be taken into consideration is the absolute accuracy (ABS), which is measured as a ratio between the number of trials with PIN found correctly and the number of all trials. However, there are two more detailed factors that may be used when analyzing results of the algorithms described in the previous section. At first, each algorithm rejects some number of trials for which it assumes that recognition is impossible. So, the first factor to be analyzed is an acceptance rate (ACR). This factor is influenced by PCF and MER parameters accordingly to the algorithm used. Then, for all remaining trials, PIN is evaluated. The number of PINs found correctly to the number of all evaluated trials is defined as a correctness rate (CRR).

Ideally, both ACR and CRR should be 100 %. However, it can be expected that both factors are dependent on each other – when the acceptance rate decreases, the remaining samples are of better quality and the correctness increases. And when the acceptance rate increases, more low quality samples are taken into account during the next step, which may result in lower correctness rate. Tuning of this two factors depends on the purpose of the trial (see **Conclusion** for examples). Obviously, the absolute accuracy (ABS) is the result of multiplication of the two above factors (ABS = ACR*CRR).

As it was presented in the previous section, the first rejection takes place after the 'extraction of fixations' step. All trials, for which it was impossible to find exactly four fixations are rejected. The next step when trials may be rejected depends on the algorithm used. For the distance based algorithm the acceptance rate depends on the proximity coefficient (PCF). As it was described in the previous section, increasing PCF decreases the number of accepted samples. For the regression based algorithm the *min_error* coefficient (MER) may be tuned to reject dubious trials. If MER is high, all samples are accepted and as it decreases, the acceptance rate (ACR) decreases as well.

To illustrate described dependency, the ACR and CRR values for trials when the regression based algorithm was used with different values of min_error (MER) parameter was presented in Fig. 1.

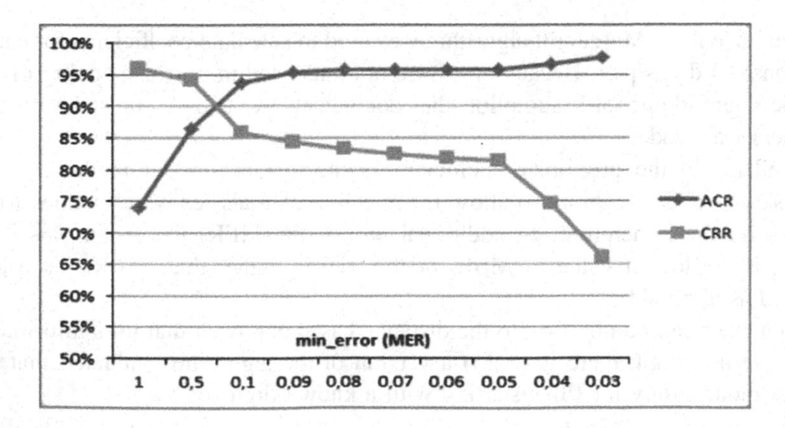

Fig. 1 ACR and CRR values for different min_error (MER) in regression based algorithm

The results obtained for both algorithms and both types of samples are shown in following tables. Table 1 presents values of the acceptance rate (ACR) and the correctness rate (CRR) for calibrated and non-calibrated trials, for the distance based (DIST) algorithm with different values of the proximity coefficient (PCF).

Table 1 CRR and ACR for different PCF for distance based algorithm

	Calibrated		Non-calibrated	
PCF	CRR	ACR	CRR	ACR
1	95 %	94 %	58 %	68 %
1.1	95 %	94 %	66 %	57 %
1.2	95 %	93 %	75 %	46 %

As it can be seen, the results for calibrated trials are quite good and stable for different values of PCF and the results are significantly better than for non-calibrated trials. As it could be expected, higher value of PCF increases the correctness (CRR) but in the same time decreases the acceptance (ACR). For PCF>=1.2 more than a half of non-calibrated trials is rejected. The best value of ABS for non-calibrated trials is only 39 % while it is about 90 % for all PCF values, when only calibrated trials are taken into consideration.

The results for the regression based algorithm (REGR) are presented in Table 2. They were calculated for different values of *min_error* (MER) parameter.

Table 2 CRR and ACR for different MER values for regression based algorithm

	Calibrated		Non-calibrated	
MER	CRR	ACR	CRR	ACR
1	74 %	96 %	56 %	94 %
0.5	86 %	94 %	66 %	91 %
0.1	94 %	86 %	74 %	80 %
0.08	96 %	83 %	81 %	73 %
0.06	96 %	82 %	85 %	61 %
0.04	97 %	75 %	90 %	44 %

It is visible that the results for calibrated trials are worse than for DIST algorithm with ABS about 80 %. However, the results for non-calibrated trials for the regression based algorithm are significantly better than for the distance based one, with ABS reaching 60 % for MER = 0.5. The algorithm is especially efficient in rejecting low quality trials. For instance, 74 % correctness (CRR) was achieved for the acceptance rate (ACR) 80 %, while for DIST algorithm the same correctness rate was achieved for ACR amounting only to 46 %.

The next research question was how the simplest possible, one point calibration can improve the results. Because there were only trials with four points available the only way to check it was to assume that the first digit of PIN is known. For

DIST algorithm it resulted in shifting fixations so that the first fixation overlapped the first (known) digit (see **Method** section for details). For REGR algorithm only PIN numbers starting with the known digit were considered as candidates (see **Method** section as well).

Table 3 Results achieved for DIST algorithm with the first fixation shift

	Calibrated		Non-calibrated	
PC	CRR	ACR	CRR	ACR
1	96 %	95 %	76 %	83 %
1.1	96 %	94 %	81 %	77 %
1.2	96 %	94 %	85 %	70 %
1.3	98 %	93 %	89 %	62 %
1.4	98 %	91 %	91 %	56 %

When considering DIST algorithm (Table 3) the results for the calibrated trials are better for the case of fixation shifting but the difference is not significant (ABS is equal about 91 % in most cases). However, the results for non-calibrated samples are significantly better with 63 % for ABS, in the best case comparing to 39 % for tests without shifting.

Table 4 Results for REGR algorithm with the first digit known

	Calibrated		Non-calibrated	
MER	CRR	ACR	CRR	ACR
0.5	87 %	94 %	69 %	90 %
0.1	95 %	85 %	83 %	74 %
0.08	96 %	83 %	88 %	69 %
0.06	98 %	80 %	91 %	58 %
0.04	98 %	74 %	92 %	43 %

The results for the regression based algorithm (Table 4) did not improve outcomes significantly when only PINs with a correct first digit were taken into account. Such condition reduced the number of PINs for which models were calculated ten times (from 7290 to 729) but it did not affect algorithms performance.

The comparison of the algorithms and the sets was presented in Fig. 2. The distance based algorithm performed very well for calibrated data (*dist_cal* and *dist_cal_shift*) while its results were very unsatisfactory for non-calibrated data (*dist_nc*). The regression algorithm was not as good as the distance based one for calibrated data (*regr_cal*) but it outperformed it for non-calibrated one (*regr_nc*). Adding information about the first digit improved outcomes of the distance based algorithm (*dist_nc_shift*) but even with this information it is not better than the normal regression based outcome (*regr_nc*). As it was shown in Table 4, adding information about the first digit of PIN did not improve significantly the results for the regression based algorithm so it was not included in Fig. 2.

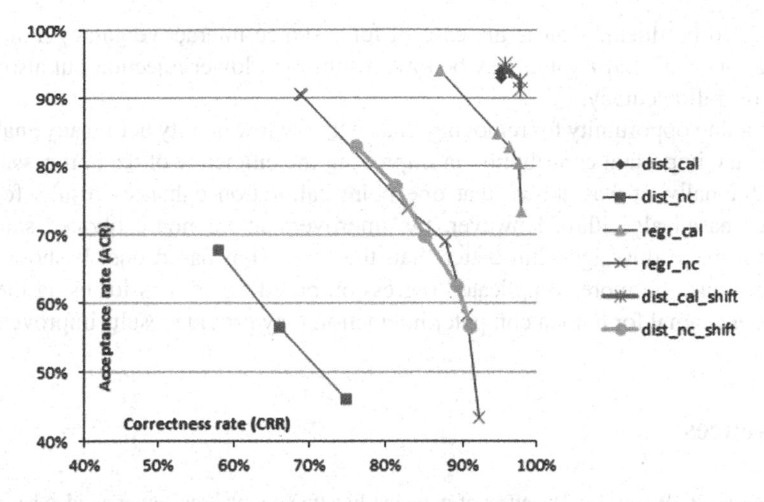

Fig. 2 ACR and CRR for different algorithms and calibrated (cal) and non-calibrated (nc) trials

5 Conclusions

The findings of the research can be divided into two groups. First of them regards trials proceeded by the per-user calibration. The results obtained for such recordings confirmed the possibility of using eyes for providing information of PIN type. Another conclusions may be drawn from outcomes obtained for various scopes of the introduced parameters - *proximity coefficient* (PCF) and *min_error* (MER). They show to what extent the size of the area of interests can be reduced not to decrease the efficiency of the method.

The second group of the findings concerns the problem of omitting a calibration process. The experiments presented in this paper showed that it is possible to use eye tracker as a pointer for simple and well defined tasks even without a prior per-user calibration. It is possible if such task does not require point to point gaze mapping, yet point to area of interests adjustment is acceptable. The studies of using eyes for PIN providing fulfil this requirement. The results obtained for the non-calibrated trials are worse than for the calibrated ones, however values of the analyzed factors indicated that in most cases proper values could be obtained.

A novel regression based algorithm was introduced and it was shown that it outperforms the distance based one for the non-calibrated samples. Additionally, it may be tuned for various types of interfaces using a *min_error* parameter. When the correctness of recognition is important, the *min_error* value may be increased and it was shown that results become more reliable (in sake of higher rejection rate). Such scenario may be useful for instance for gaze pointing of PIN at ATM when we want to be sure that PIN entered is correct even if the user is forced to enter it several times due to rejections. On the other hand there are applications in which an approximated gaze position is enough and rejections are rather undesirable for an interface to be fluent. That is the case of for instance interactive games. For such applications *min_error* value may be low, resulting in lower rejection but also with lower overall accuracy.

Providing opportunity for removing trials with the low quality before any analyses starts is the important contribution in improving the efficiency of data processing.

Additionally, it was shown that one point calibration enhances results for the distance based algorithm. However, the improvement for non-calibrated samples does not make this algorithm better than the regression based one. It shows that further studies on more complicated regression based algorithms for using the eye movement signal for human computer interaction may provide results improvement.

References

1. Blignaut, P., Wium, D.: The effect of mapping function on the accuracy of a video-based eye tracker. In: Proceedings of the 2013 Conference on Eye Tracking South Africa, pp. 39–46. ACM (2013)
2. De Luca, A., Denzel, M., Hussmann, H.: Look into my eyes!: Can you guess my password?. In: Proceedings of the 5th Symposium on Usable Privacy and Security, p. 7. ACM (2009)
3. Duchowski, A.: Eye Tracking methOdology: Theory and Practice, vol. 373. Springer, New York (2007)
4. Haręźlak, K., Kasprowski, P., Stasch, M.: Towards accurate eye tracker calibration—methods and procedures. Proceedings of the 17th International Conference in Knowledge Based and Intelligent Information and Engineering Systems—KES2014. Procedia Computer Science, vol. 35, pp. 1073–1081, Elsevier (2014) (ScienceDirect)
5. Haręźlak K., Kasprowski, P.: Evaluating quality of dispersion based fixation detection algorithm. In: 29th International Symposium on Computer and Information Sciences. Information Sciences and Systems 2014. Springer Lecture Notes on Electrical Engineering, pp. 97–104 (2014)
6. Holmqvist, K., et al.: Eye tracking: A Comprehensive Guide to Methods and Measures. Oxford University Press, Oxford (2011)
7. Jacob, R.J.K.: The use of eye movements in human-computer interaction techniques: what you look at is what you get. ACM Trans. Inf. Syst. (TOIS) **9.2**, 152–169 (1991)
8. Kasprowski, P., Haręźlak, K., Stasch, M.: Guidelines for the Eye Tracker Calibration Using Points of Regard. Information Technologies in Biomedicine, vol. 4. Advances in Intelligent Systems and Computing, vol. 284, pp. 225–236. Springer International Publishing (2014)
9. Kasprowski, P., Haręźlak, K.: Cheap and easy PIN entering using eye gaze. In: 3rd Conference on Cryptography and Security Systems. Annales UMCS Informatica, vol. 14, 1, pp. 75–84. Lublin (2014)
10. Kumar, M., et al.: Reducing shoulder-surfing by using gaze-based password entry. In: Proceedings of the 3rd Symposium on Usable Privacy and Security, pp. 13–19. ACM (2007)
11. Ramanauskas, N.: Calibration of video-oculographical eye-tracking system. Electron. Electr. Eng. **8**(72), 65–68 (2006)
12. Salvucci, D.D., Goldberg, J.H.: Identifying fixations and saccades in eye-tracking protocols. In: Proceedings of the 2000 Symposium on Eye Tracking Research and Applications, pp. 71–78. ACM Press, NY (2000)
13. Shic, F., Chawarska, K., Scassellati, B.: The incomplete fixation measure. In: Proceedings of the 2008 Symposium on Eye Tracking Research and Applications, pp. 111–114. ACM Press, NY (2008)

Argument Quality or Peripheral Cues: What Makes an Auction Deal Successful in eBay?

Min Jung Ko and Yong Jin Kim

Abstract Recently, in this online business environment, there are magnificent advancement and growth of online auction, representatively in eBay. The cause of this advancement is a kind of consumer experience that the excitement of winning a product is happier rather than simply buying it. In relation to these situations, there are many business models about online auction process. These studies were not based on actual data and focused mostly on individual cases and specific factors including the starting price of a transaction, the bidder number, bidder and seller behavior in the auction market, and the online auction strategy. This fragmented approach made understanding the whole picture of online auctions difficult. Also, there is no study addressing online auction from the perspective of the communication between bid-takers and bid-makers in which informational influence plays an important role. To fill this gap, this study takes a holistic view by defining the online auction process as an integrated persuasion process. We are going to study Argument quality and Peripheral cues that make an Auction deal successful in eBay and we got the sold data from the web crawling in eBay so that we analyzed what factors make an Auction deal successful in eBay. As a result, auction success is strong the affects of the central route than peripheral route. Our findings provide additional insights in design for internet auction system and offers findings based on actual auction data that can be compared with those of the modeling approach.

Keywords Internet auction · IS success · Information quality · ELM (Elaboration likelihood Model) · ebay

M.J. Ko
Computer Science Institute, Dongguk University, Seoul, South Korea
e-mail: mjgo@dongguk.edu

Y.J. Kim (✉)
Graduate School of Business, Sogang University, Seoul, South Korea
e-mail: yongjkim@sogang.ac.kr

R. Neves-Silva et al. (eds.), *Intelligent Decision Technologies*,
Smart Innovation, Systems and Technologies 39,
DOI 10.1007/978-3-319-19857-6_32

1 Introduction

Recently with the advancement and growth of online auction, consumers experience the excitement of winning a product at possibly cheaper price rather than simply buying it [6]. Previous studies examined traditional auctions on a single item from the game theoretic view point [37]. With the development of online auctions where multiunit auctions are common [5], many studies investigated the optimal bidding strategies of participants [5, 6], online auction as a social interaction [39], trust issues as a major success factor of the auction practice [3, 4], and the impact of online auction on business models [37]. To our knowledge, however, there is no study addressing online auction from the perspective of the communication between bid-takers and bid-makers in which informational influence plays an important role.

Auction either online or off-line can be viewed as social process [39] where information processing is influenced by the quality of information itself and the relationship between participants [36]. In online environment, the interaction between buyers and sellers, although weaker than that in traditional, face-to-face markets, creates social facilitation in which the presence of others affects the behavior of the buyer or seller [39]. In this sense, like other product purchase situations, a variety of factors in addition to trust [3, 4] and bidder strategies [5, 6] likely influence the completion of online auction deals. The growing use and importance of hypermedia systems has led researchers and designers to formulate standards in this area to provide direction to vendors and to develop hypermedia products which can be easily integrated with other. Such problems also may be avoided by using metadata to support the desired search and presentation of existing resources. This can be achieved by providing summary or other characteristics of the document separately, which might allow users to decide whether to read a document fully, thus increasing user efficiency. Therefore, upon arriving at a document, users will be able to scan the entire document before they are able to assess the contents.

These studies were not based on actual data and focused mostly on individual cases and specific factors including the starting price of a transaction, the bidder number, bidder's and seller's behavior in the auction market, and the online auction strategy [5, 6]. In addition, the previous studies often use highly abstracted concepts, which are difficult to apply to real world situations.

In order to fill these gaps, we apply an Elaboration Likelihood Model (ELM) as the overarching theory. This offers a holistic view and assumes that the auction process is a persuasion process between bidders and sellers using central or peripheral information [36]. We collected the empirical data of three products typically traded in online auctions directly from the eBay website using a crawling engine that we developed. We then categorized the collected data into central or peripheral cues to test how these cues affect the outcome of the persuasion process in a differential manner.

We found that the central and peripheral route affected online auction success in a differential manner according to the characteristics of each auction item, such as the level of personalization, formality, and complexity. This study contributes to the existing literature in two ways. Firstly, this study provides a holistic view of the online auction success by using the concept of persuasion with central and peripheral routes. Secondly, this study offers a way to collect online auction data that can be used to examine online auction success with actual transaction data. Practically, this study has implications for the design of online auction websites and the processes used to be successful.

The investigation of the sources of informational influence separately leads to the understanding of the factors that affect the success of online auction deals and may enhance the likelihood of the success of the e-commerce. Likewise, this approach may shed light to the questions such as what kinds of items should be taken into account and shown to users when a website is designed, how those items should be arranged in a website, and who will be a winner in the online auction deals.

2 Theoretical Background

2.1 IS Success

Quality can be defined as the "degree or grade of excellence, etc. possessed by a thing, used for restricted cases in which there is comparison (expressed or implied) with other things of the same kind [32]," and as "the totality of features and characteristics of a product and service that bears on its ability to satisfy given needs" [20]. It has also been defined in terms of "fitness of purpose" [21], the "conformance to requirements" [12], and "not just meeting specifications, agreed upon goals, but satisfying customers' expectations better than the competitors [11]." Consequently, information quality can be defined as a global judgment or attitude of information users relating to the superiority or excellence of the provided information.

Delone and Mclean analyzed what makes the e-trade success. So they suggested their two dimensions of information success, system quality and information quality. So these two factors can affect the matter of information system success.

For the information system success, On-line web site should have high level of system quality. System quality can be evaluated by some factors such as adaptability, availability, reliability, response time and usability. These factors are played important role in evaluating system quality. High level of system quality can work some kind of format option that can lead to information system success. On the other hand, Successful Online web site should also have high level of information quality. Information quality is important to make more user satisfaction. This information quality can be evaluated by some factors such as completeness, ease of understanding, personalization, relevance and security. High level of system quality can work some kind of details that can lead to information system success. So, these two dimensions of information factors are playing an important role in information success.

Information quality, or IQ, can also be measured in terms of its sub-dimensions or categories. In their seminal paper on information systems success, DeLone and McLean [13] established a comprehensive taxonomy of information quality based on approximately 180 articles representing the cumulative tradition of information research. Among their six dimensions of information systems success (system quality, information quality, use, user satisfaction, individual impact, and organizational impact), information quality dimension has been measured by 23 measures: importance, relevance, usefulness, informativeness, usableness, understandability, readability, clarity, format, appearance, content, accuracy, precision, conciseness, sufficiency, completeness, reliability, currency, timeliness, uniqueness, comparability, quantitativeness, and freedom from bias.

Strong, Lee, and Wang [41] define information quality as fitness for use by information consumer, with four categories and fifteen dimensions. The four categories of information quality are intrinsic, accessibility, contextual and representational information quality. Intrinsic information quality with dimensions such as accuracy, objectivity, believability, and reputation concerns the fact that information quality has quality in its own right. Contextual information quality with relevancy, value added, timeliness, completeness, and amount of data involves the requirement that information quality should be taken into account within the context of the task at hand. Accessibility with access and security dimensions and representational information quality with such dimensions as interpretability, consistency, conciseness, and ease of understanding place an emphasis on the importance of the role of information systems.

Klobas [23] defines the attributes of information quality in terms of accuracy, currency, authority, and novelty. *Authority* is the extent to which the information resource contains from authoritative sources, and *novelty* is the extent to which the resource contains information new to users.

O'Brien [31] classifies information quality into three dimensions: content, time, and form. In his framework, while the content dimension involves intrinsic quality of information, independent of form and time in which information is used, form dimension is concerned with the representational quality of the information. The time dimension deals with the order of presentation, history in the information use, and information update.

The measures of DeLone and McLean [13], Strong, Lee, and Wang [41], and Klobas [23] appear to be sufficient to evaluate information quality. However, they need conceptual integration and expansion. For example, according to [19], informativeness is comprised of such concepts as relevance, comprehensiveness, recentness, accuracy, and credibility. DeLone and McLean's [13] measures, however, contain informativeness, relevance, and accuracy. Strong et al. [41] do not address time-related issues such as order of presentation.

Using O'Brien's [31] three-dimensional information quality model consisting of content, form, and time, we are integrating and expanding DeLone and McLean's [13] and Strong, et al.'s [41] dimensions of information quality. The "content" dimension can be mapped on to seventeen of the twenty three measures of DeLone and McLean – *importance, relevance, usefulness, usableness, informativeness, content, accuracy, precision, conciseness, sufficiency, completeness, reliability, uniqueness, comparability, quantitativeness, timeliness, and freedom from bias.* This dimension can also be mapped on to the *intrinsic* and *contextual* information quality of Strong, et al. [41]. The "time" dimension consists of the *currency* measure of the DeLone and McLean model, but there is no construct in the Strong, Lee, and Wang model that captures the notion of time. It should be noted that Strong, Lee, and Wang's *timeliness* construct essentially means that information is appropriate or adapted to the times or the occasion, and has been taken as a part of the content dimension. The IQ measures of *understandability, readability, clarity, format, and appearance* in the DeLone and McLean model constitute the "form" dimension as these attributes concern the representational aspect of IQ. Similarly, Strong, Lee, and Wang's *representational* and *accessibility* IQ dimensions map on to the "form dimension" in our model.

Among the measures of the content dimension, the term relevance means the ability to retrieve material that satisfies the needs of a user, thereby it can be interpreted as the meaning of applicability, appropriateness, and usefulness [23, 44]. Relevance can thus be a representative term for such concepts as usefulness, usableness, importance, timeliness, and conciseness. Accuracy means freedom from mistake or error, or the degree of conformity of a measure to a standard and has as its synonym precision, preciseness, correctness, and exactness [27, 44]. Accuracy can be the representative measure for precision and freedom from bias. Completeness refers to the fact that there is no missing information and the availability of appropriately aggregated information, which represents such measures as sufficiency, reliability, uniqueness, and comparability [41]. Eventually the content dimension can be comprised of the measures of relevance, accuracy, and completeness.

The time dimension consists of measures of timed history and order of presentation [34] may be added to DeLone and McLean's measure: currency. While timed history is related to time series information of content and presentation, order of presentation involves synchronous or asynchronous information presentation. The effectiveness of the form dimension depends on the way of packaging information, providing interface structure and accessibility to increase understandability, readability, and clarity [31].

Through this study, we are going to analyze the eBay auction process. However, we hypothesize that eBay is successful in terms of the system quality. So we are going to research in terms of the information quality only.

2.2 ELM

The ELM (Elaboration Likelihood Model) of persuasion is a dual process theory describing how attitudes are formed and changed, developed by Richard E. Petty and John Cacioppo during the early 1980s. The model examines how an argument's position on the elaboration continuum, from processing and evaluating to peripheral issues such as source expertise or attractiveness, shapes its persuasiveness. ELM resembles the heuristic-systematic model of information processing developed about the same time by Shelly Chaiken. The ELM makes several proposals, attitudes formed under high elaboration which is the central route are stronger than those formed under low elaboration, making this level of persuasion stable and less susceptible to counter-persuasion. Attitudes formed under low elaboration which is the peripheral route are more likely to cause short-term attitude change.

Under low-elaboration conditions, a variable may act as a peripheral cue for example, the belief that experts are always right. While this is similar to the Einstein example above, this is a shortcut which unlike the Einstein example does not require thought. Under moderate elaboration, a variable may direct the extent of information processing: If an expert agrees with this position, I should really listen to what he has to say. A variable's effect on elaboration may increase or decrease persuasion, depending on the strength of the argument. If the argument is strong, enhancing elaboration will enhance persuasion; but if it is weak, thought will undermine persuasion.

Recent adaptations of the ELM have added an additional role for variables: to affect the extent to which a person trusts their thoughts in response to a message self-validation role. A person may feel if an expert presented this information, it is probably correct, and thus I can trust that my reactions to it are informative with respect to my attitude. This role, because of its Meta cognitive nature, only occurs in high-elaboration conditions.

Our study classifies a variety of information provided by eBay into central and peripheral cues. We also hypothesize that these cues affect the success of an auction depending on the characteristics of the auction items. We define the direct information as the central routes, such as the initial price, and the auction duration. We defined the indirect information as the peripheral route, such as the number of graphics, the seller's star count, and the number of selling at eBay.

2.3 Prior Auction Success Studies

With the rapid growth in the electronic commerce over the Internet, the Web became an effective alternative to traditional means of market communications, providing the transactional function of products and services and allowing to manage public relations effectively [42]. As such, Web sites have played role as a communication medium [25] and as a business enabler [33, 43]. Hence, improving

Web site quality is recognized to influence the attitude toward web site of consumers and eventually companies' bottom line.

This phenomenon has significantly raised managerial attention to what consumers experience during their visit to a corporate web site [1] and how to develop web sites responsive to user needs [38]. In particular, the design of Web sites has been considered a critical factor to satisfy users and to make users revisit a web site [28]. Poor web site design results in low user satisfaction [28] and in turn the web site failure (Buschke 1997) by causing serious web usability problems such as irrelevant information, disorientation, and cognitive overhead [10, 26]. The irrelevant information problem occurs when users reach web pages that do not contain the required information while surfing through certain links. Disorientation is the tendency to lose one's sense of location and direction in a set of non-linear documents. The cognitive overhead problem stems from additional efforts and concentration that are required of the users to manage and/or maintain several tasks or navigation trails at one time.

However, technically well-developed web sites do not guarantee the success of the web sites if they do not take into account user needs [22]. Understanding how consumers perceive web sites is very important because Internet shoppers are quite different from traditional retail shoppers [14, 35]. Internet shoppers consider convenience very important, are innovative, have propensity to variety-seeking and impulsiveness, and positive attitude toward advertising [14].

In this context, the current paper examines the impact of the amount of information quantity on perceived information quality as well as that of information quality and ease of navigation on consumer attitude toward a Web site as discussed in other Web site success studies [24, 33, 40]. Of particular interest will be the potential moderating effect of Web site visitors' interest level in determining the interrelationships among the research constructs.

Table 1 classifies the previous studies into various types according to the approach and method.

As discussed earlier, the recent studies on the electronic commerce repeatedly points to information and entertainment as the key determinants of Web site attractiveness or success. There are, however, more reasons to support a simple bidimensional attitude model. For example, Batra and Ray [2] propose that attitudes have cognitive and affective components, called "utilitarian affect" and "hedonic affect," respectively (see also [17]). In line with this bidimensional conceptualization of attitudes, consumer Web behavior also is bifurcated. For instance, Hoffman and Novak [15] identify goal-directed behavior such as online shopping (i.e., information search) and experiential, "netsurfing" behavior as "two broad categories of on-line behavior" (p. 44). Taken together, the literature indicates that practical utility (informativeness) and hedonic utility (entertainment) are the key dimensions of Web site content quality perception, which should influence post-visit attitudes. These two characteristics may influence the attitude toward Web sites regardless of the level of individual interest in the Web site (s)he visits. Therefore, the level of individual interest in the Web site, however, may hold different effects on the attitude toward the Web site, as inferred from Elaboration Likelihood Model.

Table 1 Review of prior auction success studies

Studies	Newness of auction success	Weak points
[6]	cheaper price/the optimal bidding strategies of participants	single item from the game theoretic view point
[37]	business models	Wide area concept
[3, 4]	trust issues of the auction practice	Process fairness problems
[24, 33, 40]	quantity on perceived information quality as well as that of information quality and ease of navigation on consumer attitude toward a Web site	Estimate guideline missing
[39]	a social interaction	Focusing uncertainty
[25, 33, 43]	played role communication medium a business enabler	influence the attitude toward web site
[1, 38]	raised managerial attention to what consumers experience during their visit to a corporate web site how to develop web sites responsive to user needs	Not explain between consumers experiences and site responsiveness
[42]	providing the transactional function of products and services and allowing to manage public relations effectively	define function of Services and products
[10, 26, 28], Buschke (1997), Thüring et al. (1995), Reich, Carr, De Roure, and Hall (1999), Utting and Yankelovich (1989)	a critical factor to satisfy users and to make users revisit a web site Poor web site design results in low user satisfaction by causing serious web usability problems such as irrelevant information, disorientation, and cognitive overhead	the additional efforts and concentration necessary to maintain several tasks or trails at one time
[22]	technically well-developed web sites do not guarantee the success of the web sites if they do not take into account user need	Situation of take into account user needs
[14, 35]	how consumers perceive web sites	Process perceived web sites

(continued)

Table 1 (continued)

Studies	Newness of auction success	Weak points
[36]	information processing is influenced by the quality of information itself and the relationship between participants	Impact participants
[7, 8, 36]	on dual process models of information influence	to explain the impact of message and her factors on individual information processing in the context of online auction deals

High interest consumers much more likely focus on the central message or content itself such that informativeness may carry more effects on the attitude than does entertainment. On the contrary, in the case of low interest individuals, entertainment will affect the attitude as much as informativeness.

3 Research Model and Hypotheses

Although they originated in different disciplines, IS Success and ELM having some obvious similarities. The relative advantage construct IS Success is considered to be the information process in ELM, and the complexity construct in IS Success is extremely similar to the information quality (elaboration Likelihood) concept in ELM. This suggests that IS Success and ELM reconfirm each other's findings.

In our study, we integrate the interest based on the information quality for increase the user's satisfaction in the IS Success and that is composed of consumers around the dual process for the process of persuading the factors in the ELM model. So the deal was arranged with attractive deals and directly related to the central path of the main elements of successful auctions, but, the secondary path was defined representation of the seller experience and the content representation as an important clue. Based on this rationale, IS Success and ELM were Integrated to construct the research model for our research (see Fig. 1).

The Research model adopts ELM central route and peripheral route relationship. These relationships can be stated in the following hypotheses.

H1. *The effect of* ***transaction attractiveness*** *on auction success is positively related to auction success.*
H2. *The effect of* ***content representation*** *on auction success is positively related to auction.*
H3. *The effect of* ***seller experience*** *on auction success is positively related to auction success.*

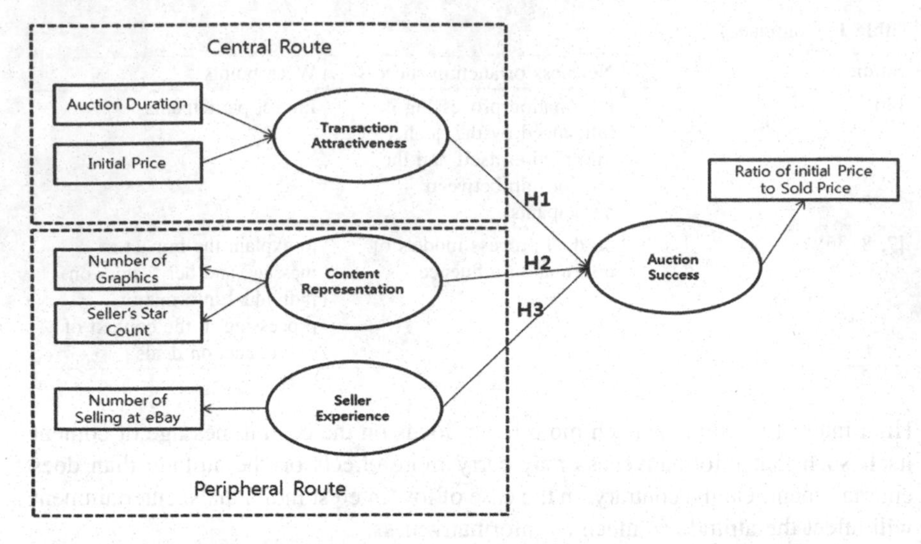

Fig. 1 Proposal research model and hypotheses

H4. ***Among the variables***, ***transaction attractiveness*** *has a* ***stronger*** *effect on auction success than the other two variables, followed by seller experience and content representation.*
H5. ***Among the variables for peripheral route***, ***seller experience*** *has a* ***stronger*** *effect on auction success than the content representation.*

4 Research Methodology

4.1 Assessment of Structural Model

Here, we provide the empirical results that focus on auction success. Figure 2 shows the results of the regressions of the auction success. The results show that three factors were uniformly associated (across all types) with auction success. In most cases, the transaction attractiveness, the content representation and seller experience affected the auction success. The model-fit value was valid affected auction success. The structural model was tested using Smart PLS and Fig. 2 presents the results of the structural model with the standardized path coefficients between constructs and coefficients associated with structural path represent standardized estimates of the parenthesized values represent standard errors. As a result, all paths were significant in $p < 0.001$. Three paths accounted for approximately 37 % of observed variance in auction success. In term of explanatory power of model, PLS draws on R^2 values for each endogenous variable and does not use Goodness-of-fit indicates utilized in covariance-based structural equation modeling [9] and approaches is the minimization of differences between the observed [18]. Accordingly, the effectiveness of a model in PLS is determined by the R^2 (0.370) values and this model can be used to predict the relationship between the construct (Table 2).

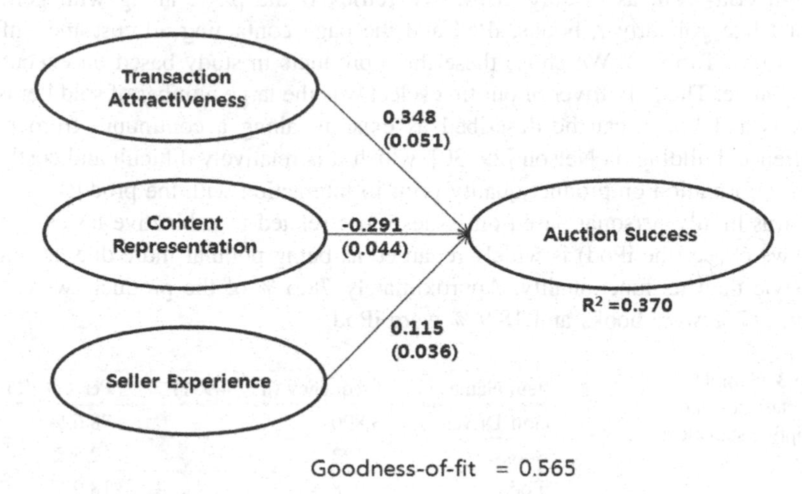

Fig. 2 The revised model

Table 2 Hypothesis testing results

Hypothesis	Path	t-test	Remarks
H1	Transaction Attractiveness → Auction Success	6.815***	Supported
H2	Content Representation → Auction Success	6.640***	Supported
H3	Seller Experiences → Auction Success	3.161***	Supported

*** $p < 0.001$(based on $t_{(100)}$, one-tailed test)

Auction success was statistically significantly related to transaction attractiveness, content representation and seller experience. The positive effect of transaction attractiveness was the path coefficients of 0.348, the other path coefficients 0.291 of content representation and seller experience was statistically positively significant at $p < 0.001$. Therefore, Hypotheses 1, 2, and 3 were supported at the 0.001 level of significance (see Table 6).

As you can see the Fig. 2, the coefficient value of transaction attractiveness is greatest influence as 0.348 than the other path on the auction success. This result lends support for H4, which proposed that the positive relationship central route rather than and peripheral route. It was also the path coefficients values in the Content Representation path is 29 %, Seller Experiences is 12 %. So we suggested H5, have not supported. That means the measured value is empirical data at eBay and has been limited in one of the measured variables corresponding to the Seller Experiences entry. This part can be solved if we include in the next experiment analyzed more items.

4.2 Data Collection

We collected empirical data for this study using online auction data available through eBay.com as of July 2014. We retrieved the page, along with general product like golf driver, books, iPod and the page containing all customer information (see Table 3). We chose these three products in study based on executive data volume. The golf driver of our first select was the large number of sold items in auctions and books can be described as existing along a continuum from pure experience, building on Nelson [29, 30], which it is relatively difficult and costly to obtain information on product quality prior to interaction with the product. Lastly, iPod was highly personal based on issues more related to subjective tastes. However, we chose (the iPod) is widely regarded as being popular more due its image and style than its functionality. Approximately 78.6 % of the products were golf driver, 2.5 % were books and 18.9 % were iPod.

Table 3 Sample Characteristics and descriptive statistics

Item Name	Frequency (n = 4,951)	Percent (%)
Golf Driver	3,890	78.6 %
Books	123	2.5 %
iPod	938	18.9 %

4.3 Operationalization of Research Variables

Table 4 shows the measurement items for constructs such as transaction attractiveness, content representation, seller experience, auction success. Each constructs measure the field name of the eBay page.

Table 4 Constructs and Measures

Construct	Definition	Source
Transaction Attractiveness	The Transaction attractiveness of using a Web-site with existing auction duration, initial price	[24, 33, 39, 40, 42]
Content Representation	The number of graphics uses a web-site and Seller's reputation like the star count of positive score in web-site	[7, 8, 10, 22, 26, 28, 36], Buschke (1997), Thüring et al. (1995), Reich, Carr, De Roure, and Hall (1999), Utting and Yankelovich (1989)
Seller Experience	The seller's experience that using a web-site will increase their performance in auction such as number of selling at eBay	[1, 3, 4, 25, 33, 36, 38, 43]
Auction Success	The winning bid of using a web-site and the ratio of initial price to sold price uses the web-site in a given time period	[4, 14, 35, 37]

4.4 Construct Reliability and Validity

We conducted confirmatory factor analysis, which specifies the links between the latent and manifest variables, factor loadings for the measurement model were computed as reported in Table 5. The criteria for acceptable level of convergent validity are individual item loadings greater than 0.7 [16]. As reported Table 4, the constructs have alpha values well above the cutoff point of 0.70 [16] but transaction attractiveness's alpha is below the guideline point cause by the collected value is real found.

Table 5 Reliability and Factor Analysis

Construct	Item	Factor loadings	Alpha
Transaction Attractiveness	Auction Duration	0.659	0.349
	Initial Price	0.874	
Content Representation	Number of Graphics	0.901	–5.107
	Seller's Star Count	0.949	
Seller Experiences	Number of Selling at eBay	1.000	1.000
Auction Success	Ratio of initial Price to Sold Price	1.000	1.000

We addressed the issue of discriminat validity which it is the square root of AVE for each construct should be greater than the correlation values of the construct with other constructs [16]. All constructs across the samples passed the guidline as reported in Table 5, providing the evidence for their discriminant validity cause by the samples exceed the threshold of 0.50 [16]. As shown in Table 6, all composite reliability exceeded the threshold value of 0.6 [16] except the content representation.

Table 6 Validity and Correlations

	Composite reliability	Average variance extracted	Transaction attractiveness	Content representation	Seller experiences	Auction success
Transaction Attractiveness	0.746	0.600	0.774			
Content Representation	0.008	0.856	–0.650	0.925		
Seller Experiences	1	1	0.126	–0.141	1.000	
Auction Success	1	1	0.551	–0.533	0.199	1.000

5 Discussion and Conclusion

Recently, because of the massive growth of e-trade based on internet, Web is a kind of new alternative way of market communication. Moreover, there is a gradual growth of online auction purchase based on e-trade.

So there are many sorts of studies about factors that affect the success of online auction. However, former studies commonly have some limits that they are only focused on technical factors like interface so that they could not accord with real online auction systems. To resolve this limits of former studies, this study focused on the real factors that affect the success of online auction and researched what makes an auction deal successful. So we collected 4,521 goods that are successful bids on e-bay. And we categorized these goods for central route which is about trade attraction and peripheral route which is about consumer experience.

The contributions of the current study are twofold. First, the current study investigates the role of consumer interest level in evaluating a Web site. Although the interest level of consumers is important to predict their behavior, it is rarely studied in the context of electronic commerce. Understanding the role of user interest with respect to the use of a Web site allows managers and system developers to design more flexible and responsive Web sites. Second, this paper investigates the effect of the perceived amount of Web site content on the perception of information quality and attitude toward the Web site based on consumer interest level. The results of this study will provide practitioners with insight regarding how to design an effective Web site from consumers' view and suggest exerting their efforts on both the main factors and the interaction effect between consumer interest level and the key constructs.

Result on this study, trade attraction in central route is the highest percent factor affecting on online auction deal success. And also in peripheral route, expression of contents is more effective than consumer experience.

This study assumed the process of online auction as a persuasion process, and defined the data as central route and peripheral route. It is important that this study is differentiated with former studies that focused on navigation or interface convenience. It will be helpful to developing high attractive web site. Also it can measure information quality and cognitive positive effects on goods. This study needs to be expended to not only successful bids but refunded bids and postponed bids.

Acknowledgments The author would like to thanks Professor Yong Jin Kim for helpful comments and ideas on earlier drafts of this paper.

References

1. Agarwal, R., Venkatesh, V.: Assessing a firm's Web presence: a heuristic evaluation procedure for the measurement of usability. Inf. Syst. Res. **13**(2), 168–186 (2002)
2. Batra, R., Michael L.R.: How advertising works at contact. In: Alwitt L.F., Mitchell, A.A. (eds.) Psychological Processes and Advertising Effect, Lawrence Erlbaum Associates,

Hillsdale, NJ (1985). Buschke, L.: The basics of building a great Web site. Training Dev. **51**(7), 46–48 (1997)
3. Ba, S., Pavlou, P.A.: Evidence of the effect of trust building technology in electronic markets: price premiums and buyer behavior. MIS Q. **26**(3), 243–268 (2002)
4. Ba, S., Whinston, A.B., Zhang, H.: Building trust in online auction markets through an economic incentive mechanism. Decis. Support Syst. **35**(3), 273 (2003)
5. Bapna, R., Goes, P., Gupta, A.: Analysis and design of business-to-consumer online auctions. Manage. Sci. **49**(1), 85–101 (2003)
6. Bapna, R., Goes, P., Gupta, A.: Replicating online Yankee auctions to analyze auctioneers' and bidders' strategies. Inform. Syst. Res. **14**(3), pp. 244–268 (2003)
7. Chaiken, S.: Heuristic versus systematic information processing and the use of source versus message cues in persuasion. J. Pers. Soc. Psychol. **39**(5), 752–766 (1980)
8. Chaiken, S., Eagly, A.H.: Communication modality as a determinant of message persuasiveness and message comprehensibility. J. Pers. Soc. Psychol. **34**, 605–614 (1976)
9. Chin, W.W.: The partial least squares approach to structural equation modeling. Mod. Methods Bus. Res. **1976**, 295–336 (1998)
10. Conklin, J.: Hypertext: an introduction and survey. IEEE Comput. Surv. **20**(9), 17–41 (1987)
11. Conti, T.: Vision 2000: Positioning the new ISO 9000 standards with respect to total quality management models. Total Qual. Manage. **10**(4/5), S454–S464 (1999)
12. Crosby, P.B.: Quality is Free. McGraw-Hill, New York (1979)
13. DeLone, W.H., McLean, E.R.: Information systems success: the quest for the dependent variable. Inf. Syst. Res. **3**(1), 61–95 (1992)
14. Donthu, N., Garcia, A.: The Internet shopper. J Advertising Res. **39**(3), 52–58 (1999)
15. Hoffman, L., Novak, T.P.: A new marketing paradigm for electronic commerce. Inf. Soc. **13** (1), 43–54 (1997)
16. Fornell, C., Larcker, D.: Evaluating structural equation models with unobservable variables and measurement error. J. Advertising Res. **39**(3), 52–58 (1999)
17. Holbrook, M.B., Hirschman, E.C.: The experiential aspects of consumption: consumer fantasies, feelings, and fun. J. Mark. Res. **18**, 39–50 (1981)
18. Hulland, J.: Use of partial least squares (PLS) in strategic management research: a review of four recent studies. Strateg. Manage. J. **20**, 195–204 (1999)
19. Iivari, J., Koskela, E.: The PIOCO model for information systems design. MIS Q. **11**(3), 401–419 (1987)
20. ISO-Document-8402: Quality Vocabulary, ISO Document 8402, International Standards Office (1986)
21. Juran, J.M.: Quality Control Handbook. McGraw-Hill, New York (1979)
22. Keeney, R.L.: The value of Internet commerce to the customer. Manage. Sci. **45**(4), 533–542 (1999)
23. Klobas, J.E.: Beyond information quality: fitness for purpose and electronic information resource use. J. Inf. Sci. **21**(2), 95–114 (1995)
24. Liu, C., Kirk P.A.: Exploring the factors associated with Web site success in the context of electronic commerce. Inf. Manage. **38**(1), 23–34 (2000)
25. Liu, C., Arnett, K.P., Capella, L.M., Beatty, R.C.: Web sites of the Fortune 500 companies: facing customers through home pages. Inf. Manage. **31**(6), 335–345 (1997)
26. Lucarella, D., Zanzi, A.: A visual retrieval environment for hypermedia information systems. ACM Trans. Inf. Syst. **14**(1), 3–29 (1996)
27. Miller, H.: The multiple dimensions of information quality. Inf. Syst. Manage. **13**(2), 79–82 (1996)
28. Nielsen, J.: Designing Web Usability: The Practice of Simplicity. New Riders Publishing, Indianapolis (2000)
29. Nelson, P.: Information and consumer behavior. J. Polit. Econ. **17**(4), 311–329 (1970)
30. Nelson, P.: Advertising as information. J. Polit. Econ. **81**(4), 729–754 (1974)
31. O'Brien, J.A.: Management Information Systems: Managing Information Technology in the Networked Enterprise. Irwin, Boston (1996)

32. OED.com: Oxford English Dictionary, http://dictionary.oed.com/ (1989)
33. Palmer, J.W.: Web site usability, design, and performance metrics. Inf. Syst. Res. **13**(2), 151–167 (2002)
34. Paulo, F.B., Masiero, P.C., de Oliveira, M.C.F.: Hypercharts: extended state charts to support hypermedia specification. IEEE Trans. Softw. Eng. **25**(1), 33–49 (1999)
35. Peterson, R.A. Sridhar, B., Bart, J.B.: Exploring the implications of the Internet for consumer marketing. Acad. Mark. Sci. J. **25**(4), 329–346 (1997)
36. Petty, R.E., Cacioppo, J.T.: The elaboration likelihood model of persuasion. Adv. Exp. Soc. Psychol. **19**, 123–205 (1986)
37. Pinker, E.J., Seidmann, A., Vakrat, Y.: Managing online auctions: current business and research issues. Manage. Sci. **49**(11), 1457–1484 (2003)
38. Price, M.: What makes users revisit a Web site? Mark. News **12,** Issue Number, March 17 1997
39. Rafaeli, S., Noy, A.: Online auctions, messaging, communication and social facilitation: a simulation and experimental evidence. Eur. J. Inf. Syst. **11**(3), 196–207 (2002)
40. Ranganathan, C. Shobha G.: Key dimensions of business-to-consumer Web sites. Inf. Manage. **39**(6), 457–465 (2002)
41. Strong, D.M., Lee, Y.W., Wang, R.Y.: Data quality in context. Commun. ACM **40**(5), 103–110 (1997)
42. Subramaniam, C., Shaw, M.J., Gardner, D.M.: Product marketing and channel management in electronic commerce. Inf. Syst. Frontiers **1**(4), 363–379 (2000)
43. Torkzadeh, G., Dhillon, G.: Measuring factors that influence the success of Internet commerce. Inf. Syst. Res. **13**(2), 187–204 (2002)
44. Webster's: Webster's New World Dictionary, Webster's New World Dictionary, New York (1991)

Performance Evaluation of Academic Research Activity in a Greek University: A DEA Approach

Gregory Koronakos, Dimitris Sotiros, Dimitris K. Despotis and Dimitris Apostolou

Abstract In this paper we present a methodology developed for assessing the research performance of the academic staff, as part of the internal evaluation of the School of Information and Communication Technologies of a Greek university. We incorporate a qualitative aspect in the assessment by categorizing the research outcomes according to their quality. Also, value judgments over the evaluation criteria are elicited from the supporting committee and incorporated in a Data Envelopment Analysis assessment framework in the form of assurance region constraints.

Keywords Higher education · Academic research assessment · Quality of academic research · Data envelopment analysis (DEA) · Assurance region

1 Introduction

Academic research is considered as one of the most important activities of academic staff in higher education. The extent and quality of academic research are determinants for the academics' appointment and advancement. However, the quality is a controversial topic because of the existence of a large volume of publications in journals of low quality. As the research activity in a university is strictly designated

G. Koronakos (✉) · D. Sotiros · D.K. Despotis · D. Apostolou
Department of Informatics, University of Piraeus, 80 Karaoli and Dimitriou, 18534 Piraeus, Greece
e-mail: gkoron@unipi.gr

D. Sotiros
e-mail: dsotiros@unipi.gr

D.K. Despotis
e-mail: despotis@unipi.gr

D. Apostolou
e-mail: dapost@unipi.gr

R. Neves-Silva et al. (eds.), *Intelligent Decision Technologies*, Smart Innovation, Systems and Technologies 39,
DOI 10.1007/978-3-319-19857-6_33

by the research activities of its staff members, the outcomes leverage its recognition and affect its position in international academic rankings (competitiveness). Moreover, there are countries where quality and performance issues play a crucial role in determining the funding that they receive from the government (e.g. in the UK and Australia). Therefore, the policy makers as well as the public draw significant attention to the results of the assessment of Higher Education Institutions (HEIs) and of their departments or faculties. Governments in many countries have already delivered policies with the aim to handle issues of accountability, cost control and enhancements of the quality of HEIs. In line with the above policies, in many countries periodical exercises are carried out by assessment bodies (committees). In the UK, for instance, the primary objective of Research Excellence Framework (REF) is the evaluation of the quality of research in publicly funded UK HEIs. It replaced the previous assessment system, last conducted in 2008, and named Research Assessment Exercise (RAE). In Australia, the Excellence in Research for Australia (ERA) initiative evaluates the quality of the research in Australian universities in order to provide advice to the Government on research matters and assist the National Competitive Grants Program (NCGP).

Beyond the aforementioned initiatives, complementary policies, such as internal assessments are often adopted in many institutions. For instance, the research development group at Helsinki School of Economics established a two-person team in order to assess the research performance and assist the administration to the allocation of the resources [1]. Recently, Greek higher education institutions have started developing diagnostic tools in order to better understand their comparative strengths and weaknesses. The goals are to provide incentives to the academics to improve the quality of their research and to enable the policy makers to allocate effectively the available resources. In [2], 20 Greek universities were evaluated by applying Data Envelopment Analysis (DEA) and econometric models. The inputs considered were: the number of academic and non-academic staff, the number of active registered students and the operating expenses, whereas the outputs considered were the number of graduates and the research income. A similar evaluation scheme was used in [3] in a department level of a Greek University. Further applications of DEA in education can be found in [4–11] among others.

In this paper we present a systematic framework developed for assessing the research performance of the academic staff, as part of the internal assessment of the ICT School in a Greek university. In this context, we focus on the efficiency assessment of 40 academic staff members undertaken by a committee of experts. Value judgments of the experts on the quality of the research outcome were incorporated in the assessment on the basis of the ERA 2010 journal classification system.

The rest of the paper unfolds as follows. Section 2 discusses the proposed methodology. Section 3 examines the research activity of the 40 academics. Finally conclusions are drawn.

2 Methods and Tools

The assessment exercise focuses on measuring the academics' research performance in the ICT School since their appointment. Therefore, the committee considered as inputs the years since the academic joined the institution and the total academic salary that s/he has received. As outputs were considered the publications in journals and the citations per year. The data of the academics are drawn from Scopus and their CVs.

Inputs

- Number of years in post.
- Total academic salary: total income received since appointment (euros).

Outputs

- Publications: number of single-author equivalent (SAE) journal papers published after the appointment.
- Citations per year: the index derives by dividing the total number of citations of journal papers by the number of years since the first publication. Self-citations are not taken into account.

In case of multiple co-authors (say N), each author is credited with the 1/N of the publication. Summing up over the publications of a particular academic, the single-author equivalent (SAE) is calculated as a measure of the quantity of his research outcome. Note here that the contribution of all the co-authors in the paper is assumed isomeric.

As long as quality matters, we made our assessments by taking into account the quality of the research records as well. First, we limited the research publications of each academic by taking into account only journal papers indexed in Scopus. Then we used the quality of the journal to classify the papers in five categories in accordance to the ERA2010 journal classification system, which ranks the journals in four tiers of quality rating (A+, A, B and C). Scopus papers not indexed by ERA2010 (unranked papers) formed a fifth category D. We note that the choice of the ERA journal classification system was made by the evaluators, which is an assumption in our assessment framework. However, other journal classification systems could be used instead. In our case though, because of the wide range of scope covered by the publications of the 40 academics of the ICT School under evaluation, a classification system that includes a wide list of journals was needed. Such a classification system is ERA2010, which comprises 20712 of a wide spectrum of scientific fields. To the best of our knowledge, there is not any other classification system so valid and thorough. For instance, the UK's Association of Business Schools (ABS) journal ranking includes a short list of journals relative to business and management science; as a result it does not meet our needs.

In many studies, the aspect of quality is incorporated in the assessment by using different weighting schemes to aggregate the different quality categories of publications. For instance in [6], the research quantity and quality were assessed by using weighted indexes of research publications. In a similar manner, in [1] weighted averages of outputs are obtained in order to shorten the outputs consisted of many factors and to account for the quality. In terms of this assessment though, the outputs are treated separately in order to obtain better insight of the quality of each academic's research activity.

2.1 Incorporating Preferences in Data Envelopment Analysis

As already mentioned, data envelopment analysis (DEA) has been widely used as an assessment framework in higher education. Indeed, a few years after the DEA was introduced [12], the technique was straightforwardly applied to the higher education sector. For instance, six DEA models used in [13] for the assessment of 20 English accounting departments. The remarks made in [11] about the attributes of higher education (absence of input and output prices, non-profit character and production of multiple outputs from multiple inputs) render it an attractive domain for DEA. In addition, DEA enables the decision makers to identify the best practice and provides methods for improvements.

For the assessment of the academic research we use the DEA model introduced in [14], under variable returns-to-scale (VRS) assumption. An output orientation is selected so as to determine the improvements for the inefficient academics by increasing the level of their outputs, given the levels of income and years in post. Assume n units, each using m inputs to produce s outputs. We denote by y_{rj} the level of the output r $(r = 1,\ldots,s)$ produced by unit j $(j = 1,\ldots,n)$ and by x_{ij} the level of the input i $(i = 1,\ldots,m)$ consumed by the unit j. The multiplier form of the output-oriented VRS DEA model for evaluating the relative efficiency of the unit j_0 is as follows:

$$
\begin{aligned}
&\min h_{j0} = \sum_{i=1}^{m} v_i x_{ij0} - w_0 \\
&s.t. \\
&\sum_{r=1}^{s} u_r y_{rj0} = 1 \\
&\sum_{r=1}^{s} u_r y_{rj} - \sum_{i=1}^{m} v_i x_{ij} + w_0 \leq 0, \quad j=1,\ldots,n \\
&u_r \geq 0 \quad (r=1,\ldots,s) \\
&v_i \geq 0 \quad (i=1,\ldots,m) \\
&w_0 \in \Re \\
&u_r \in \Omega
\end{aligned}
\tag{1}
$$

Notice that in model (1), the free in sign variable w_o corresponds to the convexity constraint of the dual (envelopment) DEA model. Also, Ω generally denotes the set of restrictions imposed on the weights that limit the freedom of the evaluated unit in selecting its optimal weights [15, 16]. In our case, these restrictions reflect the aggregated opinions of the members of the evaluation committee with respect to the intensity of differentiation in quality among the five categories of journal publications. Pairwise comparisons matrices are used to assess the priorities of the five quality categories by each individual member of the evaluation committee. The comparison scheme and scale is as in the analytic hierarchy process (AHP) c.f. [17]. Then the individual priorities are used to obtain assurance region constraints that restrain the weights given to the papers of different categories. A modified version of method proposed in [18] is used to translate the individual priorities obtained by each evaluator into assurance region constraints that define a weight space reflecting the committee as a whole. For every pair of factors (O_i, O_j), the ratio of their weights (u_i/u_j) is bounded as follows:

$$L_{ij} \leq u_i/u_j \leq U_{ij} \tag{2}$$

where the bounds are defined by the priorities w_{ki} and w_{kj} that express the priority given by the kth evaluator to the ith and the jth factor as follows:

$$L_{ij} = \min_k \frac{w_{ki}}{w_{kj}}, \ U_{ij} = \max_k \frac{w_{ki}}{w_{kj}} \tag{3}$$

It is clear from (3) that the lower and upper bounds L_{ij} and U_{ij} are defined by the extreme judgments (priorities). Thus, the preferences' central tendency is not captured adequately. We treat this issue by excluding, for each pair of factors, the ratios of priorities, which are deemed as outliers. We provide the method in detail in the next section.

3 Assessment of Academic Research Performance

In this section, we provide the results obtained from the internal assessment of the faculty members of the ICT School. Descriptive statistics for the data set (40 faculty members) are presented in the Table 1:

Table 1 Descriptive statistics of the data set

	Inputs		Outputs					
	Y*	TS*	A+*	A*	B*	C*	D*	Citations*
Max	28.50	105.68	1.33	6.42	7.42	9.20	4.26	71.75
Min	3.00	6.42	0	0	0	0	0	0.41
Average	11.73	35.55	0.14	1.07	1.57	1.73	0.79	14.94
St. Dev.	6.90	25.81	0.31	1.72	1.69	2.04	1.06	13.63
Variance	47.61	666.38	0.10	2.94	2.85	4.18	1.12	185.80

* Y: Years in post; TS: Total salary since appointment (10,000 euros); A+, A, B, C: Number of single author equivalent (SAE) papers in class A+, A, B, C journals (ERA2010); D: Number of single author equivalent (SAE) papers in unranked journals; Citations: Citations per year

The five evaluators, members of the evaluation committee, provided independently their opinion with respect to the five quality categories of journals, indicating how much one category excels or is exceled by another in terms of quality. The evaluation scheme employed was the pairwise comparisons matrix used in the context of AHP [17]. Judgments were made on the [1–9] scale proposed in [17] whereas the eigenvector method was used to calculate the priorities. The priorities are the principal right eigenvector i.e. the eigenvector corresponding to the largest eigenvalues of the reciprocal response matrix. Tables 2 and 3 exhibit the pairwise comparisons of two of the evaluators (E4 and E5).

Table 2 Pairwise comparisons matrix of evaluator E4

Categories	A+	A	B	C	D
A+	1	2	4	9	9
A	0.5	1	4	9	9
B	0.25	0.25	1	4	4
C	0.11	0.11	0.25	1	1
D	0.11	0.11	0.25	1	1

The evaluator E4 identifies three main categories (A+, A), B and (C, D) with a slight superiority of class A+ over A, whereas the evaluator E5 is more strict as he identifies only two main categories (A+, A) and (B, C, D) with a slight difference between class A+ and class A.

Table 3 Pairwise comparisons matrix of evaluator E5

Categories	A+	A	B	C	D
A+	1	3	9	9	9
A	0.333	1	7	7	7
B	0.111	0.143	1	1	1
C	0.111	0.143	1	1	1
D	0.111	0.143	1	1	1

The priorities (weights) obtained for the five categories from the pairwise comparison matrices of the evaluators E1–E5 are summarized row-wise in Table 4 and are exhibited in Fig. 1. The last column of Table 4 portrays the Consistency Ratio (CR) of each evaluator's comparison matrix. As all CRs are below 10 % the priorities are acceptable.

Table 4 Evaluators' priorities for the five journal categories

Evaluator	A+	A	B	C	D	CR
E1	0.434	0.306	0.162	0.060	0.037	0.030
E2	0.503	0.299	0.107	0.056	0.035	0.021
E3	0.570	0.250	0.076	0.076	0.027	0.1
E4	0.456	0.344	0.123	0.038	0.038	0.025
E5	0.550	0.301	0.050	0.050	0.050	0.020

As shown in Table 4 and Fig. 1, for the priorities of all the evaluators holds that $w^{A+} > w^{A} > w^{B} \geq w^{C} \geq w^{D}$. However, the intense of preference among the different journal categories is not the same. The evaluators E1 and E2 rank order the five categories with no ties (strict order), whereas E3, E4 and E5 differentiate by assuming ties for some categories.

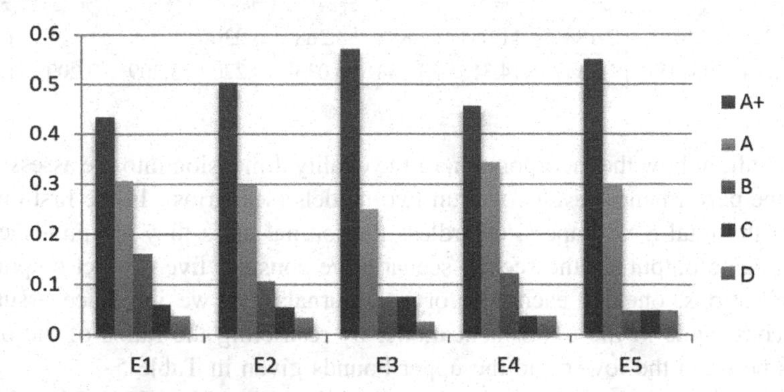

Fig. 1 Weighting preferences of five evaluators

Next we eliminate the outlier ratios of priorities across the evaluators to avoid the effect of extreme judgments. As shown in the boxplot of Fig. 2, the ratio For example, the ratio w^{A}/w^{C} of the priorities given from the evaluator E4 for class A over class C is detected as outlier and, thus, omitted before calculating the upper bound as in (3).

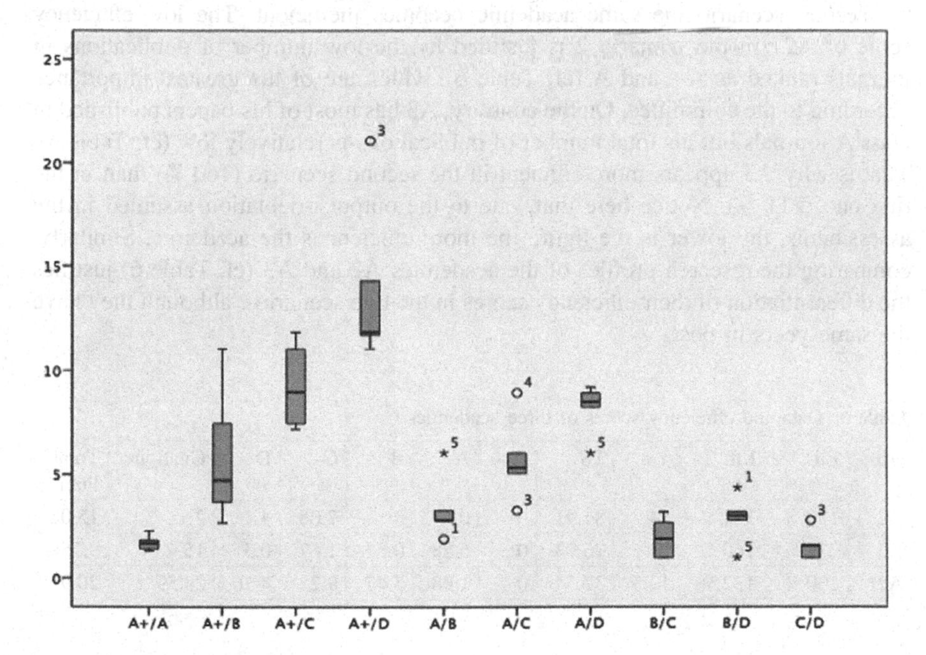

Fig. 2 Outlier detection

Table 5 contains the calculated lower and upper bounds according to Eqs. (2) and (3) after removing the upper outliers.

Table 5 Lower and upper bounds

	$\frac{u^{A+}}{u^{A}}$	$\frac{u^{A+}}{u^{B}}$	$\frac{u^{A+}}{u^{C}}$	$\frac{u^{A+}}{u^{Unr.}}$	$\frac{u^{A}}{u^{B}}$	$\frac{u^{A}}{u^{C}}$	$\frac{u^{A}}{u^{D}}$	$\frac{u^{B}}{u^{C}}$	$\frac{u^{B}}{u^{D}}$	$\frac{u^{C}}{u^{D}}$
LB	1.325	2.674	7.188	11.034	1.885	3.268	6.036	1	1	1
UB	2.284	11.034	11.847	14.315	3.268	6.036	9.220	3.209	3.209	1.629

To highlight how the incorporation of the quality dimension into the assessments affects the performance results, we run two models (scenarios). In the first one we consider the total SAE papers, regardless the journal class they are published, as one aggregate output. In the second scenario we consider five distinct outputs for the publications, one for each category of journals and we introduce assurance region constraints in the assessment model by restricting the ratios of the output weights between the lower and the upper bounds given in Table 5.

In the first scenario, 6 academics (namely A4, A18, A21, A22, A24 and A39) were deemed efficient. In the second one, only 4 of them (namely A4, A22, A24 and A39) maintained their 100 % efficiency score. For the 75 % of the academics the efficiency scores in the second scenario are lower than in the first one. This is attributed to the high number of publications in low and medium quality journals. For instance, A21 is efficient in the first scenario as he has the highest total number of publications. However, when the quality of the journals is taken into account in the second scenario the same academic becomes inefficient. The low efficiency score of A21 in the scenario 2 is justified by the low number of publications in journals ranked as A+ and A (cf. Table 6), which are of the greatest importance according to the committee. On the contrary, A3 has most of his papers published in class A journals but his total number of publications is relatively low (cf. Table 6). That is why A3 appears more efficient in the second scenario (160 %) than in the first one (211 %). Notice here that, due to the output orientation assumed in the assessments, the lower is the figure the more efficient is the academic. Similarly, comparing the research profiles of the academics A2 and A3 (cf. Table 6) justifies the differentiation of their efficiency scores in the two scenarios, although they have the same years in post.

Table 6 Data and efficiency scores of three academics

Ac.	Eff. 1	Eff. 2	Y	TS	A+	A	B	C	D	Citations	Total Publ.
A2	139 %	381 %	24	84.91	0	0	4	7.08	4	2.11	15.08
A3	211 %	160 %	24	76.73	0	6.25	0.83	1.67	0.5	15.4	9.25
A21	100 %	132 %	12.5	37.54	0	1.98	5.47	9.2	4.26	24.59	20.9

Figure 3 depicts the top 3 layers of efficient academics. The first layer comprises the efficient academics discussed above. The second layer is obtained by rerunning the scenario 2 after excluding the academics of the first layer. The third layer is obtained similarly by excluding the academics of the first two layers.

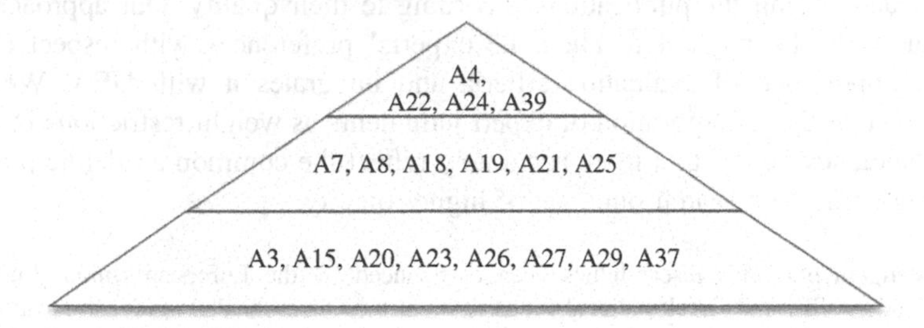

Fig. 3 The top 3 layers of efficient academics

The dual solution of the model (1) provides the necessary information to project the inefficient units on the efficient frontier. The profiles of the efficient academics in the second scenario are presented in the Table 7.

Table 7 Profile of the efficient academics in scenario 2

Ac.	Y	TS	A+	A	B	C	D	Citations
A4	16	54.749	1	4.417	3.450	7.617	1.417	41.348
A22	3	6.422	0	0.333	0	0	0	7.6
A24	7	15.607	0	0	1.450	0.500	0	71.75
A39	3.5	14.419	0	0.750	0.250	0	0	15.938

Table 8 exhibits the profile of three inefficient academics and their potential improvements so as to be rendered efficient. Specifically, it provides the levels of improvement that the academics should exhibit on each output so as to be more competitive.

Table 8 Profile and targets of three inefficient academics

	Original values							Target values					
	Eff.	A+	A	B	C	D	Cit.	A+	A	B	C	D	Cit.
A6	313 %	0.2	0.75	0.95	0.7	0.25	11	0.63	2.35	2.98	2.19	0.78	34.47
A20	199 %	0.5	0	0.8	0.33	1.67	15.8	0.99	0	1.59	0.66	3.32	31.43
A27	218 %	0	0.25	0.33	2.08	0.33	19.08	0	0.54	0.73	4.53	0.73	41.53

4 Conclusions

We evaluated the performance of research activity of 40 academics in the context of the internal assessment of a Greek ICT School. We also discussed about the crucial role of categorizing the publications according to their quality. Our approach utilizes an AHP-like method to elicit the experts' preferences with respect to the relative importance of evaluation criteria and integrates it with DEA. We also illustrated that the incorporation of expert judgments as weight restrictions in DEA yields efficiency scores that more accurately reflect the common academic practice of giving credit to research outcome of higher quality.

Acknowledgements This research has been co-financed by the European Union (European Social Fund – ESF) and Greek national funds through the Operational Program 'Education and Lifelong Learning' of the National Strategic Reference Framework (NSRF) – Research Funding Program: THALES. Investing in knowledge society through the European Social Fund.

References

1. Korhonen, P., Tainio, R., Wallenius, J.: Value efficiency analysis of academic research. Eur. J. Oper. Res. **130**, 121–132 (2001)
2. Katharaki, M., Katharakis, M.G.: A comparative assessment of Greek Universities' efficiency using quantitative analysis. Int. J. Ed. R. **49**, 115–128 (2010)
3. Kounetas, K., Anastasiou, A., Mitropoulos, P., Mitropoulos, I.: Departmental efficiency differences within a Greek University: an application of a DEA and Tobit analysis. Int. T. Oper. Res. **18**, 545–559 (2011)
4. Beasley, J.E.: Comparing university departments. Omega **18**(2), 171–183 (1990)
5. Beasley, J.E.: Determining teaching and research efficiencies. J. Oper. Res. Soc. **46**(4), 441–452 (1995)
6. Athanassopoulos, A.D., Shale, E.: Assessing the comparative efficiency of higher education institutions in the UK by the means of data envelopment analysis. Educ. Econ. **5**(2), 117–134 (1997)
7. Johnes, J., Johnes, G.: Research funding and performance in U.K. University departments of economics: a frontier analysis. Econ. Educ. Rev. **14**(3), 301–314 (1995)
8. Avkiran, N.K.: Investigating technical and scale efficiencies of Australian Universities through data envelopment analysis. Socio. Econ. Plan. Sci. **35**(1), 57–80 (2001)
9. Abbott, M., Doucouliagos, C.: The efficiency of Australian Universities: a data envelopment analysis. Econ. Educ. Rev. **22**(1), 89–97 (2003)
10. Kao, C., Hung, H.-T.: Efficiency analysis of university departments: an empirical study. Omega **36**, 653–664 (2008)
11. Johnes, J.: Data envelopment analysis and its application to the measurement of efficiency in higher education. Econ. Educ. Rev. **25**(3), 273–288 (2006)
12. Charnes, A., Cooper, W.W., Rhodes, E.: Measuring the efficiency of decision making units. Eur. J. Oper. Res. **2**, 429–444 (1978)
13. Tomkins, C., Green, R.: An experiment in the use of data envelopment analysis for evaluating the efficiency of UK University departments of accounting. Financ. Acc. Man. **4**(2), 147–164 (1988)
14. Banker, R.D., Charnes, A., Cooper, W.W.: Some models for estimating technical and scale inefficiencies in DEA. Manage. Sci. **32**, 1078–1092 (1984)

15. Charnes, A., Cooper, W.W., Huang, Z.M., Sun, D.B.: Polyhedral cone-ratio DEA models with an illustrative application to large commercial banks. J. Econometrics **46**(1–2), 73–91 (1990)
16. Thompson, R.G., Langemeier, L.N., Lee, C., Lee, E., Thrall, R.M.: The role of multiplier bounds in efficiency analysis with application to Kansas firms. J. Econometrics **46**(1–2), 93–108 (1990)
17. Saaty, T.L.: A scaling method for priorities in hierarchical structures. J. Math. Psychol. **15**(3), 234–281 (1977)
18. Takamura, Y., Tone, K.: A comparative site evaluation study for relocating Japanese government agencies out of Tokyo. Socio. Econ. Plan. Sci. **37**(2), 85–102 (2003)

Application of Bayesian Networks to the Forecasting of Daily Water Demand

Ewa Magiera and Wojciech Froelich

Abstract In this paper, we investigate the application of Bayesian Networks (BN) for the 1-step ahead forecasting of daily water demand. The water demand time series is associated with the series containing information for daily precipitation and mean temperature that play the role of the additional explanatory variables. To enable the application of the standard Bayesian network as a predictive model, all considered time series are discretized. The number of discretization intervals is assumed as a parameter of the following learn-and-test trials. To test forecasting accuracy, we propose a novel discrete type of mean absolute error measure. Then, the concept of growing window is used to learn and test several Bayesian networks. For comparative experiments, different algorithms for learning structure and parameters of the BNs are applied. The experiments revealed that a simple two-node BN outperformed all of the other complex models tested for the considered data.

Keywords Forecasting time series · Bayesian networks · Daily water demand

1 Introduction

Forecasting of daily water demand is required for the optimization of urban water supply systems. This problem has been extensively investigated in the literature [2, 3, 9]. Several predictive models have been used for this purpose: artificial neural networks and linear regression [10] as well as fuzzy logic and a genetic algorithm [20]. The problem of water demand forecasting and the application of diverse models for this task were reviewed in [11] and [5]. In addition, Bayesian networks have been used for the forecasting of time series [1, 13]. Recently, an overview of Bayesian forecasting approaches has been reported [4].

E. Magiera (✉) · W. Froelich
Institute of Computer Science, University of Silesia, Sosnowiec, Poland
e-mail: ewa.magiera@us.edu.pl

W. Froelich
e-mail: wojciech.froelich@us.edu.pl

R. Neves-Silva et al. (eds.), *Intelligent Decision Technologies*,
Smart Innovation, Systems and Technologies 39,
DOI 10.1007/978-3-319-19857-6_34

To the best of our knowledge, BNs have been never applied for the forecasting of water demand. In this paper, we are investigating just such an approach. To estimate the discretized forecasts as the outcome obtained from BNs, we propose a newly formulated forecasting measure, the discrete mean absolute forecasting error (DMAE). To obtain the highest possible forecasting accuracy, we select the optimal structure of the BN. In addition, we investigate a related problem, the dependence between equal-length discretization technique and the accuracy of forecasting. On that basis, we select the optimal number of discretization intervals for the previously chosen BN model.

The paper is organized as follows. Section 2 provides a brief introduction to Bayesian networks. In Sect. 3 basic definitions and assumptions regarding water demand time series are presented. A discrete version of mean absolute error measure for the forecasting of discrete-valued time series is proposed in Sect. 4. Section 5 describes the adaptation of Bayesian network to the forecasting of daily water demand. The results of our experiments are presented in Sect. 6. Section 7 concludes the paper.

2 Basic Notions Regarding Bayesian Networks

A Bayesian network is defined as a triple $BN = (X, DAG, P)$, where X is a set of random variables, $DAG = (V, E)$ is a directed acyclic graph, and P is a set of conditional probability distributions [7]. Each node of the graph $v_i \in V$ corresponds to the random variable $X_i \in X$ with a finite set of mutually exclusive states. Directed edges $E \subseteq V \times V$ of DAG correspond to conditional dependence between random variables. For every child node X_i, there is $P(X_i|X_{pa(i)})$, which is the conditional probability distribution for each $X_i \in X$, where: $pa(i)$ is the set of conditioning variables of the X_i. Those distributions are represented in the form of conditional probability tables CPT_i.

A Bayesian network model is usually used to calculate the posterior probability of unknown $X(t)$ given the set of evidence variables E. In terms of forecasting, BN is applied to calculate the probability distribution $P(X'(t)|E)$, where $X'(t)$ denotes the predicted value and E is the observed evidence.

The learning of Bayesian network consists of two steps: (1) structure learning and (2) learning of the conditional probability tables CPT_i. For the purpose of this study we selected the following algorithms to learn the structure of BN:

- Hill-Climbing - a greedy search on the space of the directed graphs,
- Tabu Search - a Hill-Climbing with the additional mechanism of escaping from local optima,
- Max-Min Hill-Climbing - a hybrid algorithm combining the Max-Min algorithm and the Hill-Climbing algorithm,
- Restricted Maximization - a generalized implementation of Max-Min Hill-Climbing.

Maximum Likelihood Estimation (MLE) has been selected for learning probability distributions (the parameters of BN). The details of the applied algorithms are given in [12].

3 Water Demand Time Series

Let $\{W(t)\}$ denotes the considered water demand time series, where $W \in \Re$ is a real-valued variable and $t \in [1, 2, \ldots, n]$ is a discrete time scale with the length $n \in \aleph$.

It has been reported in the literature that lower precipitation and higher temperature may lead to higher water demand [3]. As a result, those factors are considered in our study and thus the forecasting of water demand is, to a certain extent, related to the forecasting of weather time series [6, 8]. We denote $\{T(t)\}$, $\{R(t)\}$ as the time series related to exogenous variables: temperature and rain (precipitation) respectively.

Let $k \in \aleph$ be a parameter, the number of levels (bins) the time series are to be discretized into. We make the assumption that the same value of k is used for all considered variables. In consequence, equal-length intervals for the discretization are used. Such an assumption is a limitation but enables one to easily perform preliminary experiments without loss of the generality of the approach proposed in this study.

This way, for the variables $W(t)$, $T(t)$ and $R(t)$ it is possible to calculate their discretized versions $W_d(t) \in \aleph$, $T_d(t) \in \aleph$ and $R_d(t) \in \aleph$. The values of $W_d(t)$, $T_d(t)$ and $R_d(t)$ are positive integers denoting the intervals to which the actual values of $W(t)$, $T(t)$ and $R(t)$ belong.

The daily water demand $W_d(t)$ corresponds to the random variable whose value is to be predicted. The lagged $W_d(t-1)$, $W_d(t-2)$, …, $W_d(t-n)$ also correspond to the random variables within the considered BN.

To evaluate the forecasting accuracy, the time series should be partitioned into learning and testing periods. In this study, we apply the concept of a growing learning window. Following such assumption, the learning period begins at the first time step and finishes at time $t-1$. This way, the length of the growing window increases as the time flows. It is assumed that the prediction is made 1-step ahead for the step t. The BN model is retrained at every time step (day).

4 Forecasting Accuracy and Discrete Mean Absolute Error Measure

After the discretization of all considered time series it is possible to apply standard Bayesian network to the forecasting of $W_d(t)$. The forecasted discrete value is denoted as $W'_d(t)$.

Let us assume t_s and t_f as the beginning and final time steps of the forecasts obtained from the simulation. The first possibility of measuring forecasting accuracy is to use the formula (1).

$$Accuracy = \frac{\sum_{t=t_s}^{t_f} 1|W'_d(t) = W_d(t)}{t_f - t_s + 1} \tag{1}$$

This way we achieve the ratio of perfect forecasts (related to all forecasts made). The formula (1) is useful in the sense that it estimates the probability of perfect forecast assuming certain type of discretization. However, further practical applications of such a measure are limited. It is obvious that assuming a low value of k, resulting in wide discretization intervals, makes it possible to obtain high forecasting accuracy as measured by Eq. (1), independent of the applied predictive model. While Eq. (1) reflects the standard classification accuracy, it does not take into account that the discretization intervals are ordered and that they posses certain width.

To overcome that limitation, let us first calculate $d = \frac{max(W(t))-min(W(t))}{k}$ as the length of the discretization interval used for the variable $W_d(t)$. Let us analyze the situation when the forecast is perfect in terms of discrete values, i.e., $W'_d(t) = W_d(t)$. In such case the maximal possible difference between the related continuous values is equal to the length of the discretization interval d, i.e., $max|W'(t) - W(t)| = d$. In other words, although the discrete forecast $W'_d(t)$ is perfect, it's continuous value $W'(t)$ may differ from the actual continuous value $W(t)$ no more than by d. To calculate accumulated forecasting error we introduce the discrete mean absolute error (DMAE) accuracy measure that is given by the formula (2).

$$DMAE = \frac{\sum_{t=t_s}^{t_f} |W'_d(t) - W_d(t) + 1| \cdot d}{t_f - t_s + 1} \tag{2}$$

In the case of perfect discrete forecast, i.e. $|W'_d(t) - W_d(t)| = 0$, it makes a pessimistic assumption that the continuous value of error is equal to d. For higher values of $|W'_d(t) - W_d(t)|$ the related continuous value of error is accordingly increased.

5 Bayesian Approach to the Forecasting of Water Demand

It has been argued in the literature that weekly seasonality is expected for urban water demand [3]. This is related to the fact that water demand is usually lower during the weekends then during the working days [3]. In our study we investigate the time lag of 7 for the lagged variable $W_d(t = 7)$, included in the set of evidence variables E. For the same reason, we include within E the variable *Day* related to the days of the week.

Table 1 The set of random variables

Variable	Description
$W_d(t)$	water demand
$W_d(t-1)$	water demand with lag 1
$W_d(t-7)$	water demand with lag 7
$T_d(t-1)$	temperature with lag 1
$R_d(t-1)$	rain with lag 1
$Day(t)$	day of week
$Month(t)$	month

In addition, monthly seasonality is expected within water demand time series as the result of summer and winter holidays [3]. However, after numerous experiments that produced poor results, we resigned to use the lag 30 or 31 related to month for water demand. Instead the variable Month has been included in E as the exogenous variable explaining monthly seasonality.

The final set of the considered time series (related to random variables) is given in Table 1.

6 Experiments

To validate our proposed approach, real historical water demand data were used. Water demand time series has been gathered from the urban water distribution network of Sosnowiec, Poland. Weather-related time series have been collected from the weather station located in the same city. All time series spanned from 16 June 2007 to 31 December 2013 and contained 2386 time steps. The minimum length of the growing learning window was set to 20 steps (days); therefore, the first forecast was generated for the 21th day. First, the structure of the BN was learned automatically using all of the algorithms given in Sect. 2. It is important to mention here that the obtained directions of the arcs of the BNs were usually not in rapport with the direction of time flow. The structure of the BNs reflected probabilistic dependencies discovered in the data rather than by the physical precedence of events. We intentionally decline to show any of those BNs here as their structure could not be interpreted as cause-and-effect graphs. The obtained results are given in Table 2. The values of DMAE are given first with the accuracy calculated according to Eq. (1) provided in brackets.

Table 2 Forecasting accuracy of the BNs

k	Hill-Climbing	Tabu search	Max-Min HC	Restr. Max.
5	607.45(0.78)	620.02(0.76)	607.95(0.78)	607.70(0.78)
10	387.95(0.58)	405.42(0.56)	388.08(0.58)	387.82(0.58)
20	614.30(0.29)	617.06(0.29)	616.12(0.29)	615.68(0.29)
30	922.99(0.07)	924.08(0.07)	923.79(0.06)	924.67(0.07)

As shown in Table 2, the value of DMAE is quite high for all of the learning algorithms and the lowest considered number of discretization intervals $k = 5$. As the value of k grows, the DMAE decreases, achieving a much lower level for $k = 10$. After further increase of k, errors increase again, achieving very high values for $k = 30$. The obtained results suggest the need for the further optimization of k with respect to DMAE. When considering the learning algorithms, only the application of Tabu Search resulted in slightly higher DMAE in comparison to other methods.

On the other hand, as expected, the accuracy decreases when increasing the number of discretization intervals. This time, the selection of k depends on the preferences imposed by an expert. Assuming that the expert is interested only in obtaining perfect discrete forecasts, she can select the value of k on the basis of the assumed tolerance of uncertainty related to the width of the discretization interval.

In the second step of the experiments, an attempt was undertaken to design the structure of BN by trial and error. In all cases we assumed that the water demand is directly influenced by the lags and exogenous variables. This way only forward inference within the BN is required and the direction of arcs of the BN is in rapport with the physical order of events. Table 3 describes the investigated BNs and related evidence variables. The BN1 contains all evidence variables that were used previously (while learning the structure of BNs automatically). In the networks BN2 and BN3 we use $Day(t)$, $Month(t)$ and $R(t-1)$, $T(t-1)$. In BN4 we use only the calendar-related variables. In BN5 we assume that the forecasts are based on the lagged variables of water demand.

Table 3 Forecasting accuracy for the designed structures of BNs

Bayesian network	Variable
BN1	$W_d(t-1)$, $W_d(t-7)$, $Day(t)$, $Month(t)$, $R(t-1)$, $T(t-1)$
BN2	$W_d(t-1)$, $W_d(t-7)$, $R(t-1)$, $T(t-1)$
BN3	$W_d(t-1)$, $W_d(t-7)$, $Day(t)$, $Month(t)$
BN4	$Day(t)$, $Month(t)$
BN5	$W_d(t-1)$, $W_d(t-7)$

Our results are shown in Table 4. One can observe that the designed BN1 produces higher errors than all BNs with the structure learned automatically. Surprisingly, the lowest forecasting errors in terms of DMAE and accuracy were produced by the simple BN5. This result motivated our further investigations.

Table 4 Forecasting accuracy for the designed structures of BNs

k	BN1	BN2	BN3	BN4	BN5
5	743.47(0.49)	624.80(0.75)	701.23(0.60)	773.14(0.32)	594.12(0.78)
10	756.92(0.22)	448.17(0.51)	599.03(0.35)	696.22(0.20)	361.30(0.61)
20	841.12(0.14)	564.02(0.23)	661.70(0.18)	772.58(0.13)	280.91(0.40)
30	883.13(0.12)	901.16(0.15)	680.91(0.12)	815.25(0.09)	312.65(0.29)

In the following experiments we decided to test how the further elimination of explanatory variables from BN5 influences the accuracy of forecasting. For the BN6 we used only $W_d(t-1)$. For the BN7 only the $W_d(t-7)$ was applied. Moreover, the simulations were performed for the number of discretization intervals up to $k = 50$.

Table 5 Forecasting accuracy for the trivial structures of BNs

k	BN6	BN7
5	606.9486(0.7834101)	656.7314(0.7009217)
10	343.4514(0.6207373)	429.0629(0.5456221)
20	237.3486(0.4262673)	306.4914(0.3672811)
30	212.2895(0.3396313)	280.8876(0.2764977)
40	224.4943(0.2608295)	285.5600(0.2184332)
50	230.3337(0.2290323)	294.1463(0.1820276)

As can be seen in Table 5, the BN7 was outperformed by the BN6. For $k = 30$ the BN6 produced a relatively small value, i.e., DMAE = 212.2895. To observe the influence of k on DMAE we performed experiments for all k from the interval [5, 50]. These results are shown in Figs. 1 and 2.

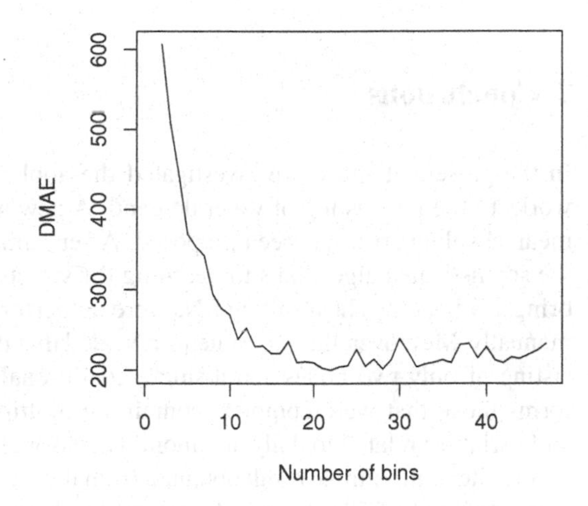

Fig. 1 Discrete mean absolute error for BN6

The series of DMAE shown in Fig. 1 demonstrates a different pattern then those generated by other BNs. The errors decrease while increasing k, but contrary the rest of the cases, they don't increase while increasing k up to 50. Starting from $k \approx 20$ they demonstrate random fluctuations. As can be seen in Fig. 2 increasing k leads to lower accuracy, i.e. the number of perfect forecasts decreases. For these reasons, we recommend using $k = 20$ when applying the proposed method to urban water demand data.

Surprisingly, the results revealed that a simple two-node BN outperforms all other complex Bayesian networks, both those with the structure learned automatically and those that were manually designed. In fact, such a simple BN represents a trivial probabilistic forecast based on a single conditional probability distribution.

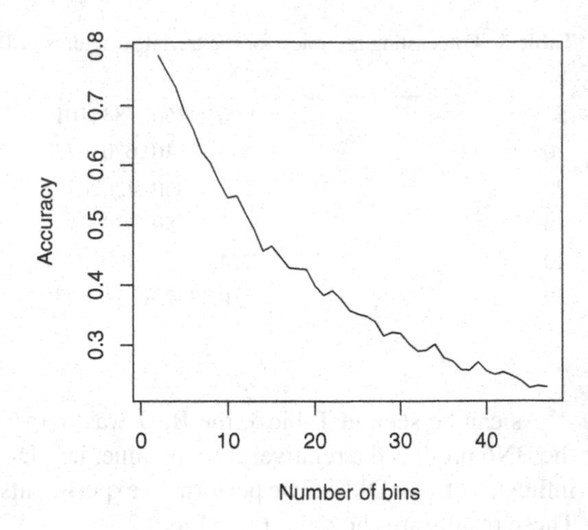

Fig. 2 Accuracy of forecasting for BN6

7 Conclusions

In the presented study, we investigated the application of standard Bayesian networks to the forecasting of water demand. A new accuracy measure called discrete mean absolute error has been proposed. After numerous trials, it became clear that the sophisticated algorithms for learning the structure of Bayesian networks did not bring any benefit. The resulting BNs were outperformed by those that were designed manually. Moreover, the experiments revealed that the simple Bayesian network consisting of only two nodes and a single conditional probability distribution outperforms those that were complex, containing multiple lags, weather-related factors, and variables related to daily and monthly seasonality.

In spite of the simple result obtained from the experiments, the proposed approach is general and will be further developed by the authors. Future directions for this research will include the application of other state-of-the-art classifiers to the forecasting of discretized time series. In addition, more sophisticated methods of discretization should be considered. Another research challenge for the future is further investigation of the relationship between the continuous forecasts based on regression techniques and those discrete-valued produced by classifiers.

Acknowledgments The work was supported by ISS-EWATUS project which has received funding from the European Union's Seventh Framework Programme for research, technological development and demonstration under grant agreement no. 619228.

The authors would like also to thank the water distribution company in Sosnowiec (Poland) for gathering water demand data and the personal of the weather station of the University of Silesia for collecting and preparing meteorological data.

References

1. Abramson, B., Brown, J., Edwards, W., Murphy, A., Winkler, R.L.: Hailfinder: a Bayesian system for forecasting severe weather. Int. J. Forecast. **12**(1), 57–71 (1996)
2. Adamowski, J., Adamowski, K., Prokoph, A.: A spectral analysis based methodology to detect climatological influences on daily urban water demand. Math. Geosci. **45**(1), 49–68 (2013)
3. Billings, R.B., Jones, C.V.: Forecasting Urban Water Demand. American Water Works Association (2008)
4. Biondi, D., Luca, D.D.: Performance assessment of a bayesian forecasting system (BFS) for real-time flood forecasting. J. Hydrol. **479**(0), 51–63 (2013)
5. Donkor, E., Mazzuchi, T., Soyer, R., Alan Roberson, J.: Urban water demand forecasting: review of methods and models. J. Water Resour. Plan. Manag. **140**(2), 146–159 (2014)
6. Froelich, W., Salmeron, J.L.: Evolutionary learning of fuzzy grey cognitive maps for the forecasting of multivariate, interval-valued time series. Int. J. Approx. Reason. **55**(6), 1319–1335 (2014)
7. Jensen, F.V.: Bayesian Networks and Decision Graphs. Springer (2001)
8. Juszczuk, P., Froelich, W.: Learning fuzzy cognitive maps using a differential evolution algorithm. Pol. J. Env. Stud. **12**(3B), 108–112 (2009)
9. Pulido-Calvo, I., Gutirrez-Estrada, J.C.: Improvedfrigation water demand forecasting using a soft-computing hybrid model. Biosyst. Eng. **102**(2), 202–218 (2009)
10. Pulido-Calvo, I., Montesinos, P., Roldn, J., Ruiz-Navarro, F.: Linear regressions and neural approaches to water demand forecasting in irrigation districts with telemetry systems. Biosyst. Eng. **97**(2), 283–293 (2007)
11. Qi, C., Chang, N.B.: System dynamics modeling for municipal water demand estimation in an urban region under uncertain economic impacts. J. Env. Manag. **92**(6), 1628–1641 (2011)
12. Scutari, M.: Bayesian network structure learning, parameter learning and inference. http://www.bnlearn.com/ (2014)
13. Vlachopoulou, M., Chin, G., Fuller, J.C., Lu, S., Kalsi, K.: Model for aggregated water heater load using dynamic Bayesian networks. In: Proceedings of the DMIN'12 International Conference on Data Mining, pp. 1–7 (2012)

SyLAR—The System for Logic and Automated Reasoning for Situation Management Decision Support Systems

Erika Matsak, Tarmo Lehtpuu and Peeter Lorents

Abstract This paper presents a modified version of forward chaining method of reasoning and a prototype implementation called Sylar – System for Logic and Automated reasoning, which can be used for situation management. The method, presented by the authors, implements Modus Ponens and Modus Tollens by Truth tables and a SAT solver to discover hidden knowledge and suggest decisions. Also, it gives the opportunity to use multi-step inferences. Proposed method is suitable for this specific kind of decision making process.

Keywords Forward chaining • Decision support system • Inferences • Situation management • Inference engine

1 Introduction

Automated reasoning methods and approaches (e.g. resolution method [26]), SAT solvers [9], matrix connection method [1], etc. can be used for development of decision support systems [31] for situation management.

Some automated reasoning methods use backward chaining for reasoning, some use forward chaining [30]. Both of them have their own advantages. Backward chaining needs the goals, and sometimes goals cannot be defined before reasoning begins, because it is important to find all possible truth values of the statements

E. Matsak (✉)
Tallinn University, Tallinn, Estonia
e-mail: erika.matsak@tlu.ee

T. Lehtpuu
Estonian Information Technology College, Tallinn, Estonia
e-mail: tlehtpuu@itcollege.ee

P. Lorents
Estonian Business School, Tallinn, Estonia
e-mail: peeter.lorents@ebs.ee

R. Neves-Silva et al. (eds.), *Intelligent Decision Technologies*,
Smart Innovation, Systems and Technologies 39,
DOI 10.1007/978-3-319-19857-6_35

from a number of axioms or from a few verified formulas. In that case forward chaining will be a better solution. However backward chaining reasoning methods have less computational problems and can be realized also with little computational resources. In spite of problems with forward chaining computational time and despite the fact, that sometimes decision can't be made in necessary quite short time interval [29], there are lot applications that use it. For example: BaseVISor 2.0 Inference Engine [21], CYC [24], Drools Business Logic Integration Platform [3]. Forward chaining starts with the available information (presented by the corresponding formulas) and uses inference rules in order to obtain more information until a desired statement, or assertion, or the answer to a question, or a necessary decision, or etc. (presented by the corresponding formula) is reached. For this purpose, we must have a suitable "machine" – an "inference engine". This engine uses forward chaining to search the suitable inference rules. Usually it is essentially one and the same rule – namely Modus Ponens – that is appropriately applied to certain specific formulas to identify the appropriate formulas, one of which is the implication, and the other is the assumption from above-mentioned implication. When these formulas are successfully found, then concrete reasoning is carried out, resulting in a formula (the "right-hand" operand in the last implication) that provides the necessary information (the answer to the question, necessary decision etc.).

In the present paper the authors present an advanced and novel version of a forward chaining reasoning method for situation management, and a prototype SyLAR (System for Logic and Automated Reasoning) implementation. The main components of the prototype system are: (I) situation descriptions from observers or sensors; (II) basic or "general" principles i.e. axioms from experts; (III) the inference rules that allow to get new useful information based on existing information; (IV) information warehouse, which contains all of the information that has been presented about this situation by experts, observers or sensors, and all information which is derived from existing information.

The new method of forward chaining reasoning, considered in this paper, gives us the opportunity to use multi-step inferences. In traditional solutions expert rules are represented in the if-then style, where the result is represented by one statement formula (f.e: $A\&B\&C\supset D$) and after applying the axiom and Modus Ponens the reasoning will stop. In the present approach axioms may have more "uncertain" results, which are represented by more complicate (not atomic) formulas, and the reasoning will continue (f.e: $A\&B\&C\supset D\vee E$). This is a main feature, which brings the difference between the system described in this paper, and many other well-known systems. In our system we use the fact that correct inference steps are related to certain true implications. Second distinction is the usage of the SAT (Propositional Satisfiability Problem) solver in the forward chaining and something a little similar to the resolution method, which is applied in some meaning in reverse way.

This approach provides the ability for the prototype SyLAR: to determine (with an additional module) whether the information, related to situation descriptions is consistent or not. It helps to highlight possible mistakes in information presented by the experts. It should be noted that SyLAR allows determining the truth value of a statement belonging to a certain collection of statements if this truth value is not initially known for some reason.

2 Theoretical Foundation for the Prototype

We note that we proceed from a classic bivalent logic (but we use below, in some places, the so-called "slightly disguised" form of a trivalent logic). We add the necessary information for situation descriptions into a database (expressed by the experts, observers, sensors, and "hidden" information inferred from the available information), using appropriate formulas [22]. Studying the descriptions of situations, provided by experts, it is possible to see that most of them are presented by formulas in the following format: X, ¬X, Y⊃X, (Y1&...&Yn)⊃X, X1∨...∨Xm, (Y1&...&Yn)⊃(X1∨...∨Xm), ¬(Y1&...&Yn), where X, X1,...,Xm, Y, Y1,...,Yn are either atomic formulas or formulas that are constructed from atomic formulas using logical operations ¬, &, ∨, ⊃, ⇔.

Note 1. In these formulas there are no individual variables and quantifiers. This, however, does not exclude the existence of individual constants; we can say that in that case we have a subset of the set of formulas of first order predicate calculus. Strictly speaking, however, mentioned subset is not the set of all propositions in first order predicate calculus (for example in the latter the formula (∀x)(x = x) occurs, that does not belong to our subset).

Note 2. The prototype implements sequents, that were introduced by Gentzen (Let us recall that the notation for the formula X is the sequent ⟶X, notation for the formula ¬X is the sequent X⟶, notation for the formula (Y1&... &Yn)⊃(X1∨...∨Xn) is the sequent Y1,...,Yn⟶X1,...,Xm etc. See [4, 10, 28], etc.). This representation of formulas using sequents is well known and easy to use for experts, who have to create axioms. But instead of natural deduction [11, 25], the SyLAR prototype uses a different kind of reasoning method, developed by authors.

The initial set of the above-mentioned formulas, used in the prototype, must be created by experts, who decide what sort or which type of information, hereinafter knowledge, is needed for the "complete picture" about a given situation. Let us call this initial set as a set of axioms, provided by experts. Logical axioms must be added to those axioms. Sequents representing axioms, given by experts, must not have empty antecedents or empty succedents; an important role is played by axioms, which contain only one statement formula in the succedent - this formula represents the decision.

The complexity of the formulas which belong to sequents (in the meaning of the "number of applications" of the logical operations ¬, &, ∨, ⊃, ⇔) is not limited in the prototype. Only those formulas are allowed, that provide descriptions of situations or the relevant axioms and are associated with knowledge of the situation (see [18]).

For inference we use only logically correct inference steps and rules (if the assumptions are true for some interpretation φ, then the conclusion is true for the same interpretation φ).

The prototype SyLAR uses classical interpretation for finding the truth value of formulas.

Next let's briefly review reasoning using natural deduction and Gentzen sequences and after that we present the same deduction using the method presented in this paper. For the rules of inference, we have taken the slightly modified versions of Modus Ponens (MP) and Modus Tollens (MT):

$$\frac{X \quad X \supset Y}{Y} \qquad \frac{X \supset Y \quad \neg Y}{\neg X} \tag{1}$$

"The slightly modified sequential forms" of the above rules are as follows:

$$\frac{\to A \quad A \to \Delta}{\to \Delta} \qquad \frac{\Sigma \to A \quad A \to}{\Sigma \to} \tag{2}$$

It is possible to demonstrate (see [10], part V) that when we select the appropriate logical axioms, then we can substitute the use of all Gentzen rules from the calculus LK (except rules for the quantifiers) to implement appropriately MP and MT.

MP is the rule we use to derive the sequence $\longrightarrow \Delta$. In other words, in order to make sure that Δ is true. MT is the rule we use to derive a sequence $\Sigma \longrightarrow$. In other words, in order to make sure that Σ is false.

In addition to the above rules, we use the rule for equivalence operator, which allows deciding the truth value of a statement, if the truth value for the equivalent statement is already known.

In the prototype we use relevant truth tables for the implementation of inference rules. Next the method, developed by authors, will be explained. Here it is important to mention, that this method is suitable for this specific kind of decision making process.

Let's agree that if we write ¬A, then we definitely know that $\varphi(A) = 0$ and otherwise if we write A, then we definitely know that $\varphi(A) = 1$. In other words if somebody provides the information, that the "event A is not present" or "signal A is absent", than we write the formula ¬A and say that $\varphi(A) = 0$. Such kind of agreement is possible in classical case, because of $\varphi(\neg A) = 1$ gets the $\varphi(A) = 0$ (that is not correct, for example, the case of intuitionistic interpretation). A similar notation is used in computers, where the binary code 0 or 1 explains whether we have electrical signal on some contact or not.

Remind that:

1. Modus Ponens needs the sequent → A (here A is a notation for any suitable logical formula, sometimes formed from few atomic formulas). The truth value of → A must be true ($\varphi(A) = 1$).

2. Modus Ponens use also the axiom (A⟶Δ). The truth value of axioms are always true (φ((A⟶Δ) = 1)). The succedent and also the antecedent of axiom may consist of some atomic formulas formed by logical operators.
3. It is possible to calculate the truth values of axiom components (here A or Δ) processing from whole axiom truth value (true).
4. Modus Ponens can be used if we get the truth value of → A (for example from observers or sensors) and if this truth value is true (signal or event is present). From above we will calculate the truth value of Δ.

Example 1 Let A be a notation of formula X1&X2 and Δ is notation of formula Y1&¬Y2. If observers or sensors (the presence of a signal can be marked with 1, the absence with 0) provide us that the φ(X1) = 1 and also the φ(X2) = 1, then we get that φ(A) = 1 and Modus Ponens can be used. Next we calculate φ(Y1) and φ(Y2) from circumstances that φ((A⟶Δ) = 1 and φ(A) = 1). We know that general truth table of implication is the following:

φ(A)	→	φ(Δ)
0	1	0
0	1	1
1	0	0
1	1	1

Fig. 1 General truth table of implication

As far as φ(A) = 1) and φ(A⟶Δ) = 1, then the truth value of Δ is also true (Fig. 1). That means that φ(Y1&¬Y2) = 1. With simple SAT solver it is possible to find that φ(Y1) = 1 and φ(Y2) = 0.

Remind also that:

1. Modus Tollens needs the sequent A → and the information about the truth value for A. The sequent A → is the "same" as formula ¬A.
2. Modus Tollens needs the axiom Σ⟶A and the truth value of axiom is always true (φ(Σ⟶A)). Also as limited before, such axiom means that φ(Σ) = 1 and φ(Λ) = 1
3. Modus Tollens can be used in two cases:
   a. if φ(A) = 0 (the event or signal is not present, the formula is ¬A) with axiom Σ⟶A;*
   b. if φ(¬A) = 0, φ(A) = 1 (event or signal is present, the formula is A) with axiom Σ⟶¬A;
4. By using the Table 1 we can find, that the φ(Σ) = 0, in other words the result formula is ¬Σ and event or signal is not present.

*. In practice there is a problem with application of Modus Ponens with axioms, where the truth value of succedent is true. Such inference allows to use both (true and false) values for antecedent Σ. Without additional information it is impossible to decide which truth value for Σ is correct.

Example 2 Let Σ be represented by formula Z1⊃Z2. If observers have provided that φ(A) = 1 (signal is present) and we use Modus Tollens with axiom Z1⊃Z2⟶¬A, then a SAT solver will search the satisfying values for φ(Z1⊃Z2) = 0. Finally, we will get as a result that the φ(Z1) = 1 and the φ(Z2) = 0.

More complex computation will be continued if the SAT solver gives the output of uncertain answer for some formulas, both truth values: true and false. In this case it is necessary to find the related axioms or/and truth values of related formulas are provided by observers (or sensors).

Theorem 1 If we have sequences A → B and A′ → B′ and A″ → B″, also know that φ(A → B) = 1 and φ(A′ → B′) = 1 and φ(A″ → B″) = 1 (the truth values of axioms is always true), then φ(A, A′, A″ → B& B′& B″) = 1. *Proof*:

$$\dfrac{\dfrac{\dfrac{\dfrac{\dfrac{A\to B}{A,A'\to B}\quad\dfrac{A'\to B'}{A,A'\to B'}}{A,A'\to B\&B'}}{A'',A,A'\to B\&B'}}{\dots\quad\dots\quad\dots}\quad\quad \dfrac{\dfrac{A''\to B''}{A',A''\to B''}}{A,A',A''\to B''}}{A,A',A''\to B\&B'\&B''} \tag{3}$$

Example 3 Let Σ be the notation of formula Z1∨Z2 and observers have provided that the signal X1 is absent. Keep in mind that A is notation of X1&X2 (see example 1) and as far as the signal X1 is absent, we get the formula ¬X1&X2 or sequence X1&X2 →. We may have an axiom Z1∨Z2 → X1&X2 and may try to use Modus Tollens. NB! According to agreement before, if truth value of succedent, calculated from observers data (here the X1) is false, then the usage of Modus Ponens in our approach means the satisfaction of φ(Z1∨Z2) = 0. In this example a SAT solver cannot solve a φ(Z1∨Z2) = 0 uniquely. We need related information. We may find, for example, the following axioms: ((X1&Z2)⊃Z1) → Y1 and X3 → Y2. Also observers may provide that the signal X3 is present (true). Now we have a system of formulas:

$$\begin{cases} X1\to \quad & Z1\vee Z2\to X1\&X2 \\ & (X1\&Z2\supset Z1)\to Y2\,. \\ \to X3 & \quad X3\to Y2 \end{cases} \tag{4}$$

Axioms will be combined to the common axiom (Gentzen sequence):

$$(Z1 \vee Z2 \rightarrow X1\&X2)\&((X1\&Z2 \supset Z1) \rightarrow Y2)\&(X3 \rightarrow Y2)) \tag{5}$$

$$(Z1 \vee Z2), \quad (X1\&Z2 \supset Z1), \quad X3 \rightarrow (X1\&X2)\& \ Y2. \tag{6}$$

$$\begin{aligned} &\text{Next we will find the Z1 and Z2 from } \varphi((Z1 \vee Z2), (X1\&Z2 \supset Z1)). \\ &X3 \rightarrow (X1\&X2)\& \ Y2) = 0 \text{ and } \varphi(X1) = 0 \text{ and } \varphi(X3) = 0. \end{aligned} \tag{7}$$

Fig. 2 Calculation of Z1 and Z2

Z1	or	Z2	and	(X1	and	Z2	then	Z1)	and	X3
0	0	0	0	0	0	0	1	0	0	0
1	1	0	1	0	0	0	1	1	0	0
0	1	1	1	0	0	1	1	0	0	0
1	1	1	1	0	0	1	1	1	0	0
0	0	0	0	0	0	0	1	0	0	0
1	1	0	1	0	0	0	1	1	0	0
0	1	1	1	0	0	1	1	0	0	0
1	1	1	1	0	0	1	1	1	0	0
0	*0*	*0*	*0*	*0*	*0*	*0*	*1*	*0*	*0*	*1*
1	1	0	1	0	0	0	1	1	1	1
0	1	1	1	0	0	1	1	0	1	1
1	1	1	1	0	0	1	1	1	1	1
0	*0*	*0*	*0*	*0*	*0*	*0*	*1*	*0*	*0*	*1*
1	1	0	1	0	0	0	1	1	1	1
0	1	1	1	0	0	1	1	0	1	1
1	1	1	1	0	0	1	1	1	1	1

As stated above, the formulas that occur in the prototype are associated with knowledge about a given situation. By finding the truth value of a formula (lets note as Z), we discover if in a given situation an event occurs (truth value of "1", which is equivalent to the fact that the formula Z is derived) or does not occur (truth value of "0", which is equivalent to the fact that the formula ¬Z is derived). Using forward chaining we are able to use previously derived statements to derive new reliable statements and obtain new knowledge (Fig. 2).

3 Knowledge Representation

In this work we use the basic concepts and results of Lorents [15, 16] to deal with knowledge.

Therefore, we call "knowledge" every such ordered pair ⟨A, B⟩, where A is the notation (symbol, sign, etc.) for B and, at the same time, B is the denotation (meaning) of A. For example:

⟨Gorby, Mikhail Sergeyevich Gorbachyov⟩, ⟨π, *the ratio of a circle's circumference to its diameter*⟩.

In that case we say that A and B have a notation-denotation relationship and we represent this as A∫B. The binary relation "∫" has several natural algebraic properties, such as well-foundedness, a specific anti-symmetry, and transitivity [17, 18].

The usage of described knowledge representation in decision support systems have been discussed for digital solutions [22]. These ideas have been used in the implementation of SyLAR. Namely: a description of the situation B is given by the formula A (or A∫B). So we get the knowledge – ⟨A, B⟩. This description corresponds to the identifier (ID) in the database, so we have the following knowledge ⟨ID, B⟩. In doing so, we consider the following constraint: each notation must have only one "meaning" or denotation, and each denotation must have only one notation.

In order to be able to describe all necessary and possible situations, the expert should be able to use natural language [23], or some suitable system, which helps to construct formulas. Such system might allow, for example, (I) to choose appropriate predicate-symbols and the individual-symbols from database, and (II) to place these symbols in the appropriate places in the predicate-logic formulas using the software.

One of the possible solutions is to use the ontology. A huge amount of ontology is prepared for the IBM Watson intelligent system and most of them are available online for downloading [27, 28].

By using OWL (Web Ontology Language) or RDF (Resource Description Framework) format files as inputs it is possible to build some special tools for selection of individuals and predicates and automatically place them to the necessary structure of logical formulas (Fig. 3).

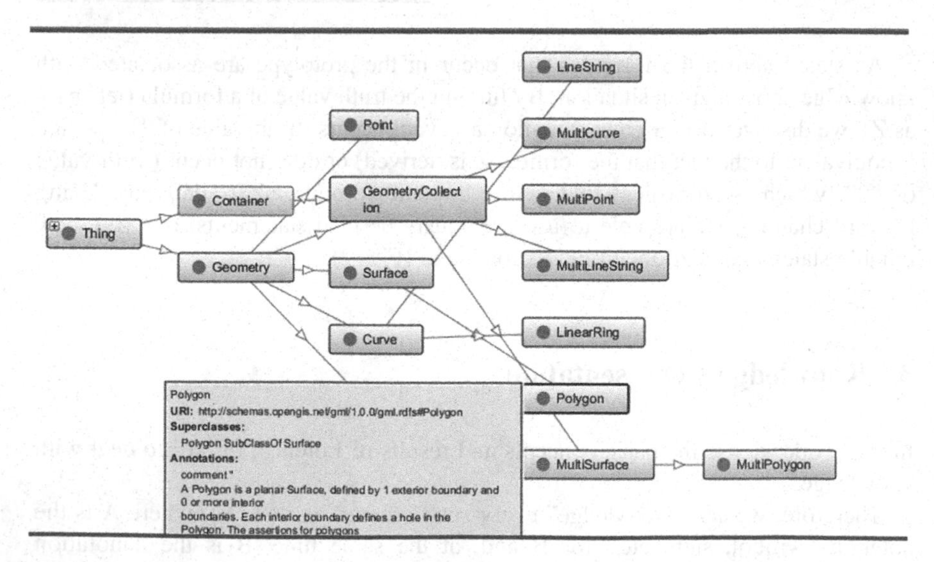

Fig. 3 The example of knowledge representation by Stanford protégé software

Here is possible to use the ideas and solutions made in SAWA assistant [31]. They use for the knowledge domain creation the OWL ontology and for the rules insertion the RuleVISor[1] SWRL editor (designed for forward chaining reasoning).

4 Feeding Knowledge to the Decision Support System

It is important to prepare a suitable *algebraic system* [6, 13, 20] so that the decision support system can operate with knowledge. For this purpose, we need to implement the system mining procedure. To implement this procedure, at first, it is necessary to identify the set of elements, or all of the objects which belong to the considered situations. Then, to identify which properties and relationships are important for situation analysis [19]. Then, using the obtained results, we create the formal descriptions of situations (knowledge), which is important for situation management. Such knowledge may be the initial basis for decision making, or the interim result, which is important for the following logical steps or the results, which are possible to get by inferences.

Note. Even in ancient times, in order to share the language between parents and children, the latter had to learn words that denote objects, as well as the properties and relationships, and then, using word components, to form sentences, including the clauses. Thus, our system also, at first, must learn the words, and second, the sentences that describe situations and are needed for communication.

5 Information from Observers and Tri-Valent Logic

In order to allow the system to make automatic conclusions, we need an initial set, i.e. inputs. We use the observers who can provide feedback on events for situation management.

Observers get the list with descriptions of things (which by nature are logic formulas), that must be assessed. Assessment of formulas gives the following boolean values: true, false, or unknown (ex. [2, 5, 12]).

Example 4 Let a traffic accident occur. The observer must indicate the cell corresponding to the area, including street or intersection as true. This gives the related logic formula, such as Crossroads (Narva mnt, Pronksi), the boolean value of true. It should also be noted, for example, whether there are some people affected (ARE (affected) boolean true), or if there are deaths (ARE (death), boolean false). Some of the information may, however, remain unknown. Let in our example be such information unknown: must the car be written-off (IS (amortizing) boolean value is unknown).

---

[1]VIStology, Inc. http://www.vistology.com/.

6 Transition from Tri-Valent to Bi-Valent

In the real decision-making process it is necessary to reduce the uncertainty (i.e. elimination of the truth value: unknown). For this purpose, we use in our prototype the "deduction philosophy", which is supported by the corresponding "truth tables in practice". In particular - if we have an amount M of formulas, which is included in the formula F, the truth value is unknown, it will explain whether from the set M – {F} can be derived the formula F or formula ¬F. If it is possible to derive F, then F is true. If it is possible to derive ¬F, then F is false.

Example 5 Let observers forward the following data:

- the statements A1 and A3 be present, in other words A1 is true and A3 is true
- the statement A2 is not present, A2 is false
- B1 is the statement whose truth value is unknown
- B2 is true
- B3 is false

Next it is important to find which axioms that are given by experts and are related to the data. Let it be, for example, A1, A2⊃A3⟶B1&B2, B3. Using this axiom (with truth value "1"!) it is possible to find the suitable truth value for B1.

In the traditional way the truth value detection for B1 is to use Gentzen's inference rules (because our fragment of the predicate calculus is complete – then the formula that has been derived from the correct formulas, must be true).

$$\dfrac{\dfrac{\dfrac{\dfrac{\to A_1}{\neg A_1 \to}}{\to \neg\neg A_1} \qquad \dfrac{\dfrac{\dfrac{\dfrac{\to A_3}{A_2 \to A_3}}{\to A_2 \supset A_3} \qquad \dfrac{A_1, A_2 \supset A_3 \to B_1 \& B_2, B_3}{A_2 \supset A_3 \to B_1 \& B_2, B_3, \neg A}}{\to B_1 \& B_2, B_3, \neg A_1}}{\neg\neg A_1 \to B_1 \& B_2, B_3}}{\dfrac{\dfrac{\dfrac{B_3 \to}{\to \neg B_3} \qquad \dfrac{\to B_1 \& B_2, B_3}{\neg B_3 \to B_1 \& B_2}}{\to B1 \& B2} \qquad \dfrac{B_1 \to B_1}{B1 \& B2 \to B1}}{\to B_1}} \tag{8}$$

The second way – as it is implemented in the prototype SyLAR– to use the Truth tables (see the Comments in "Theoretical foundation for the prototype"), which replace all intermediate inferences. At first it is necessary to define which of the rules can be used at the resulting step. This could be either MP or MT. If formula B1, which truth value is unknown, is located only in the succedent, and truth values for all other formulas (in succedent and also in antecedent) are known, then it could be possible to use the Modus Ponens at the last step. But this possibility depends of truth value of the antecedent. We assume that the information from observers is correct. This means that for using Modus Ponens, the truth value of the observers' information and the truth value of the antecedent must be the same.

Let's calculate the truth value for antecedent A1, A2⊃A3. Note. Here we recall that the sequent Y1,…,Yn⟶X1,…,Xm is the notation for the formula (Y1&…&Yn)⊃(X1∨…∨Xn)).

As the result we get the φ(A1) = 1 and φ(A2⊃A3) = 1, and also $\varphi(A_1, A_2 \supset A_3) = 1$. Modus Ponens can be used and that gives us the φ(B1&B2, B3) = 1.

Axioms have the truth value true (this means: 1) (Fig. 4). The Truth table for A1, A2⊃A3⟶B1&B2, B3:

$A_1, A_2 \supset A_3$	→	$B_1\&B_2, B_3$	
0	1	0	(a)
1	1	1	(b)
0	1	1	(c)
1	0	0	(d)

Fig. 4 The Truth table for calculation

The Truth table of implication interpretation demonstrates that the line (d) is not needed, as far as correctness of axioms must give the Truth value of axiom is 1 (true).

We have analyzed before that the truth value of antecedent must be true (1). This means that only row (b) is suitable for Modus Ponens. But row (b) shows that in this case the truth value of succedent B1&B2, B3 or formula (B1&B2)∨ B3 must be also the 1 (true).

In the present case the truth value of (B1&B2)∨ B3 must be 1 (true) (Fig. 5).

B_1	&	B_2	∨	B_3	
0	0	1	0	0	(a)
1	1	1	1	0	(b)

Fig. 5 Truth value calculation for (B1&B2)∨ B3

As result we have φ(B1) = 1 (true).

7 Inferences with Few Unknown Truth Values

Inferences that correspond to axioms with several pieces of knowledge with unknown truth values are also possible in the created prototype tables (see the example 3 in "Theoretical foundation for the prototype"). Like in system equation with several unknowns it is important to find the corresponding collection of axioms, which cover all unknown values. But in our implementation it is important that unknown knowledge values stay only in antecedent or only in succedent of axioms. In case of Modus Ponens we choose the axioms, where the unknown truth values are in succedent. Next we use the conjunction of detected axioms succedents (putting all detected succedents together) and analyze this conjunction by truth table.

8 SyLAR Prototype

SyLAR is a prototype implementation of the method described before. At the present stage it uses the randomly generated axioms. The amount of axioms is not limited and all axioms are in logical consistency. SyLAR system is an open source software and is available to the research community at the address: https://github.com/tarmolehtpuu/sylar. Next we describe the structure of the developed prototype SyLAR.

Prototype SyLAR consist of 8 packages:

- ee.moo.sylar.alphabet: Contains classes that define the alphabets for formulas and sequents
- ee.moo.sylar.formula: Contains the implementations for formulas and formula collections (sets of formulas), also contains the method for tautology and contradiction checks.
- ee.moo.sylar.generator: Contains classes for formula and sequent generation based on Markov chains. This package gives the opportunity to test the prototype without a real system and generates the fictitious axioms.
- ee.moo.sylar.inference: Contains the interface and implementation for different inference strategies:
    - InferenceEngine: implementation of an InferenceEngine, which, given a set of axioms and a set of known values, tries to infer as much as possible (until no new knowledge can be inferred).
    - InferenceException: generic exception thrown by classes in the inference package.
    - InferenceHelper: helper class for inference strategies, contains helper methods for determining if given a set of known values, all the rest of the interpretations of a formula evaluate constantly to true or false.
    - InferenceRequest: implementation of a request to perform inference, contains an axiom and a set of known values.
    - InferenceResult: contains the results of an inference strategy application, in case inference was successful it will contain all the values for new knowledge that we managed to infer.
    - InferenceStrategy: generic interface for inference strategies.
    - InferenceStrategyEQ: contains an inference strategy implementation for equivalence.

- InferenceStrategyMP: contains an inference strategy implementation for Modus Ponens.
- InferenceStrategyMT: contains an inference strategy implementation for Modus Tollens.

– ee.moo.sylar.input: contains classes related to parsing formulas and sequents.
– ee.moo.sylar.output: contains classes related to converting a formula into another format (for example SVG or .dot) for GraphViz visualization.
– ee.moo.sylar.util: contains utility classes used by other packages, such as class that provides the methods for use in SyLAR's interactive console interface, class for iterating over all possible interpretations of a formula (truth table) and the others
– ee.moo.sylar.graphical: contains the classes related to the graphical interface.

9 Discussions

The main problem of proposed the prototype is related to SAT solver computational complexity. As we know this is the first problem that has been proven to be NP-complete [7]. During many years it has been possible to participate in international SAT competition[2] and collect the new solutions and ideas. At the present time SyLAR does not use some of the most improved SAT methods. Such kind of algorithms or open source software can be implemented in the future.

Then we move, using forward chaining method of reasoning, from the data observed by observers, and the axioms given by the experts, to discovering hidden knowledge and to decision making, we can classify the area (set, class or something else) of problems. As Fig. 3 shows, a lot of available ontology are created by using subclasses. If we discover some decision, then we can start from a set of formulas related to the smallest subclass and if this subclass will not satisfy our expectations, we will move to the next level of hierarchy.

Situation Management involves interaction between people, technologies, and responses. That means that it is important to keep in focus such things like data, goals, risks, decisions etc. When the amount of data is going to be very large, or object or place of situation is going to be dangerous, then different systems made by humans can come to help in situation management. The present version of SyLAR prototype gives the improved forward chaining reasoning possibilities. User interface for situation management is under design. SAWA assistant [31], which is use the VIStology Inc software, has developed very good modules with a graphical interface and gives sufficient good possibilities for situation management. SyLAR forward chaining method is more complex comparing to traditional solution, also more complex than in BaseVISor. Also the structure of axioms in SyLAR allows

[2]The international SAT Competitions web page: http://www.satcompetition.org.

multistep inferences. Using VIStology Inc, SAWA modules and theirs implementation to/with SyLAR prototype brings wider opportunities for situation management.

10 Conclusions

The novelty and advantage of the described prototype approach involves, inter alia, the following: (I) the possibility of using a combination of inference steps and interpretations (for example, presented by boolean tables); (II) the opportunity to use SAT solvers separately in succedent and antecedent for discovering of hidden knowledge truth values and keep moving in forward chain of reasoning; (III) the possibility to apply more than one step containing inferences; (IV) the opportunity to prevent in a large number of cases, the overflow of processing capacity. (V) the opportunity to prevent in a large number of cases, the overflow of processing capacity.

For testing our distributed implementation of the system we plan to use Amazon EC2 High CPU virtual machines. This will allow us to test the performance of the algorithm in a distributed environment and test a problem on a scale that is beyond what any single general consumer computer could handle (for example 100 unknowns in one axiom).

References

1. Bibel, W.: On matrices with connections. J. Assoc. Comput. Mach. **28**(4), 633–645 (1981)
2. Birkhoff, G., von Neumann, J.: The logic of quantum mechanics. Ann. Math. **37**, 823–842 (1936)
3. Browne, P.: JBoss Drools Business Rules. Packt Publishing, Birmingham (2009)
4. Buss, S.R.: An Introduction to Proof Theory. Elsevier Science, Amsterdam (1998)
5. Church, A.: Introduction to Mathematical Logic, vol. I. Princeton University Press, Princeton, New Jersey (1956)
6. Cohn, P.M.: Universal Algebra. Harper and Row, Evanston (1965)
7. Cook, S.A.: The complexity of theorem-proving procedures. In: STOC '71 Proceedings of the third annual ACM symposium on Theory of computing, pp. 151–158. ACM, New York, NY, USA (1971)
8. d'Aquin, M., Motta, E.: Watson, more than a Semantic Web search engine. Semantic Web IOS Press (2010). (http://www.semantic-web-journal.net/sites/default/files/swj96_1.pdf)
9. Een, N., Mishchenko, A., Sörensson, N.: Applying Logic Synthesis for Speeding Up SAT, SAT (2007)
10. Gentzen, G.: Die Widerspruchsfreiheit der reinen Zahlentheorie. Math. Ann. **112**, 493–565 (1936)
11. Gentzen, G.: Investigations into logical deduction. Szabo **1969**, 68–131 (1935)
12. Goodstein, R.L.: Mathematical Logic. Leicester, England (1957)
13. Grätzer, G.: Universal Algebra, 2nd Edn. Springer, New York (2008)

14. Lorents, P.: Knowledge and information. In: Proceedings of the 2010 International Conference on Artificial Intelligence, pp. 209–215. CSREA Press (2010)
15. Lorents, P.: Formalization of data and knowledge based on the fundamental notation-denotation relation. In: International Conference on Artificial Intelligence, Proceedings, vol. III, pp. 1297–1301. CSREA Press, Las Vegas USA (2001)
16. Lorents, P.: Denotations, knowledge and lies. In: Proceedings of the International Conference on Artificial Intelligence, IC-AI' 2007, vol. II, pp. 324–329. CSREA Press, Las Vegas, USA, June 14–17 2007
17. Lorents, P.: Knowledge and taxonomy of intellect. In: Proceedings of the International Conference on Artificial Intelligence, IC-AI' 2008, vol. II, pp. 484–489. CSREA Press, Las Vegas, USA, 25–28 July 2008
18. Lorents, P.: Knowledge and logic. In: Proceedings of the International Conference on Artificial Intelligence, IC-AI' 2009: International Conference on Artificial Intelligence, IC-AI' 2009, pp. 568–570. CSREA Press, Las-Vegas (2009)
19. Lorents, P., Matsak, E.: Applying time-dependent algebraic systems for describing situations. In: 2011 IEEE Conference on Cognitive Methods in Situation Awareness and Decision Support (CogSIMA 2011), pp. 25–31. I.E.E.E. Press, Miami Beach, FL. (IEEE Catalog Number: CFP11COH-CDR) (2011)
20. Maltsev, A.I.: (Мальцев А. И.) *Алгебраические системы* (Algebraic systems). Moscow: Наука (1970)
21. Matheus, C., Dionne, B., Parent, D., Baclawski, K., Kokar, M.: BaseVISor: A forward-chaining inference engine optimized for RDF/OWL triples. In: Digital Proceedings of the 5th International Semantic Web Conference, ISWC 2006, Athens, GA (2006)
22. Matsak, E., Lorents, P.: Knowledge in digital decision support system. In: Universal Access in Human-Computer Interaction. Applications and Services: HCI International 2011, Lecture Notes in Computer Science, pp. 263–271. Springer, Orlando, Florida, USA (2011)
23. Matsak, E., Lorents, P.: Decision-support systems for situation management and communication through the language of algebraic systems. In: 2012 IEEE International Multi-Disciplinary Conference on Cognitive Methods in Situation Awareness and Decision Support (CogSIMA 2012): CogSima 2012, pp. 301–307. I.E.E.E. Press (2012)
24. Matuszek, C., Cabral, J., Witbrock, M., Deoliveira, J.: An introduction to the syntax and content of Cyc. In: Proceedings of the 2006 AAAI Spring Symposium on Formalizing and Compiling Background Knowledge and its Applications to Knowledge Representation and Question Answering (2006)
25. Prawitz, D.: Natural Deduction: A Proof Theoretical Study. Almqvist and Wiksell, Stockholm (1965)
26. Robinson, J.A.: A machine-oriented logic based on the resolution principle. J. ACM **12**(1), 23–41 (1965). doi:10.1145/321250.321253
27. Suchanek, F.M., Kasneci, G., Weikum, G.: Yago—A core of semantic knowledge. In: 16th international World Wide Web conference (2007)
28. Takeuti, G.: Proof Theory. North-Holland Publishing Company, Amstardam (1975)
29. Tomsovica, K., Liu, C.-C.: Bounding the computation time of forward-chaining rule-based systems, Data and Knowledge Engineering. Elsevier (1993)
30. Tilotma S., Navneet T., Deepali K.: Study of difference between forward and backward reasoning. Int. J. Emerg. Technol. Adv. Eng. **2**(10), ISSN: 2250–2459 (2012)
31. Tetali P.: Artificial intelligence and decision support systems. Int. J. Adv. Res. IT Eng. **2**(4), ISSN: 2278–6244 (2013)
32. Matheus, C.J., Kokar, M.M., Baclawski, K., Letkowski, J.A., Call C.: SAWA: An assistant for higher-level fusion and situation awareness. In: Proceedings of SPIE 5813, Multisensor, Multisource Information Fusion: Architectures, Algorithms, and Applications 2005, vol. 75 (05 April 2005). doi:10.1117/12.604120

Reasoning and Decision Making in an Inconsistent World

Labeled Logical Varieties as a Tool for Inconsistency Robustness

Cornelis N.J. de Vey Mestdagh and Mark Burgin

Abstract The goal of this paper is the development of foundations for robust reasoning and decision-making in pervasively inconsistent theories and deductive databases. The pervasiveness of these inconsistencies is partly inherent to the human epistemic condition (i.e. the need to make decisions on the basis of perspective bound knowledge) and partly inherent to practical limitations (e.g. incomplete knowledge). In the first case, we do not want to eliminate the inconsistencies. In the second case, we cannot eliminate them. So, in both cases, we will have to incorporate them in our descriptions of reasoning and decision making. For this reason, inconsistency handling is one of the central problems in many areas of AI. There are different approaches to dealing with contradictions and other types of inconsistency. In this paper, we develop an approach based on logical varieties and prevarieties, which are complex structures constructed from logical calculi. Being locally isomorphic to a logical calculus, globally logical varieties form a logical structure, which allows representation of inconsistent knowledge in a consistent way and provides much more flexibility and efficacy for AI than standard logical methods. Logical varieties and prevarieties are efficiently used in database theory and practice, expert systems and knowledge representation and processing. To increase efficiency, flexibility and capabilities of logical varieties and prevarieties and to model perspective bound decision making, we introduce labeling of their elements and study labeled logical varieties and prevarieties. We illustrate the viability of this by an example of a labeled logical variety, the extended Logic of Reasonable Inferences, which is and has been applied in the legal domain.

C.N.J. de Vey Mestdagh (✉)
Centre for Law & ICT, University of Groningen, Groningen, The Netherlands
e-mail: c.n.j.de.vey.mestdagh@rug.nl

M. Burgin
Department of Computer Science, UCLA, Los Angeles, USA

R. Neves-Silva et al. (eds.), *Intelligent Decision Technologies*,
Smart Innovation, Systems and Technologies 39,
DOI 10.1007/978-3-319-19857-6_36

1 Introduction

As far as we (humans) are concerned, the world *is* inconsistent. Our reproductive survival probability seems to be enhanced by our ability to perceive and process inconsistent data establishing inconsistent (alternative, mutually exclusive) projections and using them to make the more beneficial choices. If this was not the case, we would not be here to reflect upon our inconsistency handling abilities as evolution theory and the anthropic principle suggest. The argument that this ability is an independent feature of our cognition can be countered by the observation that perceiving inconsistency in a consistent world would just result in confusion and therefore be counter-*re*productive. The time dimension imposes constant change upon us. If this change *is* predetermined, then there would be only one consistent universe. Actually, this change can be brought about by humans, who are often able to foresee alternative futures caused by their actual actions. They can even experience a feeling of choice preceding the actual action because they dispose of a palette of imaginable alternative actions and, at least in their minds, they have the power to decide which one of these alternatives to select and perform. From our perspective every transition from one state to another therefore presupposes a selection of the next state from a set of accessible possibilities. To describe our universe, we therefore need a descriptive model of knowledge of the universe that incorporates the perception of the alternatives and some tentative outcomes of the selection process. Even if the selection process is predetermined, we need a description of the antecedents (that are called considerations if constituted by conscious cognitive processes) that led to the derived conclusion. It is also important to be able to explain this process to ourselves and to others, because we will have to communicate about it and to cooperate with each other in its application to be able to make collective decisions. The more we know about and share these antecedents the more we perceive them as *facts*, while the less we know about them and the less we share them the more we are inclined to call them *opinions*. An advanced scientific description of relations between antecedents and corresponding conclusions is called a theory. The situation is even more complicated because a state of mind itself can also play the role of a possible future. Indeed, when we want to describe changes of our states of mind, we consider a number of possible states of mind (e.g. opinions) to which we come, in making our decisions and in adhering to certain opinions.

In the next section we will formulate the *theoretical* requirements for the description of our inconsistency handling capabilities.

2 Reasoning and Decision Making in an Inconsistent World

Humans are able to imagine and reason about different futures (i.e. to describe their distinct characteristics and the relations between these characteristics) by perceiving relations between their (mental) actions, making selections of possible futures, reasoning about relations between themselves as individuals and some of these futures and coming into a state in which this relation is narrowed down to one possible future, which then becomes the present. Classical logic does not cover and usually denies this human capacity. Some modern logics try to incorporate this process into their scope (see Sect. 3). However, modern logics like belief revision and default reasoning still deny the fact that the state of inconsistency is not transient but permanent - there is no decisive factor which renders a perception (a fact or opinion) into the universal knowledge - while these logics consider knowledge to be inherently universal. Other modern logics like probabilistic and fuzzy logics have the same limitations and even more, they deny the presence of the context dependent factors that constitute opinions and that are used in the selection of the individual consequences of these opinions. In particular, the mean of all perceived or even possible contexts is not a realistic prediction of the future context in an actual case, while these logics consider knowledge to be valid in one common (or even universal) context. To meet these requirements for the description of our cognition, we need two things:

1. A description of our cognition that allows reasoning about (mutually exclusive) alternative actions and situations.
2. A description of our cognition that allows tagging the alternatives and situations, e.g., describing their identity and context, in order to be able to explain the connection between alternative antecedents (considerations/arguments) and alternative imagined futures (decisions for a certain perspective or opinion).

As we will see in Sect. 3, the most common forms of modeling decision making in logic (like belief revision and default reasoning; probabilistic and fuzzy logics) are not suited to realistically model reasoning and decision making in an inconsistent domain of knowledge. Logical varieties do solve the first limitation of these logics: they allow for the preservation of all individual opinions [17, 18, 25, 26] (Burgin and de Vey Mestdagh 1191). Labeled logics solve the second limitation of these logics: they allow for the identification of individual opinions and enable us to reason about these opinions. Labeled logical varieties, introduced and studied in this paper, combine advantages of both approaches essentially extending the power of logic in solving problems in all practical and theoretical areas since they both resolve the epistemic need to conserve inconsistency as the practical need to handle inconsistency even when it cannot be eliminated.

The previous argument can be illustrated most easily by taking a knowledge domain that is situated on the opinion end of the perception (fact-opinion) continuum. The legal knowledge domain is such a domain. Not only is its foundation based on the existence of different normative opinions, but it also does not presuppose the aspiration for a permanent (universal) resolution of these differences. In practice, only local and temporal legal solutions enable humans to make a move allowing them immediately to reconsider their previous decision when necessary. The rejected opinion is a part of the (context of) the selected opinions (see Sect. 7). Judicial and other social laws are not laws of nature. They allow for discretion (e.g., for balancing specific interests), qualification (e.g., they have an open texture), interpretation (e.g., they are vague) and individualization (e.g., they can be instantiated by a personal local temporal context). Even if the contents of my opinion are exactly the same as your opinion, it is still a unique opinion, which has another potential effect upon the outcome of a reasoning process. So, reasoning in the legal domain demands non explosive means of describing and processing alternatives. Furthermore it demands the preservation of opinions, even if they are locally and or temporally or even on average defeated. Finally it demands the possibility to tag each opinion and to describe its context.

The most obvious way in which the context of an opinion plays a role in reasoning is when an individual or a group tries to decide which opinion should prevail in order to be able to make an opinionated move. In other words, to make a personal local temporal decision, we need to be able to reason about opinions and their contexts. For both purposes, the opinions have to be identified. Criteria for making decisions between opinions do use this identification to compare opinions. To make things more complex: even the decision criteria are opinions and therefore alternative decision criteria compete when we make a decision. Mathematicians and logicians immediately recognize the recursion, which takes away the burden of modeling each of these levels of decision making separately. An extended example of such a decision making process including the contexts, the tags for opinions (valid, excludes, preferred) and their semantics and the decision rules (legal principles/metarules) and derivation rules (ranking principles) is given in de Vey Mestdagh and Hoepman [81]. We will develop a decision logic on the basis of the class this example belongs to and show its efficacy using inter alia this example.

In the next section we will formulate the *formal* requirements for the description of our inconsistency handling capabilities.

3 Approaches to Dealing with Inconsistent Knowledge

There is an active current research in the area of inconsistent knowledge, comprising various directions and approaches [25, 26]. Although inconsistencies bothered logicians from the time of Aristotle, the first logical systems treating contradictions appeared only in the 20th century. At first, they had the form of multivalued logics developed by Vasil'év and Łukasiewicz. Then the first relevant logics were built by Orlov. However, their work did not make any impact at the time and the first logician to have developed a formal paraconsistent logic was a student of Łukasiewicz, Jaśkowski [48]. Starting from this time, a diversity of different paraconsistent logics, including fuzzy logics, multivalued logics and relevant logics, has been elaborated (cf. [73]).

The perspective-bound character of information and information processing often results in natural inconsistency coming from different perspectives or from a faulty perception or from faulty information processing, such as processing on the basis of incomplete knowledge from a single perspective. As a result, now many understand that contradiction handling is one of the central problems in AI. Inconsistent knowledge/belief systems exist in many areas of AI, such as distributed knowledge base and databases, defeasible reasoning, dynamic expert systems, merging ontologies, ontology evolution, knowledge transition from one formalism to another, and belief revision.

There are three basic approaches to dealing with inconsistency in knowledge systems. The first one, which we call the *restoration approach*, is aimed at restoring consistency of an inconsistent knowledge system, e.g., a database [69]. The most popular direction in the first approach is called *coherence-based methodology* [60, 61]. It is performed in two steps:

1. Selection of one or several consistent subsystems in an inconsistent knowledge system;
2. Application of the classical inference on these subsystems.

Examples of systems that select one consistent subsystem are possibilistic logic [8], linear ordering [61], Adjustment and Maxi-adjustment [86, 87]. Examples of techniques that select several consistent subsystems are acceptable subbases [70], preferred subbases [14], Papini's revision function [63] and lexicographical approach [9, 50, 51].

Another way to restore consistency is to exclude those formulas (statements) that cause contradictions. This is achieved by utilizing logics that allow dynamical changes in the process of their functioning. Examples of such strategies are non-monotonic logics [53, 55, 56] and belief revision [29, 34, 39, 66, 84].

A *non-monotonic logic* is a formal logic that learning a new piece of knowledge can transform the set of what is known aimed at contradiction elimination. A non-monotonic logic can handle various reasoning tasks, such as *default reasoning* when consequences may be derived only because of lack of evidence of the contrary, *abductive reasoning* when consequences are deduced as most plausible explanations and belief revision when new knowledge may contradict old beliefs and some beliefs that cause contradictions are excluded [55, 53].

The second approach, which we call the *tolerance approach*, is to tolerate inconsistency [12] by using non-classical logics and including an inconsistent knowledge system into such a logic. As a result, inconsistency is treated at a higher level [78]. Examples of such methods are paraconsistent logics [13, 28, 67, 72] and argumentation [2, 10, 31].

The third direction, which we call the *structuring approach*, is based on implicit or explicit utilization of logical varieties, quasi-varieties and prevarieties [16–18].

In comparison with non-monotonic logics, which form the base for the first approach, logical varieties, quasi-varieties and prevarieties provide tools for preserving all points of view, approaches and positions even when some of them taken together lead to contradiction. Due to their flexibility, logical varieties, quasi-varieties and prevarieties allow treating any form of logical contradictions in a rigorous and consistent way.

Paraconsistent logics, which form the base for the second approach, are inferentially weaker than classical logic; that is, they deem *fewer* inferences valid. Thus, in comparison with paraconsistent logics, logical varieties, quasi-varieties and prevarieties allow utilization of sufficiently powerful means of logical inference, for example, deductive rules of the classical predicate calculus. Besides, Weinzierl [85] explains why paraconsistent reasoning is not acceptable for many real-life scenarios and other approaches are necessary. In addition, paraconsistent logics attempts to deal with contradictions in a discriminating way, while logical varieties, quasi-varieties and prevarieties treat contradictions and other inconsistencies by a separation technique.

Although conventional logical systems based on logical calculi have been successfully used in mathematics and beyond, they have definite limitations that often restrict their applications. For instance, the principal condition for any logical calculus is its consistency. At the same time, knowledge about large object domains (in science or in practice) is essentially inconsistent [16, 62, 79, 80]. From this perspective, Partridge and Wilks [64] write, "because of privacy and discretionary concerns, different knowledge bases will contain different perspectives and conflicting beliefs. Thus, all the knowledge bases of a distributed AI system taken together will be perpetually inconsistent." Consequently, when conventional logic is used for formalization, it is possible to represent only small fragments of the object domain. Otherwise, contradictions appear.

To eliminate these limitations in a logically correct way, logical prevarieties and varieties were introduced [16]. Logical varieties represent the natural development of logical calculi, being more advanced systems of logic, and thus, they show the direction in which mathematical logic will inevitably go. Including logical calculi as the simplest case, logical varieties and related systems offer several advantages over the conventional logic:

1. Logical varieties, prevarieties and quasi-varieties give an exact and rigorous structure to deal with all kinds of inconsistencies
2. Logical varieties allow modeling/realization of all other approaches to inconsistent knowledge. For instance, it is possible to use any kind of paraconsistent logics as components of logical varieties. In [16], it is demonstrated how logical varieties realize non-monotonic inference.
3. Theoretical results on logical varieties provide means for more efficient application of logical methods to problems in different areas (e.g. the application of the Logic of Reasonable Inferences (a logical variety) to represent and process contradicting opinions in the legal domain as described below).
4. Logical varieties allow partitioning of an inconsistent knowledge system into consistent parts and to use powerful tools of classical logic for reasoning.

5. Logical varieties allow utilization of different kinds of logic (multi-functionality) in the same knowledge system. For instance, it is possible to use a combination of the classical predicate calculus and non-monotonic calculus to represent two perspectives one of which is based on complete knowledge and the other on incomplete knowledge.
6. Logical varieties allow separation of different parts in a knowledge system and working with them separately.
7. Logical varieties provide means to reflect change of beliefs, knowledge and opinions without loss of previously existed beliefs, knowledge and opinions even in the case when new beliefs, knowledge and opinions contradict to what was before. In [18], they are applied to temporal databases.

These qualities of logic varieties are especially important for normative, in particular, legal, knowledge because this knowledge consists of a collection of formalized systems, a collection of adopted laws, a collection of existing traditions and precedents, and a collection of people's opinions. In addition, in the process of functioning, normative (legal) knowledge involves a variety of situational knowledge, beliefs and opinions. To analyze and use this diversity, it is necessary to have a flexible system that allows one to make sense of all different approaches without discarding them in an attempt to build a unique consistent system. To formalize these characteristics of normative knowledge, a form of a logical variety, the **Logic of Reasonable Inferences** (LRI) was developed [79]. The LRI was subsequently used as specification for the implementation of a *knowledge based system shell*, **Argumentator** [80]. This shell has consequently been used to acquire and represent legal knowledge. The resulting legal knowledge based system has been successfully used to test the empirical validity of the theory about legal reasoning and decision making modeled by the LRI [80].

It is interesting that several other systems used for inconsistency resolution, e.g., Multi-Context Systems [85], are also logical varieties and prevarieties. For instance, bridge rules used in Multi-Context Systems for non-monotonic information exchange are functions that glue together components of a logical variety or prevariety.

The goal of this work is the development of logical tools for working with inconsistency in knowledge systems. That is why we do not consider here specific approaches to inconsistency in data sets and databases. This is an important area and various means are used to perform data mining on inconsistent data. It was demonstrated that approaches based on *rough set theory* are especially useful for this task (cf. [42–45, 59, 65, 83]), as well as approaches based on *fuzzy set theory* (cf. [49, 71]). However, it is necessary to mention that data sets are model logical varieties and prevarieties (see Sect. 5) and it is possible to apply techniques based on model logical varieties and prevarieties to building tools for working with inconsistency in data sets and databases.

In the next sections we will describe labeling in logic (Sect. 4), logical varieties (Sect. 5) and the function of labels in logical varieties as a formal approach to handle inconsistency (Sect. 6), which overcomes the limitations of the other approaches described in this section.

4 Labeling in Logic and Named Sets

Labels are a kind of names and labeling is a specific naming. Namely, a *label* is a name of an object attached to this object. In a similar way, labeling is a process of naming studied in named set theory [22], and namely, it is a process of attaching the label to the object. This informal conception is formalized in the following way.

Let us consider a logical language L and a labeling structure A.

Definition 4.1 A *labeling* of the logical language L in the structure A is a partial mapping $l: L \to A$.

According to this definition, the mapping l assigns labels to logical formulas from L.

Treating labels in the context of names, we see that names, naming and named sets came to logic from the very beginning. To understand this, we start with the definition of a named set.

Definition 4.2 A *named set* (or a *fundamental triad*) **X** is the structure that consists of three components $\mathbf{X} = (X, f, I)$.

Each component of a named set (fundamental triad) plays its unique role and has a specific name. Namely, X is called the *support*, I is called the *component of names* or *set of names* and f is called the *naming correspondence* of the named set **X**.

This construction is prevalent in mathematics in general and in mathematical logic in particular. At the same time, *named sets* are abounding in many areas, in particular, in the legal domain. An example is elaborated on in Sect. 7.

Traditionally, mathematical logic is divided into three parts: *formal* (usually, axiomatic) *theories* (such as axiomatic set theories or formal arithmetic), *proof theory*, and *model theory*.

A formal theory includes a *logical calculus* and *logical semantics*. A *logical calculus* consists of a set A of initial statements called *axioms*, the *deductive system* as a set r of rules for correct inference, and a set T of deduced statements called *theorems*. *Logical semantics* studies inner interpretation (or truth evaluation) of logical systems for codification of the meaning of logical formulas in the form of truth-conditions, or possible truth conditions.

Proof theory studies the syntax of logic, that is, logical languages and means of logical inference, mostly, deduction. In this area, the main logical structures are propositions, predicates, logical operations (such as conjunction or disjunction), logical formulas, logical transformations (such as deduction or induction), logical calculi and syntactic or deductive logical varieties and prevarieties.

Model theory studies outer interpretations of logical systems in general structures called models.

At first, names and named sets were used only in model theory. For instance, statements of Aristotle's syllogistics contained variables and as it was demonstrated, variables are named sets.

A *variable* $\boldsymbol{x}$ is a named set (fundamental triad) $x = (x, r, D_x)$ where the support is the name of the variable, the set of names form the *domain* D_x of the variable $\boldsymbol{x}$, and correspondence r called the *interpretation mapping* connects the name x with the domain of this variable.

Names and consequently named sets play an imperative role in logic. For instance, Alonzo Church explains the importance of names for mathematical logic, contemplating the concept of name as one of the basic terms used in logic [27]. At the beginning of his book, Church clarifies the main concepts of logic. At first, he elucidates what *logic* is as a discipline. Then he makes clear what a *name* is in logic, taking into account only proper names and following the theory of (proper) names developed by Frege [33]. Only after this exposition, Church goes to such concepts as variables, constants, functions, propositions and so on. Thus, according to Church, the concept of a name is the first after the concept of logic. Shoenfield [75] also describes how names for individuals and expressions are constructed and used. To indicate importance of names in logic, Smullian [77] includes the word *name* in the title of his book, which provides a popular and informal introduction to logic.

Much later labels and corresponding named sets appeared in logical semantics with the advent of fuzzy logic. Namely, the truth value of a formula (statement) instead of taking two values {true, false} as in classical logics acquired multiple values from the interval [0, 1]. Each of these values is a number from the interval [0, 1], which means that the formula (statement) is true only to some degree. Thus, this degree c is a label, which displays the degree of truthfulness.

Fuzzy logics have been extensively studied, having many applications [5, 46, 58].

In proof theory, labels and corresponding named sets appeared in the labeled deductive systems introduced and studied by Gabbay [36]. *Labeled logics* and *labeled deductive systems* form a new and actively expanding direction in logic (cf. [37, 82]). Labeled logics use *labeled signed formulas* where labels (names of the formulas) are taken from information frames. As the result, the set of formulas in a labeled logic becomes an explicit named set, the support of which consists of logical formulas, while the set of names is an information frame, i.e., the system of labels. The derivation rules act on the labels as well as on the formulae, according to certain fixed rules of propagation. It means that derivation rules are morphisms (mappings) of the corresponding named sets.

Application of logic to programming gave birth to the direction called logic programming. Naturally, labeled logics brought forth labeled logic programs [6].

Special tools have been developed for working with names in logic, and naming always involves building new or transforming existing named sets (cf. Sect. 5.2). For instance, Gabbay and Malod [38] extend predicate modal and temporal logics introducing a special predicate $W(x)$, which names the world under consideration. Such a naming allows one to compare the different states the world (universe or individual) can be in after a given period of time, depending on the alternatives taken on the way. As Gabbay and Malod [38] remark, the idea of naming the worlds and/or time points goes back to Prior [68].

Labels play *two roles* in logic:

1. *Substitution labels* are used to represent logical formulas.
2. *Attached labels* provide additional information about logical formulas.

Substitutions are used in all kinds and types of logic. For instance, when we have a metaformula in the sense of [20] or logical schema in the sense of [54] $A \wedge B \rightarrow A$ of the propositional calculus, in it, A and B are substitution labels because it is possible to substitutes by arbitrary propositions obtaining a true formula of the classical propositional calculus. In essence, any variable in logic is a substitution label.

It is necessary to remark that a substitution is a kind of renaming [54], while renaming is one of the basic operations studied in the theory of named sets [22]. As a matter of fact, renaming is abounding in many areas, in particular, in the legal domain (see the example in Sect. 7).

There are *three modalities* of attached labels:

1. *Control labels* impact, e.g., control or direct, logical inference.
2. *Coordinated labels* are transformed in the process of inference according to deduction rules.
3. *Independent labels* and logical formulas are independently transformed in the process of inference.

Control labels have become rather popular in logic. For instance, in labeled deductive systems, consequence is defined by proof rules that work with both logical formulas and their labels [36], while in traditional logical systems, consequence is defined by proof rules only over the logical formulas. Examples of the use of different types of attached labels are given in Sect. 7.

There are also many possible functions of attached labels: Often attached labels form some structure, e.g., an algebra or information frame.

A labeled logical language L consists of pairs (φ, a) where φ is a logical formula and a is an attached label. Note that, in general, one label can consist of several other labels. For instance, if $a_1, a_2, \ldots, a_n$ are labels, then it is possible to treat the n–tuple $(a_1, a_2, \ldots, a_n)$ as another label.

Applications of logic in software technology on the one hand, e.g., in logic programming, and applications of computers to logic, e.g., in automated theorem proving, bring us to labeled inference languages. As Manna and Waldinger [54] write, proofs in formal theories that use simplistic application of deduction rules are usually far too expensive to be of practical value. So, it is necessary to build efficient proof strategies. Here labels come into the play. They can be used for labeling deduction rules in strategies, for naming proof strategies and organizing interaction between proof strategies.

Besides, proof strategies are specific types of second-order algorithms [23, 24]. As a result, labels can be used in second-order algorithms for the same purposes as in proof strategies.

In the next section we will describe the application of labels to logical varieties.

5 Structuring Inconsistency by Means of Labeled Logical Quasi-Varieties, Prevarieties, and Varieties

There are different types and kinds of logical varieties and prevarieties: *deductive or syntactic* varieties and prevarieties, *functional or semantic* varieties and prevarieties and *model or pragmatic* varieties and prevarieties [17]. Semantic logical varieties and prevarieties are formed by separating those parts that represent definite semantic units and determining logical semantics in these parts. In contrast to syntactic and semantic varieties, model varieties are essentially formalized structures. The structure of data in a distributed relational database is an example of logical model varieties.

Syntactic varieties, quasi-varieties and prevarieties are built from logical calculi as buildings are built from blocks. That is why, we, at first remind the concept of a logical calculus.

Let us consider a labeled logical language L and a labeled inference language R.

Definition 5.1 A *syntactic* or *deductive labeled logical calculus* is a triad (a named set) of the form $\mathbf{C}=(A,H,T)$ where $H\subseteq R$ and $A,T\subseteq L,A$ is the set of axioms, H consists of inference rules (rules of deduction) by which from axioms the theorems of the calculus are deduced, and the set of theorems T is obtained by applying algorithms/procedures/rules from H to elements from A.

Note that there are not only logical calculi. For instance, there are differential calculus and integral calculus in mathematical analysis.

Let $\mathbf{K}$ be a class of syntactic labeled logical calculi, K be a set of labeled inference rules, and $\mathbf{F}$ be a class of partial mappings from L to L.

Definition 5.2 A triad $\mathbf{M} = (A, K, M)$, where A and M are sets of expressions that belong to the labeled logical language L (A consists of axioms and M consists of theorems) and H is a set of inference rules, which belong to K, is called:

(1) a *labeled projective syntactic* (**K,F**)*-quasi-prevariety* if there exists a set of labeled logical calculi $\mathbf{C}_i=(A_i,H_i,T_i)$ from $\mathbf{K}$ and a system of mappings f_i: $A_i\to L$ and g_i: $M_i\to L (i\in I)$ from $\mathbf{F}$ in which A_i consists of axioms and M_i consists of some (not necessarily all) theorems of the logical calculus $\mathbf{C}_i$, i.e., $M_i\subseteq T_i$ and for which the equalities $A=\cup_{i\in I}f_i(A_i)$, $K=\cup_{i\in I}H_i$ and $M=\cup_{i\in I}g_i(M_i)$ are valid (it is possible that $\mathbf{C}_i=\mathbf{C}_j$ for some $i\neq j$).

(2) a *labeled syntactic* **K***-quasi-prevariety* if it is a projective syntactic (**K,F**)-quasi-prevariety where all mappings f_i and g_i that define $\mathbf{M}$ are inclusions, i.e., $A=\cup_{i\in I}A_i$ and $M=\cup_{i\in I}M_i$.

(3) a *labeled projective syntactic* (**K,F**)*-quasi-variety* with the depth k if it is a projective syntactic (**K,F**)-quasi-prevariety and for any $i_1, i_2, i_3, \ldots, i_k \in I$ either the intersections $\cap_{j=1^k}f_{ij}(A_{ij})$ and $\cap_{j=1^k}g_{ij}(T_{ij})$ are empty or there exists a calculus $C=(A,H,T)$ from $\mathbf{K}$ and projections $f\to\cap_{j=1^k}f_{ij}(A_{ij})$ and $g{:}\,N\to\cap_{j=1^k}g_{ij}(M_{ij})$ from $\mathbf{F}$ where $N\subseteq T$.

(4) a *labeled syntactic* **K**-*quasi-variety* with the depth k if it is a projective syntactic (**K,F**)- quasi-variety with depth k in which all mappings f_i and g_i that define **M** are bijections on the sets A_i and $M_{i,}$ correspondingly, and for all intersections (cf. Definition 4.2(3)), mappings f and g are bijections.

(5) a *(full) labeled projective syntactic* (**K,F**)-*quasi-variety* if for any $k > 0$, it is a projective syntactic (**K,F**)-*quasi-variety* with the depth k.

(6) a *(full) labeled syntactic* **K**-*quasi-variety* if for any $k > 0$, it is a **K**-quasi-variety with the depth k.

(7) a *labeled projective syntactic* (**K,F**)-prevariety if it is a projective syntactic (**K,F**)-quasi-prevariety in which $M_i = T_i$ for all $i \in I$.

(8) a *labeled syntactic* **K**-*prevariety* if it is a syntactic **K**-quasi-variety in which $M_i = T_i$ for all $i \in I$.

(9) a *labeled projective syntactic* (**K,F**)-*variety with the depth* k if it is a projective syntactic (**K,F**)-quasi-prevariety with the depth k in which $M_i = T_i$ for all $i \in I$.

(10) a *labeled syntactic* **K**-*variety with the depth* k if it is a projective syntactic (**K, F**)-quasi-variety with depth k in which $M_i = T_i$ for all $i \in I$ and for all intersections (cf. Definition 4.2(3)), $N = T$.

(11) a *(full) labeled projective syntactic* (**K,F**)-*variety* if for any $k > 0$, it is a projective syntactic (**K,F**)-variety with the depth k.

(12) a *(full) labeled syntactic* **K**-*variety* if for any $k > 0$, it is a **K**-variety with the depth k.

We see that the collection of mappings f_i and g_i makes a unified system called a prevariety or quasi-prevariety out of separate logical calculi $\boldsymbol{C}_{i,}$ while the collection of the intersections $\cap_{j=1^k} f_{ij}(A_{ij})$ and $\cap_{j=1^k} g_{ij}(T_{ij})$ makes a unified system called a variety out of separate logical calculi $\boldsymbol{C}_{i.}$ For instance, mappings f_i and g_i allow one to establish a correspondence between norms/laws that were used in one country during different periods of time or between norms/laws used in different countries.

The main goal of syntactic logical varieties is in presenting sets of formulas as a structured logical system using logical calculi, which have means for inference and other logical operations. Semantically, it allows one to describe a domain of interest, e.g., a database, knowledge of an individual or the text of a novel, by a syntactic logical variety dividing the domain in parts that allow representation by calculi.

In comparison with varieties and prevarieties, logical quasi-varieties and quasi-prevarieties are not necessarily closed under logical inference. This trait allows better flexibility in knowledge representation.

Definition 5.3 The calculi $\boldsymbol{C}_i$ used in the formation of the labeled syntactic quasivariety, prevariety or variety **M** are called *components* of **M**.

Note that it is possible that different components of a syntactic prevariety (variety) **M** have different deduction (inference) rules and even different types of deduction (inference) rules. For instance, some components have conventional, i.e., recursive, inference rules, while other components have super-recursive inference rules, which are more powerful [19]. If we consider the formal arithmetic *A*, then the famous result of Gödel asserts impossibility of proving consistency of the arithmetic *A* by recursive, inference rules [41]. However, later consistency of the arithmetic *A* was proved using more powerful superrecursive inference rules [1, 40, 74].

An example of a logical variety is a distributed database or knowledge base, each component of which consists of consistent knowledge/data. Then components of this knowledge/database are naturally represented by components of a logical variety. Moreover, very often data in a database have various labels. For example, in temporary database data elements have timestamps [21]. Besides, in one knowledge base different object domains may be represented. In these domains, some object may have properties that contradict properties of an object from another domain.

One more example of naturally formed logical varieties is the technique *Chunk and Permeate* built by [15]. This technique suggests to begin reasoning from inconsistent premises and proceeds by separating the assumptions into consistent theories (called *chunks* by the authors). These chunks are components of the logical variety shaped by them. After this, appropriate consequences are derived in one component (chunk). Then those consequences are transferred to a different component (chunk) for further consequences to be derived. This is exactly the way how logical varieties are used to realize and model non-monotonic reasoning [16]. Brown and Priest suggest that Newton's original reasoning in taking derivates in the calculus, was of this form.

An interesting type of logical varieties was developed in artificial intelligence and large knowledge bases. As Amir and McIlraith write [4, 52, 57], there is growing interest in building large knowledge bases of everyday knowledge about the world, comprising tens or hundreds of thousands of assertions. However working with large knowledge bases, general-purpose reasoning engines tend to suffer from combinatorial explosion when they answer user's queries. A promising approach to grappling with this complexity is to structure the content into multiple domain- or task-specific partitions. These partitions generate a logical variety comprising the knowledge base content. For instance, a first-order predicate theory or a propositional theory is partitioned into tightly coupled sub-theories according to the language of the axioms in the theory. This partitioning induces a graphical representation where a node represents a particular partition or sub-theory and an arc represents the shared language between sub-theories.

The technology of content partitioning allows reasoning engines to improve the efficiency of theorem proving in large knowledge bases by identifying and exploiting the implicit structure of the knowledge [4, 52, 57]. The basic approach is to convert a graphical representation of the problem into a tree-structured representation, where each node in the tree represents a tightly-connected sub-problem, and the arcs represent the loose coupling between sub-problems. To maximize the effectiveness of partition-based reasoning, the coupling between partitions is minimized, information being passed between nodes is reduced, and local inference within each partition is also minimized.

Additional advantage of partitioning is a possibility to reason effectively with multiple knowledge bases that have overlap in content [4].

The tools and methodology of content partitioning and thus, implicitly of logical varieties are applied for the design of logical theories describing the domain of robot motion and interaction [3]. Other examples are given in Burgin and de Vey Mestdagh [25, 26].

Concepts of logical varieties and prevarieties provide further formalization for local logics of Barwise and Seligman [7], many-worlds model of quantum reality of Everett [30, 32], and pluralistic quantum field theory of Smolin related to the many-worlds theory [76].

In the next section we will describe the functions of labels in logical varieties

6 Labels and Their Functions in Logical Varieties

Labels can perform multiple functions in logical varieties:

1. Labels that indicate the source of information (knowledge) represented by a component of a logical variety, assuming that each component of a logical variety is connected to a source of information (knowledge). For instance the source of a particular observation or derivation, a specific legal source (legislation, case law, expert opinion, etc.);
2. Labels that show the reliability of (the source of) information (knowledge). For instance an accepted fact finding method, a valid legal source (see Sect. 7), the reliability of a witness, etc.;
3. Labels that refer the logical formulas to the problem-knowledge which these formulas represent. For instance, items, e.g. formulas, related to the same source (a person, a body, etc.) can have the same label (see Sect. 7);
4. Labels that estimate the relevance of logical formulas to the corresponding problems;
5. Labels that estimate the importance of information (knowledge) to the related conjectures;
6. Labels that refer the logical formulas to the topics (attributes) knowledge about which these formulas represent;
7. Labels that estimate the relevance of logical formulas to the corresponding topics (attributes);
8. Labels that describe relations between objects and people involved in the process;
9. Labels that estimate the importance of information (knowledge);
10. Labels that refer the logical formulas to conjectures (see for example, Sect. 7);
11. Labels that estimate importance of information (knowledge) to the related conjectures;

12. Labels that estimate probability of conjectures represented by logical formulas in a logical variety;
13. Labels that show the degree of support of a conjecture or belief. Utilization of such labels is described in [35, 47];
14. Labels that describe connections and relations of a formula;
15. Labels that indicate where information represented by a formula is stored;
16. Labels that show time of information creation and/or acceptance (see for example, Sect. 7);
17. Labels that show time of information validity.

Labeling in information systems allows users to discern when the same information represented by a formula comes from different sources. Thus, it is possible that one and the same formula has different labels. Such a formula can belong to different components or even to the same component of the syntactic logical quasivariety, prevariety or variety. For instance, when a label indicates the reliability of information sources and the source 1 has the reliability 5, while the source 1 has the reliability 7, then a formula φ coming from the source 1 has the reliability label 5, while coming from the source 1 has the reliability label 7.

Let us consider some types of labeling.

Definition 6.1 A labeling $l: M \to A$ of a syntactic logical quasivariety (prevariety or variety) $\mathbf{M} = (A, K, M)$ is called:

(1) *structured* if formulas from different components of **M** have different labels;
(2) *discerning* if different elements from M have different labels.
(3) *univalent* if any element from M has, at most, one label.

An example of structured labeling of logical quasivarieties is a locally stratified knowledge base or database in the approach of Benferhat and Garcia [11] who suggest handling inconsistent knowledge bases by local stratification based on priorities, such as a preference relation or reliability relation, which are assigned to knowledge (data) items, allowing to provide more meaningful and reliable information to the user. Priority values are used as labels of formulas, while priorities stratify the database or its part forming a logical variety, in which a component is determined by a fixed level (value) of priority. Thus, priority labeling is structured.

Proposition 6.1. A syntactic quasivariety $\mathbf{M} = (A, K, M)$ has a structured labeling if it is discrete.

Corollary 6.1. A syntactic prevariety (variety) $\mathbf{M} = (A, K, M)$ has a structured labeling if it is discrete.

An important operation in labeled logical structures, such as logical deductive systems, logical calculi and logical varieties, is relabeling or changing labels. It is a particular case of renaming studied in the theory of named sets [22].

Labels allow logicians to connect semantic logical varieties studied in [17] and syntactic logical varieties studied here.

Let us consider a labeled logical language L and a structure Q. Note that it is possible to treat a logical language without labels as a specific labeled logical language, namely, as logical language formulas of which have empty labels.

Definition 6.2. A named set Sem $= (L, c, Q)$ where $c: L \to Q$ is a partial mapping is called a *semantics* in the language L with the scale Q.

The most popular semantic scale is the truth-scale $Q = \{\mathrm{T}, \mathrm{F}\}$ used in classical logics. Another popular semantic scale $Q = \{\mathrm{T}, \mathrm{F}, \mathrm{U}\}$ where U means *undefined* or *unknown* is used in intuitionistic logics.

Semantics in a labeled logical language allows us to define semantic calculi. Let us consider a subset E of the structure Q.

Definition 6.3. A labeled semantic E-calculus $\mathbf{C}$ is the name set $\mathbf{C} = (C, c, E)$ in which the subset C of L defined in the following way

$$C = \{\varphi \in L; c(\varphi) \in E\}$$

Let $\mathbf{K}$ be a class of semantic labeled logical calculi, Q be a semantic scale, and $\mathbf{F}$ be a class of partial mappings.

Definition 6.4. (a) A *projective labeled semantic logical* ($\mathbf{K}$, $\mathbf{F}$)*-prevariety* $\mathbf{M}$ = (M, k, E) is the named set components of which are defined in the following way:

There are semantic labeled logical calculi $\mathbf{C}_i = (C_i, c_i, E_i)$ from $\mathbf{K}$ and three systems of mappings $c_i: C_i \to E_i$, $k_i: E_i \to E$ and $g_i: M_i \to L$ from $\mathbf{F}$ such that

$$M = \cup_{i \in I} g_i(M_i), E = \cup_{i \in I} k_i(E_i) \text{ and } k = \cup_{i \in I} k_i c_i$$

(b) A *labeled semantic logical* $\mathbf{K}$*-prevariety* is a projective labeled semantic logical ($\mathbf{K}$, $\mathbf{F}$)-prevariety $\mathbf{M}$ = (M, k, E) in which all g_i and k_i are inclusions, i.e., $M = \cup_{i \in I} M_i$ and $E = \cup_{i \in I} E_i$.

Definition 6.5. A *labeled semantic logical* $\mathbf{K}$*-variety* is a labeled semantic logical $\mathbf{K}$-prevariety $\mathbf{M}$ = (M, k, E) in which all intersections $\cap_{j=1^k} M_{ij}$ are empty or there exists a calculus $\boldsymbol{C} = (\boldsymbol{C}, \boldsymbol{c}, \boldsymbol{H})$ from $\mathbf{K}$ such that $\cap_{j=1^k} M_{ij} = C, \cap_{j=1^k} E_{ij} = H$ and each restriction of c on M_{ij} coincides with c_{ij}.

Definition 6.6. Labeling $l: M \to L$ of a syntactic logical quasivariety, prevariety or variety $\mathbf{M} = (A, K, M)$ is called *semantic* if there is a semantic logical prevariety or variety $\mathbf{M}_s = (N, k, E)$ such that $N = M$ and $l(\varphi) = k(\varphi)$ for any formula φ from M.

Definitions imply the following result.

Proposition 6.2 For any syntactic logical variety $\mathbf{M} = (A, K, M)$ and any semantic logical variety $\mathbf{M}_s = (N, k, E)$, there is a semantic labeling of $\mathbf{M}$ defined by $\mathbf{M}_s$.

Control labels can essentially change properties of logical calculi and logical varieties.

By the famous Gödel theorem, the first-order formal arithmetic (Peano arithmetic) A is undecidable [41], and in particular, it has formulas that are true in the natural semantics of A but are not recursively provable (recursively deducible). An appropriate labeling can change this situation.

Proposition 6.3 There is a binary labeling l of all formulas in the language of A and a deduction operator such that all true in A formulas and only true in A formulas are recursively provable.

Proof. Let us take the standard semantic scale of labeling $Q = \{\mathrm{T}, \mathrm{F}\}$ and a formula φ from the language $L(A)$ of the arithmetic A. We attach the label T to φ when φ is true in A and the label F when φ is false in A. After doing this, we define the following deduction rule

$$(\varphi, d) \Vdash (\psi, c) \text{ if and only if } c = d = T$$

We can see that the labeled calculus $(D, \{\Vdash\}, H)$ where the set of axioms D consists of single axiom $(n = n, \mathrm{T})$ and H is the set of all theorems deduced from D by means of the deduction operator $\Vdash$ is complete and consistent assuming that A is consistent.

Proposition is proved.

In the above example, labeling extends provability making deduction more powerful because in traditional logics without labeling, it is impossible to build sufficiently powerful complete and consistent calculus as the first Gödel theorem tells us. At the same time, it is possible not only to extend power of deduction but also to restrict provability by means of an appropriate labeling as the following example demonstrates.

Let us consider a formal theory P with the classical deduction operator $\vdash$, which includes application of standard deduction rules, such as $(\varphi, \varphi \to \psi) \Rightarrow \psi$, and the set of axioms A, and take the classical semantic scale $Q = \{\mathrm{T}, \mathrm{F}\}$, We attach the label T to a formula φ from the language $L(P)$ of the theory P when φ is true in A and the label F when φ is false in A. For a set of formulas D and a formula φ from $L(P)$, we define another deduction rule

$D \Vdash (\varphi, d)$ if and only if $D \vdash \varphi$ in P and there are no formulas in D with the label F.

This gives us a syntactic calculus $\mathbf{C} = (A, \{\Vdash\}, T)$ where T are all theorems deduced from A by means of the deduction operator $\Vdash$. In this calculus, false formulas do not participate in the inference and thus, even when A or T has contradictions (false formulas), it does not cause explosion, i.e., T is not necessarily equal to L. Note that in a general case, $\mathbf{C}$ is a kind of a paraconsistent logical calculus.

Various other applications of control labels to enhancement of inference processes are described in the area of labelled deductive systems [36, 37].

In the next section we will apply a species of a labeled logic variety (the extended Logic of Reasonable Inferences) to an example from the legal domain.

7 The Logic of Reasonable Inferences, an Implementation of a Labeled Logical Variety in the Legal Domain

Our knowledge of the world is always perspective bound and therefore fundamentally inconsistent, even if we agree to a common perspective, because this agreement is necessarily local and temporal due to the human epistemic condition. The natural inconsistency of our knowledge of the world is particularly manifest in the legal domain (de Vey Mestdagh et al. 2011).

In the legal domain, on the object level (that of case facts and opinions about legal subject behavior), alternative (often contradicting) legal positions compete. All of these positions are a result of reasoning about the facts of the case at hand and a selection of preferred behavioral norms presented as legal rules. At the meta-level meta-positions are used to make a choice for one of the competing positions (the solution of an internal conflict of norms, a successful subject negotiation or mediation, a legal judgement). Such a decision based on positions that are inherently local and temporal is by definition also local and temporal itself. The criteria for this choice are in most cases based on legal principles. We call these legal principles *metaprinciples* because they are used to evaluate the relations between different positions at the object level.

To formalize this natural characteristic of (legal) knowledge we developed the Logic of Reasonable Inferences (LRI, [79]). The LRI is a logical variety that handles inconsistency by preserving inconsistent positions and their antecedents using as many independent predicate calculi as there are inconsistent positions [25, 26]. The original LRI was implemented and proved to be effective as a model of and a tool for knowledge processing in the legal domain [80]. In order to be able to make inferences about the relations between different positions (e.g. make local and temporal decisions), labels were added to the LRI. In de Vey Mestdagh et al. (2011) formulas and sets of formulas are named and characterized by labelling them in the form (A_i, H_i, P_i, C_i). These labels are used to define and restrict different possible inference relations (***Axioms*** A_i and ***Hypotheses*** H_i, i.e. ***labeled signed formulas*** and ***control labels***, see Sect. 4) and to define and restrict the composition of consistent sets of formulas (***Positions*** P_i and ***Contexts*** C_i). Formulas labeled A_i must be part of any position and context and therefore are not (allowed to be) inconsistent. Formulas labeled H_i can only be part of the same position or context if they are mutually consistent. A set of formulas labeled P_i represents a ***position***, i.e. a consistent set of formulas including all Axioms (e.g., ***a perspective on a world, without inferences about that world***). A set of formulas labeled C_i represents a ***context*** (a maximal set of consistent formulas within the (sub)domain and their justifications, c.f. ***the world under consideration***, see Sect. 4). All these labels can be used as predicate variables and if individualized to instantiate predicate variables and consequently as constants (***variables as named sets*** see Sect. 4). Certain metacharacteristics of formulas and pairs of formulas were finally described by labels (e.g., metapredicates like Valid, Excludes, Prefer) describing some of their legal source characteristics and their legal relations which could be used to rank the different positions externally. The ***semantics*** of these three Predicates (Valid, Exclude and Prefer) are described in de Vey Mestdagh et al. [81]. These three predicates describe the elementary relations between legal positions that are prescribed by the most fundamental sets of legal principles (i.e. principles regarding the legal validity of positions, principles regarding the relative exclusivity of legal positions even if they do not contradict each other and principles regarding the preference of one legal position over another). It was also demonstrated that the first requirement of Sect. 2, a *description of our cognition should allow for reasoning about (mutually exclusive) alternatives*, is attained by the LRI.

In this section we show that labels can be used formally to describe the ranking process of positions and contexts. With that the second requirement of Sect. 2 will be also satisfied, namely, *a description of our cognition should allow for local and temporal decisions for a certain alternative, which means without discarding the non-preferred alternatives like belief revision does and without using the mean of all alternatives like probabilistic logics do.* This extends the LRI from a logical variety that could be used to formalize the non-explosive inference of inconsistent contexts (opinions) and naming (the elements of) these contexts to a labeled logical variety, in which tentative decisions can be formally represented by using a labelling that allows for expressing the semantics of the aforementioned meta-predicates and prioritizing (***priority labelling***). We illustrate the use of these labels by examples.

In de Vey Mestdagh et al. 2011 (Sect. 4), a complex legal case is presented about a threat to the **tower of Pisa**. It would take too much space to repeat the informal and the full formal case description here. Just the necessary parts of the formal case description are repeated here. The full case description can be read at and downloaded from:

http://www.rug.nl/staff/c.n.j.de.vey.mestdagh/inconsistent_knowledge_as_a_natural_phenomenon_cnjdvm_def-en-reftobook.pdf (p. 29, Sect. 4).

As a part of this case four existing legal positions (P_i) consisting of Axioms (A_i) and Hypotheses (H_i) are presented:

P1	Subjectivist's view	: A1..A11; H1, H4(H1), H7(H1,H2)
P2	Objectivist's view	: A1..A11; H2, H4(H2), H5(H2), H6(H2,H1)
P3	Impressionist's view	: A1..A11; H3, H7(H3,H2)
P4	Subjectivist's + Impressionist's	: A1..A11; H1, H3, H4(H1), H7(H1,H2), H7(H3, H2)

It is shown that these positions can be formally reduced to two mutually inconsistent contexts (general perspectives C1 justified by P1, P3, P4 and C2 justified by P2):

C1	Reasonable inference	: Criminal_attempt(One_of_us,Toppling_Top)
	Justification	: P1, P3, P4
C2	Reasonable inference	: ¬Criminal_attempt(One_of_us,Toppling_Top)
		Valid(H1), Valid(H2), Excludes(R2,R1)
	Justification	: P2

There are several possible perspectives on the criteria used to decide between these two contexts. In this Section four of these tentative ***'decision models'*** and a combination of them are described. We use the term 'tentative', because the decision model does not discard contexts but just adds inferences about their preference to the contexts. The decision models are extensions of and are applied to the theorems of the case presented in de Vey Mestdagh et al. 2011.

Decision model 'Prefer the most valid position'

Prefer a minimal position which contains more *uninstantiated* valid basic formulas over a position that contains less *uninstantiated* valid basic formulas (the minimal position predicate and the uninstantiated formula predicates are left out to preserve readability).

A6 Precedent(H1), Precedent(H2)
A7 Complies_with_section_1_Penal_Code(H2)
H4 $\text{Precedent}(H_i) \rightarrow \text{Valid}(H_i)$ [*meta – rule precedent validity*]
H5 $\text{Complies\_with\_section\_1\_Penal\_Code}(H_i) \rightarrow \text{Valid}(H_i)$ [*meta – rule principle of legality*]
H8 $\neg\,(\text{Precedent}(H_i) \vee \text{Complies\_with\_section\_1\_Penal\_Code}(H_i)) \rightarrow \neg\text{Valid}(H_i)$
H9 $\text{In}(P_i, H_i) \wedge \text{Valid}(H_i) \rightarrow \text{Valid}(P_i)$
H10 $\text{In}(P_i, H_i) \wedge \neg\text{Valid}(H_i) \rightarrow \neg\text{Valid}(P_i)$
H11 $\text{Valid}(P_i) \wedge \neg\text{Valid}(P_j) \rightarrow \text{Prefer}(P_i, P_j)$

On the basis of this extended set of rules hypotheses H1 and H2 are just valid and hypothesis H3 is just invalid (all hypotheses contained in de positions P1, P2, P3 and P4 presented above). As a consequence the positions P1 (just valid uninstantiated formulas), P2 (just valid uninstantiated formulas) and P4 (valid and invalid uninstantiated formulas) are preferred over P3 (just invalid uninstantiated formulas). The double evaluation of P4 (valid and invalid) causes a split of contexts.

Decision model 'Prefer the legal position'

Prefer a position which contains more uninstantiated legal basic formulas (i.e. a formula based on legislation) over a position that contains less uninstantiated legal basic formulas:

A7 Complies_with_section_1_Penal_Code(H2)
H5 Complies_with_section_1_Penal_Code(Hi) → Valid(Hi) [*meta – rule principle of legality*]
H12 Complies_with_section_1_Penal_Code(Hi) → Legal(Hi)
H13 ¬ Complies_with_section_1_Penal_Code(Hi) → ¬Legal(Hi)

H14 $In(Pi, Hi) \wedge Legal(Hi) \rightarrow Legal(Pi)$
H15 $In(Pi, Hi) \wedge \neg Legal(Hi) \rightarrow \neg Legal(Pi)$
H16 $Legal(Pi) \wedge \neg Legal(Pj) \rightarrow Prefer(Pi, Pj)$

On the basis of this set of rules P2 (containing a legal uninstantiated formula), is preferred over P1, P3 and P4 (containing no legal uninstantiated formulas).

Decision model 'Prefer the specific position'

Prefer a position which contains the more specific uninstantiated legal basic formulas over a position that contains the more general uninstantiated legal basic formulas.

A8 $Specialis(H2, H1)$
H6 $Specialis(H_i, H_j) \rightarrow Excludes(H_i, H_j)$ [*meta – rule lex specialis*]
H18 $In(P_i, H_i) \wedge In(P_j, H_j) \wedge H6\,Excludes(H_i, H_j) \rightarrow Prefer(P_i, P_j)$

On the basis of this set of rules P2 (containing the specialis H2) is preferred over P1 and over P4 (both containing the generalis H1)

Decision model 'Prefer the position which soothes public alarm' (terrorism induced policies)

A decision model which is popular in this time of perceived terror threats is 'Prefer a position which soothes public alarm'.

A5 Recent_public_alarm(Toppling_Top)
A9 Soothes_public_alarm(H1, Toppling_Top)
A10 ¬Soothes_public_alarm(H2, Toppling_Top)
A11 Soothes_public_alarm(H3, Toppling_Top)
H7 Recent_public _alarm(y) ∧ Soothes_public_alarm(H_i, y) ∧ ¬Soothes_public_alarm(H_j, y) → Prefer(H_i, H_j) [*meta – rule terrorism induced policies*]
H19 $In(P_i, H_i) \wedge In(P_j, H_j) \wedge H7\,Prefer(H_i, H_j) \rightarrow Prefer(P_i, P_j)$

On the basis of this set of rules P1, P3 and P4 (based on one or two of the soothing rules H1 and H3) are preferred over P2 (based on the non-soothing rule H2).

Decision model 'Prefer the context with the most preferred position as a justification'

The LRI adds the inferred preference relations by definition to the contexts (cf. de Vey Mestdagh et al. 2011). The two contexts C1 and C2 described in de Vey Mestdagh et al. 2011 split into four contexts: two (equal to the originals) extended with valid formulas P4 and two (C3 and C4, equal to the originals C1 and C2) with invalid formulas P4. A simple decision model just counts the number of times a position is preferred and prefers the context(s) with the most preferred position as a justification. Position P1 is preferred 3 times in the four decision models, position

P2 6 times, position P3 once and position P4 twice. In the contexts the occurrence of the preference for position P2 is reduced to 4 times and the other preferences stay the same (less than 4), because equal inferences are just registered once (so a more elegant decision model would use the number of preferred positions based on the different decision models, but the LRI does not allow for labelling the decision models because they do not belong to the formal part of the logic).

The following decision rule will conclude to prefer C2 (and C4) over C1 (and C2):
Most_preferred(Pi) ∧ In_justification(Ci,Pi) ∧ ¬ In_justification(Cj,Pi) → Prefer (Ci,Cj)

The reasonable inference in both C2 and C4 is that it is no criminal attempt if one of us tries to topple the Tower of Pisa just by pushing it, which is actually the dominant (but not the only!) opinion of the majority of Dutch courts. Note that this preference is just added to (in this case) all contexts and that none of the contexts is discarded because of the preference. Actually the logic does not decide but just suggests a decision (that a human or computer can process further). This is exactly what we want. Interestingly enough in a more complex case the LRI can derive multiple preference relations between contexts including contradictory ones. In that case metapredicates can be added to evaluate the preference relation between these preference relations (if we had the brains this would end up in an infinite recursion, but fortunately we lawyers as all humans are fairly limited in our capacity to think new metalevels of decision making through).

8 Conclusion

In this paper, we have described the theory and application of labelled logical varieties. It was demonstrated that labelled logical varieties provide means for overcoming some severe restrictions of common logical approaches to inconsistency. It allows to preserve all knowledge positions instead of discarding them (like belief revision) and instead of replacing them by an aggregate such as a mean (probabilistic approaches).

Labelled logical varieties allow us to formally describe and to model the underlying reality of:

1. *Reasoning about (mutually exclusive) alternatives without discarding or adapting them*
2. *Deciding locally and temporally for a certain alternative*

The preservation of all alternatives does satisfy the epistemic need to conserve inconsistency (natural inconsistent perspectives) and in doing that can en passant satisfy the practical need to handle unwanted inconsistency when it cannot be eliminated.

We think that this is in accord with the human epistemic condition, which does not know and does not accept universal and infinite closure under any circumstance.

Further research will focus on the extension and application of labelled logical varieties in knowledge systems, particularly in the legal domain. It would also be useful to develop techniques based on model logical varieties and prevarieties for building tools that would allow working with inconsistency in data sets and databases.

References

1. Ackermann, W.: Zur Widerspruchsfreiheit der Zahlentheorie. Math. Ann. **117**, pp. 162–194 (1940)
2. Amgoud, L., Cayrol, C.: On the acceptability of arguments in preference-based argumentation. In: Proceedings of the 14th Conference on Uncertainty in Artificial Intelligence (UAI'98), pp. 1–7 (1998)
3. Amir, E.: Dividing and Conquering Logic. Ph.D. thesis, Stanford University, Computer Science Department, 2002
4. Amir, E., McIlraith, S.: Partition-based logical reasoning for first-order and propositional theories. Artif. Intell. **162**(1/2), 49–88 (2005)
5. Bandemer, H., Gottwald, S.: Fuzzy Sets, Fuzzy Logic, Fuzzy Methods With Applications. Wiley, New York (1996)
6. Barker, C.: Continuations in natural language. In: Thielecke, H. (ed.), Proceedings of the 4th Continuations Workshop, pp. 55–64. Technical report CSR-04-1, School of Computer Science, University of Birmingham (2004)
7. Barwise, J., Seligman, J.: Information Flow: The Logic of Distributed Systems. Cambridge Tracts in Theoretical Computer Science 44, Cambridge University Press (1997)
8. Benferhat, S., Dubois, D., Prade. H.: Representing default rules in possibilistic logic. In: Proceedings of the 3rd International Conference of Principles of Knowledge Representation and Reasoning (KR'92), pp. 673–684 (1992)
9. Benferhat, S., Cayrol, C., Dubois, D., Lang, J., Prade, H.: Inconsistency management and prioritized syntax-based entailment. In: Proceedings of the 13th International Joint Conference on Artificial Intelligence (IJCAI'93), pp. 640–645 (1993)
10. Benferhat, S., Dubois, D., Prade, H.: How to infer from inconsistent beliefs without revising? In: Proceedings of the 14th International Joint Conference on Artificial Intelligence (IJCAI'95), pp. 1449–1455 (1995)
11. Benferhat, S., Garcia, L.: Handling locally stratified inconsistent knowledge bases. Stud. Logica. **70**, 77–104 (2002)
12. Bertossi, L.E., Hunter, A., Schaub, T. (eds.): Inconsistency Tolerance. LNCS, vol. 3300, Springer, Heidelberg (2005)
13. Besnard, P., Hunter, A.: Quasi-classical logic: non-trivializable classical reasoning from inconsistent information. In: Proceedings of ECSQARU'95, LNAI, vol. 946, pp. 44–51 (1995)
14. Brewka, G.: Preferred subtheories: an extended logical framework for default reason. In: Proceedings of the 11th Int Joint Conference on Artificial Intelligence (IJ CAI'89), pp. 1043–1048 (1989)

15. Brown, B., Priest, G.: Chunk and permeate: a paraconsistent inference strategy, part I: the infinitesimal calculus. J. Philos. Logic **33**(4), 379–388 (2004)
16. Burgin, M.: Logical methods in artificial intelligent systems (in Russian). In: Vestnik of the Computer Society, No. 2, pp. 66–78 (1991)
17. Burgin, M.: Logical varieties and covarieties. (in Russian) In: Methodological and Theoretical Problems of Mathematics and Information and Computer Sciences, pp. 18–34. Kiev (1997)
18. Burgin, M.: Logical Tools for Program Integration and Interoperability. In: Procedings of the IASTED International Conference on Software Engineering and Applications, pp. 743–748, MIT, Cambridge (2004)
19. Burgin, M.: Super-recursive algorithms. Monographs in computer science. Springer, New York (2005). ISBN 0-387-95569-0
20. Burgin, M.: Languages, Algorithms, Procedures, Calculi, and Metalogic. Preprint in Mathematics LO/0701121, 31 pp. (electronic edition: http://arXiv.org) (2007)
21. Burgin, M.: structural organization of temporal databases. In: Proceedings of the 17th International Conference on Software Engineering and Data Engineering (SEDE-2008), ISCA, pp. 68–73, Los Angeles, California (2008)
22. Burgin, M.: Theory of Named Sets. Nova Science Publishers, New York (2011)
23. Burgin, M., Debnath, N.: Reusability as design of second-level algorithms. In: Proceedings of the ISCA 25th International Conference "Computers and their Applications" (CATA-2010), ISCA, pp. 147–152. Honolulu, Hawaii (2010)
24. Burgin, M., Gupta, B.: Second-level algorithms, superrecursivity, and recovery problem in distributed systems. Theor. Comput. Syst. **50**(4), 694–705 (2012)
25. Burgin, M., de Vey Mestdagh, C.N.J.: The Representation of Inconsistent Knowledge in advanced knowledge based systems. In: Koenig, A., Dengel, A., Hinkelmann, K., Kise, K., Howlett, R.J., Jain, L.C. (eds.) Knowlege-Based and Intelligent Information and Engineering Systems, vol. 2, pp. 524–537. Springer, ISBN 978-3-642-23862-8 (2011)
26. Burgin, M., de Vey Mestdagh, C.N.J.: Consistent structuring of inconsistent knowledge. In: Journal of Intelligent Information Systems, pp. 1–24. Springer US, Sept 2013
27. Church, A.: Introduction to Mathematical Logic. Princeton University Press, Princeton (1956)
28. Da Costa, N.C.A.: Calcul propositionnel pour les systemes formels inconsistants. Compte Rendu Academie des Sciences (Paris) **257**, 3790–3792 (1963)
29. Dalal, M.: Investigations into a theory of knowledge base revision: preliminary report. In: Proceedings of the Seventh National Conference on Artificial Intelligence (AAAI'88), pp. 475–479 (1988)
30. DeWitt, B.S.: The Many-Universes interpretation of quantum mechanics. In: Foundations of Quantum Mechanics, pp. 167–218. Academic Press, New York (1971)
31. Dung, P.M.: On the acceptability of arguments and its fundamental role in non-monotonic reasoning, logic programming and n-person games. Artif. Intell. **77**, 321–357 (1995)
32. Everett, H.: 'Relative State' formulation of quantum mechanics. Rev. Mod. Phys. **29**, 454–462 (1957)
33. Frege, F.L.G.: Über Sinn und Bedeutung. In: Zeitschrift für Philosophie und philosophische Kritik, pp. 25–50 (1892)
34. Friedman, N., Halpern, J.Y.: A knowledge-based framework for belief change, part II: revision and update. In: Proceedings of the Fourth International Conference on the Principles of Knowledge Representation and Reasoning (KR'94), pp. 190–200 (1994)
35. Gabbay, D.M.: Labelled deductive systems and the informal fallacies. In: Van Eemeren F.H. et al. (eds.). Proceedings of the 3rd International Conference on Argumentation, vol. 2: Analysis and Evaluation, Sponsored by ISSA, International Society for the Study of Argumentation, pp. 308–319 (1994)
36. Gabbay, D.M.: Labelled deductive systems. Oxford Logic Guides, vol. 33, Clarendon Press/Oxford Science Publications, Oxford (1996)

37. Gabbay, D.M., D'Agostino, M.: A generalization of analytic deduction via labelled deductive systems part 1: basic substructural logics. J. Autom. Reasoning **13**, 243–281 (1994)
38. Gabbay, D.M., Malod, G.: Naming worlds in modal and temporal logic. J. Log. Lang. Inf. **11**, 29–65 (2002)
39. Gärdenfors, P., Rott, H.: Belief revision. In: Handbook of Logic in Artificial Intelligence and Logic Programming, vol. 4, pp. 35–132. Oxford University Press (1995)
40. Gentzen, G.: Die Widerspruchfreiheit der reinen Zahlentheorie. Math. Ann. **112**, 493–565 (1936)
41. Gödel, K.: Über formal unentscheidbare Sätze der Principia Mathematica und verwandter Systeme I. Monatsh. Math. Phys. **38**(1), 173–198 (1931–1932)
42. Gogoi, P., Das, R., Borah, B., Bhattacharyya, D.K.: Efficient rule set generation using rough set theory for classification of high dimensional data. Int. J. Smart Sens. Ad Hoc Netw. (IJSSAN) **1**(2), 13–20 (2011)
43. Greco, S., Matarazzo, B., Slowinski, R., Stefanowski, J.: Variable consistency model of dominance-based rough sets approach. In: Rough Sets and Current Trends in Computing. Lecture Notes in Computer Science, vol. 2005, pp. 170–181 (2001)
44. Grzymala-Busse, J.W.: Knowledge acquisition under uncertainty-A rough set approach. J. Intell. Rob. Syst. **1**, 3–16 (1988)
45. Grzymala-Busse, J.W.: Rough set theory with applications to data mining. Real World Appl. Comput. Intell. Stud. Fuzziness Soft Comput. **179**, 221–244 (2005)
46. Hájek, P.: Metamathematics of Fuzzy Logic. kluwer Academic publishers, Dordrecht (1998)
47. Hunter, A., Liu, W.: Knowledge base stratification and merging based on degree of support. In: Quantitative and Qualitative Approaches to Reasoning and Uncertainty (ECSQARU'09), LNCS, vol. 5590, pp. 383–395. Springer (1994)
48. Jaśkowski, S.: Rachunek zdań dla systemów dedukcyjnych sprzecznych. Studia Societatis Scientiarun Torunesis (Sectio A) **1**(5), 55–77 (1948)
49. Kong, H., Xue, G., He X., Yao, S.: A proposal to handle inconsistent ontology with fuzzy OWL. In: 2009 WRI World Congress on Computer Science and Information Engineering, pp. 599–603 (2009)
50. Lehmann, D.: Another perspective on default reasoning. Ann. Math Artif. Intell. **15**, 61–82 (1995)
51. Lehmann, D.: Belief revision, revised. In: Proceedings of the Fourteenth International Joint Conference on Artificial Intelligence (IJCAI'95), pp. 1534–1540 (1995)
52. MacCartney, B., McIlraith, S.A., Amir, A., Uribe, T.: Practical partition-based theorem proving for large knowledge bases. In: Proceedings of the Eighteenth International Joint Conference on Artificial Intelligence (IJCAI-03), pp. 89–96 (2003)
53. Makinson, D.: Bridges from Classical to Nonmonotonic Logic. College Publications (2005)
54. Manna, Z., Waldinger, R.: The Deductive Foundations of Computer Programming. Addison-Wesley, Boston/New York/Toronto (1993)
55. Marek, W., Truszczynski, M.: Nonmonotonic Logics: Context-Dependent Reasoning. Springer, New York (1993)
56. McDermott, D., Doyle, J.: Non-monotonic logic I. Artif. Intell. **25**, 41–72 (1980)
57. McIlraith, S., Amir, E.: Theorem proving with structured theories. In: Proceedings of the 17th International Joint Conference on Artificial Intelligence, (IJCAI'01), pp. 624–631 (2001)
58. McNeill, D., Freiberger, P.: Fuzzy Logic. Simon and Schuster (1993)
59. Mollestad, T., Skowron, A.: A rough set framework for data mining of propositional default rules. In: Proceedings of the 9th International Symposium on Foundations of Intelligent Systems, pp. 448–457 (1996)
60. Nebel, B.: Belief revision and default reasoning: syntax-based approaches. In: Proceedings of the Second International Conference on the Principles of Knowledge Representation and Reasoning (KR'91), pp. 417–428 (1991)

61. Nebel, B.: Base revision operations and schemes: semantics, representation and complexity. In: Proceedings of the Eleventh European Conference on Artificial Intelligence (ECAI'94), pp. 341–345 (1994)
62. Nguen, N.T.: Inconsistency of knowledge and collective intelligence. Cybern. Syst. **39**(6), 542–562 (2008)
63. Papini, O.: A complete revision function in propositional calculus. In: Proceedings of the 10th European Conference on Artificial Intelligence (ECAI'92) (1992)
64. Partridge, D., Wilks, Y.: The Foundations of Artificial Intelligence. Cambridge University Press, Cambridge (1990)
65. Pawlak, Z.: Rough set and intelligent data analysis. Int. J. Inform. Sci. **147**, 1–12 (2002)
66. Pollock, J.L., Gillies, A.: Belief revision and epistemology. Synthese **122**, 69–92 (2000)
67. Priest, G., Routley, R., Norman, J. (eds.): Paraconsistent Logic: Essays on the Inconsistent. Philosophia Verlag, München (1989)
68. Prior, A.N.: Past, Present and Future. Clarendon Press, Oxford (1967)
69. Rescher, N., Manor, R.: On inference from inconsistent premisses. Theor. Decis. **1**(2), 179–217 (1970)
70. Rescher, N.: Plausible reasoning: an introduction to the theory and practice of plausibilistic inference (1976)
71. Resconi, G., Hinde, C.J.: Active sets, fuzzy sets and inconsistency. In: IEEE International Conference on Fuzzy Systems (FUZZ), pp. 1–8, Barcelona, Spain (2010)
72. Ross, T.J.: Fuzzy Logic with Engineering Applications. McGraw-Hill P. C. (1994)
73. Routley, R., Plumwood, V., Meyer, R.K., Brady, R.T.: Relevant Logics and Their Rivals. Atascadero, Ridgeview, CA (1982)
74. Schütte, K.: Beweistheorie. Springer, Berlin (1960)
75. Shoenfield, J.R.: Mathematical Logic. Addison-Wesley, Reading (2001)
76. Smolin, L.: The Bekenstein bound, topological quantum field theory and pluralistic quantum field theory. Penn State preprint CGPG-95/8-7; Los Alamos Archives preprint in physics, gr-qc/9508064, electronic edition: http://arXiv.org (1995)
77. Smullian, R.: What is the Name of this Book?. Prentice Hall, Englewood Cliffs (1978)
78. Toulmin, S.: The Uses of Argument. Cambridge University Press (1956)
79. de Vey Mestdagh, C.N.J., Verwaard, W., Hoepman, J.H.: The logic of reasonable inferences. In: Breuker, J.A., de Mulder, R.V., Hage, J.C. (eds.) Legal Knowledge Based Systems, Model-based legal reasoning, Proceedings of the 4th annual JURIX Conference on Legal Knowledge Based Systems, pp. 60–76. Vermande, Lelystad (1991)
80. de Vey Mestdagh, C.N.J.: Legal expert systems. Experts or Expedients? The Representation of Legal Knowledge in an Expert System for Environmental Permit Law. In: Ciampi, C., Marinai, E. (eds.) The Law in the Information Society, Conference Proceedings on CD-Rom, Firenze, 8 pp (1998)
81. de Vey Mestdagh, C.N.J., Hoepman, J.H.: Inconsistent knowledge as a natural phenomenon: the ranking of reasonable inferences as a computational approach to naturally inconsistent (legal) theories. In: Dodig-Crnkovic, G. & Burgin, M. (eds.) Information and Computation, pp. 439–476. World Scientific, New Jersey (2011)
82. Viganò, L., Volpe, M.: Labeled natural deduction systems for a family of tense logics. In: Demri, S., Christian S., Jensen, C.S. (eds.) 15th International Symposium on Temporal Representation and Reasoning (TIME 2008), pp. 118–126, University of Quebec, Montreal, Canada, 16–18 June 2008. IEEE Computer Society (2008)
83. Wang, G., Liu, F.: The inconsistency in rough set based rule generation. In: Rough Sets and Current Trends in Computing, Lecture Notes in Computer Science, vol. 2005, pp. 370–377 (2001)
84. Wassermann, R.: An algorithm for belief revision. In: Proceedings of the 7th International Conference of Principles of Knowledge Representation and Reasoning (KR'2000) (2000)
85. Weinzierl, A.: Comparing inconsistency resolutions in multi-context systems. In: Slavkovik, M. (ed.) Student Session of the European Summer School for Logic, Language, and Information, pp. 17–24 (2010)

86. Williams, M.A.: Transmutations of knowledge systems. In: Proceedings of the 4th International Conference of Principles of Knowledge Representation and Reasoning (KR'94), pp. 619–629 (1994)
87. Williams, M.A.: A practical approach to belief revision: reason-based change. In: Proceedings of the 5th International Confeence of Principles of Knowledge Representation and Reasoning (KR'96), pp. 412–421 (1996)

A Study on Composition of Elements for AHP

Takafumi Mizuno and Eizo Kinoshita

Abstract We discuss how to estimate evaluation values of composed elements for the Analytic Hierarchy Process (AHP). If there are two elements whose evaluation values are known, this study provides an estimation value of their composition without precise evaluation process. In this paper, we deal with three composing operations: "and", "or", and "not". We describe properties which these estimation values must satisfy, and provide concrete estimation calculations.

Keywords AHP (Analytic Hierarchy Process) · Relational database

1 Introduction

AHP (the Analytic Hierarchy Process) [1, 2] is a decision making process which selects best one from a set of alternatives based on multi criteria to achieve an ultimate purpose. Through the process, decision makers also obtain scores, which are positive real numbers, of alternatives.

AHP consists of three phases: making a layered structure, evaluation of importance values, and synthesis of the evaluation values.

In the first phase, decision makers construct a hierarchical structure (Fig. 1). The structure consists of three layers. The top layer represents an ultimate purpose of the decision making, the second layer consists of criteria, and the bottom layer consists of alternatives.

In the next phase, decision makers evaluate importance of every element. The main procedure of the phase is the pairwise comparison method. Let $\{s_1, \cdots, s_n\}$ be a set of elements. Decision makers show a ratio of importance of elements s_i and s_j in positive real number (ratio r_{ij} = importance of s_i / importance of s_j) for all pair

T. Mizuno (✉) · E. Kinoshita
Faculty of Urban Science, Meijo University, Gifu, Japan
e-mail: tmizuno@urban.meijo-u.ac.jp

E. Kinoshita
e-mail: kinoshit@urban.meijo-u.ac.jp

R. Neves-Silva et al. (eds.), *Intelligent Decision Technologies*,
Smart Innovation, Systems and Technologies 39,
DOI 10.1007/978-3-319-19857-6_37

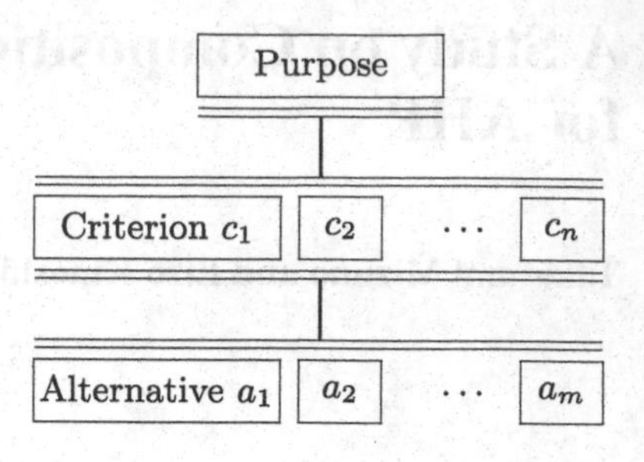

Fig. 1 Three layered structure of AHP

(i,j) of elements, and arrange the ratio into a pairwise comparison matrix W which has a value r_{ij} in the i-th row in the j-th column. The method outputs an evaluation value of the element of s_i as i-th element of the prime eigenvector of the matrix W. Decision makers calculate weights of criteria by using the pairwise comparison methods. These weights are positive real number and a sum of them is 1. For each criterion, decision makers evaluate alternatives in respect of the criterion by using pairwise comparison methods.

In the last phase, scores of alternatives are calculated. A score of an alternative is a sum of evaluation values multiplied by weights of corresponding criteria. Let $Alt = \{a_1, \cdots, a_m\}$ be a set of alternatives, $Cr = \{c_1, \cdots, c_n\}$ be a set of criteria, a value u_{ac} be an evaluation value of an alternative a in respect of a criterion c, and a value w_c be a weight of the criterion c. Then a score of the alternative a is calculated as

$$S_a = \sum_{c \in Cr} u_{ac} w_c. \tag{1}$$

Or, with a normalization,

$$s_a = \frac{S_a}{\sum_{\xi \in Alt} S_\xi}. \tag{2}$$

1.1 A Motivation of Our Study

In this study, we have a question: "How do we estimate scores of composed elements without doing full evaluation processes?" In the evaluation process of AHP, comparisons for all pairs of elements cause increase in cost of decision makings. Then we want to reduce the cost rationally by estimating evaluation values of elements when they consist of elements whose values are known.

Applicable area of our study is design. In Web page design, Web pages consist of many widgets which are easy to compose and to decompose. Many properties such as colors and font sizes can be composed easy too. And designers can give options or choices of interfaces to users. Users can change designs when browsing them. Costs of evaluations will be huge if decision makers evaluate all widgets, properties, and their possible combinations. Similar applicable area is evaluations of beverages or cooking.

The study also can apply AHP with blank data. Pairwise comparison matrices often have blank elements. Harker [3] suggests a method that proceeds pairwise comparisons in this case, and this is a standard method which treats the case. The method replaces blanks of a matrix to zero, and adds number of blanks in a row to a value of corresponding diagonal element. Validity of the method stands on transitivity rule of ratio of importance values, and decision makers proceed with the method by looking only ratios on the matrix. Our approach can look into relations or structures of elements, and fill blanks appropriate values which corresponds to the structures.

Of course, decision makers have to evaluate precisely by regular processes even if an element is composed of other well-known elements. But we want rough estimation of the composed element in advance of the precise evaluation for verification of the evaluation. If the evaluation value greatly differs from the estimation value, decision makers may fail to look anything in the process. Or the composition varies any qualities unexpectedly. In either case, decision makers must analyze the irrationality.

Other motivation of the study is automation of evaluation processes.

Notations and Terminologies In this paper, a word "element" means any alternative or any criterion of decision makings. Elements are represented in letters a, b, c, d, A. And positive real numbers are represented in letters u, r, x, y, ρ.

We represent "importance value" of an element a as $E(a)$. This value is positive real number (> 0), and indicates importance of the element. Note that $E(a)$ is not any function which has concrete calculation process. We suppose that decision makers can acquire only ratios of importance values. Decision makers directly show ratios such as $E(a)/E(b)$, and they obtain proportional values of $E(a)$ or $E(b)$ with the pairwise comparison method. We refer to the proportional values as "evaluation values". With any value $\rho(> 0)$, an evaluation value of an element a is represented as $e(a) = \rho E(a)$. In short, we suppose that these values are represented in ratio scales. We write values of an element a in $e_c(a)$ and $E_c(a)$, when we emphasize that evaluations focus on a criterion c.

2 Estimation of Evaluation Values of Composed Elements

In this paper, we deal with three composing operations: "and", "or", and "not". They have corresponding abstract semantics. Operation "and" applies an element to base element, or picks common properties of them up. Operation "or" adds an element to base element, or provides the element as options. Operation "not" cancels effects of an element.

The purpose of the study is to make expressions of $e(a \text{ and } b)$, $e(a \text{ or } b)$, and $e(\text{not } a)$, when $e(a)$ and $e(b)$ are known.

2.1 Required Properties

We consider properties that the operations must have, and itemize them here.

Comparable property. Composed elements will be compared with any element, or their estimation values are greater than zero.

$$e(a \text{ and } b) > 0, \qquad e(a \text{ or } b) > 0, \qquad e(\text{not } a) > 0. \tag{3}$$

Identity property. For operations "and" and "or", if a composed element consists of same element, then the composition does not change its evaluation value.

$$e(a \text{ and } a) = e(a), \qquad e(a \text{ or } a) = e(a). \tag{4}$$

Commutative property.

$$e(a \text{ and } b) = e(b \text{ and } a), \qquad e(a \text{ or } b) = e(b \text{ or } a). \tag{5}$$

Note that we do not suppose commutativity of composing more than two elements.
Modest property. Because these are just estimation without precise measurements, we do not expect that compositions drastically change evaluation values. Based on this demand, operation "and", at least, must suppress estimation value under evaluation values of elements. Operation "or", at least, must make estimation value be larger than minor evaluation value of elements, since the operation increases options (If the operation shrinks the value, we can ignore the option and can take bare element).

$$e(a \text{ and } b) \le \max\{e(a), e(b)\}, \qquad e(a \text{ or } b) \ge \min\{e(a), e(b)\}. \tag{6}$$

Marginal property. We put a property when an evaluation value of one element is sufficiently large. If $e(A) > e(a)$ then

$$\frac{e(a \text{ and } b)}{e(a)} > \frac{e(A \text{ and } b)}{e(A)}, \tag{7}$$

$$\frac{e(A)}{e(a \text{ or } A)} > \frac{e(a)}{e(a \text{ or } A)}. \tag{8}$$

Equation (7) means that increasing an evaluation value of an element decreases its effects of composing other elements into it. Equation (8) means that increasing an evaluation value of an element increases its share of an estimation value of composed elements.

Relation of operations "and" and "or". Because operation "or" extends options, we suppose that its estimation value is larger than value of operation "and".

$$e(a \text{ or } b) \geq e(a \text{ and } b). \tag{9}$$

2.2 An Algebraic Model

First, we consider a model that estimates evaluation values of composing elements by arithmetic calculations of evaluation values. We represent operations "and", "or", and "not" in arithmetic operations "$\odot$", "$\oplus$", and "$\diamond$" respectively.

$$e(a \text{ and } b) \equiv e(a) \odot e(b), \tag{10}$$

$$e(a \text{ or } b) \equiv e(a) \oplus e(b), \tag{11}$$

$$e(\text{not } a) \equiv \diamond e(a). \tag{12}$$

Table 1 An example database which consists of one table "Coffees"

Coffees

Voter ID	Bean	Flavor	Sweetness	Aftertaste
1	Java	medium	medium	medium
2	Blue Mountain	high	sour	medium
3	Kilimanjaro	high	medium	smooth
4	Java	medium	medium	smooth
5	Kilimanjaro	medium	sour	smooth

Next work is to suggest concrete calculation of them.

In this paper, we propose simple calculations as follows.

$$x \odot y \equiv \sqrt{xy}, \tag{13}$$

$$x \oplus y \equiv x + y - \sqrt{xy}. \tag{14}$$

Estimation values calculated by Eqs. (13) and (14), and they satisfy above properties. Equation (14) stands on an assumption of relation of $\odot$ and $\oplus$.

$$x \oplus y = x + y - x \odot y. \tag{15}$$

We can regard $\odot$ and $\oplus$ as functions whose domain are the set of positive real numbers. Graphs of these functions are depicted in Fig. 2.

The operation $\diamond$ is treated in Sect. 3.

2.3 A Relational Model

Besides the algebraic model, we describe a model in which we can analyze importance values and structures of elements. In the model, we suppose that all elements of AHP can be expressed in restrictions of queries for relational databases.

Relational database is a set of table. Table 1 is a simple example for an explanation. This is a single table database, and the table records votes for favorite coffees. Candidates are Java, Blue Mountain, and Kilimanjaro. There are five voters and they write their impressions of coffee's properties (flavor, sweetness, and aftertaste). Restrictions of alternative for the decision making are "bean=Java", "bean=Blue Mountain", and "bean=Kilimanjaro". The first alternative is written in relational algebra form as

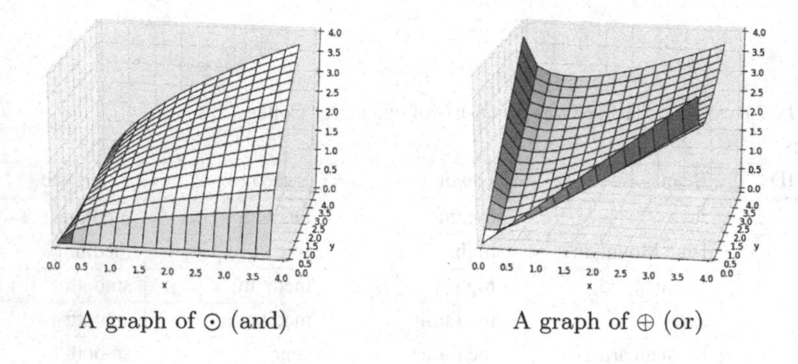

Fig. 2 Graphs of functions of operations ("and" and "or") in our algebraic model

$$a = \sigma_{\text{bean=Java}}, \tag{16}$$

or is written in SQL query as

```
SELECT * FROM Coffees WHERE bean="Java";
```

We also can express criteria as restrictions of queries. For examples, let us consider a criterion "smoothness", and let us define the criterion as "sweetness is medium and aftertaste is smooth". The criterion can be expressed in relational algebra form

$$c = \sigma_{\text{sweetness=medium} \wedge \text{aftertaste=smooth}}, \tag{17}$$

or is expressed in SQL query as

```
SELECT * FROM Coffees
    WHERE sweetness="medium" AND aftertaste=''smooth'';
```

In the model, we suppose that we can directly calculate an importance value of an element a using results of corresponding query.

$$E(a) \equiv f(\sigma_a). \tag{18}$$

Decision makers prepare appropriate map f for their decision making. In this paper, we use a simple map.

$$E(a) \equiv \#(\sigma_a) + 1. \tag{19}$$

A map # returns number of result tuples of the query. We put a term "+1" on ending for cases in which there are no results of the query.

In the examples of Coffees table, an importance value of the criterion "smoothness" is

$$E(c) = E(\sigma_{\text{sweetness=medium}\wedge\text{aftertaste=smooth}}) = 2 + 1 = 3. \tag{20}$$

An importance value of an alternative "bean=Java" based on the criterion "smoothness" is

$$E_c(a) = E(\sigma_{\text{bean=Java}\ \wedge c}) = 1 + 1 = 2. \tag{21}$$

We can calculate importance values of composed elements by combining restrictions of queries.

$$E(a \text{ and } b) \equiv \#(\sigma_{a\wedge b}) + 1, \tag{22}$$

$$E(a \text{ or } b) \equiv \#(\sigma_{a\vee b}) + 1, \tag{23}$$

$$E(\text{not } a) \equiv \#(\sigma_{\neg a}) + 1. \tag{24}$$

Evaluation values are calculated by using ratios of pair of these importance values. Intuitively, operators $\wedge$, $\vee$, $\neg$ correspond to set operation $\cap$, $\cup$, *not* respectively. So they hold submodular law.

$$\#(\sigma_{a\vee b}) = \#(\sigma_a) + \#(\sigma_b) - \#(\sigma_{a\wedge b}). \tag{25}$$

And

$$\begin{aligned} E(a \text{ or } b) &= \#(\sigma_a) + \#(\sigma_b) - \#(\sigma_{a\wedge b}) + 1 \\ &= \#(\sigma_a) + 1 - 1 + \#(\sigma_b) + 1 - 1 - \#(\sigma_{a\wedge b}) - 1 + 1 + 1 \\ &= E(a) + E(b) - E(a \text{ and } b). \end{aligned} \tag{26}$$

3 Discussions

We define the operator $\diamond$ (not) in the algebraic model, at first, as

$$\diamond x \equiv \frac{1}{x}. \tag{27}$$

This is natural definition because the operation "not" means canceling effects.

But the division has an inconvenience to proceed with AHP. In AHP, there are only ratio scales, and it is nonsense without factors of proportionality on importance values. An evaluation value can be expressed in $\rho E(a)$, and the division cancels the coefficient ρ from the estimation when we apply a to the operated element "not a". This means that we cannot lump the estimation value with other evaluation values.

So we obtain a property for the algebraic model; estimation values of composed elements retain the coefficients. Operators $\odot$ and $\oplus$ hold this property.

$$\rho x \odot \rho y = = \sqrt{\rho x \rho y} = \rho \sqrt{xy} = \rho(x \odot y) \tag{28}$$

$$\rho x \oplus \rho y = \rho x + \rho y - \rho \sqrt{xy} = \rho(x \oplus y) \tag{29}$$

where ρ is any coefficient greater than zero (> 0). While

$$\diamond \rho x = \frac{1}{\rho x} \neq \rho(\diamond x) \tag{30}$$

We express the case of the Eq. (30) as its both sides have different dimensions.

Then we revise the semantics of operation "not", and improve its operator $\diamond$. New semantics of "not" is that the operation "not" cancels effects of the element and restores to standard. In other word, by preparing a standard elements d,

$$e(a \text{ and } (\text{not}_d\, a)) = e(d). \tag{31}$$

The new operation "not" depends on the standard element d. And new definition of $\diamond$ is

$$e(\text{not}_d\, a) \equiv \diamond_d e(a), \qquad \diamond_d x \equiv \frac{(e(d))^2}{x}. \tag{32}$$

This definition holds the property of dimension.

In the relational model, the importance value "1" means importance of null set or contradiction. In the algebraic model, an evaluation value of contradiction, which is defined under a standard element d, is expressed as $e(d) = e(a \text{ and } \text{not}_d\, a)$.

Dominant AHP [4] is an improved AHP to avoid a rank reversal problem. The problem was pointed out by Belton and Gear [5]. In AHP, each evaluation values of alternatives have different scales on each criterion, and this causes the rank reversal problem. A main idea of Dominant AHP is normalization of scales of evaluation values by remarking on standard alternative; each evaluation value $e_c(a)$ of any alternative a based on a criterion c is divided by $e_c(d)$. With interpretations based on this paper, in Dominant AHP, evaluation values are normalized by the evaluation value of contradiction. If decision makers accept estimation values of composed elements, they can use these composed elements as standard in the process.

4 Conclusions and Future Works

In this paper, we use a term "synthesis" when we discuss combining weights of criteria and evaluation values of alternatives, and we use a term "composition" when we discuss combining among alternatives or combining among criteria. Former studies on AHP treat syntheses, while this study treats composition.

In this paper, we describe properties that estimation values of composed elements must satisfy, and we provide simple estimation methods that satisfy the properties. We must validate the methods or calculations by quantitative psychological experiments.

References

1. Saaty, T.L.: A scaling method for priorities in hierarchical structures. J. Math. Psychol. **15**, 234–281 (1977)
2. Saaty, T.L.: The Analytic Hierarchy Process. McGraw-Hill, New York (1980)
3. Harker, P.T.: Alternative modes of questioning in the analytic hierarchy process. Math. Model. **9**, 353–360 (1987)
4. Kinoshita, E., Nakanishi, M.: Proposal of new AHP model in light of dominant relationship among alternatives. J. Oper. Res. Soc. Jpn. **42**, 180–197 (1999)
5. Belton, V., Gear, T.: On a shortcoming of Saaty's method of analytic hierarchies. OMEGA Int. J. Manag. Sci. **11**, 228–236 (1983)

Probabilistic Forecasting of Solar Power: An Ensemble Learning Approach

Azhar Ahmed Mohammed, Waheeb Yaqub and Zeyar Aung

Abstract Probabilistic forecasts account for the uncertainty in the prediction helping the decision makers take optimal decisions. With the emergence of renewable technologies and the uncertainties involved with the power generated through them, probabilistic forecasts can come to the rescue. Wind power is a mature technology and is in place for decades now, various probabilistic forecasting techniques are used here. On the other hand solar power is an emerging technology and as the technology matures there will be a need for forecasting the power generated days ahead. In this study, we utilize some of the probabilistic forecasting techniques in the field of solar power forecasting. An ensemble approach is used with different machine learning algorithms and different initial settings assuming normal distribution for the forecasts. It is observed that having multiple models with different initial settings gives exceedingly better results when compared to individual models. Getting accurate forecasts will be of great help where the large scale solar farms are integrated into the power grid.

Keywords Solar power · Probabilistic forecasting · Pinball loss function · Ensemble learning

A.A. Mohammed · W. Yaqub · Z. Aung (✉)
Institute Center for Smart and Sustainable Systems (iSmart),
Department of Electrical Engineering and Computer Science,
Masdar Institute of Science and Technology, Abu Dhabi, UAE
e-mail: zaung@masdar.ac.ae

A.A. Mohammed
e-mail: amohammed@masdar.ac.ae

W. Yaqub
e-mail: wyaqub@masdar.ac.ae

R. Neves-Silva et al. (eds.), *Intelligent Decision Technologies*,
Smart Innovation, Systems and Technologies 39,
DOI 10.1007/978-3-319-19857-6_38

1 Introduction

Renewable energy sources have gained popularity in the past decade because of the increasing global warming. The environmental impact of the average barrel of oil is much more today than it was in 1950. The production of fossil fuels has some environmental impact as it releases large amounts of carbon dioxide into the atmosphere. The impact is only going to increase with the depletion of resources as we try to access less accessible resources which results in more costs on transport etc. [5]. According to the U.S. Energy Information Administration, the energy consumption across the world is going to increase by 56 percent between 2010 and 2040. Renewable energy along with nuclear power is the fastest growing energy resource, the growth rate is estimated at 2.5 percent every year [14].

As the world looks for more environmental friendly energy resources, solar energy is viewed as an important clean energy source. The amount of solar energy striking the earth's surface every hour is more than sufficient to meet the energy needs of the entire human population for one year [9]. Solar energy is one of the most abundant resources available and it can help reduce the ever increasing dependency on energy imports. It can also protect us from the price fluctuations of the fossil fuels and help stabilize the electricity generation costs in the long run. Solar photovoltaic (PV) power plants do not emit any green house gases and they use little water, with the rapid increase in air pollution this benefit of solar PV becomes even more important. The technology roadmap for solar PV envisions a 4600 GW of PV capacity by 2050 this will result in the reduction of 4 gigatonnes (Gt) of carbon dioxide per year [15].

When large solar farm(s) are integrated into the power grid, solar power forecasting becomes essential for grid operators who make decisions about the power grid operations and as well as for electric market operators [20, 23]. The output of solar farms varies with every season, month, day and even hour. This can be challenging for grid operators and they have to turn on and off the power plants to balance the grid. Forecasting the power generated ahead of time can help avoid these problems. It also helps in the optimal use of resources [18]. With large solar farms coming into the forefront, short-term ramp events will have a higher impact. Ramp events occurs when the power generation suddenly stops because of a cloud cover and also when the cloud cover lifts and the power generation ramps up again [21].

No forecast is perfect. Every forecast carries along with it a certain amount of inaccuracy and this can be attributed to the errors in real-time measurements and model uncertainty. "Probabilistic forecasting" [8] is the forecast that takes probability distributions over future events. Probabilistic forecasts are preferred to point (a.k.a. single-valued) forecasts as they take into account the uncertainties in the predicted values which helps in taking effective decisions. Probability forecasts are being used for quite some in the case of predicting binary events, events like "what is the probability that it rains today?" and other similar events. But the focus is shifting towards applying them in more general events. Some examples include flood risk assessment, weather prediction, financial risk management among others [8].

Traditionally, in the domain of solar forecasting, point forecasting methods [2, 12, 13, 19] has widely been used. Nonetheless, recently, probabilistic forecasting was used for forecasting of solar irradiance by employing stochastic differential equations [16]. In the electricity market high asymmetric costs are associated with the need to continuously balance the grid to avoid grid failure. These costs arise due to the intrinsic uncertainty associated with the emerging renewable power sources such as wind and solar. Hence a thorough understanding of the uncertainty associated with the prediction is necessary to effectively manage the grid [16]. Thus, we believe that probabilistic forecasting can become a power trend in solar power forecasting.

In this paper, as our **research contributions**, we have:

- Proposed a probabilistic forecasting method for solar power forecasting using an ensemble of different machine learning models, and
- Demonstrated that the use of ensemble learning approach offers significantly better results when compared to individual models.

2 Related Work

Bacher et al. [2] introduce a new two stage model for online short-term solar power forecasting. In the first stage, clear sky model is used to normalize the solar power and in the second stage linear time series models are used to forecast. The clear sky model gives the solar power estimation at any given time. The normalized values are obtained by taking the ratio of solar power in clear sky and the observed solar power. This is done to ensure the resulting values are more stationary, the values must be stationary to apply the statistical smoothing algorithms which assume stationarity. It takes the observed power as input and uses statistical smoothing techniques and quantile regression to give the estimated power values. The linear time series models use adaptive recursive least squares (RLS) method. The coefficients of the model are estimated by minimizing the weighted residual sum of errors, the new values are given a higher weight. This is necessary as the models need to adapt to the changing conditions.

Marquez et al. [19] use Artificial Neural Network (ANN) model to forecast the global and direct solar irradiance with the help of National Weather Service's (NWS) database. Eleven input variables are used in total, nine from the NWS database i.e. the meteorological data and two additional variables solar zenith angle and normalized hour angle are also added. To extract the relevant features, Gamma Test (GT) is used and to efficiently search the feature space, Genetic Algorithm (GA) is used. After extracting the relevant features, ANN model is used to generate the forecasts. The most important features are found to be the solar geometry variables, probability of precipitation, sky cover, minimum and maximum temperature. The models are accurate during summer rather than winter as there are more days with clear skies in summer.

Hossain et al. [12] propose a hybrid intelligent predictor for 6 hour ahead solar power prediction. The system uses an ensemble method with 10 widely-used regression models namely Linear Regression (LR), Radial Basis Function (RBF), Support Vector Machines (SVM), Multi-Layer Perceptron (MLP), Pace Regression (PR), Simple Linear Regression (SLR), Least Median Square (LMS), Additive Regression (AR), Locally Weighted Learning (LWL), and IBk (an implementation of k-th Nearest Neighbor, KNN, Algorithm). Their results shows that, with respect to mean absolute error (MAE) and mean absolute percentage error (MAPE), the top most accurately performing regression models are LMS, MLP, and SVM.

E. B. Iversen et al. [16] propose a framework for calculating the probabilistic forecasts of solar irradiance using stochastic differential equations (SDE). They construct a process which is limited to a bounded state space and it gives zero probability to all the events outside this state space. To start with, a simple SDE model is used which tracks the solar irradiance from the numerical weather prediction model. This basic model is further improved by normalizing the predicted values using the maximum solar irradiance, this helps in capturing seasonality and trend. The proposed model outperformed some of the complex benchmarks.

3 Data Set

The data used in this study is taken from the Global Energy Forecasting Competition (GEFCOM) 2014 [7, 11]. This is the first competition on probabilistic forecasting in the power and energy industry. The competition lasted for 16 weeks which included 15 tasks. Each week the task is to forecast the power values for the next period. For the first task, hourly data is provided for each of the 3 zones from April 1, 2012 until April 1, 2013. For each of the remaining tasks, the data for the next month is provided. By the end of the last task, the data contained 56,953 records ranging from April 1, 2012 until May 31, 2014.

Table 1 shows the installation parameters for the three solar farms (one in each zone). Table 2 shows the independent variables (a.k.a features or attributes) that are given as part of the training and testing data sets. The objective is to predict the probabilistic distribution of the solar power generation values. The power values are normalized to range between 0 and 1 as the nominal power value for each of the solar farms is different. The location and time zones of the solar farms are not disclosed.

Table 1 Installation parameters for the solar farms. (Note: the actual names and locations of the zones and solar farms are not disclosed by the GEFCOM organizers)

Zone	Type	Number	Power	Orientation	Tilt
1	Solarfun SF160-24-1M195	8	1,560 W	38° clockwise from North	36°
2	Suntech STP190S-24/Ad+	26	4,940 W	327° clockwise from North	25°
3	Suntech STP200-18/ud	20	4,000 W	31° clockwise from North	21°

4 Methodology

This section describes the approach that we use to obtain probabilistic forecasts. The below sections describe how the data is grouped and the models used in the process followed by the methods to generate probabilistic forecasts.

4.1 Grouping the Data

The data provided contains no missing values and hence there is no need to handle missing values. The solar power values in the data are normalized to scale appropriately across all the three solar farms. No other data preprocessing techniques are used. All the variables provided have an impact on the solar power generated. However the data in each zone is grouped based on hour. Hence for each zone 24 different models are used.

Table 2 Description of independent variables

Sr.	Description	Unit
1	Total column liquid water	kg/m^2
2	Total column ice water	kg/m^2
3	Surface pressure	Pa
4	Relative humidity at 1000 $mbar$	%
5	Total cloud cover	$0-1$
6	10 metre U wind component	m/s
7	10 metre V wind component	m/s
8	2 metre temperature	K
9	Surface solar rad down	J/m^2
10	Surface thermal rad down	J/m^2
11	Top net solar rad	J/m^2
12	Total precipitation	m

Figure 1 shows the solar power values for the month of April 2012. A clear trend can be seen with the values going to zero during the time and slowly peaking in the afternoon, there are more fluctuations in the data and it is widely spread. Whereas Fig. 2 shows the hourly data for a particular time (1 am) observed for the year 2013. This shows that the values are not as dispersed as in the case of daily values. This will help avoid the problem of outliers while fitting the models.

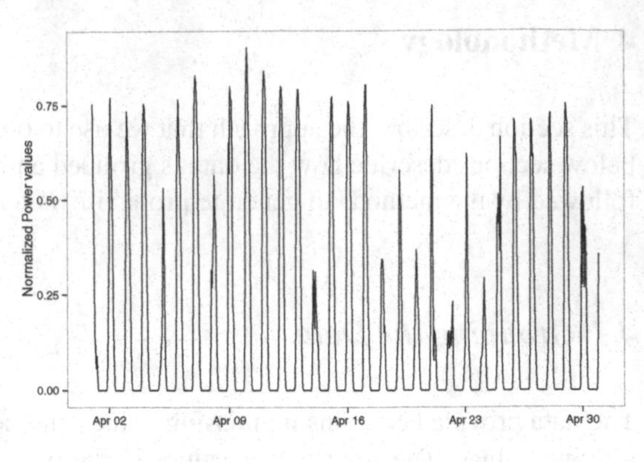

Fig. 1 Observed solar power values for the month of April, 2012 in Zone 1

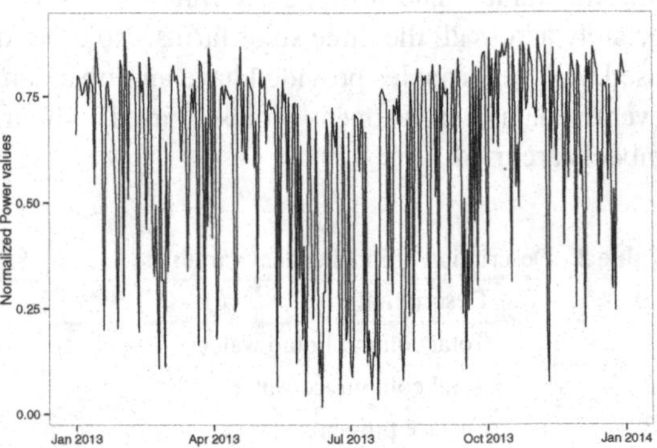

Fig. 2 Solar power values recorded at 1 am across the year 2013 in Zone 1

4.2 Base Models Used

The following seven individual machine learning methods are used as base models for ensemble learning in this study.

- **Decision Tree Regressor:** A model is fit using each of the independent variables. For each of the individual variables, mean squared error is used to determine the best split. Maximum number of features to be considered at each split is set to the total number of features [4].
- **Random Forest Regressor:** An ensemble approach that works on he principle that a group of weak learners when combined would give a strong learner. The weak learners used in random forest are decision trees. Breiman's bagger is used in which at each split all the variables are taken into consideration [3].
- **KNN Regressor (Uniform):** The output is predicted using the values from the k-nearest neighbors (KNNs) [1]. In the uniform model, all the neighbors are given an equal weight. Five nearest neighbors are used in this models i.e. the 'k' value is set to five. Distance metric "Minkowski" is used in finding the neighbors.
- **KNN Regressor (Distance):** In this variant of KNN, the neighbors closer to the target are given higher weights. The choice of *k* and distance metric are same as above.

- **Ridge Regression:** Penalizes the use of large number of dimensions in the dataset using linear least squares to minimize the error [10].
- **Lasso Regression:** A variation of linear regression that uses shrinkage and selection method. Sum of squares error is minimized but with a constraint on the absolute value of the coefficients [22].
- **Gradient Boosting Regressor:** An ensemble model usually using the decision trees as weak learners, it builds the model in stage-wise manner by optimizing the loss function [6].

4.3 Generating Probabilistic Forecasts Using Ensemble Learning

Three ensemble learning approaches are used to generate the probability forecasts using the values generated from the models mentioned above.

- **Naive model:** A cumulative probability distribution where the first quantile is the lowest among the values and 99th quantile was the highest.
- **Normal distribution:** The mean and standard deviation of the point forecasts from above seven individual models are used to generate 99 quantiles assuming normal distribution.
- **Normal distribution with different initial settings:** This method is similar to the above normal distribution method but the models are run with two different initial settings, including the month as a variable and taking only the values for the 30 most recent days as inputs.

5 Evaluation Metric

Pinball loss function [17] is used as an evaluation metric as we are dealing with probabilistic forecasts and not point forecasts. Let the 99 quantiles generated 0.01, 0.02,..., 0.99 be defined as $q_1, q_2 \dots, q_{99}$ respectively and $q_0 = -\infty$ the natural lower bound and $q_100 = +\infty$ the natural upper bound. Then the score L for q_i is defined as:

$$L(q_i, y) = \begin{cases} (1 - i/100)(q_i - y) & \text{if } y < q_i \\ i/100(y - q_i) & \text{if } y \geq q_i \end{cases}$$

where y is the observed value and $i = 1, 2, \dots, 99$. To evaluate the overall performance, this score is averaged across all target quantiles. Lower scores indicate better forecasts.

6 Results and Discussions

The results when using only the individual models are shown in Table 3. The average pinball loss in the table refers to average value across 15 months starting from April 2013 for the 3 zones. Random forests and gradient boosting regression gave the best results. A small part of the results for a 24 hour period is shown in Fig. 3 for better resolution. The point forecasts are very close to the observed values but they do not take into account the error associated with them and hence we need probabilistic forecasts. There is a significant improvement in the results when the ensemble methods are used as shown in Table 4.

Table 3 Individual performance of different models

Individual model	Average pinball loss
Decision Tree	0.0249
Random Forest	0.0194
KNN (Uniform)	0.0223
KNN (Distance)	0.0224
Ridge Regression	0.0206
Lasso Regression	0.0218
Gradient Boosting	0.0193

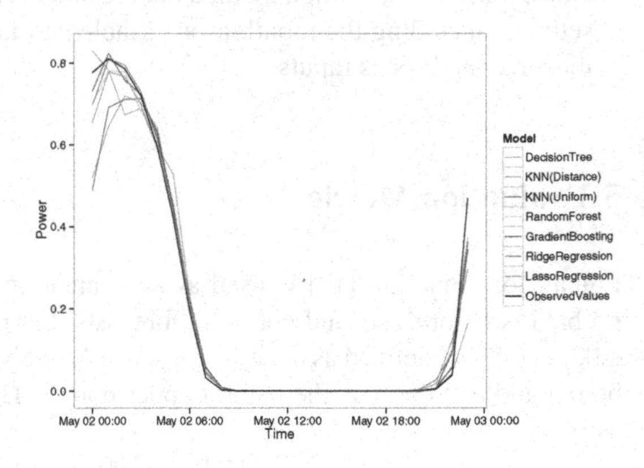

Fig. 3 Point forecasts of different models for 24 hour period on May 2nd, 2013

Table 4 Ensemble performance of the three methods

Ensemble model	Average pinball loss
Naive model	0.0157
Normal Distribution	0.0152
Normal Distribution with different initial settings	0.0148

Among the three zones, Zone1 had the best results. For example the average error value in Zone1 for Naive model is 0.019673 whereas it is 0.022617 and 0.022531 for Zone2 and 3 respectively. The hourly error values also varied significantly with hours 11 through 18 showing a zero error value since the power generated during that times is zero. The highest error values are observed during the hours zero through four. The monthly error values are shown in Fig. 4. Very low error are observed in the months of May and June whereas August has the highest error rate. These fluctuations in the error rates are caused by the cloud cover. In general better forecasts are achieved in summer because of clear sky and there are high error rates during the winter with more cloud cover during the daytime.

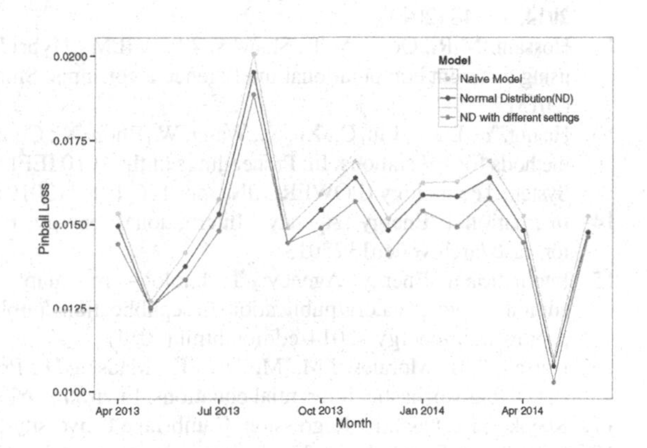

Fig. 4 Pinball loss values for different months

7 Conclusion

In this study, we generated the probabilistic forecasts using ensemble methods assuming normal distribution for the point forecasts obtained from the individual models. For each hour in each zone, a different model is used to avoid problems with outliers. There is a significant improvement in the performance of ensemble model when compared to individual models. However, these models can be further improved as the assumption that the forecasts follow normal distribution is too restrictive.

References

1. Altman, N.S.: An introduction to kernel and nearest-neighbor nonparametric regression. Am. Stat. **46**(3), 175–185 (1992)
2. Bacher, P., Madsen, H., Nielsen, H.A.: Online short-term solar power forecasting. Sol. Energy **83**(10), 1772–1783 (2009)
3. Breiman, L.: Random forests. Mach. Learn. **45**(1), 5–32 (2001)

4. Breiman, L., Friedman, J., Stone, C., Olshen, R.A.: Classification and regression trees. Taylor & Francis (1984)
5. Davidson, D.J., Andrews, J.: Not all about consumption. Science **339**(6125), 1286–1287 (2013)
6. Friedman, J.H.: Greedy function approximation: a gradient boosting machine. Ann. Stat. 1189–1232 (2001)
7. GEFCOM: Global energy forecasting competition 2014. http://www.drhongtao.com/gefcom (2014)
8. Gneiting, T., Katzfuss, M.: Probabilistic forecasting. Ann. Rev. Stat. Appl. **1**, 125–151 (2014)
9. Goldemberg, J., Johansson, T.B., Anderson, D.: World energy assessment: overview: 2004 Update. United Nations Development Programme, Bureau for Development Policy (2004)
10. Hoerl, A.E., Kennard, R.W.: Ridge regression: biased estimation for nonorthogonal problems. Technometrics **12**(1), 55–67 (1970)
11. Hong, T.: Energy forecasting: past, present, and future. Foresight: Int. J. Appl. Forecast. Winter **2014**, 43–48 (2014)
12. Hossain, M.R., Oo, A.M.T., Shawkat Ali, A.B.M.: Hybrid prediction method for solar power using different computational intelligence algorithms. Smart Grid renew. Energy **4**(1), 76–87 (2013)
13. Huang, Y., Lu, J., Liu, C., Xu, X., Wang, W., Zhou, X.: Comparative study of power forecasting methods for PV stations. In: Proceedings of the 2010 IEEE International Conference on Power System Technology (POWERCON), pp. 1–6. IEEE (2010)
14. International Energy Agency: International energy outlook 2013. http://www.eia.gov/forecasts/archive/ieo13 (2013)
15. International Energy Agency: Technology roadmap: solar photovoltaic energy—2014 edition. www.iea.org/publications/freepublications/publication/technology-roadmap-solar-photovoltaic-energy--2014-edition.html (2014)
16. Iversen, E.B., Morales, J.M., Møller, J.K., Madsen, H.: Probabilistic forecasts of solar irradiance using stochastic differential equations. Environmetrics **25**(3), 152–164 (2014)
17. Koenker, R.: Quantile Regression. Cambridge University Press, New York (2005)
18. Letendre, S.E.: Grab the low-hanging fruit: use solar forecasting before storage to stabilize the grid. http://www.renewableenergyworld.com/rea/news/article/2014/10/grab-the-low-hanging-fruit-of-grid-integration-with-solar-forecasting (2014)
19. Marquez, R., Coimbra, C.F.M.: Forecasting of global and direct solar irradiance using stochastic learning methods, ground experiments and the NWS database. Sol. Energy **85**(5), 746–756 (2011)
20. Perera, K.S., Aung, Z., Woon, W.L.: Machine learning techniques for supporting renewable energy generation and integration: a survey. In: Data Analytics for Renewable Energy Integration—Second ECML PKDD Workshop, DARE 2014, Lecture Notes in Computer Science, vol. 8817, pp. 81–96 (2014)
21. Runyon, J.: Transparency and better forecasting tools needed for the solar industry. http://www.renewableenergyworld.com/rea/news/article/2012/12/transparency-and-better-forecasting-tools-needed-for-the-solar-industry (2015)
22. Tibshirani, R.: Regression shrinkage and selection via the lasso. J. Roy. Stat. Soc.: Ser. (Methodol.) **58**(1), 267–288 (1996)
23. Wikipedia: Solar power forecasting. http://en.wikipedia.org/wiki/Solar_power_forecasting (2015)

Using Argumentation to Develop a Set of Rules for Claims Classification

Jann Müller and Tobias Trapp

Abstract The first step in insurance claims processing is the classification of accident types, for example work-related or domestic accidents, using a set of business rules. In some cases the rules are ambiguous, so a claim is assigned more than one possible classification. If the process is to be automated, the rule set should allow for as few ambiguities as possible, because every ambiguous match requires human intervention. In this paper, we present a technique based on argumentation theory for minimising the number of ambiguous matches in the development of a set of decision rules. We evaluate our approach with a case study and some theoretical results.

Keywords Argumentation · Classification · Insurance claims processing

1 Introduction

Enterprise resource planning (ERP) software usually works with nonambiguous data, which is necessary for the automation of business processes. If there is ambiguous data, additional checks are necessary to determine the target system in the corporate IT landscape, the relevant business process, and what kind of rules have to be applied. Sometimes even manual post-processing is required.

Consider an example from statutory health insurance. Insurance give information about accidents by filling out questionnaires, which are scanned and used in IT-supported processes. One of the first steps is to determine the accident type, for example, whether it was a household accident or a workplace accident. More than 40 rules are used to discern ten different accident types. However, the current rule set often produces ambiguous results: In a case study, the automated classification

J. Müller (✉)
SAP UK Ltd, The Concourse, Queen's Road, Belfast BT3 9DT, UK
e-mail: j.mueller.11@ucl.ac.uk

T. Trapp
AOK Systems GmbH, Breitlacher Str. 90, 60489 Frankfurt, Germany
e-mail: tobias.trapp@sys.aok.de

R. Neves-Silva et al. (eds.), *Intelligent Decision Technologies*,
Smart Innovation, Systems and Technologies 39,
DOI 10.1007/978-3-319-19857-6_39

of 1305 questionnaires resulted in 799 (61 %) cases where more than one accident type was assigned.

Ambiguous classifications may be caused by several factors: Insufficient data (illegible hand-writing, erroneous reports), contradictory data, complex domain models and compliance with legislation. Since every ambiguous classification has to be examined manually, the rule set should be designed in a way that minimises the number of ambiguous outcomes.

This leads to two following challenges for the design of rule sets. (1) How can rule sets be described so that the rationale behind classifications is easily accessible both for designers (subject matter experts) and for implementers of the rule sets? In particular, (2) how can possible ambiguities in a rule set be detected? We propose to address these challenges using ontologies and argumentation, two methods for formally describing and analysing knowledge.

2 Ontologies

We use ontologies [12] to formally describe rule sets for classification. Ontologies can be interpreted in various description logics such as $\mathcal{ALC}$, $\mathcal{SHOIN}$(D) or $\mathcal{SHOIQ}$(D) [3, 9] in which questions such as the subsumption of one rule by another rule can be decided efficiently. This makes them useful for the design, analysis and documentation of rule sets regardless of the implementation in a business management rule system (BMRS), even if the BMRS only supports procedural rules or decision tables.

In this paper we use the following, simplified example.[1] We formalise different accident types and their relationships in $\mathcal{ALC}$, as found eg. in the Web Ontology Language (OWL) [4].

$$\begin{aligned}
&\texttt{Accident}\\
\texttt{StaircaseAccident} &\sqsubseteq \texttt{Accident}\\
\texttt{OccupationalAccident} &\sqsubseteq \texttt{Accident}\\
\texttt{HouseholdAccident} &\sqsubseteq \texttt{Accident}\\
\texttt{SpecialCareAccident} &\sqsubseteq \texttt{Accident}
\end{aligned}$$

`Accident` is the most general concept, subsuming the more specialised classes such as `StaircaseAccident` and `HouseholdAccident`. Some accidents potentially belong to several classes: One could for example imagine a special care accident in a household. In this case, some accident types take preference over others, because for

[1]The concept model of accident types is in reality quite complex and can even involve requests for additional (master and application) data from other IT systems, the modelling of which is a challenge in itself.

legal and business reasons it is crucial that every accident is assigned exactly one type.

Every piece of information that can possibly be extracted from questionnaires is represented as an individual in the `SelfReport` class. For example, an accident on a dark, slippery staircase is modeled as

$$\texttt{SelfReport(slipperyStaircase)}$$
$$\texttt{SelfReport(darkStaircase)}$$

We then define rules to relate this information to accident types. The rule "if an incident involved a slippery staircase, then it is a staircase accident" is represented as

$$\exists \texttt{reported(Incident, slipperyStaircase)} \sqsubseteq \texttt{StaircaseAccident} \quad (1)$$

For each accident we have a set of additional information from the scanned questionnaire (named `ds`):

$$\texttt{Incident(ds); OccupationalAccident(ds)}$$
$$\texttt{reported(ds, slipperyStaircase)}$$

We use a reasoner to classify the information of `ds`. In this example, the accident is both a member of `OccupationalAccident` (from the assertion) and of `StaircaseAccident` (from the rule).

In statutory health insurance, the questionnaires and the set of business rules are standardised within an organisation for compliance reasons. As a result, the TBox of the ontology is fixed, as well as assertions for rules such as (1). The set of all individuals of type `SelfReport` is not fixed, as in a concrete questionnaire not all possible assertions will occur. The more individuals of type `SelfReport` are reported for a certain incident the more likely it is that the classification assigns more than one accident type. This behaviour is correct for a classification service but it can lead to ambiguity. To avoid this we will show how to reduce the number of classifications using argumentation theory.

3 Formal Methods of Argumentation

Formal methods of argumentation [5] are a method for dealing with conflict in a knowledge base. At the heart of argumentation lies the notion of an argument as a claim paired with evidence to support it. Because of this metaphor, argumentation can provide intuitive explanations for conflicts.

There are various proposals for the structure of arguments [5]. While individual argument are usually consistent, the same is not always true for sets of arguments, where conflicts may arise from mutually exclusive conclusions. Such sets of arguments and their conflicts can be given a graph structure [8] which allows one to compute consistent subsets, or "extensions", of arguments.

Before we formalise arguments about classification rules, we briefly recall the relevant definitions introduced by Dung [8].

Definition 1 (Argument graph). *An argument graph is a pair* $(\mathcal{A}, \mathcal{R})$, *where* $\mathcal{A}$ *is a set of arguments and* $\mathcal{R}$ *is an attack relation* $\mathcal{R} \subseteq \mathcal{A} \times \mathcal{A}$.

The intuitive meaning of $A\mathcal{R}B$ is that A is incompatible B. In our setting, attacks will arise from ambiguous classifications. Given an argument graph $G = (\mathcal{A}, \mathcal{R})$, a set of arguments $A \subseteq \mathcal{A}$ is **conflict-free** iff there are no $a, b \in A$ such that $(a, b) \in \mathcal{R}$. Extensions are conflict-free subsets of $\mathcal{A}$ that meet certain conditions:

Definition 2 (Semantics). *Let* $\mathcal{F} = (\mathcal{A}, \mathcal{R})$ *be an argument graph and* $E \subseteq A$. *E is an* admissible extension *iff it is conflict-free and defends all its elements. E is a* complete extension *iff it is admissible and contains all the arguments it defends. E is a* grounded extension *iff it is the minimal complete extension. E is a* stable extension *if E attacks all arguments in* $A \backslash E$.

The grounded extension is unique and it is the smallest consisting of arguments containing all elements it defends.

3.1 Accident Classification with Abstract Arguments

We will express classifications as an argument graph $(\mathcal{A}, \mathcal{R})$ in which the internal structure of the arguments in $\mathcal{A}$ is not specified. This argument graph arises from a preference relation over accident types: If type s is strictly preferred over type t, then we model it as an attack of s on t.

Definition 3 (Preference-based argument graph). *Let* $(S, \leq)$ *be a set with a partial order. The* preference-based argument graph *of S is given by* $(\mathcal{A}, \mathcal{R})$ *where*

1. $\mathcal{A} = S$
2. $\mathcal{R} = \{(s, t) \in S \times S \mid s \geq t \text{ and } s \neq t\}$

Example 1 Figure 1 shows a preference-based argument graph with one argument for each accident type. `OccupationalAccident` and `SpecialCareAccident` attack all other accident types since they are all specialisations of the former. So the classification result is either `OccupationalAccident` or `SpecialCareAccident` and we could reduce the number of possible ambiguous outcomes. The grounded extension of the argument graph in Fig. 1 consists of `OccupationalAccident` and `SpecialCareAccident`.

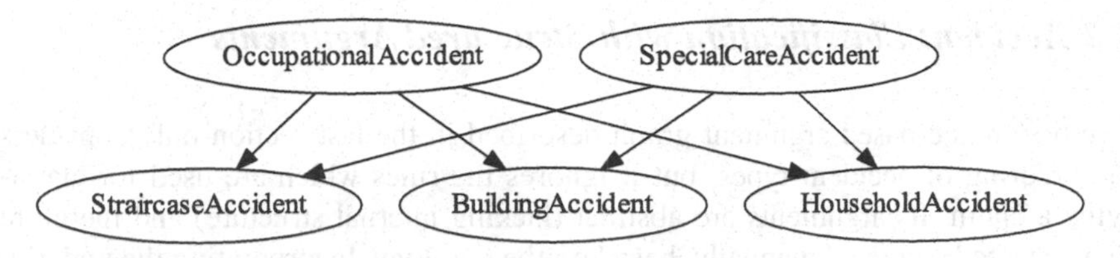

Fig. 1 Attack relations between accident types

If a partial order is antisymmetric then its preference-based argument graph is free of cycles. This is because a cyclic attack can only result from an equality arising from $a \leq b$ and $b \leq a$ (or a transitive extension of that).

Proposition 1 *The preference-based argument graph of a poset* $(S, \leq)$ *is acyclic if and only if* $\leq$ *is antisymmetric.*

Apart from being acyclic (see [13]), this graph has the property that no extension contains more than one class belonging to the same hierarchy.

Proposition 2 *Let* $(S, \leq)$ *be a set with a partial order and let* $(\mathcal{A}, \mathcal{R})$ *be its preference-based argument graph. Let* $\mathcal{E} \subseteq \mathcal{A}$ *be conflict free. For all* $A_1, A_2 \in \mathcal{E}$*:* $A_1 \nleq A_2$ *or* $A_1 = A_2$.

Proof Suppose $\mathcal{E}$ conflict-free with $A_1, A_2 \in \mathcal{E}$ and $A_1 < A_2$. Then A_2 attacks A_1 (by Def. 3), which contradicts the assumption that $\mathcal{E}$ is conflict-free.

Proposition 3 *Let* $(S, \leq)$ *be a bounded join semi-lattice and let* $(\mathcal{A}, \mathcal{R})$ *be its preference-based argument graph. Let* $s \in S$ *be the maximum element under* $\leq$. *Then* $\{s\}$ *is the grounded extension of* $(\mathcal{A}, \mathcal{R})$.

Proof Let $\vee$ be the join operation of $(S, \leq)$. Let $s = \bigvee_{s \in S} s$ be the maximum element of S. Since the semi-lattice is bounded, s exists and is unique. Let $s' \in S$. Then $s \vee s' = s$ so $s' \leq s$ and, if $s \neq s'$, s attacks s' (by Def. 3 Cond.3) and s' does not attack s. So s attacks every argument except itself and it is not attacked, so the grounded extension of $(\mathcal{A}, \mathcal{R})$ is $\{s\}$.

To classify a claim using abstract argumentation, we first construct a set of assertions (ABox) from the data of an incident report, then calculate the classifications of the incident using the ontology and finally, if there is more than one possible classification, accept only those classes that are in the grounded extension of the preference-based argument graph.

We can use this method to detect ambiguous classifications (resulting in symmetric attacks). However, due to the abstract nature of arguments, we do not gain anything over directly using the preference relation of arguments. Furthermore, we cannot reuse the rules from the ontology in the argumentation process.

While abstract argumentation gives us the tools to describe conflicts between classifications, it does not explain what causes the conflicts. However, as rule designers this information is exactly what we are interested in. In the next section we will embellish the arguments with structure based on classification rules.

3.2 Accident Classification with Structured Arguments

The preference-based argument graph described in the last section only considers the ordering of accident types, but it ignores the rules which are used for classifying a claim. Its arguments are abstract (lacking internal structure) and therefore they have to be created manually, based on the ontology. Incorporating the underlying ontology into the argument graph has two advantages: First, it ensures that the set of arguments and the set of possible claims are coherent. Secondly, arguments with structure (the facts and rules they are based on) can be used to explain *why* a classification is ambiguous.

The claims of arguments are assertions about accident types, for example `StaircaseAccident(ds)`. The evidence or support of arguments are assertions together with classification rules. Each argument therefore represents the application of a rule to a particular case, as shown in Example 2.

The following definition of arguments captures allows us to use the full expressiveness of whatever description logic we choose as the foundation of our particular ontology Δ, since it only refers to the subconcept relationship, which is part of every description logic.

Definition 4 (Δ-Argument). *Let Δ be an ontology, let $C \sqsubseteq D$ be a GCI and let i be an individual. Then $\langle i, C \sqsubseteq D\rangle_\Delta$ is an* argument *if and only if*

1. $\Delta \models C \sqsubseteq D$ *and*
2. $\Delta \models C(i)$

We omit the subscript Δ whenever possible. If $A = \langle C \sqsubseteq D, i\rangle$ is an argument then $D(i)$ is the **claim** of A, $C \sqsubseteq D$ is its **support** and the individual i is its **witness**. For every Δ-Argument $\langle i, C \sqsubseteq D\rangle_\Delta$, the claim $D(i)$ can also be inferred from the ontology, that is, $\Delta \models D(i)$. If two arguments A, B relate to the same individual and the claim of argument A is a subconcept of argument B's support, we say that A **leads to** B, short $A \rightsquigarrow B$:

Definition 5 *Let $A = \langle i, C \sqsubseteq D\rangle_\Delta$, $B = \langle i, E \sqsubseteq F\rangle_\Delta$ be two arguments. A leads to B, short $A \rightsquigarrow B$, if and only if $\Delta \models D \sqsubseteq E$.*

Since $\sqsubseteq$ is reflexive and transitive, $\rightsquigarrow$ is reflexive, transitive and antisymmetric (wrt. equivalence of concepts under $\equiv_\Delta$). Every classification is captured by a corresponding argument.

Example 2 For the report `ds` we get two arguments A_1 and A_2. For A_1 we have the following components:

$$support(A_1) = \exists\texttt{reported(Incident, slipperyStaircase)} \sqsubseteq \texttt{StaircaseAccident}$$
$$witness(A_1) = \texttt{ds}$$
$$claim(A_1) = \texttt{StaircaseAccident(ds)}$$

and for A_2 we get

$support(A_2) = \texttt{OccupationalAccident} \sqsubseteq \texttt{OccupationalAccident}$
$witness(A_2) = \texttt{ds}$
$claim(A_2) = \texttt{OccupationalAccident(ds)}$

Note that we use the fact that every concept is a sub-concept of itself to model "atomic" arguments, i.e. arguments whose support and claim are identical. Atomic arguments don't involve a reasoning step to get from support to conclusion, but they are helpful for assertions that can be derived directly from questionnaires, for example if the accident type is stated explicitly on the form.

Now that we can express rules as arguments, we need to define attacks in order to identify inconsistencies in a rule set, the most important aspect of practical applications of argumentation (cf. [1]).

In our case, the "attacks" relation should reflect ambiguous classifications of accidents. We have a set C_{acc} of concepts of accident types with $C_{acc} = \{\texttt{Accident}, \texttt{StaircaseAccident}, \ldots\}$. If there are two arguments that classify the same individual into two different classes $C, C' \in C_{acc}$, then there should be an attack between them. If we further know that C is strictly preferred over C', the attack should be asymmetric, defeating only the "weaker" argument for C'. As rule designers we are interested in cases where no preference can be established, that is, in symmetric attacks (resulting in multiple preferred extensions).

To avoid any problems with the consistency of the graph's extensions [7] we define attacks with respect to $\sqsubseteq$ for a given ontology Δ, and we will require the set of arguments to be well formed:

Definition 6 *A set of Δ-arguments $\mathcal{A}$ over an ontology Δ is* ***well-formed*** *iff for every pair of arguments $\langle i, C \sqsubseteq D\rangle_\Delta \in \mathcal{A}$, $\langle i, E \sqsubseteq F\rangle_\Delta \in \mathcal{A}$: If $\Delta \models D \sqsubseteq E$ then there is an argument $\langle i, C \sqsubseteq F\rangle_\Delta \in \mathcal{A}$*

Any set of arguments is the subset of well-formed set of arguments, which can be computed by adding the appropriate arguments.

Proposition 4 *For every set of Δ-arguments S, there exists a set of Δ-arguments S' such that S' is well-formed and $S \subseteq S'$.*

Proof S' is the closure of S under Def. 6.

We will first define an attacks relation based on the requirements outlined above. We then formalise the concept of "ambiguous classification" and show that ambiguous classifications lead to symmetric attacks as required.

Definition 7 **(Attack).** *Let C be a set of concepts with a partial order $\leq_C$, let Δ be an ontology and let A_1, A_2 be Δ-arguments with $support(A_1) = \texttt{C} \sqsubseteq \texttt{D}$ and $support(A_2) = \texttt{C}' \sqsubseteq \texttt{D}'$. A_1* ***Δ-attacks*** *A_2 iff*

1. *$witness(A_1) = witness(A_2)$ and $\Delta \not\models \texttt{D} \equiv \texttt{D}'$ and $\texttt{D}' \not<_C \texttt{D}$ or*
2. *There exists an argument A_3 such that A_1 attacks A_3 and $A_3 \rightsquigarrow A_2$*

Condition 2 of Def. 7 ensures that attacks on sub-arguments are propagated upwards in the concept hierarchy of Δ. In our definition of Δ-attacks we do not have to worry about the rationality postulates by Caminada and Amgoud [7] since the knowledge represented by our arguments is always derived from a consistent ontology. We cannot obtain an inconsistent set of formulae from a consistent set of arguments because *any* set of arguments stands for a consistent set of formulae. Attacks in our argumentation system are thus an expression of underspecification rather than overspecification (causing direct inconsistency).

According to this definition, argument A_2 attacks A_1 (p.) if we assume that `OccupationalAccident` is ranked strictly higher than `StaircaseAccident`. Arguments with an ontology and a partial order of concepts (classes) form an ontology-based argumentation system, short OAS.

Definition 8 *An* ***ontology-based argumentation system*** *(OAS) is a triple* $(\Delta, \mathcal{A}, \leq_C)$ *where*

1. Δ *is an ontology*
2. $\mathcal{A}$ *is a well-formed set of* Δ*-arguments*
3. $\leq_C$ *is a partial order of concepts*

Every OAS gives rise to an argument graph whose attacks are determined by $\leq_C$, according to Def. 7.

Definition 9 *The* ***ontology argument graph*** *of an OAS* $(\Delta, \mathcal{A}, \leq_C)$ *is given by* $(\mathcal{A}, \mathcal{R}_{\leq_C})$ *where* $\mathcal{R}_{\leq_C}$ *is the* Δ*–attacks relation induced by* $\leq_C$*.*

The argumentation formalism developed above can be used to express the problem of ambiguous classifications.

Definition 10 *An* ***ambiguous classification*** *in an OAS* $(\Delta, \mathcal{A}, \leq_C)$ *is a pair of arguments* $a = \langle i, C \sqsubseteq D \rangle_\Delta$, $a' = \langle i', C' \sqsubseteq D' \rangle_\Delta$ *with* $a, a' \in \mathcal{A}^*$ *such that*

1. $\Delta \not\models D \equiv D'$ *and*
2. $i = i'$ *and*
3. *There is no strict preference on* $\leq_C$ *under* $\{D, D'\}$

It follows from Def. 7 that a and a' mutually attack each other. In fact, all mutual attacks in the argument graph arise from ambiguous classifications:

Proposition 5 *Let* $(\Delta, \mathcal{A}, \leq_C)$ *be an OAS, let* $G = (\mathcal{A}^*, \mathcal{R}_{\leq_C})$ *be its argument graph and let* $a, a' \in \mathcal{A}^*$. $\{a, a'\}$ *is an ambiguous classification if and only if* $\{(a, a'), (a', a)\} \subseteq \mathcal{R}_{\leq_C}$*.*

Proof ($\Rightarrow$) Let $a = \langle i, C \sqsubseteq D \rangle_\Delta$, $a' = \langle i', C' \sqsubseteq D' \rangle_\Delta$ be an ambiguous classification in G. Then, by Def. 10 Cond. 3, $D' \not<_C D$, so a attacks a'. By Def. 10 Cond. 3 again, $D \not<_C D'$, so a' attacks a. Therefore, $\{(a, a'), (a', a)\} \subseteq \mathcal{R}_{\leq_C}$

($\Leftarrow$) Let $a = \langle i, C \sqsubseteq D \rangle_\Delta$, $a' = \langle i', C' \sqsubseteq D' \rangle_\Delta$ be arguments with $\{(a, a'), (a', a)\} \subseteq \mathcal{R}_{\leq_C}$. By Def. 7 Cond. 1, $a' \not<_C a$ and $a \not<_C a'$, so by Def. 10 a, a' is an ambiguous classification.

Ambiguous classifications can be resolved by adding a strict preference, that is, by adjusting the preference relation $\leq_C$. If the classification of an incident report is non-ambiguous, then the grounded extension consists of exactly one argument and it is a stable extension, too:

Proposition 6 *Let $S = (\Delta, A, \leq_C)$ be an OAS such that there is only one individual i in Δ, let G be its argument graph and let $\mathcal{E}$ be the grounded extension of G. Then either $\mathcal{E} = \emptyset$ or $|\mathcal{E}| = 1$ and $\mathcal{E}$ is stable*

Proof Let S, G and $\mathcal{E}$ as stated. ($|\mathcal{E}| \leq 1$) Assume $A, B \in \mathcal{E}$ with $A \neq B$. Then either the conclusions of A, B are unrelated in $\leq_C$ - in which case A and B mutually attack each other, or $A >_C B$ or $B >_C A$ - in which case there is an asymmetric attack, so $\{A, B\}$ is not conflict-free (contradiction - $\mathcal{E}$ conflict-free).

(2 $|\mathcal{E}| = 1$ implies $\mathcal{E}$ stable): Assume $\mathcal{E} = \{A\}$ is the grounded extension and let $B \neq A$. If $A \leq_C B$, then B attacks A. Because $A \in \mathcal{E}$, there must be an argument $A' \in \mathcal{E}$ such that A' defends A and $A' \neq A$. This contradicts the assumption that $|\mathcal{E}| = 1$. Otherwise, if A and B unrelated in $\leq_C$, then A and B mutually attack each other and the same argument applies. Therefore $A > B$, so A attacks B.

3.3 Acting on Classifications

Once a claim has been classified, several actions such as sending an e-mail or adding an item to an analyst's work list are triggered. If the accident type is ambiguous, those actions can only be performed after a the classification has been confirmed manually. However, our argumentation system allows us to identify some actions that may be performed even before the accident type is resolved.

Suppose we want to send an e-mail (`SendEmail`) for any accidents regardless of their type (`Accident`), and create a work-item `WorkItem` only for accidents involving staircases. We can express this with two rules, `Accident` $\sqsubseteq$ `SendEmail` and `StaircaseAccident` $\sqsubseteq$ `WorkItem`.

Given the incident report from p. , we get two additional arguments A_3 and A_4. The conclusion of A_3 is `SendEmail(ds)` and the conclusion of A_4 is `WorkItem(ds)`. Recall that A_2 attacks A_1 (p.). Consequently, A_2 attacks A_4 as A_4 uses the conclusion of A_1. The grounded extension of the argument graph is therefore A_2, A_3, so `SendEmail` is sceptically acceptable.

If we did not strictly prefer `OccupationalAccident` over `StaircaseAccident`, then A_1 and A_2 would mutually attack each other. The resulting graph then has two preferred extensions, one that contains A_2 and A_3 and one that contains A_1, A_3 and A_4. In this case, the argument for `WorkItem` is only credulously acceptable and an analyst has to review the case before the action can be triggered. `SendEmail` however is part of the grounded extension and thus is acceptable in any case, so the action can therefore be performed immediately.

4 Conclusion

In this paper we described the problem of ambiguous classifications of accident reports in the insurance industry and proposed a method for improving the set of classification rules. Our approach utilises the explanatory power of argumentation to draw out the causes of ambiguous classifications. We discussed some formal properties of our system and demonstrated its use with a case study based on real-world data.

Argumentation has been applied to the merging of ontologies [6, 10] and to handling inconsistent ontologies [11], two problems arising from overspecification. The classification problem on the other hand is one of underspecification.

The paper [2] applies argumentation to a classification problem, by using it to enhance a statistical (machine learning) classifier. A similar direction was taken in [14] where several agents, each with their own local training data, use argumentation to reconcile classifications. The fundamental difference between [2, 14] and our approach is that our argumentation system is used to analyse and improve a set of hand-written classification rules, whereas the other two are applications of argumentation to machine-learning techniques.

There are two avenues for future work. First, we would like to investigate the use of argumentation for identifying actions as outlined in Sect. 3.3. Secondly, we want to use the argumentation system to automatically suggest improvements to the preference order of accident types, instead of just pointing out the sources of ambiguity.

References

1. Amgoud, L., Besnard, P.: Logical limits of abstract argumentation frameworks. J. Appl. Non-Class. Logics **23**(3), 229–267 (2013)
2. Amgoud, L., Serrurier, M.: Agents that argue and explain classifications. Auton. Agent. Multi-Agent Syst. **16**(2), 187–209 (2008)
3. Baader, F., Horrocks, I., Sattler, U.: Description Logics, International Handbooks on Information Systems, vol. 9: Handbook on Ontologies, chap. 1, pp. 3–28, 1st edn. Springer (2004)
4. Bechhofer, S., van Harmelen, F., Hendler, J., Horrocks, I., McGuinness, D.L., Patel-Schneider, P.F., Stein, L.A.: OWL Web Ontology Language (2004)
5. Besnard, P., Hunter, A.: Elements of Argumentation, 1st edn. The MIT Press, Cambridge (2008)
6. Black, E., Hunter, A., Pan, J.Z.: An argument-based approach to using multiple ontologies. In: Third International Conference On Scalable Uncertainty Management (SUM'09), pp. 68–79. No. 5785 in LNCS. Springer (2009)
7. Caminada, M., Amgoud, L.: On the evaluation of argumentation formalisms. Artif. Intell. **171**(5–6), 286–310 (2007)
8. Dung, P.: On the acceptability of arguments and its fundamental role in nonmonotonic reasoning, logic programming and n-person games. Artif. Intell. **77**(2), 321–357 (1995)
9. Eiter, T., Ianni, G., Krennwallner, T., Polleres, A.: Rules and ontologies for the semantic web. In: Baroglio, C., Bonatti, P.A., Maluszynski, J., Marchiori, M., Polleres, A., Schaffert, S. (eds.) Reasoning Web. Lecture Notes in Computer Science, vol. 5224, pp. 1–53. Springer (2008)
10. Flouris, G., Huang, Z., Pan, J.Z., Plexousakis, D., Wache, H.: Inconsistencies, negations and changes in ontologies. In: 21st AAAI Conference, pp. 1295–1300 (2006)

11. Gómez, S.A., Nevar, C.I.C., Simari, G.R.: Reasoning with inconsistent ontologies through argumentation. Appl. Artif. Intell. **24**(1–2), 102–148 (2010)
12. Gruber, T.R.: Toward principles for the design of ontologies used for knowledge sharing. Int. J. Hum.-Comput. Stud. **43**(5–6), 907–928 (1995)
13. Kaci, S., van der Torre, L.W.N., Weydert, E.: Acyclic argumentation: Attack = conflict + preference. In: Brewka, G., Coradeschi, S., Perini, A., Traverso, P. (eds.) ECAI. Frontiers in Artificial Intelligence and Applications, vol. 141, pp. 725–726. IOS Press (2006)
14. Wardeh, M., Coenen, F., Bench-Capon, T.J.M.: Multi-agent based classification using argumentation from experience. Auton. Agent. Multi-Agent Syst. **25**(3), 447–474 (2012). http://dx.doi.org/10.1007/s10458-012-9197-6

Journal Impact Factor Revised with Focused View

Tetsuya Nakatoh, Hayato Nakanishi and Sachio Hirokawa

Abstract The evaluation of an article is an important issue for scientific research. A journal impact factor is used widely to evaluate the impact of the journals. An impact factor is considered as an influence measure of articles. Articles in a high impact factor journal tend to have strong impact on the wide research fields. However, this does not imply that they have big influence in a particular research area. Measuring the speciality and the generality of research results is not trivial task. The present paper proposes a generalized method to evaluate the influence of a journal with respect to a focused view. An empirical evaluation was conducted on "bibliometrics" related 10,186 articles.

Keywords Bibliometrics · Scientometrics · Impact factor

1 Introduction

Literature review is a basic activity of scientific research. Databases of scientific articles are useful to search for related work. However, the evaluation of the search results is more difficult compared to search them. The citation count of an article that represents the number of articles that refer the article is used as an objective evaluation.

T. Nakatoh (✉) · S. Hirokawa
Research Institute for Information Technology, Kyushu University,
Hakozaki 6-10-1, Fukuoka 812-8581, Japan
e-mail: nakatoh@cc.kyushu-u.ac.jp

S. Hirokawa
e-mail: hirokawa@cc.kyushu-u.ac.jp

H. Nakanishi
Graduate School of Integrated Frontier Sciences, Kyushu University,
Hakozaki 6-10-1, Fukuoka 812-8581, Japan

R. Neves-Silva et al. (eds.), *Intelligent Decision Technologies*,
Smart Innovation, Systems and Technologies 39,
DOI 10.1007/978-3-319-19857-6_40

Evaluation of an journal is widely used for approximated evaluation of an article. It is based on an assumption that the content of articles on a good journal should have hight quality. The journal impact factor [3, 4] is one of the most known measures. Given a journal and a year, the journal impact factor evaluates how the recent articles of the journal are referred in the present year. The following Eq. (1) formally defines the journal impact factor.

$$IF(j, y) = \frac{C(P(\{j\}, \{y-1, y-2\}), P(J, \{y\}))}{|P(\{j\}, \{y-1, y-2\})|}, \quad (1)$$

where $P(J, Y)$ represents the set of all articles published in the journals J in the years Y and $C(X, Z)$ denotes the number of citations of the articles in X in the set of articles Z.

The journal impact factor is considered as the most up to date evaluation of the journal and reliable measure assuming that the quality of articles on a journal does not change drastically. The journal impact factor is used for assessment of researcher's achievement as well as the measure of articles.

The journal impact factor does not mean that the articles in a hight scored journal match the purpose of a user who searches for an appropriate article in a particular field or with a keyword. When we search for related articles, we consider the following two kinds of qualities.

(A) Relatedness with respect to query words
(B) Reputation based on the journal

A journal covers several research genres. The impact factor does not always evaluate the different genres equally. For example, "Nature" is known as one of the best scientific journal with a high impact factor. There are some good articles on "bibliometrics" in Nature. But, it is not a journal focused only on "bibliometrics" but covers a wide range genres.

The present paper proposes a combination of the impact factor and the relatedness with respect to the search genre. We consider the process of literature review consists of the first step where related articles are retrieved by appropriated keywords and the second step where they are narrowed down according to the quality evaluation. A difficulty is that we cannot combine this two process at the same time. Imagine that two articles X and Y are obtained by a keyword and that X is published in Nature and that Y is published in a journal A. If we believe only in the impact factor, then we would choose X first for detailed inspection even though Y matches best with the keyword. Particularly, when the impact factor of A is comparatively low, we would miss the chance to consider Y. We solve this dilemma to choose either the relatedness or the impact factor, by restricting the set of articles in counting citation for the focused genre.

We are aiming at the search technique for a literature survey and construction of a search engine for it. The present paper considers genre designation using a key word and proposes the focused impact factor. We show that the focused impact factor is suitable as a combination of relatedness and the impact factor.

2 Related Work

The most crucial task in the literature review is in selecting the articles with respect to the search purpose. If we think of appropriate query keywords, search engine is useful to find those articles that match our requests. However, in the next step of search, we need to read the abstracts and the contents to make an evaluation of the articles we get. This second process is much harder and time consuming compared to the search process by search engine. The natural language processing is not yet matured to evaluate the quality of the article objectively nor subjectively. Citation information represents some kind of reputation among the research community. It is widely used as a surrogate measure for objective evaluation. The best known one is the journal impact factor.

Journal Impact Factor (JIF) [3, 4] is one of the most popular evaluation measure of scientific journals. Thomson Reuters updates and provides the score of journals in Journal Citation Reports (JCR) every year. The JIF of a journal describes the citation counts of an average article published in the journal. JIF is considered as a defacto standard to evaluate not only a journal, but also a researcher and a research organization as well.

Pudovkin and Garfield [1] pointed out that JIF is not appropriate to be used as a measure to compare the different disciplines. They chose several genres to compare JIF of top ranked journals of each genre. They revealed that JIF of journals in "Physics, Multidisciplinary" and that of "Genetics & Heredity" have a big difference. Modification of JIF by normalization has been studied as one of the key issues [1, 2].

The purpose of the present paper and our goal is not in the evaluation of journals but in considering a better method for literature review. The impact factor is helpful to select good journals, however, it does not guarantee that an article in high JIF journal is strongly related to the query a user is looking for. We have to face a choice of JIF and the relatedness of the contents. The present paper proposes to combine the relatedness of a journal to the user's query with JIF.

Bergstrom proposed EigenFactor [5] to get rid of the problem of impact factor. It is listed in JCR as well. EigenFactor adjusts the weight for each citation, where JIF treats all citations equally. EiginFactor of a citation of an article of a journal A from another article of another journal B is increased when A has a smaller citation count than B does. Our proposal is similar to EigenFactor in the sense that both care the quality of each citation. The difference is that EigenFactor uses the weight of the citing journal and that we consider the contents of the citing journal. We restrict the citing articles with a keyword in calculation of the impact factor. Thus our method is much concerned with the contents of the articles.

3 Proposal of Focused Impact Score

We propose Focused Impact Score in order to evaluate journals. It gives the evaluation value of the journal for the field for the purpose of investigation.

The Eq. (1) is a formula of Impact Factor [3, 4]. The journals J of the Eq. (1) is "Web of Science Core Collection" provided by Tomson Roiter. Therefore, the definition of Impact Factor contains not only a formula but a database.

First, we propose Impact Score (*IS*) as a valuation method of journals. The database of the *IS* is not limited to "Web of Science Core Collection." Moreover, the *IS* limits the investigation scope using keywords. The Eq. (2) is the definition of the *IS*.

$$IS(j, W, y) = \frac{C(P'(\{j\}, W, \{y-1, y-2\}), P'(J, \emptyset, \{y\}))}{|P'(\{j\}, W, \{y-1, y-2\})|}, \tag{2}$$

where $P'(J, W, Y)$ represents the set of all articles published in the journals J in the years Y and contains the keywords W, and $C(X, Z)$ denotes the number of citations of the articles in X in the set of articles Z.

Furthermore, we define Focused Impact Score(F-IS). F-IS restricts the scientific field of the cited article. This classifies two sorts of articles as follows.

1. When the cited article is not related to a query keyword.
2. When the cited article is related to a query keyword.

By using only 2, the citation by the article of a field with few relations is not used for calculation of F-IS.

In this paper, restriction by the same keyword as *IS* is used as a method of restricting the field of cited articles. The Eq. (3) is the calculation method of Focused Impact Score by this proposal.

$$F\text{-}IS(j, W, y) = \frac{C(P'(\{j\}, W, \{y-1, y-2\}), P'(J, W, \{y\}))}{|P'(\{j\}, W, \{y-1, y-2\})|}, \tag{3}$$

where $P'(J, W, Y)$ is a set of a article which published in journal J at year Y, and which has the query keywords W. $C(X, Z)$ is the total number by which a article in X was cited in a article in Z.

4 Evaluation of Focused Impact Score

4.1 Gathering Article Data

The data of articles was gathered from Scopus.[1] In this experiment, "bibliometrics" was chosen as a query keyword. 10186 articles published from 1976 to 2015 were gathered using search API.

4.2 Basic Analysis

In this section, we conduct basic analysis of the obtained articles. Although some articles have no citation at all, there is also an article with 2,977 citation. Table 1 shows a part of number of citations of articles. It has the number of citations of the top 10 articles, and of bottom 10 articles.

Figure 1 plots the data of Table 1 as a log-log graph. In this graph, the frequency of citations seems to follow power law.

4.3 Comparison of Impact Score and Focused Impact Score

We selected ten journals with many articles as journals to analyze. Table 2 shows the selected journals and the number of articles.

We will analyze in detail the data of 10 journals published in five latest years (from 2010 to 2014). Since the data in 2015 would still be imperfect, it was not included in the objects for analysis.

Table 1 Number of articles

Cited by	Number of articles	Cited by	Number of articles
0	3024	:	:
1	1236	1003	1
2	759	1003	1
3	569	1003	1
4	483	1003	1
5	422	1003	1
6	325	1003	1
7	284	1229	1
8	241	1441	1
9	178	1630	1
:	:	2977	1

[1]http://www.scopus.com/.

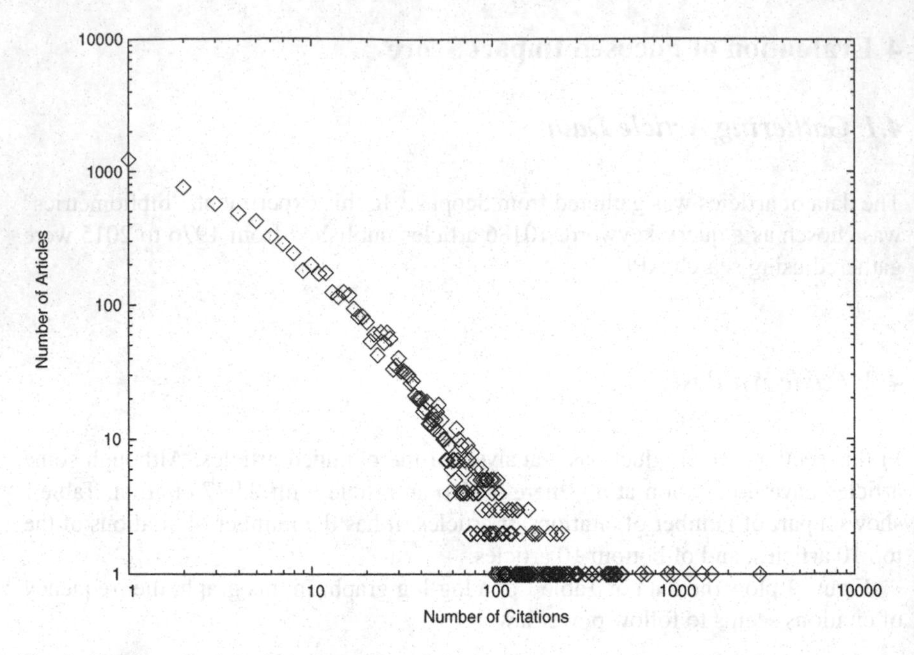

Fig. 1 Number of articles in each number of citations

Table 2 Number of articles on each journal

Number of articles	Journal name
726	Scientometrics
226	JASIS
178	Nature
138	Journal of Informetrics
100	PLoS ONE
84	Information Processing and Management
84	Research Evaluation
81	Technological Forecasting and Social Change
79	Journal of Information Science
59	Research Policy
58	Journal of Documentation

First, Impact Score and Focused Impact Score for every year were calculated. Figure 2 and Fig. 3 made transitions of the value the line graph. A clear difference cannot be read in two graphs. However, both Scores in 2014 are low in the graphs. It may be the reason that capture of new information is not perfect (Fig. 4).

We decided that the data in 2012 was stable from the graph. Then, the data in 2012 was decided to be an object of detail analysis. Impact Score and Focused Impact Score were calculated about 10 journals in 2012. Figure 5 plots them to two dimensions. Some correlation is found by two values. However, there are also some journals with a different tendency.

The straight line in Fig. 5 is a mean line of plots. The journals of conformity above this straight line increase by the focus. Conversely, the journals of conformity below this straight line decrease by the focus. As an example, the graph shows that the conformity of "Scientometrics" and "JAIST" is high. Moreover, "LNCS", "Nature", "PLoS ONE", and "TechnologicalForecasting and Social Change" have low conformity.

Now, we introduce *Focus-Ratio* as a ratio of two scores. The definition is shown in Eq. 4.

$$Focus\text{-}Ratio = \frac{F\text{-}IS(j, W, y)}{IS(j, W, y)} \tag{4}$$

Figure 6 expresses the relation between the *Focus-Ratio* and the number of articles. The graph shows "Scientometrics" has many articles and conforms to "bibliometorics." Although there are not many articles which journal "Research Evaluation" has relatively, it has high relation to "bibliometorics".

The relation between *Focus-Ratio* and the number of citations was expressed to Fig. 7. This graph is easier to distinguish. "Nature" and "Technological Forecasting and SocialChange" have many citations or the relation to "bibliometorics" is not high.

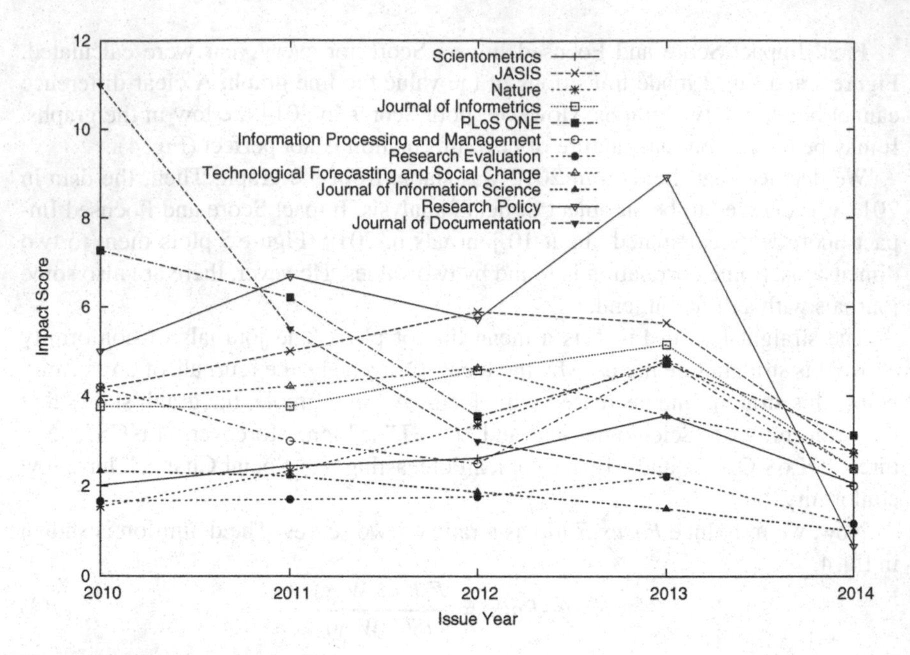

Fig. 2 Secular change of impact score

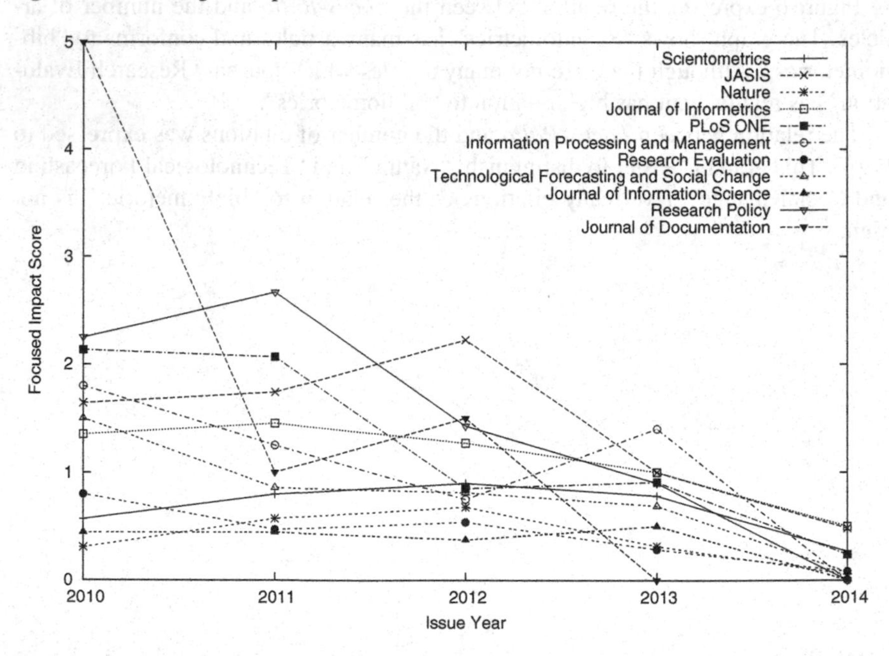

Fig. 3 Secular change of focused impact score

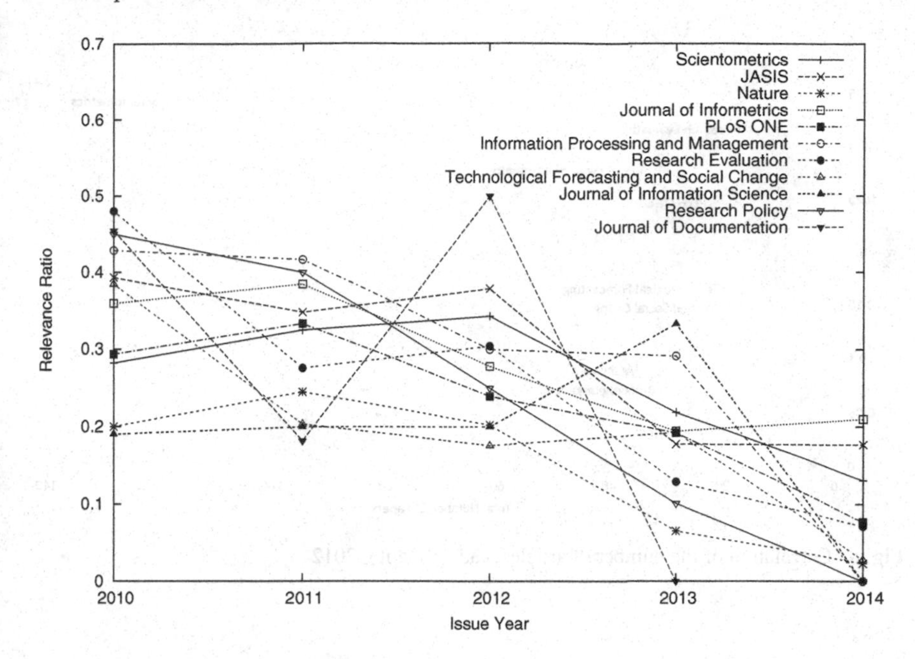

Fig. 4 Secular change of relevance ratio

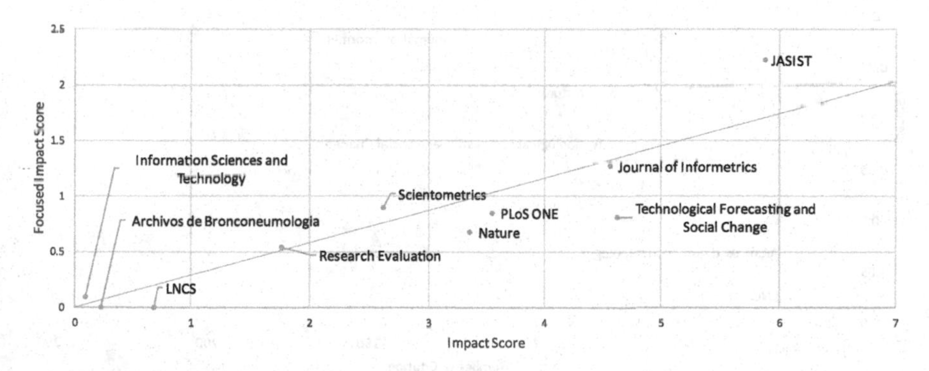

Fig. 5 Comparison of impact score & focused impact score, 2012

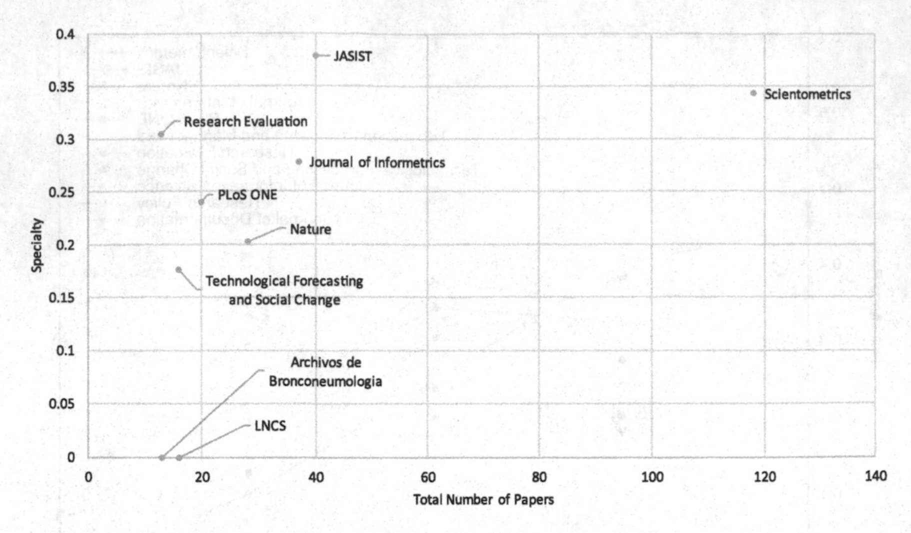

Fig. 6 Correlation of the number of articles, and speciality, 2012

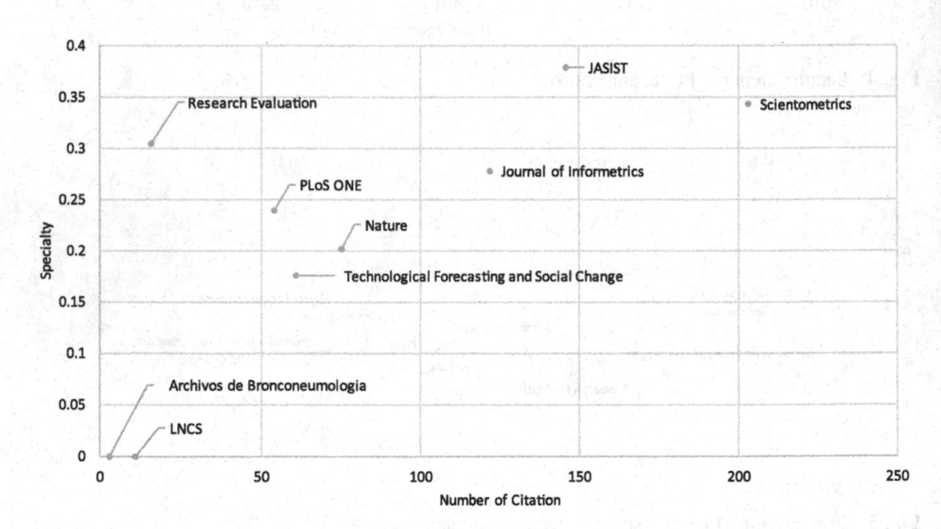

Fig. 7 Correlation of the number of citations, and speciality, 2012

5 Conclusion

Literature review quires selection of articles among a large candidate set of articles obtained the query words. Objective measures are crucial in evaluating the contents and the quality of articles. The journal impact factor is one of the standard measure to evaluate journals. However, it does not distinguish the relatedness of the articles that appear in the same journal. The present paper proposes the focused impact factor that combines the citation analysis with relatedness to the query. Empirical study is conducted to confirm the usefulness of the proposed measure.

We plan to estimate the effectiveness of the proposed method in cases using more than one key word. Construction of the literature survey system based on the method is a further work.

Acknowledgments This work was partially supported by JSPS KAKENHI Grant Number 24500176.

References

1. Pudovkin, A.I., Garfield, E.: Rank-normalized impact factor: a way to compare journal performance across subject categories. In: Proceedings of the ASIST Annual Meeting, vol. 41, pp. 507–515 (2004)
2. Marshakova-Shaikevich, I.: The standard impact factor as an evaluation tool of science fields and scientific journals. Scientometrics **35**(2), 283–290 (1996)
3. Garfield, E.: Citation indexes for science. Science **122**(3159), 108–111 (1955)
4. Garfield, E.: The history and meaning of the journal impact factor. J. Am. Med. Assoc. **295**(1), 93–99 (2006)
5. Bergstrom, C.: Eigenfactor: Measuring the value and prestige of scholarly journals. **68**(5) (2007)

The Improvement of Pairwise Comparison Method of the Alternatives in the AHP

Kazutomo Nishizawa

Abstract The aim of this study was the improvement of the evaluation method of pairwise comparison for the Analytic Hierarchy Process (AHP) and the Analytic Network Process (ANP). In the traditional AHP and the ANP, the evaluations of pairwise comparison are carried out using a nine-stage evaluation process, and then the same process in inverse. However, in fact, the maximum evaluation value of 9 has been discovered not to have been very useful. This paper proposes the improvement of the pairwise comparison method for the AHP and a procedure that does not perform pairwise comparison for each criterion. The proposed method hypothetically offered two beneficial results. One is the ability to take into account the importance of the criteria for the evaluation of each alternative without the pairwise comparison of each criterion. The other is that the overall evaluation of the alternatives is easily obtained by the row sum of the principal eigenvector for each criterion.

Keywords AHP · ANP · Pairwise comparison · Criteria · Alternatives

1 Introduction

This study focuses on improvement of the evaluation method of pairwise comparison for the Analytic Hierarchy Process (AHP) and the Analytic Network Process (ANP). This improvement method is related to the normalization method of alternatives and the overall evaluation in the AHP and the ANP.

In the traditional AHP and the ANP, the pairwise comparisons were carried out using a nine-stage evaluation process. However, in fact, the maximum evaluation value 9 seems to not to have been very useful. In the traditional AHP [9], $\boldsymbol{w}$, the overall evaluation vector of alternatives is obtained using Eq. (1),

$$\boldsymbol{w} = \boldsymbol{W}\boldsymbol{v}, \tag{1}$$

K. Nishizawa (✉)
Nihon University, 1-2-1 Izumicho, Narashino, Chiba 275-8575, Japan
e-mail: Kazutomo@nihon-u.ac.jp

R. Neves-Silva et al. (eds.), *Intelligent Decision Technologies*,
Smart Innovation, Systems and Technologies 39,
DOI 10.1007/978-3-319-19857-6_41

where $\boldsymbol{W}$ is the weight matrix of alternatives which were evaluated by each criterion, and $\boldsymbol{v}$ is the weight vector of criteria. In the traditional method, the sum of the column vector of matrix $\boldsymbol{W}$ is normalized to 1. This normalization method entails some problems, for example rank reversal [1]. The Dominant AHP was proposed by Prof. Kinoshita et al. [3, 4], and other improved normalization methods were also proposed [2, 11]. Furthermore, in the traditional ANP [10], the super matrix $\boldsymbol{S}$ is constructed and the converged matrix $\boldsymbol{S}^{\infty}$ is calculated. The form of the super matrix of the two-cluster ANP, which is equivalent to the AHP, is shown in Eq. (2),

$$\boldsymbol{S} = \left[\begin{array}{c|c} \boldsymbol{0} & \boldsymbol{V} \\ \hline \boldsymbol{W} & \boldsymbol{0} \end{array}\right], \tag{2}$$

where sub-matrix $\boldsymbol{W}$ is the evaluation weight of the alternatives, sub-matrix $\boldsymbol{V}$ is the evaluation weight of the criteria and $\boldsymbol{0}$ is a zero matrix. Sub-matrix $\boldsymbol{W}$ in Eq. (2) and $\boldsymbol{W}$ in Eq. (1) are equivalent. Sub-matrix $\boldsymbol{V}$ is not constructed through the traditional AHP. From the result of $\boldsymbol{S}^{\infty}$, we can obtain the weights of criteria and the weights of alternatives using the traditional ANP method. (Note that $\boldsymbol{S}$ must be a stochastic matrix, so we need the normalization operation of $\boldsymbol{S}$. In the traditional ANP, each sum of the column vector of $\boldsymbol{S}$ is normalized to 1.)

However, in the pairwise comparisons of the AHP and the ANP, it is not easy to compare each criterion. Therefore, in the previous studies of the ANP, an improved super matrix, consisting of the sub-matrix $\boldsymbol{W}$ and the transposed sub-matrix $\boldsymbol{W}^T$, was proposed [5, 6]. This method was called Transposed Matrix Method (TMM). As shown below, $\boldsymbol{W}^T$ is used instead of $\boldsymbol{V}$ in Eq. (2), and the resulting form of the TMM super matrix is shown in Eq. (3).

$$\boldsymbol{S} = \left[\begin{array}{c|c} \boldsymbol{0} & \boldsymbol{W}^T \\ \hline \boldsymbol{W} & \boldsymbol{0} \end{array}\right] \tag{3}$$

Meanwhile, the normalization method in the AHP and the ANP is very important. In previous studies, improved normalization methods for the AHP and the TMM were proposed [8]. However, from results obtained using the proposed method, the improvements were insufficient to take into account the weight of criteria in the overall evaluation of alternatives [7, 8].

To reiterate, this study proposes an improved pairwise comparison method for the AHP. In the proposed method, in the pairwise comparison between alternatives, the maximum evaluation value for each criterion is introduced. And a procedure not employing the pairwise comparison for each criterion is also proposed. In this paper, the proposed method is applied to the AHP. As a result, the calculation process of the proposed method has been simplified compared with the traditional AHP.

Section 2 of this paper describes the proposed method. Section 3 explains the verification of the proposed method, starting with the example in which the traditional AHP method is used. Then the result of the example using the proposed method is shown. Finally, in Sect. 4, we discuss the results obtained through the proposed method and conclude this study.

2 The Proposed Method

In the traditional AHP and the ANP, the evaluations of pairwise comparisons were carried out through a nine-stage evaluation process, and then this same process in inverse. Therefore, in this study, the maximum evaluation value is introduced in the pairwise comparison of the alternatives for each criterion. In the proposed method, the pairwise comparisons between each criterion are not performed in the calculation of the the overall evaluation of the alternatives.

Procedure of the proposed method:

P1: Determine the maximum evaluation value of the pairwise comparison for each criterion.
P2: Perform the pairwise comparison of alternatives based on the maximum evaluation value in P1.
P3: Based on the comparison matrix in P2, calculate the row sum of the principal eigenvector of each alternative.

This procedure allows for greater ease in obtaining overall evaluation of alternatives compared with the traditional method.

In this section, the concept of the proposed method is explained.

First, let's consider the simplest 2 × 2 comparison matrix. Assuming that the maximum evaluation value is 2, then the comparison matrix $\boldsymbol{A}$ and the corresponding principal eigenvector $\boldsymbol{w}$ (w_1 and w_2) are obtained in Eq. (4) using the power method. The principal eigenvector $\boldsymbol{w}$ is not normalized.

$$\boldsymbol{A} = \begin{bmatrix} 1 & 2 \\ 1/2 & 1 \end{bmatrix}, \qquad \boldsymbol{w} = \begin{bmatrix} 1.000000 \\ 0.500000 \end{bmatrix} \tag{4}$$

Similarly, assuming that the maximum evaluation value is 9, the results obtained are shown in Eq. (5).

$$\boldsymbol{A} = \begin{bmatrix} 1 & 9 \\ 1/9 & 1 \end{bmatrix}, \qquad \boldsymbol{w} = \begin{bmatrix} 1.000000 \\ 0.111111 \end{bmatrix} \tag{5}$$

As a result, from Eq. (4) and Eq. (5), the value range of the principal eigenvector is different from the maximum evaluation values.

Next, let's consider the 4 × 4 comparison matrix. The example matrix $\boldsymbol{A}$ and the principal eigenvector $\boldsymbol{w}$ (w_1 to w_4) obtained are shown in Eq. (6). The calculated vector $\boldsymbol{w}$ is not normalized and therefore, in the power method, the maximum value of $\boldsymbol{w}$ is 1. Since, all elements of the upper triangular of $\boldsymbol{A}$ are greater than 0, the order of elements of $\boldsymbol{w}$ is $w_1 > w_2 > w_3 > w_4$.

$$\boldsymbol{A} = \begin{bmatrix} 1.0000 & 2.0000 & 4.0000 & \underline{5.0000} \\ 0.5000 & 1.0000 & 2.0000 & 3.0000 \\ 0.2500 & 0.5000 & 1.0000 & 2.0000 \\ 0.2000 & 0.3333 & 0.5000 & 1.0000 \end{bmatrix}, \qquad \boldsymbol{w} = \begin{bmatrix} 1.000000 \\ 0.521180 \\ 0.281770 \\ 0.170338 \end{bmatrix} \tag{6}$$

In Eq. (6), the maximum element of $\boldsymbol{A}$ is $\boldsymbol{a_{14}}$, and the value of this element is 5 (underlined). We intentionally convert the value of $\boldsymbol{a_{14}}$ to the arbitrary maximum evaluation value, and the other elements of $\boldsymbol{A}$ are proportionally converted.

In the first example, we convert the value of $\boldsymbol{a_{14}}$ in $\boldsymbol{A}$ to 2. The converted comparison matrix $\boldsymbol{A_2}$ and the corresponding principal eigenvector $\boldsymbol{w_2}$ are shown in Eq. (7). The converted maximum evaluation value of $\boldsymbol{A_2}$ is underlined in Eq. (7).

$$\boldsymbol{A_2} = \begin{bmatrix} 1.0000 & 1.2500 & 1.7500 & \underline{2.0000} \\ 0.8000 & 1.0000 & 1.2500 & 1.5000 \\ 0.5714 & 0.8000 & 1.0000 & 1.2500 \\ 0.5000 & 0.6667 & 0.8000 & 1.0000 \end{bmatrix}, \quad \boldsymbol{w_2} = \begin{bmatrix} 1.000000 \\ 0.765055 \\ 0.601115 \\ 0.496700 \end{bmatrix} \tag{7}$$

In the next example, we convert the value of $\boldsymbol{a_{14}}$ in $\boldsymbol{A}$ to the maximum evaluation value 9 (underlined), and $\boldsymbol{A_9}$ and $\boldsymbol{w_9}$ are obtained in Eq. (8).

$$\boldsymbol{A_9} = \begin{bmatrix} 1.0000 & 3.0000 & 7.0000 & \underline{9.0000} \\ 0.3333 & 1.0000 & 3.0000 & 5.0000 \\ 0.1429 & 0.3333 & 1.0000 & 3.0000 \\ 0.1111 & 0.2000 & 0.3333 & 1.0000 \end{bmatrix}, \quad \boldsymbol{w_9} = \begin{bmatrix} 1.000000 \\ 0.399678 \\ 0.167058 \\ 0.079910 \end{bmatrix} \tag{8}$$

Further, for the various maximum evaluation values of $\boldsymbol{a_{14}}$ in $\boldsymbol{A}$, the principal eigenvectors are calculated by the power method, respectively. The results are summarized in Table 1. Each principal eigenvector is not normalized.

Table 1 The principal eigenvector for various maximum evaluation values

Maximum evaluation value	w_1	w_2	w_3	w_4
2	1.000000	0.765055	0.601115	0.496700
3	1.000000	0.648587	0.434414	0.315254
4	1.000000	0.574281	0.341423	0.224111
5	1.000000	0.521180	0.281770	0.170338
6	1.000000	0.480672	0.240153	0.135360
7	1.000000	0.448414	0.209429	0.111051
8	1.000000	0.421930	0.185801	0.093322
9	1.000000	0.399678	0.167058	0.079910

From Table 1, in the case where the maximum evaluation value is 2, the maximum element of eigenvector is 1 and the minimum element of eigenvector is 0.496700. In case the value is 9, the minimum value of the eigenvector will be 0.079910. As a result, if the maximum evaluation value of the comparison matrix is small (large) then the difference value between the maximum element of the principal eigenvector and the minimum element is small (large).

We propose the use of these properties for the improvement of the pairwise comparison method, thus facilitating the obtention of the overall evaluation vector of alternatives.

3 Verification of the Proposed Method

In this section, verification of the proposed method is explained.

At first, the example of comparison matrices and the corresponding eigenvectors are illustrated below. The overall evaluation vector of alternatives is calculated using the traditional AHP method. Next, based on this example, the overall evaluation vector of alternatives is calculated using the proposed method. Finally, the overall evaluations of alternatives, obtained by the proposed method and the traditional method, are compared and considered.

3.1 The Example Using the Traditional AHP

The example in this study consists of the five criteria and the three alternatives.

First, $\boldsymbol{A}$, the comparison matrix of the five criteria in this example, and $\boldsymbol{v}$, the corresponding principal eigenvector, are shown in Eq. (9). The principal eigenvector $\boldsymbol{v}$ is not normalized.

$$\boldsymbol{A} = \begin{bmatrix} 1 & 3 & 2 & 1/2 & 1 \\ 1/3 & 1 & 5 & 1/4 & 1/2 \\ 1/2 & 1/5 & 1 & 1/4 & 1/3 \\ 2 & 4 & 4 & 1 & 2 \\ 1 & 2 & 3 & 1/2 & 1 \end{bmatrix}, \quad \boldsymbol{v} = \begin{bmatrix} 0.567933 \\ 0.365566 \\ 0.179137 \\ 1.000000 \\ 0.533289 \end{bmatrix} \tag{9}$$

Next, for these five criteria, $\boldsymbol{A_1}$ to $\boldsymbol{A_5}$, the comparison matrices of three alternatives, and $\boldsymbol{w_1}$ to $\boldsymbol{w_5}$, the corresponding principal eigenvectors, are shown below. These principal eigenvectors are not normalized.

$$\boldsymbol{A_1} = \begin{bmatrix} 1 & 2 & 6 \\ 1/2 & 1 & 4 \\ 1/6 & 1/4 & 1 \end{bmatrix}, \quad \boldsymbol{w_1} = \begin{bmatrix} 1.0000 \\ 0.5503 \\ 0.1514 \end{bmatrix} \tag{10}$$

$$\boldsymbol{A_2} = \begin{bmatrix} 1 & 2 & 2 \\ 1/2 & 1 & 1 \\ 1/2 & 1 & 1 \end{bmatrix}, \quad \boldsymbol{w_2} = \begin{bmatrix} 1.0000 \\ 0.5000 \\ 0.5000 \end{bmatrix} \tag{11}$$

$$\boldsymbol{A_3} = \begin{bmatrix} 1 & 5 & 3 \\ 1/5 & 1 & 1/3 \\ 1/3 & 1/3 & 1 \end{bmatrix}, \quad \boldsymbol{w_3} = \begin{bmatrix} 1.0000 \\ 0.1644 \\ 0.4054 \end{bmatrix} \tag{12}$$

$$A_4 = \begin{bmatrix} 1 & 2 & 8 \\ 1/2 & 1 & 7 \\ 1/8 & 1/7 & 1 \end{bmatrix}, \qquad w_4 = \begin{bmatrix} 1.0000 \\ 0.6025 \\ 0.1037 \end{bmatrix} \tag{13}$$

$$A_5 = \begin{bmatrix} 1 & 1/3 & 3 \\ 3 & 1 & 4 \\ 1/3 & 1/4 & 1 \end{bmatrix}, \qquad w_5 = \begin{bmatrix} 0.4367 \\ 1.0000 \\ 0.1907 \end{bmatrix} \tag{14}$$

Then, based on the obtained vector in Eq. (10) to Eq. (14), $\boldsymbol{W}$, the weight matrix of the alternatives, is constructed in Eq. (15).

$$\boldsymbol{W} = \begin{bmatrix} 1.0000 & 1.0000 & 1.0000 & 1.0000 & 0.4367 \\ 0.5503 & 0.5000 & 0.1644 & 0.6025 & 1.0000 \\ 0.1514 & 0.5000 & 0.4054 & 0.1037 & 0.1907 \end{bmatrix} \tag{15}$$

Multiplying $\boldsymbol{W}$, in Eq. (15), and $\boldsymbol{v}$, in Eq. (9), in the traditional AHP, the vector is obtained in Eq. (16).

$$\boldsymbol{Wv} = \begin{bmatrix} 2.345571 \\ 1.660606 \\ 0.546892 \end{bmatrix} \tag{16}$$

By normalizing to 1 the sum of each column of the obtained vector in Eq. (16), $\boldsymbol{w}$, the overall evaluation vector of the traditional AHP is obtained.

$$\boldsymbol{w} = \begin{bmatrix} 0.515163 \\ 0.364722 \\ 0.120115 \end{bmatrix} \tag{17}$$

The proposed method is verified in the next subsection based on the vector $\boldsymbol{w}$, in Eq. (17).

3.2 The Result of the Example Obtained Through the Proposed Method

In this subsection the result obtained through the proposed method is shown. Based on the example, illustrated in the previous subsection, the overall evaluation vector of alternatives is obtained through steps P1 to P3.

Firstly, the maximum evaluation values are determined using the proposed procedure P1.

In fact, the criteria vector $\boldsymbol{v}$ is unknown. Therefore the maximum evaluation values are determined by the decision makers. However in this paper, to compare with the result of an example obtained using the traditional method, the maximum evaluation value of each criterion is determined based on the criteria vector $\boldsymbol{v}$ in Eq. (6). In order to correspond to a nine-stage evaluation, the maximum evaluation values are obtained by multiplying v by 9 times and rounded. Obtained maximum evaluation values are shown in Eq. (18).

$$v = \begin{bmatrix} 0.567933 \\ 0.365566 \\ 0.179137 \\ 1.000000 \\ 0.533289 \end{bmatrix} \rightarrow \begin{bmatrix} 5 \\ 3 \\ 2 \\ 9 \\ 5 \end{bmatrix} \tag{18}$$

In Eq. (18), the maximum evaluation value of A_1 is 5, A_2 is 3, and so on.

Next, using the proposed procedure P2, A_1 to A_5, the comparison matrices of the example, are converted. In the comparison matrix A_1 in Eq. (10), the maximum value of elements is 6 and the maximum evaluation value in Eq. (18), is 5, and we then convert 6 to 5. At the same time the other elements of A_1 are proportionally converted. Converted matrix A_1' and the corresponding principal eigenvector w_1' are shown in Eq. (19). The maximum evaluation value of A_1' is underlined in Eq. (19). The principal eigenvector w_1' is not normalized.

$$A_1' = \begin{bmatrix} 1.000000 & 1.800000 & \underline{5.000000} \\ 0.555556 & 1.000000 & 3.400000 \\ 0.200000 & 0.294118 & 1.000000 \end{bmatrix}, \quad w_1' = \begin{bmatrix} 1.0000 \\ 0.5942 \\ 0.1869 \end{bmatrix} \tag{19}$$

Similarly, in A_2 in Eq. (10), the maximum value of elements is 2 and the maximum evaluation value is 3. First 2 is converted to 3, then 5 is converted to 2 in A_3, 8 to 9 in A_4, and 4 to 5 in A_5. As a result, converted matrices A_2' to A_5' and the corresponding principal eigenvector, w_2' to w_5', are shown in Eq. (20) to Eq. (23). The maximum evaluation value of each matrix is underlined in each matrix. The principal eigenvectors, w_1' to w_5', are not normalized.

$$A_2' = \begin{bmatrix} 1.000000 & 3.000000 & \underline{3.000000} \\ 0.333333 & 1.000000 & 1.000000 \\ 0.333333 & 1.000000 & 1.000000 \end{bmatrix}, \quad w_2' = \begin{bmatrix} 1.0000 \\ 0.3333 \\ 0.3333 \end{bmatrix} \tag{20}$$

$$A_3' = \begin{bmatrix} 1.000000 & \underline{2.000000} & 1.500000 \\ 0.500000 & 1.000000 & 0.666667 \\ 0.666667 & 1.500000 & 1.000000 \end{bmatrix}, \quad w_3' = \begin{bmatrix} 1.0000 \\ 0.4807 \\ 0.6933 \end{bmatrix} \tag{21}$$

$$A_4' = \begin{bmatrix} 1.000000 & 2.142857 & \underline{9.000000} \\ 0.466667 & 1.000000 & 7.857143 \\ 0.111111 & 0.127273 & 1.000000 \end{bmatrix}, \quad w_4' = \begin{bmatrix} 1.0000 \\ 0.5750 \\ 0.0901 \end{bmatrix} \tag{22}$$

$$A_5' = \begin{bmatrix} 1.000000 & 0.272727 & 3.666667 \\ 3.666667 & 1.000000 & \underline{5.000000} \\ 0.272727 & 0.200000 & 1.000000 \end{bmatrix}, \quad w_5' = \begin{bmatrix} 0.3792 \\ 1.0000 \\ 0.1438 \end{bmatrix} \tag{23}$$

Then, from the obtained vector in Eq. (19) to Eq. (23), the improved weight matrix of the alternatives $\boldsymbol{W'}$ is constructed in Eq. (24).

$$\boldsymbol{W'} = \begin{bmatrix} 1.0000 & 1.0000 & 1.0000 & 1.0000 & 0.3792 \\ 0.5942 & 0.3333 & 0.4807 & 0.5750 & 1.0000 \\ 0.1869 & 0.3333 & 0.6933 & 0.0901 & 0.1438 \end{bmatrix} \quad (24)$$

Finally, through the proposed procedure P3, the improved overall evaluation vector of alternatives is obtained. In the proposed method, contrary to the traditional AHP method, the improved overall evaluation vector is obtained by calculating each row sum of $\boldsymbol{W'}$ in Eq. (24). Further by normalizing to 1 the sum of each column of the obtained vector, the improved overall evaluation vector $\boldsymbol{w'}$ is obtained in Eq. (25).

$$\boldsymbol{w'} = \begin{bmatrix} 0.497061 \\ 0.338624 \\ 0.164315 \end{bmatrix} \quad (25)$$

By comparing the overall evaluation vector $\boldsymbol{w}$ in Eq. (17) with $\boldsymbol{w'}$ in Eq. (25), the values from the traditional AHP method and from the proposed method are almost same.

4 Conclusion

In this paper, for the AHP, the improvement of pairwise comparison method was proposed. In this proposed method, the maximum evaluation value was introduced and the procedure used to obtain the overall evaluation vector of the alternatives was improved. This procedure omits the pairwise comparisons for each criterion.

Comparing the results calculated using the proposed method to those obtained through the traditional method, following was established.

1. The results of the overall evaluation of alternatives by the proposed method show that the values are almost same as those of the traditional method.
2. By using the maximum evaluation value of pairwise comparison of each alternative omitting the pairwise comparison of each criterion, it is possible to take into account the importance of the criteria to the alternative evaluation.
3. In the proposed procedure, the overall evaluation vector of alternatives was obtained through the row sum of the principal eigenvector for each criterion.
4. In the proposed method the procedure of the overall evaluation of alternatives has been simplified compared with the traditional procedure.

Future study will focus on a method of determining the maximum evaluation value of each criterion.

References

1. Belton, V., Gear, T.: On a Short-coming of Saaty's method of Analytic Hierarchies. Omega **11**, 228–230 (1983)
2. Belton, V., Gear, T.: The legitimacy of rank reversal—a comment. Omega **13**, 143–145 (1985)
3. Kinoshita, E., Nakanishi, M.: Proposal of new AHP model in light of dominant relationship among alternatives. J. Oper. Res. Soc. Japan **42**, 180–197 (1999)
4. Kinoshita, E., Sugiura, S.: A comparison study of dominant AHP and similar dominant models. J. Res. Inst. Meijo Univ. **7**, 115–116 (2008)
5. Nishizawa, K.: Simple AHP based on three-level evaluation and two-cluster ANP for the decision makers. J. Jpn. Symp. Anal. Hierarchy Process **1**, 97–104 (2007) (In Japanese)
6. Nishizawa, K.: Two-cluster ANP which consists of the evaluation sub-matrix of the alternatives and its transposed matrix instead of pair-wise comparisons among the criteria. J. Jpn. Symp. Anal. Hierarchy Process **2**, 59–67 (2010) (In Japanese)
7. Nishizawa, K.: Normalization method based on dummy alternative with perfect evaluation score in AHP and ANP. Intell. Decis. Technol. **1**(SIST 15), 253–262 (2012)
8. Nishizawa, K.: Improving of the weight normalization method on alternatives in AHP and ANP. In: Smart Digital Futures 2014, pp. 155–163. IOS Press (2014)
9. Saaty, T.L.: The Analytic Hierarchy Process. McGraw-Hill, New York (1980)
10. Saaty, T.L.: The Analytic Network Process. RWS Publications, Pittsburgh (1996)
11. Schoner, B., Wedley, W.C., Choo, E.U.: A unified approach to AHP with linking pins. Eur. J. Oper. Res. **13**, 384–392 (1993)

Using Super Pairwise Comparison Matrix for Calculation of the Multiple Dominant AHP

Takao Ohya and Eizo Kinoshita

Abstract We have proposed a super pairwise comparison matrix (SPCM) to express all pairwise comparisons in the evaluation process of the dominant analytic hierarchy process (D-AHP) or the multiple dominant AHP (MDAHP) as a single pairwise comparison matrix. This paper shows the example of using SPCM with the application of the logarithmic least squares method for calculation of MDAHP.

Keywords Super pairwise comparison matrix • The dominant AHP • The multiple dominant AHP • Logarithmic least square method

1 Introduction

The analytic hierarchy process (AHP) proposed by Saaty [1] enables objective decision making by top-down evaluation based on an overall aim.

In actual decision making, a decision maker often has a specific alternative (regulating alternative) in mind and makes an evaluation on the basis of the alternative. This was modeled in the dominant AHP (D-AHP), proposed by Kinoshita and Nakanishi [2].

If there are more than one regulating alternatives and the importance of each criterion is inconsistent, the overall evaluation value may differ for each regulating alternative. As a method of integrating the importances in such cases, the concurrent convergence method (CCM) was proposed. Kinoshita and Sekitani [3] showed the convergence of CCM.

Meanwhile, Ohya and Kinoshita [4] proposed the geometric mean multiple dominant AHP (GMMDAHP), which integrates weights by using a geometric mean

T. Ohya (✉)
School of Science and Engineering, Kokushikan University, Tokyo, Japan
e-mail: takaohya@kokushikan.ac.jp

E. Kinoshita
Faculty of Urban Science, Meijo University, Gifu, Japan
e-mail: kinoshit@urban.meijo-u.ac.jp

R. Neves-Silva et al. (eds.), *Intelligent Decision Technologies*,
Smart Innovation, Systems and Technologies 39,
DOI 10.1007/978-3-319-19857-6_42

based on an error model to obtain an overall evaluation value. Herein, such methods of evaluation with multiple regulating alternatives will be generically referred to as the multiple dominant AHP (MDAHP).

Ohya and Kinoshita [5] proposed a super pairwise comparison matrix (SPCM) to express all pairwise comparisons in the evaluation process of D-AHP or MDAHP as a single pairwise comparison matrix.

Ohya and Kinoshita [6] showed, using the error models, that in the dominant AHP an evaluation value resulting from the application of the logarithmic least squares method (LLSM) to a SPCM necessarily coincide with that of the evaluation value resulting obtained by using the geometric mean method to each pairwise comparison matrix. Ohya and Kinoshita [6] also showed, using the error models, that in MDAHP an evaluation value resulting from the application of the logarithmic least-squares method (LLSM) to a SPCM does not necessarily coincide with that of the evaluation value resulting from the application of the geometric mean multiple dominant AHP (GMMDAHP) to the evaluation value obtained from each pairwise comparison matrix by using the geometric mean method.

Ohya and Kinoshita [7] showed the treatment of hierarchical criteria in D-AHP with super pairwise comparison matrix.

This paper shows the example of using SPCM with the application of LLSM for calculation of MDAHP.

2 MDAHP and SPCM

This section explains D-AHP, GMMDAHP and a SPCM to express the pairwise comparisons appearing in the evaluation processes of D-AHP and MDAHP as a single pairwise comparison matrix. Section 2.1 outlines D-AHP procedure and explicitly states pairwise comparisons, and Sect. 2.2 outlines GMMDAHP. Section 2.3 explains the SPCM that expresses these pairwise comparisons as a single pairwise comparison matrix.

2.1 Evaluation in D-AHP

The true absolute importance of alternative $a(a=1, \ldots, A)$ at criterion $c(c=1, \ldots, C)$ is v_{ca}. The final purpose of the AHP is to obtain the relative value (between alternatives) of the overall evaluation value $v_a = \sum_{c=1}^{C} v_{ca}$ of alternative a. The procedure of D-AHP for obtaining an overall evaluation value is as follows:

D-AHP

Step 1: The relative importance $u_{ca} = \alpha_c v_{ca}$ (where α_c is a constant) of alternative a at criterion c is obtained by some kind of methods. In this paper, u_{ca} is obtained by applying the pairwise comparison method to alternatives at criterion c.

Step2: Alternative d is the regulating alternative. The importance u_{ca} of alternative a at criterion c is normalized by the importance u_{cd} of the regulating alternative d, and u_{ca}^{d} $(=u_{ca}/u_{cd})$ is calculated.
Step3: With the regulating alternative d as a representative alternative, the importance w_c^d of criterion c is obtained by applying the pairwise comparison method to criteria, where, w_c^d is normalized by $\sum_{c=1}^{C} w_c^d = 1$.
Step4: From u_{ca}^{d}, w_c^d obtained at Steps 2 and 3, the overall evaluation value $t_a = \sum_{c=1}^{C} w_c^d u_{ca}^d$ of alternative a is obtained. By normalization at Steps 2 and 3, $u_d = 1$. Therefore, the overall evaluation value of regulating alternative d is normalized to 1.

2.2 GMMDAHP

In this paper, the regulating alternatives are assumed to be alternative 1 to alternative D. This assumption can generally be satisfied by renumbering alternatives.

Let $\hat{w}_c^{d(d')}$ be the unknown evaluation value of the criterion c from the alternative $r \neq r'$. Then, Kinoshita and Nakanishi [2] proposed a following evaluation rule under their assumption.

$$\hat{w}_c^{d(d')} = \frac{w_c^{d'} u_{cd}/u_{cd'}}{\sum_{c'} w_{c'}^{d'} u_{c'd'}/u_{c'd'}} \tag{1}$$

MDAHP requires the decision maker to evaluate criteria from the viewpoint of each regulating alternative d. Let w_c^d be the evaluation value of the criterion c from the regulating alternative d, w_c^d are normalized as $\sum_c w_c^d = 1$ for all d. If the evaluation values are consistent, all $\widehat{w}_c^{d(d')}$ are same for all $d' \in D$. But, almost all evaluations involve inconsistency.

GMMDAHP calculate the weights for the criteria with the geometric means from multiple regulating alternatives as the following formula.

$$\hat{w}_c^d = \frac{\left(\prod_{d' \in D} w_c^{d'} u_{cd}/u_{cd'}\right)^{\frac{1}{D}}}{\sum_{c'} \left(\prod_{d' \in D} w_{c'}^{d'} u_{c'd}/u_{c'd'}\right)^{\frac{1}{D}}} \tag{2}$$

$\hat{w}_c^d$ is the geometric mean of $w_c^{d(d')}$ and multiplied constant of normalization as $\sum_c \hat{w}_c^d = 1$.

All $\hat{w}_c^d$ always have following relations.

$$\hat{w}_c^d = \frac{\hat{w}_c^{d'} u_{cd}/u_{cd'}}{\sum_{c'} \hat{w}_{c'}^{d'} u_{c'd}/u_{c'd'}} \tag{3}$$

So, all $\hat{w}_c^d$ are consistent.

2.3 SPCM

The relative comparison values $r_{c'a'}^{ca}$ of importance v_{ca} of alternative a at criteria c as compared with the importance $v_{c'a'}$ of alternative a' in criterion c', are arranged in a (CA × CA) or (AC × AC) matrix. This is proposed as the SPCM $R = (r_{c'a'}^{ca})$ *or* $(r_{a'c'}^{ac})$.

In a (CA × CA) matrix, index of alternative changes first. In a (CA × CA) matrix, SPCM's $(A(c-1)+a,\ A(c'-1)+a')$ th element is $r_{c'a'}^{ca}$.

In a (AC × AC) matrix, index of criteria changes first. In a (AC × AC) matrix, SPCM's $(C(a-1)+c,\ C(a'-1)+c')$ th element is $r_{a'c'}^{ac}$.

In a SPCM, symmetric components have a reciprocal relationship as in pairwise comparison matrices. Diagonal elements are 1 and the following relationships are true:

If $r_{c'a'}^{ca}$ exists, then $r_{ca}^{c'a'}$ exists and

$$r_{ca}^{c'a'} = 1/r_{c'a'}^{ca}, \tag{4}$$

$$r_{ca}^{ca} = 1. \tag{5}$$

Pairwise comparison at Step 1 of D-AHP consists of the relative comparison value $r_{ca'}^{ca}$ of importance v_{ca} of alternative a, compared with the importance $v_{ca'}$ of alternative a' at criterion c.

Pairwise comparison at Step 3 of D-AHP consists of the relative comparison value $r_{c'd}^{cd}$ of importance v_{cd} of alternative d at criterion c, compared with the importance $v_{c'd}$ of alternative d at criterion c', where the regulating alternative is d.

SPCM of D-AHP or MDAHP is an incomplete pairwise comparison matrix. Therefore, the LLSM based on an error model or an eigenvalue method such as the Harker method [8] or two-stage method is applicable to the calculation of evaluation values from an SPCM.

3 Numerical Example of Using SPCM for Calculation of MDAHP

Three alternatives from 1 to 3 and four criteria from I to IV are assumed, where Alternative 1 is the regulating alternative.

As the result of pairwise comparison between alternatives at criteria c ($c = \mathrm{I}, \ldots, \mathrm{IV}$), the following pairwise comparison matrices R_c^A, $c = \mathrm{I}, \ldots, \mathrm{IV}$ are obtained:

$$R_{\mathrm{I}}^A = \begin{pmatrix} 1 & 1/3 & 5 \\ 3 & 1 & 3 \\ 1/5 & 1/3 & 1 \end{pmatrix}, \quad R_{\mathrm{II}}^A = \begin{pmatrix} 1 & 7 & 3 \\ 1/7 & 1 & 1/3 \\ 1/3 & 3 & 1 \end{pmatrix},$$

$$R_{\mathrm{III}}^A = \begin{pmatrix} 1 & 1/3 & 1/3 \\ 3 & 1 & 1/3 \\ 3 & 3 & 1 \end{pmatrix}, \quad R_{\mathrm{IV}}^A = \begin{pmatrix} 1 & 3 & 5 \\ 1/3 & 1 & 1 \\ 1/5 & 1 & 1 \end{pmatrix}$$

With regulating alternatives 1 to 3 as the representative alternatives, importance between criteria was evaluated by pairwise comparison. As a result, the following pairwise comparison matrix R_1^C, R_2^C, R_3^C is obtained:

$$R_1^C = \begin{pmatrix} 1 & 1/3 & 3 & 1/3 \\ 3 & 1 & 3 & 1 \\ 1/3 & 1/3 & 1 & 1/3 \\ 3 & 1 & 3 & 1 \end{pmatrix}, \quad R_2^C = \begin{pmatrix} 1 & 9 & 1 & 5 \\ 1/9 & 1 & 1/3 & 1 \\ 1 & 3 & 1 & 1 \\ 1/5 & 1 & 1 & 1 \end{pmatrix},$$

$$R_3^C = \begin{pmatrix} 1 & 1/3 & 1/9 & 3 \\ 3 & 1 & 1/5 & 5 \\ 9 & 5 & 1 & 5 \\ 1/3 & 1/5 & 1/5 & 1 \end{pmatrix}$$

The (CA × CA) order SPCM for this example is

$$R_{(CA\times CA)} = \begin{pmatrix}
1 & 1/3 & 5 & 1/3 & & & 3 & & & 1/3 & & \\
3 & 1 & 3 & & 9 & & & 1 & & & 5 & \\
1/5 & 1/3 & 1 & & & 1/3 & & & 1/9 & & & 3 \\
3 & & & 1 & 7 & 3 & 3 & & & 1 & & \\
 & 1/9 & & 1/7 & 1 & 1/3 & & 1/3 & & & 1 & \\
 & & 3 & 1/3 & 3 & 1 & & & 1/5 & & & 5 \\
1/3 & & & 1/3 & & & 1 & 1/3 & 1/3 & 1/3 & & \\
 & 1 & & & 3 & & 3 & 1 & 1/3 & & 1 & \\
 & & 9 & & & 5 & 3 & 3 & 1 & & & 5 \\
3 & & & 1 & & & 3 & & & 1 & 3 & 5 \\
 & 1/5 & & & 1 & & & 1 & & 1/3 & 1 & 1 \\
 & & 1/3 & & & 1/5 & & & 1/5 & 1/5 & 1 & 1
\end{pmatrix}$$

For pairwise comparison values in an SPCM, an error model is assumed as follows:

$$r_{c'a'}^{ca} = \varepsilon_{c'a'}^{ca} \frac{v_{ca}}{v_{c'a'}}. \tag{6}$$

Taking the logarithms of both sides gives

$$\log r_{c'a'}^{ca} = \log v_{ca} - \log v_{c'a'} + \log \varepsilon_{c'a'}^{ca} \tag{7}$$

To simplify the equation, logarithms will be represented by overdots as $\dot{r}_{c'a'}^{ca} = \log r_{c'a'}^{ca}$, $\dot{v}_{ca} = \log v_{ca}$, $\dot{\varepsilon}_{c'a'}^{ca} = \log \varepsilon_{c'a'}^{ca}$. Using this notation, Eq. (7) becomes

$$\dot{r}_{c'a'}^{ca} = \dot{v}_{ca} - \dot{v}_{c'a'} + \dot{\varepsilon}_{c'a'}^{ca}, \quad c, c' = 1, \ldots, C, \ a, a' = 1, \ldots, A. \tag{8}$$

From Eqs. (4) and (5), we have

$$\dot{r}_{ca}^{c'a'} = -\dot{r}_{c'a'}^{ca} \tag{9}$$

$$\dot{r}_{ca}^{ca} = 0 \tag{10}$$

If $\dot{\varepsilon}_{c'a'}^{ca}$ is assumed to follow an independent probability distribution of mean 0 and variance σ^2, irrespective of c, a, c', a', the least squares estimate gives the best estimate for the error model of Eq. (8) according to the Gauss Markov theorem

There are two types of pairwise comparison in the dominant AHP: $r_{ca'}^{ca}$ at Step 1 and $r_{c'd}^{cd}$ at Step 3. Then Eq. (8) comes to following Eq. (11) by vector notation.

$$\dot{\mathbf{Y}} = \mathbf{S}\dot{\mathbf{x}} + \dot{\varepsilon} \tag{11}$$

where

$$\dot{\mathbf{x}} = (\dot{v}_{\mathrm{I2}} \ \dot{v}_{\mathrm{I3}} \ \dot{v}_{\mathrm{II1}} \ \dot{v}_{\mathrm{II2}} \ \dot{v}_{\mathrm{II3}} \ \dot{v}_{\mathrm{III1}} \ \dot{v}_{\mathrm{III2}} \ \dot{v}_{\mathrm{III3}} \ \dot{v}_{\mathrm{IV1}} \ \dot{v}_{\mathrm{IV2}} \ \dot{v}_{\mathrm{IV3}})^T,$$

$$\dot{\mathbf{Y}} = \begin{pmatrix} \dot{r}_{\mathrm{I2}}^{\mathrm{I1}} \\ \dot{r}_{\mathrm{I3}}^{\mathrm{I1}} \\ \dot{r}_{\mathrm{II1}}^{\mathrm{I1}} \\ \dot{r}_{\mathrm{III1}}^{\mathrm{I1}} \\ \dot{r}_{\mathrm{IV1}}^{\mathrm{I1}} \\ \dot{r}_{\mathrm{I3}}^{\mathrm{I2}} \\ \dot{r}_{\mathrm{II2}}^{\mathrm{I2}} \\ \dot{r}_{\mathrm{III2}}^{\mathrm{I2}} \\ \dot{r}_{\mathrm{IV2}}^{\mathrm{I2}} \\ \dot{r}_{\mathrm{II3}}^{\mathrm{I3}} \\ \dot{r}_{\mathrm{III3}}^{\mathrm{I3}} \\ \dot{r}_{\mathrm{IV3}}^{\mathrm{I3}} \\ \dot{r}_{\mathrm{II2}}^{\mathrm{II1}} \\ \dot{r}_{\mathrm{II3}}^{\mathrm{II1}} \\ \dot{r}_{\mathrm{III1}}^{\mathrm{II1}} \\ \cdot \\ \cdot \\ \cdot \\ \dot{r}_{\mathrm{IV2}}^{\mathrm{IV1}} \\ \dot{r}_{\mathrm{IV3}}^{\mathrm{IV1}} \\ \dot{r}_{\mathrm{IV3}}^{\mathrm{IV2}} \end{pmatrix} = \begin{pmatrix} \log(1/3) \\ \log 5 \\ \log(1/3) \\ \log 3 \\ \log(1/3) \\ \log 3 \\ \log 9 \\ \log 1 \\ \log 5 \\ \log(1/3) \\ \log 9 \\ \log 3 \\ \log 7 \\ \log 3 \\ \log 3 \\ \cdot \\ \cdot \\ \cdot \\ \log 3 \\ \log 5 \\ \log 1 \end{pmatrix}, \quad \mathbf{S} = \begin{pmatrix} -1 & & & & & & & & & & \\ & -1 & & & & & & & & & \\ & & -1 & & & & & & & & \\ & & & & & -1 & & & & & \\ & & & & & & & & -1 & & \\ 1 & -1 & & & & & & & & & \\ 1 & & & -1 & & & & & & & \\ 1 & & & & & & -1 & & & & \\ 1 & & & & & & & & & -1 & \\ & 1 & & & -1 & & & & & & \\ & 1 & & & & & & -1 & & & \\ & 1 & & & & & & & & & -1 \\ & & 1 & -1 & & & & & & & \\ & & 1 & & -1 & & & & & & \\ & & 1 & & & -1 & & & & & \\ & & & & & \cdot & & & & & \\ & & & & & \cdot & & & & & \\ & & & & & \cdot & & & & & \\ & & & & & & & & 1 & -1 & \\ & & & & & & & & 1 & & -1 \\ & & & & & & & & & 1 & -1 \end{pmatrix}$$

To simplify calculations, $v_{11} = 1$ that is $\dot{v}_{11} = 0$. The least squares estimates for formula (11) are calculated by $\hat{\mathbf{x}} = (\mathbf{S}^{\mathrm{T}}\mathbf{S})^{-1}\mathbf{S}^{\mathrm{T}}\dot{\mathbf{Y}}$.

Table 1 shows the evaluation values obtained from the SPCM for this example.

Table 1 Evaluation values obtained by SPCM+LLSM

	Criterion I	Criterion II	Criterion III	Criterion IV	Overall evaluation value
Alternative 1	1	2.2582	0.5466	1.8840	5.6883
Alternative 2	1.9261	0.3242	1.0529	0.4767	3.7793
Alternative 3	0.4022	0.9306	2.6479	0.2986	4.2787

4 Conclusion

SPCM of MDAHP is an incomplete pairwise comparison matrix. Therefore, the LLSM based on an error model or an eigenvalue method such as the Harker method or two-stage method is applicable to the calculation of evaluation values from an SPCM. In this paper, we showed the way of using SPCM with the application of LLSM for calculation of MDAHP with the numerical example.

References

1. Saaty, T.L.: The Analytic Hierarchy Process. McGraw-Hill, New York (1980)
2. Kinoshita, E., Nakanishi, M.: Proposal of new AHP model in light of dominative relationship among alternatives. J. Oper. Res. Soc. Japan **42**, 180–198 (1999)
3. Kinoshita, E., Sekitani, K., Shi, J.: Mathematical properties of dominant AHP and concurrent convergence method. J. Oper. Res. Soc. Japan **45**, 198–213 (2002)
4. Ohya, T., Kinoshita, E.: Proposal of super pairwise comparison matrix. In: Watada, J., et al. (eds.) Intelligent Decision Technologies, pp. 247–254. Springer-Verlag, Berlin (2011)
5. Ohya, T., Kinoshita, E.: Super pairwise comparison matrix in the multiple dominant AHP. In: Watada, J. et al (eds.) Intelligent Decision Technologies, vol. 1, Smart Innovation, Systems and Technologies 15, pp. 319–327. Springer-Verlag, Berlin, 2012
6. Ohya, T., Kinoshita, E.: Super pairwise comparison matrix with the logarithmic least-squares method. In: Neves-Silva, R. et al. (eds.) Intelligent Decision Technologies, Volume 255 Frontiers in Artificial Intelligence and Applications, pp. 390–398. IOS press, (2013)
7. Ohya, T. Kinoshita, E.: The treatment of hierarchical criteria in dominant AHP with super pairwise comparison matrix. In: Neves-Silva, R. et al. (eds.) Smart Digital Futures, pp. 142–148. IOS press (2014)
8. Harker, P.T.: Incomplete pairwise comparisons in the analytic hierarchy process. Math. Model. **9**, 837–848 (1987)

Fuzzy Cognitive Maps and Multi-step Gradient Methods for Prediction: Applications to Electricity Consumption and Stock Exchange Returns

Elpiniki I. Papageorgiou, Katarzyna Poczęta, Alexander Yastrebov and Chrysi Laspidou

Abstract The paper focuses on the application of fuzzy cognitive map (FCM) with multi-step learning algorithms based on gradient method and Markov model of gradient for prediction tasks. Two datasets were selected for the implementation of the algorithms: real data of household electricity consumption and stock exchange returns that include Istanbul Stock Exchange returns. These data were used in learning and testing processes of the proposed FCM approaches. A comparative analysis of the two-step method of Markov model of gradient, multi-step gradient method and one-step gradient method is performed in order to show the capabilities and effectiveness of each method and conclusions are based on the obtained MSE, RMSE, MAE and MAPE errors.

Keywords Fuzzy cognitive map · Multi-steps algorithms · Gradient method · Markov model of gradient · Electricity consumption predcition · Stock exchange returns prediction

---

E.I. Papageorgiou (✉)
Department of Computer Engineering, Technological Educational Institute (T.E.I.) of Central Greece, Lamia, Greece
e-mail: epapageorgiou@iti.gr

E.I. Papageorgiou · C. Laspidou
Information Technologies Institute, Center for Research and Technology Hellas, CERTH, 6th Km Charilaou-Thermi Rd., 57001 Thermi, Greece

K. Poczęta · A. Yastrebov
Department of Computer Science Applications, Kielce University of Technology, Kielce, Poland
e-mail: k.piotrowska@tu.kielce.pl

A. Yastrebov
e-mail: a.jastriebow@tu.kielce.pl

C. Laspidou
Department of Civil Engineering, University of Thessaly, Pedion Areos, 38334 Volos, Greece
e-mail: laspidou@iti.gr

R. Neves-Silva et al. (eds.), *Intelligent Decision Technologies*,
Smart Innovation, Systems and Technologies 39,
DOI 10.1007/978-3-319-19857-6_43

1 Introduction

Fuzzy cognitive map (FCM) is an effective tool for modeling decision support systems. It describes real objects as a set of concepts (nodes) and connections (relations) between them [9]. There are various extensions of FCM; for example: augmented FCM [16], fuzzy grey cognitive map [13, 17], and relational FCM [19]. Each one of them improves some aspects of the operation of traditional maps.

The task of prediction of stock exchange returns is difficult due to chaotic and dynamic nature of the stock market processes. In [1, 2] Hybrid Radial Basis Function Neural Networks was proposed as a predictive modeling tool to study the daily movements of stock indicators. The application of FCMs for stock market modeling has been investigated in [4].

The application of Fuzzy Cognitive Maps to time series modeling has been previously discussed by Homenda et al. in [6–8] and Lu et al. [10]. Methodologies presented in the previous work of Homenda et al. as well as in the last article of Lu et al. [10] are fundamentally different from ours. Homenda et al. in [6] proposed, for the first time, a new modeling methodology that joins FCMs and moving window approach to model and forecast time series, focused on the case of univariate time series forecasting. Also in [7, 8] nodes selection criteria for FCM designed to model time series were proposed. Furthermore, some simplification strategies by posteriori removing nodes and weights were presented.

Alternative approaches to FCM-based time series modeling are related to classification. Published works in this stream of research include for example: [11]. FCM was also applied for the effective prediction of multivariate interval-valued time series [3].

The previously proposed multi-step learning algorithms based on gradient method [15, 20] and Markov model of gradient [18, 21] are applied for multivariate prediction of Istanbul Stock Exchange National 100 Index [1, 2] and the electric power consumption in one household with a one-minute sampling rate [5]. Simulation analysis is accomplished with the use of ISEMK (Intelligent Expert System based on Cognitive Maps) software tool [14], in which the learning algorithms have been implemented.

The objectives of this paper are the following: (1) to investigate multi-step learning algorithms of FCM for multivariate time series prediction, (2) to present experimental evaluation of the proposed methods.

In this article, we briefly introduce fuzzy cognitive maps and we present the multi-step supervised learning based on gradient method and Markov model of gradient. After describing the basic features of the ISEMK software tool, we present selected results of experiments on real data. Finally, we outline conclusions on the suitability of these methodologies.

2 Fuzzy Cognitive Maps

The structure of FCM is based on a directed graph:

$$< X, R >, \tag{1}$$

where $X = [X_1, ..., X_n]^T$ is the set of the concepts; $i = 1, 2, ..., n$, n is the number of concepts; $R = \{r_{j,i}\}$ is relations matrix.

When all nodes are activated by an initial value, they influence each other. The new values of concepts are calculated according to the formula:

$$X_i(t+1) = F\left(X_i(t) + \sum_{j \neq i} r_{j,i} \cdot X_j(t)\right), \tag{2}$$

where t is discrete time, $t = 0, 1, 2, ..., T$, T is end time of simulation, $X_i(t)$ is the value of the i-th concept, $i = 1, 2, ..., n$, n is the number of concepts, $r_{j,i}$ is the relation weight between the j-th concept and the i-th concept, value from the interval $[-1, 1]$, $F(x)$ is stabilizing function, which can be chosen in the form:

$$F(x) = \frac{1}{1 + e^{-cx}}, \tag{3}$$

where $c > 0$ is a parameter.

3 Multi-step Learning Algorithms

In the article, multi-step learning algorithms for fuzzy cognitive maps based on gradient method [15, 20] and Markov model of gradient [18, 21] are described.

3.1 Gradient Method

Multi-step supervised learning based on gradient method is described by the equation [20]:

$$\begin{aligned} r_{j,i}(t+1) = P_{[-1,1]}(&\sum_{k=0}^{m_1} \alpha_k \cdot r_{j,i}(t-k) + \\ &\sum_{l=0}^{m_2} (\beta_l \cdot \eta_l(t) \cdot \left(Z_i(t-l) - X_i(t-l)\right) \cdot h_{j,i}(t-l))), \end{aligned} \tag{4}$$

where $\alpha_k, \beta_l, \eta_l$ are learning parameters, which are determined using experimental trial and error method, $k = 1, ..., m_1$; $l = 1, ..., m_2$, m_1, m_2 are the number of the steps of the method; t is a time of learning, $t = 0, 1, ..., T$, T is end time of learning; $h_{j,i}(t)$ is a sensitivity function; $X_i(t)$ is the value of the i-the concept (predicted value), $Z_i(t)$

is the reference value of the i-th concept (monitored value); $P_{[-1,1]}(x)$ is an operator of design for the set $[-1, 1]$. Sensitivity function $y_{j,i}(t)$ is described as follows:

$$h_{j,i}(t+1) = \left(h_{j,i}(t) + X_j(t)\right) \cdot F'(X_i(t) + \sum_{j \neq i} r_{j,i} \cdot X_j(t)), \tag{5}$$

where $F'(x)$ is the derivative of the stabilizing function.

Learning parameters α_k, β_l, η_l have to satisfy the conditions (6)–(9) to reach the convergence of the multi-step gradient method [20].

$$\sum_{k=0}^{m_1} \alpha_k = 1, \tag{6}$$

$$0 < \eta_l(t) < 1, \tag{7}$$

$$\eta_l(t) = \frac{1}{\lambda + t}, \tag{8}$$

$$\lambda > 0, \beta_l \geq 0. \tag{9}$$

A special case of multi-step gradient method is the one-step algorithm, which modifies the relations matrix according to the formula:

$$r_{j,i}(t+1) = P_{[-1,1]}(r_{j,i}(t) + \beta_0 \cdot \eta_0(t) \cdot \left(Z_i(t) - X_i(t)\right) \cdot h_{j,i}(t)). \tag{10}$$

3.2 Markov Model of Gradient

Another example of multi-step algorithms is the two-step method based on Markov model of gradient [18]. Modification of the relation weight is described by the formula [21]:

$$r_{j,i}(t+1) = P_{[-1,1]}(r_{j,i}(t) - y_{j,i}(t)), \tag{11}$$

where: $y_{j,i}$ is the Markov model of gradient, described as follows:

$$y_{j,i}(t+1) = a \cdot y_{j,i}(t) - \beta_0 \cdot \eta_0(t) \cdot \left(Z_i(t) - X_i(t)\right) \cdot h_{j,i}(t))). \tag{12}$$

where a is a learning parameter, $a \in (0, 1)$.

Termination criterion for the presented methods can be expressed by the formula:

$$J(t) = \frac{1}{n} \sum_{i=1}^{n} (Z_i(t) - X_i(t))^2 < e, \tag{13}$$

where $J(t)$ is a learning error function; e is a level of error tolerance.

4 ISEMK System

ISEMK is a computer software, which is a universal tool for modeling phenomena based on FCM. ISEMK realizes [14]:

- the implementation of fuzzy cognitive map based on expert knowledge or historical data;
- multi-step supervised learning based on gradient method or Markov model of gradient with the use of historical data;
- testing of learned FCMs based on historical data;
- exporting data of learning and FCM analysis to .csv files;
- proper visualizations of done research.

Figure 1 present sample results of the learned FCM testing.

5 Simulation Results

The comparative analysis of two-step method of Markov model of gradient, multi-step gradient method and one-step gradient method was performed. Mean Squared Error (14), Root Mean Square Error (15), Mean Absolute Percentage Error (16) and Mean Absolute Error (17) were used to estimate the performance of the FCM learning algorithms.

$$MSE = \frac{1}{n_D(T-1)} \sum_{t=1}^{T-1} \sum_{i=1}^{n_D} \left(Z_i(t) - X_i(t) \right)^2, \tag{14}$$

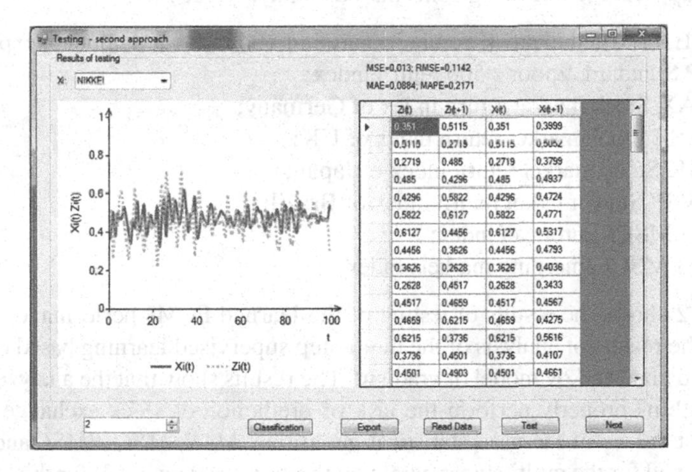

Fig. 1 Sample results of FCM testing

$$RMSE = \sqrt{\frac{1}{n_D(T-1)} \sum_{t=1}^{T-1} \sum_{i=1}^{n_D} \left(Z_i(t) - X_i(t)\right)^2}, \tag{15}$$

$$MAPE = \frac{1}{n_D(T-1)} \sum_{t=1}^{T-1} \sum_{i=1}^{n_D} \left| \frac{Z_i(t) - X_i(t)}{Z_i(t)} \right|, \tag{16}$$

$$MAE = \frac{1}{n_D(T-1)} \sum_{t=1}^{T-1} \sum_{i=1}^{n_D} \left| Z_i(t) - X_i(t) \right|, \tag{17}$$

where: t – time of testing, $t = 0, 1, ..., T-1$, T – the number of the test records, $i = 1, ..., n_D$, n_D – the number of the output conepts, $X_i(t)$ – the value of the i-th output concept, $Z_i(t)$ – the reference value of the i-th output concept.

The performance of the chosen supervised learning algorithms and fuzzy cognitive map was evaluated based on real multivariate time series for prediction of stock exchange returns and electricity consumption.

5.1 Prediction of Stock Exchange Returns

The aim of the analysis is to determine relations between the Istanbul Stock Exchange National 100 Index (ISE100) and seven other international stock market indicators. The used dataset includes returns of Istanbul Stock Exchange with seven other international indeces from January 5, 2009 to February 22, 2011 [1, 2]. The dataset was divided into two groups: 425 records were used in learning process and 100 other records were used to evaluate the performance of the analyzed algorithms.

The map with the following concepts was analyzed [1, 2]:

- X_1 – ISE100 Istanbul stock exchange national 100 index – output concept;
- X_2 – SP Standard &poor's 500 return index;
- X_3 – DAX Stock market return index of Germany;
- X_4 – FTSE Stock market return index of UK;
- X_5 – NIK Stock market return index of Japan;
- X_6 – BVSP Stock market return index of Brazil;
- X_7 – EU MSCI European index;
- X_8 – EM MSCI emerging markets index.

Figure 2 shows the results of testing of the learned FCMs performance. Table 1 presents the results of analysis of the multi-step supervised learning based on gradient method and Markov model of gradient. The results show that the analyzed learning algorithms properly perform the task of prediction of stock exchange returns. The lowest values of the used statistical quantities: *MAE*, *MSE*, *RMSE* and *MAPE* were obtained for the multi-step gradient method ($m_1 = 0, m_2 = 2$) for the following parameters: $\alpha_0 = 1$, $\beta_0 = 5$, $\beta_1 = 3$, $\beta_2 = 2$, $\lambda = 10$, $e = 0.001$, $c = 2$.

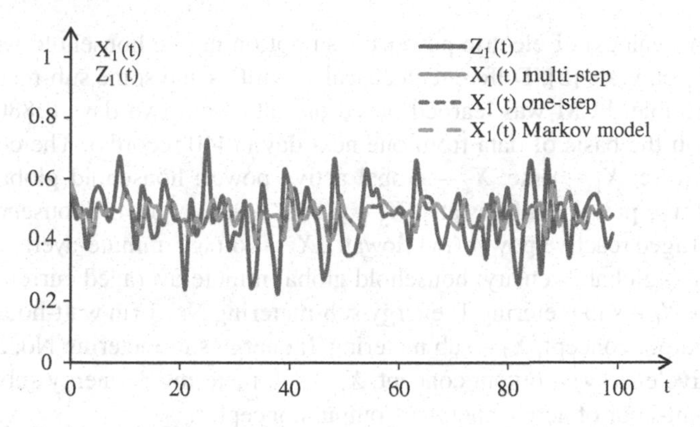

Fig. 2 Obtained values of $X_1(t)$ during testing

Table 1 Chosen results of the analysis (for the following parameters: $e = 0.001$, $c = 2$)

Type	Learning parameters	*MAE*	*MSE*	*RMSE*	*MAPE*
one-step	$\alpha_0 = 1, \beta_0 = 50, \lambda = 100$	0.0884	0.013	0.1142	0.2171
multi-step	$\alpha_0 = 1, \beta_0 = 30, \beta_1 = 20, \lambda = 100$	0.0878	0.0129	0.1134	0.2154
multi-step	$\alpha_0 = 1, \beta_0 = \beta_1 = 20, \beta_2 = 10, \lambda = 100$	0.0876	0.0128	0.1132	0.2149
Markov model	$a = 0.2, \beta_0 = 50, \lambda = 100$	0.0886	0.0131	0.1146	0.2179
one-step	$\alpha_0 = 1, \beta_0 = 10, \lambda = 10$	0.0872	0.0126	0.1122	0.2109
multi-step	$\alpha_0 = 1, \beta_0 = 7, \beta_1 = 3, \lambda = 10$	0.087	0.0125	0.1109	0.2103
multi-step	$\alpha_0 = 1, \beta_0 = 5, \beta_1 = 3, \beta_2 = 2, \lambda = 10$	**0.0869**	**0.0125**	**0.1118**	**0.21**
Markov model	$a = 0.1, \beta_0 = 10, \lambda = 10$	0.0873	0.0126	0.1124	0.2114

5.2 Prediction of Electricity Consumption

On-going urbanization and technology advances, coupled with resource depletion and emissions prove that city populations exert great pressures with increased consumptions and pollution. The need to increase energy efficiency is really urgent. Societies are urged to be innovative, which includes to turn to renewable energy sources [22, 23], or to use relevant Information and Communication Technologies (ICT) and social computing to raise awareness of stakeholders on their consumption [23]. The ability to predict electricity consumption is at the heart of exploiting ICT to make electricity use more efficient and to lead to new savings in smart cities; this article develops suitable methodologies that show that prediction of electricity consumption is possible and should be applied to promote savings.

The simulations of multi-step learning algorithms for FCMs were performed on the provided dataset with measurements (different electrical quantities and some sub-metering values) of electric power consumption in one household with a one-minute sampling rate [5]. Different electrical quantities and some sub-metering values are available. FCM was learned based on data from two days (2880 records) and tested on the basis of data from one next day (1440 records). The concepts of the problem are: X_1 – time, X_2 – global active power: household global minute-averaged active power (in kilowatt), X_3 – global reactive power: household global minute-averaged reactive power (in kilowatt), X_4 – voltage: minute-averaged voltage (in volt), X_5 – global intensity: household global minute-averaged current intensity (in ampere), X_6 – sub metering 1: energy sub-metering No. 1 (in watt-hour of active energy) – output concept, X_7 – sub metering 2: energy sub-metering No. 2 (in watt-hour of active energy) – output concept, X_8 – sub metering 3: energy sub-metering No. 3 (in watt-hour of active energy) – output concept.

The structure of the initialized map is presented in Fig. 3. The most important concepts are X_6, X_7, and X_8. Sub metering 1 corresponds to the kitchen, containing mainly a dishwasher, an oven and a microwave. Sub metering 2 corresponds to the laundry room, containing a washing-machine, a tumble-drier, a refrigerator and a light. Sub metering 3 corresponds to an electric water-heater and an air-conditioner. Figure 4 illustrates the results of testing of the learned FCMs performance.

Table 2 presents the results of analysis of the multi-step supervised learning based on gradient method and Markov model of gradient.

The results show that the analyzed learning algorithms properly perform the task of prediction of electricity consumption. The lowest values of the *MSE*, *RMSE* and *MAPE* were obtained for the multi-step gradient method ($m_1 = 0$, $m_2 = 1$) for the following parameters: $\alpha_0 = 1$, $\beta_0 = 30$, $\beta_1 = 20$, $\lambda = 100$, $e = 0.001$, $c = 5$. The lowest value of the *MAE* was obtained for the multi-step gradient method ($m_1 = 0$, $m_2 = 2$) for the following parameters: $\alpha_0 = 1$, $\beta_0 = \beta_1 = 20$, $\beta_2 = 10$, $\lambda = 100$, $e = 0.001$, $c = 5$.

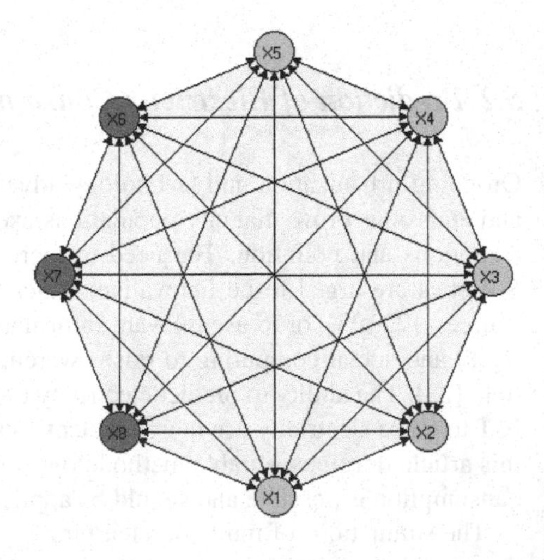

Fig. 3 Structure of the initialized map

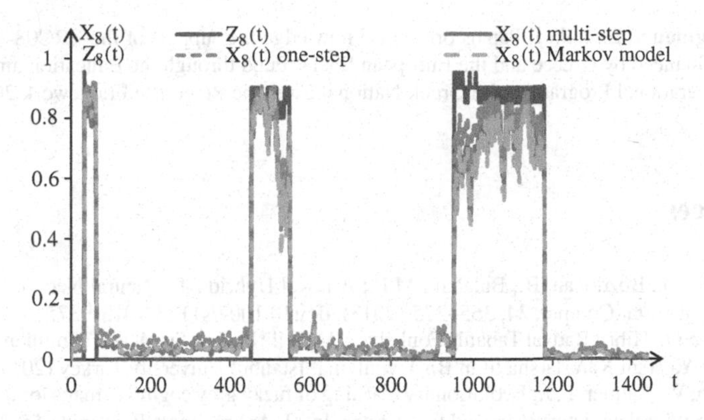

Fig. 4 Obtained values of $X_8(t)$ during testing

6 Conclusion

This paper contains the description of use of FCMs with efficient FCM learning algorithms to predict time series of different datasets concerning electricity consumption and stock exchange returns. The main contribution of this research study is the analysis of FCM learning algorithms of the multi-step learning based on gradient method and Markov model of gradient. Their prediction capabilities are investigated by simulations performed with the use of ISEMK software tool. For statistical quantities *MAE*, *MSE*, *RMSE* and *MAPE* were used to assess model performance. Our analysis shows that FCM with multi-step supervised learning algorithms can be used for multivariate time series prediction. The multi-step gradient method seems most efficient with respect to the accuracy of prediction expressed by the used errors.

Table 2 Chosen results of the analysis (for the following parameters: $e = 0.001$, $c = 5$) with the use of data from January

Type	Learning parameters	*MAE*	*MSE*	*RMSE*	*MAPE*
one-step	$\alpha_0 = 1, \beta_0 = 50, \lambda = 100$	0.0484	0.0117	0.1083	0.1494
multi-step	$\alpha_0 = 1, \beta_0 = 30, \beta_1 = 20, \lambda = 100$	0.0487	**0.0099**	**0.0997**	**0.129**
multi-step	$\alpha_0 = 1, \beta_0 = \beta_1 = 20, \beta_2 = 10, \lambda = 100$	**0.0473**	0.0103	0.1017	0.13
Markov model	$a = 0.1, \beta_0 = 50, \lambda = 100$	0.0613	0.0143	0.1197	0.163
one-step	$\alpha_0 = 1, \beta_0 = 30, \lambda = 100$	0.0531	0.0109	0.1046	0.1656
multi-step	$\alpha_0 = 1, \beta_0 = 20, \beta_1 = 10, \lambda = 100$	0.0492	0.0119	0.1093	0.1341
multi-step	$\alpha_0 = 1, \beta_0 = \beta_1 = \beta_2 = 10, \lambda = 100$	0.0504	0.0113	0.1064	0.1406
Markov model	$a = 0.5, \beta_0 = 30, \lambda = 100$	0.0515	0.0135	0.116	0.1328

Acknowledgments Elpiniki I. Papageorgiou acknowledge the support of the ERC08- RECITAL project, co-financed by Greece and the European Social Fund through the Education and Lifelong Learning Operational Program of the Greek National Strategic Reference Framework 2007–2013.

References

1. Akbilgic, O., Bozdogan, H., Balaban, M.E.: A novel Hybrid RBF Neural Networks model as a forecaster. Stat.Comput. **24**, 365–375 (2013). doi:10.1007/s11222-013-9375-7
2. Akbilgic, O.: Hibrit Radyal Tabanlš Fonksiyon Alarš ile Deiken Seimi ve Tahminleme: Menkul Kšymet Yatšršm Kararlaršna likin Bir Uygulama. Istanbul University,Turkey (2011)
3. Froelich, W., Salmeron, J.: Evolutionary learning of fuzzy grey cognitive maps for the forecasting of multivariate, interval-valued time series. Int. J. Approximate Reasoning **55**, 1319–1335 (2014)
4. Froelich, W., Wakulicz-Deja, A.: Learning fuzzy cognitive maps from the web for stock market decision support system. In: Węgrzyn-Wolska, K.M., Szczepaniak, P.S. (eds.) Advance Intelligence Web, ASC 43, pp. 106–111. Springer-Varlag, Heidelberg (2007)
5. Hébrail G., Bérard A.: UCI machine learning repository, EDF R&D, Clamart, France. http://archive.ics.uci.edu/ml (2012)
6. Homenda, W., Jastrzębska, A., Pedrycz, W.: Modeling time series with fuzzy cognitive maps. In: 2014 IEEE International Conference on Fuzzy Systems (FUZZ-IEEE), pp. 2055–2062, Beijing, China, 6–11 July 2014
7. Homenda, W., Jastrzębska, A., Pedrycz, W.: Nodes selection criteria for fuzzy cognitive maps designed to model time series. Adv. Intell. Syst. Comput. **323**, 859–870 (2015)
8. Homenda, W., Jastrzębska, A., Pedrycz, W.: Time series modeling with fuzzy cognitive maps: Simplification strategies. The Case of a Posteriori Removal of Nodes and Weights. LNCS 8838, pp. 409–420 (2014)
9. Kosko, B.: Fuzzy cognitive maps. Int. J. Man-Mach. Stud. **24**, 65–75 (1986)
10. Lu, W., Pedrycz, W., Liu, X., Yang, J., Li, P.: The modeling of time series based on fuzzy information granules. Expert Syst. Appl. **41**, 3799–3808 (2014)
11. Papageorgiou, E.I., Froelich, W.: Application of evolutionary fuzzy cognitive maps for prediction of pulmonary infections. IEEE Trans. Inf. Technol. Biomed. **16**(1), 143–149 (2011)
12. Papageorgiou, E.I., Froelich, W.: Multi-step prediction of pulmonary infection with the use of evolutionary fuzzy cognitive maps. Neurocomputing **92**, 28–35 (2012)
13. Papageorgiou, E.I., Salmeron, J.L.: Learning fuzzy grey cognitive maps using nonlinear hebbian-based approach. Int. J. Approximate Reasoning **53**, 54–65 (2012)
14. Piotrowska, K. (Poczęta, K.): Intelligent expert system based on cognitive maps. Studia Informatica **33**(2A 105), 605–616 (2012) (in Polish)
15. Poczęta K., Yastrebov, A.: analysis of fuzzy cognitive maps with multi-step learning algorithms in valuation of owner-occupied homes. In: 2014 IEEE International Conference on Fuzzy Systems (FUZZ-IEEE), pp. 1029–1035. Beijing, China (2014)
16. Salmeron, J.L.: Augmented fuzzy cognitive maps for modelling LMS critical success factors. Knowl.-Based Syst. **22**(4), 275–278 (2009)
17. Salmeron, J.L.: Modelling grey uncertainty with fuzzy cognitive maps. Expert Syst. Appl. **37**, 7581–7588 (2010)
18. Shilman, S.V., Yastrebov, A.I.: Convergence analysis of some class of multi-step stochastic optimization algorithms. Autom. Telemechanics **8**, (1976) (in Russian)
19. Słoń, G.: The use of fuzzy numbers in the process of designing relational fuzzy cognitive maps. Lecture Notes in Artificial Intelligence LNAI 7894/Part1, pp. 376–387. Springer-Verlag (2013)
20. Yastrebov, A., Piotrowska, K., Poczęta, K.: Simulation analysis of multistep algorithms of relational cognitive maps learning. In: Yastrebov, A., Kuźmińska-Sołśnia, B., Raczyńska, M. (eds.) Computer Technologies in Science, Technology and Education, pp. 126–137. Institute for Sustainable Technologies, National Research Institute, Radom (2012)

21. Yastrebov, A., Poczęta, K.: Application of fuzzy cognitive map with two-step learning algorithm based on Markov model of gradient for time series prediction. Logistyka 6/2014 (2014) (in Polish)
22. Laspidou C., Kalliantopoulos, V.C.: Design and technoeconomic analysis of a photovoltaic system installed on a house in Xanthi, Greece. Fresen. Environ. Bull. **21**(8c), 2494–2498
23. Laspidou, C.: ICT and stakeholder participation for improved urban water management in the cities of the future. Water Util. J. **8**, 79–85 (2014)

Decision Making Support Technique Based on Territory Wellbeing Estimation

Tatiana Penkova

Abstract This paper presents the technique of comprehensive decision making support in territory management based on estimation of current state of the territory and formation of control actions. In order to provide the decision preparation support the author proposes a method of wellbeing level estimation that allows to calculate a wellbeing index taking into account various aspects of vital activity. In order to provide the intellectual support of decision formation the author comes up with a model of generation of the control recommendations.

Keywords Decision making support • Wellbeing estimation • Control recommendations • Territory management

1 Introduction

Automation of territory management requires the development of methods and algorithms that provide the decision making support. In the course of working out a reasonable decision, there is a need in a comprehensive data analysis about the current state of the territory. A generalized representation of information about processes in various sectors of socio-economic vital activity is a critical condition for effective territory management. Based on estimation results there should be generated a number of control actions aimed at the target level of wellbeing.

The challenges of analytical decision making support of the territory management have been investigated by many scientists and a lot of researches have accumulated a great methodological base [1–4]. However, as a rule, the available methods consider separate aspects of decision making process. The development of comprehensive decision making support based on territory state estimation and decision formation still remains a topical problem. A key stage in the decision

T. Penkova (✉)
Institute of Computational Modelling SB RAS, Krasnoyarsk, Russia
e-mail: Penkova_t@icm.krasn.ru

R. Neves-Silva et al. (eds.), *Intelligent Decision Technologies*,
Smart Innovation, Systems and Technologies 39,
DOI 10.1007/978-3-319-19857-6_44

making process is the estimation of the control object condition [4–8]. The quality of decisions depends largely on the system of applied indicators [1, 3, 4, 7–10] and methods of their identification [5, 6, 10]. The method should allow an analyst to detect the risk factors and identify the priority directions of territory evolution. Moreover, to estimate the state of complex social and economic objects the method should provide the analysis of heterogeneous indicators and assessment of their changes according to geographically-oriented standard. Apart from this, the formation of control decisions requires expert knowledge and should be founded on the analysis of current situation with identification of particular problems and their causes. It is necessary to take into account the depth of the problem and generate a set of control recommendations.

This paper presents the technique of comprehensive decision making support in territory management. In order to provide the decisions preparation support the author proposes a method of estimation of the wellbeing level that allows to calculate the wellbeing index taking into account various aspects of vital activity. In order to provide the intellectual support of decisions formation the author proposes a model of generation of the control recommendations. The solutions are considered using examples of social monitoring.

The outline of paper is as follows. Section 2 presents the technique of comprehensive decision making support in territory management. Section 3 describes the method of territory wellbeing level estimation. Section 4 speaks about the model of generation of the control recommendations. Section 5 draws the conclusion.

2 Technique of Comprehensive Decision Making Support

The technique of decision making support in territory management is based on estimation of the territory wellbeing and generation of the control recommendations. Figures 1, 2 and 3 show the IDEF0 diagrams of these processes. IDEF0 methodology provides a function modelling using the language of Structured Analysis and Design Technique (SADT) [11]. An IDEF0 model describes the functions (e.g. activities, actions, processes or operations); inputs and outputs in form of data needed to perform the function and the data that is produced as a result of the function respectively; controls which constrain or govern the function and mechanisms which can be thought of as a person or device which performs the function.

According to Fig. 1 the decision making support process consists of two basic stages: estimation of the territory wellbeing and generation of the control recommendations.

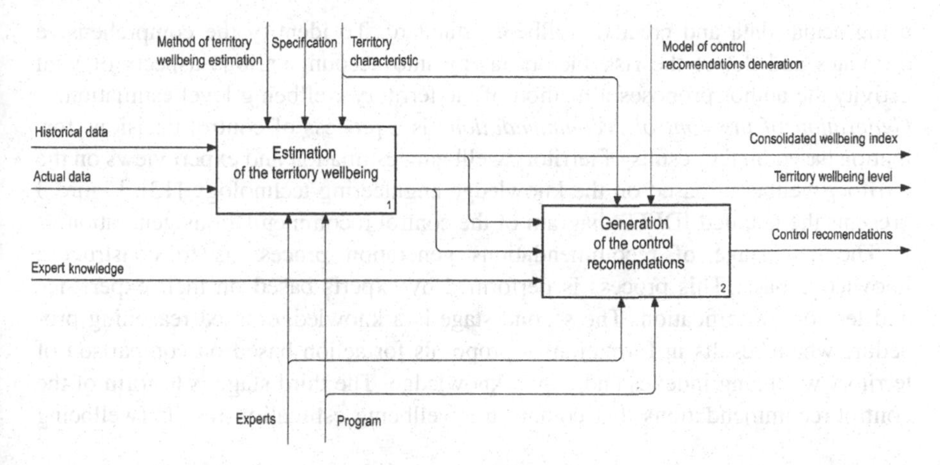

Fig. 1 IDEF0 diagram of the decision making support based on estimation of the territory wellbeing and generation of the control recommendations

Estimation of the territory wellbeing is based on creation of the territory wellbeing standard and estimation of the territory wellbeing level. Figure 2 presents the detailed IDEF0 diagram of the territory wellbeing estimation.

The first stage of wellbeing estimation process is to develop the geographically-oriented wellbeing standard which is a target level of wellbeing and it is required for correct estimation of current state of the territory. The creation of the territory wellbeing standard includes: identification of hierarchy of indicators (i.e. the set of primary indicators and levels of their aggregation), identification of significance coefficients of indicators, identification of normative values and estimation scale. This process is performed by experts using historical data based on territory characteristic and specification [12]. The second stage of wellbeing estimation process is to calculate the wellbeing index and identify the territory wellbeing level

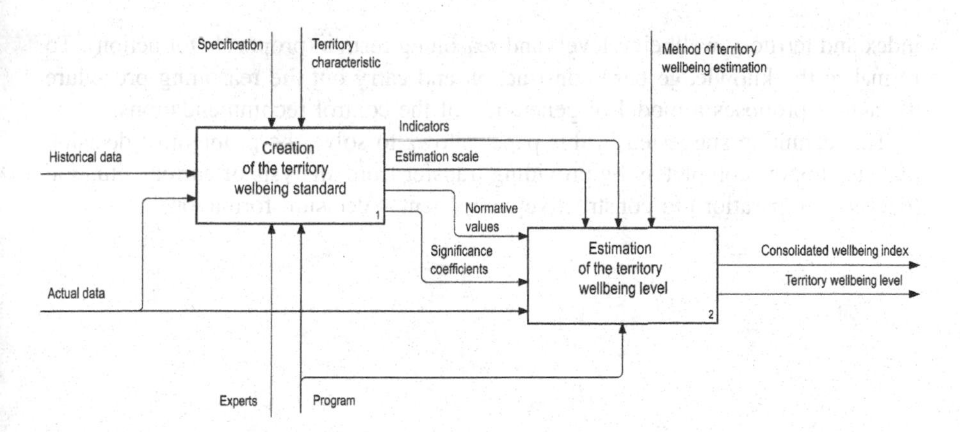

Fig. 2 IDEF0 diagram of estimation of the territory wellbeing

using actual data and created wellbeing standard. To identify the comprehensive estimates and detect the risk factors taking into account various aspects of vital activity the author proposes a method of the territory wellbeing level estimation. *Generation of the control recommendations* is a process of control decisions formation based on the results of territory wellbeing estimation and expert views on the territory wellbeing based on the knowledge engineering technology [13]. Figure 3 presents the detailed IDEF0 diagram of the control recommendations generation.

The first stage of recommendations generation process is to construct a knowledge base. This process is performed by experts based on their experience and territory specification. The second stage is a knowledge-based reasoning procedure which results in formation of proposals for action based on comparison of territory wellbeing indexes and expert knowledge. The third stage is to form of the control recommendations that contain the wellbeing estimation results (wellbeing

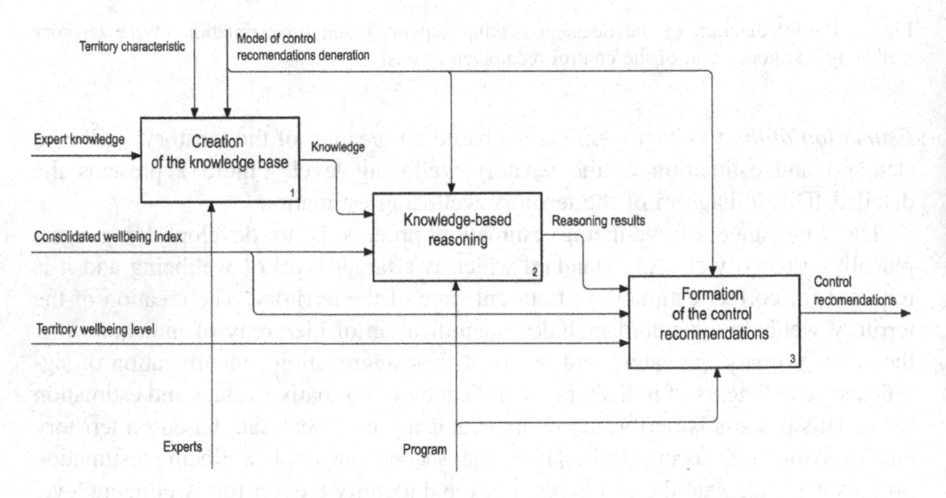

Fig. 3 IDEF0 diagram of generation of the control recommendations

index and territory wellbeing level) and reasoning results (proposals for action). To formalize the knowledge base construction and carry out the reasoning procedure the author proposes a model of generation of the control recommendations.

The technique suggested in this paper allows to solve the problem of decision making support completely by providing transfer from analysis of current situation (decision preparation) to constructive component – decision formation.

3 Method of Estimation of the Territory Wellbeing Level

The method of estimation of the territory wellbeing level is an improvement to the approach to estimation of complex socio-economic objects [8]. The method provides the calculation of individual and consolidated wellbeing indexes based on integration of heterogeneous indicators and interpretation of index values.

This method includes three basic stages: *estimation of the primary indicator* calculates the individual wellbeing index using actual data and normative values of the primary indicators; *estimation of the comprehensive indicator* calculates the consolidated wellbeing index using individual indexes and significance coefficients of indicators; *identification of the wellbeing level* is a semantic interpretation of index values based on estimation scale using fuzzy logic.

Estimation of the primary indicator

The set of primary indicators contains heterogeneous indicators: indicators which should be increased to improve wellbeing level (e.g. birth rate, family income and housing) and indicators which should be decreased to improve wellbeing level (e.g. death rate, morbidity and crime rate).

The individual wellbeing index is calculated by:

$$i_k = 1 + \Delta P_k \cdot S_k, \quad (1)$$

where i_k – is a individual index of k-th indicator; ΔP_k – is a compliance coefficient showing how the actual values of k-th indicator correspond with standard; $S_k = \pm 1$ – is a coefficient which characterizes the "polarity" of k-th indicator, $S_k = 1$ when the change of indicator is proportional to index and $S_k = -1$ when the change of indicator is inversely proportional to index.

Compliance coefficient is calculated by:

$$\Delta P_k = \begin{cases} 0, \text{ if } P_k \in [N_k, Z_k] \\ \frac{P_k - Z_k}{Z_k - N_k}, \text{ if } P_k > Z_k \\ \frac{P_k - Z_k}{Z_k - N_k}, \text{ if } P_k < N_k \end{cases}, \quad (2)$$

where ΔP_k – is a compliance coefficient showing how the actual values of k-th indicator correspond with standard; P_k – is a actual value of k-th indicator; $[N_k, Z_k]$ – is a range of normative value of k-th indicator, N_k – is a lower limit of the range, Z_k – is a upper limit of the range.

The value of individual index can be identified as $i_k > 1$ that demonstrates significant improvement of indicator or as $i_k < 1$ that demonstrates significant degradation of indicator. This feature of index is extremely important for estimation of wellbeing and understanding the current state of the territory.

In contrast to existing methods [5, 6, 10] the author suggests using the ranges of the normative values as a wellbeing standard instead of the single value and estimating the significance of indicator change relative to the standard range.

Estimation of comprehensive indicator

Consolidated wellbeing index is calculated by:

$$I = \sum_{k=1}^{n} u_k \cdot i_k, \tag{3}$$

where I – is a consolidated wellbeing index; u_k – is a significance coefficient of k-th indicator, $u_k > 0$ and $\sum u_k = 1$, n – number of indicators. The value of consolidated index can also be identified as $I > 1$ or $I < 1$ that demonstrates significant improvement or significant degradation of the territory wellbeing respectively.

Identification of wellbeing level

Each index value corresponds to a value of linguistic variable "Wellbeing level" with expert confidence level [13]. The linguistic variable "Wellbeing level" is identified by the following set of values: "Critical", "Very low", "Low", "Middle", "Satisfactory", "Acceptable", "Perfect" and membership functions $\mu_j(x)$, where x – is a value of estimate, $x \in \{I\}$, $j = \overline{1,7}$ – is a number of wellbeing level:

$$\mu_1 = \begin{cases} 1, & x < 0 \\ 0, & x \geq 0 \end{cases}. \quad \mu_2 = \begin{cases} 1, 0 \leq x \leq 0,18 \\ 25(0,22 - x), 0,18 < x < 0,22 \\ 0, 0,22 \leq x \end{cases}.$$

$$\mu_3 = \begin{cases} 0, 0 \leq x < 0,18 \\ 25(x - 0,18), 0,18 < x < 0,22 \\ 1, 0,22 \leq x \leq 0,38 \\ 25(0,42 - x), 0,38 < x < 0,42 \\ 0, 0,42 \leq x \end{cases}. \quad \mu_4 = \begin{cases} 0, 0 \leq x < 0,38 \\ 25(x - 0,38), 0,38 < x < 0,42 \\ 1, 0,42 \leq x \leq 0,58 \\ 25(0,58 - x), 0,58 < x < 0,62 \\ 0, 0,62 \leq x \end{cases}.$$

$$\mu_5 = \begin{cases} 0, 0 \leq x < 0,58 \\ 25(x - 0,58), \ 0,58 < x < 0,62 \\ 1, 0,62 \leq x \leq 0,78 \\ 25(0,82 - x), 0,78 < x < 0,82 \\ 0, 0,82 \leq x \end{cases}. \quad \mu_6 = \begin{cases} 0, 0 \leq x < 0,78 \\ 25(x - 0,78), 0,78 < x < 0,82 \\ 1, 0,82 \leq x \leq 1 \end{cases}.$$

$$\mu_7 = \begin{cases} 1, & x > 1 \\ 0, & x \leq 1 \end{cases}.$$

Table 1 Example of indicators for estimation of the social wellbeing level

Indicators	u_k	$[N_k, Z_k]$		P_k	ΔP_k	Index
Social wellbeing	–	–	–	–	–	0.69
Population structure	0.15	–	–	–	–	0.61
Labor market	0.15	–	–	–	–	0.80
Housing facilities	0.19	–	–	–	–	0.75
Standard of living	0.19	–	–	–	–	0.60
Psycho-emotional tension	0.17	–	–	–	–	0.53
Children's drug addiction	0.21	4.20	7.20	7.16	0	1.00
Teenage drug addiction	0.20	120.00	150.00	180.56	1.02	−0.02
Mortality from suicide	0.17	32.00	35.00	35.00	0	1.00
Drug addiction	0.15	200.00	250.00	265.38	0.31	0.69
Alcoholism	0.15	1100.00	1300.00	1333.63	0.17	0.83
…	…	…	…	…	…	…
Medical provision	0.16	–	–	–	–	0.85

Let us consider an example of wellbeing level estimation for indicators of social monitoring. A fragment of hierarchy of indicators, initial data and estimation results are shown in Table 1.

The interpretation of index values for comprehensive indicators gives the following wellbeing level: "Medical provision" – "Acceptable", "Housing facilities" – "Satisfactory" and "Psycho-emotional tension" – "Middle" with 100 % expert confidence; "Labour market" – between "Acceptable" and "Satisfactory", "Standard of living" – between "Middle" and "Satisfactory" with 50 % expert confidence that reveals the equal possibility of improvement or degradation; "Population structure" – "Middle" with 25 % and "Satisfactory" with 75 % that points out a positive tendency. The general social wellbeing level is identified as "Satisfactory".

Thus, the method proposed in this paper provides the estimation of the current territory wellbeing level by analyzing heterogeneous indicators and assessing their changes according to geographically-oriented standard. Hierarchical structure of indicators allows an analyst to investigate the various fields in detail and detect the risk factors.

4 Model of Generation of the Control Recommendations

The generation of control recommendations is based on cognitive analysis of the situation with identification of particular problems and their causes [2, 14]. Figure 4 presents the model of generation of the control recommendations that can be used for any situation in decision making process.

In Fig. 4 P_i – is a "problem" indicator, that characterizes the analyzed situation; $S_{1j}(S_{l-1j}(\ldots S_{1j}(P_i)))$ – is a "cause" indicator, that characterizes the cause of the

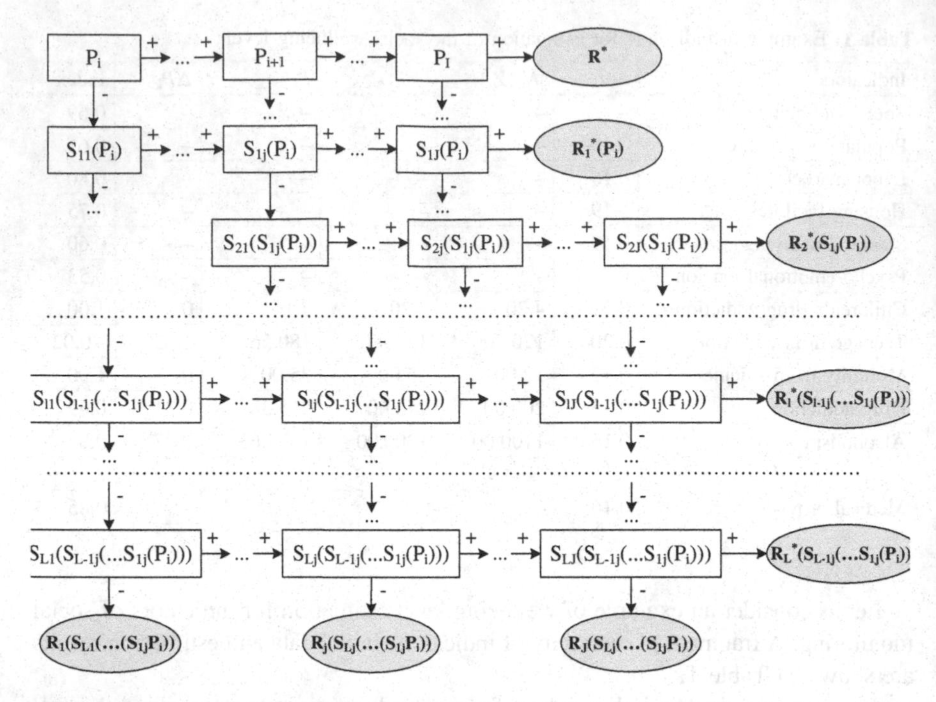

Fig. 4 Model of generation of the control recommendations based on cognitive map of the situation

problem situation; $i = 1,\ldots,I$ – is a number of "problems"; $l = 1,\ldots,L$ – is a number of "causes"; $j = 1,\ldots,J$ – is a number of "causes" on the l-th level of hierarchy; (+)/(−) – is a type of relation that corresponds to the state of indicator – "normal"/"abnormal" accordingly; R^* – is a general recommendation when all "problem" indicators are "normal"; $R_l^*(S_{l-1}(\ldots S_{lj}(P_i)))$ – is a general recommendation when all "cause" indicators of l-th level are "normal"; $R_j^*(S_{Lj}(\ldots S_{lj}(P_i)))$ – is a recommendation when "cause" indicator of L-th level is "abnormal".

While modelling, firstly, the "problem" indicators which characterize the analyzed situation are identified. Then, for each "problem" indicator a set of "cause" indicators is identified. This model is unique in the way that the hierarchical relations between "cause" indicators are applied in accordance with hierarchical structure of wellbeing estimation indicators.

In a cognitive model, there are two types of cause-end-effect relations: positive (+) and negative (−) which corresponds to "normal" and "abnormal" state of indicators accordingly and which are taken into account for generation of recommendations. Unlike previously suggested models based on cognitive map [15], the author proposes to analyze the estimates of comprehensive indicators instead of actual values of primary indicators. This way allows to operate with generalized information and, in case of nonconformance of wellbeing target level, to go deeper to detect the particular causes and generate the set of control recommendations.

In order to formalize the expert knowledge and carry out the reasoning procedure according to proposed model we have developed the rules with the following construction:

$$T\!:P\!:R\!:\ IF < (x_1 \sim a_i)\&(x_2 \sim a_i)\&\ldots\&(x_n \sim a_i) > \\ THEN < Q_1, Q_2, \ldots, Q_n > \quad (4)$$

where T – is an identifier of unique rule name in knowledge base; P – is an identifier of affected zone (e.g. rules for control of “problem” concepts or rules for control of “cause” concepts); R – is an identifier of rule priority that describes the privilege of rule for conflict resolution; $x_1, x_2, \ldots, x_n$ – are estimates of comprehensive indicators; a_i – is a value of estimate established by expert as a “normal”; “~” symbol – is a comparative operation including $\geq$, $<$; $Q_1, Q_2, \ldots, Q_n$ – are operations, Q_i: C→C′, where C, C′ – are concepts of model, “→” symbol – is a operation of moving from one concept of model to another.

For the suggested model the following types of rules have been developed:

$$P1\!:\ P\!:\forall\, i \in \{1, 2, \ldots, I-1\}\!:\ IF < P_i \geq a_i > THEN < P_i \rightarrow P_{i+1} >;$$

$$P2\!:\ P\!:\exists\, i \in \{1, 2, \ldots, I\}\!:\ IF < P_i < a_i > THEN < P_i \rightarrow S_{11}(P_i) >;$$

$$P3\!:\ P\!:\forall\, i \in \{1, 2, \ldots, I\}\!:\ IF < P_i < a_i > THEN < P_i \rightarrow R^* >;$$

$$P4\!:\ S\!:\forall l \in \{1, 2, \ldots, L\}, \forall j \in \{1, 2, \ldots, J-1\}\!:\ IF < S_{lj}(S_{l-1j}(\ldots S_{1j}(P_i))) \geq a_i > \\ THEN < S_{lj}(S_{l-1j}(\ldots S_{1j}(P_i))) \rightarrow S_{lj+1}(S_{l-1j}(\ldots S_{1j}(P_i))) >;$$

$$P5\!:\ S\!:\forall\, l \in \{1, 2, \ldots, L-1\}, \exists\, j \in \{1, 2, \ldots, J\}\!:\ IF < S_{lj}(S_{l-1j}(\ldots S_{1j}(P_i))) < a_i > \\ THEN < S_{lj}(S_{l-1j}(\ldots S_{1j}(P_i))) \rightarrow S_{l+1j}(S_{lj}(\ldots S_{1j}(P_i))) >;$$

$$P6\!:\ S\!:\forall\, l \in \{1, 2, \ldots, L\}, \forall j \in \{1, 2, \ldots, J\}\!:\ IF < S_{lj}(S_{l-1j}(\ldots S_{1j}(P_i))) \geq a_i > \\ THEN < S_{lj}(S_{l-1j}(\ldots S_{1j}(P_i))) \rightarrow R^*_l(S_l(\ldots S_{1j}(P_i))), \\ R^*_l(S_l(\ldots S_{1j}(P_i))) \rightarrow S_{l-1j+1}(S_{l-2j}(\ldots S_{1j}(P_i))) >;$$

$$P7\!:\ S\!:\ l = L, \forall j \in \{1, 2, \ldots, J-1\}\!:\ IF < S_{lj}(S_{l-1j}(\ldots S_{1j}(P_i))) < a_i > \\ THEN < S_{lj}(S_{l-1j}(\ldots S_{1j}(P_i))) \rightarrow R_j(S_{Lj}(\ldots S_{1j}(P_i))), \\ R_j(S_{Lj}(\ldots S_{1j}(P_i))) \rightarrow S_{lj+1}(S_{l-1j}(\ldots S_{1j}(P_i))) >;$$

$$P8\!:\ S\!:\ l = L, \forall j = J\!:\ IF < S_{lj}(S_{l-1j}(\ldots S_{1j}(P_i))) < a_i > \\ THEN < S_{lj}(S_{l-1j}(\ldots S_{1j}(P_i))) \rightarrow R_j(S_{Lj}(\ldots S_{1j}(P_i))) >;$$

$$P9\!:\ S\!:l = 1, \forall j \in \{1, 2, \ldots, J\}\!:\ IF < S_{lj}(S_{l-1j}(\ldots S_{1j}(P_i))) \geq a_i > \\ THEN < S_{lj}(S_{l-1j}(\ldots S_{1j}(P_i))) \rightarrow R_1^*(P_i) >.$$

The reasoning procedure based on rules is a sequential comparison of antecedent of rule with actual data and execution of relevant consequent. When all “causes” have been analyzed and the set of recommendations has been created for one “problem” then it is time to start the analysis of “causes” of other “problems”.

So, for example in social monitoring, the “problem” is that “social wellbeing index in reporting period is lower than one in previous reporting period”. This “problem” according to hierarchy of indicators is corresponds to the following “causes”: “Labor market”, “Housing facilities”, “Psycho-emotional tension” and etc. Each of them, in turn, is indentified by subordinate “causes”. For instance, “Labor market” is identified by: “jobless rate”, “working conditions”; “Psycho-emotional tension” is indentified by: “drug addiction”, “alcoholism” and etc. In addition, “normal” state of indicators is identified as 0.7 that match the “Satisfactory” wellbeing level. The following rules are possible:
P6.4: S: IF $< I_{population\ structure} \ \& \ I_{labor\ market} \ \& \ I_{housing\ facilities} \ \& \ I_{standard\ of\ living} \ \& \ I_{psycho\text{-}emotional\ tension} \ \& \ I_{medical\ provision} >$ *THEN < Make the standard tougher. Check the correction of initial data, probably, there is an unrecorded “causes” indicator which has an impact on this “problem” indicator >;*
P4.4: S: IF $< I_{labor\ market} > 0.7 >$ THEN $< I_{housing\ facilities} >$;
P5.4: S: IF $< I_{labor\ market} < 0.7 >$ THEN *<Probably, the indicators of hard working conditions are unrecorded; should change the system of estimating indicators. Should provide the usage of modern manufacturing equipment, medical and treatment provision. Should carry out the additional working process audit>.*

The first rule in example forms the general recommendation when there is “problem” but all “causes” are normal. The second rule implements the moving from one “causes” to another at the same level of hierarchy. The third rule forms particular recommendation when “cause” indicator at the last level of hierarchy is abnormal.

Thus, the model of control recommendations generation gives an opportunity to analyze the existing situation with identification of particular problems and their causes and generate of decisions using expert knowledge.

5 Conclusion

This paper presents the technique that provides the comprehensive decision making support in territory management based on estimation of the territory wellbeing and formation of control actions. To estimate the territory wellbeing level the author has proposed a method that is based on analysis of heterogeneous indicators and assessment of their changes according to standard. This method allows a decision maker to detect the risk factors and identify the priority directions of territory development to achieve the target level of wellbeing. To provide the intellectual support of decision making, formalization of knowledge base constructing and implementation of reasoning procedure the author proposes a model of generation of the control recommendations. This model allows a decision maker to analyze the current situation with identification of particular problems and their causes taking into account the depth of the problem and generate the set of control recommendations using expert knowledge. The solutions suggested in this paper can be used for a wide range of tasks in territory-industry management in the different fields.

References

1. Tsybatov, V.: Technology of forecasting of socio-economic activity of region based on methods of balances. In: 13th International Conference on the Application of Artificial Intelligence in Engineering, pp. 69–72 (1998)
2. Trahtengerts, E.A.: Computer-Aided management decision-making support systems. J. Problemy Upravleniya **1**, 13–28 (2003)
3. Nozhenkova, L.F.: Information-analytical technologies and systems of region management support. J. Comput. Technol. **14**(6), 71–81 (2009)
4. Makarov A.M., Galkina N.V., Savenkov B.V.: Criteria of social and economic efficiency in the estimation of territory viability. J. Chelyabinsk State Univ. **1**(2), 57–62 (2003)
5. Kalinina V.V.: Modern approaches to the estimation of the industrial complex of region. J. Volglgrad Nat. Univ. **2**(19), 62–69 (2011)
6. Murias, P., Martinez, F., Miguel C.: An economic wellbeing index for the spanish provinces: A data envelopment analysis approach. Soc. Ind. Res. **77**, 395–417 (2006)
7. Becht, M.: The theory and estimation of individual and social welfare measures. J. Econ. Surv. **9**(1), 53–87 (1995)
8. Borisova, E.V.: Index method of comprehensive quantitative estimation of the quality of complex objects. In: XII International Conference "Mathematics, Computer, Education" Izhevsk, Issue 1, pp. 249–159 (2005)
9. Helliwell, J.F., Putnam R.D.: The social context of well-being. J. Philos. Trans. R. Soc. **359** (1449), 1435–1446
10. Diamantopoulos, A., Winklhofer, H.M.: Index construction with formative indicators: An alternative to scale development. J. Mark. Res. **38**(2), 269–277 (2001)
11. Marca D.A., McGowan C.L.: SADT: Structured Analysis and Design Technique, 392 pp. McGraw-Hill Book Co., New York (1987)
12. Penkova, T.: Method of wellbeing estimation in the territory management. LNCS **8582**, 57–68 (2014)
13. Ueno H., Icidzuka M.: Presentation and application of knowledge, 220 pp. Mir, Moscow (1989)
14. Zhilina N.M., Chechenin G.I. et al.: The Automated System of Social-hygiene Monitoring of the Health and Human Environment – The Decision Making Tool, 192 pp. Novokuznetsk (2005)
15. Axelrod R.: Structure of Decisions: The Cognitive Maps of Political Elites, 404 pp. Princeton University Press, Princeton (1976)

An Approach for Text Mining Based on Noun Phrases

Marcello Sandi Pinheiro, Hercules Antonio do Prado, Edilson Ferneda and Marcelo Ladeira

Abstract The use of noun phrases as descriptors for text mining vectors has been proposed to overcome the poor semantic of the traditional bag-of-words (BOW). However, the solutions found in the literature are unsatisfactory, mainly due to the use of static definitions for noun phrases and the fact that noun phrases per se do not enable an adequate relevance representation since they are expressions that barely repeat. We present an approach to deal with these problems by (*i*) introducing a process that enables the definition of noun phrases interactively and (*ii*) considering similar noun phrases as a unique term. A case study compares both approaches, the one proposed in this paper and the other based on BOW. The main contribution of this paper is the improvement of the preprocessing phase of text mining, leading to better results in the overall process.

Keywords Text mining · Natural language processing · Preprocessing · Noun phrases

M.S. Pinheiro
Brazilian Army CDS – QGEx Setor Militar Urbano, Brasília, DF 70.630-904, Brazil
e-mail: msandipinheiro@gmail.com

H.A. do Prado (✉) · E. Ferneda
Graduate Program on Knowledge and IT Management Catholic, University of Brasilia, SGAN 916 Av. W5, Brasília, DF 70.790-160, Brazil
e-mail: hercules@ucb.br

E. Ferneda
e-mail: eferneda@pos.ucb.br

H.A. do Prado
Embrapa - Management and Strategy Secretariat Parque Estação Biológica – PqEB S/N°, Brasília, DF 70.770-90, Brazil

M. Ladeira
Edifício CIC/EST, Universidade de Brasília – IE - CIC Campus Universitário Darcy Ribeiro, Brasilia, DF 70.910-900, Brazil
e-mail: mladeira@unb.br

R. Neves-Silva et al. (eds.), *Intelligent Decision Technologies*, Smart Innovation, Systems and Technologies 39,
DOI 10.1007/978-3-319-19857-6_45

1 Introduction

Usually, the development of Text Mining (TM) applications involves (*i*) preprocessing and representing a given corpus in a vectorial space, (*ii*) running a learning algorithm on the corpus vectorial representation and (*iii*) evaluating the results on the basis of appropriate metrics [1, 2]. A crucial task to assure the quality of the results is the process of building a vector that represents the considered corpus. In this sense, the traditional *bag-of-words* (BOW) has limitations imposed by the way it is defined. By using only the words from a corpus, excluding the stopwords, to build the vector it is unlikely to keep a reasonable semantic from it, since the grammar structures from the texts are not captured.

To face this problem, some approaches have been presented, involving the use of dictionaries, thesaurus, or specific ontologies [3, 4], that contribute to capture relevant characteristics, more complex than simple isolated words, to enable the achievement of better results from the TM process. Although the effectiveness of these approaches, they are restricted to certain knowledge domains, beyond involving high costs to develop the support software.

Alternatively, methods based on Computational Linguistics (CL) [5] have shown to be promising, since they explore the peculiar nature of textual sources that, unlikely the word based approaches, are able to capture semantic structures from texts.

A thread explored in CL is the use of noun phrases as basic information units extracted from the corpus to compose a BOW [6]. In this approach, all noun phrases that satisfy a given structure are included in the vector - from now called *bag-of-noun-phrases* (BON), along with their relevance in each document. However, two problems arise from this approach [7–9]: (*i*) once defined the structural types for the noun phrases, they are not allowed to change, what leads to the inclusion of many irrelevant noun phrases in the BON, and (*ii*) as a noun phrase repeats very rarely, the calculation of its relevance based in the same relation TF*IDF used for the BOW showed to be inadequate to the BON.

In this paper some approaches to deal with these problems are discussed. Aiming at enriching the BON with noun phrases really relevant under the point of view of a user, an environment to support the BON generation interactively was developed. In this environment, the extracted noun phrases pass on human judgment to decide if it should remain in the BON or not. During a sequence of cycles, the BON is built incrementally, producing a refined structure. Regarding the relevance issue, it was searched an alternative to increase the frequency of the information units. The solution adopted was to treat each set of similar noun phrases as one unique. To evaluate this similarity a matching algorithm based in the Levenshtein Distance [10] was applied. However, only the aggregation of similar noun phrases did not enable an adequate expression of relevance, as observed in the experiments performed in this work. So, in accordance with the literature [1, 2, 10–14], it was tested some modifications in the traditional formula for relevance calculation (TF*IDF) in order to meet a formula with the desired characteristics.

2 Previous Works

According to Kuramoto [8], noun phrases can be very useful as texts descriptors and in many situations that require a set of information units to represent a text, like the BON for TM. The early approaches for TM presented limited effectiveness due to the use of simple words it the corpus representation. Usually, words were considered as loose symbols collected from texts, with no reference to objects or facts in the real world. Conversely, noun phrases exhibit a logic and semantic structure inside the text they represents. Kuramoto [8] emphasizes that the importance of noun phrases as descriptors in TM comes from the fact they are information units one can extract from a text. Noun phrases can have many syntactic configurations. For example, "O estudo da economia da informação" is a complex noun phrase, since it is composed by two simpler noun phrases: "A economia da informação" and "A informação".

Additionally, noun phrases can be represented hierarchically in a taxonomy. In the example above, three hierarchical levels can be identified: "A informação" as the most general noun phrase, an intermediate level with the noun phrase "A economia da informação", and the more specific "O estudo da economia da informação".

In short, unlike words, noun phrases keep their meaning when extracted from a text. This way, the central question is how to use this rich structure adequately for TM.

Some previous works [7–9] did not succeed in adopting a BON instead of a BOW because no previous selection of noun phrases were done, leading to a BON much bigger than necessary. An alternative to overcome this problem was suggested by Kuramoto [8] and consists in the adoption of the noun phrases taxonomy to drive the selection of descriptors.

For obtaining the hierarchical levels of noun phrases, Kuramoto [8] proposed a set of syntactic rules to describe the structures of noun phrases in Portuguese. He also defined a general structure for noun phrases. Noun phrases are described by the rule

$$N'' = D' + N', \tag{1}$$

where N'' is the symbol that represents the noun phrase, D' is a determinant, and N' represents a related predicate or, at least, a noun. A determinant D' can be an article, a possessive pronoun, a demonstrative pronoun, or an empty determinant. An empty determinant is usual in Portuguese. A related predicate N' can be any expression in the forms: *(i)* N' = noun, or $N + EP$, or $N + SP$, or N; *(ii)* N represents a noun; *(iii)* $EP = P + N$ means a prepositional expansion (e.g., from fire, from system); *(iv)* $SP = P + N''$ means a noun phrase prepositional (e.g., from the information, from the business); P stands for a preposition.

According to Kuramoto [8], the general rule belongs to a bigger set of rules that describes the syntax of a noun phrase, pointing out that it does not describe all classes of noun phrase. However, it represents near 90 % of the noun phrases in Portuguese. Noun phrases are descriptors that refer to objects or facts from the real world, independently the way they were extracted or assigned to the texts. The findings of Kuramoto [8] were used by Bräscher [15], that applied the notion of noun phrase to solve ambiguity problems related to synonymy, polysemy, and homonymy in Information Retrieval.

Silva and Vieira [9, 16] applied noun phrases as descriptors for TM and compared the results with the traditional approach. The error rate of the traditional approach was smaller than the method based in noun phrases, contradicting the conjecture of Kuramoto [8].

We believe that the unsatisfactory results of using noun phrases as descriptors in the TM process [9] relies in the way they were extracted and treated in the previous works. In the next sessions it is presented and discussed an approach to deal with the found flaws.

3 The Proposed Approach

The proposed approach comprises the following six steps: *(i)* Stopwords removal, *(ii)* Part-of-Speech tagger (POS), *(iii)* Noun phrases retrieval, *(iv)* Matching, *(v)* Weighting, and *(vi)* Mining.

Stopwords removal involves the elimination of punctuation, abbreviations, dates, numbers, special characters, and any kind of unwanted word. All other classes of words (e.g. articles and conjunctions propositions) are terms that help in the formation of a noun phrase and, so, cannot be excluded in this process.

In POS tagger, all words are marked in the text. MXPost tagger [17] was chosen by having presented the higher level of accuracy (90.25 %) for Portuguese [18]. This accuracy was later improved to 96 % by applying the corpora Mac-Morpho [19]. Table 1 shows the list of MacMorpho tags.

Noun phrases retrieval starts with an analysis of the tagged documents in order to detect the more adequate structures in face of the mining objective using the software prototype specially constructed to these applications. A human judgment is conducted by means of interactions between the TM expert and the domain user. After some interactions, the more appropriate structures to define the noun phrases to be retrieved are specified. Finally, the noun phrases considered as descriptors are retrieved.

As example, a fragment of document from the corpus adopted for the case study is shown below, along with their tags, noun phrases structures, and descriptors:

- *Original text*: "Venho por meio desta tão conceituada empresa solicitar a vossa atenção e ação para os referidos postes na Rua Celidônia, na altura do nº 355, Guratiba, Rio de Janeiro, que não têm mais condições de sustentar a rede elétrica. Por ser um poste de baixa renda, suas condições são precárias. Pedimos a Vossa Senhoria a intervenção para este assunto!"

- *Tagged text*: [VENHO_V POR_PREP MEIO_NUM DESTA_N TÃO_ADV CONCEITUADA_PCP EMPRESA_N SOLICITAR_V A_ART VOSSA_N ATENÇÃO_N E_KC AÇÃO_N PARA_PREP OS_ART REFERIDOS_PCP POSTES_N NA_PREP| + RUA_NPROP CELIDÔNIA_NPROP NA_NPROP ALTURA_N DO_NPROP N_NPROP GURATIBA_NPROP RIO_NPROP DE_NPROP JANEIRO_NPROP QUE_PRO-KS-REL NÃO_ADV TÊM_V MAIS_N CONDIÇÕES_N DE_PREP SUSTENTAR_V A_ART REDE_N ELÉTRICA_ADJ POR_PREP SER_V UM_ART POSTE_N DE_PREP| + BAIXA_ADJ RENDA_V SUAS_PROADJ CONDIÇÕES_N SÃO_V PRECÁRIAS_ADJ PEDIMOS_V A_ART VOSSA_NPROP SENHORIA_NPROP A_ART INTERVENÇÃO_N PARA_PREP ESTE_PROADJ ASSUNTO_N]
- *Noun phrases structures*: {<ART N> , <N KC N> , <N PREP|+NPROP NPROP> , <ART N ADJ>}
- *Retrieved descriptors*: {A-VOSSA, ATENÇÃO-E-AÇÃO, POSTES-NA-RUA-CELIDÔNIA, A-REDE-ELÉTRICA, UM-POSTE, A-INTERVENÇÃO}

Table 1 MacMorpho tags used in MXPost

<table>
<tr><th>Part of speech</th><th>Tags</th><th>Part of speech</th><th>Tags</th></tr>
<tr><td>Adjetivo</td><td>ADJ</td><td>Palavra denotativa</td><td>PDEN</td></tr>
<tr><td>Advérbio</td><td>ADV</td><td>Preposição</td><td>PREP</td></tr>
<tr><td>Advérbio conectivo subordinativo</td><td>ADV-KS</td><td>Pronome relativo conectivo subordinativo</td><td>PRO-KS-REL</td></tr>
<tr><td>Advérbio relativo subordinativo</td><td>ADV-KS-REL</td><td>Pronome conectivo subordinativo</td><td>PRO-KS</td></tr>
<tr><td>Artigo (definido ou indefinido)</td><td>ART</td><td>Símbolo de moeda corrente</td><td>CUR</td></tr>
<tr><td>Conjunção coordenativa</td><td>KC</td><td>Pronome adjetivo</td><td>PROADJ</td></tr>
<tr><td>Conjunção subordinativa</td><td>KS</td><td>Pronome substantivo</td><td>PROSUB</td></tr>
<tr><td>Interjeição</td><td>IN</td><td>Verbo</td><td>V</td></tr>
<tr><td>Nome</td><td>N</td><td>Verbo auxiliar</td><td>VAUX</td></tr>
<tr><td>Nome próprio</td><td>NPROP</td><td>Pronome pessoal</td><td>PROPESS</td></tr>
<tr><td>Numeral</td><td>NUM</td><td>Contrações e ênclises</td><td>|+</td></tr>
<tr><td>Particípio</td><td>PCP</td><td>Mesóclises</td><td>|!</td></tr>
<tr><td>Estrangeirismos</td><td>|EST</td><td>Números de Telefone</td><td>|TEL</td></tr>
<tr><td>Apostos</td><td>|AP</td><td>Datas</td><td>|DAT</td></tr>
<tr><td>Dados</td><td>|DAD</td><td>Horas</td><td>|HOR</td></tr>
</table>

Traditional approaches apply stemming techniques to both reduce the amount of words in a BOW and to obtain terms with higher frequency [11]. In fact, this technique improves the classification accuracy [1, 2, 18, 20]. On the other hand, while this approach is adequate for a BOW, it is not used for a BON, since stemming is not applicable. So, an alternative for shrinking similar noun phrases, based on a matching algorithm, were adopted [21, 22]. The Levenshtein Distance

(LD), a.k.a. Edit-Distance [10], algorithm allows to find similar noun phrases that can be represented in a unique form. This algorithm assesses the similarity between two strings A and B, based in the amount of deletions, insertions, or replacements required to transform A into B. The bigger the distance LD, the more different they are. A slight modification was done in the algorithm in order to have as output the percentage of modifications since this was the metric used for similarity.

Weighting refers to the attribution of a relevance value to the retrieved descriptors. In general, this problem is approached by adopting variations of TF*IDF metric [1, 2, 9–12, 23]. There is a consensus on the necessity of experimentation to adjust this metric for specific cases [1, 12, 13, 24]. In the present case, a data set with artificial data representing the very nature of noun phrases, in terms of frequency was created. It was also considered that, the simpler the noun phrase the bigger the probability of its appearance in the text, and vice versa. On the basis of the experiments results, the following formula for traditional TF*IDF was obtained:

$$tfidf'(j) = \frac{1}{tf(j)} \times idf'(j) \tag{2}$$

where tf represents the frequency of a noun phrase j in a document; idf is defined as:

$$idf'(j) = \frac{\ln\left(\frac{N}{df(j)}\right)}{\ln(10)} \tag{3}$$

In this case, the inverse document frequency is a measure of how much information the noun phrase provides, that is, whether the noun phrase is common or rare across all documents. It is the logarithmically scaled fraction of the documents that contain the noun phrase, obtained by dividing the total number of documents by the number of documents containing the noun phrase, and then taking the logarithms above of that quotient and denominator.

The rationale for this formula comes from the necessity to enhance the weight of low frequency elements, as is the case of noun phrases in the BON. The first noun phrases inverts the local importance of noun phrases, enhancing the weight of them; the second noun phrases acts as a scale reductor leading the values for *tfidf* to remain in a well behaved range. This can be observed in Fig. 1.

The three denoted ranges of frequency, *(a)*, *(b)* and *(c)*, represent operative intervals of frequency. It means the ranges in which the interesting noun phrases appear. In range *(a)* low frequency noun phrases (that is the common case) have their weight enhanced. Range *(b)* includes intermediately frequent noun phrases that usually not occur. Range *(c)* is, practically a non-operative range, since even occurring a few high frequency noun phrases, they are not of interest by having simple (general) structures.

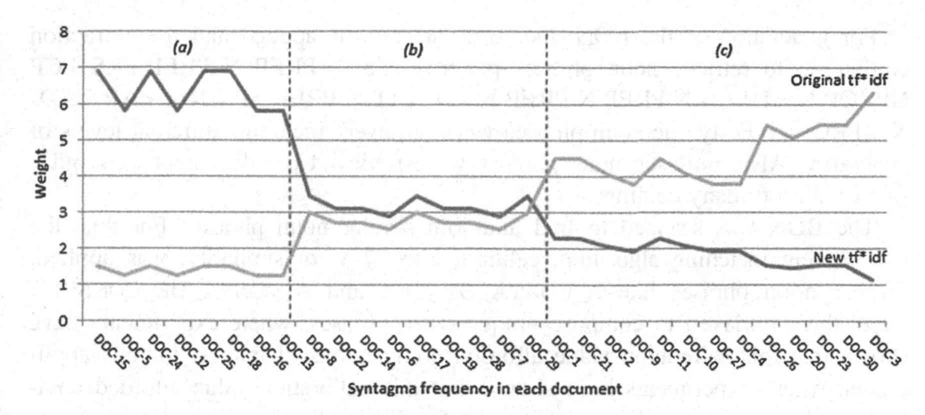

Fig. 1 Frequency of a noun phrase for the two forms of TF*IDF

A procedure to perform the text preprocessing and BON creation was implemented, including: *(i)* standardization of words; *(ii)* stopwords removal; *(iii)* tagging of words using MXPost Tagger; *(iv)* descriptors retrieval according the defined noun phrases structure; *(v)* generation of BON. Remember that an interaction with the user to refine the BON is underlying to the process. As output, a BON in CSV format is generated.

From a BON in CSV format, usually any TM software can be applied.

4 Case Study and Performance Analysis

An analysis of customers' opinions from an electric company was carried out as case study. The idea was to characterize groups of customers in order to guide the managers in coping with critical relationships.

The corpus used in the case study comprises 2,000 documents related to e-mails from corporate clients of an electric company, received in 2007 and 2008. The mining objective is to figure out categories of opinions regarding the services from the company. The mining task considered most appropriate for this application was clustering analysis, since no previous classes were known. For the sake of comparison, the traditional method and the approach here proposed were applied. So, both a BOW and a BON were created.

The BOW was generated using the traditional formula TF*IDF and the extraction was configured to retrieve nouns, adjectives, verbs and verbs in past perfect tense. A threshold of ten noun phrases per document were used based on [5, 6, 25, 26], in order to avoid an uncontrolled increase of the BOW.

For generation of the BON, the formula (2) was applied and the extraction configured to retrieve noun phrases patterns (e.g., <PREP N PREP|+KS PCP NPROP N ADJ>, <N PREP N PREP N>, <ART N PREP ART N>, <ART ADJ N NPROP>). Forty one noun phrases were retrieved, including different levels of generality. Also, only ten noun phrases were retrieved per document to avoid a BON with too many columns.

The BON was scanned to find and join similar noun phrases. For this, the Levenshtein matching algorithm, calibrated to 77 % of similarity, was applied. Similar noun phrases like `A_CONTA_DE_LUZ` and `A_CONTA_DE_CONSUMO` were, then, retrieved to compose unique noun phrases, where experiments have shown that this calibration value afforded comparisons of interest for this application, where experiments have shown that this calibration value afforded comparisons between noun phrases of interest for this application.

Next, the *k*-Means algorithm was applied, both over the BOW and the BON, under an overflow process designed to allow iterations using different values for *k*. The iterative process produced the configurations specified by the values of *k* that was evaluated for cohesion of groups and distance among them. The same process to identify the optimal *k* was applied to BOW and BON, reaching $k = 6$ for BOW and $k = 4$ for BON. It was used the Davies-Bouldin measure [27] to evaluate the compactness of groups and their mean distances.

After the clusters generation, the documents were labeled according the clusters they belong to and the characterization of them was performed by applying an algorithm for identifying associative rules. This option was considered adequate since all descriptors are, in each case (BOW or BON), from the same nature, words or noun phrases. The algorithm Tertius was applied because it allows one to define an invariable right side for the rules. So, it was possible to characterize each cluster.

The threshold of generated rules was tuned in order to have rules for all clusters. The rules were evaluated on the basis of the expert experience. Some examples of rules for the BOW, using the traditional TF*IDF [28], are shown below:

- Cluster 0: {BR, LEONARDO, BR and MORAIS, BR and FERNANDO}
- Cluster 1: {BR and CLIENTE}
- Cluster 2: {VENCIMENTO, CRESCIMENTO and VENCIMENTO, ATENDIMENTO and VENCIMENTO, ATENDIMENTO and CRESCIMENTO and VENCIMENTO}
- Cluster 3: {}
- Cluster 4: {BR and FERNANDO, BR and MORAIS, FERNANDO and MORAIS, LEONARDO and MORAIS}
- Cluster 5: {}

For example, according the expert, cluster 2 can be considered as well formed, since it represents a set of clients that complained against increases in their bills and about the due date. No good characterization was observed in the other groups.

Examples of associations found in the BON are shown below:

- Cluster 0 – Requirements for relationship updating for big clients: {MOTIVO_DA_SOLICITACAO, ATUALIZACAO_DO_RELACIONAMENTO, ATRAVES_DA_SOLICITACAO, CLIENTES_EMPRESARIAIS_TEL, ASSUNTO_ENC_ATUALIZACAO, ATUALIZACAO_DO_RELACIONAMENTO}
- Cluster 1 – Claims for improvement the customer treatment: {LIGHT_GRANDES_CLIENTES_LIGHT, A_MEDICAO_DE_ACOMPANHAMENTO, A_IMPOSSIBILIDADE_DE_GERENCIAMENTO, ENERGIA_ELETRICA_TEL, A_CONCESSISONARIA_DE_ENERGIA, O_TEMPO_DE_ATENDIMENTO, VERIFICACAO_NO_LOCAL}
- Cluster 2 – Request for technical assistance and installation: {A_POSSIBILIDADE_DE_INSTALACAO, O_PROCEDIMENTO_DE_INSTALACAO, UM_PEDIDO_DE_LIGACAO_NOVA, AS_CONTAS_DE_FORNECIMENTO, MUDANCA_DE_MEDIDOR, UM_DESLIGAMENTO_A_PEDIDO}
- Cluster 3 – Complaints regarding the consumption of energy: {MOTIVO_DA_SOLICITACAO, ENERGIA_ELETRICA_TEL, O_CONSUMO_DE_ENERGIA_ELETRICA, AS_CONTAS_DE_ENERGIA_ELETRICA}

Notice that the explanation concatenated with the cluster name was assigned by the expert on the basis of the noun phrases belonging to each cluster. This explanation was made possible due to the richer semantic of noun phrases in comparison to words alone.

The results with the BOW could be improved with the expansion of the stopwords list. However, the semantic issue would not be improved only by increasing words in the stopwords list. The results found with the BON showed to be more understandable, on the basis of the human judgment provided by the expert. In this sense, it is arguable that formula (2) succeeded in capturing the relevant noun phrases that represent each cluster.

5 Conclusion and Further Efforts

A flexible and effective proposal for improving TM process based on noun phrases was presented and discussed. The main advantages of this proposal are: *(i)* independence of dictionaries or thesaurus as occurs in other approaches; *(ii)* preservation of the words structures by avoiding stemming procedures; *(iii)* the potential to be used in other languages than Brazilian Portuguese.

Some important contributions for TM were introduced by: *(i)* identifying the drawbacks related to the use of noun phrases as descriptors; *(ii)* introducing an approach for the effective use of noun phrases as descriptors; *(iii)* designing a frequency calculation process able to attribute relevance for items with a limited range of frequency; *(iv)* improving TM process in comparison to the traditional method.

The use of Levenshtein matching algorithm has shown to be efficient and flexible, allowing a tune of the matching level according to the mining objective. However, the same general rule of thumb (*garbage in, garbage out*) applies to the proposed approach: the quality of the corpus is determinant for the mining results.

References

1. Weiss, S.M., Indurkhya, N., Zhang, T., Damerau, F.L.: Text Mining: Predictive Methods for Analyzing Unstructured Information. Springer, New York (2005)
2. Zanasi, D. (ed.): Text Mining and Its Applications to Intelligence, CRM and knowledge management. Advances in Management Information, WIT Press, Southampton (2007)
3. Berry, M.W., Castellanos, M. (eds.): Survey of Text Mining: Clustering, Classification, and Retrieval, 2nd Edn. Springer, New York (2007)
4. Konchady, M.: Text mining application programming. Cengage Learning (2006)
5. Lopes, M.C.S.: Textual data mining using clustering techniques for Portuguese language. Ph.D. Thesis, COPPE/UFRJ, Rio de Janeiro, Brazil (2004) (in Portuguese)
6. Furtado, M.I.V.: Business Intelligence for the private higher education: a text mining approach. Ph.D. Thesis, COPPE/UFRJ, Rio de Janeiro, Brazil (2004) (in Portuguese)
7. Gelbukh, A.: Computational linguistics and intelligent text processing. In: Second International Conference—CICLing, Mexico City, Mexico, Springer (2001)
8. Kuramoto, H.: Proposing a system of information retrieval by computer assisted with application to Portuguese. PhD thesis, University Lumière, Lyon, France (1999) (in French)
9. Silva, C.F., Vieira, R.: Grammatical and syntactic groups in automatic categorization with support vector machines. In: XXV SBC, V ENIA (2005) (in Portuguese)
10. Levenshtein, V.: Binary codes capable of correcting spurious insertions and deletions of ones. Probl. Inf. Transm. **1**, 8–17 (1965)
11. Navarro, G.: A guided tour to approximate string matching. ACM Comput. Surv. **33**(1), 31–48 (2001)
12. Baeza-Yates, R., Ribeiro-Neto, B.: Modern Information Retrieval. Addison-Wesley Longman Publishing Co., Inc., Boston (1999)
13. Feldman, R., Sanger, J.: The Text Mining Handbook: Advanced Approaches in Analyzing Unstructured Data. Cambridge University Press, Cambridge (2007)
14. Manning, C.D., Raghavan, P., Schütze, H.: Introduction to Information Retrieval. Cambridge University Press, Cambridge (2008)
15. Bräscher, M.: Automatic processing of ambiguity in information retrieval. Ph.D. Thesis, UnB, Brasília, Brazil (1999) (in Portuguese)
16. Silva, C.F., Vieira, R.: Categorization of Portuguese texts with decision trees, SVM and linguistic information. In: XXVII SBC, TIL—V Workshop em Tecnologia da Informação e da Linguagem Humana. Rio de Janeiro, Brazil (2007) (in Portuguese)
17. Ratnaparkhi, A.: A maximum entropy part-of-speech tagger. In: Conference on Empirical Methods in Natural Language Processing, pp. 133–142. University of Pennsylvania (1996)
18. Aires, R.V.X.: Implementation, adaptation, combination and evaluation of taggers for Portuguese of Brazil. MsC Thesis, USP, São Carlos, Brazil, 2000 (in Portuguese)
19. Mac-Morpho: Corpora in Portuguese for tagger MXPost (in Portuguese)
20. Matsunaga, L.A.: An automatic text categorization methodology for distribution of bills to standing committees of the legislative chamber of the Brazilian Federal District. Ph.D. Thesis, COPPE/UFRJ, Rio de Janeiro, RJ, Junho, 2007 (in Portuguese)
21. Feldman, R., Dagan, I.: Knowledge discovery in textual databases (KDT). In: KDD-95, Montreal, pp. 112–117. AAAI Press, Canada, 20–21 Aug 1995

22. Hearst, M.A.: Untangling text data mining. In: 37th Annual Meeting of the Association for Computational Linguistics, pp. 3–10 (1999)
23. Silva, C.F., Vieira, R., Osório, F.S.: Use of linguistic information in text categorization using neural networks. In: VIII SBRN, pp. 1–6 (2004) (in Portuguese)
24. Metzler, D.: Generalized inverse document frequency. In: 17th ACM Conference on Information and Knowledge Management—CIKM. Napa Valley, California, USA (2008)
25. Sirmakessis, S. (ed.): Text mining and its applications: Results of the NEMIS launch conference, vol. 138. Studies in Fuzziness and Soft Computing, Springer (2004)
26. Meij, J. (ed.): Dealing with the data flood: mining data, text and multimedia. STT/Beweton, The Hague (2002)
27. Davies, D.L., Bouldin, D.W.: A cluster separation measure. IEEE Trans. Pattern Anal. Mach. Intell. **1**(2), 224–227 (1979)
28. Papineni, K.: Why inverse document frequency? In: Second Meeting of the North American Chapter of the NAACL. Association for Computational Linguistics (2001)

Reinforcement Learning Strategy for Solving the MRCPSP by a Team of Agents

Piotr Jędrzejowicz and Ewa Ratajczak-Ropel

Abstract In this paper the strategy for the A-Team with Reinforcement Learning (RL) approach for solving the Multi-mode Resource-Constrained Project Scheduling Problem (MRCPSP) is proposed and experimentally validated. The MRCPSP belongs to the NP-hard problem class. To solve this problem a team of asynchronous agents (A-Team) has been implemented using multiagent system. An A-Team is the set of objects including multiple agents and the common memory which through interactions produce solutions of optimization problems. These interactions are usually managed by the static strategy. In this paper the dynamic learning strategy is suggested. The proposed strategy based on reinforcement learning supervises interactions between optimization agents and the common memory. To validate the proposed approach computational experiment has been carried out.

Keywords Multi-mode resource-constrained project scheduling · MRCPSP · Optimization · Agent · A-Team · Reinforcement learning

1 Introduction

The Multi-mode Resource Constrained Project Scheduling Problem (MRCPSP) has attracted a lot of attention and many exact, heuristic and metaheuristic solution methods have been proposed in the literature in recent years [12, 15, 18]. The current approaches to solve these problems produce either approximate solutions or can only be applied for solving instances of the limited size. Hence, searching for more effective algorithms and solutions to the problems is still a lively field of research. One of the promising directions of such research is to take advantage of the parallel and

P. Jędrzejowicz · E. Ratajczak-Ropel (✉)
Chair of Information Systems, Gdynia Maritime University,
Morska 83, 81-225 Gdynia, Poland
e-mail: ewra@am.gdynia.pl

P. Jędrzejowicz
e-mail: pj@am.gdynia.pl

R. Neves-Silva et al. (eds.), *Intelligent Decision Technologies*,
Smart Innovation, Systems and Technologies 39,
DOI 10.1007/978-3-319-19857-6_46

distributed computation solutions, which are the common feature of the contemporary multiagent systems.

Modern multiagent system architectures are an important and intensively expanding area of research and development. There is a number of multiple-agent approaches proposed to solve different types of optimization problems. One of them is the concept of an A-Team, originally introduced by Talukdar et al. [14]. The idea of the A-Team was used to develop the software environment for solving a variety of computationally hard optimization problems called JABAT [1, 8]. JADE based A-Team (JABAT) system supports the construction of the dedicated A-Team architectures.

A-Team is a system composed of the set of objects including multiple agents and the common memory which through interactions produce solutions of optimization problems. Several strategies controlling the interactions between agents and memories have been recently proposed and experimentally validated. The influence of such strategy on the A-Team performance was investigated by Barbucha at al. [1]. In [2] the reinforcement learning based strategy for the synchronous team of agents has been considered. The similar topics were also considered by other authors for different multi-agent systems, e.g. [6].

Reinforcement Learning (RL) [3] belongs to the category of unsupervised machine learning algorithms. It can be described as learning which action to take in a given situation (state) to achieve one or more goal(s). The learning process takes place through interaction with an environment. Reinforcement Learning is usually used in solving combinatorial optimization problems at three levels [17]: direct level, metaheuristic level and hyperheuristic level.

In this paper the RL based on utility values to learn interaction strategy for the A-Team solving the MRCPSP is proposed and experimentally validated. The concept of using the RL to control the strategy parameters instead of the parameters required by the respective metaheuristics for RCPSP has been proposed in [10]. In this paper the additional rules have been proposed, the approach has been extended to MRCPSP and experimentally validated. It is expected that introducing the proposed RL will result in obtaining high quality solutions in an efficient manner.

The paper is constructed as follows: Sect. 2 contains the MRCPSP formulation. Section 3 gives some information on JABAT environment. Section 4 provides details of the proposed RL based approach. Section 5 describes settings of the computational experiment carried-out with a view to validate the proposed approach and contains a discussion of the computational experiment results. Finally, Sect. 6 contains conclusions and suggestions for future research.

2 Problem Formulation

A single-mode Resource-Constrained Project Scheduling Problem (RCPSP) consists of a set of n activities, where each activity has to be processed without interruption to complete the project. The dummy activities 1 and n represent the beginning and the end of the project. The duration of an activity $j, j = 1, \dots, n$ is denoted by d_j where $d_1 = d_n = 0$. There are r renewable resource types. The availability of each resource type k in each time period is r_k units, $k = 1, \dots, r$. Each activity j requires r_{jk} units of resource k during each period of its duration, where $r_{1k} = r_{nk} = 0, k = 1, \dots, r$. All parameters are non-negative integers. There are precedence relations of the finish-start type with a zero parameter value (i.e. $FS = 0$) defined between the activities. In other words activity i precedes activity j if j cannot start until i has been completed. The structure of a project can be represented by an activity-on-node network. SS_j (SP_j) is the set of successors (predecessors) of activity $j, j = 1, \dots, n$. It is further assumed that $1 \in SP_j, j = 2, \dots, n$, and $n \in SS_j, j = 1, \dots, n-1$. The objective is to find a schedule S of activities starting times $[s_1, \dots, s_n]$, where $s_1 = 0$ and precedence relations and resource constraints are satisfied, such that the schedule duration $T(S) = s_n$ is minimized.

In case of the MRCPSP each activity $j, j = 1, \dots, n$ may be executed in one out of M_j modes. The activities may not be preempted and a mode once selected may not change, i.e., a job j once started in mode m has to be completed in mode m without interruption. Performing job j in mode m takes d_{jm} periods and is supported by a set R of renewable and a set N of non-renewable resources. Considering the time horizon, that is, an upper bound T on the project's makespan, one has the available amount of renewable (doubly constrained) resource as well as certain overall capacity of the non-renewable (doubly constrained) resource. The objective is to find a makespan minimal schedule that meets the constraints imposed by the precedence relations and the limited resource availability.

The MRCPSP is NP-hard as a generalization of the RCPSP [5]. Moreover, if there is more than one nonrenewable resource, the problem of finding a feasible solution for the MRCPSP is NP-complete [11].

3 The JABAT Environment

JABAT is the software environment facilitating the design and implementation of the A-Team architecture for solving various combinatorial optimization problems using population-based approach.

JABAT produces solutions to combinatorial optimization problems using a set of optimization agents, each representing an improvement algorithm. The initial population of solutions (individuals) is generated or constructed. Individuals from the population are, at the following computation stages, improved by independently acting optimization agents. The main functionality of the proposed environment includes organizing and conducting the process of search for the best solution.

The behavior of the A-Team is controlled by the, so called, interaction strategy defined by the user. An A-Team uses a population of individuals (solutions) and a number of optimization agents. All optimization agents within the A-Team work together to improve individuals from its population in accordance with the interaction strategy.

To adapt the proposed architecture for solving the particular problem, the following classes of agents need to be designed and implemented:

SolutionManager - represents and manages the process of solving the problem instance by the A-Team.

OptiAgent - represents a single improving algorithm (e.g. local search, simulated annealing, genetic algorithm etc.).

To describe the problem Task class representing the instance of the problem and Solution class representing the solution is used.

JABAT has been designed and implemented using JADE (Java Agent Development Framework) [4]. More detailed information about the JABAT can be found in [1, 8].

4 A-Team with Interaction Strategy Controlled by RL

JABAT was successfully used by the authors for solving the RCPSP and MRCPSP problems [9, 10]. The approach proposed in this paper is based on the earlier JABAT adaptation for MRCPSP [9] and RCPSP [10]. The most important differences include extending the RL controlled strategy and adapting it to MRCPSP. The mechanisms of remembering the environment state and sending and receiving individuals have been rebuild. Agent interaction strategy has been made dependent on current solutions characteristics.

To adapt JABAT to solving MRCPSP using RL strategy two sets of classes and agents were implemented. The first set includes classes describing the problem. They are responsible for reading and preprocessing of the data and generating random instances of the problem.

The second set includes classes describing the optimization agents. Each of them includes the implementation of an optimization algorithm used to solve the MRCPSP. All of them are inheriting from the OptiAgent class. The prefix Opti is assigned to each agent with its embedded algorithm:

OptiLSA - implementing the Local Search Algorithm (LSA),

OptiTSA - implementing the Tabu Search Algorithm (TSA),

OptiCA - implementing the Crossover Algorithm (CA),

OptiPRA - implementing the Path Relinking Algorithm (PRA).

In our earlier approaches the different static interaction strategies were used [1]. The proposed RL interaction strategy extends the most promising approach identified in [1] - Blocking2 strategy. Now the interaction strategy is based on the following assumptions and rules:

- All individuals in the initial population of solutions are generated randomly and stored in the common memory.
- Individuals for improvement are selected by the solution manager from the common memory randomly and blocked which means that once selected individual (or individuals) cannot be selected again until the OptiAgent to which it has been sent returns its improved solution.

- The returning individual, which represents a feasible solution, replaces its original version before eventual further attempted improvements take place.
- A new feasible solution is generated with fixed probability P_{rg} to replace randomly selected worse one from the population.
- For each level of learning the environment state is remembered and used in the learning scheme. This environment state includes: the best individual and the population average diversity. The state is calculated every fixed number of iterations $itNS = \lfloor PopulationSize/AgentsNumber \rfloor$. To reduce the computation time, average diversity of the population is evaluated by comparison with the best solution only. Diversity of two solutions is evaluated as the sum of differences between activities starting times in the projects.
- The A-Team stops computations where the average diversity in the population is less then fixed threshold (e.g. 0.01 %).

Four learning rules (RL rules), as well as their combinations, are formulated and adopted to the proposed basic strategy. As a result the following interaction strategies are considered:

- RL1 - controls replacement of one individual from the population by another, randomly generated one.
- RL2 - controls the method of selecting an individual for replacement.
- RL3 - controls the process of clustering individuals in the population according to certain features, and next selecting a cluster.
- RL4 - controls the procedure of selecting individuals to be forwarded to optimization agents from the cluster chosen by RL3.
- RL123 - RL1 and RL2 and RL3 are used together.
- RL1234 - RL1 and RL2 and RL3 and RL4 are used together.

The first three strategies and their combinations have been proposed and tested for RCPSP in [10]. The most effective has proven to be combination of strategies denoted as RL123. In this approach the additional RL4 strategy is proposed.

The considered RL strategies are based on utility values reinforcement proposed in [2, 13]. The SolutionManager is the agent that manages the population of solutions in accordance with the chosen strategy. It sends the individuals to optimization agents (OptiAgents) and receives improved solutions (individuals) from them. The SolutionManager calculates and preserves environment state and uses the RL rules to manage its actions.

In case of the RL1 the probability of randomly generating a new solution P_{rg} is calculated as $P_{rg} = w_{rg}/PopulationSize$. Initially the weight w_{rg} is set to 0 to intensify the exploitation within an initial population. The w_{rg} is increased where the average diversification in population decreases. The w_{rg} is decreased in three cases: where average diversity in the population increases, where the best solution is found and where the new solution is randomly generated.

In case of the RL2 different methods of adding the new randomly generated individual to the population are considered. For each method the selection probability is calculated. Three methods are possible: the new solution replaces the random one, the random worse one or the worst solution in the population. The weight w_m for each method is calculated, where $m \in M$, $M = \{random, worse, worst\}$. The w_{random} is increased where the population average diversity decreases and it is decreased in the opposite case. The w_{worse} and w_{worst} is decreased where the population average diversity decreases and they are increased in the opposite case. The probability of selection the method m is calculated as

$$P_m = \frac{w_m}{\sum_{i \in M} w_i}.$$

In case of the RL3 strategy, RL utility values are introduced for clusters (groups) of solutions and for different kinds of algorithms. The RL is applied during solving a single instance of the problem. In this case the additional parameters of environment state are remembered as the features of each individual (solution) in the population. These are the average number of unused resources in each period of time $\overline{f_{URP}}$ and the average lateness of activities $\overline{f_{LA}}$. For each algorithm the matrix of weights allocated to each solution feature is remembered. The weight is increased where the optimization agent implementing given kind of algorithm returns better or the best solution. The weight is decreased where the optimization agent returns non better solution. The probability of choosing the algorithm is calculated as

$$P^{Alg}_{\overline{f_{URP}}\,\overline{f_{LA}}} = \frac{w^{Alg}_{\overline{f_{URP}}\,\overline{f_{LA}}}}{\sum_{i \in F_{URP}, j \in F_{LA}} w^{Alg}_{ij}},$$

where $i \in F_{URP}$ and $j \in F_{LA}$. F_{URP} and F_{LA} denote the sets of average unused resources in each period of time and average activity lateness, respectively.

The RL4 strategy is complementary to the RL3. For each individual from the population the matrix of weights is remembered. It reflects results of the subsequent improvement attempts. In the considered case the matrix of four weights is used, one for each OptiAgent. The initial weights are idetical. After receiving an "improved" individual from the OptiAgent the respective weight is decreased when the value of solution is worse and increased when this value is better or equal to the one before the attempted improvement took place. The probability of selecting an individual for improvement is calculated as

$$P^{Alg}_S = \frac{w^{Alg}_S}{\sum_{i \in G_{RL3}} w^{Alg}_i},$$

where G_{RL3} is the group of individuals indicated by RL3.

The rules RL1 and RL2 are used consecutively just before adding the new solution to the population after receiving it by the SolutionManager from the OptiAgent. The RL3 and RL4 strategy is used just before reading the solution from the population. The RL3 and RL4 act asynchronously to the RL1 and RL2.

5 Computational Experiment

5.1 Settings

To evaluate the effectiveness of the proposed approach the computational experiment has been carried out using benchmark instances of MRCPSP from PSPLIB.[1]. The test sets include: mm10 (multi mode, 10 activities), mm12, mm14, mm16, mm18, mm20. Each set includes 640 problem instances.

In the experiment the following parameters have been used:

- *PopulationSize* = 30,
- *itNSnoImp* = *PopulationSize* (it is the number of iteration in which the state is calculated and no improvement has been detected)
- *AgentsNumber* = 4 (the proposed A-Team includes four optimization agents, representing the LSA, TSA, CA and PRA algorithms)
- $itNS = \lfloor PopualtionSize/AgentsNumber \rfloor = 7$,
- initial values for probabilities and weights:
  - $w_{rg}0 = 0$,
  - w_m: $w_{random} = 30$, $w_{worse} = 60$, $w_{worst} = 10$,
  - $w^{Alg}_{f_{URP}, f_{LA}} = 1$,
  - $w^{Alg}_{S} = 50$.

Values of parameters have been chosen experimentally based on the earlier experiments in JABAT and the preliminary experiments for the RL strategies.

In case of the positive reinforcement the additive adaptation for the weights is used: $w = w + 1$, and in case of the negative reinforcement the additive $w = w - 1$ or root adaptation $w = \sqrt{w}$ is used. If the utility value falls below 1, it is reset to 1; if the utility value exceeds a certain max_w, it is reset to max_w if the utility value is assigned a non-integer value, it is rounded down. These update schemes proved effective in [2, 13].

The experiment has been carried out using nodes of the cluster Holk of the Tricity Academic Computer Network built of 256 Intel Itanium 2 Dual Core 1.4 GHz with 12 MB L3 cache processors and with Mellanox InfiniBand interconnections with 10 Gb/s bandwidth.

5.2 Results

During the experiment the following characteristics of the computational results have been calculated and recorded: Mean Relative Error (MRE) calculated as the deviation from the optimal solution, Mean Computation Time (MCT) which has been

[1]See PSPLIB at http://www.om-db.wi.tum.de/psplib/.

needed to find the best solution and Mean Total Computation Time (MTCT) which has been needed to stop all optimization agents and the whole system. Each instance has been solved five times and the results have been averaged over these solutions.

The computational experiment results are presented in Tables 1, 2 and 3. The results obtained for the proposed approach are good and very promising. The mean relative error for RL123 and RL1234 below 1 % in case of 10, 12, 14, 16 and 18 activities and below 1.6 % in case of 20 activities have been obtained.

It can be seen that in case of using a single rule the most effective is the RL1 strategy, and the least effective the RL3 but the differences are very small. After conjunction of three and four RL rules coactively the effectiveness increases. The most effective proves to be the conjunction of all four considered rules in the RL1234 strategy. Simultaneously, it can be observed that introducing more RL rules to the strategy insignificantly influences computation time. Hence additional and other learning parameters or methods could be considered.

The presented results are comparable with the results reported in the literature [7, 15, 16], see Table 4. However in case of the agent based approach it is difficult to compare computation times as well as the numbers of schedules, which is another widely used measure of the algorithm efficiency in project scheduling. In the proposed agent-based approach computation times as well as numbers of schedules differ between nodes and optimization agent algorithms working in parallel. The results obtained by a single agent may or may not influence the results obtained by the other agents. Additionally the computation time includes the time used by agents to prepare, send and receive messages.

Table 1 MRE from the optimal solution for benchmark test sets mm10 - mm20

Strategy	mm10	mm12	mm14	mm16	mm18	mm20
Blocking2	0.41 %	0.47 %	0.69 %	0.81 %	0.95 %	1.80 %
RL1	0.38 %	0.48 %	0.61 %	0.80 %	0.98 %	1.58 %
RL2	0.38 %	0.48 %	0.58 %	0.78 %	0.96 %	1.64 %
RL3	0.39 %	0.49 %	0.58 %	0.78 %	0.97 %	1.70 %
RL34	0.36 %	0.47 %	0.56 %	0.77 %	0.95 %	1.63 %
RL123	0.32 %	0.43 %	0.55 %	0.75 %	0.94 %	1.58 %
RL1234	0.28 %	0.41 %	0.54 %	0.75 %	0.92 %	1.55 %

Table 2 MCT [s] for benchmark test sets mm10 - mm20

Strategy	mm10	mm12	mm14	mm16	mm18	mm20
Blocking2	2.23	2.23	3.31	3.52	4.13	5.01
RL1	1.45	1.47	3.08	3.28	4.22	4.62
RL2	1.34	1.33	3.09	3.25	4.04	4.57
RL3	1.27	1.20	3.11	3.12	3.59	4.57
RL34	1.31	1.35	3.24	3.34	4.01	4.23
RL123	1.27	1.29	3.23	3.45	4.37	4.17
RL1234	1.29	1.43	3.25	3.56	4.45	4.21

Table 3 MTCT [s] for benchmark test set mm10 - mm20

Strategy	mm10	mm12	mm14	mm16	mm18	mm20
Blocking2	35.12	25.12	27.02	29.30	28.56	36.09
RL1	14.21	20.71	23.43	26.15	26.15	32.12
RL2	13.49	23.08	24.15	26.47	26.47	32.14
RL3	12.76	22.34	25.14	26.39	26.45	32.21
RL34	13.31	21.97	25.96	26.06	26.56	32.34
RL123	12.42	22.85	25.50	26.78	27.21	32.23
RL1234	13.14	23.12	24.56	25.89	27.84	32.29

The experiment results show that the proposed implementation is effective and using reinforcement learning based on the utility values to control different elements of the interaction strategy in A-Teams architecture is beneficial.

6 Conclusions

The computational experiment results show that the proposed dedicated A-Team architecture supported by Reinforcement Learning to control the interaction strategy is an effective and competitive tool for solving instances of the MRCPSP. Presented results are comparable with solutions known from the literature and in some cases outperform them. It can be also noted that they have been obtained in a comparable time. However, in this case time comparison may be misleading since the algorithms are run using different environments, operating systems, numbers and kinds of processors. In case of the agent-based environments the significant part of the time is used for agent communication which has an influence on both - computation time and quality of the results.

The presented experiment should be extended to examine different and additional parameters of the environment state and solutions as well as iteration numbers, probabilities and weights. On the other hand the additional or other RL rules should be examined.

Table 4 Literature reported results [7, 15, 16]

Set	Algorithm	Authors	MRE	MCT [s]	Computer
mm10	Distribution algorithm	Wang & Fang	0.01 %	1	2.2 GHz
	Genetic algorithm	Van Peteghem & Vanhoucke	0.01 %	0.12	2.8 GHz
	Hybrid genetic algorithm	Lova et al.	0.04 %	0.1	3 GHz
	our approach		*0.28* %	*1.29*	*1.4* GHz
mm12	Distribution algorithm	Wang & Fang	0.02 %	1.8	2.2 GHz
	Genetic algorithm	Van Peteghem & Vanhoucke	0.09 %	–	–
	Hybrid genetic algorithm	Lova et al.	0.17 %	–	–
	our approach		*0.41* %	*1.29*	*1.4* GHz
	Hybrid scatter search	Ranjbar et al.	0.65 %	10	3 GHz
mm14	Distribution algorithm	Wang & Fang	0.03 %	1	2.2 GHz
	Genetic algorithm	Van Peteghem & Vanhoucke	0.22 %	0.14	2.8 GHz
	Hybrid genetic algorithm	Lova et al.	0.32 %	0.11	3 GHz
	our approach		*0.54* %	*3.25*	*1.4* GHz
	Hybrid scatter search	Ranjbar et al.	0.89 %	10	3 GHz
mm16	Distribution algorithm	Wang & Fang	0.17 %	1	2.2 GHz
	Genetic algorithm	Van Peteghem & Vanhoucke	0.32 %	0.15	2.8 GHz
	Hybrid genetic algorithm	Lova et al.	0.44 %	0.12	3 GHz
	our approach		*0.75* %	*3.56*	*1.4* GHz
	Hybrid scatter search	Ranjbar et al.	0.95 %	10	3 GHz

(continued)

Table 4 (continued)

Set	Algorithm	Authors	MRE	MCT [s]	Computer
mm18	Distribution algorithm	Wang & Fang	0.19 %	1	2.2 GHz
	Genetic algorithm	Van Peteghem & Vanhoucke	0.42 %	0.16	2.8 GHz
	Hybrid genetic algorithm	Lova et al.	0.63 %	0.13	3 GHz
	our approach		*0.92 %*	*4.45*	*1.4* GHz
	Hybrid scatter search	Ranjbar et al.	1.21 %	10	3 GHz
mm20	Distribution algorithm	Wang & Fang	0.32 %	1	2.2 GHz
	Genetic algorithm	Van Peteghem & Vanhoucke	0.57 %	0.17	2.8 GHz
	Hybrid genetic algorithm	Lova et al.	0.87 %	0.15	3 GHz
	our approach		*1.55 %*	*4.21*	*1.4* GHz
	Hybrid scatter search	Ranjbar et al.	1.64 %	10	3 GHz

References

1. Barbucha, D., Czarnowski, I., Jędrzejowicz, P., Ratajczak-Ropel, E., Wierzbowska, I.: Influence of the working strategy on A-team performance, smart information and knowledge management. In: Szczerbicki, E., Nguyen, N.T. (eds.) Studies in Computational Intelligence, vol. 260, pp. 83–102 (2010)
2. Barbucha, D.: Search modes for the cooperative multi-agent system solving the vehicle routing problem. Intell. Auton. Syst. Neurocomput. **88**, 13–23 (2012)
3. Barto, A.G., Sutton, R.S., Anderson, C.W.: Neuronlike adaptive elements that can solve difficult learning control problems. IEEE Trans. Syst. Man Cybern. **SMC-13**, 835–846 (1983)
4. Bellifemine, F., Caire, G., Poggi, A., Rimassa, G.: JADE. A White Paper Exp. **3**(3), 6–20 (2003)
5. Błażewicz, J., Lenstra, J., Rinnooy, A.: Scheduling subject to resource constraints: Classification and complexity. Discrete Appl. Math. **5**, 11–24 (1983)
6. Cadenas, J.M., Garrido, M.C., Muñoz, E.: Using machine learning in a cooperative hybrid parallel strategy of metaheuristics. Inf. Sci. **179**(19), 3255–3267 (2009)
7. Lova, A., Tormos, P., Cervantes, M., Barber, F.: An efficient hybrid genetic algorithm for scheduling projects with resource constraints and multiple execution modes. Int. J. Prod. Econ. **117**(2), 302–316 (2009)
8. Jędrzejowicz, P., Wierzbowska, I.: JADE-based A-team environment. Comput. Sci.–ICCS. Lect. Notes Comput. Sci. **3993**, 719–726 (2006)
9. Jędrzejowicz, P., Ratajczak-Ropel, E.: New generation A-Team for solving the resource constrained project scheduling. In: Proceedings of the Eleventh International Workshop on Project Management and Scheduling, pp. 156–159. Istanbul (2008)
10. Jędrzejowicz, P., Ratajczak-Ropel, E.: Reinforcement learning strategies for A-team solving the resource-constrained project scheduling problem. Neurocomputing **146**, 301–307 (2014)

11. Kolisch, R.: Project scheduling under resource constraints–Efficient heuristics for several problem classes. Ph.D. thesis, Physica, Heidelberg (1995)
12. Liu, S., Chen, D., Wang, Y.: Memetic algorithm for multi-mode resource-constrained project scheduling problems. J. Syst. Eng. Electron. **25**(4), 609–617 (2014)
13. Nareyek, A.: Choosing Search Heuristics by Non-Stationary Reinforcement Learning Metaheuristics: Computer Decision-Making. Academic Publishers, Kluwer (2001)
14. Talukdar, S., BaerentzenL, G.A, De Souza, P.: Asynchronous teams: Co-operation schemes for autonomous, computer-based agents. In: Technical Report EDRC 18–59-96, Carnegie Mellon University, Pittsburgh (1996)
15. Van Peteghem, V., Vanhoucke, M.: A genetic algorithm for the preemptive and non-preemptive multi-mode resource-constrained project scheduling problems. Eur. J. Oper. Res. **201**(2), 409–418 (2010)
16. Ranjbar, M., Reyck, B., De Kianfar, F.: A hybrid scatter search for the discrete time/resource trade-off problem in project scheduling. E. J. Oper. Res. **193**(1), 35–48 (2009)
17. Wauters, T.: Reinforcement learning enhanced heuristic search for combinatorial optimization, Doctoral thesis, Department of Computer Science, KU Leuven (2012)
18. Węglarz, J., Józefowska, J., Mika, M., Waligora, G.: Project scheduling with finite or infinite number of activity processing modes–a survey. Eur. J. Oper. Res. **208**, 177–205 (2011)

Comparison of Classical Multi-Criteria Decision Making Methods with Fuzzy Rule-Based Methods on the Example of Investment Projects Evaluation

Bogdan Rębiasz and Andrzej Macioł

Abstract In the process of investment decision making, next to financial indicators many other aspects of investment projects are increasingly often considered. This leads to the multi-criteria evaluation of a project. In the work one compared results of multi-criteria evaluation of the investment projects realized by using *TOPSIS* and *AHP* methods with results obtained at the use of rule-based methods, especially fuzzy reasoning techniques. To comparisons were used chosen investments in the metallurgical industry. The work finish conclusions defined on the basis carried out calculations.

Keywords Evaluation of investment projects · Multi-criteria methods · Hybrid data · Fuzzy reasoning

1 Introduction

A commonly used criterion for selecting investment projects are financial criteria. Very often non-financial criteria are taken into account. This leads to the multi-criteria evaluation of a project. Furthermore one can say that the multi-criteria decision-making process in the evaluation of investments has some specific characteristics [1, 2]: it has to take into consideration either quantitative or qualitative effects, naturally occurring competitiveness and even contradiction of criteria, it has to take into consideration both the uncertainty of each alternative and the uncertainty originating from the difficulty to establish the importance of every goal.

The aim of our study was to determine whether different methods based on the fuzzy logic theory allow for effective evaluation of investment projects and as may

B. Rębiasz (✉) · A. Macioł
AGH University of Science and Technology, Faculty of Management, Kraków, Poland
e-mail: brebiasz@zarz.agh.edu.pl

A. Macioł
e-mail: amaciol@zarz.agh.edu.pl

R. Neves-Silva et al. (eds.), *Intelligent Decision Technologies*,
Smart Innovation, Systems and Technologies 39,
DOI 10.1007/978-3-319-19857-6_47

be applicable in the longer term in economic field. Under the attention we took two groups of methods. The first group contains fuzzy version of the most classical *multi-criteria* methods. One can here mention Fuzzy *Analytical Hierarchy Process* (*FAHP*) and *Fuzzy Technique for Order of Preference by Similarity to Ideal Solution* (fuzzy *TOPSIS*). The second group is the modelling techniques based on fuzzy inference systems (*FIS*). Results of our research can contribute to the indication of necessary actions to the dissemination methods based on fuzzy logic theory in economics.

2 Investment Projects Evaluation with Uncertain Knowledge – State of the Art

AHP is an approach to decision making that involves structuring multiple choice criteria into a hierarchy, assessing the relative importance of these criteria, comparing alternatives for each criterion and determining an overall ranking of the alternatives [3, 4]. Many studies have explored the field of the fuzzy extension of Saaty's theory [1, 5].

TOPSIS method is based on the idea that the chosen alternative should have the shortest distance from the *Positive Ideal Solution* (*PIS*) and on the other hand the largest distance from the *Negative Ideal Solution* (*NIS*).

Some of *AHP* applications include technology choice [6], vendor selection of a telecommunications system [7], project selection, budget allocation [1]. Liu et al. introduces an evaluation method based on an uncertain linguistic weighted operator to the risk evaluation of the high-tech project investment [8]. Rębiasz et al. [9] present the usage of the *AHP* and *TOPSIS* methods for the evaluation of projects and selection of the most profitable project from the steel industry.

An alternative to the classic (but also fuzzy) method of investment projects evaluation based on comparing of cases are rule-based methods and especially fuzzy reasoning techniques. The concept of fuzzy models was originally created for the needs of *Fuzzy Logic Control* (*FLC*) [10]. The two types of *FIS*, namely Mamdani [11] and Sugeno *FIS* [12] are widely accepted and applied to many real-world problems. Although these methods have been developed for the *FLC*, you can find more and more frequent reports of using them to solve problems in other fields. As examples can be mentioned a ranking method on the basis of fuzzy inference system (*FIS*) proposed for supplier selection problem presented in [13], risk evaluation in electricity market [14], traffic flow prediction using Mamdani and Sugeno Fuzzy Inference Systems [15].

The other way to solve the problem occurrence of uncertainty can be solved by using evidential reasoning approach. Premises and conclusions of the rules of such system can be expressed in various ways (linguistic variables, probability distributions, crisp values). A way to write the rules in such system as well as the inference methodology was proposed by Yang et al. [16]. The rule-base inference methodology proposed by Yang uses the evidential reasoning (*RIMER*). Yang presents this approach on simple technical example but evidential reasoning is more often used in different areas in economic problems too. As example one can read about use of evidential reasoning for financial audit support [17]. The same approach was presented by the authors in [18].

3 Numerical Examples

A numerical examples, which are based on investment projects analysis in metallurgical industry, is studied in this section. The example aims to determine the utility of the following investment projects: modernization of the shape mill (*A1*), construction of the cold rolled sheet mill (*A2*), construction of the hot-dip galvanizing sheet plant (*A3*), construction of the sheet organic coating plant (*A4*), Construction of the drawing mill (*A5*). For simplification the input indicators defined in the process of projects evaluation include only the following:

- The financial criterion (*C1*):
  - profitability Index - *PI* (*C11*),
  - return on equity - *ROE%* (*C12*),
- The market criterion (*C2*):
  - forecasted dynamics of the market (C21),
  - competitiveness of products (*C22*):
    - quality of products (*C221*),
    - price of products (*C222*),
- Impact on the environment (*C3*).

In the Table 1 is presented the original characteristics of analysed projects.

The characteristics of projects were defined as follows. Continuous variable *PI* and *ROE%* was defined in the form of normal probability density function, forecasted dynamics of the market was defined by experts as linguistic variable transformed to triangular fuzzy number. The remaining parameters were determined by experts in the form of belief structure (*H*,a; *M*,b; *L*,c), where *L* means low, *M* medium and *H* high. The defined above characteristics were next transformed to the form required by every of the methods used for project evaluation. The way of the conversion is described below for every method separately.

Below one presented the evaluation of chosen projects using the multi-criterion methods (*FAHP* and fuzzy *TOPSIS*) and fuzzy inference systems (*FIS*).

Table 1 The characteristics of analysed projects

Projects	Input indicators					
	C11[1]	C12[1]	C21[2]	C221	C222	C3
A1	(1,40; 0,2)	(8,0; 1,6)	*approx 3,5 % (2,2; 3,5; 4,8)*	*(H, 02; M, 0,8)*	*(M, 1)*	*(M, 03; L, 0,7)*
A2	(1,12; 0,15)	(7,2; 1,1)	*approx 2, 5 % (1,8; 2,5; 3,2)*	*(H, 05; M, 0,5)*	*(M, 1)*	*(M, 03; L, 0,7)*
A3	(2,12; 0,22)	(12,1; 2,1)	*approx 4, 5 % (3,2; 4,5; 5,8)*	*(H, 07; M, 0,3)*	*(M, 0,2; H, 0,8)*	*(M, 0,8; L0,2)*
A4	(2,81; 0,26)	(15,3; 2,2)	*approx 6,0 % (4,7; 6,0; 7,3)*	*(H, 07; M, 0,3)*	*(M, 0,2; H, 0,8)*	*(M, 0,9; L,0,1)*
A5	(1,13; 0,17)	(6,2; 1,1)	*approx 2,5 % (1,8; 2, 5; 3,2)*	*(L, 0,8; M,0,1; L, 0,1))*	*(L, 0,9; M,0,1)*	*(L, 1)*

[1](m, s) – average, standard deviation
[2](a, b, c) – triangular fuzzy number

3.1 Evaluation of Chosen Projects Using FAHP

Let $v_1, v_2, \ldots, v_l$ denote the weights of criteria and p_{ij} ($i = 1,2,\ldots, k, j = 1, 2,\ldots, l$) the performance of ith alternative on jth criterion. Now the global priority (P_i) of the ith alternative can be obtained as the weighted sum of performances [5, 19]:

$$P_i = \sum_{j=1}^{l} v_j p_{ij} \tag{1}$$

The higher the value P_i, the more preferred the alternative.

In order to construct pairwise comparison of alternatives under each criterion or about criteria a triangular fuzzy comparison matrix is defined as follows [19]:

$$\tilde{A} = \left(\tilde{a}_{ij}\right)_{n\times n} = \begin{bmatrix} (1,1,1) & l_{12}, m_{12}, u_{12} & \cdots & (l_{1n}, m_{1n}, u_{1n}) \\ (l_{21}, m_{21}, u_{21}) & (1,1,1) & \cdots & (l_{2n}, m_{2n}, u_{2n}) \\ \vdots & \vdots & \vdots & \vdots \\ (l_{n1}, m_{n1}, u_{n1}) & (l_{n2}, m_{n2}, u_{n2}) & \cdots & (1,1,1) \end{bmatrix} \tag{2}$$

where:
$\tilde{a}_{ij} = \left(l_{ij}, m_{ij}, u_{ij}\right)$ is the triangular fuzzy number representing uncertain comparison judgment.

To the calculation $v_1; v_2,\ldots,v_l$ and p_{ij} ($i = 1, 2,\ldots,k, j = 1, 2,\ldots, l$) we used *Fuzzy Extent Analysis* proposed by Chang [19].

In the Table 2 is presented the ranking scale for criteria and alternatives. In the Table 3 is presented pairwise comparison of criteria and in Table 4 pairwise comparison of subcriteria prepared by experts. When defining these magnitudes, one used the scale presented in Table 2. In the Table 5 is presented final priority of sub-criterion.

For example in Table 6 is presented the fuzzy evaluation matrixes relevant to the criterion *C11*. This table arose across the data transformation contained in the Table 1 and the use of the scale from the Table 2. For example, if *C11* for *A3* is equal to (2,12; 0,22) and for *A1* is equal to (1,40; 0,20), then *A3* according to the criterion *C11* is "Very strongly more import" than *A1* and we choose (6,7,8).

Table 2 Ranking scale for criteria and alternatives

Definition	Fuzzy intensity of importance	Explanation
Equally important	(1,1,1)	Two factors contribute equally to the objective
Moderately more important	(2,3,4)	Experience and judgement slightly favour one over the other
Strongly more import	(4,5,6)	Experience and judgement strongly favour one over the other
Very strongly more import	(6,7,8)	Experience and judgement very strongly favour one over the other; its importance is demonstrated in practice
Extremely more import	(8,9,10	The evidence favouring one over the other is of the highest possible validity

Table 3 Pairwise comparison of criteria

C1	(1,1,1)	(2,5,6)	(6,7,8)
C2	(0.167,0.2,0.5)	(1,1,1)	(1,3,5)
C3	(0.125,0.143,0.167)	(0.2,0.333,1)	(1,1,1)

Table 4 Pairwise comparison subcriteria

	C11	*C12*		*C21*	*C22*		*C221*	*C222*
C11	(1,1,1)	(1,3,5)	*C21*	(1,1,1)	(0.2,0.333,1)	*C221*	(1,1,1)	(1,1,1)
C12	(0.2,0.333,1)	(1,1,1)	*C22*	(1,3,5)	(1,1,1)	*C222*	(1,1,1)	(1,1,1)

Table 5 Final priority of sub-criterion

The sub-criterion	Final priority of sub-criterion
C11	0.330
C12	0.213
C21	0.111
C221	0.110
C222	0.034
C3	0.203

Table 6 The fuzzy evaluation matrix relevant to the criterion *C11*

	A1	*A2*	*A3*	*A4*	*A5*
A1	(1,1,1)	(1,1,1)	(0.125,0.143,0.167)	(0.125,0.143,0.167)	(1,1,1)
A2	(1,1,1)	(1,1,1)	(0.125,0.143,0.167)	(0.125,0.143,0.167)	(1,1,1)
A3	(6,7,8)	(6,7,8)	(1,1,1)	(0.2,0.333,1)	(6,7,8)
A4	(6,7,8)	(6,7,8)	(1,3,5)	(1,1,1)	(6,7,8)
A5	(1,1,1)	(1,3,5)	(0.125,0.143,0.167)	(0.125,0.143,0.167)	(1,1,1)

In Table 7 is presented the global priority obtained for each investment project.

Table 7 The global priority obtained for each candidate

	A1	*A2*	*A3*	*A4*	*A5*
Final score	0.07	0.062	0.372	0.531	0.027

According to the data in Table 7 projects *A4* and *A3* are characterized by the highest utility.

3.2 Evaluation of Chosen Projects Using Fuzzy TOPSIS

The foundation for the solution ranking in *TOPSIS* method is the closeness coefficient. This coefficient is calculated in accordance with the equation [20]:

$$\overline{CC_i^*} = \frac{\overline{d}_i^-}{\overline{d}_i^* + \overline{d}_i^-}, \quad i = 1, \ldots, m \tag{3}$$

where d_i^* and d_i^- adequately are the distance of each alternative (investment project) $i = 1, 2, \ldots, m$ from *Fuzzy Positive Ideal Solution* (*FPIS*) and the *Fuzzy Negative Ideal Solution* (*FNIS*). Distances are calculated with the use of weights defining the importance of individual criteria. *FPIS* consists of all best criteria values available. *FNIS* consists of all worst criteria values achievable. The closer the value $\overline{CC_i^*}$ to one, the better the solution.

In the Table 8 is presented the characteristics of analysed projects in the suitable form for the fuzzy *TOPSIS* method. This table arose across the data transformation contained in the Table 1 and the use of the scale of alternative assessment from the Table 9. For example, if *C11* for A1 is equal to (1,40; 0,20) we choose “Fair” – (3,5,7) and for *A3* is equal to (2,12; 0,22) we choose “Very good” – (7,9,9).

Table 8 The characteristics of projects in the suitable form for the fuzzy TOPSIS method

	A1	A2	A3	A4	A5
C1.1	(3,5,7)	(3,5,7)	(5,7,9)	(7,9,9)	(3,5,7)
C1.2	(1,3,5)	(3,5,7)	(5,7,9)	(7,9,9)	(1,3,5)
C2.1.1	(1,3,5)	(3,5,7)	(5,7,9)	(5,7,9)	(1,1,3)
C2.1.2	(3,5,7)	(3,5,7)	(1,3,5)	(1,3,5)	(5,7,9)
C2.2	(3,5,7)	(1,3,5)	(5,7,9)	(7,9,9)	(1,3,5)
C3	(3,5,7)	(3,5,7)	(1,3,5)	(1,3,5)	(5,7,9)

Table 9 Fuzzy ratings for linguistic variables

Fuzzy number	Alternative assessment	Weights of the criterion
(1, 1, 3)	Very poor (VP)	Very low (VL)
(1, 3, 5)	Poor (P)	Low (L)
(3, 5, 7)	Fair (F)	Medium (M)
(5, 7, 9)	Good (G)	High (H)
(7, 9, 9)	Very good (VG)	Very high (VH)

In calculations was used weights of criterions elaborated in the *FAHP* method presented in the Table 5. In the Table 10 is presented closeness coefficients distances for analyzed investment projects.

Table 10 Distances from *FPIS* and *FNIS* for alternatives and closeness coefficients

	A1	*A2*	*A3*	*A4*	*A5*
Closeness coefficients	0.137	0.144	0.164	0.190	0.118

Projects *A4* and *A3* according to data in the Table 10 are the best.

3.3 Evaluation of Chosen Projects Using Mamdani Fuzzy Inference Method

Mamdani model is a rule set, each of which defines a so called fuzzy point. The rules are as follows:

$$R_k:\ IF\ (x_1\ is\ A_{k,1})\ AND\ (x_2\ is\ A_{k,2})\ AND\ldots AND\ (x_m\ is\ A_{k,m,})\ THEN\ (y_k = B_k) \quad (4)$$

where: x_i are crisp values of current input, $A_{k,i}$ and B_k are linguistic values (represented by fuzzy sets) of the variables x_i and y_k in the respective universes.

In *FLC* applications inference is performed in four steps: fuzzification, rules evaluation, aggregation of the rule outputs, defuzzification. Further, the standard Mamdani *FLC* system will be extended inter alia to work as inputs with linguistic terms. In this case fuzzification phase is skipped. Depending on the use of the model may also be omitted defuzzification phase. In the case of use the Mamdani model for the implementation of rule-based systems in the analysis of the relationship between qualitative factors that describe certain phenomena is particularly useful to offer input variables as linguistic terms. The inputs are applied to the antecedents of the fuzzy rules. The result of the antecedent evaluation, which in turn determines the value of conclusion requires a suitable fuzzy implication operator. In our investigation we used the operator algebraic product PROD.

Now the result of the antecedent evaluation can be applied to the membership function of the consequent. The most common method is to cut the consequent membership function at the level of the antecedent truth. The membership functions of all rule consequents are then combined into a single fuzzy set.

In our example the rules are hierarchically organized in sub-rule bases. The excerpt of knowledge model in form of *if-then rules* is presented in Table 11.

Table 11 The knowledge model in form of *if-then rules*

Number	Antecedent	Consequent
1	(X_2 is *L*)	X_1 is *L*
2	(X_2 is *M*)	X_1 is *M*
…	…	…
14	(X_5 is *L*)∧ (X_6 is *M*)	X_4 is *L*
15	(X_1 is *L*)	X_{10} is *L*
…	…	…

Mamdani method requires that all input variables are either directly presented in the form of linguistic variables or transformed into this form. In our example, we have to deal with input variables presented in different forms. That is why it is necessary to transform a part of them into the form of linguistic variables. Each of input variables and the utility of investment (projects are defined as having values of high (H), medium (M), or low (L) with the exception of the harmful impact on the environment (low L, medium M, heavy H).

The recapitulation of the results of the evaluation of projects is presented in the Table 12.

Table 12 The results of projects evaluation by Mamdani method

	A1	A2	A3	A4	A5
Low	0.067	0.447	0.000	0.000	0.430
Middle	0.932	0.553	0.311	0.025	0.570
High	0.001	0.000	0.689	0.975	0.000

3.4 Evaluation of Chosen Projects Using Takagi-Sugeno Model

The structure of the knowledge base in case of the Takagi-Sugeno method is the same as when using the Mamdani method, different is just a formula of the conclusions:

$$R_k: \; IF\;(x_1\; is\, A_{k,1})\; AND\;(x_2\; is\, A_{k,2})\; AND \ldots AND\;(x_m\; is\, A_{k,m,})\; THEN\;(y_k = f_k(X)) \tag{5}$$

where: $f_k(X)$ is a function of the input vector X.

The excerpt of knowledge model in form of *if-then rules* is presented in Table 13.

Table 13 The excerpt of knowledge model in form of if-then rules

Number	Antecedent	Consequent
1	(X_2 *is L*)	$X_1 = X_2$
2	(X_2 is *M*)	$X_1 = X_2$
…		
14	(X_7 is *H*)∧(X_8 is *H*)	$X_6 = 0{,}713X_7 - 0{,}080X_8 + 0{,}500$
15	(X_5 is *L*)∧ (X_6 is *L*)	$X_4 = 0{,}350X_5 + 0{,}350X_6 + 0{,}300$
…		

The attributes of investment projects are presented in numerical scale. It was assumed that for all attributes this scale ranges from 0 to 10. It is also assumed that the values scoring in the range from 0 to 2 correspond to the linguistic variable low, from 4 to 6 the variable middle and in the range from 8 to 10 value high. For a combination of all of these variables were determined functions addictive next conclusions of the relevant premises.

The input data in the form suitable to Takagi-Sugeno method for the previously presented investment projects are presented in the Table 14.

Table 14 The characteristic of investment projects in the form suitable to Takagi-Sugeno method

Projects	Input indicators					
	C11	C12	C21	C221	C222	C3
A2	4.04	4.60	4.00	7.00	5.00	2.20
A3	6.27	7.05	6.00	7.80	8.20	4.20
A4	7.80	8.65	4.81	7.80	8.20	1.40
A5	4.07	4.10	4.00	2.20	5.00	2.20

The recapitulation of the results of the evaluation of projects by Takagi-Sugeno method is presented in the Table 15.

Table 15 The results of projects evaluation by Takagi-Sugeno method

Projects	Evaluation in range of 0 to 10
A1	4.282
A2	4.113
A3	6.922
A4	7.282
A5	3.583

3.5 Evaluation of Chosen Projects Using Modified Evidential Reasoning Approach

In terms of occurrence of hybrid data premises and conclusions of the rules of inference system should be expressed in various ways (linguistic variables, probability distributions, crisp values). A way to write the rules in such system as well as the inference methodology was proposed by Yang et al. [16]. The rule-base inference methodology proposed by Yang uses the evidential reasoning (*RIMER*). The essence of this methodology is to combine a generic knowledge representation scheme using a belief structure with evidential reasoning (*ER*) approach. In [18] the modified *RIMER* method was presented.

The generalized form of a rule in this system can be presented as follows:

$$R_k : \textit{if } A_1^k \wedge A_2^k \wedge \cdots \wedge A_{T_k}^k, \textit{ then } \left\{ (D_1, \beta_{1k}), \cdots, (D_N, \beta_{Nk}) \right\}, \sum_{i=1}^{k} \beta_{ik} \leq 1, \quad (6)$$

with a rule weight Θ_k and attribute weights $\sigma_{k1}, \sigma_{k2}, \ldots, \sigma_{kT_k}$, $k = 1, 2 \ldots L$ where $\beta_{ik}(i = 1,2 \ldots N)$ is the belief degree to which D_i is believed to be the consequent, in the *kth* packet rule, the input satisfies the packet antecedents A^k.

The definitions of the extended rules with the consequents having the dedicated degrees of belief are given in Table 16. The numbers shown in the second column represent the weights assigned to the rules.

Table 16 Rule base with the belief structure

Number	θ	Antecedent	Consequent
1	1	(X_2 is *L*)	X_1 is (*L*, 1)
2	1	(X_2 is *M*)	X_1 is (*M*, 1)
…	…	…	…
14	1	(X_7 is *H*)∧(X_8 is *H*)	X_6 is (*H*, 0,2), (*M*, 0,5), (*L*, 0,1)
15	0, 9	(X_5 is *L*)∧ (X_6 is *L*)	X_4 is (*L*, 1)
…			

In this method the input data of different form have to be transformed into structure. In our case we used the same approach to hybrid data transformation into belief structure as by Mamdani and Takagi-Sugeno methods.

The reasoning algorithm was presented in [16, 18]. The recapitulation of the results of the evaluation of projects by Modified Evidential Reasoning Approach method is presented in the Table 17 where β_i is the belief degree to which the finally evaluation of investment project is to be low, middle or high and β_D represents the remaining belief degrees unassigned to any of this values.

Table 17 Results of the evaluation of investment projects

Projects	$\beta_{1\ (utlity\ L)}$	$\beta_{2\ (utlity\ M)}$	$\beta_{3\ (utlity\ H)}$	β_D
A1	0.075	0.627	0.004	0.293
A2	0.096	0.550	0.002	0.352
A3	0.001	0.149	0.317	0.532
A4	0.000	0.159	0.576	0.266
A5	0.113	0.571	0.000	0.316

4 Comparison of Results of the Evaluation of Investment Projects

Results of carried out research indicate that fuzzy methods based on multi-criteria evaluation of projects as and the modelling techniques based on fuzzy inference systems (*FIS*) give similar results of the comparison of analysed projects. In all methods one fixed that most effective were projects *A3* and *A4* and remaining were evaluated on average level. However this does not mean that in every case one can these methods treat complementarily. Restrictions and advantages of techniques analysed in our publication one can present as follows: (1) for obvious reasons, multi-criteria evaluation method (based on *AHP* and *TOPSIS*) cannot be used to assess individual projects, unless we have the suitable patterns; (2) methods of the multi-criteria evaluation are a lot more sensitive to experts judgments; (3) Mamdani method is least expressive from among examined alternatives due to the lack of the possibility of the differentiation of the description of the conclusion what leads to eliminating of the influence of less significant premises; (4) Takagi-Sugeno method requires the application of suitable techniques of knowledge acquisition (without them the settlement of functions describing conclusions is unusually difficult) and this in turn requires the access to historical recordings what in case of the evaluation of the investment not always is possible; (5) the more expressive *FIS* method there is undoubtedly the method based on evidential reasoning approach – does not require also certainly the application of methods of knowledge acquisition on the base of historical recordings.

5 Conclusions and Further Research

Analyzed method give similar results of the comparison of projects. However the more general application of fuzzy methods requires further research works. In our opinion the most suitable method of multi-criteria evaluation of projects is the method based on evidential reasoning approach. The evaluation of the project is defined here in the form of the belief structure characterizing the utility of the project. Thereby necessary is the elaboration of the method enabling the unambiguous arrangement of projects due to their utility.

Furthermore a key-problem is currently the knowledge acquisition particularly in case of the lack of historic adequate data.

References

1. Enea, M., Piazza, T.: Project selection by constrained fuzzy AHP. Fuzzy Optim. Decis. Mak. **3**, 39–62 (2004)
2. Tarimcilar, M.M., Khaksari, S.Z.: Capital budgeting in hospital management using the analytic hierarchy process. Socioecon. Plann. Sci. **25**, 27–34 (1991)
3. Olson, D.L.: Decision Aids for Selection Problems. Springer Verlag, New York (1996)
4. Souder, W.: Project Selection and Economic Appraisal. Van Nostrand Reinhold, New York (1984)
5. Van Laarhoven, P.J.M., Pedrycz, W.: A fuzzy extension of Saaty's priority theory, Fuzzy Sets and Syst. **11**, 229–241 (1983)
6. Chan, F.T.S., Chan, M.H., Tang, N.K.H.: Evaluation methodologies for technology selection. J. Mater. Process. Technol. **107**, 330–337 (2000)
7. Wand, T.: Fuzzy discounted cash flow analysis. In: Karwowski W., Wilhelm, M. R. (eds.) Applications of Fuzzy Sets Methodologies in Industrial Engineering, pp. 91–102 (1989)
8. Liu, P., Zhang, X., Liu, W.: A risk evaluation method for the high-tech project investment based on uncertain linguistic variables. Technol. Forecast. Soc. Change. **78**, 40–50 (2011)
9. Rebiasz, B., Gawel, B., Skalna, I.: Fuzzy multi-attribute evaluation of investments. In: 2013 Federated Conference on Computer Science and Information Systems (FedCSIS), pp. 977–980 (2013)
10. Iancu, I.: Reasoning system with fuzzy uncertainty. Fuzzy Sets Syst. **92**(1), 51–59 (1997)
11. Mamdani, E.H., Assilian, S.: An experiment in linguistic synthesis with a fuzzy logic controller. Int. J. Man-Mach. Stud. **7**, 1–13 (1975)
12. Sugeno, M.: Industrial Applications of Fuzzy Control. Elsevier science Publisher, Amsterdam (1985)
13. Amindoust, A., Ahmed, S., Saghafinia, A., Bahreininejad, A.: Sustainable supplier selection: A ranking model based on fuzzy inference system, Appl. soft Comput. **12**(6), 1668–1667 (2012)
14. Medina, S., Moreno, J.: Risk evaluation in Colombian electricity market using fuzzy logic. Energy Econ. **29**, 999–1009 (2007)
15. Kaur, A., Kaur, A.: Comparison of Mamdani-type and Sugeno-type fuzzy inference systems for air conditioning system. Int. J. Soft Comput. Eng. **2**, 323–325 (2012)
16. Yang, J.B., Liu, J., Wang, J., Sii, H.S., Wang, H.W.: Belief rule-base inference methodology using the evidential reasoning approach—RIMER. IEEE Trans. Syst. Man, Cybern. Part A Syst. Humans. **36**, 266–285 (2006)

17. Mock, T.J., Sun, L., Srivastava, R.P., Vasarhelyi, M.: An evidential reasoning approach to Sarbanes-Oxley mandated internal control risk assessment. Int. J. Account. Inf. Syst. **10**, 65–78 (2009)
18. Rębiasz, B., Macioł, A.: Hybrid data in the multiobjective evaluation of investments. Procedia Comput. Sci. **35**, 624–633 (2014)
19. Chang, D.-Y.: Applications of the extent analysis method on fuzzy AHP. Eur. J. Oper. Res. **95** (3), 649–655 (1996)
20. Saghafian, S., Hejazi, S.R.: Multi-criteria Group Decision Making Using A Modified Fuzzy TOPSIS Procedure. Int. Conf. Comput. Intell. Model. Control Autom. Int. Conf. Intell. Agents, Web Technol. Internet Commer. **2** (2005)

Leveraging Textual Information for Improving Decision-Making in the Business Process Lifecycle

Rainer Schmidt, Michael Möhring, Ralf-Christian Härting, Alfred Zimmermann, Jan Heitmann and Franziska Blum

Abstract Business process implementations fail, because requirements are elicited incompletely. At the same time, a huge amount of unstructured data is not used for decision-making during the business process lifecycle. Data from questionnaires and interviews is collected but not exploited because the effort doing so is too high. Therefore, this paper shows how to leverage textual information for improving decision making in the business process lifecycle. To do so, text mining is used for analyzing questionnaires and interviews.

Keywords Decision-making • Bpm • Process interviews • Text mining • Context data

1 Introduction

Business process management has been identified as critical success factor for the implementation of large applications systems [1]. Therefore, it does not support, that proper modelling of business processes is crucial for the success of business process implementations [1]. Research shows, that the improper integration of stakeholders may cause an incomplete elicitation of requirements [2]. The root cause is the expert-driven approach used for business process modelling [3]. The business process models are created by experts who decide about to include or not information provided by other stakeholders. Unfortunately, this selection is often driven by inappropriate selection criteria [4].

R. Schmidt (✉)
Munich University of Applied Sciences, Munich, Germany

M. Möhring · R.-C. Härting · J. Heitmann · F. Blum
Aalen University of Applied Sciences, Aalen, Germany

A. Zimmermann
Reutlingen University, Reutlingen, Germany

R. Neves-Silva et al. (eds.), *Intelligent Decision Technologies*,
Smart Innovation, Systems and Technologies 39,
DOI 10.1007/978-3-319-19857-6_48

On the other hand, plenty of information is contained in interviews, questionnaires etc. during all phases of the business process lifecycle [5, 6]. During process design, interviews and questionnaires deliver both semi- and unstructured data defining requirements for the process to be implemented. Comments and suggestions collected in parallel to process execution may give important hints for improving an existing process. In the evaluation phase, interviews and questionnaires are very important means to evaluate process performance. Furthermore, it is important to talk with process owners and other persons related to the implemented business processes, because of the continuous changes of business processes (e.g. cultural change [2]).

Unfortunately, the knowledge contained in interviews and questionnaires rarely leveraged to support business process management. Instead, they are used by the business process managements as raw material for their experts-driven approach. Therefore, this paper demonstrates how to use unstructured data from interviews and questionnaires for improving business processes. Text mining is used to compare the process descriptions (e.g. process interviews) with process models (e.g. Event-Driven-Process Chains (EPC) [7]) in order to improve the quality of business process modelling. Up to now, there is a lack of research in leveraging unstructured data from interviews and questionnaires for improving the quality of newly designed or already implemented business process models.

The paper is structured as follows. First, the methodology of design of the prototype and Text Mining algorithms are introduced. In the next section, we develop our prototype for evaluation and test it. Finally, a related work section as well as an outlook and the conclusion are given.

2 Prototype Design: Text Mining for Business Process Model Check

There are a number of approaches to represent business processes in a (semi-) formal way. Petri-Nets were identified as advantageous means for representing business process models [8]. In the beginning, Petri-Nets were used for process modelling directly. Over the time several approaches using Petri-Nets such as INCOME [9] were developed. In addition, the widely used ARIS approach is using Petri-Nets as basis for its EPC. In parallel, extensions to object-oriented approaches such as activity diagram were developed [10]. Typically, experts use these approaches to represent the intrinsic information on the business processes. However, in parallel, a lot of extrinsic information exists, that contains extrinsic information the business processes. Unfortunately, this information is only partially used for validating business process models and quality improvement.

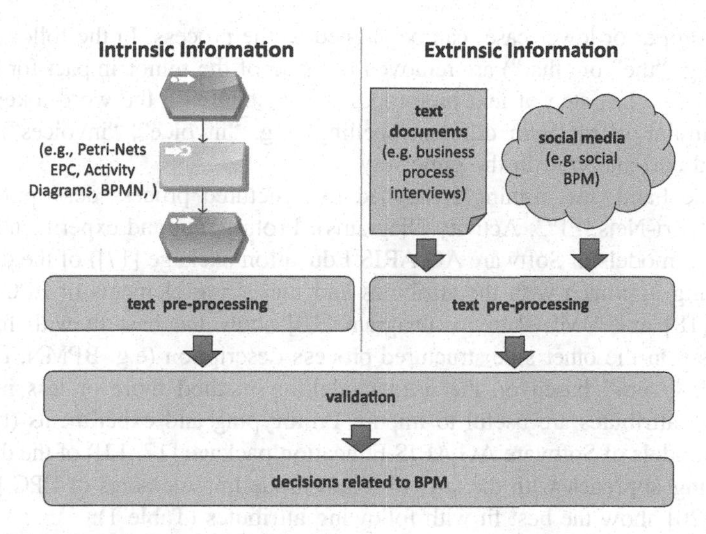

Fig. 1 Using intrinsic and extrinsic information for business process check

Our approach is based on a comparison between a (semi-) formal process model containing intrinsic information and extrinsic information collected from interviews and questionnaires, as shown in Fig. 1. To leverage both intrinsic and extrinsic information text mining is used. Text Mining can be defined as the process of extracting interesting patterns or knowledge from (unstructured) text files or documents [11]. Therefore, Text Mining is an extension of traditional data mining [11–13]. Text documents are particularly difficult to analyze because of the lack of structure. To manage this lack, Text Mining provides different technical possibilities (e.g. Text classification algorithms [14]).

Text documents (e.g. Interviews of process owner) are frequently an input and foundation of business process modelling projects [15]. Based on this documents business processes can be modelled, reviewed etc. To handle and review these modelling operations Text Mining can be one source of quality checks and can so improve BPM projects. To process text documents with Text Mining several steps are needed for pre-processing (see Fig. 2) [16].

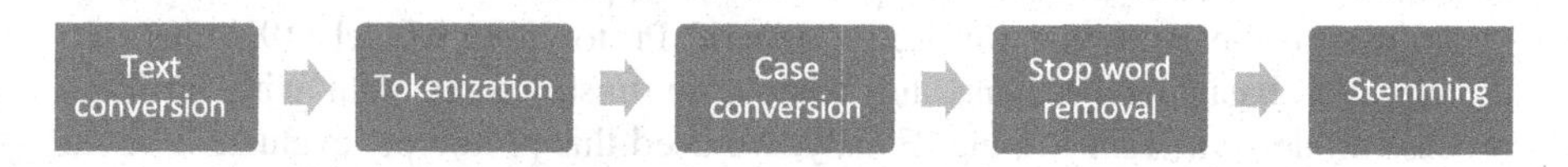

Fig. 2 Text pre-processing according to [16]

First the text (e.g. Business process interview) must be converted to a standard text document file and irrelevant sentences can be removed [16]. The next step is tokenization of the text, while each word (separated by a space character) of the text is extracted to a single token. Then a case conversion (to lower case) is implemented, to get a consistent form of tokens. Therefore, same words with different spelling (upper or lower case) can be aligned to the process. In the following stop words (e.g. “the” or “that”) are removed because of the minor impact for this text analytics. The last step of text pre-processing is stemming the word tokens. As a result, similar words with different spelling (e.g. “invoice”, “invoices”) can be processed and analyzed in the same way.

On one hand, text mining is applied to structured process descriptions (e.g. BPMN, Petri-Nets, EPC, Activity Diagrams). Prototyping and experiments (based on process models of Software AG ARIS Education package [17]) of the described text mining approach with the attributes and modelling elements of EPC [7] and BPMN [18] and UML Activity Diagram [10] show the best fit with following attributes (On the other side structured process description (e.g. BPMN, EPC) are imported. Overall based on the used modelling method more or less important modelling attributes are useful to import. Prototyping and experiments (based on process models of Software AG ARIS Education package [17, 12]) of the described text mining approach with the attributes and modelling elements of EPC [19] and BPMN [20] show the best fit with following attributes (Table 1):

Table 1 **Useful** business process modelling method attributes for text-mining

EPC	BPMN	Activity diagram (UML)
• Name of events • Name of functions • Name of organization units • Name of information, resource or material objects	• Name of events • Name of activity's • Name of data objects • Name of swim lanes and pools	• Name of actions • Name of decisions

On the other side, unstructured process descriptions (e.g. text documents or posts in social media) are imported to a text-mining environment to gather knowledge for business process improvement. In general, only the text document or post with a process description must be imported to the text-mining environment (e.g. customized text-mining algorithm in text mining software like Rapid Miner or R Project). Other information (e.g. time of generating the post or text document) can be removed. Then the text pre-processing steps (described above) with steps e.g. tokenization, stemming, remove stop words are applied.

To design the prototype we used a general Prototyping Model [19]. First, we identify the basic requirements for comparing unstructured data with business process models via Text Mining. Finally, we used this prototype, evaluate it based on quality criteria, and give an outlook for the next improved version of it. Depending on the use case and the adopted modelling methods a variation and evaluation of the attributes of the modelling and is needed. Only the relevant attributes of a process model are exported from the modelling environment software to a standard text document file (e.g. PDF, TXT, DOC), the text pre-processing steps are applied.

After text pre-processing of the unstructured and structured process descriptions a validation algorithm checks the congruence of the two descriptions. Based on the validation algorithm the result can be a percentage or absolute value of congruence (respectively accordance) of the two process descriptions (unstructured and structured). Classification algorithms for validation can be for instance artificial neuronal networks (ANN), KNN or SVM based algorithms [20].

The result of the validation step is the congruence of the unstructured and structured process description. Based on this results implications for improving the business process model can be applied. For instance, if the congruence of the unstructured process description (e.g. A process interview) is very low (e.g. <20 %), the process may be invalid and must be revised. If the congruence is very high (>80 %) the process may be modelled very well. Therefore, a better quality of process modelling can be implemented. However, some limitations and use cases specific adoptions must be considered. If different classification algorithms are implemented to improve the robustness of the system, the congruence of the data must be calculated during aggregating the different classification results. This aggregating can be done via e.g. calculating the arithmetic mean [21] of the different classification results [22]:

$$\text{Process\_fit} = \frac{\sum_1^n confidence_i}{n} \quad (1)$$

Equation 1: Calculating the fit of structured and unstructured process data.

We implement this kind of a system for checking unstructured and structured process data in a prototype according to general prototype model [19]. After identifying the described basic requirements, we develop and test our prototype in the next section.

3 Text Mining Prototype for Business Process Management Using Process Interviews

The concept developed so far shall now be applied to a sample interview. Therefore, we develop and use a prototype implemented in Rapid Miner 5 [23].

The next section shows a short unstructured description of the business process "important credit application processing" as a sample:

An important credit application received by Mr. Right and he creates the credit data set in the creditIS system based on the document credit application. Then the creditIS system checks if the integrity of the data set is okay.

If it is, the credit agreement and a credit agreement document are made by Ms. Willson based on the data in the creditIS system. If the integrity check is not okay, Ms. Luck updates the data set in the creditIS.

Based on this interview one business process modelling team creates the following EPC process model (Fig. 3) Another business process modelling team modelled the following EPC process model.

Our text mining approach can now be applied to e.g. (pre-)check whether the business process models are modelled correctly. To test it the text mining approach is implemented as a prototype in Rapid Miner 5 [24, 25].

First, the interviews associated with the processes are loaded into the environment and pre-processed (see Sect. 4). They contain the unstructured process descriptions. Then a validation of the interviews is made to separate it. In this example, the validation is made with an SVM (support vector machines), artificial neuronal network (ANN) and k-nn based classification algorithm, because of the recommend use of these algorithms in the literature [20].

The Support Vector Machine (SVM) can be defined as a supervised algorithm for (text) classification [20], where a hyperplane between the positive and negative examples of the training data set is created [26, 27]. This algorithm calculates the result based on the best-fit best in relative values. The confidence values can be between 0 and 1 (0: no confidence; 1 high confidence). For example, if the confidence value is 0.75, there is a high congruence and accordingly support of the validation between the models and data. If the value is for instance 0.10, there is a very little support.

The k-nn (K-Nearest Neighbor Method) is a classification algorithm based on the concept of group elements according to the short distance to the classified document [20]. The confidence values can be here only 1 and 0 (high confidence; no confidence). Artificial neuronal networks (ANN) simulate input values to output values (in general over a hidden layer) based on a structure according to the human neurons [21, 23]. Confidence values can be between 0 and 1 (similar to SVM classification).

This factor can be seen as the evaluation of the design of our prototype. Furthermore, an aggregation of the different classification results can be done by calculating Eq. 1.

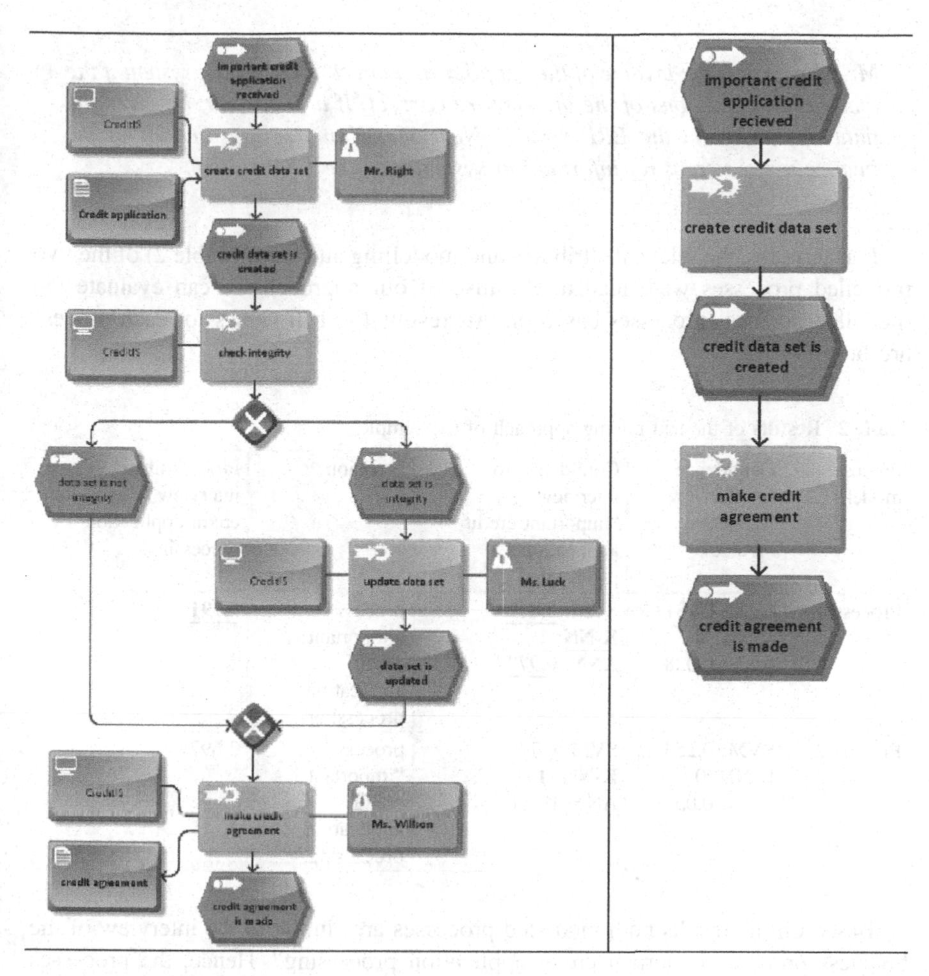

Fig. 3 Sample Processes 1 & 2

The result is a validation model to identify the specific of each business process interview. Then the modelling attributes and elements (Table 1) of the structured process description are exported from the modelling environment (in this example from ARIS) and pre-processed via Text Mining algorithms in Rapid Miner. The Apply Model step process and validate the pre-processed, structured process description with the model of the unstructured process descriptions (process interviews). In this example, the interviews from the described process above and one another interview about the process "handle invoices" are loaded. The interview of the business process "handle invoices" is defined as the following:

Mr. Mayer gets the invoice of the supplier and check via *the ERP system if the amount and positions of the invoice are correct. If it is correct, he gives the final approval* via *the ERP system. Next Mr. Han pay the amount of the invoice* via *the treasury information system and the ERP system.*

Furthermore, the relevant attributes and modelling attributes (Table 2) of the two modelled processes were loaded. Because of our approach we can evaluate the modelled business processes based on this results (best fit values for each process are bold):

Table 2 **Results** of the text mining approach of the sample

Process model	Confidence to interview "handle invoice"	Confidence to interview "important credit application processing"	Decision	Process fit to interview "important credit application processing"
Process 1	SVM: 0.181 K-NN: 0 ANN: 0.028	SVM: **0.819** **K-NN: 1** **ANN: 0.972**	process "important credit application processing"	**2.791**
Process 2	SVM: 0.253 K-NN: 0 ANN: 0.050	SVM: 0.747 **K-NN: 1** ANN: 0.950	process "important credit application processing"	2.697

Based on the results both modelled processes are similar to the interview of the business process "important credit application processing". Hence, the processes are correct classified. However, the best fit of the modelled process has Process 1 (based on Process_fit value according to Eq. 1: 2.791 > 2.697). In comparison with both models, (manually) Process 1 has also the best fit. Therefore, the Text Mining approach shows quality differences of the process models and can so improve the modelling process by correcting quality problems. Results may be better with a huge amount of process documents, where the classification algorithm can learn the differences. Based on this metrics, the design can be evaluated. The best fit shows a Text Mining based approach under use of SVN and ANN classification algorithms, because of the non dichotomic confidence values as well as the fit of the tested sample data (refer to Table 2 and sample case).

However, there are some limitations. This approach only based on the occurrence of words. If the model is modelled bad, but with the same steps in an incorrect order, this approach will calculate wrong results and wrong decisions. Furthermore, forks, gateways, etc. are not interpreting very well. For quality pre-Check this approach is very simple to adopt and can improve so huge modelling projects.

4 Related Work

There are a number of other approaches for gathering knowledge from semi- and unstructured context data have been developed. Process mining [15, 28] collects semi-structured history information from event-logs in order to discover processes, verify the compliance of process execution etc. A first approach to automatically extract process information from policy documents is described in [29]. General aspect of Business Process Intelligence is described in [30]. In [31] a approach for process models from natural language text is described. In contrast to our approach, this approach focuses on the generation of business process models from text documents and does not validate unstructured process descriptions (especially process interviews or audits) with structured process descriptions. Furthermore, this approach is more focusing on grammar and consistent sentences. The inverse functionality, the generation of natural language form business process models is described in [32]. Several other interesting applications of natural language process in conjunction with business process models are described in [33]. Examples are the detection and correction of guideline violations or the derivation of services from process models. There are a number of other approaches for extracting models from natural language such as the NL approach [34], that is still in an early stage. The extraction of workflow models from maintenance manuals is described in [35]. Furthermore, SBVR [36] is a standard for defining business rule and business vocabulary semantic and can be used to define and analyze the semantic structure of current process models. SBVR uses a deductive approach contrary to the inductive approach presented here. The detection of process fragments of clinical documents is investigated in [37]. However, these approaches do not take care about the validation and crosscheck between unstructured (e.g. Interview) and structured (e.g. EPC process model). There are a number of other approaches for gathering knowledge from semi- and unstructured context data have been developed. Process mining [10, 15, 24, 29] collects semi-structured history information from event-logs in order to discover processes, verify the compliance of process execution etc. A first approach to automatically extract process information from policy documents is described in [29]. General aspect of Business Process Intelligence is described in [30]. In [31] a approach for process models from natural language text is described. In contrast to our approach, this approach focuses on the generation of business process models from text documents and does not validate unstructured process descriptions (especially process interviews or audits) with structured process descriptions. Furthermore, this approach is more focusing on grammar and consistent sentences. The inverse functionality, the generation of natural language form business process models is described in [32]. Several other interesting applications of natural language process in conjunction with business process models are described in [33]. Examples are the detection and correction of guideline violations or the derivation of services from process models. There are a number of other approaches for extracting models from natural language such as the NL approach [34], that is still in an early stage. The extraction of workflow models from maintenance manuals is described in [35]. The detection of process fragments of clinical documents is investigated in [37]. However, these approaches do not take care about the validation and crosscheck between unstructured (e.g. Interview) and structured (e.g. EPC process model).

5 Conclusion and Outlook

Our approach can e.g. pre-check the quality of process models based on unstructured descriptions of the business processes (e.g. Process interviews or audits, discussions in social BPM). Therefore, problems in the quality of modes can be exposed automatically and very fast. Enterprises can save costs by detecting modelling problems in an early time and improve decision-making in the business process management lifecycle. Furthermore, more and more unstructured data (e.g. through social BPM like Processlife (Software AG)) is generated. Therefore, an automatic approach is needed. Academics can profit from a new approach of Text Mining and Business Process Modelling in an early stage and can so adopt and improve current approaches. Opportunities to explore and correlate unstructured with structured data of business process models exist. Furthermore, techniques of quality management of BPM and integration of social software can be improved. However, there are some limitations and grand possibilities for future research. Our approach analyses, processes the occurrences of words from the unstructured process descriptions, and compares it with information from structured process descriptions. The order of the occurrences in the descriptions and forks/gateways, etc. is not interpreted completely; Future work will also use them. Furthermore, adoption and empirical applications of this approach for improvement is needed. There might be interesting context-based differences (e.g. of industry sector specific process modelling), which can be explored in the future and so adopt the approaches.

References

1. Reijers, H.A.: Implementing BPM systems: The role of process orientation. Bus. Process Manag. J. **12**(4). pp. 389–409 (2006)
2. Baumöl, U.: Cultural change in process management. In: Handbook on Business Process Management 2, pp. 487–514. Springer (2010)
3. Erol, S., Granitzer, M., Happ, S., Jantunen, S., Jennings, B., Johannesson, P., Koschmider, A., Nurcan, S., Rossi, D., Schmidt, R.: Combining BPM and social software: Contradiction or chance?. J. Softw. Maint. Evol. Res. Pract. **22**(6–7), pp. 449–476 (2010)
4. Schmidt, R., Nurcan, S.: Augmenting BPM with Social Software. In: Business Process Management Workshop, pp. 201–206 (2010)
5. Kettinger, W.J., Teng, J.T., Guha, S.: Business process change: a study of methodologies, techniques, and tools. MIS Q, pp. 55–80 (1997)

6. Trkman, P.: The critical success factors of business process management. Int. J. Inf. Manag. **30** (2), 125–134 (2010)
7. Scheer, A.-W., Nüttgens, M.: ARIS architecture and reference models for business process management. In: van der Aalst, W., Desel, J., Hrsg, O.A.: Business Process Management, pp. 376–389. Springer, Berlin, Heidelberg (2000)
8. van Der Aalst, W.M.: Three good reasons for using a Petri-net-based workflow management system. In: Proceedings of the International Working Conference on Information and Process Integration in Enterprises (IPIC'96), pp. 179–201 (1996)
9. Oberweis, A., Schätzle, R., Stucky, W., Weitz, W., Zimmermann, G.: INCOME/WF: A Petri net based approach to workflow management. Wirtschaftsinformatik **97**, pp. 557–580
10. Russell, N., van der Aalst, W.M., Ter Hofstede, A.H., Wohed, P.: On the suitability of UML 2.0 activity diagrams for business process modelling. In: Proceedings of the 3rd Asia-Pacific Conference on Conceptual Modelling-Volume 53, pp. 95–104 (2006)
11. Tan, A.-H.: Text mining: The state of the art and the challenges. In: Proceedings of the PAKDD 1999 Workshop on Knowledge Discovery from Advanced Databases, pp. 65–70 (1999)
12. Fayyad, U., Piatetsky-Shapiro, G., Smyth, P.: From data mining to knowledge discovery in databases. AI Mag. **17**(3), pp. 37 (1996)
13. Simoudis, E.: Reality check for data mining. IEEE Intell. Syst. **11**(5), 26–33 (1996)
14. Forman, G.: An extensive empirical study of feature selection metrics for text classification. J. Mach. Learn. Res. **3**, 1289–1305 (2003)
15. Van der Aalst, W.M., Weijters, A.: Process mining: a research agenda. Comput. Ind. **53**(3), 231–244 (2004)
16. Tan, P.-N., Blau, H., Harp, S., Goldman, R.: Textual data mining of service center call records. In: Proceedings of the Sixth ACM SIGKDD International Conference on Knowledge Discovery and Data Mining, pp. 417–423 (2000)
17. Fayyad, U., Piatetsky-Shapiro, G., Smyth, P., Uthurusamy, R.: Advances in Knowledge Discovery and Data Mining. The MIT Press, Cambridge (1996)
18. Chinosi, M., Trombetta, A.: BPMN: An introduction to the standard. Comput. Stand. Interfaces **34**(1), pp. 124–134 (2012)
19. Naumann, J.D., Jenkins, A.M.: Prototyping: The new paradigm for systems development. Mis Q, pp 29–44 (1982)
20. Hotho, A., Nürnberger, A., Paaß, G.: A brief survey of text mining. Ldv Forum **20**, 19–62 (2005)
21. Watson, R.T., Pitt, L.F., Kavan, C.B.: Measuring information systems service quality: Lessons from two longitudinal case studies. MIS Q, pp. 61–79 1(998)
22. Kohler, U., Kreuter, F., Data analysis using Stata. Stata Press (2005)
23. Ertek, G., Tapucu, D., Arın, İ., Text mining with rapid miner. Rapid Miner Data Min. Use Cases Bus. Anal. Appl. p. 241 (2013)
24. Rapid Miner Studio, Rapid Miner. Verfügbar unter: http://rapidminer.com/products/rapidminer-studio/. [Zugegriffen: 24 Feb 2014]
25. Miner, G., IV, J.E., Hill, T., Nisbet, R., Delen, D., Fast, A.: Practical Text Mining and Statistical Analysis for Non-structured Text Data Applications. Academic Press (2012)
26. Leopold, E., Kindermann, J.: Text categorization with support vector machines. How to represent texts in input space? Mach. Learn. **46**(1–3), pp. 423–444 (2002)
27. Joachims, T.: Text Categorization with Support Vector Machines: Learning with Many Relevant Features. Springer, Heidelberg (1998)
28. van der Aalst, W., Adriansyah, A., de Medeiros, A.K.A., Arcieri, F., Baier, T., Blickle, T., Bose, J.C., van den Brand, P., Brandtjen, R., Buijs, J.: Process mining manifesto. In: Business process management workshops, pp. 169–194 (2012)
29. Li, J., Wang, H.J., Zhang, Z., Zhao, J.L.: A policy-based process mining framework: mining business policy texts for discovering process models. Inf. Syst. E-Bus. Manag. **8**(2), 169–188 (2010)

30. Linden, M., Felden, C., Chamoni, P.: Dimensions of business process intelligence. In Business process management workshops, pp. 208–213 (2011)
31. Friedrich, F., Mendling, J., Puhlmann, F.: Process model generation from natural language text. In: Advanced Information Systems Engineering, pp. 482–493 (2011)
32. Leopold, H., Mendling, J., Polyvyanyy, A.: Generating natural language texts from business process models. In: Advanced Information Systems Engineering, pp. 64–79 (2012)
33. Leopold, H.: Natural Language in Business Process Models. Springer, Heidelberg (2013)
34. Ackermann, L., Volz, B.: Model [NL] generation: Natural language model extraction. In: Proceedings of the 2013 ACM workshop on Domain-specific modeling. 45–50 (2013)
35. Schumacher, P., Minor, M., Schulte-Zurhausen, E.: Extracting and enriching workflows from text. In: 2013 IEEE 14th International Conference on Information Reuse and Integration (IRI), pp. 285–292 (2013)
36. Team, S. et al.: Semantics of Business Vocabulary and Rules (SBVR), Technical Report dtc/06–03–02 (2006)
37. Thorne, C., Cardillo, E., Eccher, C., Montali, M., Calvanese, D.: Process fragment recognition in clinical documents. In: AI* IA 2013: Advances in Artificial Intelligence, pp. 227–238. Springer (2013)

FCA for Users' Social Circles Extension

Soumaya Guesmi, Chiraz Trabelsi and Chiraz Latiri

Abstract Nowadays, more and more users share real-time information and news on social networks and micro-blogging such as Flickr, BlogCatalog and Livejournal. With the increasing number of registered users in this kind of sites, find relevant and useful information becomes difficult. As the solution to this information glut problem, recommendation technologies emerged to help the exchanging of reliable sources of information, products or documents. One of the main mechanisms for organizing users online social networks is to categorize their friends into what we refer to as *social circles*. Currently, users identify and extend their social circles manually, or in a basic way by identifying friends sharing common activities. Neither approach is really satisfactory since they are time consuming and may function poorly as a user adds more friends. In this context, we develop a new algorithm called $Recom_{Friend}$ for users' social circles extension and friends recommendation based on a set of selected Communities Of Shared Interests (COSI) using Formal Concept Analysis (FCA) techniques. Carried out experiments over real-world datasets emphasize the relevance of our proposal and open many issues.

Keywords FCA · Users' interests · Social circle extension · Social networks

1 Introduction

Online social networks and micro-blogging sites allow users to follow streams of posts generated by hundreds of their friends and acquaintances. Users' friends generate overwhelming volumes of information and to cope with the "information overload" they need to organize their personal social networks. One of the main mechanisms for users of social networking sites, to organize their networks and the content generated by them, is to categorize their friends into what we refer to as *social circles*. Currently, all major social networks and micro-blogging sites provide such functionality, for example on Flickr, Facebook and Livejournal. Generally, users in such sites

S. Guesmi (✉) · C. Trabelsi · C. Latiri
Faculty of Sciences of Tunis, University Tunis El-Manar, Tunis, Lipah

R. Neves-Silva et al. (eds.), *Intelligent Decision Technologies*,
Smart Innovation, Systems and Technologies 39,
DOI 10.1007/978-3-319-19857-6_49

seek for identifying and extending their social circles either manually, or in a naive manner by identifying friends sharing common attributes. Neither approach fulfills users' expectations or needs: the former is time consuming and does not help users to add 'interesting' friends, while the latter fails to capture individual aspects of users' communities and may function poorly when profile information is missing.

In this paper, we investigate the problem of the users' social circles extension. In particular, given a single user with his personal social network, our goal is to identify and extend his social circles, subset of his friends with which he shares same interests. Currently, users identify and extend their social circles manually, or in a basic way by identifying friends sharing common activities. Hence, we formulate the problem of social circles discovering as a clustering problem on the social network of friendships. For this purpose, we made use of the Formal Concept Analysis (AFC) techniques and especially the notion of communities coverage based on a new selection criteria, *aka*, the relevance score to identify the minimal set of Communities of Shared Interests (COSI). Given these identified communities, we proceed by applying a new proposed algorithm, called $Recom_{Friend}$.

The remainder of the paper is organized as follows: Sect. 2 thoroughly scrutinizes the related work. In Sect. 3, we present our users' social circles extension approach and we introduce the $Recom_{Friend}$ algorithm for users' social circles extension. The experimental study of the overall approach is illustrated in Sect. 4. Finally, Sect. 5 concludes this paper and sketches avenues for future work.

2 Related Work

With the rapid growth of social networks, finding relevant and useful information become essential. Therefore, the recommendation of better friend is the essential factor to find relevant and useful information. Hence, recent years witnessed an overwhelming number of publications on different aspects of communities' detection and friends' recommendation. In this respect, Silva *et al.*, [14] introduced a new clustering index using Genetic Algorithm to suggest friends. They combined Knowledge of the topology and structure of networks with quantitative properties such as density, size, average path length or clustering coefficient. Differently, Kwon et *al.*, [7] propose to firstly measure the strength friendship and then suggests friends from a list using spiritual and social context.

Actually, many friends' recommendation approaches are based on communities discovering such as in [13], which proposed the use of a previously detected communities from different dimensions of social networks for capturing the similarities of these different dimensions and help recommendation systems to work based on the found similarities. Yan et *al.*, [16] used the K-means algorithm for clustering. Indeed, this approach combines the interests of users and their social relationships to discover and extend communities describing a given order to check their social network interaction. Nevertheless, the use of such technique requires the authors to perform several experiments to determine the number of communities (clusters) introduced as input to the K-means algorithm. Paolillo and Wright in [10], also used the k-means algorithm as a clustering technique. They built 2 graphs from the social network LiveJournal.com, one for the property "FOAF:knows" and another one for the property "FOAF:interest". This separation induces loss of knowledge while it could be used to filter their sources, improve their relevance and accuracy and customize their results.

Actually, the problem of discovering communities for social networks has been reduced in several works to a clustering problem. Several approaches are based on the disjoint clustering to define a model used in the communities' discovery [2, 5]. Many studies focused on optimizing the Girvan and Newman [5] approach [1, 6, 15]. In [8], the authors proposed the OSLOM algorithm (Order Statistics Local Optimization Method) aiming to test the statistical significance of a cluster with respect to a global null model during cluster extension. To extend a cluster, a value r is computed for each neighbor vertex and to check whether it is included in the current cluster. However OSLOM algorithm often leads to a significant number of outliers or singletons. In fact, disjoint and overlapping approaches have mostly focused on topological properties of social networks, regardless the semantic information. To overcome this drawback, recent approaches made use of Formal Concept Analysis (FCA) for a conceptual clustering of social networks. Hence, the proposed approach in [3] attempts to extract communities, preserving knowledge shared in each community where the inputs are bipartite graphs and the outputs a Galois hierarchy that reveals communities semantically defined with their shared knowledge or common attributes, where vertices are a lattice extents and edges are labeled by lattice intents (*i.e.*, shared knowledge). As described in the survey conducted by Plantié and Crampes [12], discovering communities based on FCA techniques is the most accurate, because it extracts communities using their precise semantics. Discovering communities in social networks is an important topic, which focuses on analyzing the relationships between social actors. Traditional researches have mostly focused on topological properties of social networks, ignoring the semantic information. However, real social networks are often embedded in particular social contexts defined by specific information carriers. To overcome these aforementioned limitations, we introduce in this paper, a new approach based on Formal Concept Analysis (*FCA*) techniques which combines the extraction frequent concepts techniques [11] with the social data for users' social circles extension.

3 Users' Social Circles Extension Approach

Users' Social Circles Extension approach aims to discover the similar interests between a target user and other social users and then suggests users with the most similar interests to the target user. Hence, before the step of Users' Social Circles Extension, a selected set of Communities of Shared Interests must be firstly detected. The overall approach is based on two main steps:

1. The Communities of Shared Interests selection based on Formal Concepts analysis techniques;
2. The Users' social circles extension using a new algorithm called $Recom_{Friend}$ which takes as input the selected communities to produce a set of recommended Friends.

3.1 Communities of Shared Interests Selection

The first step of our proposed approach for users' social circles extension is to select the 'pertinent' set of communities of users sharing similar interests. To this aim, we firstly build a Communities Tree formally represented by the Galois lattice [11]. This Communities Tree represents a set of Communities of Shared Interests (*COSI*) hierarchies, each node represents a community and an edge represents the subsumption relation (≤) based on [11]. As input data, we consider a binary relation between a set of users(U) and a set of interests(I) represented by User_Interest_Context (UI_Context), defined as following. Let U = $\{u_1, u_2, \ldots, u_n\}$ be a set of users and I = $\{i_1, i_2, \ldots, i_n\}$ be the set of interests.

Definition 1 *A User Interest Context UI_Context=(U, I, R) consists of two sets U and I and a relation R between U and I. We write uRi or (u, i)∈R and read it as "the user u has the interest i".*

Definition 2 *A Community of shared interests (COSI) of a UI_Context=(U, I, R) is a set of pair (U, I) with U is the extent and I is the intent of the Communities of shared interests.*

Definition 3 *A Communities Tree is a set of Communities of shared interests (COSI) defined by UI_Context=(U, I, R) and hierarchically organized by the Subsumption relation (≤).*

Indeed, the subsumption relation of a communities Tree, is defined between communities (*COSI*), $COSI_1 \leq COSI_2$. Let $COSI_1 = (U_1, I_1)$ and $COSI_2 = (U_2, I_2)$ ($COSI_1$ is subsumed by $COSI_2$ or $COSI_2$ subsumes $COSI_1$), if $U_1 \subseteq U_2$ and dually $I_2 \subseteq I_1$.

Example: Let as consider the UI_context=(U,I,R), where U= {Tom, Ken, Bob, Jon, Ben} is the set of users and I= {Art, Music, Sport, Dance} is the set of interests as it's represented in the following Table 1. The Communities Tree of this context(UI_Context) is sketched in Fig. 1. It is formed by eight Communities Of Shared

Table 1 A toy example of a user interest context

	Art	Music	Sport	Dance
Tom	×	×		
Ken	×	×		
Bob		×	×	
Jon		×	×	×
Ben			×	×

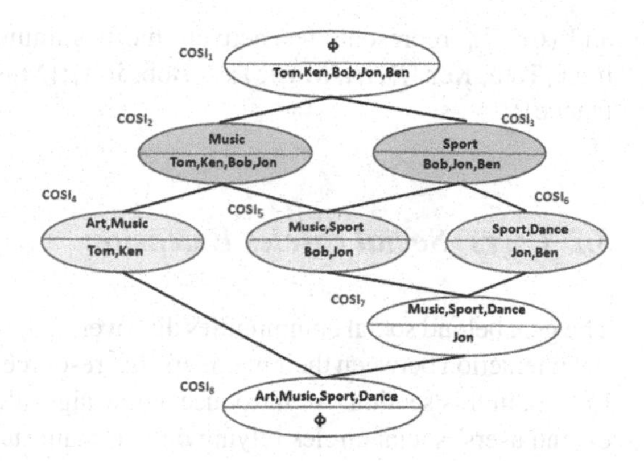

Fig. 1 Example of a communities tree related to the UI_context given in Table 1

Interests (*COSI*) extracted by the Norris algorithm [9]. As the set of all (*COSI*) belonging to the Communities Tree is notably large compared to the number of users or interests, we will in particular consider a specific condensed representation called Communities Coverage (CV). In this respect, to extract the Communities Coverage (CV), we introduce a new selection criteria, called the relevance score, based on the definition of [4]. Indeed, Communities Coverage represents the minimum set of communities containing both a maximum set of users sharing a maximum set of similar interests. To apply the notion of the Communities Coverage to our Tree Communities we give the following definitions:

Definition 4 *Let* $U=\{u_1, u_2, \ldots, u_n\}$ *be a set of users and* $I=\{i_1, i_2, \ldots, i_n\}$ *be the set of users' interests. A Communities Coverage of a UI_context= (U, I, R) is defined as a set of Communities* $CV=\{(COSI)_1, (COSI)_2, \ldots, (COSI)_n\} \subset K$ *such that any couple (u, i) in the UI_context=(U, I, R) is included in at least one community of CV.*

Consequently, a powerful selection process should be based on both the maximum number of users (cardinality of extent) and the maximum number of interests (cardinality of intent) of its community. To formalize the new criterion, we give the following definitions:

Definition 5 *Let* $(COSI)_j=(U_j, I_j)$. *We define:*
$\mathcal{L}(COSI)_j$= *Length of* $(COSI)_j$: *the number of interests in the intent* I_j.
$\mathcal{W}(COSI)_j$= *Width of* $(COSI)_j$: *the number of users in the extent* U_j.
Relevance (RL) of a community $(COSI)_j$: *is a function of the width and the length defined as follows:* $RL(COSI)_j = [\mathcal{W}(COSI)_j \times \mathcal{L}(COSI)_j] - [\mathcal{W}(COSI)_j + \mathcal{L}(COSI)_j]$

Definition 6 *Communities Of Shared Interests* $(COSI)_j=(U_j, I_j) \in$ *Communities Coverage(CV) if it maximizes the Relevance score.*

Example: Let as consider the User Interest Context (UI_context) depicted in Table 1 to illustrate the selection criteria of Communities Coverage. A Communities Coverage of the UI_context=(U,I,R), shown in grey color in Fig. 1, can be constituted by three communities: S $=\{(COSI)_4; (COSI)_5; (COSI)_6\}$ where: $(COSI)_4$, $(COSI)_5$

and $(COSI)_6$ represents respectively the Community Of Shared Interests containing ({Tom, Ken};{Art, Music}); ({Bob, Jon};{Music, Sport}); ({Jon, Ben};{Sport, Dance}).

3.2 Users' Social Circles Extension

The goal behind social communities discovering is to enrich networks and to increase the interaction between their members, *ie.*, resource sharing, exchange messages, etc. Indeed, in this section, we introduce a new algorithm, called $Recom_{Friend}$, aiming to extend users' social circles relying on two main steps:

Algorithm 1: RECOM_FRIEND

```
Input: - Communities Coverage CV, Communities Tree and u_cv.
Output: - The set of recommended Friends Friend(u_cv)
begin
    Relevance(COSI_cv):= 0; Friend(u_cv) := 0; BestRelevance := 0;
    foreach Communities Of Shared Interests COSI_cv from CV do
        if u_s ∈ Ext(COSI_cv) and i_s ∈ int(COSI_cv) then
            if Relevance (COSI_cv) > BestRelevance then
                BestRelevance := Relevance(COSI_cv);
                CSIC := COSI_cv;
    while BP(CSIC) != null do
        if Friend(u_cv) ∉ Ext(BP) then
            Friend(u_cv) := Ext(BP) ∪ Friend(u_cv);
    if Friend(u_cv) != null then
        return Friend(u_cv);
end
```

- Given a user u_{cv} who has a set of interests $\{i_{cv}\}$, find the community which has the 'BestRL' and contain the user u_{cv} and the set of interests $\{i_{cv}\}$. This amounts to study communities of the Coverage CV, noted by $COSI_{cv}$. For this aim, we compute the Relevance score (given in definition 5) of all communities $COSI_{cv}$ to find the community that has the 'BestRL'(maximum Relevance), called the most Confident Shared Interest Community noted *CSIC*.
- Extract a set of BigParents(BP) of *CSIC*. The BigParents are the communities' parents of $COSI_{cv}$ belonging to the level1 of the Communities Tree.

The pseudo code of the $Recom_{Friend}$ algorithm is sketched by Algorithm 1. $Recom_{Friend}$ takes as input the Communities Coverage CV ($COSI_{cv}$) as well as the Communities Tree (*COSI*) and the user u_{cv}. $Recom_{Friend}$ algorithm outputs a set of recommended friends Friend (u_{cv}) to the user u_{cv}. $Recom_{Friend}$ starts by handling all $COSI_{cv}$ to determine the most Confident Shared Interest Communities (*CSIC*), *i.e.*, having the best relevance, containing the user u_{cv}. Then, $Recom_{Friend}$ extracts BigParents(BP) of the most *CSIC* and recommends the set of users BigParents (BigParents extensions (Ext(BP)) to the user u_{cv} (Friend(u_{cv}) = Ext(BP)).

$Recom_{Friend}$ algorithm relies on the Communities Coverage CV to suggest effectively, to a user u_{cv}, a set of friends with whom he shares a set of interests i_{cv}. Social users circles are then enriched by adding the set of recommended friends to the confident Shared Interest Community noted *CSIC* that contain the user u_{cv}.

Example: As it's represented in Communities Tree depicted in Fig. 1, the extracted Communities Coverage(CV) contains three communities: CV = $\{COSI_4, COSI_5, COSI_6\}$. For example, the user 'Bob' belonging to the community $COSI_5$ (Friend(Bob)), has two BigParents $COSI_3$ and $COSI_2$ (communities colored with pink). By applying our $Recom_{Friend}$ algorithm, we obtain a set of Friend(Bob)={Tom, Ken, Ben}. So the set of users {Tom,Ken,Ben} are recommended to the user 'Bob'. Therefore, we extend the community $COSI_5$ by the set of users {Tom,Ken,Ben}.

4 Results and Discussion

In this section, we show through extensive carried out experiments the assessments of our approach performances *vs.* those of Louvain [1], OSLOM [8], CM [15] and CONGA [6], respectively. We have applied our experiments on three real-world datasets collected from BlogCatalog,[1] Flickr [2] and FOAF.[3] BlogCatalog is a social blog directory where the bloggers can register their blogs under pre-specified categories which we used as users' interests. Flickr is a content sharing platform, with a focus on photos where users can share their contents, upload tags and subscribe to different interest groups, we use this interest groups as users' interests. FOAF, *i.e.*, Friend of a Friend, is a machine-readable ontology which describes people and their interests. We use the "foaf:interest" property that represents the user's interest in the real world. The characteristics of the 3 datasets are reported in Table 2.

[1] http://www.blogcatalog.com.

[2] http://www.flickr.com.

[3] http://www.foaf-project.org.

Table 2 Datasets characteristics

Data sets	BlogCatalog	Flickr	FOAF
Interests	119	100	1307
Nodes	358	436	333
Links	11048	86841	28882
Network density	0.17	0.911	0.522
Clustering coefficient	0.693	0.941	0.797

4.1 Semantic Evaluation

As mentioned in Sect. 3.1, the extraction of the 'best' community should not only be based on the number of users but on both the maximum number of users and the maximum number of interests of its community. Actually, the best approach is the one which extracts the maximum number of vertices and maximum number of interests from the network. Hence, for evaluating the effectiveness of our approach in communities detection, we compare it with Louvain, OSLOM, CM and CONGA approaches, by computing the semantic coverage (Sem_Coverage) over the 3 datasets. According to the definition 5, we define the Sem_Coverage as follows : $Sem\_Coverage = \frac{IC \times VC}{IT \times VT}$. It indicates how many vertices and interests are assigned to communities, *i.e.,* the number of assigned vertices(VC) and interests(IC) divided by the total number of vertices(VT) and interests(IT) containing in the network. As depicted in Table 3, our approach cover a significant number of interests and users (45,9 %, 73,7 % and 44,81 % of interests and users on BlogCatalog, Flickr and FOAF datasets, respectively). The obtained results highlight that our approach is efficient for covering the maximum number of users and the maximum number of interests over different datasets. Moreover, we can report that our approach has the highest Sem_Coverage comparing to the other approaches, over the 3 datasets. In fact, we can explain the low accuracy of other approaches over all measures, by the fact that it tend to detect the entire graph as a community and try essentially to maximize the ratio of intra-community links to inter-community links without taking into account the semantic relation between users.

Table 3 Semantic Coverage of our approach *vs.* that of Louvain, CM, CONGA and OSLOM above FOAF, BlogCatalog and Flickr datasets

Data sets	FOAF	BlogCatalog	Flickr
Our approach	**44.81**	**45.9**	**73.7**
Louvain	3.09	7.68	8.14
CM	2.02	6.52	29.47
CONGA	4.05	10.86	
OSLOM		12.09	

4.2 Statistical Evaluation

The statistical evaluation of our approach is carried out by computing the Stat_Coverage, the recall and the F_β score before and after applying the $Recom_{Friend}$ algorithm on the 3 datasets. The Stat_coverage indicates how many vertices are assigned to communities (i.e., the number of assigned vertices(VC) divided by the total number of vertices(VT) containing in the network).

To evaluate the Social Circles Extension via ground truth before and after extension, we consider each explicit interest (categories) of Flickr, BlogCatalog and FOAF networks as a ground truth community (GT) (Each interest corresponds to an individual GT community) [17]. Given a set of algorithmic communities C and the ground truth communities S, precision indicates how many vertices are actually in the same GT community, recall indicates how many vertices are predicted to be in the same community in a retrieved community and F_β score is the harmonic mean of precision and recall. Precision and Recall are defined as follows : $Precision = \frac{|C \cap S|}{|C|}$, $Recall = \frac{|C \cap S|}{|S|}$. To consider both precision and recall, we use F_β score to measure the test accuracy. We choose $\beta = 2$ in order to weigh recall higher than precision because giving good recommendations is more important than excluding irrelevant ones. F_β score is computed as follows: $F_\beta = (1 + \beta^2) \times \frac{Precision \times Recall}{\beta^2 \times Precision + Recall}$.

As depicted in Table 4, the Stat_Coverage is increased above BlogCatalog and Flickr datasets compared to the baseline communities (discovered communities with our community discovery approach). We can note that the $Recom_{Friend}$ algorithm enables to discover new users which are not discovered by our community discovery approach. No Stat_Coverage increasing is noted for FOAF dataset, we can explain that by the fact that all nodes are covered by applying our community Discovery approach (the Stat_Coverage is 100 % before applying the extension algorithm).

An interesting observation is that the $Recom_{Friend}$ algorithm improves the matching with the ground truth communities over all cases, all results are increased in terms of Recall and F_2 score. We can note that the FOAF dataset is the most beneficiary of this $Recom_{Friend}$ extension algorithm, it is increased by 73.8 % in terms of recall and 58.98 % in terms of F_2 score. It's clear that communities are changed and extended by adding new users belonging to another discovered communities (belong to the communities Coverage). The lowest results are obtained by applying this algorithm on Flickr dataset, it is increased only by 3.34 % on recall measure and 1.7 % in F_2 score. On BlogCatalog the recall and F_2 score show a significant improvement by 13.9 % and 10.88 % respectively. we can note that the $Recom_{Friend}$ algorithm not only discovers new users which are not discovered by the community discovery approach but also it respects the initial structure of baseline communities and enhance the quality of communities.

Table 4 Stat_Coverage, Recall and F_2 score of $Recom_{Friend}$ before (BF) and after (AF) communities extension and the improvement (Improve) above FOAF, BlogCatalog and Flickr datasets

Measures	Stat_Coverage			Recall			F_2 score		
Data sets	BF	AF	Improve	BF	AF	Improve	BF	AF	Improve
FOAF	100 %	100 %	0	26.74	46.48	+73.8 %	29.74	47.28	+58.98 %
BlogCatalog	68,11 %	71 %	+4.24 %	31.49	35.88	+13.9 %	31.79	35.25	+10.88 %
Flickr	93 %	98,8 %	+6.23 %	36.75	37.98	+3.34 %	40,16	40.84	+1.7 %

5 Conclusion and Future Work

In this paper, we have introduced a novel approach for users' social circles extension through the proposition of a new algorithm called $Recom_{Friend}$. We tackle the challenge of discovering a 'pertinent' set of communities of shared interests based on semantic aspects. We have evaluated our approach on real-world datasets through several metrics. Our future avenues for future work mainly address the focus on other users' attribute such as their working place, their organization, their localization and so on. It remains to be seen whether more sized datasets further increase the performance bar for users' social circles extension.

References

1. Blondel, V.D., Guillaume, J.L., Lambiotte, R., Lefebvre, E.: Fast unfolding of communities in large networks. J. Stat. Mech.-Theory Exp. **10** (2008)
2. Clauset, A., Newman, M.E.J., Moore, C.: Finding community structure in very large networks. Phys. Rev. E **70**(066111), 1–6 (2004)
3. Crampes, M., Plantié, M.: Détection de communautés chevauchantes dans les graphes bipartis. In: MARAMI 2012: conférence sur les modèles et l'analyse des réseaux: Approches mathématiques et informatiques (2012)
4. Ganter, B., Wille, R.: Formal Concept Analysis: Mathematical Foundations. Springer, Berlin (1999)
5. Girvan, M., Newman, M.E.J.: Community structure in social and biological net- works. In: Proceedings of the National Academy of Sciences, vol. 99, pp. 7821–7826 (2002)
6. Gregory, S.: A fast algorithm to find overlapping communities in networks. In: Proceedings of the 2008 European Conference on Machine Learning and Knowledge Discovery in Databases - Part I. pp. 408–423 (2008)
7. Kwon, J., Kim, S.: Friend recommendation method using physical and social con- text. Int. J. Comput. Sci. Netw. Secur. **10**, 116–120 (2010)
8. Lancichinetti, A., Radicchi, F., Ramasco, J.J., Fortunato, S.: Finding statistically significant communities in networks. CoRR abs/1012.2363 (2010)
9. Norris, E.M.: An algorithm for computing the maximal rectangles in a binary relation. Revue Roumaine de Mathématiques Pures et Appliquées **23**, 243–250 (1978)
10. Paolillo, J., Wright, E.: Social network analysis on the semantic web: Techniques and challenges for visualizing foaf. In: Visualizing the Semantic Web: XML-Based Internet and Information Visualization pp. 229–241 (2006)
11. Pasquier, N., Bastide, Y., Taouil, R., Lakhal, L.: Efficient mining of association rules using closed itemset lattices. Inf. Syst. **24**, 25–46 (1999)
12. Planti, M., Crampes, M.: Survey on social community detection. In: Book Chapter, Social Media Retrieval, Computer Communications and Networks. pp. 65–85 (2013)
13. Sahebi, S., Cohen, W.: Community-based recommendations: a solution to the cold start problem. In: Workshop on Recommender Systems and the Social Web (RSWEB), held in conjunction with ACM RecSys11 (October 2011)
14. Silva, N.B., Tsang, I.R., Cavalcanti, G.D.C., Tsang, I.J.: A graph-based friend recommendation system using genetic algorithm. In: IEEE Congress on Evolutionary Computation. pp. 1–7. IEEE (2010)

15. Yan, B., Gregory, S.: Detecting communities in networks by merging cliques. CoRR (2012)
16. Yan, F., Jiang, J., Lu, Y., Luo, Q., Zhang, M.: Community discovery based on social actors' interests and social relationships. In: Proceedings of the 2008 Fourth International Conference on Semantics. Knowledge and Grid, pp. 79–86. IEEE Computer Society, Washington, DC, USA (2008)
17. Yang, J., Leskovec, J.: Defining and evaluating network communities based on ground-truth. In: ICDM. pp. 745–754. IEEE Computer Society (2012)

Mining Historical Social Issues

Yasunobu Sumikawa and Ryohei Ikejiri

Abstract This paper presents a framework for identifying human histories that are similar to a modern social issue specified by a learner. From the historical data, the learner can study how people in history tried to resolve social issues and what results they achieved. This can help the learner consider how to resolve the modern social issue. To identify issues in history similar to a given modern issue, our framework uses the characteristics and explanation of the specified modern issue in two techniques: clustering and classification. These techniques identify the similarity between historical and the modern issues by using matrix operation and text classification. We implemented our proposed framework and evaluated it in terms of analysis time. Experimental results proved that our framework has practical usage with an analysis time of only about 0.7 s.

Keywords History education · Transfer of learning · Authentic learning · Clustering · Text classification · Semi-supervised learning

1 Introduction

Study and analysis of historical data can provide numerous benefits, including an enhanced perception of the legacies of the past in the present and enabling learners to make valuable connections through time. One of the goals of imparting modern history education at high schools is to enable students to study how people in history tried to solve social issues. Students can then apply this knowledge to consider creative solutions to current social issues [1, 2]. This process of applying

Y. Sumikawa (✉)
Tokyo University of Science, Tokyo, Japan
e-mail: yas@cs.is.noda.tus.ac.jp

R. Ikejiri
The University of Tokyo, Tokyo, Japan
e-mail: ikejiri@iii.u-tokyo.ac.jp

R. Neves-Silva et al. (eds.), *Intelligent Decision Technologies*,
Smart Innovation, Systems and Technologies 39,
DOI 10.1007/978-3-319-19857-6_50

knowledge obtained in one context, such as studying history, to another, such as considering solutions to current social issues, is called *transfer of learning* [3]. Although some past history education research studies proposed effective methods to enhance transfer of learning, each method uses only one theme selected by researchers [4, 5]. Modern research in education fields has suggested that teachers or the learning environment should provide a selectively suitable theme for each student. For example, if a Ph.D./postdoctoral student, or a university professor studies the history of labor laws, and a newspaper reports that politicians are contemplating modifications to the salary of university researchers, using this news as an introduction to teaching the history may be a suitable introduction for the learner; however, this is may not be useful for many high-school students. A learning method that connects what students learn to real-world issues is called *authentic learning* [3].

In this paper, we present a framework that can serve as a tool for authentic learning, and can be implemented for history education. We call this framework *human history miner* (HHM). If a learner inputs a current issue's characteristics and explanation, HHM detects relevant histories that are similar to the current issue by using clustering and classifying. Clustering determines the similarity by matrix operation using characteristics of the historical and the current issues. Classifying uses the explanation to assign a history name label to the modern issue based on traditional text classification.

The contributions of HHM can be summarized as follows:

1. HHM detects similarities between past and modern social issues. This contribution is useful for both history education and text mining. With regard to the former, studying similar histories can promote the transfer of learning, known as analogy [6]. For the latter research field, text mining, HHM presents a new method to detect similarities. By considering a time axis, HHM detects similarities between objects at a distance from each other along the time axis. In contrast, past techniques detect similarities between objects at the same point on the axis.
2. HHM presents a study environment to impart authentic learning for history education. Using this environment, learners can study the history relevant to every current issue. To the best of our knowledge, HHM is the first framework that provides a learning environment of this type for history education.

The remainder of this paper is organized as follows: Sect. 2 provides summaries of several related works. We describe the data representation in our framework in Sect. 3, and detail the two techniques that we use to detect similarities between historical social issues and current social problems in Sect. 4. Section 5 details experimental results that show the analysis time for determining the similar histories. Section 6 contains our conclusions.

2 Related Work

2.1 History Education

Drie and Boxtel reviewed previous studies on historical reasoning [7], and found six components of historical reasoning: asking historical questions, using sources, contextualization, argumentation, using substantive concepts, and using metaconcepts. However, previous studies do not take into account the students' ability to use learning about the past as a means of solving present and future problems.

Mansilla examined in depth how students apply history to current problems successfully [8]. She used the Rwandan Genocide of 1994 as an example of a modern social problem and the Holocaust as a similar event in the past. As a result of her research, students were able to successfully use history to build an informed comparison base between both cases of genocide, recognize historical differences between them, appropriately apply historical modes of thinking to examine the genocide in Rwanda, and generate new questions and hypotheses about the genocide. However, Mansilla did not investigate what themes could enable other histories to connect with other modern problems.

Lee insists that a *usable historical framework* can make connections between events in the past and potentially between events in the past and present [1]. This framework must be an overview of long-term patterns of change, and be an open structure, capable of being modified, tested, improved, and even abandoned. This would enable students to assimilate new history to the existing framework. In addition, its subject should be human history, not some privileged sub-set of it. This is the underlying concept of a theoretical framework for using learning about the past as a means to solve present and future problems. However, Lee did not explain how to create concrete themes for the usable historical framework and learning environment.

Ikejiri designed a competitive card game for high-school students studying world history, where players can construct causal relations in the modern age by using historical causal relations [4]. This educational material includes a staged learning method for identifying causal relationships within modern societal problems by using references to historical causal relations. The effectiveness of this game was tested, and the results proved that this educational tool is effective in improving the ability to both associate past and current events with similar characteristics and to analyze causal relations in modern problems by referencing historical problems. Further, Ikejiri et al. designed another competitive card game that high-school students can use to apply learning from the policies created in world history to create new policies that would revitalize Japan's economy [5]. The effectiveness of this game was tested, and it revealed that the number of policies proposed to invigorate Japan's economy increased after the high-school students' use of this card game. These two studies provide effective methods for high-school students to use history to analyze causal relations in modern problems by referencing historical problems and to cultivate alternative solutions to confront contemporary social problems. However, the theme that can connect historical contexts with modern society is set by a researcher each time manually. Thus, these learning methods are not automated according to modern context.

HHM can dynamically connect history with modern society, and present a learning environment where students can apply history to authentic modern social problems immediately.

2.2 *Mining History*

Some prevalent technologies, such as lifelogging [9], versioning [10], and malware protection [11], use history as a convenience feature. These technologies collect several types of data about each user, such as video and audio data, positions, documents, pages viewed, and items purchased, etc.

Au Yeung and Jatowt studied how current society remembers the past by analyzing how many events from history are referenced in news articles [12]. Odijk et al. developed a tool to help historians involved in searching historical data [13]. Although these studies may be useful for implementing traditional history education and for history researchers, they are not designed as tools to enhance the transfer of learning.

3 Data Representation

As mentioned in Sect. 2.1, history education should use the usable historical framework proposed by Lee [1]. In this study, we use histories included in Lee's framework. Moreover, we organize these history data explanations in the order of cause, issue, solution, and results because this order has been proved to promote transfer of learning [6]. We assume that explanations of modern social issues are provided in news articles.

4 Algorithms to Determine Similarity Between Historical and Modern Social Issues

This section first provides an overview of HHM, and then details each technique.

4.1 *System Overview*

Figure 1 shows an overview of the functionality of HHM. When a user inputs the characteristics and explanation of a modern social issue, the processes of

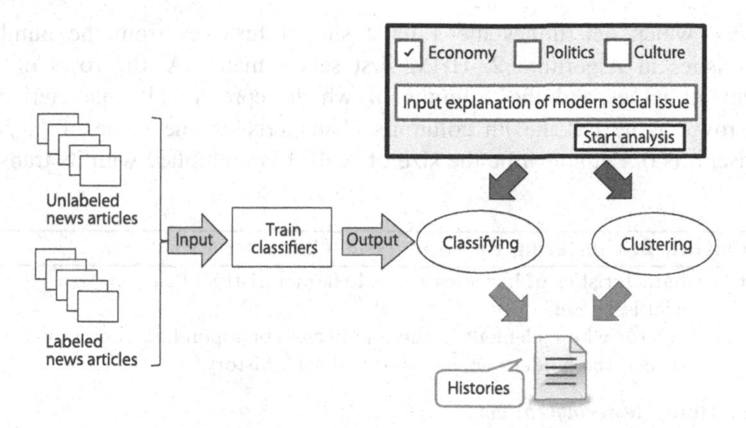

Fig. 1 System overview

Algorithm 1: Discovering histories similar to a modern issue

Input : Characteristics of the modern social issue *cm* and a text file *tf* explaining the modern social issue
Output: Histories

Function: *hhm*(*cm*, *tf*)
1: Let *ch* be characteristics of histories.
2: Call *clustering*(*ch*, *cm*) and *classifying*(*tf*) in parallel, and let *clu* and *cla* be results of *clustering* and *classifying*, respectively.
3: **return** histories included in *clu* and *cla*.

---

clustering and classifying, respectively, are applied to the inputed data. This process is defined in the function *hhm* described in Algorithm 1. HHM first loads histories and their characteristics *ch*. It then calls two functions, *clustering* and *classifying*, in parallel. The *clustering* takes *ch* and the characteristics of a modern social issue as its arguments. The *classifying* takes the explanation of the modern social issue as its argument. Note that the *classifying* function uses a model that is already learned before the user inputs the two kinds of information to retrieve histories as quickly as possible. Finally, HHM displays the histories based on these results.

4.2 Clustering Based on Characteristics of Issues

Every social issue generally has some defining characteristics related to economy, politics, industry, culture, and so on. If two issues have the same characteristics, we can reasonably claim that they are similar. Further, the probability of a past and a current issue being similar increases with the increase in the number of characteristics they have in common. We show the pseudo code of the function *clustering*, which determines the k most similar histories from the number of characteristics in Algorithm 2. HHM first sets a matrix A, the rows of which represent all issues and the columns of which represent all characteristics. If the ith row's issue has the jth column's characteristics, the value of (i, j) is 1; otherwise, it is 0. Let $m \times n$ be the size of A. If A is multiplied with its transposed

Algorithm 2: Clustering Based on Issue Characteristics

Input : Characteristics of histories ch and characteristics of the modern social issue cm

Output: A vector whose element is the number of corresponding characteristics between the modern social issue and each history

Function: *clustering*(ch, cm)

1: Create a feature matrix A from the feature vectors of ch and cm.
2: Create a transposed matrix ${}^tA$ from A.
3: **return** a row of A^tA corresponding to the modern social issue.

Table 1 Example data used in clustering

Name	Economy	Politics	Industry
History 1	0	0	1
History 2	1	1	1
Current issue	1	1	0

$$A \stackrel{def}{\Leftrightarrow} \begin{pmatrix} 0 & 0 & 1 \\ 1 & 1 & 1 \\ 1 & 1 & 0 \end{pmatrix}, \qquad A^tA = \begin{pmatrix} 1 & 1 & 0 \\ 1 & 3 & 2 \\ 0 & 2 & 2 \end{pmatrix}$$

Fig. 2 Determining the similarities in Table 1 by performing matrix operation

matrix ${}^tA$, we get an $m \times m$ matrix whose rows and columns represent issues and each element (i,j) represents the number of corresponding characteristics between ith row's issue and jth row's issue; therefore, HHM can return the most similar k histories from the result matrix.

For example, consider the history data set shown in Table 1. HHM determines the feature matrix A as shown in Fig. 2. Following the multiplication of A by ${}^tA$, we see that the current issue has two characteristics similar to those of with History 2 but none to those of History 1. From this result, HHM determines that History 2 is more similar to the current issue than History 1.

4.3 Classifying Histories into Categories

Motivation for Using Classification. Because HHM only deals with social issues iteratively occurring in the long-term period, we can use news articles regarding current social issues for determining the similarity between historical and modern issues. For example, the issue of Trans-Pacific Partnership (TPP) that has been written about in several news articles is similar to the free trade policy that George Canning worked on in the 19th century because TPP relates to free trade. Therefore, we can determine the similarity by text classifiers with history and news articles. However, there are two issues. First, training classifiers using only history data involves a high cost for making training data sets because the explanation of history data must be presented in a suitable order as mentioned in Sect. 3. Second, every word has its own history, and its meaning can change over time. In contrast, we can easily obtain numerous news articles. Given these observations, if we consider each name of history data as a label, and news articles are labeled by the history name, then it is possible to train classifiers based on a few labeled news articles and many unlabeled ones. This is known as *semi-supervised learning*. Because this method only uses news articles written in modern words, both of the above mentioned issues can be simultaneously resolved. Moreover, using only news articles enables us to use any text classification algorithm based on semi-supervised learning in order to

---

Algorithm 3: Training Classifiers

Function: *trainingClassifier*()
1: Train classifiers.
2: Let *nuun* be the number of unlabeled news articles.
3: **while** *nuun* is greater than zero **do**
4: Use the classifiers to assign history name labels to unlabeled news articles.
5: Let *new_nuun* be the number of unlabeled news articles.
6: **if** (*nuun* > *new_nuun*) **then**
7: // some news articles were labeled.
8: Train classifiers again.
9: Update *nuun* by the number of unlabeled news articles of new data set.
10: **else**
11: // no news articles can be labeled.
12: **break**
13: **end if**
14: **end while**

Algorithm 4: Assigning a history name label to unlabeled news articles

---

Input : Text file *tf* explaining the modern social issue
Output: Names of similar histories

Function: *classifying*(*tf*)
1: For *tf*, make a bag-of-words *bow*, and remove stop words.
2: Recall a model learned in *trainingClassifier*().
3: **return** result of labeling to *bow* by the model.

---

detect similar histories. Text classification based on semi-supervised learning has been extensively studied [14–16], and HHM can benefit from this.

Algorithm. HHM's classification consists of two phases. The first phase is to train classifiers based on semi-supervised learning. The second phase is to label a news article that the user inputs.

Algorithm 3 depicts a pseudo code that defines a function *trainingClassifier*. This function uses the Expectation-Maximization algorithm, which is one of the most popular algorithms in semi-supervised classification [16]. HHM first trains classifiers with labeled news articles (line 1). The classifiers assign each unlabeled news article a suitable label (line 4). If some news articles are labeled, the classifiers are trained again with the set of all newly labeled news articles (line 8). This process continues until the number of labeled news articles is zero, or if the classifiers cannot assign any labels to the unlabeled news articles (line 10).

Algorithm 4 defines the function *classifying* that uses classifiers trained by Algorithm 3 to assign a history name label to the inputed modern social issue that has been entered. Because the inputed issue is text based, a bag-of-words (BoW) is first created from it. Morphological analysis is then applied to divide words if the text is written in a language that does not use spaces between words, such as Japanese, Chinese, and Korean. Furthermore, the processes to remove stop words, facilitate stemming, and using term frequency-inverse document frequency, are also applied at this stage. The BoW is then used by classifiers trained in Algorithm 3 to assign a suitable history name label.

5 Experimental Results

We measured the analysis time taken by HHM on a laptop computer with a system configuration similar to mobile phones in order to check HHM can be practically used in school classrooms, even though most schools currently either prohibit the use of the Internet or do not have an Internet server. We used a computer with an Intel Core2Duo U9600 CPU with processing speed of 1.6 GHz on the CentOS operating system. The system parameters of the cache memories are shown in Table 2.

Table 2 System parameters of cache memories

Parameters	L1D	L2
Total size (KB)	32	3,072
Line size (bytes)	64	64
Number of lines	512	49,152
Associativity	8	12

We implemented the *clustering*, *classifying*, and *hhm* functions in Java, and the *trainingClassifier* function in Python. For clustering functionality, we used an 800 × 50 matrix as matrix *A* of Algorithm 2 because we considered these numbers are sufficient for the purpose of our study. For example, the National Council for the Social Studies provides a guideline for classroom teachers [17], which defines 10 social themes and 24 terms. Therefore, we thought that 50 characteristics were sufficient for our purposes. For classifying functionality, we implemented a naive Bayes classifier, one of the most popular probabilistic classifiers based on Bayes' theorem. To train the classifier, we used six histories, and then for each history, we used 10 labeled and 1,500 unlabeled news articles included in the Mainichi newspaper articles published in 2012.[1] To test the functionality of HHM, we used a Japanese newspaper article with a file size of 708 bytes. Because the newspaper was written in Japanese, we performed morphological analysis using MeCab [18], and then removed stop words in *classifying*.

Table 3 displays the analysis times taken by the *clustering*, *classifying*, and *hhm* functions. All functions returned results within 0.7 s. Therefore, in practice HHM can be used on any computer or mobile device. In general, when a user inputs information, HHM should use three threads, one each for entering data, loading the history data used in the clustering process, and loading the model for the function *classifying*. Given this process, the analysis times provided in Table 3 do not include the loading times because we assumed that the input times were longer than these loading times.

Table 3 Results of analyzing time (second)

Clustering	Classifying	Line 2 of hhm
0.362	0.651	0.651

HHM ignores similarity between characteristics, and thus can miss detecting a few similar issues. If we define two characteristics-national debt and trade-an issue related to national debt *ind* can be considered similar to one related to trade *it* because both characteristics relate to money. However, HHM would determine that *ind* is not similar to *it*. This missed detection can be corrected by forming a hierarchy of characteristics. If two issues belong to different characteristics on the same level in the hierarchy, but these characteristics have the same characteristics on an upper level, HHM determines that the issues may be similar. This solution is one of the most important future works of this research.

As mentioned in Sect. 4.3, we can use any classification algorithm with HHM's classifiers. Further, we can use any corpus of news articles because HHM is simply a classification framework based on semi-supervised learning. To obtain better classification accuracy, it is necessary to carefully choose a combination of classifier algorithm and corpus.

[1] CD-Mainichi Newspapers 2012 data, Nichigai Associates, Inc., 2012 (Japanese).

6 Conclusions

In this paper, we proposed a learning framework called HHM that provides an authentic learning environment for history education. HHM presents issues in recorded human history that are similar to a modern social issue by using clustering and classification to promote transfer of learning for history education.

To show that HHM is a practical tool, we implemented a prototype, and measured the analysis time of the framework on a laptop computer. The results proved that HHM is practical because its analysis time was only about 0.7 s. This time was based on a test case using 800 histories and 50 characteristics for clustering, a 708 bytes input text file, and a model learned by analyzing 9,060 news articles. In future work, we intend to address the following issues: (1) developing an algorithm that determines similarity based on the characteristics hierarchy, (2) developing an interface to displaying the results of a similarity search, and (3) evaluating if HHM can promote transfer of learning by testing it on high-school students.

Acknowledgments This work was supported by JSPS KAKENHI Grant Number 26750076.

References

1. Lee, P.: Historical literacy: theory and research. Int. J. Hist. Learn. Teach. Res. **5**(1), 25–40 (2005)
2. Ministry of Education, Culture, S.S., Technology: Japan course of study for senior high schools (2009)
3. Donovan, M.S., Bransford, J.D., James W. Pellegrino, E.: How people learn: bridging research and practice. National Academy Press (1999)
4. Ikejiri, R.: Designing and evaluating the card game which fosters the ability to apply the historical causal relation to the modern problems. Japan Soc. Educ. Technol. **34**(4), 375–386 (2011) (in Japanese)
5. Ikejiri, R., Fujimoto, T., Tsubakimoto, M., Yamauchi, Y.: Designing and evaluating a card game to support high school students in applying their knowledge of world history to solve modern political issues. In: International Conference of Media Education, Beijing Normal University (2012)
6. Gick, L.M., Holyoak, J.K.: Analogical problem solving. Cogn. Psychol. **12**, 306–355 (1980)
7. van Drie, J., van Boxtel, C.: Historical reasoning: Towards a framework for analyzing students' reasoning about the past. Educ. Psychol. Rev. **20**(2), 87–110 (2008)
8. Boix-Mansilla, V.: Historical understanding: beyond the past and into the present. Knowing, Teaching, and Learning History: National and International Perspectives, pp. 390–418 (2000)
9. Byrne, D., Kelliher, A., Jones, G.J.: Life editing: third-party perspectives on lifelog content. In: Proceedings of the SIGCHI Conference on Human Factors in Computing Systems. CHI '11, pp. 1501–1510. New York, NY, USA, ACM (2011)
10. Zimmermann, T., Weissgerber, P., Diehl, S., Zeller, A.: Mining version histories to guide software changes. IEEE Trans. Softw. Eng. **31**(6), 429–445 (2005)
11. Rieck, K., Trinius, P., Willems, C., Holz, T.: Automatic analysis of malware behavior using machine learning. J. Comput. Secur. **19**(4), 639–668 (2011)

12. Au Yeung, C.m., Jatowt, A.: Studying how the past is remembered: towards computational history through large scale text mining. In: Proceedings of the 20th ACM International Conference on Information and Knowledge Management. CIKM '11, 1231–1240. New York, NY, USA, ACM (2011)
13. Odijk, D., de Rooij, O., Peetz, M.H., Pieters, T., de Rijke, M., Snelders, S.: Semantic document selection: historical research on collections that span multiple centuries. In: Proceedings of the Second International Conference on Theory and Practice of Digital Libraries. TPDL'12, pp. 215–221. Berlin, Heidelberg, Springer (2012)
14. Cong, G., Lee, W., Wu, H., Liu, B.: Semi-supervised text classification using partitioned em. In: Lee, Y., Li, J., Whang, K.Y., Lee, D. (eds.) Database Systems for Advanced Applications. Lecture Notes in Computer Science, vol. 2973, pp. 482–493. Springer, Berlin (2004)
15. Ghani, R.: Combining labeled and unlabeled data for multiclass text categorization. In: Proceedings of the Nineteenth International Conference on Machine Learning. ICML '02, , pp. 187–194. San Francisco, CA, USA, Morgan Kaufmann Publishers Inc. (2002)
16. Nigam, K., McCallum, A.K., Thrun, S., Mitchell, T.: Text classification from labeled and unlabeled documents using em. Mach. Learn. **39**(2–3), 103–134 (2000)
17. Golston, S.: The revised NCSS standards: ideas for the classroom teacher. Soc. Educ. **7**(4), 210–216 (2010)
18. Kudo, T., Yamamoto, K., Matsumoto, Y.: Applying conditional random fields to japanese morphological analysis. In: Proceedings of EMNLP, pp. 230–237 (2004)

Evolving an Adaptive Artificial Neural Network with a Gravitational Search Algorithm

Shing Chiang Tan and Chee Peng Lim

Abstract In this paper, a supervised fuzzy adaptive resonance theory neural network, i.e., Fuzzy ARTMAP (FAM), is integrated with a heuristic Gravitational Search Algorithm (GSA) that is inspired from the laws of Newtonian gravity. The proposed FAM-GSA model combines the unique features of both constituents to perform data classification. The classification performance of FAM-GSA is benchmarked against other state-of-art machine learning classifiers using an artificially generated data set and two real data sets from different domains. Comparatively, the empirical results indicate that FAM-GSA generally is able to achieve a better classification performance with a parsimonious network size, but with the expense of a higher computational load.

Keywords Adaptive resonance theory · Artificial neural network · Supervised learning · Gravitational search algorithm · Data classification

1 Introduction

Artificial neural networks (ANNs) learn a collection of information from a database through a learning process. An ANN consists of an arbitrary number of hidden neurons (nodes) which contain important information for solving a particular problem. One approach to constructing an ANN structure is an incremental learning scheme, which learns and accumulates information from a database continuously during a training process. Examples of incremental learning ANNs include unsupervised models such as self-organizing map [1], growing neural gas [2]; semi-supervised

S.C. Tan (✉)
Multimedia University, Cyberjaya, Malaysia
e-mail: sctan@mmu.edu.my

C.P. Lim
Deakin University, Geelong, Victoria
e-mail: chee.lim@deakin.edu.au

R. Neves-Silva et al. (eds.), *Intelligent Decision Technologies*,
Smart Innovation, Systems and Technologies 39,
DOI 10.1007/978-3-319-19857-6_51

models such as growing self-organizing map [3]; and, supervised models such as radial basis function [4] and adaptive resonance theory (ART) [5] networks. However, an incremental learning ANN is most likely not fully optimized, therefore resulting in compromised generalization.

On the other hand, a number of algorithms inspired by natural phenomena, which include physical (such as simulated annealing [6] and gravitational search algorithm [7]) and biological (such as genetic algorithm [8], artificial immune system [9], ant colony optimization [10], particle swarm optimization [11], and bacterial foraging algorithm [12]) processes, have been introduced in the literature for solving optimization problems. These heuristic algorithms are useful for tackling different types of search and optimization problems in various areas, e.g. medical data classification pertaining to breast cancer [13] and production schedule optimization pertaining to job shop assembly [14]. In this paper, the main aim is to optimize the information learned by an ART network using a gravitational search algorithm (GSA). In particular, a supervised ART network, namely, Fuzzy ARTMAP (FAM) [15], is adopted for integration with the GSA. During the training process, FAM performs incremental learning with the capability of overcoming the *stability-plasticity* dilemma [15]. The dilemma addresses the challenge on how a system could preserve previously learned information (stability) while continuously accommodating newly learned information (plasticity). The proposed FAM-GSA model inherits the advantages of its constituents originated from different computing paradigms. It first adapts the information from a supervised ART learning process. Then, such information continues to evolve by applying the GSA for search and optimization operations.

The organization of this paper is as follows. In Sect. 2, both FAM and GSA are described before explaining the proposed FAM-GSA model. In Sect. 3, the effectiveness of FAM-GSA is evaluated using three benchmark studies. The results from the experiment are compared and analyzed. A summary of the work is presented in Sect. 4.

2 Method

2.1 Fuzzy ARTMAP

FAM [15] is a supervised version of the ART network. It learns information from a set of training samples incrementally. New information is adapted into the existing knowledge base through a self-organizing and self-stabilizing learning mechanism. An FAM structure is composed by two fuzzy ART modules (i.e., ART_a, which is an input module and ART_b, which is an output module) that are linked through a map field, F^{ab}. Each ART module is composed of three layers of nodes, i.e., $F_0^a(F_0^b)$ denotes the normalization layer that complement-codes [15] an $M-$ dimensional input vector, $\boldsymbol{a}$, to a $2M-$ dimensional vector, $\boldsymbol{A}$; $F_1^a(F_1^b)$ denotes the input layer that receives the complement-coded input vector; $F_2^a(F_2^b)$ denotes a the recognition layer that contains prototype nodes that represent the input samples. During the incremental learning process, additional recognition nodes can be created on-the-fly to encode new information into $F_2^a(F_2^b)$.

On presentation of a pair of input-target data sample, the learning procedure of FAM ensues, as follows. The input sample, $\boldsymbol{a}$, is assigned to ART$_a$ whereas its corresponding target output, c, (where $c \in [1, 2, \ldots, C]$, C is the number of target classes is assigned to ART$_b$. In F_1^a, $\boldsymbol{a}$ is complement-coded as pattern $\boldsymbol{A}$, and then transmitted to F_2^a. In F_2^a, the activation level of each recognition node is determined by a choice function [15], as follows:

$$T_j = \frac{\left|\boldsymbol{A} \wedge \boldsymbol{w}_j^a\right|}{\chi_a + \left|\boldsymbol{w}_j^a\right|} \tag{1}$$

where χ_a is the choice parameter that is close to zero [15]; $\boldsymbol{w}_j^a$ represents the weight of the j-th recognition node. The node with the highest activation level is identified as the winning node (denoted as node J) under the *winner-take-all* competition scheme. A vigilance test is performed to measure the similarity between the winning prototype, $\boldsymbol{w}_J^a$, and $\boldsymbol{A}$, and it is compared against a threshold (vigilance parameter, $\rho_a \in [0, 1]$, as follows [15]:

$$\frac{\left|\boldsymbol{A} \wedge \boldsymbol{w}_J^a\right|}{\left|\boldsymbol{A}\right|} \geq \rho_a \tag{2}$$

If the vigilance test is not satisfied, a new round of search for another winning node is initiated. Such search process for finding a winning node is repeated until the selected node is able to pass the vigilance test. However, when none of the existing recognition nodes can fulfill the vigilance test, a new node is inserted into F_2^a to encode the input sample. At ART$_b$, the output module also undergoes the same pattern-matching cycle as in ART$_a$. In this case, the target vector c is assigned to ART$_b$ to identify a winning node that represents the target class.

After the winning nodes in F_2^a and F_2^b have been identified, a prediction is made from F_2^a to F_2^b through F^{ab}. A map-field vigilance test is carried out to determine the prediction [15], as follows:

$$\frac{\left|\boldsymbol{y}^b \wedge \boldsymbol{w}_J^{ab}\right|}{\left|\boldsymbol{y}^b\right|} \geq \rho_{ab} \tag{3}$$

where $\boldsymbol{y}^b$ represents the output vector of $\boldsymbol{y}^b$; $\boldsymbol{w}_J^{ab}$ represents the weight vector from F_2^a to F^{ab}; and ρ_{ab} denotes the map-field vigilance parameter. If the map-field vigilance test fails, it indicates that the winning node in F_2^a has made a wrong prediction of the target class in F_2^b. As such, a match-tracking process [15] is carried out. In this case, ρ_a, which is initially set to a user-defined baseline vigilance parameter $\bar{\rho}_a$, is raised to

$$\rho_a = \frac{|\boldsymbol{A} \wedge \boldsymbol{w}_J^a|}{|\boldsymbol{A}|} + \delta \tag{4}$$

where δ is a small positive value. Increasing ρ_a results in failure of the ART_a vigilance test. Therefore, a new search cycle in ART_a is carried out with a higher level of ρ_a. This search process is reiterated until a correct prediction is made between the winning nodes in F_2^a and F_2^b. At this juncture, FAM goes into a learning phase, whereby $\boldsymbol{w}_J^{ab}$ is updated [15] as follows:

$$\boldsymbol{w}_J^{a(new)} = \beta_a \left(\boldsymbol{A} \wedge \boldsymbol{w}_J^{a(old)} \right) + (1 - \beta_a) \boldsymbol{w}_J^{a(old)} \tag{5}$$

where $\beta_a \in [0, 1]$ is the learning rate of ART_a. Equations (1)–(5) indicate the learning process in ART_a. The same learning process occurs in ART_a, where all equations with superscript or subscript a are replaced by b.

2.2 Gravitational Search Algorithm (GSA)

The GSA is a heuristic search and optimization algorithm, which is inspired from Newton's laws of gravitation and motion, on a population of agents (objects) [7]. The interaction among the objects is affected by the gravity force and the object mass. An object with a smaller mass is attracted to move towards objects with larger mass. All objects under the GSA undergo both exploration and exploitation processes. An object with a heavier mass indicates a better solution. Therefore, it moves slower than lighter objects, and this exemplifies the exploitation process of the GSA.

In GSA, a population of N potential solutions to a problem is represented by the positions of N objects, as follows:

$$\boldsymbol{X}_i = \left(x_i^1, \ldots, x_i^d, \ldots, x_i^n \right) \quad \text{for} \quad i = 1, 2, \ldots, N \tag{1}$$

where x_i^d indicates the position of the i-th object in the d-th dimension. At an arbitrary time step, t, the force exerting on object i from object j is defined as follows:

$$F_{ij}^t(t) = G(t) \frac{M_{pi}(t) \times M_{aj}(t)}{R_{ij}(t) + \epsilon} \left(x_j^d(t) - x_i^d(t) \right) \tag{2}$$

where M_{aj} denotes the active gravitational mass related to object j; M_{pi} denotes the passive gravitational mass related to object i; $G(t)$ denotes a gravitational constant at time t; ϵ denotes a small positive constant; and, $R_{ij}(t)$ denotes the Euclidean distance between objects i and j, as follows:

$$R_{ij}(t) = \left\| \boldsymbol{X}_i(t), \boldsymbol{X}_j(t) \right\|_2 \quad (3)$$

Initially, $G(t_0) = G_0$, where G_0 denotes the initial value of G at the first cosmic quantum-interval of time t_0. During the search process, $G(t)$ is gradually reduced with time to govern search accuracy, as follows:

$$G(t) = G(t_0) \times \left(\frac{t_0}{t}\right)^{\beta}, \quad \beta < 1 \quad (4)$$

To impart a stochastic characteristic into the GSA, the total force exerting on object i in dimension d comprises a randomly weighted sum of the d-th component of the forces applied from other objects, i.e.,

$$F_i^d(t) = \sum\nolimits_{j=1, j \neq i}^{N} rand_j F_{ij}^d(t) \quad (5)$$

where $rand_j$ is a random number between 0 and 1. According to the law of motion, the acceleration of object i at time t in the d-th direction, $a_i^d(t)$, is defined as follows:

$$a_i^d(t) = \frac{F_i^d(t)}{M_{ii}(t)} \quad (6)$$

where M_{ii} denotes the inertia mass of the i-th object. On the other hand, the velocity of an object at time step $t + 1$ is a fraction of its current velocity and acceleration at t, i.e.,

$$v_i^d(t+1) = rand_i \times v_i^d(t) + a_i^d(t) \quad (7)$$

where $rand_i$ is a uniformly distributed random number between 0 and 1. Then, the position of object i is updated as follows:

$$x_i^d(t+1) = x_i^d(t) + v_i^d(t+1) \quad (8)$$

The performance of an object is measured by a fitness function, which is expressed in terms of gravitational and inertia masses. A larger object has a higher efficacy. This also indicates that the object has a higher degree of attraction, and it moves more slowly. Assuming

$$M_{ai} = M_{pi} = M_{ii} = M_i, \quad i = 1, 2, \ldots, N \quad (9)$$

the gravitational and inertial masses are updated as follows:

$$m_i(t) = \frac{fit_i(t) - worst(t)}{best(t) - worst(t)} \quad (10)$$

$$M_i(t) = \frac{m_i(t)}{\sum_{j=1}^{N} m_j(t)} \quad (11)$$

where $fit_i(t)$ is the fitness value of object i at time t. For a maximization problem,

$$best(t) = \max_{j \in \{1,\ldots,N\}} fit_j(t) \quad (12a)$$

$$worst(t) = \min_{j \in \{1,\ldots,N\}} fit_j(t) \quad (12b)$$

2.3 Integration Between FAM and GSA

The proposed FAM-GSA model undergoes an adaptation process in two phases. In the learning phase, FAM learns information directly from a set of training data. Information is updated and organized in terms of recognition nodes. In the search and optimization phase, the GSA is used to evolve the connection weights of FAM. In such a way, FAM-GSA constructs a knowledge base directly from a set of training data, and such knowledge base evolves and adapts continuously to achieve an optimized solution. The training procedure of FAM-GSA is described as follows:

Phase 1: Network training. FAM undergoes a supervised learning process in one training epoch to learn information from a training data set. The acquired information is kept in terms of connections weights of the ART_a recognition nodes, and new information can be accumulated continuously owing to the incremental learning capability of FAM.

Phase 2: Search and adaptation with the GSA
Step 1: Initialize the population. All connection weights of the recognition nodes from ART_a are concatenated to form an object, as follows:

$$l_0 = [w_1^a \, w_2^a \, w_3^a \ldots w_S^a] \quad (12)$$

where S denotes the number of recognition nodes in ART_a. A population of the objects is formed by using a modified function [16]:

$$l_k = l_0 + \text{RMF} \times \text{rand}(1, n) \quad (13)$$

where l_k is the k-th (for $k = 1, 2, \ldots, K-1$) replicated chromosome of l_0 from a perturbation process; RMF $\in [0, 1]$ denotes the range multiplication factor; rand(1,n) is a uniformly distributed random vector between 0 and 1, and n is the length of l_0.
Step 2: Perform the fitness evaluation. The fitness of each object is its accuracy rate in classifying the training samples correctly.
Step 3: Update the GSA parameters, i.e., Eqs. (4) and (9)–(12).
Step 4: Compute the total force in different directions, i.e., Eq. (5).
Step 5: Compute the acceleration (Eq. (6)) and velocity (Eq. (7)) rates.
Step 6: Revise the objects' position, i.e., Eq. (8).
Step 7: Repeat Step 2 to Step 6 until a stopping criterion (e.g., the maximum number of the search cycle) is met.

3 Experiment

The efficacy of the proposed FAM-GSA model is assessed using three benchmark data sets, which are the syntactic Ripley data set [17] and two real-world data sets from the UCI machine-learning repository [18] (i.e., the iris flowers and steel plate faults). In each experiment, FAM-GSA was trained using a set of default parameters in two adaptation phases. In the first phase, FAM was established based on one training epoch with $\beta = 1$ and $\bar{\rho}_a = 0.0$. In the second phase, the GSA performed the gravitational search and adaptation process within 10 cycles with RMF = 0.30 and $K = 10$. Unless otherwise stated, the experiment was repeated 100 times, each with a different sequence of training samples on a random basis. The results were analyzed and compared statistically.

3.1 The Ripley Data Set

The Ripley data set [17] is a nonlinear binary classification problem. Each class represents a bimodal distribution of input features, which are generated from two Gaussian mixture distributions with an equal covariance. In this experiment, 250 samples were allocated for training whereas 1000 samples were used for test. The bootstrapping method [19] was employed to estimate the average results (with 95 % confidence intervals) of FAM-GSA. FAM was trained with both single-epoch and multi-epoch learning processes.

Table 1 shows the results. The average accuracy rate of FAM-GSA is 85.07 % whereas those from single-epoch and multi-epoch FAM models are 81.56 % and 83.32 %, respectively. The 95 % confidence interval indicates that FAM-GSA outperforms both models of FAM statistically. Such finding indicates the advantage of incorporating the gravitational search and optimization process as an additional adaptation mechanism to improve the generalization capability of FAM. On the other hand, the network size of FAM-GSA, which is in terms of the average number of recognition nodes, is similar to that of single-epoch FAM. FAM-GSA is obviously smaller in terms of the network size as compared with multi-epoch FAM. The training time of FAM-GSA is the longest. This is expected because FAM-GSA undergoes both supervised ART learning and GSA search and optimization processes. In summary, FAM-GSA is a more accurate model in classifying the non-linear Ripley data samples at the expense of a longer training time as compared to the original FAM network (Table 2).

Table 1 Classification results of the FAM-based models using the Ripley data set (the numbers in parenthesis indicate the 95 % confidence intervals of the results)

Algorithm	Classification correct rate (%)	Network size	Training time (s)
FAM-GSA	85.07 [84.58 85.54]	9.9 [9.3 10.4]	15.11 [14.65 15.60]
FAM (single-epoch)	81.56 [80.83 82.22]	10.2 [9.8 10.8]	0.45 [0.44 0.46]
FAM (multi-epoch)	83.40 [82.64 84.09]	23.4 [20.7 27.0]	9.27 [9.12 9.41]

Table 2 Performance comparison between FAM-GSA and other metaheuristic ANNs using the iris data set [20]

Algorithm	Classification correct rate (%)	Network size	Running time (s)
RBF-1	90.91	50	0.12
RBF-2	91.80	50	0.75
GA-RBF-1	86.71	32	1.81
GA-RBF-2	95.04	32	1.81 + 0.07
GA-RBF-3	93.42	32	1.81 + 0.49
FAM (single-epoch)	94.22	5.1	0.15
FAM-GSA	94.98	5.1	4.33

3.2 The Iris Data Set

The proposed FAM-GSA model is also compared with other metaheuristic ANNs [20], which include the RBF network trained by a genetic algorithm (GA-RBF-1) and its variants (GA-RBF-2 and GA-RBF-3) using the popular iris data set. The data set records the measurements of the petal width and length of three types of iris flowers. In this experiment, 70 % of data samples were used for training, while the remaining samples were used for test. Table 3 shows the classification performances. FAM-GSA has a smaller network size than all RBF-based networks [20]. In general, FAM-GSA is a compact model that is able to perform with a higher classification accuracy rate than the RBF networks (except for GA-RBF-2) at the cost of a longer training time.

3.3 The Steel Plate Data Set

FAM-GSA is applied to solve a fault detection problem in the steel manufacturing industry. The steel plate data set contains 1941 records that indicate seven types of faulty condition of a stainless steel leaf [18]. Each record presents a set of numeric based on 27 input attributes of a steel-plate image, which include geometric shape, luminosity, orientation, edges, and contour. The target output is one of seven faults, which are pastry, Z-scratch, K-scratch, stains, dirtiness, bumps, and others.

A stratified two-fold cross validation technique [21] was employed. The data set was divided into a training set of 1457 samples and a test set of 484 samples. The number of samples in both training and test sets approximately followed the output class distribution of the original data set. The results were averaged across 30 runs, which was the same as the experimental setup in [22]. Table 3 shows the classification performance of FAM-GSA, along with an evolutionary ANN (EANN) and other machine-learning techniques, i.e., two ANN classifiers (i.e., MLP and RBF), one statistical classifier (i.e., SVM), two rule-based classifiers (i.e., C4.5 and PART) and an instance-based classifier (i.e., 1-NN) [22]. In [22], four different feature selection techniques were utilized to create a two-stage EANN (i.e., TSEAFS). The structure and weights of TSEAFS were identified separately in two successive evolutionary processes. TSEAFS was compared with other state-of-art classifiers using the original data set and another data set with reduced input features. For clarity, only the results from the original data set as reported in [22] were used for comparison. From Table 3, it can be observed that FAM-GSA outperforms other classifiers. This finding again shows the advantage of integrating the supervised ART classifier with a gravitational search capability to form the proposed FAM-GSA model.

Table 3 Performance comparison with other machine-learning techniques using the steel plate data set [22]

Algorithm	Classification correct rate (%)
FAM-GSA	65.40
FAM	64.84
C4.5	39.05
1-NN	49.17
SVM	57.02
PART	46.69
MLP	53.50
RBF	59.94
TSEAFS	51.46

4 Summary

This paper presents a novel metaheuristic ANN, which involves the integration of a supervised fuzzy ART classifier and a heuristic algorithm based on the laws of Newtonian gravity. In FAM-GSA, the network structure is formed incrementally, and the connection weights are evolved in two modes of adaptation. The classification performance of FAM-GSA is evaluated using three benchmark studies. The first is a benchmark problem showing the effectiveness of FAM-GSA in classifying a set of syntactic data samples with bimodal distributions in the data space. The second and third experiments investigate the performance of FAM-GSA in classifying two real benchmark data samples from different application domains. The former involves a data classification task for identifying three different types of the iris flowers whereas the latter is a benchmark problem for classifying steel faults. The performance of FAM-GSA is compared against its original FAM network and also other recently proposed machine-learning classifiers in the literature. In general, the results from these benchmark studies indicate that FAM-GSA is a compact ANN capable of producing high classification accuracy rates.

Further work will be devoted to develop a more dynamic FAM network wherein its connection weights and hidden nodes could be evolved simultaneously. The effectiveness and applicability of the proposed FAM-GSA model will be evaluated using additional benchmark data sets as well as real-world data sets from different domains. Besides that, the use of FAM-GSA in solving multi-objective classification problems will be investigated.

References

1. Shen, F., Ouyang, Q., Kasai, W., Hasegawa, O.: A general associative memory based on self-organizing incremental neural network. Neurocomputing **104**, 57–71 (2013)
2. Fišer, D., Faigl, J., Kulich, M.: Growing neural gas efficiently. Neurocomputing **104**, 72–82 (2013)
3. Allahyar, A., Yazdi, H.S., Harati, A.: Constrained semi-supervised growing self-organizing map. Neurocomputing **147**, 456–471 (2015)
4. Reiner, P., Wilamowski, B.M.: Efficient incremental construction of RBF networks using quasi-gradient method. Neurocomputing **150**, 349–356 (2015)
5. Zhang, Y., Ji, H., Zhang, W.: TPPFAM: Use of threshold and posterior probability for category reduction in fuzzy ARTMAP. Neurocomputing **124**, 63–71 (2014)
6. Kirkpatrick, S., Gelatto, C.D., Vecchi, M.P.: Optimization by simulated annealing. Science **220**, 671–680 (1983)
7. Rashedi, E., Nezamabadi-pour, H., Saryazdi, S.: GSA: A gravitational search algorithm. Inf. Sci. **179**, 2232–2248 (2009)
8. Mitchell, M.: An Introduction to Genetic Algorithms. MIT Press, Cambridge (1996)
9. Farmer, J.D., Packard, N., Perelson, A.: The immune system, adaptation and machine learning. Phys. D **2**, 187–204 (1986)
10. Dorigo, M., Maniezzo, V., Colorni, A.: Ant system: optimization by a colony of cooperating agents. IEEE Trans. Syst. Man, Cybern. Part B **26**, 29–41 (1996)

11. Kennedy, J., Eberhart, R.C.: Particle swarm optimization. Proc. IEEE Int. Conf. Neural Netw. **4**, 1942–1948 (1995)
12. Passino, K.M.: Biomimicry of bacterial foraging for distributed optimization and control. IEEE Trans. Control Syst. Mag. **22**, 52–67 (2002)
13. Karakış, R., Tez, M., Kılıç, Y.A., Kuru, Y., Güler, İ.: A genetic algorithm model based on artificial neural network for prediction of the axillary lymph node status in breast cancer. Eng. Appl. Artif. Intell. **26**, 945–950 (2013)
14. Wong, T.C., Ngan, S.C.: A comparison of hybrid genetic algorithm and hybrid particle swarm optimization to minimize make span for assembly job shop. Appl. Soft Comput. **13**, 1391–1399 (2013)
15. Carpenter, G.A., Grossberg, S., Markuzon, N., Reynolds, J.H., Rosen, D.B.: Fuzzy artmap: a neural network architecture for incremental supervised learning of analog multidimensional maps. IEEE Trans. Neural Networks **3**, 698–713 (1992)
16. Baskar, S., Subraraj, P., Rao, M.V.C.: Performance of hybrid real coded genetic algorithms. Int. J. Comput. Eng. Sci. **2**, 583–602 (2001)
17. Ripley, B.D.: Neural networks and related methods for classification. J. Roy. Stat. Soc.: Ser. B (Methodol.) **56**, 409–456 (1994)
18. Asuncion, A., Newman, D.J.: UCI Machine Learning Repository [http://www.ics.uci.edu/~mlearn/MLRepository.html]. University of California, School of Information and Computer Science, Irvine, CA (2007)
19. Efron, B.: Bootstrap methods: another look at the jackknife. Ann. Stat. **7**, 1–26 (1979)
20. Ding, S., Xu, L., Su, C., Jin, F.: An optimizing method of rbf neural network based on genetic algorithm. Neural Comput. Appl. **21**(2012), 333–336 (2012)
21. Kohavi, R.: A study of cross validation and bootstrap for accuracy estimation and model selection. In: Proceedings of the 14th International Joint Conference Artificial Intelligence (IJCAI), pp. 1137–1145. Morgan Kaufmann (1995)
22. Tallón-Ballesteros, A.J., Hervás-Martínez, C., Riquelme, J.C., Ruiz, R.: Feature selection to enhance a two-stage evolutionary algorithm in product unit neural networks for complex classification problems. Neurocomputing **114**, 107–117 (2013)

Improving Model-Based Mobile Gaze Tracking

Miika Toivanen and Kristian Lukander

Abstract Mobile, wearable gaze tracking provides flexible opportunities for extending gaze tracking research outside of laboratory environments. Wearable trackers are predominantly video based and fall in to two categories: pupil-corneal reflection methods and physical model-based methods. A number of error sources affect the feature extraction and gaze mapping and therefore the accuracy and precision of both systems. Here, we present two methods for improving tracking results: an advanced user calibration procedure for estimating gaze vectors applicable with any model-based method and a Bayesian tracker for tracking any number of corneal reflections and the pupil center, applicable with both types of trackers. The results show clear improvements over the stability and robustness of recognizing and tracking features in the eye image and, ultimately, estimating the gaze vector.

Keywords Mobile gaze tracking · Bayesian tracking · User calibration

1 Introduction

Mobile, wearable gaze tracking provides flexible opportunities for extending gaze tracking research outside of laboratory environments. Mobile trackers can be divided into two broad categories: so called pupil-corneal reflection (PCR) trackers and physical model-based (PMB) trackers which utilize a simplified model of the human eye and multiple corneal reflections. Generally, PCR trackers suffer from calibration problems due to movement of the camera(s) in relation to the calibrated eye while utilizing simple, planar calibration schemes. While the user calibration for PMB devices can be quite straightforward, they may in turn require more tedious device calibration and a more complex tracking scheme while providing robustness toward camera/frame movement.

M. Toivanen (✉) · K. Lukander
Brain Work Research Center, Finnish Institute of Occupational Health, Helsinki, Finland
e-mail: miika.toivanen@ttl.fi

R. Neves-Silva et al. (eds.), *Intelligent Decision Technologies*,
Smart Innovation, Systems and Technologies 39,
DOI 10.1007/978-3-319-19857-6_52

A number of error sources, such as measurement noise, parallax difference between the tracked eye and the scene camera, and calibration and mapping errors, can produce inaccuracies in tracking results (see, e.g., [10]). Additionally, physiological phenomena like the changing pupil diameter have been shown to significantly affect the accuracy of point of gaze (POG) tracking [1]. Even if a gaze tracker would perform perfectly, in natural, binocular viewing conditions, where the distance to the gaze target and therefore vergence varies, intra-eye fixation disparities of up to 2° can be common while fixating a target [2] and this can further be modulated by cognitive effects like time-on-task and vigilance [11]. Hence, a plausible target accuracy of around one degree can be thought of being attainable with a head-mounted gaze tracker.

We present two methods for improving tracking results: an advanced user calibration procedure for estimating the gaze vectors and a Bayesian tracker for the (multiple) corneal reflections and the pupil center. The user calibration method can be used with any PMB tracker while the Bayesian tracking scheme is applicable with both types of trackers. The methods are presented as extensions of our previously published mobile gaze tracker [13]. The results show clear improvements over the stability and robustness of recognizing and tracking features in the eye image and, ultimately, over the accuracy and precision of the estimated gaze vector.

2 The Method

2.1 The Gaze Tracking Model

The basic principle of the method is presented in [13] as part of the gaze tracking glasses (see Fig. 1 for their prototype) and outlined here in brief. The glasses contain one camera (standard web camera modified to work with near-infrared light), looking at the eye via a hot mirror acting as the "lens" of the glasses, and one scene camera pointing forward. The method uses three-dimensional models of the cameras relative to each other and of the human eye. 3D centers of the corneal sphere and the pupil are estimated from the eye image and the optical axis (OA) is taken to traverse through these (see Fig. 1). The OA somewhat diverges from the "real" gaze vector (GV) which traces from the fovea through the pupil center and ultimately to the real point of gaze (POG); the angle between these is sometimes called kappa angle.

For computing the corneal center, reflections of point light sources on the first corneal surface and an assumption about the spherical shape of the cornea is utilized. Taken that the radius of a sphere is known, at least two light sources are needed to compute the center of the sphere [6, 15]. The system here uses six LED point light sources located around the frame to make the system of equations highly overdetermined and to ensure at least two good quality reflections regardless of the direction of gaze. The pupil center is estimated with the starburst algorithm [12] and elliptical fitting [6], using the estimated corneal center.

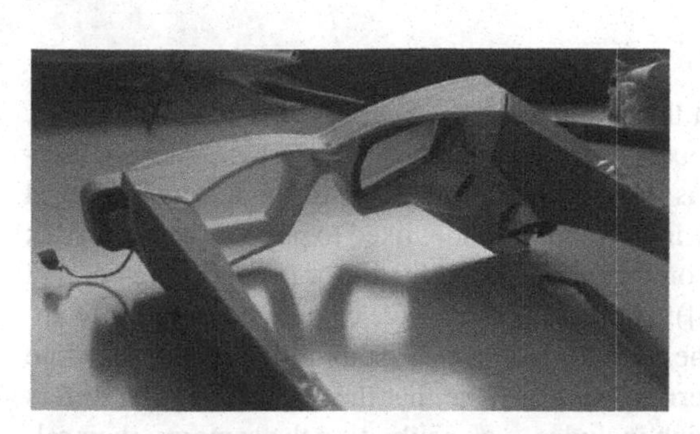

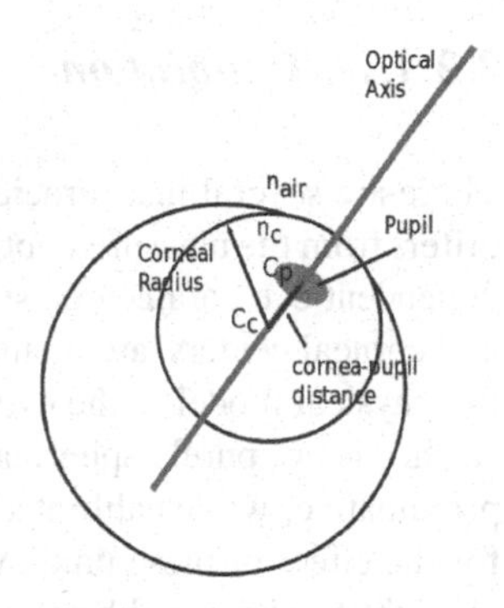

Fig. 1 Left: Prototype of the gaze tracking glasses of [13]. Right: A simplified model of the human eye. C_c and C_p refer to corneal and pupil centers; n_{air} and n_c refer to refractive indices of air and cornea

In order to have a reliable estimate of the corneal and pupil centers, the eye camera as well as the locations of the LEDs in relation to the eye camera must be carefully calibrated. The vector OA can be estimated in eye camera coordinates as starting from the estimated corneal center and pointing forward through the pupil center. This can then be mapped through an affine transformation to the scene camera coordinates through the calibrated relation between the eye and scene cameras. Finally, the POG can be defined as the intersection of the gaze vector and a virtual, orthogonal plane on an assumed distance d from the scene camera origin. With monocular tracking, gaze depth, i.e., the focal distance of the viewer, is unknown and must be set manually.

Therefore, if a 3×3 matrix **A** denotes the linear part of the transformation and a 3×1 vector **a** the translation part, the POG in scene camera coordinates is:

$$\mathbf{p} = \mathbf{A}(\mathbf{C}_{\mathrm{c}} + d\mathbf{L}) + \mathbf{a} \tag{1}$$

where **p** denotes the 3D POG vector in scene camera coordinates, $\mathbf{C}_{\mathrm{c}}$ denotes the 3D corneal center, and $\mathbf{L} \equiv \mathbf{P}_{\mathrm{c}} - \mathbf{C}_{\mathrm{c}}$ is the optical axis where $\mathbf{P}_{\mathrm{c}}$ denotes the 3D pupil center. Vectors $\mathbf{P}_{\mathrm{c}}$, $\mathbf{C}_{\mathrm{c}}$, and hence **L** are all in eye camera coordinates.

In the original method, as described in [13], the corneal radius was input to the system manually, based on values obtained with, e.g., a keratometer. Other eye parameters, such as the refractive index of the cornea and the depth of the pupil from the corneal surface, were input based on population averages from the literature. The system allowed for adjusting these values interactively if the estimated gaze point was clearly off of the actual gaze target. Also, the original system only included a simple translation of the gaze point in the scene camera coordinates to compensate for some of the inherent errors such as the kappa angle. Here we present a systematic calibration model which maps the evaluated optical axis into a POG vector, improving accuracy. As a second improvement, for robustly detecting the glints and the pupil in the eye image we present a remedying tracking scheme which takes into account the temporal behaviour of the eye.

2.2 User Calibration

There are several inaccuracies in the outlined model Sect. 2.1: the real gaze vector differs from the measured optical axis (and the difference varies between users); user-dependent eye parameters, such as the corneal radius or the distance between pupil and corneal centers, are manually input (and probably fixed to population averages); the physical model of the eye is somewhat over-simplified (for instance, the corneal surface is not purely spherical [4]); and the used pin-hole camera model is only approximative, while calibrated. The purpose of the user calibration is to compensate for the effect of these unknown error sources. Note that the user calibration differs from the hardware calibration which includes, e.g., calibrating the cameras, their relative position, and the LED locations related to the eye camera. The user calibration in gaze tracking systems typically aims at translating (often somewhat heuristically) the estimated POG in some 2D coordinate system according to the 2D error vectors measured during the calibration procedure ([6, 10]). In our solution, we correct the measured 3D vector, which can be done either in the eye camera coordinates or in the scene camera coordinates, resulting thus in two different calibration models.

Correcting the Gaze Vector in Eye Camera Coordinates In this calibrating scheme the gaze vector is calibrated in the eye camera coordinates, that is, the optical axis $\mathbf{L}$ in (1) is multiplied by a coefficient matrix $\mathbf{K}_g$. This linear map can be thought of as responding to, e.g., rotation of the visual axis and hence to the kappa angle. The POG formula (1) can be rewritten to contain a calibrated optical axis:

$$\mathbf{p}_g = \mathbf{A}(\mathbf{C}_c + d\mathbf{K}_g\mathbf{L}) + \mathbf{a} \tag{2}$$

where $\mathbf{p}_g$ is the corrected POG. To estimate $\mathbf{K}_g$, the subject needs to focus her gaze on several pre-defined locations which are extracted from the scene camera video, either manually or automatically, and whose 3D vectors are computed using the scene camera calibration. Let us gather these N test gaze points as $3 \times N$ sized matrix $\mathbf{P}_t^{cal}$ and the corresponding measured corneal centers and optical axes as $\mathbf{C}_c^{cal}$ and $\mathbf{L}^{cal}$. Equation (2) can then be solved for $\mathbf{K}_g$ using the pseudoinverse:

$$\mathbf{K}_g = \frac{1}{d}\mathbf{A}^{-1}(\mathbf{P}_t^{cal} - \hat{\mathbf{a}} - \mathbf{A}\mathbf{C}_c^{cal})\mathbf{L}^{cal^T}(\mathbf{L}^{cal}\mathbf{L}^{cal^T})^{-1} \tag{3}$$

where $\hat{\mathbf{a}}$ is a matrix with N $\mathbf{a}$ vectors: $\hat{\mathbf{a}} \equiv [\mathbf{a}\ \mathbf{a}\ ...\ \mathbf{a}]$. As the matrix equation (3) contains three scalar equations and the matrix K contains nine unknowns, at least three independent measurements are needed to solve $\mathbf{K}_g$, i.e., to invert the matrix $\mathbf{L}^{cal}\mathbf{L}^{cal^T}$.

Correcting the Point of Gaze in Scene Camera Coordinates The problem with the previous calibration model is that apart from the kappa angle, it may not capture well the other inaccuracies of the model. Here, we present another calibration model in which we correct the gaze vector in scene camera coordinates. In order to capture a wider family of transformations than what is possible with linear mapping, the correction map is taken to be affine (i.e. a linear map and a translation). Let us denote with $\mathbf{p}_s$ the corrected POG and with $\tilde{\mathbf{p}}_o$ the observed POG, augmented by adding "1" in the end: $\tilde{\mathbf{p}}_o \equiv [\mathbf{p}_o{}^T \; 1]^T$, so that the affine mapping can be linearized:

$$\mathbf{p}_s = \mathbf{K}_s \tilde{\mathbf{p}}_o \tag{4}$$

where the coefficient $\mathbf{K}_s$ is a 3×4 matrix. By combining the augmented POG vectors, observed during the calibration process with N test samples, as $4 \times N$ sized matrix $\mathbf{P}_o^{cal}$ and the "true" gaze points as $\mathbf{P}_t^{cal}$, the K matrix can be computed with pseudoinverse:

$$\mathbf{K}_s = \mathbf{P}_t^{cal} \mathbf{P}_o^{cal\,T} (\mathbf{P}_o^{cal} \mathbf{P}_o^{cal\,T})^{-1}. \tag{5}$$

As the matrix K_s contains 12 unknowns, at least four independent measurements are needed to solve $\mathbf{K}_s$.

The real mapping between $\mathbf{p}_t$ and $\mathbf{p}_o$ may not be affine but its true nature is a difficult topic and likely unattainable. Here, we settle with using the affine model in this calibration scheme. We also tested using a linear model but with worse results. A problem with this calibration scheme is that the results can be assumed to be accurate only for the single gaze distance used for the calibration.

2.3 Glint Tracking

This subsection presents a probabilistic framework for tracking glints in the eye video. The underlying approach is applicable to basically any method where the gaze is computed based on some features that need to be located in the eye image.

In [13], the glints are detected by thresholding the eye image and applying heuristic, geometric rules to assign the observed glints to their corresponding light sources. However, noise in the images, and the topography of the true eye surface causes variation in forms of missed and false detection which often results in severe fluctuation in the estimated POG. As glint localization is performed independently for each video frame, the method neglects the temporal dimension which would provide useful information: for most of the time (that is, during fixations), the glints stay approximately stationary. Using the prior information about glint locations from the previous frame can be used to enhance and stabilize the detection. For this, we use a Bayesian approach (see, e.g., [5]). The idea is to locate the glints within a frame sequentially, one at a time, and to use the location of the glints in the previous frame and the expected shape of the glint grid as prior information.

The Bayesian Model Let us denote the tth (grayscale) eye video frame as $\mathbf{I}^t$ and the location of the glint grid in the frame as $\mathbf{x}^t$ whose size equals the number of LEDs, N_L. Hence, the observed data is $\mathbf{I}^t$ and the unknown parameter of which we are interested is $\mathbf{x}^t$. Following Bayes' formula we get that the unnormalized posterior probability of $\mathbf{x}^t$ is

$$p(\mathbf{x}^t|\mathbf{I}^t, \mathbf{x}^{1:t-1}) \sim p(\mathbf{I}^t|\mathbf{x}^t)p(\mathbf{x}^t|\mathbf{x}^{1:t-1}) \tag{6}$$

where $\mathbf{x}^{1:t-1}$ denotes the past glint grid locations of which the joint likelihood is independent and where the value of the unknown normalization constant is irrelevant. The conditional posterior for any component $i \in [1, ..., N_L]$ of $\mathbf{x}^t$ can be written as

$$p(x_i^t|\mathbf{I}^t, \mathbf{x}^{1:t-1}, \mathbf{x}_{1:i-1}^t) \sim p(\mathbf{I}^t|x_i^t)p(\mathbf{x}_i^t|\mathbf{x}^{1:t-1}, \mathbf{x}_{1:i-1}^t), \; i \in [1, ..., N_L] \tag{7}$$

where $\mathbf{x}_{1:i-1}^t$ denotes the already estimated components of the parameter vector of which the likelihood is independent. The likelihood function is taken to be

$$p(\mathbf{I}^t|x_i^t) = e^{\beta \mathbf{I}_s{}^t(x_i^t)} \tag{8}$$

where $\mathbf{I}_s$ is a scaled eye image: $\mathbf{I}_s(x) \in [0, 1] \; \forall x$, and β is a pre-fixed parameter. The prior distribution is a Gaussian:

$$p(x_i^t|\mathbf{x}^{1:t-1}, \mathbf{x}_{1:i-1}^t) = \mathcal{N}(\mu, \mathbf{C}) \tag{9}$$

where the covariance is a diagonal 2×2 matrix, $\mathbf{C} = \sigma_i \mathbb{1}$, and where the mean of the prior distribution is

$$\mu = \begin{cases} x_i^{t-1} & \text{when } i = 1 \\ x_{i-1}^t + x_i^{t^*} - x_{i-1}^{t^*} & \text{when } i > 1 \end{cases} \tag{10}$$

where t^* is explained below (initially, $t^* = t_1$). Every past grid locations ($\mathbf{x}^{1:t-1}$) needs thus not be stored but only the previous estimate ($\mathbf{x}^{t-1}$) and $\mathbf{x}^{t^*}$. The variance σ of the prior distribution is a pre-defined constant which it is set four times larger for the first component : $\sigma_i = 4\sigma_0, i = 1$ and $\sigma_i = \sigma_0, i > 1$. This ensures that the search space is larger for the first component, enabling correctly locating the grid also after saccades, and narrower after this "initialization". Thus, the first component is searched near the location of the corresponding glint in the previous frame with a wide search window. The subsequent components are searched using a narrow search window near the location of the glint just estimated in the same frame, added with the presumed distance vector between the two glints. For speeding up the computations, the likelihood and prior distributions are only evaluated around the prior mean within a window whose width and height is six times the variance of the prior distribution. Figure 2 gives an example of the Bayesian approach for locating a glint.

Estimating the Maximum of the Posterior We search a value for $\mathbf{x}^t$ that maximizes the posterior (6) by maximizing (7) one component at the time. Instead of searching just one such maximum a posteriori (MAP) estimate, many estimates are searched independently and parallelly by using different ordering for the components of $\mathbf{x}^t$. As this approach resembles that of Particle filtering (also known as Sequential Monte

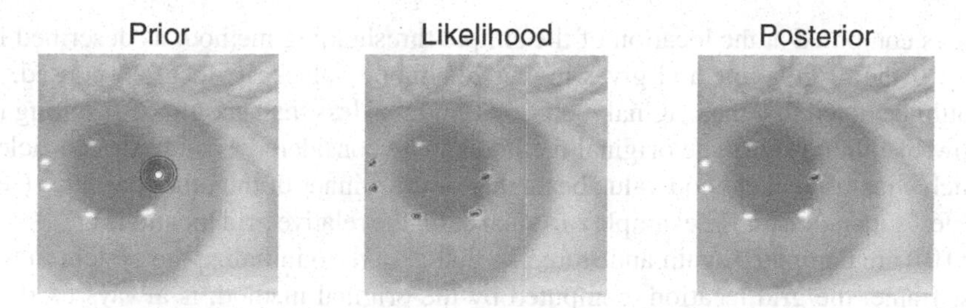

Fig. 2 An example of searching the rightmost glint with the Bayesian framework. Left panel superimposes as contour plot the prior distribution, the middle panel the likelihood distribution, and the right panel the (unnormalized) posterior distribution for the glint location

Carlo sampling [3]) methods, we call the independent searches "particles" which may be considered as candidates for the parameter values. For examples on using Particle filtering for locating image features, see [16, 18]. In our implementation, we use six particles, each with a different component order. Identifying the glints as in Table 1, the order for the first particle could be $\{A, ..., F\}$, for the second particle $\{B, ..., F, A\}$, and so on. The idea behind using different orders for different particles is that some particles may be initialized in a false location if, e.g., there are additional confusing reflections in the eye surface. After each particle has searched their MAP estimates, they are compared and the one with the largest average evidence, i.e., the integral of the unnormalized conditional posterior (7), is the "winner" whose parameter values are used.

To prevent the method from identifying false locations – which may happen after a large saccade, moving all glint locations a considerable amount – the sum of the likelihood values of the winner particle is compared to the sum of the likelihood

Table 1 An illustrative example of the particle based scheme for estimating the MAP value of (6). The table shows the order of the components i of each six particle, denoted as $P_1, ..., P_6$, and the variance σ of the prior distribution (σ_0 is a pre-defined constant). The letters A,...,F refer to the identities of the glints as in the left figure. P_{orig} refers to the location, found with the original thresholding method [13]. Log(evi) and log(lkh) refer to the logarithm of the evidence and likelihood values, summed over each component. The values here are illustrative. In this case, the winner would be particle P_3 whose log(evi) value is underlined. The location of P_3 would also be more likely to correspond to the true glint locations than the location of the original method; hence, its location would be used

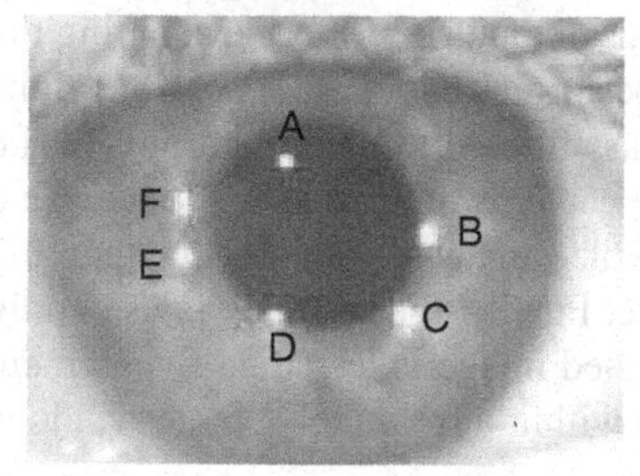

i	σ_i	P_1	P_2	P_3	P_4	P_5	P_6	P_{orig}
1	$4\sigma_0$	A	B	C	D	E	F	
2	σ_0	B	C	D	E	F	A	
3	σ_0	C	D	E	F	A	B	
4	σ_0	D	E	F	A	B	C	
5	σ_0	E	F	A	B	C	D	
6	σ_0	F	A	B	C	D	E	
	log(evi)	5.1	4.4	<u>7.2</u>	5.9	6.5	3.8	
	log(lkh)			0.5				0.2

values computed at the location of the simple thresholding method, as described in [13]. If the original method gives higher likelihood values, its location is used. It should be noted that the original method often finds less than six glints, resulting in a low likelihood sum. The original method can be considered as an "extra particle" which wins if its likelihood value beats that of the winner of the other particles (see Table 1 for an illustrative example). In such case, the relative grid locations ($x_i^{t^*} - x_{i-1}^{t^*}$ in (10)) are computed again and stored so that $t^* = t$. To initialize the system in the first frame, the grid location, computed by the original method, is always used so that $t^* = t_1$.

The differences between the presented framework and the Sequential Monte Carlo sampling is that we only use six particles and instead of sampling we use the deterministic (MAP) value for each particle. This is mainly for the sake of computation speed – using a "real" sampling scheme would probably mean abandoning the real-time property of the system, unless possibly implementing the algorithm in some special computing unit such as a graphics card. However, the possibility for sampling is left for future development. For instance, having $6! = 720$ particles, covering all possible permutations, would probably lead to even more robust tracking as at least one of such particles would locate the glints in the most salient order.

2.4 Pupil Tracking

A problem with the pupil detection scheme of [13] is the same as with glint detection: it is sensitive to noise and does not utilize temporal information. Therefore, pupil detection would likely benefit from a similar approach to the one used for glint tracking in Subsect. 2.3. The unnormalized posterior probability for pupil center location $\mathbf{P}_c{}^t$ in the tth eye image frame is

$$p(\mathbf{P}_c{}^t | \mathbf{I}, \mathbf{P}_c{}^{t-1}) \sim p(\mathbf{I} | \mathbf{P}_c{}^t) p(\mathbf{P}_c{}^t | \mathbf{P}_c{}^{t-1}) \,. \tag{11}$$

The likelihood is similar to (8):

$$p(\mathbf{I}^t | x_i^t) = e^{\beta(1 - \mathbf{I}_s{}^t(x_i^t))} \tag{12}$$

and the prior distribution is again a Gaussian with the mean value being the previous estimate and the covariance being a diagonal constant matrix. Before evaluating the likelihood, morphological opening operation is performed to get rid of possible glints inside the pupil and the image is mildly low-pass filtered to smoothen noise. The MAP estimate of (11) is computed and the pupil center is defined as the mean value of a pupil segment whose pixel values in the scaled image $\mathbf{I}_s$ are smaller than 0.2 (i.e. dark) and which contains the MAP estimate. Finally, the estimate is used only if the sum of the pupil segment pixels in the scaled image is less than five percent of the size of the segment and if the location is within a (relatively) large window, set in the beginning around the first found pupil location – otherwise, the location of the original method is used.

The estimated pupil center is utilized only when the original method seems to fail to locate the pupil correctly. The mislocalization is assessed by the size of the found pupil; often, a falsely detected pupil is notably larger or smaller than a typical pupil. For computing the 3D pupil center, needed for computing OA, we settle with a heuristic solution of using the previous "decently" fitted pupil ellipse with the estimated 2D center (see [13] for details). Improving the estimation of the 3D pupil is left for future development.

3 Results

3.1 Experimental Setup

In this section we report the performance of the presented methods. We used five test subjects (age 30-37 years, one male) with normal vision. The experiments were performed in a dark room where the subjects were looking at the monitor from a distance of 80 cm while wearing the gaze tracking glasses. The subjects were asked to hold their head still. The eye and scene camera recorded 30 frames per second in VGA resolution. The presented algorithms were implemented in Matlab. The original method is written in C++. The parameters of the presented model were set as follows: in the glint tracking algorithm, the β value in likelihood (8) was 20 and the constant variance in the prior model (9) was set to $\sigma_0^2 = 5$ pixels; in the pupil tracking algorithm, the corresponding values are 30 and 50. The method is not very sensible about these values; they were chosen based on the prior experience on previous experiments. For the eye parameters, population averages were used as we are aiming for minimal user calibration: the distance between pupil and corneal centers was fixed to $r_d = 3.5$ mm, the corneal radius to $R_c = 7.7$ mm, and the refractive index of the cornea to 1.336.

For the user calibration, the subjects watched a video where a dot, whose size was one degree, changed its location after each 3 seconds. The locations formed a regular 3×3 grid, shown in Fig. 3. To reduce the number of spontaneous blinks, and therefore possibly corrupt calibration samples, after every three locations the dot was fixed for three more seconds and its color changed from black to brown, instructing the subject to blink her eyes.

For testing the performance, the subjects looked at a similar video with the same intervals for dot location changing and blinking pauses but with a denser 8×8 grid where the distance between adjacent dots was approximately 2.5 degrees in vertical direction and 3.5 degrees in horizontal direction (see Fig. 3) – the grid thus covered a visual angle of approximately 18 degrees in horizontal direction and 25 degrees in vertical direction. As an error measure, an angle between the measured gaze vector and the real gaze vector in scene camera coordinates was computed. The real gaze vector was computed by detecting the dot in the frames with simple computer vision algorithms and mapping the 2D location into a 3D vector using the camera calibration. After each change in the dot location, 10 samples (corresponding to 333 milliseconds) were discarded due to latency in a saccade related to stimulus onset. Initially, one experiment was conducted for each test subject.

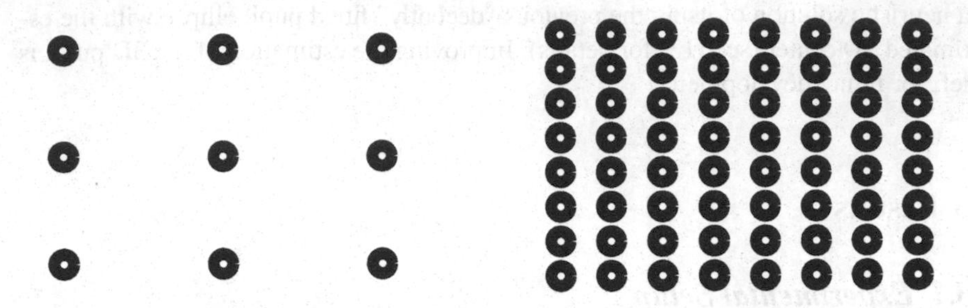

Fig. 3 Left: The grid used in the calibration procedure. Right: the grid used in the testing phase

The gaze tracking field still lacks a standard way for reporting tracker performance. For progress here, Holmqvist et al. have reported definitions in [8, 9] and Tobii has released a specification for remote eye trackers [17]. They define *accuracy* as "the average angular offset (distance) (in degrees of visual angle) between measured fixation locations and the corresponding locations of the fixation targets" [8] and *precision* as "the ability of the eye tracker to reliably reproduce the same gaze point measurement" [17], computed as the root-mean-square (RMS) of successive samples. Here, we will report these error metrics.

3.2 Performance

Accuracy The mean and median values of the angles in each experiment as well as over all the experiments are tabulated in Table 2. The presented methods seem to work relatively well apart from the measurement of subject S5 whose calibration data was poor; it seemed that the eye camera failed to see all the glints properly, resulting in imprecise calibration matrices. Especially, the table shows a clear improvement in the performance when using the presented algorithms over the original method. The two different user calibration schemes seem to perform equally well. The distribution of the angles are skewed, as revealed by the difference between the mean and median values. For instance, the 10 % percentile values over all the angles are 0.71, 0.73, and 1.41 for K_g, K_s, and *Orig* methods. To make the comparison fair, the original method was modified a bit from the method presented in [13] to include the rejection of a misdetected pupil and using the previous decent pupil instead. The user calibration method for the original method was the one that uses the K_g matrix; hence, the first and last rows in Table 2 are comparable as they have the same user calibration. The system failed to compute the gaze (at least within area of the scene camera view) for all the frames. An obvious reason for this is blinking but the pupil or the glints might sometimes be misdetected in spurious locations resulting in the POG outside the scene camera view. The percentages of the frames containing successful measurements are 84 % for the original method and 90 % with the presented glint and pupil tracking algorithms (independent of the user calibration scheme).

Table 2 *Accuracy.* The angle errors between the computed gaze vectors and the true gaze vectors, in degrees. Shown are the mean and median values for each test subject (SN) and for all the measurements. K_g refers to calibration of the gaze vector in eye camera (Eq. (2)) and K_s refers to calibration of the gaze point in scene camera (Eq. (4)). *Orig* refers to the method of [13] which was here calibrated with the K_g matrix

	Mean					Median					All	
	S1	S2	S3	S4	S5	S1	S2	S3	S4	S5	Mean	Med.
K_g	2.4	2.2	3.4	1.4	7.5	1.9	1.7	2.1	1.2	7.1	3.3	1.9
K_s	1.6	2.2	3.6	1.8	10.8	1.4	1.8	2.3	1.6	7.5	3.9	2.0
Orig	6.5	11.7	7.1	4.5	9.8	5.7	8.8	4.7	3.4	9.3	7.9	5.9

Precision Table 3 shows the precision, i.e., the RMS values of successive gaze locations for each measurement. The RMS values were computed separately for each of the 64 fixations. It can be seen that the blink and pupil tracking algorithms stabilize the system considerably as compared to the original method.

Table 3 *Precision.* Root-mean-square values of the angle errors, computed separately for each fixation, in degrees. Shown are the mean and median values for each test subject (SN) and for all the measurements. See the caption of Table 2 for definitions of each row

	Mean					Median					All	
	S1	S2	S3	S4	S5	S1	S2	S3	S4	S5	Mean	Med.
K_g	0.6	0.7	1.5	0.4	0.4	0.5	0.5	0.9	0.3	0.3	0.7	0.4
K_s	0.5	0.6	1.4	0.4	0.9	0.5	0.4	0.9	0.3	0.4	0.7	0.5
Orig	2.9	6.3	4.1	2.2	3.6	2.2	5.9	3.5	1.7	2.4	3.8	2.6

Error divergence Figure 4 illustrates the divergence of the errors between the grid points, for single measurement and for all the measurements. In the left panel, the target points fail to form a regular grid due to slight head motions. The deviations for each grid point is small as also revealed in Table 3. In the right panel, a median of the angles was computed for each grid point and measurement and these were averaged over the measurements to get a proportionality factor for the radius' in the figure. It seems that the rightmost edge causes most severe problems although also the errors for some of the points in the middle of the grid are somewhat large.

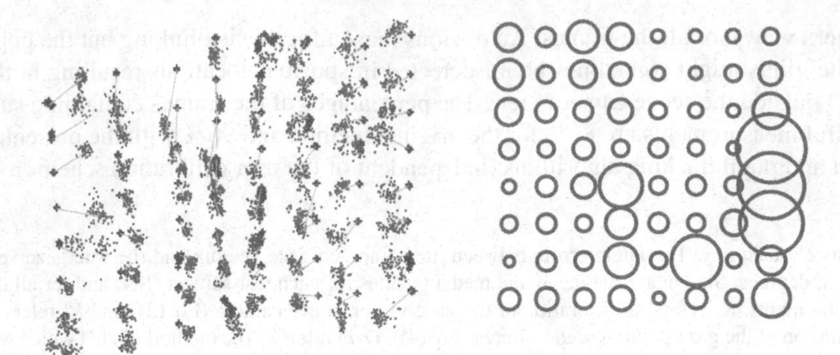

Fig. 4 Left: All the measured gaze points (red dots) and target points (blue crosses) for test subject S4. The lines are drawn between the mean values of each cluster. Note that the distance between two adjacent dots was approximately 2.5 degrees in vertical direction and 3.5 degrees in horizontal direction. Right: The dispersion of the errors between the grid points, computed from all the measurements. The radius of a circle is proportional to the average median error measure for the corresponding grid point (color figure online)

Within-subject variability In order to assess the variability of errors in terms of repeated experiments with the same test subject, we conducted another measurement session (on a separate day) for subject S4 with four additional measurements. These results were compared across each other and with data from the original experiment, presented in Tables 2 and 3. We calibrated the gaze vector in eye camera, using two different K_g matrices: the original matrix used in the original measurement and a recalibrated matrix computed from data recorded before the repeated measurements. Hence, this also serves to test the re-usability of a subject-specific calibration between sessions. The calibration and test stimuli were identical to those of the original experiment.

Figure 5 presents the accuracy and precision of the four repeated measurements together with the original measurement for S4. It seems that the recalibration brings some benefit in terms of better accuracy but does not, as assumed, improve precision. In addition, Fig. 5 shows that the variability across repeated measurements is fairly good although there is a slight trend of decreasing accuracy in repeated measurements when using the original user calibration.

4 Discussion

In this paper, we have presented two improvements for our original physical model-based gaze tracking glasses system [13]. The first improvement is a user calibration methodology which should compensate for the inaccuracies that arise, e.g., from using the simplified physical model and population averages as eye parameters. The calibration process involves recording various training samples and solving the

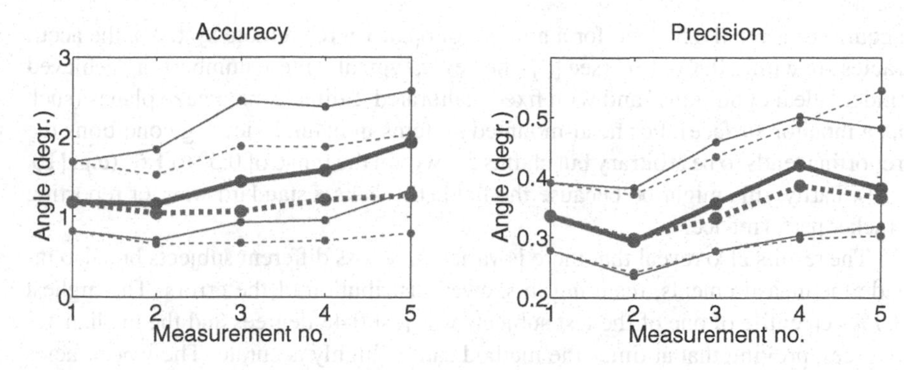

Fig. 5 Accuracy (left panel) and precision (right panel) of the repeated experiments of subject S4. Measurement number 1 refers to the original experiment and measurement numbers 2,...,5 to the four additional experiments. The thick lines refer to the median values and the thin lines to the 25th and 75th percentiles. The solid blue lines refer to using the original user calibration (corresponding to measurement no. 1) and the dashed red lines to the user calibration of the repeated session (conducted just before measurement no. 2) (color figure online)

calibration matrix which can act on the 3D gaze vector in eye camera coordinates or on the 3D point of gaze, mapped to the scene camera coordinates. Of these two possibilities, the former has the advantage of being independent of the gaze distance (i.e. distance between the eye and point of gaze) while the latter mapping might capture better some inaccuracies due to its affine nature. Another improvement is the algorithm for tracking the corneal glints and pupil center in the eye image by considering also the estimates of the previous frame. For this, we presented an advanced Bayesian framework and a semi-heuristic method for estimating the MAP value; independent particles maximize the posterior, conditional on the already estimated values, each with different order for the glints, slightly akin to a Sequential Monte Carlo method.

The promise of the methods were tested in experiments where five test subjects fixated on a regularly moving dot on a display. For a fair comparison, the original method was calibrated using the presented method. The results show that the glint and pupil tracking method clearly improves the accuracy of the original system: the errors are up to five times smaller. The original system is sensitive to noise, causing it to mislocate pupil and glints and to locate only a subset of the glint grid [1]; these lead to false estimates for POG. Additionally, the original method is very subject-specific and would perform better for persons whose eye parameters are close to the population averages and whose eye surface gives good reflections for the LED light sources – the gain of the presented methods can thus be seen as making the original method to work for more users in various conditions.

Fixation precision has not been rigorously reported in evaluations of head-mounted gaze tracking systems. Most commercial manufacturers report

[1]Theoretically any subset of the glint grid with at least two glints should suffice to compute the POG accurately but due to the inaccurate LED calibration the accuracy improves with more detected glints.

accuracies around 0.5°, and for a number of open-source remote systems, the accuracies are within 0.8° - 1.0° (see [7]), however, typically these numbers are achieved under "ideal conditions" and with fixed, calibrated 2-dimensional gaze planes (such as a monitor surface). For head-mounted systems in natural viewing conditions the reporting tends to be arbitrary but claims are within the range of 0.5° to 1.6° (e.g. [10, 14]). Partly, this might be because the field still lacks a standard way for reporting tracker performance.

The results also reveal that there is variation across different subjects but also inside the measurements, resulting in skewed distribution of the errors. The highest 10 % percentile of one of the test subjects was just 0.48 degrees and the median 1.2 degrees, proving that at times the method can be highly accurate. The inaccuracies mainly stem from mislocalization of the glints and/or pupil, imperfect LED calibration, and insufficient user calibration model. The contribution of each of these to the error is undetermined and left for future studies, as well as improving the component(s) that cause most of the error. Also, in some user studies it may be possible to discard the portion of the data with inaccurate results.

We propose that the presented glint tracking algorithm is usable within any gaze tracking method that is based on tracking some eye features. For instance, the simpler pupil-corneal reflection trackers are often based on locating a single glint in the eye image. The Bayesian framework also has the nice feature of being modular, making it possible to modify the likelihood function and the prior model independently; one could, e.g., incorporate prior knowledge about the motion of the pupil. In addition, the suggested user calibration models can be utilized in any mobile gaze tracking method.

The presented algorithms were implemented in Matlab. In a Linux Ubuntu with a 2.60 GHz processor, the glint and pupil tracking takes about 50 ms. As the Matlab algorithms run parallelly with the C++ implementation, our software package with the implementation and the original C++ software is capable of processing about 20 frames per second. The original software is available as open source at https://github.com/bwrc/gaze_tracker_glasses.

References

1. Choe, K.W., Blake, R., Lee, S.H.: Pupil size dynamics during fixation impact the accuracy and precision of video-based gaze estimation. Vis. Res. (2015)
2. Cornell, E.D., Macdougall, H.G., Predebon, J., Curthoys, I.S., et al.: Errors of binocular fixation are common in normal subjects during natural conditions. Optom. & Vis. Sci. **80**(11), 764–771 (2003)
3. Doucet, A., De Freitas, N., Gordon, N.: Sequential Monte Carlo Methods in Practice. Springer (2001)
4. Dubbelman, M., Sicam, V., Van der Heijde, G.: The shape of the anterior and posterior surface of the aging human cornea. Vis. Res. **46**(6), 993–1001 (2006)
5. Gelman, A., Carlin, J.B., Stern, H.S., Rubin, D.B.: Bayesian Data Analysis. Chapman & Hall/CRC, Boca Raton (2004)

6. Hennessey, C., Noureddin, B., Lawrence, P.: A single camera eye-gaze tracking system with free head motion. In: Proceedings of the 2006 Symposium on Eye Tracking Research & Applications, pp. 87–94. ACM (2006)
7. Hennessey, C., Noureddin, B., Lawrence, P.: Fixation precision in high-speed noncontact eye-gaze tracking. IEEE Trans. Syst. Man Cybern. Part B: Cybern. **38**(2), 289–298 (2008)
8. Holmqvist, K., Nyström, M., Andersson, R., Dewhurst, R., Jarodzka, H., Van de Weijer, J.: Eye Tracking: A Comprehensive Guide to Methods and Measures. Oxford University Press, Oxford (2011)
9. Holmqvist, K., Nyström, M., Mulvey, F.: Eye tracker data quality: what it is and how to measure it. In: Proceedings of the Symposium on Eye Tracking Research and Applications, pp. 45–52. ACM (2012)
10. Kassner, M., Patera, W., Bulling, A.: Pupil: an open source platform for pervasive eye tracking and mobile gaze-based interaction. arXiv preprint arXiv:1405.0006 (2014)
11. Lavine, R.A., Sibert, J.L., Gokturk, M., Dickens, B.: Eye-tracking measures and human performance in a vigilance task. Aviat. Space Env. Med. **73**(4), 367–372 (2002)
12. Li, D., Babcock, J., Parkhurst, D.J.: Openeyes: a low-cost head-mounted eye-tracking solution. In: Proceedings of the 2006 Symposium on Eye Tracking Research & Applications, pp. 95–100. ACM (2006)
13. Lukander, K., Jagadeesan, S., Chi, H., Müller, K.: Omg!: a new robust, wearable and affordable open source mobile gaze tracker. In: Proceedings of the 15th International Conference on Human-Computer Interaction with Mobile Devices and Services, pp. 408–411. ACM (2013)
14. Noris, B., Keller, J.B., Billard, A.: A wearable gaze tracking system for children in unconstrained environments. Comput. Vis. Image Underst. **115**(4), 476–486 (2011)
15. Shih, S.W., Liu, J.: A novel approach to 3-D gaze tracking using stereo cameras. IEEE Trans. Syst. Man Cybern. Part B: Cybern. **34**(1), 234–245 (2004)
16. Tamminen, T., Lampinen, J.: Sequential Monte Carlo for Bayesian matching of objects with occlusions. IEEE Trans. Pattern Anal. Mach. Intell. **28**(6), 930–941 (2006)
17. Tobii: Accuracy and precision test method for remote eye trackers. Referenced 23.01.2015. http://www.tobii.com/es/eye-tracking-research/global/about-tobii-pro/eye-tracking/test-method (2011)
18. Toivanen, M., Lampinen, J.: Incremental object matching and detection with bayesian methods and particle filters. Comput. Vis. IET **5**(4), 201–210 (2011)

Artificial Intelligence Based Techniques for Rare Patterns Detection in the Industrial Field

Marco Vannucci and Valentina Colla

Abstract This work presents a review of methods used for improving the performances of classifiers when coping with uneven dataset, focusing on the problem of the detection of rare events. The paper introduces and analyses the problem and outlines the ideas behind classical and advanced methods used for facing the problem. The achievements of the discussed approaches are shown on some real world problems coming from the industrial fields.

1 Introduction

Many real world tasks deal with the problem of the detection of rare patterns. The correct identification of these patterns is often a key issue for the related applications because, for instance, they represent particular situations of interest: it is the case of machinery malfunctions or defective products in the manufacturing field, of fraudulent transactions in the financial framework, of rare diseases in the medical fields. In all these examples, the correct and reliable identification of the interesting patterns is far more important than the identification of other patterns. Moreover, when pursuing this main objective, the eventual generation of *false alarms* (i.e. false positive) does not represent a big issue. These classification problems are characterized by a not uniform distribution of the observations in terms of the *class* variable which is also the one to be predicted by the classifier. Due to the class-unbalance, this problem in literature is commonly known as classification of uneven/unbalanced datasets. In the industrial field, many practical tasks belong to this class of problems since, for instance, when monitoring the status of a manufacturing process for fault detection purposes, most observations are related to the normal situations, while the abnormal

M. Vannucci (✉) · V. Colla
TeCIP Institute, Scuola Superiore Sant'Anna, Via G. Moruzzi, 1, 56124 Pisa, Italy
e-mail: mvannucci@sssup.it

V. Colla
e-mail: colla@sssup.it

R. Neves-Silva et al. (eds.), *Intelligent Decision Technologies*,
Smart Innovation, Systems and Technologies 39,
DOI 10.1007/978-3-319-19857-6_53

ones are usually a few. Similarly, in quality control tasks, the rate of defective products is far lower than the rate of non defective ones.

The above introduced unfrequent patterns are hard to be detected by using standard classifiers which are designed to achieve *globally* optimal performances throughout the whole dataset. For this reason standard techniques, when coping with unbalanced datasets, pursue the optimal performance by classifying nearly all samples as belonging to the most frequent class [1]. The little number of samples corresponding to infrequent events prejudices their correct characterization and makes the separation of the classes difficult for the classifier. Unbalanced datasets are difficult to handle for classification purpose independently on the kind of the technique employed, as shown in [2].

This papers describes the most relevant techniques utilized for the efficient and reliable detection of rare patterns within real world classification tasks and introduces some advanced techniques developed *ad-hoc* for handling this problem. The paper is organised as follows: in Sect. 2 a review of most widely used *classical* methods for unbalanced datasets classification are depicted, subsequently, in Sect. 3 more advanced approaches are described in detail and their employ on real world problems is shown and their performances compared to other classical approaches in order to put into evidence their efficiency; in Sect. 4, some final remarks on the use of the described techniques and some future perspectives are outlined.

2 Classical Approaches

Class imbalance degrades the classification performance of standard methods by drastically limiting the rate of detected rare events. The factors reducing these performance indexes are manifold and interacting and may include: little number of rare samples, low separability, the presence of noise and outliers of patterns which compromise the characterization of rare patterns. All these factors are very often observed in real world datasets and specially in the industrial ones. This consideration, together to the practical relevance of this topic, led to the development and improvement of many techniques for unbalanced dataset classification handling.

Among literature approaches, two different families of methods, described more in detail in the subsequent sections, can be distinguished:

External methods use unmodified versions of existing algorithms but they operate on the distribution of the dataset samples used for the training, by re-sampling the data to reduce the impact of data unbalance.

Internal methods include all those approaches where new algorithms are expressly designed for facing the problems related to unbalance in datasets. This group includes also the standard algorithms which have been modified for this purpose. The main drawback of this approach is related to the portability since these methods are often *problem-specific*.

2.1 Review on External Methods

External methods operate on the training dataset by re-sampling it and varying the unbalance rate aiming at the balancing of the classes in order to avoid the classification problems encountered by standard methods. This method, independent on the problems and employed algorithms, has the main advantage of portability.
There are two possible approaches to the re-sampling process, both operating in the direction of the reduction of the unbalance rate:

- **under-sampling**, which operates by removing from the dataset samples belonging to the majority classes until the desired ratio between positive and negative samples is reached;
- **over-sampling**, which appends minority class samples to the dataset.

Resampling techniques normally do not completely balance the classes frequencies but increases the frequency of the less represented one. A fix unbalance ratio which optimizes the performance of any classifier for any problem does not exist, but it is rather dependent on the problem and original dataset characteristics and, in practice, this quantity must be determined experimentally.

Under-sampling techniques operate by removing frequent samples from the training dataset. The simplest way to select the samples to be removed is random but it risks to remove samples with high informative content; for this reason more sophisticated techniques perform a focused removal of frequent samples by selecting only those located in the extreme regions of the input space, in order to reduce the size of input space areas which the classifier would assign to frequent patterns and to create a more specific concept of the corresponding class [4]. Oversampling methods perform the resampling task by adding samples belonging to the unfrequent class to the training dataset. As for undersampling, this operation can be done by randomly replicating patterns that are present in the dataset or by replicating only those positive samples which lie on the boundary zones of the input space between positive and negative samples. This latter method is applied in order to spread the regions of the input space which the classifier will put into correspondence with the minority class and to limit eventual classification conflicts that standard classifiers would solve in favour of the majority class [1].

Among the oversampling approaches it is worth to note the SMOTE algorithm, proposed in [5]. The main element of novelty of this approach is that the new rare samples which are added to the dataset are not taken from the existing ones but they are created synthetically. In facts, one undesired effect of the replication of existing rare patters is the creation of smaller and more specific input space areas devoted to the minority class. Replication does not add any further information to the dataset but only overbalance the decision metrics of the classifier. On the other hand, the creation of new unfrequent samples broadens the zones of the input space associated to them. Synthetic data are created by SMOTE and placed in the input space where they *likely* could be; in particular, new minority samples are located along the lines connecting existing minority samples. Depending on the number of synthetic samples required to rebalance the dataset, different numbers of couples of rare patterns are selected for the generation of new samples.

Although undersampling is more used in practice, no one of these two techniques is preferable in general with respect to the other: their performances are related to the specific problems, classifiers and datasets. A combination of the two approaches has been attempted in [6] where focused undersampling and oversampling are put into practice: the results achieved by this method are comparable to those achieved by single approaches.

2.2 Review on Internal Methods

Internal methods do not perform any modification of the data used for learners training but they focus on the development of new algorithms expressly designed to handle uneven datasets and detect rare patterns within them on the basis of the extraction of salient information from the scarce amount of observations belonging to rare patterns available. In many cases these approaches are so oriented to the identification of rare patterns that, as a counter effect, they generate an high rate of false positive type misclassification: this kind of errors are often tolerated due to the characteristics of the practical applications.

The main reason for traditional methods do not perform well when coping with unbalanced datasets is that ignoring rare samples is as penalized as ignoring frequent ones, thus classifier tend to focus their prediction to most probable events.

The most basic internal approach exploits this intuition in order to award the detection of infrequent patterns by altering the misclassification cost matrix which determines the cost of any kind of possible classification error. For promoting the detection of rare patterns, an improvement of the missed detection of a rare sample cost can be put into practice. Costs can be adjusted so as to be inversely proportional to the associated class frequencies or on the basis of application driven criteria which take into account users requirements. The adoption of an altered cost matrix is a general concept which can be used within the training of most common types of learners such as neural networks [7] and decision trees [4].

The use of other techniques such as Support Vector Machines (SVM) for unbalanced dataset classification problems has been deeply investigated in [8]. From the results, this method seems to be less affected by the training dataset imbalance than the others and for this reason SVM have been often used as a starting point for the development of more complex approaches. The v-SVM method [9] belongs to this class of approaches: it is based on the use of a parametric SVM where the so-called v-parameter is suitably tuned in order to favour the detection of patterns belonging to a specific class. Due to its characteristic v-SVM can be employed to detect whether a pattern belongs to the rare ones by training it only with rare patterns; on the other hand the number of false positive could be quite high.

A more complex learner architecture based on SVM was presented in [8] where the SVM is coupled with SMOTE and a cost matrix penalizing misclassification of rare samples. This method forces the boundary hyper-planes separating positive and negative patterns to be regular and to give more importance to the rare patterns by shifting the boundary far from them. This combination of methods, according to the performed tests on popular datasets, leads to satisfactory results improving the results obtained by other methods.

Another popular approach which has been re-designed in order to efficiently cope with unbalance datasets is the Radial Basis Function (RBF) network. An RBF architecture based on the use of rectangular activation functions in the hidden layer was proposed in [10] and addressed as Rectangular Basis Function Network (RecBF). In this framework, a particular basis function is put into relation to each class in the training dataset: the use of rectangles allows more precision in the detection of the boundary the input space regions reserved to each class. Two regions are associated to each basis function: a core region which identifies the full membership degree of the zone to a certain class, and a support region which surrounds and contains the core region and corresponds to a lower membership to the class. During the RecBF training process several rectangular basis functions are created for each target class so that at the end of the training several functions will be devoted to the characterization of the minority class.

3 Advanced Methods

In this section some advanced approaches expressly developed by the authors for coping with the classification of unbalanced datasets are presented. These approaches, which have been designed to be general purpose, include both internal and external methods. The basic ideas underneath these methods are described in order to point out the characteristics of each method and the main results achieved by these methods on exemplar unbalanced datasets coming from the industrial fields are presented.

3.1 Thresholded Neural Networks

As mentioned in Sect. 1, the effect of class unbalance in the training datasets is the generation of classifiers biased toward the classification of frequent patterns. In [11] a method based on the use of a Multi-layer Perceptron Feed-Forward Neural Network (MLP-FFNN) combined to a threshold operator aiming to the mitigation of this effect is presented. This method, called Thresholded Artificial Neural Network (TANN), employs a two layers feed-forward neural network with one single output neuron: the output of the network is a continuous value in the range [0;1]: output values lower than a fixed threshold are associated to frequent class and the others to the rare class. The TANN aims to increase the sensitivity to rare patterns by means of the tuning of the mentioned threshold in order to maximize the performance of the classifier on the basis of several criteria involving the detection of the rare events and the overall classification accuracy. The training algorithm of the TANN is divided into two subsequent steps: the first one trains the neural network sub-system while the second one selects the optimal threshold value among a set of candidate thresholds (t) according to the merit function Eq. 1:

$$E(t) = \frac{\gamma Det(t) - FA(t)}{Corr(t) + \mu FA(t)} \tag{1}$$

where *Det* represents the rate of main events detected, *FA* the rate of false alarms, *Corr* the overall rate of correct classifications while γ and μ are two empirical parameters. The employed merit function Eq. 1 formalizes the requirements of the obtained learner in terms of general performance and detection of rare events.
The TANN method has been used for classification purpose within two industrial problems:

Occlusion Refers to the problem of the detection of casting nozzles during the steel manufacturing process. The occlusion of these nozzles is highly detrimental and its detection is a key issue of the problem. Within the employed datasets 3 % of the observations belong to occlusion situations.

Metal sheet quality (MSQ) Concerns the automatic grading of metal sheets quality on the basis of the information provided by a set of sensors which inspect sheets surface. The grading, according to the number and type and characteristics of the reported defects decides whether a product can be put into market or not. In this problem the minority class includes 10 % of the observations within the available dataset.

The results achieved by TANN on these problems are shown in tables Table 1 and Table 2 for the *Occlusion* and MSQ problems respectively. Results are presented in terms of overall correct classification, rare patterns detected and false alarms generated and compared to other approaches.

Table 1 Results obtained by the TANN method on the *Occlusion* problem

Method	Correct %	Detected %	False alarms %
Decision tree	98	17	2
FFNN	98	17	2
SVM	97	0	0
TANN	77	67	21

Table 2 Results obtained by the TANN method on the metal sheet quality problem

Method	Correct %	Detected %	False alarms %
Decision tree	83	68	7
FFNN	72	29	4
SVM	87	65	1.5
TANN	83	77	3

From the results reported in the tables above, the performance improvement related to the use of TANN with respect to the other methods stands out: an higher rate of rare patterns is detected and the overall classification performance are kept high. In the case of the *Occlusion* dataset the TANN produces an higher rate of false alarms with respect to other approaches but, within the context of the problem, it does not represent an issue provided an high rate of occlusions is properly detected.

3.2 The LASCUS Method

The LASCUS method [12] was designed for facing uneven dataset classification in presence of rare patterns whose correct identification is particularly important within the problem context (i.e. diseased within the medical field or machine malfunctions in manufacturing). In these cases it is very important to detect as much rare patterns as possible, without paying too much attention to the generation of false alarms which are not harmful in the application framework. The basic idea of LASCUS is to partition efficiently the input space and to suitably label the so-formed regions in order to promote the method goal. In particular, regions where the concentration of rare patterns is *high enough* will be assigned to the rare class in order to favour its classification. The LASCUS training process consists of a first phase which determines the input space partition and is performed by means of the use of a Self Organizing Maps (SOM); subsequently the rare and frequent patterns are assigned to the so-formed clusters according to the distance of the patterns to the centroids corresponding to each cluster and rare patterns concentrations are calculated for each cluster. The determination of the critical rare events concentration which determines the assignment of a cluster to the rare class is obtained by calculating the rates of false alarms, rare detected events and overall accuracy for the set of rare events concentrations reported by all the clusters: these features are fed to a fuzzy inference system which implements a human driven criterium for the LASCUS performance evaluation. The higher rated threshold is then selected.

LASCUS has been tested on a wide set of datasets both coming from the industrial framework and the UCI dataset repository. The complete achievements of this method can be consulted in [12]. In this work its performance on the *Occlusion* and *MSQ* datasets previously introduced are reported in Table 3. This table puts into evidence that LASCUS further achieves better results than the classical approaches and the TANN method.

Table 3 Results obtained by the LASCUS method on the *MSQ* and *Occlusion* datasets

Dataset	Correct %	Detected %	False alarms %
MSQ	84.4	85.8	4.1
Occlusion	80.2	76.9	19.5

3.3 The SUNDO Resampling Method

The SUNDO method [13] belongs to the family of external methods and combines oversampling and undersampling in order to create a suitable rebalanced training dataset from an original unbalanced dataset. SUNDO requires the user to specify the target unbalance rate. Given the required unbalance rate, the number of synthetic samples to be created through oversampling (n_{over}) and to be eliminated through undersampling (n_{under}) is determined. Oversampling creates $k \cdot n_{over}$ new *rare* patterns on the basis of their original distribution, subsequently, n_{over} samples are selected among the created ones according to their distance from frequent observations: the closest samples to frequent observations are eliminated in order to limit inter-class interferences during the training. Undersampling, on the other hand, eliminates n_{under} samples on the basis of the relative distances among couples of frequent observations: once these distances are calculated, the n_{under} closest couples are selected and one sample is eliminated from each of these couples. This procedure reduces the redundancy among frequent samples. The SUNDO methods has been tested on the *Occlusion* and MSQ datasets and its performance have been compared to those of SMOTE, the most widely used external methods which uses oversampling. The obtained datasets have been fed to SVM based classifiers. The results reported in Table 4 for different unbalance rates show that SUNDO outperforms SMOTE and that its performance are comparable to the other advanced internal approaches.

Table 4 Results obtained by the SUNDO method for different unbalance rate on the *MSQ* and *Occlusion* datasets

Occlusion				
	Unb. Rate %	Correct %	Detected %	False alarms %
SMOTE	30	91.9	55.9	7.7
	50	76.3	83.3	23.7
SUNDO	30	91.2	70	8.5
	50	79	83.3	21
MSQ				
	Unb. Rate %	Correct %	Detected %	False alarms %
SMOTE	30	90.5	66.8	2
	50	89	62.6	2.6
SUNDO	30	90.5	67	2
	50	89.9	66.6	2.8

3.4 Dynamic Resampling

An original external method was presented in [14] where the concept of *Dynamic resampling* was introduced. The basic idea behind this approach is to use a different resampled dataset for the distinct phases of a FFNN training process and to build the training dataset so that it contains at each phase all the rare patterns but only a varying subset of frequent patterns. The resampling rate is defined by the user. The training process is divided into *blocks of epochs* and for each of them a different training dataset is set up. At each block of epochs the dataset is formed by all the rare samples and the same amount of frequent samples which are probabilistically selected according to two criteria: probability is higher for previously less frequently selected samples; probability is proportional to the classification performance achieved by the FFNN during the training within the blocks of epochs involving each sample. This approach allows to train the FFNN by using balanced datasets and, in the meantime, not to loose the informative content related to any frequent pattern. Moreover those samples whose use is mostly advantageous are selected with an higher probability with respect to the others. The performance achieved by the Dynamic resampling method on the *Occlusion* and MSQ datasets are reported in Table 5 for different unbalance rates. The method obtains encouraging results as it is able to increase the rate of detected rare patterns maintaining acceptable the rate of false alarms.

Table 5 Results obtained by the Dynamic resampling method for different unbalance rate on the *MSQ* and *Occlusion* datasets

Occlusion				
	Unb. Rate %	Correct %	Detected %	False alarms %
Occlusion	50	74	72	26
	40	82	65	18
MSQ	50	88	71	3
	25	89	75	5

4 Conclusions

In this paper a review of most common and some advanced approaches for the improvement of classifiers performance when coping with uneven datasets was presented. The focus of the paper was on the detection of rare patterns in those cases where they represent particular situations whose detection is fundamental such as in the industrial field.

The presented advanced methods belong both to the internal and external methods. The performance achieved on the tested industrial problems and the results which can be found in literature, put into evidence the advantages which can be drawn from the use of specific techniques expressly developed for handling unbalanced datasets. Unfortunately, among the tested methods, no one outperforms the others, thus no optimal solution exists and users have to investigate on the basis of the handled problems.

These techniques can be combined according to specific problems requirements in order to maximize these advantages. The future perspectives within this field include in fact the fusion of the internal and external approaches and the creation of hybrid techniques.

References

1. Estabrooks, A.: A combination scheme for inductive learning from imbalanced datasets. MSC thesis. Faculty of computer science, Dalhouise university (2000)
2. Estabrooks, A., Japkowicz, N.: A multiple resampling method for learning from imbalanced dataset. Comput. Intell. **20**(1) (2004)
3. Japkowicz, N.: The class imbalance problem: significance and strategies. In: Proceedings of the 2000 International Conference on Artificial Intelligence (IC-AI'2000), Las Vegas, Nevada
4. Chawla, N.V.: C4.5 and imbalanced data sets: investigating the effect of sampling method, probabilistic estimate, and decision tree structure. In: Workshop on Learning from Imbalanced Dataset II, ICML, Washington DC (2003)
5. Chawla, N.V., Bowyer, K.W., Hall, L.O., Kegelmeyer, W.P.: SMOTE: synthetic minority over-sampling technique. J. Artif. Intell. Res. (16), 321–357 (2002)
6. Ling, C., Li, C.: Data mining for direct marketing problems and solutions. In: Proceedings of the Fourth International Conference on Knowledge Discovery and Data Mining, New York (1998)
7. De Rouin, E., Brown, J., Fausett, L., Schneider, M.: Neural network training on unequally represented classes. In: Intelligent Engineering Systems Through ANNs, pp. 135–141. ASME Press, New York (1991)
8. Akbani, R., Kwek, S., Japkowicz, N.: Applying support vector machines to imbalanced datasets. In: Proceedings of 15th European Conference on Machine Learning, Pisa, Italy, 20–24 September 2004
9. Scholkopf, B., et al.: New support vector algorithms. Neural Comput. **12**, 1207–1245 (2000)
10. Berthold, M.R., Huber, K.P.: From radial to rectangular basis functions: a new approach for rule learning from large datasets. Technical report, University of Karlsruhe (1995)
11. Vannucci, M., Colla, V., Sgarbi, M., Toscanelli, O.: Thresholded neural networks for sensitive industrial classification tasks. Lecture Notes in Computer Science, 5517 LNCS, pp. 1320–1327 (2009)
12. Vannucci, M., Colla, V.: Novel classification method for sensitive problems and uneven datasets based on neural networks and fuzzy logic. Appl. Soft Comput. **11**(2), 2383–2390 (2011)
13. Cateni, S., Colla, V., Vannucci, M.: A method for resampling imbalanced datasets in binary classification tasks for real-world problems. Neurocomputing **135**, 32–41 (2014)
14. Vannucci, M., Colla, V., Vannocci, M., Reyneri, L.M.: Dynamic resampling method for classification of sensitive problems and uneven datasets. Commun. Comput. Inf. Sci. **298** CCIS, 78–87 (2012)

Strategic Decision-Making from the Perspective of Fuzzy Two-Echelon Supply Chain Model

Junzo Watada and Xian Chen

Abstract Game theory is applied widely to solving various problems to get optimal decisions. This paper analyzes the Stackelberg behaviors between a manufacturer and two retailers. As the market demand used to be uncertain, in order to describe the market demand fitting to real situations, this paper applys fuzzy variable in the two-echelon supply chain problem. Then, a numerical example is used to illustrate the result in comparison between fuzzy variable and crisp number. Fuzzy variable and crisp number are compared with mean absolute percentage difference rate of fuzzy demand (MAPDR-FD).

Keywords Two-echelon supply chain · Fuzzy variable · Optimal decision · Stackelberg

1 Introduction

Game theory becomes more and more common in the research on two-echelon supply chain and plays a pivotal role in optimal decision making.

We used to face how much amount to assign each of branches or of sections as a target amount in their region. The most important is the whole company's selling volume but on the other hand we have to encourage each of branches or of sections to work hard and obtain the best result.

In this paper, let us make a problem simplified. We have the manufacturing section M and two retailing Sects. 1 and 2. These three sections build a supply chain.In this case, there is a competition between the two retailing Sects. 1 and 2.

J. Watada (✉) · X. Chen
The Graduate School of Information, Production and Systems, Waseda University, Kitakyushu, Fukuoka 808-0135, Japan
e-mail: junzow@osb.att.ne.jp

X. Chen
e-mail: chenxian@fuji.waseda.jp

R. Neves-Silva et al. (eds.), *Intelligent Decision Technologies*, Smart Innovation, Systems and Technologies 39,
DOI 10.1007/978-3-319-19857-6_54

In game situation, these two retailers 1 and 2 are Stackelberg situation. Therefore, when we give a leading position to 1, Sect. 2 will follows section A's initial decision. The objective here is to obtain their own best result in the Stackelberg situation.

Then, according to the best results of the two reailers got before, The manufacturer can know the decision made by retailers and decide the optimal assignment of price, amount, etc.

In order to solve this problem, let us consider the uncertain situation such as uncertain demand. This situation resulted in building a fuzzy two-echelon supply chain model.

In this paper, fuzzy variable is used in building model of two-echelon supply chain.

The remaining consist of the following sections. In Sect. 2, past research works will be shortly reviewed and Sect. 3 provides prereminary mathematical preparation for building a fuzzy two-echelon supply chain model. In Sect. 4 we define fuzzy two-echelon supply chain model and its solution based on Stackelberg game. Then, Sect. 5 discusses the numerical real-life problem of the stratgic decision-making of how to obtain the optimal assignment for each of the two retailer sections. At the end in Sect. 6 we summarize the paper as conclusions and explain the remaining problems.

2 Previous Research Results

In two-echelon supply chain, quantity, cost and price are key factors in order to get optimal decision. For example, Pui-Sze CHOW et al. [1] analyze the effect when taking a minimum order quantity (MOQ) in a two-echelon supply chain. Govindan KANNAN [2] uses multi-objective models in a two-echelon supply chain to optimize the total cost. Ilham SLIMANI and Said ACHCHAB [3] concerned with the optimization of the inventory and transportation in a supply chain. Y.A. HIDAYT et al. [4] analyze an inventory problem with only a single-supplier and a single-buyer. Cai Jian-hu and Wang Li-ping [5] analyze the influence of providing adequate pricing incentives before demand information is revealed. Geon Cho et al. [6] take a quantitative model when there is a partnership in supply chain. Daogang Qu and Ying Han [7] consider a single manufacturer and a single retailer model which has dual distribution. Longfei He and Jianyong Sun [8] consider a single vendor and a single retailer originated from China market background. CAI Jian-hu et al. [9] proposed an inventory model in a two-echelon supply chain. Yihong Hu and Jianghua Zhang [10] consider that when there is a price-only contract to a retail, how will be the manufacturer's reaction to a new-vendor. Fangxu Ren and ZhongYuan [11] analyze pricing and advertising model under equilibrium. Tiezhu Zhang and Longying Li [12] consider some contracts in order to get the win-win situation. Huilin Chen and Kejing Zhang [13] take the full-return coordination mechanism in consider. Fangxu Ren [14] considers the relation between the size of slotting allowance with the members' co-operation in a two-echelon supply chain.

Moreover, there are also many researches studying the demand function. Junping Wang and Shengdong Wang [15] study the effect to the chain members when the different forms of demand function are used. Jiang Meixian et al. [16] consider a model of coordinating pricing and express that the market demand is effected by retailer price and promotion expenses. Lau and Lau [17] analyze the difference of the optimal decision when facing the different demand curve in Stackelberg game situation.

3 Preparations

As the demand of the market is always uncertain. This paper takes the demand of the market as fuzzy variable so as to make the value more close to real situation. To illustrate the difference between the fuzzy variable and crisp number, Mean Absolute Percentage Different Rate of the Fuzzy Demand (MAPDR-FD) will be used.

3.1 Fuzzy Variable

Fuzzy Variable is defined as a function which has a possibility space $(\Theta, p(\Theta), Pos)$, where Θ denote a non-empty set, $p(\Theta)$ is a power set of Θ, and Pos is a possibility measure. The possibility measure satisfies the following conditions:

(1) $Pos\{\emptyset\} = 0$
(2) $Pos\{\Theta\} = 1$
(3) $Pos\{\bigcup_{i=1}^{m} A_i\} = \sup_{1 \leq i \leq m} Pos\{A_i\}$

By taking the triangular fuzzy variable $D = (a, b, c)$, the membership function is given as follows:

$$\mu(x) = \begin{cases} \dfrac{x-a}{b-a} & ; a \leq x \leq b \\ \dfrac{c-x}{c-b} & ; b \leq x \leq c \\ 0 & ; otherwise \end{cases} \tag{1}$$

Supposing A^c is the complement set of A, then the necessity measure of A is given as follows:

$$Nec(A) = 1 - Pos(A^c) \tag{2}$$

The credibility measure is denoted as

$$Cr\{A\} = \frac{1}{2}[1 + Pos(A) - Pos(A^c)]; \tag{3}$$

$$Cr\{A\} = \frac{1}{2}[Pos(A) + Nec(A)]; \tag{4}$$

Let D be a fuzzy variable. The expected value of D is give as follows:

$$E[D] = \int_0^{\infty} Cr\{D \geq r\}dr - \int_{-\infty}^{0} Cr\{D \leq r\}dr \tag{5}$$

When D is a triangular fuzzy number (a, b, c).

$$E[D] = \frac{a + 2b + c}{4} \tag{6}$$

3.2 Model Assumptions and Notations

This paper is to analyze the behavior of a supply chain with a manufacturer and two retailers.The assumptions and notations are given as follows:

Q_i: quantity ordered by retailer i, $(i = 1, 2)$;
c: unit cost of manufacturing;
w: the wholesale price per unit ;
p_i: the sale price by retailer i, $(i = 1, 2)$;
D_i: the demand of retailer i if prices are zero
Π_i: retailer i's profit $(i = 1, 2)$;
Π_M: manufacture's profit;
Π_i^*: retailer i's maximum profit $(i = 1, 2)$;
Π_M^*: manufacture's maximum profit

The downward-sloping demand function is defined as follows:

(McGuire and Staelin (1983) and Ingene and parry (1995) have used this type of demand curve with $a_i = 1$ successively). In this paper let $a_i = 1$. So the demand function is:

$$Q_i = (D_i - p_i + \theta p_j) \qquad (i, j = 1, 2) \tag{7}$$

where θ is the degree of substitutability between retailers.

4 Two-Echelon Supply Chain Model Based on Stackelberg game

In this Stackelberg game, we assume retailer 1 as a Stackelberg leader and retailer 2 as a Stackelberg follower. Then the profit function of retailer 1 and retailer 2 will be given as:

$$\begin{aligned} \Pi_{r1} &= (p_1 - w)Q_1 \\ &= (p_1 - w)(D_1 - p_1 + \theta p_2) \end{aligned} \tag{8}$$

$$\begin{aligned} \Pi_{r2} &= (p_2 - w)Q_2 \\ &= (p_2 - w)(D_2 - p_2 + \theta p_1) \end{aligned} \tag{9}$$

The reaction function of the retailer 2 is obtained by solving $(\frac{d\Pi_{r2}}{dp_2} = 0)$:

$$P_2 = \frac{(D_2 + w + \theta p_1)}{2} \tag{10}$$

Then retailer 1 knows the reaction function (4) of retailer 2 and substitute it into his profit function. Then the profit function of retailer 1 is as follows:

$$\Pi_{r1} = \frac{(p_1 - w)(2D_1 + \theta D_2 - 2p_1 + \theta^2 p_1 + \theta w)}{2} \tag{11}$$

To get the optimal price of retailer 1, setting $\frac{d\Pi_{r1}}{dp_1} = 0$, the function is obtained as follows:

$$P_1^* = \frac{(2D_1 + \theta D_2 - \theta^2 w + 2w + \theta w)}{[2(2 - \theta^2)]} \tag{12}$$

Then the maximum profit of retailer 1 is written as

$$\Pi_{r1}^* = \frac{(2D_1 + \theta D_2 + \theta w + \theta^2 w - 2w)^2}{[8(2 - \theta^2)]} \tag{13}$$

To get the optimal sale price and the maximum profit of retailer 2, the optimal sale price of retailer 1 can be substituted into reaction function of retailer 2. Then the function can be given as:

$$\begin{aligned} P_2^* = \ & \left(2\theta D_1 + 4D_2 - \theta^2 D_2 + 4w + 2\theta w \right. \\ & \left. -\theta^2 w - \theta^3 w\right) / [4(2 - \theta^2)] \end{aligned} \tag{14}$$

$$\begin{aligned}\Pi_{r2}^* &= \left(\theta^2 D_2 - 2\theta D_1 - 4D_2 + 4w - 2\theta w\right. \\ &\quad \left.-3\theta^2 w + \theta^3 w\right)^2 / [16(2-\theta^2)^2] \end{aligned} \quad (15)$$

The profit function of manufacturer can be expressed as:

$$\Pi_M = (w - c)(Q_1^* + Q_2^*) \quad (16)$$

$$\Pi_M = (w - c)[D_1 + D_2 - (p_1^* + p_2^*) + \theta(p_1^* + p_2^*)] \quad (17)$$

Then the optimal price can be obtained by solving . $\frac{d\Pi_M}{dw} = 0$,

$$w^* = \frac{c}{2} + \frac{B}{(2J)} \quad (18)$$

where

$$B = 2[\theta + 2(D_1 + D_2)] - \theta[-2D_1 + \theta^2 D_2 + \theta(2D_1 + D_2)] \quad (19)$$

$$J = (8 - 4\theta - 7\theta^2 + 2\theta^3 + \theta^4) \quad (20)$$

So the maximum profit of the manufacturer can be given as follows:

$$\Pi_M^* = \frac{(8 + 4\theta + 3\theta^2 - \theta^3)(c - c\theta - D_1 D_2)}{16(\theta - 1)(2 - \theta^2)} \quad (21)$$

5 Numerical Examples

The numerical examples let us compare the difference of results between the fuzzy and non-fuzzy data sets. Setting $\theta = 0.5$ represents the moderate degrees of competition between two retailers (Table. 1).

Table 1 Assumed parameters

c	θ
2	0.5

5.1 Mean Absolute Percentage Difference Rate of the Fuzzy Demand (MAPDR-FD)

To find the difference between fuzzy and non-fuzzy data sets, the mean absolute percentage difference rate of the fuzzy demand is introduced. The formula is expressed as follows:

$$MAPDR - FD = \frac{100\,\%}{n}\sum_{k=1}^{n}\left\|\frac{\pi_k^l - \pi_{Fk}^l}{\pi_k^l}\right\| \tag{22}$$

where π_k^l is non-fuzzy value and π_{Fk}^l is fuzzy value $l = i, j, M$ and $k = 1, 2, \ldots, n$.

5.2 Fuzzy Variables

Let X_{D_1} be a fuzzy variable of demand faced by retailer 1. Assume $D_1 = (9, 19, 32)$

Let X_{D_2} be a fuzzy variable of demand faced by retailer 2. Assume $D_2 = (7, 18, 28)$

The possibility distribution of the triangular fuzzy variable are written as:

$$\mu_{D_1}(x) = \begin{cases} \frac{x-9}{10} & ; 9 \le x \le 19 \\ \frac{32-x}{13} & ; 19 \le x \le 32 \\ 0 & ; otherwise \end{cases} \tag{23}$$

$$\mu_{D_2}(x) = \begin{cases} \frac{x-7}{11} & ; 7 \le x \le 18 \\ \frac{28-x}{10} & ; 18 \le x \le 28 \\ 0 & ; otherwise \end{cases} \tag{24}$$

According to the possibility, necessity and credibility measures, the expected value can be calculated by using the following equation.

$$E[\xi] = \frac{a + 2b + c}{4} \tag{25}$$

The expected value of a triangular fuzzy variable can be calculated as:

Table 2 Fuzzy demand of market faced by supplier 1 and 2

	Fuzzy Triangular Number	Expected Value
D_1	(9, 19, 22)	19.75
D_2	(7, 18, 28)	17.75

5.3 Comparing the Results Between Fuzzy and Non-Fuzzy Data Sets

From Tables 2 and 3, it is easy to find that by using fuzzy data set, the profit of manufacturer is higher than the one without fuzzy data set. The quantity ordered by retailers using fuzzy number is more than the one using crisp number. On the other hand, the saleprices setted by retailers 1 and 2 using fuzzy number are lower than these using crisp number. When the fuzzy demand is less than crisp demand, the profit of retailer with fuzzy variable is lower than that without using fuzzy data set. On the other hand, when the fuzzy demand is greater than crisp demand, the profit of retailer using fuzzy data set is higher than that without using fuzzy data set.

In addition, the MAPRD-FD shows the difference between the fuzzy data set and crisp number data set.

As a result, in this situation, when the retailer sections are competitive with each other, the retailer 1 and 2 can get the optimal price and amount based on the stackelberg game analysis. According to the result, the manufacturer can also decide the optimal assignment of target price and amount. As the market demand is uncertain, we emloyed fuzzy variables in this calculation. Then we got the result that the quantity ordered by each of retailers 1 and 2 should be 0.09. The optimal profit of retailers 1 ,2 and manufacturer will be 0.483, 0.393 and 2.653. Then the profit shown in Table 3 each section can get the price and amount.

Table 3 The difference between fuzzy number and no-fuzzy data sets

	Fuzzy Data Set ($\times 10^2$)	Non-Fuzzy Data Set ($\times 10^2$)	MAPRD-FD
Π^*_{r1}	0,483	0.481	+4.30 %
Π^*_{r2}	0.393	0.451	−13.2 %
Π^*_M	2.653	2.588	+2.53 %
P^*_1	0.200	0.248	−19.2 %
P^*_2	0.187	0.240	−22.1 %
Q^*_1	0.090	0.060	+45.6 %
Q^*_2	0.090	0.070	+42.2 %

5.4 Real Life Problem

The data of Coca-Cola firm is analyzed to explain the different profit by using fuzzy and crisp number data sets.

Let X_{D_1} be a fuzzy variable of demand D_1 faced by LanZhou market.

$$D_1 = (1.174 \times 10^4, 2.098 \times 10^4, 5.269 \times 10^4) \quad (26)$$

Let X_{D_2} be a fuzzy variable of demand D_2 faced by XiNing market.

$$D_2 = (0.584 \times 10^4, 1.574 \times 10^4, 2.214 \times 10^4) \quad (27)$$

Then the possibility distribution of the triangular fuzzy variable can be written as:

$$\mu_{D_1}(x) = \begin{cases} \dfrac{x - 1.173 \times 10^4}{0.924 \times 10^4} & ; 1.174 \times 10^4 \le x \le 2.098 \times 10^4 \\ \dfrac{5.268 \times 10^4 - x}{3.171 \times 10^4} & ; 2.098 \times 10^4 \le x \le 5.269 \times 10^4 \\ 0 & ; otherwise \end{cases} \quad (28)$$

$$\mu_{D_2}(x) = \begin{cases} \dfrac{x - 5844}{9897} & ; 5844 \le x \le 15741 \\ \dfrac{22140 - x}{6399} & ; 15741 \le x \le 22140 \\ 0 & ; otherwise \end{cases} \quad (29)$$

The same as above, according to the possibility, necessity and credibility, we can get the cxpect value. the results are shown in Tables 4 and 5.

Table 4 Fuzzy demand faced by LanZhou and XiNing market

	Fuzzy Triangular number ($\times 10^4$)	Expected Value value ($\times 10^4$)
D_1	(1.174,2.098,5.269)	2.660
D_2	(0.584,1.574,2.214)	1.487

From Table 5, we can get a conclusion that by using fuzzy number the profit of the manufacturer is higher than using crisp numbers. When the fuzzy demand is less than crisp demand, the profit of retailer using fuzzy variable is lower than use crisp

Table 5 The difference between the fuzzy number and no-fuzzy number

	Fuzzy Data Set ($\times 10^6$)	Non-Fuzzy Data Set ($\times 10^6$)	MAPRD-FD
Π^*_{r1}	97.0	60.0	+60.0 %
Π^*_{r2}	28.0	34.0	–16.1 %
Π^*_M	300	251	+19.7 %
P^*_1	0.290	0.320	–8.00 %
P^*_2	0.240	0.290	–18.1 %
Q^*_1	0.092	0.038	+142 %
Q^*_2	0.053	0.020	+162 %

numbers. On the other hand, when the fuzzy demand is greater than crisp demand, the profit of retailer using fuzzy variable is higher than use crisp numbers.

The saleprices setted by retailer1 and 2 using fuzzy data set are lower than these using crisp number. On the contrary, under the fuzzy demand environment, the quantities ordered by retailers1 and 2 are more.

From the stackerberg game analysis, we can get the optimal assignment of price and amount. As the demand is not uncertained, we employ fuzzy variable to describe it. It is easy to find that though saleprices setted by retailers using fuzzy data set are lower than these using crispe number, the quantities ordered by retailers under fuzzy demand enviroment are more than using crisp number. As the profit by using fuzzy data set is higher, it is more benifit for manufacturer to use fuzzy data set. We can also find that the the more the profit is, the bigger the difference between using the fuzzy data set and crisp number.

To simplify the real problem, this paper analyse the bahaviors in a duopolostic situation. The result showed in the table is not affected by other manufacutures. So this result is different from the real environment, but we can surely know that by using fuzzy data set , the profit of manufacuturer is more than using crisp number.

Moreover,according to the MAPRD-FD, the degree of difference by using the fuzzy number and no-fuzzy number can be calculted.

In conclution, we regard retailer 1 as the leader and retailer 2 as the follower, after doing the analysis based on stackelberg game, then we can get the optimal amount and price shown in Table 5. The quantity ordered by retailer 1 and 2 should be 9,214 and 2,027 respectivly. Also, the optimal decision of manufacture can be gotten. As the market demand is uncertain mentioned before, fuzzy variables will be employed in this real life problem. Then we can finally get the result that the profits of retailer 1 2 and manufacturer are 97, 28 and 300 million USD. In this paper, many factors are ignored. By simplifying the real situation, we discussed the effect of fuzzy demand and non-fuzzy demand under the simplified situation as portion 3 mentioned. In order to get their largest profits, the manufacturer and retailers will make decisions as above. So under the fuzzy demand conditions, they will get more profit. However, in the real situation, without any contract between the retailers, the retailers will be competitive with each other. So in order to expand the market demand, the retailers always reduce the price until the Nash Equilibrium is gotten. In this paper, we only take the first situation into consideration. And as a result, we can know the difference between the fuzzy demand and non-fuzzy demand.

6 Conclusions and Future Work

In this paper, the competitive between a manufacturer and two retailers are shown. The stackbelberg game always be used to solve this problem in order to make a best decision. By using the fuzzy variable, the profit of the manufacture will be higher than the one which use crisp number.

This paper analyse a manufacturer and two retailers. As the future work, to analyse the market more accurately, there are more retailers need to be considered. Facing the market demand using fuzzy variable, the best decision of the manufacture and retailers should be discussed.

References

1. Chow, P.-S., Choi, T.-M., Cheng, T.C.E.: Impacts of minimum order quantity on a quick response supply chain. IEEE Trans. Syst. Man Cybern. Part A: Syst. Hum. **42**(4), 868–879 (2012)
2. Kannan, G.: A multi objective model for two-echelon production distribution supply chain. In: International Conference on Computers & Industrial Engineering, CIE 2009, pp. 624–627, 6–9 July 2009
3. Slimani, I., Achchab, C.S.: Game theory to control logistic costs in a two-echelon supply chain. In: 2014 International Conference on Logistics and Operations Management (GOL), pp. 168–170, 5–7 June 2014
4. Hidayat, Y.A., Suprayogi, S., Liputra, D.T., Islam, S.N.: Two-echelon supply chain inventory model with shortage, optimal reorder point, and controllable lead time. In: 2012 IEEE International Conference on Management of Innovation and Technology (ICMIT), pp. 163–167, 11–13 June 2012
5. Cai J., Wang, L.: Study on dynamic game models in a two-echelon supply chain. In: 2009 International Conference on Computational Intelligence and Security, pp. 138–141, 11–14 Dec 2009
6. Cho, G., So, S.-H., Kim, J.J., Park, Y.S., Park, H.H., Jung, K.H., Noh, C.-S.: An Optimal decision making model for supplier and buyer's win-win strategy in a two echelon supply chain. In: Proceedings of the 41st Annual Hawaii International Conference on System Sciences, p. 69, 7–10 Jan 2008
7. Qu, D., Han, Y.: Optimal Pricing for dual-channel supply chain with fairness concerned manufacturer. In: 2013 25th Chinese Control and Decision Conference (CCDC), pp. 1723–1727, 25–27 May 2013
8. He, L., Zhao, D., Sun, J.: Channel coordination on supply chain with unbalanced bargaining power structure by exogenous-force impact: China market background. In: 2008 IEEE International Conference on Service Operations and Logistics, and Informatics, IEEE/SOLI 2008, pp. 2018–2023, 12–15 Oct 2008

9. Cai, J., Wang, L., Shao, Z.: Study on an advance order tactic in a two-echelon supply chain. In: 2009 International Conference on Multimedia Information Networking and Security, pp. 613–617, 18–20 Nov 2009
10. Hu, Y., Zhang, J., Xu, Z.: Asymmetric demand information's impact on supply chain performance and relationship under price-only contract. In: 2007 IEEE International Conference on Automation and Logistics, pp. 2891–2896, 18–21 Aug 2007
11. Ren, F.: Dynamic Pricing decision in the two-echelon supply chain with manufacturer's advertising and dominant retailer. In: 2011 Chinese Control and Decision Conference (CCDC), pp. 391–395, 23–25 May 2011
12. Zhang, T., Li, L.: Study on a combined Supply Chain contracts of target rebate and quantity discount. In: 2009 Second International Conference on Intelligent Computation Technology and Automation, pp. 988–992, 10–11 Oct 2009
13. Chen, H., Zhang, K.: Stackelberg Game in a two-echelon supply chain under buy-back coordination contract. In: 2008 IEEE International Conference on Service Operations and Logistics, and Informatics, IEEE/SOLI 2008, pp. 2191–2196, 12–15 Oct 2008
14. Ren, F.: Research on slotting allowance decision in the two-echelon supply chain with retailer-led. In: 2010 International Conference on Logistics Systems and Intelligent Management, pp. 25–29, 9–10 Jan 2010
15. Wang, J., Wang, S.: Optimal manufacturing, ordering, pricing and advertising decisions in a two-echelon supply chain system. In: Proceedings of the Sixth International Conference on Management Science and Engineering Management Lecture Notes in Electrical Engineering, vol. 185, pp. 589–597 (2013)
16. Jiang, M., Yan, L., Jin, S., Feng, D., Coordinating price model for retailer dominant in a two-echelon supply chain. In: 2009 16th International Conference on Industrial Engineering and Engineering Management. IE&EM '09. pp. 1474–1477, 21–23 Oct 2009
17. Thisana, W., Rosarin, D., Watada, J.: A game approach to competitive decision making in logistics, ISME (2012)
18. Yang, S.-L., Zhou, Y.-W.: Two-echelon supply chain models: considering duopolistic retailers' different competitive behaviors. Int. J. Prod. Econ. 104–116 (2006)
19. Global Task Force 'Game Theory' Comprehensive Law Publishing (2003)
20. Lau, A., Lau, H.: Some two-echelon supply-chain games: improving from deterministic symmetric-information to stochastic asymmetric formation models. Eur. J. Oper. Res. **161**(1), 203–223 (2005)
21. Lau, A., Lau, H.S.: Effects of a deman-curve's shape on the optimal solutions of a multi-echelon inventory/pricing model. Eur. J. Oper. Res. **147**, 530–548 (2003)
22. Lau, H., Zhao, A.: Optimal ordering policies with two suppliers when lead times and demands are all stochastic. Eur. J. Oper. Res. **68**(05), 120–153 (1993)
23. Ding, H.H.: Research of supplies chain management in enterprise cooperation based on game theory. J. Shijiazhuang Railway Inst. (Social Sciences) **3**(3), 20–24 (2009)
24. Gao, Y.J., Wu, Ch.G., En. Wang, Y.: Enterprises cooperation game analysis on supply chain management based on strategy. J. Hefei Univ. Technol. (Social Sciences), **19**(04), 40–43 (2005)
25. Li, Y., Wang, L.: A Study on manufacturer and suppliers for product innovation coordinating to avoid the risk based on game theory. In: 2010 IEEE 17th International Conference on Industrial Engineering and Engineering Management (IE&EM)
26. Gumrukcu, S., Rossetti, M.D., Buyurgan, N.: Quantifying the costs of cycle counting in a two-echelon supply chain with multiple items. Int. J. Prod. Econ. **116**, 263–274 (2008)
27. SUa, J.: The Stability analysis of manufacturer -stackelberg process in two-echelon supply-chain. Procedia Eng. **24** 682–688 (2011)

Dynamic-Programming–Based Method for Fixation-to-Word Mapping

Akito Yamaya, Goran Topić, Pascual Martínez-Gómez and Akiko Aizawa

Abstract Eye movements made when reading text are considered to be important clues for estimating both understanding and interest. To analyze gaze data captured by the eye tracker with respect to a text, we need a noise-robust mapping between a fixation point and a word in the text. In this paper, we propose a dynamic-programming–based method for effective fixation-to-word mappings that can reduce the vertical displacement in gaze location. The golden dataset is created using FixFix, our web-based manual annotation tool. We first divide the gaze data into a number of sequential reading segments, then attempt to find the best segment-to-line alignment. To determine the best alignment, we select candidates for each segment, and calculate the cost based on the length characteristics of both the segment and document lines. We compare our method with the naïve mapping method, and show that it is capable of producing more accurate fixation-to-word mappings.

Keywords Eye tracking research · Error correction · Fixation-to-word mapping · Segment-to-line alignment · Dynamic programming · FixFix

---

This work was supported by JSPS KAKENHI Grant Number 24300062.

A. Yamaya (✉) · A. Aizawa
The University of Tokyo, 7-3-1 Hongo, Bunkyo-ku, Tokyo 113-8654, Japan
e-mail: yamaya@nii.ac.jp

A. Aizawa
e-mail: aizawa@nii.ac.jp

G. Topić · P. Martínez-Gómez · A. Aizawa
National Institute of Informatics, 2-1-2 Hitotsubashi Chiyoda-ku, Tokyo 101-8430, Japan
e-mail: goran_topic@nii.ac.jp

P. Martínez-Gómez
e-mail: pascual@nii.ac.jp

R. Neves-Silva et al. (eds.), *Intelligent Decision Technologies*,
Smart Innovation, Systems and Technologies 39,
DOI 10.1007/978-3-319-19857-6_55

1 Introduction

Understanding the cognitive process of reading is important for a variety of applications (e.g., document layout optimization). To analyze this cognitive process, we need an accurate mapping of each fixation point to the corresponding word in text. However, accurate fixation-to-word mapping is difficult, because gaze data captured from eye trackers often contain noise. This may be caused by inaccurate calibration, head movement, and other sources. Thus, the measured gaze location often does not correspond to the spot that the reader is actually looking at, and analysis of such data may produce misleading conclusions. When studying reading gaze data, the vertical displacement of the gaze location contributes more to the noise than the horizontal displacement. Horizontal displacement is relatively tolerable because the mapping will still usually lead to the same (or at least neighboring) word. However, vertical displacement of just one line is equivalent to a jump of perhaps 10 words, which might completely invalidate experimental data. Therefore, in this study, we focus on vertical drift adjustment. Though several means of addressing fixation-to-word mismatches have been reported [3, 4], they are mostly designed for a particular task, such as translation, or are focus on a specific conditions, such as measured gaze data at the beginning of a new line often being correct. If we measure the gaze data during the sequential reading process, independent of a specific task or condition, different implementation is required to find the best fixation-to-word mapping.

In this paper, we propose a dynamic-programming–based method for effective fixation-to-word mapping that can reduce the vertical displacement in gaze location. The golden dataset used in this paper is created using FixFix, our web-based manual annotation tool, which provides an easy-to-use interface for adjusting the fixation positions. We first divide the gaze data into a number of sequential reading segments based on the regular pattern of the horizontal saccade length. We then attempt to find the best segment-to-line alignment using the line length. To determine the best alignment path from the start to the end of the gaze data, we select candidates for each segment, and calculate the cost based on both segment and document line length characteristics. We compare our method with the naïve mapping method, and show that our method can estimate the fixation-to-word mapping with a higher degree of accuracy.

2 Related Work

The stream of gaze data points captured using an eye-tracker is influenced by two main factors: precision and accuracy of the capturing device. In case of low precision, there are variable errors – random fluctuations in the captured gaze location, that can often be compensated for by clustering gaze points into their gravity center, forming detected fixation. Low accuracy is reflected in systematic errors – consistent deviations in a single direction at a certain screen location and specific period of time; these errors are more difficult to compensate. There have been several recent attempts to address this problem.

In [1], it was noted that some tasks require subjects to fixate on certain locations to accomplish the task objective. The authors suggested measuring the disparity at these required fixation locations, and so detecting when the calibration of an eye tracker has degraded beyond a certain threshold, or even compensating for systematic errors in a post-processing stage. While these required fixation locations (RFLs) could be found in some psycholinguistic studies, they are not likely to be found in experiments involving natural reading tasks.

When recording measurements, it is relatively common to see a horizontal progression of the reading path (ascending x-coordinates of gaze data points) and slight systematic vertical deviations (ascending or descending y-coordinates). Naïve heuristics would then map the first few fixations to a certain line, and the rest of the fixations to lines above or below. However, this is not likely to work in a real-world reading situation. In [2], the authors used this assumption to correct fixation-to-word mappings. Their method first recognizes the most likely beginning of a reading line, and then re-maps fixations in the same horizontal progression on the same reading line.

In the above strategy, determining the most likely initial point of a reading line is critical, but the first fixation after a return sweep may contain inaccuracies in gaze landing locations. To increase the robustness of heuristics that recognizes the beginning of reading lines, the authors of [3] proposed using the first N fixations of a horizontal progression. Although the results displayed improvements in fixation re-mapping, it does not overcome accumulated systematic errors.

In [4], the authors further generalized fixation-to-word mapping to a cost minimization problem. For each captured fixation, the method considers mappings to the closest word and the closest words on the lines above and below. However, the mapping decision is not made at that stage, and the possible mappings are kept as alternative hypotheses. Every mapping decision has an associated cost function, which is calculated based on the distance between the fixation and the mapped word, and the distance between the current and previous fixations. Such a cost function introduces the intuition that the next fixated word should be the most likely to follow the previous fixated word, if the text is to be read naturally. Thus, all possible fixation-to-word mappings are encoded in a lattice of hypotheses, and the optimal solution is obtained using dynamic programming.

The methods described above may not perform well in the presence of long-range regressions or skimming events. For this reason, the authors in [5] proposed a feature-based image-registration method that spatially transforms the image representation of fixations to match the image representation of the text. This method only considers vertical transformations, and uses several optimization strategies to search for the optimal scaling and translation parameters. However, for reasons of computational complexity, only spatial transformations that remain constant across the screen can be considered, and this method cannot correct systematic errors that are different in different locations.

3 Dataset Construction

3.1 FixFix: Manual Annotation Tool

FixFix is our easy-to-use web-based manual annotation tool that will be publicly available.[1] Figure 1 illustrates the appearance of FixFix. Uploaded gaze data consists of black points to represent fixations and black arrows to represent the saccade path. Return sweeps, eye movements from the end of a specific line to the start of the next line, are shown by narrow green lines. We can select any fixation and drag-and-drop it to the intended position. Two modes are implemented for moving fixations, *single mode* and *non-single mode*. In *single mode*, we can move fixations one-by-one. This mode is effective when there is a specific spike in fixations because of systematic errors in the eye tracker. In *non-single mode*, we can move multiple fixations at the same time. For example, if a fixation between return sweeps is moved, the remaining fixations are dynamically adjusted along with it. This mode is effective when whole sets of fixations are shifted by head movements. FixFix has a function to focus on a range of fixations by selecting the two end points. The range's extent can be moved backwards and forwards in time using arrow keys. After the manual annotation, the corrected gaze data are stored on the server, and can be downloaded as a TSV file (format information and more details will be described on our repository).

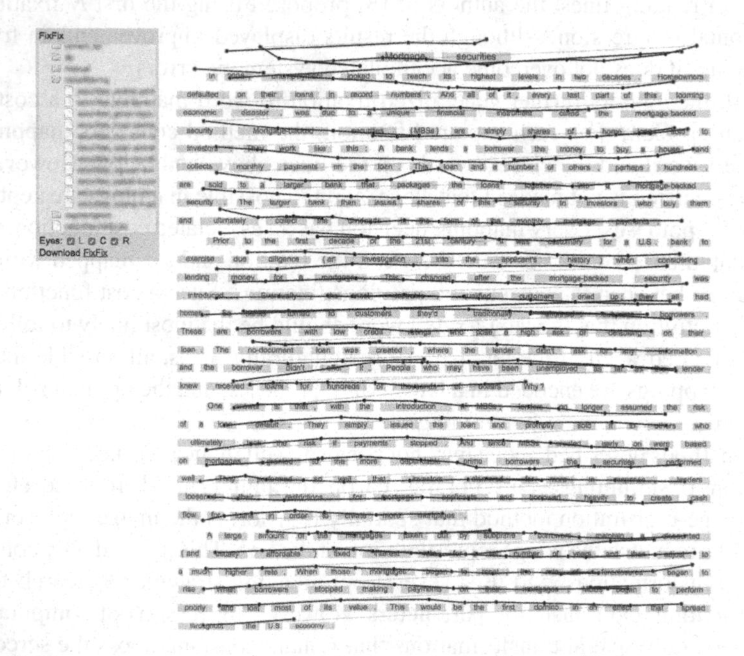

Fig. 1 FixFix: our easy-to-use web-based manual annotation tool

[1]https://github.com/KMCS-NII/fixfix/.

3.2 Golden Dataset

We use a part of the dataset in [5] for our experiments. For data collection, Tobii TX300 was used to capture gaze samples at a rate of 300 Hz. We used Dispersion-Threshold Identification (I-DT) algorithm [6] for detecting fixations from the raw input. The dispersion threshold was set to 53 px, and the duration threshold to 60 μs. Each reading session started immediately after the text was displayed on the screen, and finished when the subject signaled the operator with a hand gesture. Eleven sets of gaze data were obtained from eleven participants reading the text shown in Fig. 1. According to [5], participants read the text naturally, and did not use any equipment to fix their head position, so this dataset is very noisy.

Using FixFix, we manually annotate the gaze data obtained from the eye tracker and create the golden dataset for our experiments. To guarantee the quality of manual annotation, two annotators manually re-map each fixation to its most likely position. Using four of the gaze data file, we confirmed that 84.4 % of fixation-to-word mappings agreed within two annotators: evaluation measure of its agreement is percentage of exactly matching of two annotators' fixation-to-word mappings in the total number of fixations. For this reason, we assessed our annotation results as feasible.

4 Proposed Method

Our method for fixation-to-text alignment consists of two steps: *reading path segmentation* and *segment-to-line alignment*. In the reading path segmentation, we first convert captured gaze data from the eye tracker into a series of valid fixation points. The fixations are then segmented into a set of sequential reading segments. In the segment-to-line alignment, we search for a minimum cost alignment between the reading segments and text lines using dynamic programming. The cost of the alignment considers both the horizontal segment length and vertical movement gaps as penalties.

4.1 Reading Path Segmentation

First, we exclude fixations whose duration is less than 10 ms or more than 600 ms, though commonly used cutoff points of fixation duration is ranging from 70–100 ms according to [7]. The main reason for filtering out the fixations shorter than 10 ms (and not the customary 50–70 ms) was to allow for the potentially shorter fixations to be used as a return sweep detection feature. On the other hand, we pruned out the overly long fixations because they no longer represent normal reading behavior. Also, our method aims to detect return sweeps from a set of fixations with short duration time in the beginning and end of a line. We also eliminate bursty vertical spikes using the saccade distance. Fixations are regarded as spikes, and removed, if the vertical inward- and backward-saccade distances are longer than the space between lines in the text.

Next, we divide the fixation sequence into a set of sequential reading segments corresponding to reading lines. Although the classification of the reading and skimming scanpath has been studied in the past [8], these methods are not directly applicable here, because our objective is to identify a scanpath that belongs to a single line of text, with backward saccades allowed if they occur on the same line. Instead, we apply a simple return sweep detection method based on the horizontal backward saccade length [9]. Similarly, we can detect the regression point based on the vertical backward saccade length. The threshold value is determined based on the periodical spikes observed during reading.

4.2 Segment-to-Line Alignment

Let $S = \{s_1, s_2, \ldots, s_N\}$ be a sequence of reading segments obtained from the gaze data, and $T = \{t_1, t_2, \ldots, t_M\}$ be the lines of text displayed on the screen. Suppose the mapping function σ assigns a text line to the reading segment: i.e., $\forall s_i \in S$, $\exists t_j$ such that $\sigma(s_i) = t_j \in T$. Then, our segment-to-line alignment problem can be formulated as the problem of finding the optimal σ between S and T.

We selected the Needleman–Wunsch algorithm [10], which is commonly used in bioinformatics to align protein or nucleotide sequences. The optimal alignment is found by completing a scoring matrix. The initial matrix is created with N+1 columns and M+1 rows. The extra row and column are needed to align with any gap at the start of the matrix. Cell (0, 0) is assigned a value of zero. The first row and column are initially filled with the gap score. This gap score is then added to the previous cell of the row or column. After creating the initial matrix, the scoring function is introduced for cells (1, 1) to (N, M). A detailed definition of the scoring function is given in Sect. 4.3. To find the maximum score of each cell (i, j), we require the three neighboring scores from cells (i-1, j-1), (i, j-1), and (i-1, j). From the assumed values, we add the match/mismatch score to the diagonal value (i-1, j-1), and add the insertion or deletion gap score to the left-hand cell (i, j-1) and top cell (i-1, j). The maximum score is selected from these three values. When a score is calculated from the top or from the left, this represents an insertion or deletion in our alignment. When we calculate scores from the diagonal cell, this represents the alignment of either a match or mismatch sequence. The optimal alignment is obtained by following the immediate predecessors from the last cell (N, M) to the original cell (0, 0). The left part of Fig. 2 shows an example of the scoring matrix with match, mismatch, and gap scores of 5, -3, -4, respectively.

	-	t_1	t_2	t_3	t_4	t_5	t_6
-	0	-4	-8	-12	-16	-20	-24
s_1	-4	-3	-7	-3	-7	11	-15
s_2	-8	1	-3	-7	-6	-2	-6
s_3	-12	-3	6	2	-2	-6	-5
s_4	-16	-7	2	3	-1	3	-1
s_5	-20	-11	2	-1	0	-1	8

D(i-1, j-1)	D(i-1, j)
D(i, j-1)	D(i, j)

Fig. 2 An example of the scoring matrix

4.3 Scoring Functions

The score of cell (i, j), denoted as $D(i,j)$, is given by the following formula (see also the right-hand part of Fig. 2):

$$D(i,j) = max \begin{cases} D(i-1,j-1) + Sim(s_i,t_j) \\ D(i,j-1) + inPen \\ D(i-1,j) + delPen \end{cases} \quad (1)$$

where $Sim(s_n,t_m)$ represents a similarity score to measure the degree of match between the reading segment and the text line, and *inPen*, *delPen* are gap penalties for insertion and deletion, respectively.

Given a reading segment s_i, the set of neighboring text lines, denoted as $Neighbor(s_i)$, is defined based on the average Y-coordinate of that segment. In our case, we consider three neighbors: (1) the nearest line of s_i, (2) one line above the nearest line, and (3) one line below the nearest line. Figure 3 shows a sequential reading segment (the upper part of Fig. 3) and its neighboring text lines (the bottom part of Fig. 3).

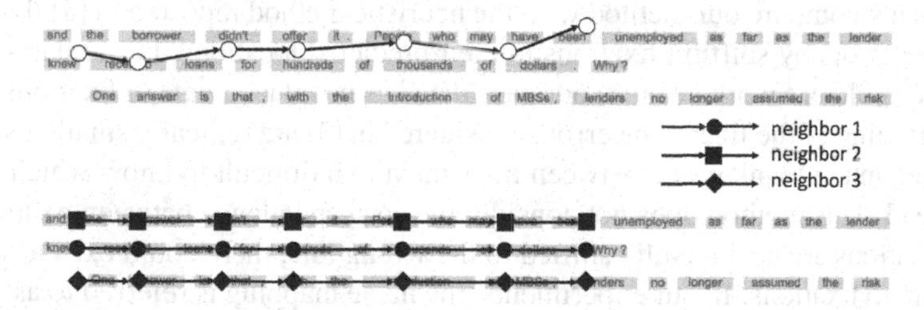

Fig. 3 Sequential reading segment and its neighboring text lines

In normal sequential reading, we assume that the segment length should be roughly equal to the corresponding line length. Thus, we set the following similarity score:

$$Sim(s_i, t_j) = \begin{cases} \dfrac{1 - Dist(s_i, t_j)}{Len(t_j)} & Dist(s_i, t_j) < \alpha \wedge t_j \in Neighbor(s_i) \\ -\dfrac{1 - Dist(s_j, t_j)}{Len(t_j)} & \text{otherwise} \end{cases} . \quad (2)$$

$$Dist(s_i, t_j) = abs(SegLen(s_i) - LineLen(t_j))/LineLen(t_j) \quad (3)$$
$$Len(t_j) = LineLen(t_j)/MedLen \quad (4)$$

where $SegLen(s_i)$ and $LineLen(t_j)$ represent the length of a segment and text line, respectively. *MedLen* is the median of the entire text line. α is a parameter for determining whether the segment-to-line alignment is a match or mismatch. In our experiments, we used $\alpha = 0.2$. $Len(t_j)$ is used to place extra weight on shorter lines at the end of a paragraph. This is motivated by our observation that such lines often serve as a strong clue for the alignment.

We set *inPen* and *delPen* to zero and β, respectively, because we consider segment insertion to be less important, whereas line deletion is harmful because skipping lines is unnatural during sequential reading. We empirically decide to use $\beta = 100$ in our experiments.

5 Experiments

5.1 Experimental Settings

We compare our output with the naïve mapping. In the naïve mapping, no fixations are moved, and each fixation corresponds to the vertically closest word in the text. We does not compare our method with the heuristic method reported in [3] that aims to reduce errors by shifting fixations to their predicted locations. In our dataset, the heuristic could not work because our gaze data has misplacements of fixations in the first beginning of the lines. The errors considered in [3] are typically smaller, such as a fixation on the blank space between lines making it difficult to know which line is being read. This method does not consider the correspondence between fixations. If these fixations are accidentally shifted to the wrong line, there could exist a solitary line with no fixations. In our experiments, the naïve mapping is referred to as *Naïve*.

To assess the accuracy of the fixation-to-word mapping, we calculated the average normalized distance between the offset position of the mapped word in the reference file (wr_i) and the test file (wt_i). We refer to this evaluation index as the *Average Word Distance AWD*. This is the enhanced version of the *Average Character Distance* reported in [4]. The value of N given by the offset distances was normalized by the fixation duration *dur*. The *AWD* reflects the distance between the fixation-to-word mappings of two files:

$$AWD = \frac{1}{N} \sum_{i=0}^{n} dur_i * |wr_i - wt_t|. \quad (5)$$

5.2 Results and Discussion

Table 1 shows the average word distance for individual participants using *Naïve* and the proposed method (*Proposed*). As we can see, for most subjects, our method results in a large decrease in average word distance (lower error) compared to the baseline method. This is because our method successfully tracks the spatial relationship between reading segments, and is able to leverage information from different reading and document line lengths to find the best alignment. In Fig. 4, we have plotted the distribution of correctly and incorrectly mapped fixations. Blue circles represent correct mappings, and red crosses represent incorrect mappings. We can see that most fixations are correctly mapped to the appropriate line, except for subjects D and J.

Figure 5 shows an example of the gaze data before the application of our method. This example shows three sequential reading segments. The second segment in Fig. 5 is comparatively short with respect to the other segments, but its nearest document line is relatively long. In Fig. 6, we show the fixations after being re-mapped using the proposed method. Our method shifts the segment one line down, achieving a more plausible alignment.

From further qualitative observations in other reading sessions, we observed that our method incorrectly mapped some portions of gaze data. This tends to occur where there is a larger density of regressions than in the rest of the gaze data. As regressions are a natural part of reading behavior, it is important to handle these regressions appropriately in future work. Our method has been applied to not only the text shown in Fig. 1, but also other types of text with different spatial information, no visual information are presented here because of space limitations. We observed a degree of robustness to different types of text, although incorrect mappings were apparent when there was a larger density of regressions.

Table 1 AWD for individual participants

Participant	Naïve	Proposed	Participant	Naïve	Proposed
A	0.839	0.022	G	1.139	0.009
B	1.662	0.407	H	2.756	0.342
C	0.724	0.404	I	0.421	0.026
D	0.394	2.904	J	1.174	2.300
E	1.981	0.838	K	1.233	0.422
F	1.179	0.373	average	1.227	0.732

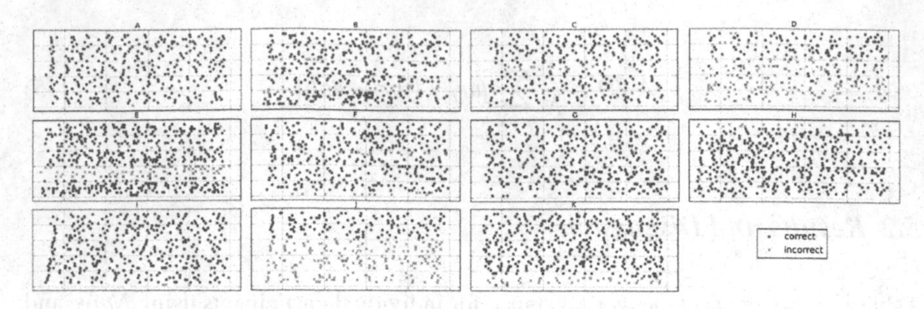

Fig. 4 Correct–Error distribution

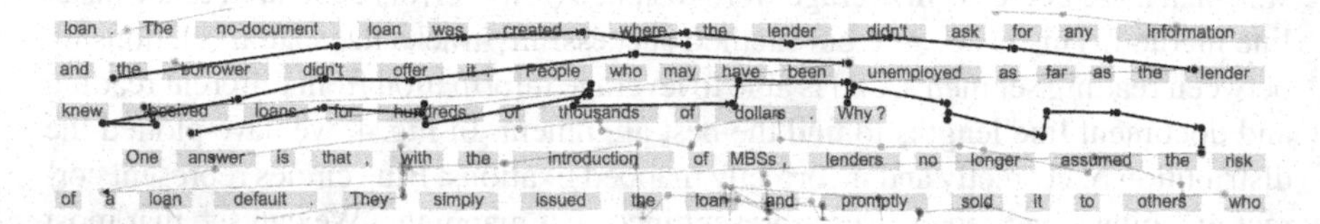

Fig. 5 An example of gaze data before applying the proposed method

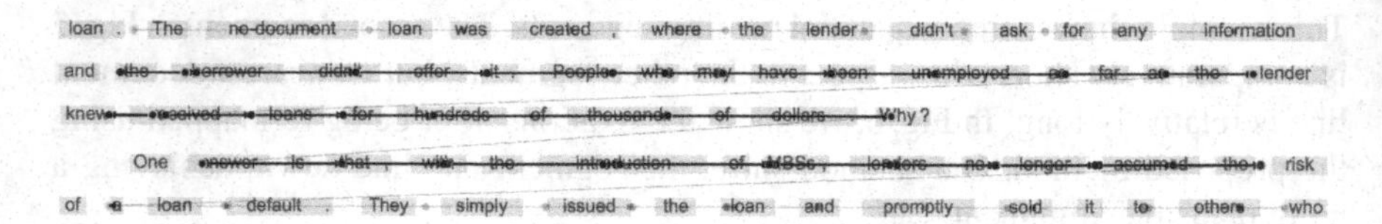

Fig. 6 An example of gaze data after applying the proposed method

6 Conclusion

In this paper, we have proposed an effective dynamic programming method to provide fixation-to-word mappings. Our method uses a cost function based on the length of reading segments and document text lines to determine whether a reading line matches a document line. We use dynamic programming to determine the best global alignment across the entire stream of gaze data samples. We have compared our method with the naïve mapping, and demonstrated that our approach gives more accurate fixation-to-word mappings. The specific characteristics of our solution mean that the proposed method may be more effective if the text consists of paragraphs whose last line is significantly shorter than the average line length in the same paragraph. In future work, we plan to refine our method to make it more robust when users read specific lines more than once. Moreover, we will consider horizontal systematic errors, as well as vertical errors, to produce better overall fixation-to-word mappings.

References

1. Hornof, A., Halverson, T.: Cleaning up systematic error in eye-tracking data by using required fixation locations. Behav. Res. Methods **34**, 592–604 (2002)
2. Hyrskykari, A.: Utilizing eye movements: overcoming inaccuracy while tracking the focus of attention during reading. Comput. Hum. Behav. **22**, 657–671 (2005)
3. Abhijit, M., Michael, C., Pushpin, B.: A heuristic-based approach for systematic error correction of gaze data for reading. In: 24th International Conference on Computational Linguistics, pp. 71–80 (2012)
4. Michael, C.: Dynamic programming for re-mapping noisy fixations in translation tasks. J. Eye Mov. Res. **6**(2), 1–11 (2013)
5. Pascual, M.G., Akiko, A.: Recognition of understanding level and language skill using measurements of reading behavior. In: 2014 International Conference on Intelligent User Interfaces (IUI-2014), pp. 95–104 (2014)
6. Salvucci, D.D., Joseph, H.G.: Identifying fixations and saccades in eye-tracking protocols. In: Proceedings of the 2000 Symposium on Eye Tracking Research & Applications. ACM (2000)
7. Holmqvist, K., et al.: Eye tracking: a comprehensive guide to methods and measures. Oxford University Press (2011)
8. Ralf, B., Jörn, H., Andreas, D., Georg, B.: A robust realtime reading-skimming classifier. In: The Symposium on Eye Tracking Research and Applications (ETRA'12), pp. 123–130 (2012)
9. Ryohei, T., Tadayoshi, H., Akiko, A.: Text width optimization based on gaze information. In: The 27th Annual Conference of the Japanese Society for Artificial Intelligence (2013) (in Japanese)
10. Needleman, S.B., Christian, D.W.: A general method applicable to the search for similarities in the amino acid sequence of two proteins. J. Mol. Biol. 443–453 (1970)

A Bilevel Synthetic Winery System for Balancing Profits and Perishable Production Quality

Haiyu Yu, Junzo Watada and Jingru Li

Abstract Recent research works show an increasing trend in analyzing agriculture problem because of its good applicability and usability. Also, the rapid development of mathematical model mitigates the difficulty to solve such abstract and uncertain problem. In this paper, the questions are discussed about a wine system, which contains some grape blocks and winery together. In the paper, a bilevel model is developed for obtaining the balance between profits and production quality.

Keywords Bilevel programming · Harvest route · Profit · Quality

1 Introduction

Recent research results show an increasing trend in analyzing agriculture problem because of its good applicability and usability. Also, the rapid development of mathematical model lightens the hardness to tackle such abstract and uncertain problem. In this paper, wine related issues are well discussed and a profitable model is developed based on the provided background.

In the wine industry, as grapes are most characterized by its perishability after harvested, it is a standard practice for wine companies to develop a system which contains the grape blocks and winery together to reduce the loss in transportation process. Besides wine productivity equipment, larger wineries may also feature warehouses, bottling lines, laboratories and large expanses of tanks known as tank

H. Yu (✉) · J. Watada · J. Li
1-15-D210, HIbikino, Wakamatsu-Ku, Kitakyushu-Shi, Fukuoka-Ken, Japan
e-mail: yuhaiyu@ruri.waseda.jp

J. Watada
e-mail: junzo.watada@gmail.com

J. Li
e-mail: jingruli@fuji.waseda.jp

R. Neves-Silva et al. (eds.), *Intelligent Decision Technologies*,
Smart Innovation, Systems and Technologies 39,
DOI 10.1007/978-3-319-19857-6_56

blocks. In the harvesting process, the grape blocks of the same wine company are interrelated and interdependent, which enables the company to get the best quality ingredients to produce wine by designing a favorable harvesting route among the blocks. That is to say, as the harvest dates differ in blocks, then a favorable harvesting route will contribute to obtain the optimum quality in the whole system. A wine unit is defined as "case" in actual production process. For standard bottle (750 ml), one case contains 12 bottles.

Many research works are published on wine field in recent 30 years. The early work of Ashenfelter et al. [1] focuses on the relationship between wine quality and weather in Bordeaux, illustrating quality-weather relationship graph by linear regression. In 2003, Helene Bombrun et al. [2] collect statistical analysis using data from the prices of 8,460 individual California wines and consider how information about grape characteristics and wine assessments affects wine prices. Finally the conclusion is drawn that the wine price hinges on grape variety, grape location, label designation and grape vintage. The objective of Folwell et al. [3] is to develop an accurate depiction of current investment costs of constructing, owning, and operating a winery in the state of Washington. The result is useful in evaluating the expected profitability, cash flows, and potential risk associated with investing in a winery. Recently, Ferrer et al. [4] built a single level optimum model to minimize harvesting cost through the optimal allocation of resources (optimal allocation of labor and machinery), as well as providing the winery with the required grapes. The model also considers another operational aspect: the routing of the harvesting operations, which is relevant in terms of the high costs or time induced by the need of moving the operations from one block to another. In 2012, Taster' Bias et al. [5] make a supplementary for the theory proposed by Helene Bombrun et al. [6], getting a clear perspective on the conclusion that the weather during a grape growing season is a fundamental and easily measured determinant of the quality of the mature wines. As a result, analysts can utilize the resulted conclusion to calculate the predictability of the quality of new vintages accumulates. Finally, Principal component analysis and fuzzy evaluation mathematical methods are applied in analyzing the wine quality by Qiuye et al. 6. This study grades the wine grape based on the data of physical and chemical indexes of wine grape and the quality of wine in 2012 CUMCM. According to the international standard of wine grade, a fuzzy comprehensive evaluation model can be set up for the grading of wine grape.

The above research content mainly concentrated in two aspects–profits and wine quality. The relationship between these two aspects is not only association but also confliction. Since it is standard practice for wine companies to develop a system which contains the grape blocks and winery together, the set background for this paper is that a wine system contains two wineries as well as several blocks. Although several determining factors influence on wine quality as elaborated in the review part, for the blocks in the same area, harvest route among blocks still plays the determinant role. Usually, a good harvest scheduling among the blocks is due to improve the wine quality and reduce labor costs in the meanwhile. However, when taking the capacity of winery into consideration, it is unwarranted to harvest all the grapes totally on the optimal harvest date. In addition, if severe weather comes without expectation, more serious problems will ensue. The core issue is how to strike a balance of contradiction by using mathematical methods.

In the early 1960s, an optimum mathematical model called bilevel programming came studied. Bilevel optimization is a special kind of optimization where one problem is embedded (nested) within another. The outer optimization task is commonly referred to as the upper-level optimization task, and the inner optimization task is commonly referred to as the lower-level optimization task. These problems involve two kinds of variables, referred to as the upper-level variables and the lower-level variables. [7] In this mathematical programming model, the two decision makers will usually have conflict in constraint resources, and both want to reach his own optimization objective. Thus, they need to take their counterpart's choice into consideration. An important feature of the bilevel programming problem is the hierarchical relationship between two autonomous, and possibly conflictual decision makers. Therefore, such structure is in good agreement with the profit-quality problem presented above and the conflict points can be well described in the constraints. If the amount of profits is assumed as the upper level, while the wine quality accounts for the lower level, then the balanced result can be calculated by using mathematical method.

The paper is structured as follows. In Sect. 2 we state the formulation of the bilevel model and discuss how the harvest route is realized. In Sect. 3 the application is worked out based on the bilevel model. At last, the conclusion will be drawn on after calculation.

2 Bilevel Formulation

The general form of bilevel programming problem can be mathematically stated as

$$\begin{aligned} \min_{\mathbf{x}} \quad & F(\mathbf{x}, \mathbf{y}) \\ s.t. \quad & (\mathbf{x}, \mathbf{y}) \leq 0 \\ \min_{\mathbf{y}} \quad & f(\mathbf{x}, \mathbf{y}) \\ s.t. \quad & \mathbf{g}(\mathbf{x}, \mathbf{y}) \leq 0, \end{aligned} \tag{1}$$

where $\mathbf{x} \in \mathbb{R}^{n_1}$ and $\mathbf{y} \in \mathbb{R}^{n_2}$ are respectively the upper level variables and lower level variables. Similarly, the function $F: \mathbb{R}^{n_1} \times \mathbb{R}^{n_2} \to \mathbb{R}$ is the upper level objective function and $f: \mathbb{R}^{n_1} \times \mathbb{R}^{n_2} \to \mathbb{R}$ is the lower level objective function, while the vector-valued functions $\mathbf{G}: \mathbb{R}^{n_1} \times \mathbb{R}^{n_2} \to \mathbb{R}^{m_1}$ and $\mathbf{g}: \mathbb{R}^{n_1} \times \mathbb{R}^{n_2} \to \mathbb{R}^{m_2}$ are respectively called the upper level and lower level constraints.

As discussed in the introduction part, the goal functions of upper level and lower level are respectively to maximize the profits and maximize the quality of harvested grapes. As the profits are defined as the difference between wine sales volume and expenditures, to increase sales volume and to reduce expenditures are both effective measures to realize the goal of upper level. For the upper level of the model, unit wine price is a key determinant of wine sales volume. Besides, expenditures are consist of productive resource cost, labor hiring cost, transportation cost among the blocks and initial equipment investment. Actually, for the same wine company, unit wine price directly depends on the grape quality, which is associated with harvest scheduling. Under normal circumstances, appropriate harvest date paves the way for maintaining the high quality of grapes and deviation from the optimum date will lead to the decline in quality of grapes. In addition, harvest scheduling will influence labor allocation, that is to say, when the harvest time are only concentrated in a few days, much more labors should be allocated and harvest mode includes manual harvest and mechanical harvest. Accordingly, the expenditure in labor aspect will increase.

For the lower level, the quality of wine is measured by the deviation from the optimum date and the penalty for harvest delay. There is a clear agreement that some deterioration is produced if we harvest ahead of time. Thus, delaying harvesting seems to be less dangerous, but it exposes the crop to an increasing danger of rain [8].

3 Application Model

Based on the description in Sect. 2, we make use of the following notation described in Table 1.

Table 1 Definition of notation involved in the bilevel model

1.1 sets of indexes	
B_1	set of blocks suitable for manual harvest
B_2	set of blocks suitable for mechanical harvest
	$B = B_1 \cup B_2$
K	set of harvest modes, $k \in K = \{1, 2\}$
	K = 1 is manual harvest
	K = 2 is mechanical harvest
1.2 model variables	
M_r	set of actual total amount of r rating grape harvested, $r \in R = \{1, 2, 3\}$
x_{jtkb}	kilograms harvest in block j, at period t, in mode k, routed to winery b
y_{jtkb}	binary variable with value 1 when block j initiate harvesting at period t, in harvest mode k, routed to winery b, and 0 otherwise.
D_{itb}	the actual harvest *nth* date in harvest period when the quality is the best in block i, during period t and with destination winery b.
v_{jtk}	binary variable indicating if there is harvest or not in block j, during period t, and in harvest mode k

(continued)

Table 1 (continued)

1.1 sets of indexes	
z_{ijtb}	binary variable that is 1 when an operation moves from block i to block j in period t and with destination winery b, 0 otherwise
τ_{jtb}	variable used to eliminate subtours in the MTZ formulation. It presents the position inside the cycle of block j, at period t, routed to winery b
w_t^H	number of workers hired at period t
w_t^F	number of workers hired at period t
u_{jtb}	productive resources used in block j in period t. It is a continuous variable such that for k = 1 it is machine hours and for k = 2 it is number of workers
1.3 model parameters	
p_r	price for different wine ratings, $r \in R = \{1, 2, 3\}$
α_r	wine juice yield for rating r, $r \in R$
β_r	maximum allowable production deviation rate from planning amount each day; for rating r
c_m	cost of initial equipment investment for different factory scale m, $m \in M = \{1, 2, 3, 4, 5\}$
c_k	unit cost of productive resource k. For k = 1 it is cost per hour of machine and for k = 2 it is cost per worker
H	hiring cost of adding one unit of labor
F	saving cost for reducing one unit of labor
T	days available to produce wine among one year
Q_{jt}	quality cost for block $j \in B$ in period t
D_{ij}	distance between block i and j, $i, j \in B$
P_{kj}	productivity of a unit of productive resource k in kg/day, $j \in B$
G_{jkb}	estimated kg of grapes in block $j \in B$ harvest in mode $k \in K$, with destination to winery b
W_{jt}	harvest window parameter indicating whether block j can be harvested in period t. It is 1 if possible, and 0 otherwise
L_{kbt}	processing capacity of winery b for grapes harvest in mode k in period t
B_{jtk}	maximum volume that can be harvested in block j at time t using harvested mode k
d_{ib}	the *nth* date in harvest period when the quality is the best in block i and with destination winery b
A_t	machine availability, in hours, in period t
M	a "big M" value
M_{Tr}	total planning grape production amount each year of rating r
R	operations relocation cost in $ per kilometer
V_k	minimum volume to be harvested from a block using k harvest mode
λ	penalty parameter used for the relative balance between operational and quality cost
N	minimum number of contractors needed *card*(*B*) number of elements in the set B, number of blocks

To build the bilevel model, the preparatory steps are written in the following.

(1) Determine the investment cost according to the production scale of winery.
(2) How to plan the harvest routing among blocks.
(3) Determine the grape quality by harvest date.

In the first step, initial investment cost is mainly associated with winery scale. The annual production planning is set up based on many factors. This approach has the disadvantage that the number of pairwise comparisons to be made, may become very large and thus become a lengthy task. However, in our planning, the size of options is acceptable and AHP is simple and clear to solve a option problem based on existed uncertain and subjective information. Therefore, Analytic Hierarchy Process (AHP) is the optimal policy to decide which winery scale is the best based on historical data.

The analytic hierarchy process (AHP) is a structured technique for organizing and analyzing complex decisions, based on mathematics and psychology. It was developed by Thomas L. Saaty in the 1970s and has been extensively studied and refined since then. [10]

The core objective of AHP is to determine the importance value after comparing every two factors when making the decisions.

And historical data of investment cost divided by scale of winery in the following is collected from a winery in Washington. (Table 2) [3].

Table 2 Total cost by equipment category and winery size ($)

Winery size(cases)					
Cost category	2000	5000	10000	15000	20000
Equipment	5.609×10^5	8.101×10^5	1.331×10^6	1.822×10^6	2.339×10^6
Labor	4.136×10^4	1.144×10^5	1.632×10^5	3.122×10^5	4.079×10^5
Insurance and maintenance	1.135×10^5	2.002×10^5	3.341×10^5	5.042×10^5	6.613×10^5
Cash flow protections	1.150×10^4	3.030×10^4	5.011×10^4	6.907×10^4	8.517×10^4
Total cost	7.273×10^6	1.151×10^6	1.878×10^6	2.708×10^6	3.493×10^6
Total average cost	3.030×10	1.918×10	1.565×10	1.504×10	1.456×10

As described, when making the decision, there are 4 treatment options to be compared—(*equipment* × *labor, equipment* × *insurance and maintenance, equipment* × *cash flow protections, labor insurance and maintenance, labor* × *cash flow protections, insurance and maintenance* × *cash flow protections*),which can be shortly written as ($R \times A$, $R \times B$, $R \times C$, $R \times D$, $B \times C$, $B \times D$, $C \times D$). Under the assumption that sufficient funds are available, we can make subjective comparison judgments results as Table 3 shows:

Then, the propose of Consistency Ratio (CR) is proposed by Saaty to value if our opinions are reasonable. Consistency Ratio is defined as

Table 3 Full matrix based on paired comparisons

	$R\times A$	$R\times B$	$R\times C$	$R\times D$
$R\times A$	1	$\frac{1}{7}$	$\frac{1}{9}$	$\frac{1}{9}$
$R\times B$	7	1	$\frac{1}{5}$	$\frac{1}{5}$
$R\times C$	9	5	1	$\frac{1}{2}$
$R\times D$	9	5	2	1

$$CR = CI/RI$$

If the value of Consistency Ratio is smaller or equal to 10 %, the result is said consistent. If the Consistency Ratio is greater than 10 %, we need to consider revising our subjective judgments [11].

In the formula *RI* is accessible when the comparison matrix is known. It is only connected with the number of items compared in a matrix (Table 4).

Table 4 Random Consistency Index – RI

n	2	3	4	5	6	7	8	9	10
RI	0.00	0.58	0.9	1.12	1.24	1.32	1.41	1.45	1.51

From the comparison matrix, it can be easily seen that when the size of comparison matrix n = 4, RI is supposed to be 0.9. While, at the same time, CI is defined as

$$\mathrm{CI} = \frac{\lambda_{max} - n}{n - 1}$$

Later we calculate the matrix in matlab. As our calculation result of Consistency Ratio is less than 10 %, then this result is acceptable. And the scale is assumed to be 10,000 cases, and total investment cost $c_m = 1.878 \times 10^6$.

In the second step, as discussed, harvest route plays the dominant role in the model. Supposed that several blocks are in the wine system, harvest operations are relevant to grape quality and transportation costs inside the blocks. It can be reasonable to harvest some blocks earlier, if they are nearby and quality effect are not significant. Therefore, it is necessary to specify the routing in sequence. Conventionally, respect to routing problem,the formulation works in a graph G = (N, A) with n nodes. The Miller-Tucker-Zemlin formulation (MTZ) refers to a complete graph G = (N, A) where $N = \{n_1, \ldots, n_m\}$ is the vertex set and $A = \{(n_i, n_j): n_i, n_j \in N, i \neq j\}$ is the arc set in our application. u_{ij} is a binary variable, assumed to be 1 if we go from node i to j in the circuit. [9] Under normal circumstances, this model starts at node 1. The constraints of MTZ is formed as following.

$$\sum_{j \neq i} u_{ij} = 1, \forall i = 1, \ldots, n \tag{2}$$

$$\sum_{i \neq j} u_{ij} = 1, \forall j = 1, \ldots, n \tag{3}$$

$$\gamma_i - \gamma_j + 1 \leq (n-1)(1-u_{ij}), \forall i, j, i \neq 1, j \neq 1 \tag{4}$$

$$\gamma_1 = 1; 2 \leq \gamma_i \leq n. \tag{5}$$

However, when this model is given a richer application background, some minor changes are made to satisfy the demands. Following is the modified routing model applied in the wine system.

$$\sum_{j \neq i} z_{ijtb} = y_{it2b}, \quad \forall i, t, b \tag{6}$$

$$\sum_{i \neq j} z_{ijtb} = y_{jt2b}, \quad \forall j, t, b \tag{7}$$

$$\tau_{itb} - \tau_{jtb} + 1 \leq (N-1)(1-z_{ijtb}) \tag{8}$$

$$\forall \mathrm{i}, \mathrm{j}; \mathrm{i} \neq 1, \mathrm{j} \neq 1, \forall \mathrm{t}, \tag{9}$$

$$2y_{it2b} \leq \tau_{itb} \leq N, \forall i \geq 2, \forall t, b \tag{10}$$

$$\tau_{1tb} = y_{it2b}; z_{ijtb} \in \{0, 1\}$$

$$\tau_{itb} \geq 0 \quad \forall i, t, b \tag{11}$$

The modified takes more information into consideration. For example, the rich information in subscript of each parameter creates the conditions for removing the barrier of starting at the first farm. In addition, we use additional parameters in subscript. Specifically, let z_{ijtb} be equal to 1 if block j is harvested after block i, beginning in day t, with destination b, and τ_{itb} be the order in the sequence of block i. y_{itkb} is the replacement for u_{ij}, will be equal to 1 only when the harvest process begins on day t.

In the third step, during the harvest window, the optimal harvest date assumes the best quality raw materials in the same production area. Then, the deviation between the optimal harvest date and the actual harvest date leads to the quality degradation. Therefore, it is a suggested way to use the distance between optimal date and actual date to value the quality of raw materials. Generally, the distance is represented by the absolute value. However, as absolute value is not suitable for calculation, it can be converted to square to improve its calculability.

In the respect of wine value, European wine scoring method is preferred. This method is totally 20 scores, in which 10 scores is basic. The scoring and quality are interrelated of each other (Table 5).

Table 5 The relationship of wine scoring and evaluation

Score	Evaluation
17–20	superior
13–17	accepted
10–13	under the criterion

In order to acquire the actual data of wine quality, we take Bordeaux of France as an example. In 2000, Gregory V. Jones et al. [12] sort out the collected data in harvest period.

Near harvest time, the key vintage quality characteristics are the chemical composition of the grapes. [12] Two of the chief determinants of crop ripeness and quality are the relative amounts of sugar and acid found in the berries leading up to harvest. [13] Therefore, the quality of wine can be judged by measuring the composition. By means of a comprehensive analysis of data in the two tables, the relationship between harvest date and evaluation can be set up after calculation.

Based on the previously defined parameters $|d_{ib} - D_{ib}|$ is feasible to indicate the distance between the optimal harvest date and actual harvest date. From the table Bordeaux reference vineyard production and quality record in harvest window, we can obtain the conclusions as following.

$$\begin{cases} |d_{ib} - D_{ib}| = 0, \textit{ the quality is superior} \\ 0 < |d_{ib} - D_{ib}| \leq 28, \textit{ the quality is accepted} \\ 28 < |d_{ib} - D_{ib}| \leq 39, \textit{ the quality is under the criterion} \end{cases}$$

in order to make it easier to calculate, we would better convert the absolute value i to square. Then, it is converted to

$$\begin{cases} (d_{ib} - D_{ib})^2 = 0, \textit{the quality is superior} \\ 0 < (d_{ib} - D_{ib})^2 \leq 784, \textit{the quality is accepted} \\ 784 < (d_{ib} - D_{ib}) \leq 1521, \textit{the quality is under the criterion} \end{cases}$$

As a result, the whole model is

upper level

$$\max_{\substack{M_r, w_t^H, w_t^F, u_{jtk} \\ v_{jtk}, z_{ijtb}, \tau_{jtb}}} \sum_{r=1,2,3} p_r M_r \alpha_r - \sum_{k,j,t} c_k u_{jtk} - H \sum_t w_t^H + F \sum_t w_t^F$$

$$- R \sum_{i,j,t,b:\, i \neq j} D_{ij} z_{ijtb} - c_m$$

$$s.t. \, (1 - \beta_r) \frac{M_T}{T} \leq M_r \leq (1 + \beta_r) M_r$$

$$c_m = \begin{cases} c_1, M_T = 2000\,case \\ c_2, M_T = 5000\,case \\ c_3, M_T = 10000\,case \\ c_4, M_T = 15000\,case \\ c_5, M_T = 20000\,case \end{cases}$$

$$R_r = \begin{cases} R_1 (d_{ib} - D_{ib})^2 = 0, p_1 = 3.229 \times 10^2 \$/750ml \\ R_2 0 < (d_{ib} - D_{ib})^2 \leq 784, p_2 = 2.021 \times 10^2 \$/750ml \\ R_3 784 < (d_{ib} - D_{ib})^2 \leq 1521, p_3 = 1.179 \times 10^2 \$/750ml \end{cases}$$

$$\sum_t \sum_k v_{jtk} \geq 1,$$

$$\sum_{j \in B_2} u_{jt2} = \sum_{j \in B_2} u_{j(t-1)2} + w_t^H - w_t^F,$$

$$\sum_{j \in B_1} u_{jt1} \leq A_t,$$

$$u_{jt2} \geq N v_{jt2},$$

$$\sum_{j \neq i} z_{ijtb} = y_{iy2b}, \forall t, b, j \in B: W_{it} \neq 0; W_{jt} \neq 0,$$

$$\tau_{jtb} \geq \tau_{itb} + 1 - (card(B) - 1)(1 - z_{ijtb}),$$

lower level

$$\min_{x_{jtkb}, y_{jtkb}, D_{itb}} \sum_i (d_{ib} - D_{ib})^2 + \lambda \sum_{j,t,k,b} Q_{jt} x_{jtkb}$$

$$s.t. \quad \forall t, b; i, j \in B_2; i \neq j; W_{it} \neq 0; W_{jt} \neq 0,$$

$$\sum_{j \in B} x_{jtkb} \leq L_{kbt}, \forall k, b, t,$$

$$\sum_t x_{jtkb} = G_{jkb}, \forall k, b, j,$$

$$x_{jtkb} \leq G_{jkb} v_{jtk}, \forall t, b, k, j,$$

$$x_{jtkb} \geq V_k v_{jtk}, \forall t, b, k, j,$$

$$\sum_b x_{jtkb} \leq B_{jtk},$$

$$\sum_b y_{jtkb} \leq M v_{jtk},$$

$$y_{jtkb} \leq \sum_{s \leq t: W_{js} = 1} v_{jsb},$$

$$\sum_b \sum_{t: W_{jt} = 1} y_{jtkb} \leq 1,$$

$$y_{jtkb} \leq W_{jt},$$

$$\sum_b x_{jtkb} \leq P_{kj} u_{jtk},$$

$$\tau_{1tb} = y_{1t2b}, \forall t\colon W_{1t} \neq 0, \forall b,$$
$$\tau_{jtb} \geq 2y_{jt2b}, \forall j, t \in B\colon W_{jt} \neq 0, \forall b,$$
$$\tau_{jtb} \leq card(B),$$
$$\forall j, t \in B\colon W_{jt} \neq 0, \forall b$$
$$z_{ijtb}, v_{jtk}, y_{jtkb} \in \{0, 1\}, \forall t, b, k, i, j$$
$$w^H_{tj}, w^F_{ti}, u_{jtk}, x_{jtkb}, \tau_{jtb} \geq 0, \forall t, b, k, i, j$$

In the model, the harvest route is in the constraint part of the lower level while the promised grape rating criteria is in the constraint part of upper level. The first constraint in upper level also gives the limit for daily production limit according to the annual production planning.

4 Results

After calculating the data in matlab, the final results in upper level harvest is as following (Table 6).

Table 6 Results in upper level

Total amount of grapes with superior quality	2134.1706 kg
Total amount of grapes with accepted quality	111868.1831 kg
Total amount of grapes that are under the criterion	23197.64627 kg
Total sales volume	21439251.35$
Cost when harvest (manual mode and machine mode)	55945.23$
Operations relocation cost	14648.83$
Investment cost	1878000$
Profits	19470978.19$

From the lower level, we can also get the actual kilograms by harvest date both in manual mode and machine mode. Since the harvest window seems a little bit big, we show the period only from Sep.9 to Sep.18 (10 days).

5 Conclusions

Bilevel model was applied in this paper to work out the optimum result when considering with profits and wine quality. In the model presented, several methods such as principal component analysis, routing strategy were efficiently used to calculate some portions. Upper level and lower level are well connected by routing strategy and the evaluation method of wine quality. The final result shows the real routing and the maximum profits this year in the system.

References

1. Ashenfelter, O., Ashmore, D., Lalonde, R.: Bordeaux wine vintage quality and the weather. Chance **8**(4), 7–14 (1995)
2. Bombrun, H., Sumner, D.A.: What determines the price of wine. The value of grape characteristics and wine quality assessments. AIC Issues Brief 18 (2003)
3. Folwell, R.J., Ball, T., Clary. C.: Small Winery Investment and Operating Costs. Washington State University Extension (2005)
4. Ferrer, J.C., et al.: An optimization approach for scheduling wine grape harvest operations. Int. J. Prod. Econ. **112**(2) 985–999 (2008)
5. http://www.wine-economics.org/data/(2012)
6. Qiuye, Q., et al.: The application of grape grading based on PCA and fuzzy evalua. Adv. J. Food Sci. Technol. **5**(11) (2013)
7. Sinha, A., Malo, P., Deb, K.: Efficient evolutionary algorithm for single-objective bilevel optimization. arXiv preprint arXiv:1303(3901) (2013)
8. Bordeu, E., Mira, P., Rivadeneria, R.: Madurez fenologica: experiencia en chile. Topicos de Actualizacion en Viticultura y Enologia **2002**, 190–192 (2002)
9. Desrochers, M., Laporte, G.: Improvements and extensions to the Miller-Tucker-Zemlin subtour elimination constraints. Oper. Res. Lett. **10**(1), 27–36 (1991)
10. Chen Z., Zhu, S.: The research of mobile application user experience and assessment model. In: International Conference on Computer Science and Network Technology (ICCSNT) IEEE, vol. 4, pp. 2832–2835 (2011)
11. Teknomo, K.: Analytic hierarchy process (AHP) tutorial. Retrieved January 2006 (11) (2011)
12. Jones, G.V., Davis, R.E.: Climate influences on grapevine phenology, grape composition, and wine production and quality for Bordeaux, France. Am. J. Enology Viticulture **51**(3), 249–261 (2000)
13. Araujo, F, Williams, L.E., Matthews, M.A.: A comparative study of young 'Thompson Seedless' grapevines under drip and furrow irrigation. II. Growth, water use efficiency and nitrogen partitioning. Sci. Hortic. **60**(3), 251–265 (1995)

Building a Sensitivity-Based Portfolio Selection Models

Huiming Zhang, Junzo Watada, Ye Li, You Li and Bo Wang

Abstract Sensitivity Analysis is a method to evaluate the influence of each variable change. In portfolio selection model, it is essential to evaluate the sensitivity of each stock or security return rate in investment decision making. Investors look for selecting stable stocks or securities. For this purpose, sensitivity analysis should play a pivotal role. It is important for the decision-making to get both sensitivity and stability of each selection. This paper proposes a new portfolio-selection model (PSM) called the sensitivity-based portfolio selection models (SPSM). The SPSM model will focus on the sensitivity of the selected portfolio. In order to analyze the sensitivity of portfolio selection models, a sensitivity analysis will be introduced for calculating out insensitive stocks or securities with maximum return and minimum risk. *Abstract* environment.

Keywords Sensitivity analysis · Portfolio selection model · Value at risk (VaR)

H. Zhang (✉) · J. Watada · Y. Li
Graduate School of Information, Production and Systems,
Waseda University, Kitakyushu, Fukuoka 808-0135, Japan
e-mail: huimingde@gmail.com

J. Watada
e-mail: junzow@osb.att.ne.jp

Y. Li
e-mail: liye@akane.waseda.jp

Y. Li
College of Finance, Nanjing University of Finance & Economics,
No. 3, Wenyuan Road, Xianlin College Town, Nanjing 210008, People's Republic of China
e-mail: ruby1025@gmail.com

B. Wang
School of Management and Engineering, Nanjing University,
Cang Ping Xiang 5, Gulou District, Nanjing 210008, People's Republic of China
e-mail: bwangips@gmail.com

R. Neves-Silva et al. (eds.), *Intelligent Decision Technologies*,
Smart Innovation, Systems and Technologies 39,
DOI 10.1007/978-3-319-19857-6_57

1 Introduction

Portfolio selection model has been well developed based on a mean-variance approach proposed by Markowitz [1]. More than 50 years have been spent to the portfolio under uncertain environment random selection development. The key principle of the mean-variance model is to use the expected returns of the portfolio as the investment risk [2]. It is accepted as a principle that a portfolio is designed according to the risk tolerance, time frame and investment objectives of investor. Each asset's monetary value may affect the risk and reward ratio of the portfolio and is referred to as the asset allocation of the portfolio. When determining a proper asset allocation, we target to maximize the expected return and minimize the risk.

In finance, security returns in conventional models are always determined by historical data. However, such data are not always available. On the other hand, securities are always affected by infinitive factors.

It is impossible to simulate the influence of all factors well. Therefore, in today's stock markets, it is not reasonable to evaluate future security returns only based on the historical data. In order to forecast more accuracy, the expert's opinions from related fields need to be considered [3]. The model SA-PSM is mainly used in the stock market, analyzing and finding the balance between the value and risks.

The sensitivity analysis has been used in a range of purposes [4]. Including (1) Testing the robustness of the results or the uncertainty of the system (2) Increased understanding of the relationships between input and output variables in a system or model (3) Reducing the uncertainty: identifying model inputs (4) Searching for errors in the model (by encountering unexpected relationships between inputs and outputs) (5) Model simplification-fixing model inputs which have no effect on the output, or identifying and removing redundant parts of the model structure (6) Enhancing communication from modelers to decision makers (e.g. by making recommendations more credible, understandable, compelling or persuasive) (7) Finding regions in the space of input factors for which the model output is either maximum or minimum or meets some optimum criterion (see optimization and Monte Carlo filtering).

For example, in a decision problem, the analyst may want to identify cost drivers as well as other quantities for which we need to acquire better knowledge in order to make an informed decision. On the other hand, some quantities have no influence on the predictions, so that we can save resources at no loss in accuracy by relaxing some of the conditions. However, the sensitivity analysis cannot analyze the uncertainty and the instability.

In this study, proposing a new portfolio-selection model (PSM) called the Sensitivity Analysis based on portfolio-selection-models (SA-PSM). This model use both types risk measure methods: variance is utilized to evaluate the portfolio stability. So this model can improve the stableness of the capital allocation. Therefore, the obtained selection result could be more helpful in decision making [5].

The remainder of this paper is organized as follows. Section 2 provides some basic mathematical preparation on portfolio selection model. In the Sect. 3, we explain the method of sensitivity analysis and sensitivity-based portfolio selection model. Section 4 illustrates a numerical example and Finally, conclusions are drawn in Sect. 5.

2 Portfolio Selection Model

In condition of high uncertainty, financial companies that fewer risks and are more stable. Sensitivity analysis plays a pivotal role in uncertainty quantification and propagation of uncertainty.

This paper introduces two steps of the sensitivity analysis. Let us define the sensitivity analysis model.

Model 1 *(Portfolio Selection Model)*

In order to construct the sensitivity, let us consider the following models. Function $F(x)$ is expectation,meaning the expected return. Function $G(x)$ is variance, meaning the risk.

$$\begin{cases} \max F(x) & \{=\sum_{i=1}^{n} x_i a_i\} \\ \min G(x) & \{=\sum_{i=1}^{n}\sum_{j=1}^{n} x_i \sigma_{i,j} x_j\} \\ Subject\ \ to & \\ & \sum_{i=1}^{n} x_i = 1 \\ & x_i \geq 0. \quad i = 1, 2, \cdots, n \end{cases} \tag{1}$$

3 Sensitivity Analysis

In order to minimize the sensitivity, let us consider the following values.

Let us define the sensitivity of the linear function $O(x)$ of two objectives $F(x)$ and $G(x)$ at variable x_i using F_{x_i} and G_{x_i} denoting the partial differentiation of the objectives $F(x)$ and $G(x)$ by variable x_i, respectively. Then, the sensitivity of $O(x)$ by t can be calculated as

$$\frac{dO(x)}{dt} = \sum_{i=1}^{n} \{\omega_F F_{x_i} + \omega_G G_{x_i}\} \frac{dx_i}{dt} \tag{2}$$

where

$$O(x) = \omega_F F(x) + \omega_G G(x) \tag{3}$$

and ω_F, ω_G denote the weight of the objectives, respectively and $\omega_F + \omega_G = 1$.

Definition 1 *(Sensitivity of the total objective) Let us define the sensitivity TS of the total objective O(x) as the differentiation by t in the following:*

$$
\begin{aligned}
TS &= \frac{dO(x)}{dt} \\
&= \sum_{i=1}^{n} \frac{\partial O(x)}{\partial x_i}\frac{dx_i}{dt} \\
&= \omega_F \frac{dF(x)}{dt} + \omega_G \frac{dG(x)}{dt} \\
&= \sum_{i=1}^{n} (\{\omega_F \frac{\partial F(x)}{\partial x_i} + \omega_G \frac{\partial G(x)}{\partial x_i}\} \frac{dx_i}{dt}) \\
&= \sum_{i=1}^{n} (\{\omega_F F_{x_i} + \omega_G G_{x_i}\} \frac{dx_i}{dt}) \qquad (4) \\
&= \sum_{i=1}^{n} (w_F a_i + w_G 2 \sum_{j=1}^{n} \sigma_{ij} x_j) \frac{dx_i}{dt} \qquad (5)
\end{aligned}
$$

where

$$
\begin{aligned}
\alpha_i = Fx_i &= \frac{\partial \sum_{i=1}^{n} x_i a_i}{\partial x_i} \\
&= a_i \qquad (6) \\
\beta_i = G_{x_i} &= \frac{\partial \sum_{i=1}^{n} \sum_{j=1}^{n} \sigma_{ij} x_i x_j}{\partial x_i} \\
&= 2 \sum_{j=1}^{n} \sigma_{ij} x_j \qquad (7)
\end{aligned}
$$

After processing Eqs. (3)–(5) we can get the sensitivity of each objective function, which are defined as F_{x_i}, G_{x_i}

Note that from here ω_F, ω_G are set to 1.

Take this ten stocks as the return, we can calculate out the sensitivity of each variable, and also, we can calculate the total sensitivity which will be shown below:

Table 2 shows the result of sensitivity analysis.

Then we can understand which variables are more sensitive and which variables are more stable. But we would better select portfolio with more stable investment, that is, less sensitive variables (stable stocks). Then we formulate the problem including the sensitivity in the objective, The total sensitivity *TS* is defined as Eqs. (4) and (5).

Table 1 Data employed from NYSE [4]

Stock	STV x_1	CNEP x_2	CNTF x_3	CBEH x_4	CHL x_5	ACH x_6	CAST x_7	DANG x_8	CNAM x_9	CBAK x_{10}
a_i	-0.01503	0.00380	0.01031	0.02711	0.00371	-0.00903	-0.00118	-0.01145	-0.03755	-0.02906
$\sigma(i,j)$	1	-0.9505	-0.5349	-0.756	-0.5535	-0.3448	0.2036	-0.8166	-0.6358	0.6679
	-0.9505	1	0.3157	0.5492	0.7762	0.6183	-0.3065	0.8517	0.6734	-0.8331
	-0.5349	0.3157	1	0.9504	-0.3257	-0.4419	0.6381	0.5863	-0.2216	-0.1812
	-0.756	0.5492	0.9504	1	-0.0977	-0.2783	0.3737	0.6746	0.0917	-0.296
	-0.5535	0.7762	-0.3257	-0.0977	1	0.9609	-0.5993	0.5361	0.6883	-0.8017
	-0.3448	0.6183	-0.4419	-0.2783	0.9609	1	-0.4833	0.4668	0.4863	-0.7962
	0.2036	-0.3065	0.6381	0.3737	-0.5993	-0.4833	1	0.2212	-0.8821	0.0197
	-0.8166	0.8517	0.5863	0.6746	0.5361	0.4668	0.2212	1	0.1862	-0.9011
	-0.6358	0.6734	-0.2216	0.0917	0.6883	0.4863	-0.8821	0.1862	1	-0.2832
	0.6679	-0.8331	-0.1812	-0.296	-0.8017	-0.7962	0.0197	-0.9011	-0.2832	1

Let us solve the following problem including the sensitivity objective:

$$Return : \max \sum_{i=1}^{n} x_i a_i \tag{8}$$

$$Risk : \min \sum_{i} \sum_{j} \sigma_{ij} x_i x_j \tag{9}$$

$$Sensitivity : \min \sum_{i=1}^{n} \{a_i \frac{dx_i}{dt} + 2 \sum_{j=1}^{n} (\sigma_{ij} \frac{dx_i}{dt} x_j)\} \tag{10}$$

Subject to

$$\sum_{i=1}^{n} x_i = 1 \tag{11}$$

$$x_i \geq 0 \quad i = 1, 2, \cdots, n \tag{12}$$

We can solve the same problems from Table 1.

On the other hand, when we consider (13)–(15), these sensitivities can be defined by the changing rate of $a_j \frac{dx_j}{dt}$ in (13), $2 \sum_{j=1}^{n} x_j \sigma_{ij} \frac{dx_i}{dt}$ in (14).

4 Building Sensitivity-Based VaR Portfolio Selection Model

Definition 2 *Sensitivity*

The sensitivity TS can be defined as follows:

$$TS = \frac{dO(x)}{dt} = \sum_{i=1}^{n} \left(\frac{\partial \{\omega_F F(x) + \omega_G G(x))\}}{\partial x_i} \frac{dx_i}{dt} \right) \tag{13}$$

That means, the total sensitivity is calculated using α_i, β_i and the change of return rate per time.

Using this definition, we can define the sensitivity-based VaR portfolio selection model as follows:

Model 2 *(the sensitivity-based portfolio selection model)*

$$Return : \max \sum_{i=1}^{n} x_i a_i \tag{14}$$

Table 2 Computation results

Stock	STV x_1	CNEP x_2	CNTF x_3	CBEH x_4	CHL x_5	ACH x_6	CAST x_7	DANG x_8	CNAM x_9	CBAK x_{10}
PSM Capital Allocation				50.4 %	40.8 %			9.0 %		
$x_i(23-29)$	0.0000	0.0000	0.0504	0.4522	0.4079	0.0000	0.0000	0.0896	0.0000	0.0000
$x_i(22-28)$	0.0000	0.4354	0.0000	0.0961	0.0000	0.0000	0.2978	0.0000	0.0000	0.1707
$x_i(21-27)$	0.1630	0.0000	0.0076	0.0000	0.0000	0.0000	0.4623	0.0000	0.0000	0.3670
$x_i(20-26)$	0.4428	0.0000	0.0000	0.0000	0.0000	0.0000	0.2029	0.3543	0.0000	0.0000
$x_i(19-25)$	0.1012	0.0000	0.0000	0.0000	0.0000	0.0000	0.3807	0.1207	0.3974	0.0000
$x_i(16-22)$	0.3022	0.0000	0.0971	0.0000	0.0000	0.1319	0.4688	0.0000	0.0000	0.0000
$x_i(15-21)$	0.0525	0.0000	0.3099	0.1877	0.0000	0.0000	0.4499	0.0000	0.0000	0.0000
$\frac{dx_i}{dt}$	-0.0250	0.0311	-0.0345	0.0352	0.0437	-0.0094	-0.0575	0.0053	-0.0142	0.0253
$\alpha_i = a_i \frac{dx_i}{dt}$	0.0004	0.0001	-0.0004	0.0010	0.0002	0.0001	0.0001	-0.0001	0.0005	-0.0007
$\beta_i = 2 \sum_{j=1}^{n} \sigma_{ij} \frac{dx_j}{dt}$	-0.1475	0.1658	-0.0459	0.0201	0.1754	0.1382	-0.1769	0.0732	0.2072	-0.1018
PSM Total $(TS_{x_i}) = (\alpha_i + \beta_i x_i)$	0.0004	0.0001	-0.0004	0.0111	0.0717	0.0001	0.0001	0.0065	0.0005	-0.0007
SPSM Capital Allocation	32.7 %			31.2 %	27.9 %					8.2 %
SPSM Total $(TS_{x_i}) = (\alpha_i + \beta_i x_i)$	-0.0479	0.0001	-0.0004	0.0072	0.0491	0.0001	0.0001	-0.0001	0.0005	-0.0091

$$Risk : \min \sum_i \sum_j \sigma_{ij} x_i x_j \tag{15}$$

$$Sensitivity : \min \sum_{i=1}^{n} \{a_i \frac{dx_i}{dt} + 2 \sum_{j=1}^{n} (\sigma_{ij} \frac{dx_i}{dt} x_j)\}$$

$$= \sum_{i=1}^{n} \{\alpha_i + \beta_i x_i\} \tag{16}$$

Subject to

$$\sum_{i=1}^{n} x_i = 1 \tag{17}$$

$$x_i \geq 0 \quad i = 1, 2, \cdots, n \tag{18}$$

where ω_F, ω_G are given as a weight for each objective function. In this discussion, let us set them to 1.

Let us approximate $\frac{dx_i}{dt}$ by the following:

$$\frac{dx_i(t)}{dt} \simeq \frac{x_i(t + \triangle t) - x_i(t)}{t + \triangle t - t} = \frac{x_i(t + \triangle t) - x_i(t)}{\triangle t} \tag{19}$$

If we want to know the change rate of return on one week, we approximately calculate it in the following:

$$Change \ rate \ of \ return = \frac{x_i(t)}{dt} \simeq \frac{x_i^{max} - x_i^{min}}{7} \tag{20}$$

where x_i^{max} and x_i^{min} are the maximum and minimum values in a week.

The calculation can be performed based on Table 2. We may prepare the changing rate of return rate x_i, that is, $\frac{dx_i}{dt}$.

5 Illustrative Application

Let us analyze the stocks of NYSE(New York Stock Exchange) in December 2011. Table 2 shows the result obtained by the portfolio selection model 1 and the sensitivity-based portfolio selection model 2. In the portfolio selection model 1,

CBEH and CHL were allocated 50.4 % and 40.8 % of the total investment. But in the sensitivity-based selection model 2 these are rather high sensitive variables with 0.0072 and 0.0491, respectively comparing with other stocks. Therefore, these stocks are not stable.

The sensitivity-based portfolio selection model 2 changed the allocation for stocks STV, CBEH and CH with 32.7 %, 31.2 % and 27.9 %. So the total investment allocation is distributed to the less sensitive stock STV with -0.0479 as well.

6 Conclusions

In this paper, we have introduced a sensitivity analysis for the portfolio selection model, which shows both the sensitivity of each variance and the total sensitivity of the model. To the existing PSMs, the approach is more acceptable for general investors, so the calculation of the sensitivity will be quite necessary. To calculate out sensitivity of each variance and the total sensitivity, we have improved the basic PSM models to get the definition of sensitivity which can be expressed in a more clear mathematic way. Also from the experiment result we can find out that, to the PSMs, different variance has different sensitivity which will make influence to the analysis of the whole model, to the total sensitivity, it will show a overall description of whether the model is stable or sensitive. To the proposed PSM, the proposed sensitivity analysis is proved useable, for calculating out both the sensitive variances and the stable variances. In the further research, we will introduce the proposed model into some other applications, and also, we will do some more analysis not only in the value of sensitivity but also the relationship between different variance's sensitivity.

References

1. Markowitz, H.: Portfolio selection. J. Financ. **7**(1), 77–91 (1952)
2. Lin, P.-C., Watada, J., Wu, B.: Statistic test on fuzzy portfolio selection model. In: 2011 IEEE International Conference on Fuzzy Systems, Taipei, Taiwan, 27–30 June 2011
3. Li, Y., Wang, B., Watada, J.: Building a fuzzy multi-objective portfolio selection model with distinct risk measurements. In: 2011 IEEE international Conference on Fuzzy Systems, Taipei, Taiwan, 27–30 June 2011
4. Pannell, D.J.: Sensitivity analysis of normative economic models: theoretical framework and practical strategies. Agric. Econ. **16**, 139–152 (1997)
5. Wang, B., Wang, S., Watada, J.: Fuzzy-portfolio-selection models with value-at-risk. In: IEEE Transactions on Fuzzy Systems, August

Author Index

R. Neves-Silva et al. (eds.), *Intelligent Decision Technologies*,
Smart Innovation, Systems and Technologies 39,
DOI 10.1007/978-3-319-19857-6

Y

Z

Printed in the United States
By Bookmasters